11A9$5415

HUMAN FACTORS: UNDERSTANDING PEOPLE-SYSTEM RELATIONSHIPS

HUMAN FACTORS: UNDERSTANDING PEOPLE-SYSTEM RELATIONSHIPS

BARRY H. KANTOWITZ
Purdue University
Professor of Psychological Sciences and Industrial Engineering

ROBERT D. SORKIN
Purdue University
Professor of Psychological Sciences

JOHN WILEY & SONS
NEW YORK CHICHESTER BRISBANE TORONTO SINGAPORE

Copyright © 1983, by John Wiley & Sons, Inc.

All rights reserved. Published simultaneously in Canada.

Reproduction or translation of any part of
this work beyond that permitted by Sections
107 and 108 of the 1976 United States Copyright
Act without the permission of the copyright
owner is unlawful. Requests for permission
or further information should be addressed to
the Permissions Department, John Wiley & Sons.

Library of Congress Cataloging in Publication Data:

Kantowitz, Barry H.
Human factors.

Includes bibliographical references and indexes.
1. Human engineering. 2. Man-machine systems.
3. Psychology, Applied. I. Sorkin, Robert D., 1937–
II. Title.

TA166.K365 1983 620.8'2 82-19967
ISBN 0-471-09594-X

Printed in the United States of America

10 9

PREFACE

As teachers of human factors are well aware, the demand for human factors specialists has ballooned in recent years with a concomitant growth in student enrollment in beginning human factors courses. Students in such diverse areas as psychology, industrial engineering, business management, and other disciplines are seeking instruction in human factors even though most of them will not become full-time human factors professionals. This increased demand has inevitably created new texts, whereas years ago only one text dominated the field.

We believe that it is crucial to distinguish between a text and a handbook. Many traditional human factors texts overwhelm the student with a wealth of details and rules of thumb that students are expected to follow automatically. While details are necessary for the human factors professional, they are often lost in a beginning human factors course. This is especially true for the student who seeks only to learn the important general principles of human factors. Far too often the amount of detail incorrectly leads the student to believe that human factors is just "knobs and dials" with simple solutions that can be found by looking up the proper table. We believe that understanding how the human factors professional goes about solving problems is far more important than learning how to look up simple solutions in tables, especially since such solutions are never valid for all situations. This text teaches the student to think like a human factors specialist but deliberately omits, simplifies, or relegates to appendices much of the details that are more appropriate for handbooks.

This does not mean that the text has been watered down to a low level. Indeed, the text is intended for college seniors and first-year graduate students. Concepts, rather than details, are stressed and some students will undoubtedly find this text more difficult than others of a more traditional nature. They will also find it more challenging and more thought provoking. Answers cannot always be found in convenient tables.

Some of the most important concepts stressed in this text are theoretical models. There is a new look in human factors today with practitioners realizing that theory helps to solve practical problems. The student of human factors must become acquainted with such theoretical concepts as the theory of signal detection, information theory, feedback and control models, and decision models—to name only a few. Human factors specialists can be crudely arranged along a spectrum ranging from psychologists—who are so concerned with theory that they never solve any problems—to engineers—who are so anxious to solve problems that they solve the wrong one. We balance these two extremes while believing that judicious application of theory offers the best hope for rapid progress in human factors. This text reflects that belief with chapters written by psychologists, engineers, and a computer scientist.

We have tried very hard to produce a text that looks to the future of human factors rather than merely surveying its past. To do this, we have emphasized important new topics such as human computer interaction, human information processing, environmental aspects of human factors, and legal issues in human factors. We have also emphasized the most recent research findings in human factors, especially from the last five years. Human factors is a vigorous field and only a text that stresses the cutting edge of recent findings can do it justice.

Certain propaedeutic devices have been used to maintain student interest without sacrificing scholarly accuracy. Foremost among these are the inclusion in every chapter of boxed material that illustrates how even the most abstract theoretical topics are intimately related to real-world events. Human factors is, if anything, a pragmatic discipline and authors have a responsibility to relate its models and methods to real-world examples that students will find meaningful. Other boxes relate to endemic human factors problems—such as defining and achieving system goals—that cross specific content areas. These boxes try to integrate the text so that students will not think that issues in one chapter (e.g., macroenvironments) are unrelated to other chapters (e.g., visual displays). In fact, both these chapters feature a discussion of link analysis that can be applied both to design of displays and to crime prevention. We have stressed the system approach to human factors with many cross references among chapters to show the student that general principles of people-system relationships are more important than the specific applications used to illustrate these principles.

We note with pleasure that more and more women are appearing in our human factors classes and profession. In order to encourage this trend, we have gone to great lengths to avoid sexist language. Thus, the traditional phrase "man-machine system" has been replaced by "person-machine system." This may sound awkward to those accustomed to the traditional phrase but we believe that the improvement in accuracy justifies some negative transfer. Indeed, there are places in the text where elegance of English has given way to awkward phrases like "his/her" in the interest of avoiding even the appearance of sexist language. However, we have refused to edit or delete reports of research that some reviewers found sexist. We hope that readers can distinguish between honest reporting of research versus any sexist opinions held by the authors.

There is a limit to how much human factors can be learned by reading any text. We have prepared an accompanying Workbook with exercises for each chapter and encourage students and instructors to use this important resource.

It takes far more than two authors to produce a book and we are pleased to acknowledge the assistance of many others. First, we thank Professors James Buck and Hubert Dunsmore for contributing chapters. Human factors is interdisciplinary and their help has created a text that reflects the expert knowledge of industrial engineering and computer science. Two sturdy reviewers, William Howell and Irving Goldstein, struggled with early versions of each chapter and offered constructive and helpful criticism. We thank them for their Herculean efforts and are pleased that the final product meets with their approval. Several anonymous reviewers commented on individual chapters and we gratefully acknowledge their assistance. The staff at John Wiley & Sons provided expert professional aid. Our first editor, Jack Burton, and his assistant, Connie Rende, were responsible for getting this project started and his enthusiasm and ideas guaranteed a good beginning. His duties were assumed by Mark Mochary and his assistant, Maryellen Costa, who kept the project on track and on time. It is risky for authors to list publishing staff other than their editor(s), since so many people worked on the book that some might be inadvertently omitted. However, the staff at Wiley was so helpful that we take this risk and apologize in advance for any accidental omissions. Jan Lavin skillfully supervised production, juggling the many components of a book so that the whole emerged intact. Our designer, Loretta Saracino, worked closely with us to produce an attractive format for the ideas expressed in the text. Our picture editor, Kathy Bendo, looked long and hard for suitable photographs. Susan Giniger edited the manuscript, removing any inadvertent sexist phrasing. Finally, we thank our wives and families who tolerated long hours and some neglect while we were busy preparing this book. We look forward to your reactions and suggestions.

Barry H. Kantowitz
Robert D. Sorkin

CONTENTS

PART ONE

INTRODUCTION

1 SYSTEMS AND PEOPLE

It is dusk. The jet plane descends swiftly toward the ground. Inside, passengers stretch and prepare to gather up their belongings, anxious to leave the aluminum cocoon. Wisps of grey clouds float by the windows as the controlled descent continues. The touchdown comes sooner than expected. It takes the passengers a few minutes to discover that their plane has made a perfect landing in the bay. No one is harmed but the passengers are very upset. Many fear that the plane will sink before they can escape. The crew tries to maintain order but the crowd is on the verge of hysteria. Some passengers mistakenly inflate their life vest before leaving the plane, impeding progress toward the exits. The plane is evacuated without loss of life but the passengers have been badly frightened. It has been a terrifying experience. Some will never get on another airplane for the rest of their lives.

The operators on the 11 to 7 shift on March 28, 1979 have the plant at 97% capacity. The Integrated Control System is in full automatic mode. The shift supervisor and two auxiliary operators are transferring resin from condensate polisher tank No. 7 to the regeneration tank. However, a resin blockage is hampering transfer so attempts are made to clear the line. Suddenly at 04:00:37 there is a total loss of feedwater and a turbine trip. All emergency pumps start automatically and the reactor continues to run at full power for another eight seconds. Then the electromatic relief valve opens as the pressure reaches its setpoint and the reactor trips. Three Mile Island has begun.

You are taking a shower. In order to adjust the water temperature you must reach through the spray. The water is too cold so you rotate the left-hand control counterclockwise. Too far. A scalding stream strikes a particularly delicate portion of your anatomy. &#"$#*&%%:!!!

Human factors, like a vacant seat in class, is most noted when it is absent. When systems function properly, few congratulate the human factors specialist for a job well done. But when disaster strikes there is a sudden interest in using the knowledge of the human factors expert to apply a quick fix. The foregoing examples refer to errors in design and operation that could have been prevented by sound human factors analyses. But such analyses must be performed as part of the system design from the very inception of the design process.

This chapter introduces the basic principles of human factors. Human factors is the discipline that tries to optimize the relationship between technology and the human. Any person-machine system is fair game for human factors. But technology goes beyond machines. Buildings, rooms, noises, and heat also affect how the human interfaces with technology. Anywhere you find technology and people interacting, there is a need for human factors.

PERSON-MACHINE SYSTEMS

Although it is possible to define a system formally with mathematical symbols and set theory (Hall & Fagen, 1956), we will use an informal verbal definition. A person-machine system is an arrangement of people and machines interacting within an environment in order to achieve a set of system goals. The human factors specialist tries to optimize the interaction between people and machine elements of the system, while taking the environment into account. A representative schematic of a person-machine system is shown in Figure 1-1. The right half of the diagram represents the machine subsystem as it appears to the human factors specialist. Visual and other displays (see Chapters 7 and 8) represent the internal equipment status in a form the human can understand. Controls

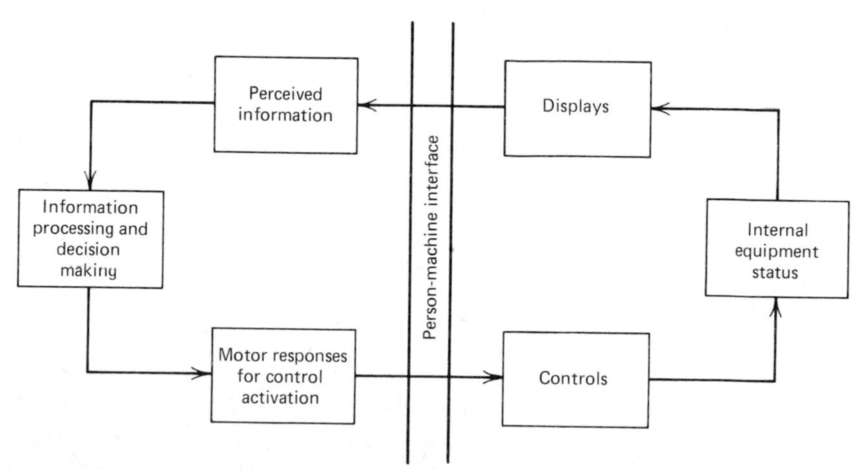

Figure 1-1. Representation of the person-machine system. (From Meister, 1971.)

(see Chapter 10) allow the human operator to make changes in the internal equipment. These two boxes represent some of the most important human factors aspects of the machine and everything else is bunched into a single box labeled internal equipment status. The engineers who designed this internal equipment spent months arranging detailed subsystems that are not represented directly in Figure 1-1. From their viewpoint this is a gross oversimplification of the machine subsystem and it is almost insulting to represent all their effort by a single box. This oversimplification does not mean that the human factors specialist fails to appreciate the importance of this engineering accomplishment. It does mean that to a large extent the detailed inner workings of the machine—the gears and cogs, integrated circuits, and digital logic—are irrelevant. The human factors specialist must design the specifications for displays and controls and also help the engineering team to ensure that system dynamics are compatible with human skills. Other specialists are responsible for implementing these specifications and the human factors specialist does not particularly care whether a necessary time delay is achieved by a bucket brigade circuit or by a delay line.

The human subsystem is represented by the left side of Figure 1-1. Information from the displays is perceived (see Chapters 3 and 4). This information is processed (Chapter 6) and decisions are made (Chapter 14). Motor responses are made (Chapter 5) to alter control settings. Figure 1-1 is not a complete representation of the human. There is nothing to show the workings of the brain and central nervous system directly. While the brain is important, the human factors specialist does not need to know what is happening in individual neurons in order to optimize the system. Figure 1-1 highlights those portions of the person-machine system that are of greatest importance to human factors. It neglects other important aspects of machines and people that are less crucial to human factors.

Perhaps the most important part of Figure 1-1 is the vertical line separating the machine subsystem from the human subsystem. This line represents the interface between human and machine. Information flows across this interface in both directions: from the machine to the person and from the person to the machine. Thus, Figure 1-1 represents a closed-loop system (Chapter 12) since you can start at any point in the diagram and trace your way around the complete system, eventually returning to the starting point. Much of this book is concerned with getting information across this interface. This can be accomplished more readily by changing the machine than by changing the human. Displays and controls must be designed to match human capabilities (see Figure 1-2).

Defining System Goals

Before anyone can start to design a person-machine system, they must know what the system is supposed to do. This seems obvious at first, but we will soon see that systems have several goals and that some of these goals compete with other goals. It is often impossible to accomplish all

If the Soap Falls Out of the Bathtub, Try This

Man in bathtub snaps fingers and pet Golf-Beaked Soap Hawk (**A**) hits soap (**B**) into baseball glove (**C**)–Rebound of spring (**D**) causes glove to throw soap past cat (**E**) into trough (**F**)–Breeze from flying soap causes cat to catch cold–She sneezes, blowing soap off trough–Soap hits string (**G**), which pulls trigger of pistol (**H**), shooting ram (**I**) against small car (**J**), into which soap has meantime fallen–Car carries soap up platform (**K**), dumping it back into tub and man can continue with his bath.

Figure 1-2. Person-machine system with no human factors analysis. (From *The Best of Rube Goldberg,* compiled by Charles Keller, Prentice-Hall, 1979.)

the desirable system goals. When this happens, it is better to discover it at the start of the design process rather than after the system goes out the door to customers. In many firms, systems goals are specified by management (or by a customer) and the human factors team is simply given a set of goals to accomplish. Trade-offs among goals are seldom clearly specified. The team may produce a system that realizes the ambiguous goals they were given, only to be castigated later on when management realizes those goals were not what it really wanted. Thus, a prudent human factors team will spend some time analyzing a list of system goals and even going back to management for clarification when goals conflict.

An example will make this conflict more visible. Let us suppose we have been hired to create a list of system goals for the postal service. We can start by writing down several desirable goals.

1. Speedy service.
2. Frequent mail delivery.
3. Home delivery to all U.S. residents.
4. Low cost.
5. Guaranteed transit time.

Already we can see some problems with this very short list. The first problem is ambiguity. Precisely how speedy is "speedy service?" Does delivery to "all U.S. residents" include hermits living in caves on the slopes of Mount Hood? These goals must be refined to be more specific. For example, frequent mail delivery could be defined as twice a day. In Europe at one time housewives mailed in their daily grocery orders that were delivered the same afternoon. Or frequent mail delivery could mean once per day seven days a week. Is 20 cents a letter low cost?

Even after these ambiguities are cleared up, the problem of conflict among goals remains. A system that delivers mail to 95% of U.S. residents will incur very high costs in rural or other low-density areas. This conflicts with the goal of low cost. Should residents in these areas pay a surcharge for daily delivery, should they have less frequent delivery than urban areas, or should these rural delivery costs be absorbed by the entire system? Management may lack the information required to decide. Thus, instead of immediately setting off to design a system, the human factors team may first have to specify the trade-offs among conflicting system goals. For example, home delivery to 80% of residents may cost 18.5 cents per letter while delivery to 95% might raise the cost to 35 cents. Without knowledge of these trade-offs a rational decision (see Chapter 14) cannot be made. The problems caused by conflicting system goals are so important that later chapters will occasionally contain a special "box" called Conflicting System Goals that illustrates the problem within a particular system context.

Levels of Technology

There are two schemes for classifying levels of technology. First, we can organize technology by its general level of complexity and detail. One such scheme (Carson, 1970) has three levels of complexity. The first and lowest level concerns relatively simple responses to components of the system such as its knobs and dials. The focus of interest is the individual element rather than general system properties. The human factors specialist working at this level must keep track of many small details. At the second level the flow of behavior is more important than individual components. This would include tasks like tracking (see Chapter 5) and driving a car along a winding road. Here the specialist is concerned with more general human capabilities, such as avoiding an overload of information. The third and highest level involves the total system and decisions as to what tasks should be performed by people and what other tasks by machines. Machines are broadly defined and this third level also includes relationships between people and their environments (see Chapters 16 to 19). For example, the famous architect Le Corbusier defined a house as "a machine to live in."

A second classification scheme creates levels of technology by focusing on the contributions of people and machines to the total system. At the lowest level, the person supplies both power and control of the system

Figure 1-3. Power Mechanic by Lewis Hines. This famous photograph illustrates the first level of technology where the human supplies power and control. (George Eastman Housc.)

Figure 1-4. This welder illustrates the second level of technology. The tool supplies power while the human controls it. (Michael C. Hayman/Photo Researchers.)

Figure 1-5. A paper-making plant illustrates the third level of technology. The plant supplies power and information but the human is still in control. (Courtesy Westvaco Corporation.)

Figure 1-6. The cockpit of a Boeing 757 jet illustrates the highest level of technology. When the autopilot is engaged the airplane supplies power, control, and information while the human monitors the operation. (Courtesy of Boeing.)

(Figure 1-3). A human wielding a shovel is functioning at this lowest level of technology. The first improvement has the machine supplying power, while the human still exercises control (Figure 1-4). A human operating a punch press is one example of this level of technology. At the next level the machine supplies power and information while the human still controls (Figure 1-5). Any factory where people read displays and turn controls functions at this level. Finally, the highest level of technology has the machine supplying power, information and control while the human monitors the operation (Figure 1-6). An automated plant run by computers is an example of this level.

At which of the four levels is human factors most needed? Most people would select the fourth level, since it uses the most sophisticated technology. People tend to associate human factors with high technology. However, this is incorrect. Human factors is needed at all four levels. As long as people are required to function in systems at all four levels, society gains improved productivity by using human factors. This text contains examples from every level of technology showing how human factors helps.

It is true that human factors is more difficult to apply at higher levels, simply because great system complexity implies more choices for system designers. This large number of choices takes longer to analyze. Since there is a trend to more sophisticated technology, where the human is a monitor of system behavior rather than an active controller, more human factors efforts will be aimed at this level.

But before the human factors specialist can get down to work, a more general philosophical question must be raised. One approach to the human as monitor is to realize that most of the time the system works quite well without human intervention. Indeed, the system often will work better if the human keeps hands off. For example, in the Three Mile Island accident the automatic safety systems worked as designed and turned on the emergency pumps. The human operators goofed and manually turned off the pumps. Therefore, one possible conclusion is that the human should be eliminated entirely from systems, or if that is not possible—for example, there are legal requirements that operators be present in nuclear power plants—then the person's role should be minimized. This will, on the average, decrease the opportunity for human error and so increase system reliability. Furthermore, it will lower operating costs by substituting machines for people.

Another philosophy is to keep the human operator involved as much as possible in the system, even if this means creating artificial tasks, such as logging in display readings by hand, to keep the person busy. Thus, should the machine part of the system fail, the human will be able to leap in and fix the problem. (The minimization philosophy assumes that the human probably won't be able to fix the problem anyway.) Thus, this philosophy argues that if you want to land a probe on the moon, it is best to have human astronauts along in case something unexpected goes wrong. The human is clever and adaptive and can often solve unanticipated problems.

We can't tell you which philosophy to believe or which your company should adopt. However, our own personal beliefs bias us against including a human in a system when there is little for the person to contribute. In fact, such an undemanding job is quite stressful (see Chapter 19) and unpleasant. We believe that the human factors expert has an obligation to provide meaningful work for people. This leads us to the next topic, dividing work between people and machines.

Allocation of Function

Allocation of function is the process whereby the designer decides which tasks (or functions) should be given (or allocated) to the machine subsystem and which to the human subsystem. Many handbooks and texts have tables telling the reader which kinds of tasks are performed better by people and which by machines. For example, a machine is better at performing 100 arithmetic calculations within 10 seconds or lifting 500 pound castings. A person is better at recognizing patterns embedded in noise. While such lists are fun to read (and write), they offer little assistance in practical situations (Chapanis, 1970). In real life the designer tries to delegate as many functions as possible to machines. The human gets stuck with the leftover functions. This approach appears to be considerably less elegant than the rational deliberative process where the designer carefully decides if the person or the machine is better suited to some specific task. But, it is more sensible. The reliability of machines can be increased less expensively than can the reliability of people (see Chapter 2) by putting extra components in parallel. At one time, the machine subsystem was almost entirely hardware so that subsequent changes were difficult and expensive. Now, software (computer programs, see Chapter 13) is as important as hardware in many machine subsystems. Software can be changed fairly easily. The human was allocated some functions in older systems to allow flexibility for change. Now this flexibility can be achieved through software modification making it practical for the designer to allocate even more functions to the machine.

Therefore, the major decision in allocation of function involves checking that the human is left with a reasonable set of tasks. These tasks should neither overload nor underload the operator's capabilities. In the case of overload it may be necessary to add a second human operator if the machine cannot assume more tasks. In the case of underload, it may be advantageous to either take some tasks away from the machine or to have the human perform them in parallel with the machine. In this latter case, the parallel tasks must be structured so that the human feels they are necessary and important.

Some systems give the human the option of changing allocation of function on-line (in real time as the system is operating). A simple example is the cruise control on your car. Engaging the cruise control allocates part of the workload, maintaining a constant vehicle speed, to the machine subsystem. There are times when such an allocation is desirable and other times when it makes more sense for the human to control vehicle speed. The human controls allocation of function by deciding whether or not to engage the cruise control.

Achieving Systems Goals

In order to achieve system goals, designers must proceed systematically. The following convenient set of seven relevant questions (Meister, 1971) can serve as a checklist for human factors specialists.

1. What inputs and outputs must be provided to satisfy system goals? To answer this question, it helps to think of the entire system as a big "black box." Inputs enter the box and outputs are emitted from the box. In the postal service example, inputs would be letters or parcels that are to be mailed. Outputs would be mail that has been delivered. Note that subsystem inputs and outputs—for example, mail to be placed in a delivery truck—lie within the general system box and so do not qualify as system inputs and outputs.

2. What operations are required to produce system outputs? In the postal example, some necessary operations would be sorting mail and transporting it.
3. What functions should the person perform within the system? This falls under the category of allocation of function. In the postal example, mail could be sorted by people or by machines or by both together. At one time postal clerks were required to memorize sorting schemes and mail was sorted entirely by people tossing mail into bins. Now in many larger post offices people sit at machines and key in zip codes that the machines use to send mail to the appropriate bin or location. It is also technologically possible to have machines recognize typed characters and sort some mail without human assistance. However, people are still better than machines at reading handwritten addresses.
4. What are the training and skill requirements of human operators? More complicated machine subsystems do not necessarily imply lower personnel requirements. For example, the newer nuclear power plants with computer-driven control rooms require additional training because the operator has many more choices in calling up information on the CRT (cathode ray tube) display.
5. Are the tasks demanded by the system compatible with human capabilities? As mentioned earlier, the designer must avoid tasks that create either overload or underload for the human operator.
6. What equipment interfaces does the human need to perform the job? This is a traditionally important area of human factors. The specialist will devote a great deal of time to determining optimal displays and controls, operating procedures, and informational diagnostics.
7. Does the person help or hurt the machine subsystem and vice versa? This is the final check. For example, computers operate much faster than people so that the machine subsystem may not allow the human subsystem enough time to make decisions. Similarly, the slow human may force the machine to wait for a human response instead of doing other tasks.

Unless the system is very simple, the human factors specialist will not be able to answer all seven questions in a single pass. Instead, the best answers for now will be listed and then as more information is acquired, the list will be gone through again and again. This iterative procedure, whereby the same design process is repeated, is typical of human factors methods. We seldom know enough to get all the answers right the first time. Indeed, it is not unheard of for a system to go out the door with some questions unanswered due to lack of time or other resources. The human factors specialist often cannot provide an optimal solution within the resources allocated to human factors in a project and must settle for finding a workable solution. This may be frustrating when the specialist feels that a better answer could be provided if he or she only had another month, more computer time, more technicians, more research dollars, etc. But human factors is a pragmatic discipline and resources are always finite and limited. Human factors experts who insist on finding the best answer regardless of practical constraints wind up teaching human factors instead of doing it.

HONOR THY USER

The first commandment of human factors is "Honor Thy User." If there is one thing you should remember 25 years from now about human factors it is "Honor Thy User." Everything else merely embellishes this dominant theme.

Before you can honor the user, you must first know who the user will be. Different user populations have different human factors requirements. When Japan first started manufacturing automobiles they were designed for Japanese citizens. This created a problem when the cars were exported to the United States. American men were unable to depress the brake pedal without simultaneously depressing either the clutch or accelerator pedals. In hindsight, the reason for this design error was clear. Americans are physically larger than Japanese and have bigger feet. While pedal spacing was satisfactory for the smaller Japanese foot, the pedals were too close together for the larger American foot. Since American feet had not been specified in the original design characteristics, the first commandment of human factors had been violated.

This same error can occur in more sophisticated ways. A large American telecommunications company, highly regarded for its human factors efforts, was designing a system for presenting prerecorded messages by telephone at times specified by the user. Since the hardware for this effort was specified, the human factors designers had to come up with the equivalent of a miniature computer language where the user could enter appropriate commands. So computer scientists were selected to create this new minilanguage. After several months of effort they came up with a powerful language that could efficiently control prerecorded messages. Furthermore, being prudent designers they tested their new system by using secretaries as test subjects. Since this device was intended for use by secretaries, this was the ideal test population. Alas, none of the secretaries were able to master the new language. After some thought, the designers decided they needed a greatly improved instruction manual and so spent several more months perfecting a tutorial manual with many examples and illustrations. A second test found a few secretaries able to use the device but most were still baffled. The designers were very unhappy and questioned the successful secretaries carefully. It turned out that all of them had prior experience with computer programming. Now the problem was clear.

The designers, being computer programmers themselves, had created a system that was optimal for anyone who could think like a computer programmer. But the typical user lacked computer skills. The designers created a new inefficient language that required many more instructions to complete any given recording schedule. But this new language was easy to learn and easy to use. A final test showed that the great majority of secretaries could now accomplish the task. The designer's training, which had stressed efficiency in minimizing the length of a

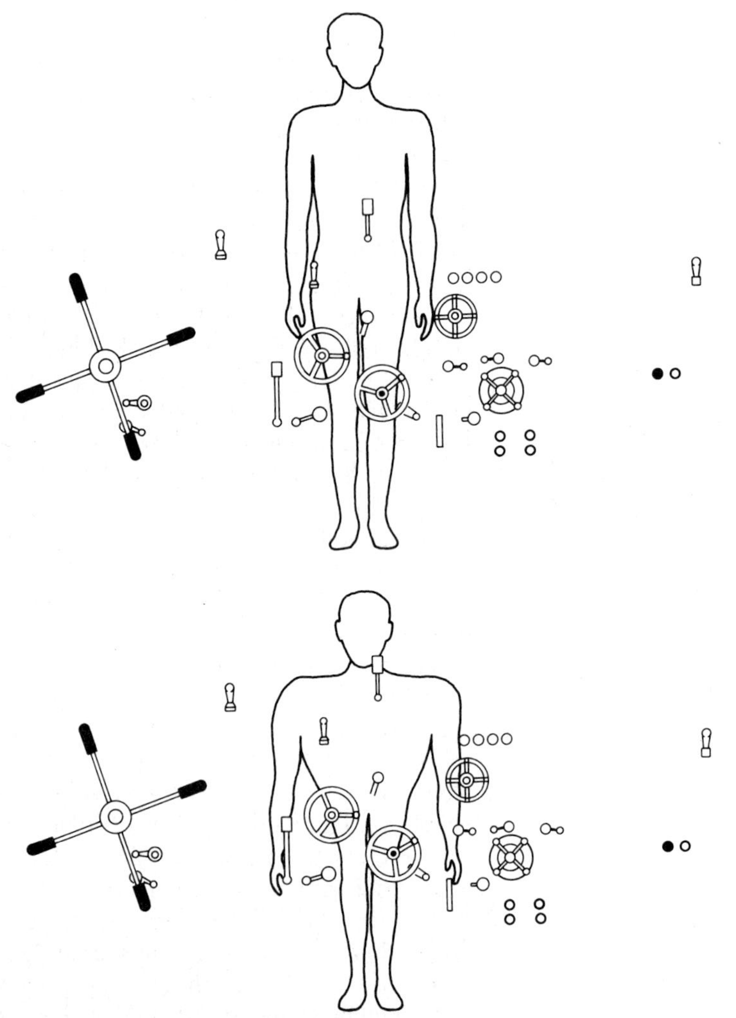

Figure 1-7. The controls of a lathe in current use are not within easy reach of the average man. They are placed so that the ideal operator should be 1372 mm ($4\frac{1}{4}$ ft) tall, 640 mm (2 ft) across the shoulder, and have a 2348 mm (8 ft) arm span. (From Applied Ergonomics, Guildford, Surrey, United Kingdom: IPC Science and Technology Press, Ltd., 27, 1969, Vol. 1.)

program, had tricked them into violating the first commandment. They had designed for themselves rather than for the user. But disaster was prevented because the designers were wise enough to test their product before sending it out the door. So an important aspect of honoring the user is to test using the same population that will eventually operate the system.

The last violation to be discussed is the design of a lathe (Figure 1-7). Here the user population is known. But, as Figure 1-7 shows, the lathe was designed for a human with a very short trunk and extremely long arms. Few if any humans have these body dimensions. This example demonstrates that it is easier to change the machine than to change the person who must operate it.

MODELS

If the human factors specialist could find an appropriate data table for any design situation, there would be little need for models. Since this will never happen, there must be a way to bridge the gap between design problems and solutions. One very important use of models allows the designer to make a reasonable guess in the absence of data.

Models are abstract representations of systems or subsystems. They can be physical, mathematical, verbal or combinations of these. A scale model airplane in a wind tunnel is a physical model as is a set of electronic components wired together to represent a tornado. Almost any equation can be a mathematical model of a system. When quantitative models are not accurate or must be too complex, verbal descriptions are substituted. This is especially true when the human is the subsystem of interest since the human is so adaptive and learns so quickly that a quantitative model that applies to beginners may not work for trained professionals. In general, quantitative models are preferable to qualitative models even if the quantitative model is not completely accurate.

Models vary in degree of abstraction. Least abstract models are based on observation of real world events. Figure 1-8 shows an operational sequence diagram (Kurke, 1961) for getting up in the morning. Special symbols are used to classify events: a circle represents received information, a rectangle represents operator action and so forth. As you work your way through the diagram the model seems so realistic (and so concrete), you may not realize that it is a model rather than a set of data. However, the structure imposed by the various symbols transforms the observations on which the information flow is based into an elementary model. This model represents the minimal level of abstraction but nevertheless, it still goes beyond a literal and unorganized repetition of data.

Most of the models you will encounter in this text are more abstract than operational sequence diagrams. Many involve some mathematics, although we have tried to keep the necessary equations as simple as pos-

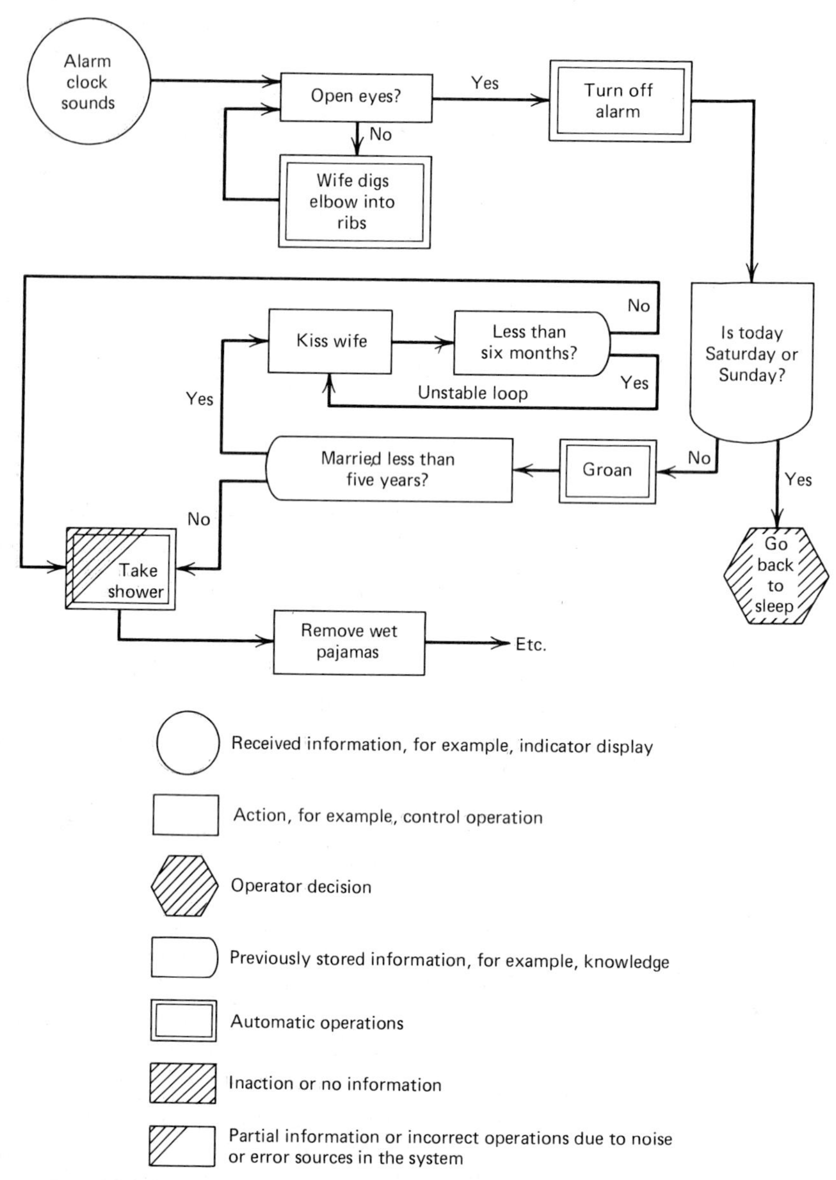

Figure 1-8. Operational sequence diagram for awakening in the morning.

sible. Some of the models you will meet include the theory of signal detection (Chapter 3), information theory (Chapter 5), the limited-capacity channel (Chapter 6), feedback and control models (Chapter 12), and decision models (Chapter 14). So, as this preview demonstrates, models are a rather important aspect of human factors. Models perform three important functions. First, they allow us to make predictions in situations where relevant data are not available. Second, they guide research by suggesting pertinent experiments. Third, they provide a framework that helps to organize data.

Most models attempt to describe and predict the behavior of systems under normal operating conditions. But there is a special class of models that specify the upper limit of performance for a system. These are called optimal, or normative, models. They are not models of behavior and especially do not represent human performance (Kantowitz & Hanson, 1981). Instead, they state how well a system can perform under ideal conditions using optimum strategies. Thus, we can compare actual system performance to the performance of an optimal model to discover how much room there is for improvement. For example, information theory tells us if a code is optimal but does not tell us how to achieve the best coding. Comparing the actual code to the mathematical prediction helps us to decide if additional effort at increasing coding efficiency will be worthwhile. Some other optimal models you will encounter are the theory of ideal observers, which offers a normative basis for signal detection, and Bayes' theorem, which offers a normative basis for decision making. You will find that human performance falls far short of optimal performance.

ENGINEERING ECONOMICS

It has often been said that the last symbol in any proper set of engineering equations is the dollar sign. The human factors specialist must be aware of the economic implications of human factors decisions. One issue that often arises is based on a comparison of original system costs (hardware plus installation) and operating costs (including maintenance, updating system software, etc.) Human factors decisions affect both original and operating costs. The naive designer will weigh both kinds of costs equally by adding them to arrive at a total cost for the life of the system. As we will soon demonstrate, this simple approach is incorrect because it neglects the time value of money.

It is an elementary economic fact that a dollar in your pocket now is worth more than the promise of a dollar in your pocket tomorrow, even if that promise is 100% reliable. If you don't need the dollar in your pocket today you can rent it to someone who does and they will pay you interest (rent) for the privilege of using your dollar. If you rent your dollar for one year at an interest rate of 10% at the end of the year you will have \$1.10. Renting for a second year will get you interest on the new amount of \$1.10 of \$.11. (This is called compounding.) There is a simple equation that relates the future values (FV) of a sum of money N years from now to its present value (PV):

$$FV = PV(1 + i)^N$$

where i is the interest rate paid. Solving this equation for PV yields $PV = (1/(1 + i)^N)FV$.

TABLE 1-1 *10.00 Percent Compound Interest Factors*

	Single Payment	
N Periods	Compound Amount Factor $(1+I)^{**}N$	Present Worth Factor $\frac{1}{(1+I)^{**}N}$
1	1.1000000	.9090909
2	1.2100000	.8264463
3	1.3310000	.7513148
4	1.4641000	.6830135
5	1.6105100	.6209213
6	1.7715610	.5644739
7	1.9487171	.5131581
8	2.1435888	.4665074
9	2.3579477	.4240976
10	2.5937425	.3855433
11	2.8531167	.3504939
12	3.1384284	.3186308
13	3.4522712	.2896644
14	3.7974983	.2633313
15	4.1772482	.2393920
16	4.5949730	.2176291
17	5.0544703	.1978447
18	5.5599173	.1798588
19	6.1159090	.1635080
20	6.7274999	.1486436
21	7.4002499	.1351306
22	8.1402749	.1228460
23	8.9543024	.1116782
24	9.8497327	.1015256
25	10.8347059	.0922960
26	11.9181765	.0839055
27	13.1099942	.0762777
28	14.4209936	.0693433
29	15.8630930	.0630394
30	17.4494023	.0573086
31	19.1943425	.0520987
32	21.1137767	.0473624
33	23.2251544	.0430568
34	25.5476699	.0391425
35	28.1024368	.0355841
36	30.9126805	.0323492
37	34.0039486	.0294083
38	37.4043434	.0267349
39	41.1447778	.0243044
40	45.2592556	.0220949

Note. $^{**}N$ is exponent N.

The factor $1/(1 + i)^N$ that multiples FV is called the present worth factor. Present worth factors for an interest rate (i) of 10% are given in Table 1-1. At a 10% interest rate the present value of $1 20 years from now is 15 cents.

Knowing how present value is calculated we can return to the foregoing example and determine the lifetime cost of a system based on original cost and operating costs. We wish to compare two airplane navigation aids, which we cleverly call NAVAID A and NAVAID B. NAVAID A costs $20,000 and B costs only $12,000. But B has a useful life of only 3 years versus 6 years for A. Assumed operating costs for the two systems are as follows (Morlok, 1978 p. 358):

Year	NAVAID A	NAVAID B
1	4000	4000
2	4200	4250
3	4400	4550
4	4600	–
5	4900	–
6	5400	–

Incorrectly summing these columns and the original cost gives totals of $47,500 for NAVAID A and $49,600 for two of NAVAID B. Therefore, a naive designer would conclude that NAVAID A is less expensive over the life of the system. But this neglects the time value of money. The correct analysis, using present worth factors from Table 1-1, for NAVAID A is as follows:

$$\$20{,}000 + \$4000 + \$4200(.909) + \$4400(.826) + \$4600(.751) + \$4900(.683) + \$5400(.621) = \$41{,}610.$$

For NAVAID B where a second unit must be purchased in the fourth year we get:

$$\$12000 + \$4000 + \$4250(.909) + \$4550(.826) + (\$1200 + 4000)(.751) + \$4250(.683) + \$4550(.621) = \$41{,}373$$

Thus, NAVAID B is slightly less expensive over the life of the system. While this example demonstrates the importance of present worth factors, it is somewhat oversimplified since it neglects important aspects of the world like inflation, tax laws, and depreciation. For a more complete treatment of the economic aspects of engineering see texts by Parks (1973) and Newman (1980).

One important and neglected topic in engineering economics was the subject of a presidential address to the Human Factors Society (Parsons, 1970) on the value of life and death. Placing an economic value on death may seem crass to some, but it is done everyday when manufacturers decide how much safety equipment should be included in their products. (See Chapter 20 for a discussion of legal aspects of product safety.) The economic utility of some feature, say safety belts in vehicles, cannot be established without estimating the value of a human life. Estimates of the economic value of a life range from 30 pieces of silver to million-dollar insurance policies. Another possibility is the sum of a person's lifetime earnings. Jul (1969 cited by Parsons, 1970) suggested a comparative approach based on expenses used to prevent death from various causes:

Space exploration	$1,000,000,000
Civil aviation	$1,000,000
Automobile	$1,000
Malnutrition	$1

Parsons (1970) estimated that by choosing to return the Apollo astronauts to earth, U. S. taxpayers have spent about half a billion dollars per astronaut compared to a one-way mission where the astronauts expired in space after completing their job. While Parsons was not advocating a kamikaze mission, it is clear that the life of an astronaut is much more highly valued than the life of an ordinary citizen.

Parsons offered several suggestions for taking life and death into account more directly in human factors research and practice. He advocated research into "caution" behavior (e.g., highway accidents and drivers hitting trains at railway crossings), selection of medical patients for lifesaving treatments, lives saved as a criterion for military defense systems, population control, and the amount of pain accompanying death. Not all these suggestions have been acted on vigorously. The dollar sign is still the ultimate engineering symbol. But we believe that Parsons was correct. The ultimate goal of human factors must be to balance human welfare and the dollar sign. We ask you to keep this goal in mind as you read this book.

SUMMARY

Human factors is the discipline that tries to optimize the relationship between people and technology. Information flows in two directions across the human-machine system interface: from the human to the machine and from the machine to the human. Since the design of the human is fixed, the human factors specialist must alter the design of the machine to optimize the flow of information in both directions across the human-machine system interface.

Before any person-machine system can be designed, the goals of the system must be clearly defined. Analysis of system goals often reveals conflicts that need be resolved before a design can be formulated. The prudent human factors specialist clarifies trade-offs among system goals before continuing the design process rather than after a design is finished.

Allocation of function refers to the division of tasks between people and machines. While it is tempting to allocate as many functions as possible to the machine, this approach works only if the human is left with a reasonable and coherent set of tasks. Some systems allow the operator to decide on-line how much work will be performed by the machine subsystem and how much by the human.

The basic principle of human factors is "Honor thy user." The best systems are those that meet the needs of the person who will use the system. This goal can be accomplished by using models to help design the system. Finally, the designer is limited to finite resources and must take engineering economics as a part of system goals. A "perfect" system that costs 100 times more than those sold by competitors may never be used. The designer has the difficult tasks of balancing economic goals and human welfare.

APPENDIX

Sources of Human Factors Information

Your first source of human factors information is obvious: you are holding it now. In particular, the References at the end of this book can lead you to more knowledge. We have been careful to include both specific references (mostly journal articles) as well as some general references (books and review articles) in every chapter. We have also tried to use recent references because technology in human factors changes rapidly. Since technical reports are difficult to obtain, we have kept references to them to a minimum and instead have used published journal articles you will be able to find. An additional advantage of published articles lies in the review process that articles must survive before reaching publication. Quality control for published articles is generally better than for technical reports.

Ergonomics Abstracts, a journal published by Taylor & Francis Ltd (4 John Street, London WCIN 2ET, England), available in most libraries, is the single best source for additional information. Ergonomics is a field that includes human factors and industrial psychology. Table A-1 shows the classification scheme used by *Ergonomics Abstracts;* you can see that it covers everything in this text and then some. Furthermore, each issue contains a List of Applications (Table A-2) classifying abstracts by industry. The numbers refer to abstracts contained in the journal (Table A-3). This system offers a convenient and thorough way to find new information about ergonomics and we recommend using it. (Indeed, there is a Workbook project based on using *Ergonomics Abstracts.*) This is the most efficient way to keep up to date in human factors.

Another good way to get information is to browse through the major human factors journals. Many journals publish articles related to human factors as you can tell by skimming through the References section of this text and/or an issue of *Ergonomics Abstracts*. However, there are three important journals that are dedicated to human factors research. In the United States, the most important journal is appropriately called *Human Factors* – the official journal of the Human Factors Society – and can be obtained from the Society, Box 1369 Santa Monica, California 90406.

(*Text continued on page 29*)

TABLE A-1 *Abbreviated Outline of Classification*[a]

ERGONOMICS

1.0.0 General References

MAN AS A SYSTEMS COMPONENT

2.0.0 Perceptual Input Processes

- Basic Visual Data
- Basic Auditory Data
- Basic Kinaesthetic Data
- Basic Tactile Data
- Basic Vestibular Data
- Basic Olfactory and Gustatory Data
- Basic Thermal Perceptual Data
- Basic Pain Data
- Inputs: Choice of Modalities

3.0.0 Central Processes

- Short Term Memory
- Long Term Memory, Learning
- Problem Solving
- Decision Making
- Attention and Set
- Search and Scanning
- Time Perception

4.0.0 Basic Motor Processes [Psychological Components]

- Positioning Movements
- Repetitive Movements
- Tracking Movements
- Serial Movements
- Preferred Limb
- Manual Dexterity
- Involuntary Reflexes
- Time and Effector Movements
- Organization of Effector Processes

5.0.0 Perceptual-Motor Performance

- Reaction Time
- Processing and Transmission of Information

6.0.0 Factors Affecting Perceptual-Motor Performance

- Arousal and Vigilance
- Fatigue
- Perceptual (Mental) Load
- Knowledge of Results
- Monotomy and Boredom
- Sensory Deprivation
- Sleep Deprivation
- Anxiety and Fear
- Isolation
- Ageing
- Drugs and Alcohol
- Circadian Rhythm

7.0.0 Basic Physiological Processes

- Cardiac Processes
- Respiratory Processes
- Basic Metabolic Processes
- Body Temperature Regulation Processes

8.0.0 Work Capacity

- Static Work Load
- Dynamic Work Load

9.0.0 Basic Anthropometric Data

- Data on Human Body Measurement

10.0.0 Basic Data on Body Mechanics

- Muscular Strength
- Posture
- Limb Movements

TABLE A-1 (*Continued*)

11.0.0 Basic Data on Sensory Physiology

- Human Visual Physiological Processes
- Human Auditory Physiological Processes
- Human Kinaesthetic Physiological Processes
- Human Vestibular Physiological Processes
- Human Thermal Sensitivity Physiological Processes
- Human Central Processes

12.0.0 Factors Affecting Physiological and Biomechanical Functions

- Fatigue
- Knowledge of Results
- Stress
- Ageing
- Drugs and Alcohol
- Diet, Food and Nutrition
- Exercise and Physical Training
- Circadian Rhythm
- Sleep Deprivation

THE DESIGN OF THE MAN-MACHINE INTERFACE

13.0.0 Visual Displays

- Analogue
- Digital
- Qualitative
- Pictorial and Symbolic
- Static Graphic
- Dynamic Graphic

14.0.0 Auditory Displays

- Non-Verbal
- Verbal (Speech)
- Transmission Equipment

15.0.0 Kinaesthetic and Tactile Displays

- Kinaesthetic Discrimination of Control Position and Load through Movement Extents
- Tactile Discrimination by Shape and Texture Coding

16.0.0 Controls

- Rotary Movement Controls – Knobs, Cranks, Wheels, etc.
- Linear Movement Controls – Levers, Pushbuttons, Pedals, Handles
- Special Purpose Controls – Mobile Remote Handlers, Manipulator Tongs, etc.

17.0.0 Specialized Input Devices

- Keyboards
- Computer Interfaces

18.0.0 Display-Control Dynamics

- Display-Control Movement Ratios
- Display-Control Compatibility
- Aided Controls
- Quickened Displays

19.0.0 Workspace Layout

- Limits of Working Areas, Access, Angle of Orientation, etc.
- Location of Components
- Furniture Specifications

20.0.0 Equipment Design

- Hand Tools Design
- Machine Design

TABLE A-1 (*Continued*)

21.0.0 Physical Environmental Factors Affecting Human Performance and Processes

- Illumination
- Noise
- Vibration
- Motion
- Atmosphere
- Thermal Conditions
- Altitude and Depth
- Space
- Other Factors

22.0.0 Specialized and Protective Clothing and Equipment

- Specialized and Protective Clothing
- Specialized and Protective Equipment
- Personal Equipment
- Prosthetics

SYSTEMS DESIGN AND ORGANIZATION

23.0.0 Assignment of Functions to Men and Machines

- Task Complexity

24.0.0 Work Design and Organization

- Paced and Unpaced Performance
- Working Hours and Rest Pauses
- Shift Work

25.0.0 Training

- Psychomotor
- Adaptive
- Other

26.0.0 Selection

27.0.0 Motivation and Attitudes

- Incentives
- Morale
- Job Satisfaction
- Job Enlargement
- Absenteeism and Turnover

METHODS, TECHNIQUES AND EQUIPMENT IN ERGONOMICS

28.0.0 Perceptual-Motor Processes

- Visual
- Auditory
- Other Perceptual Input Processes
- Short-Term Memory
- Long-Term Memory, Learning
- Problem Solving
- Other Central Processes
- Effector Processes
- Reaction Time
- Information Processing and Transmission
- Factors Affecting Perceptual-Motor Performance

29.0.0 Basic Physiological, Anthropometric and Biomechanical Data

- Cardiac Processes
- Respiratory Processes
- Basic Metabolic Processes
- Body Temperature Regulation
- Work Capacity
- Body Mechanics
- Sensory Physiology
- Factors Affecting Physiological and Biomechanical Functions

TABLE A-1 (*Continued*)

30.0.0 Data Presentation
- Visual Displays
- Auditory Displays
- Kinaesthetic and Tactile Displays

31.0.0 Input Facilities
- Controls
- Specialized Input Devices
- Display Control Dynamics

32.0.0 Workspace and Equipment Design

33.0.0 Physical Environmental Factors Affecting Performance
- Illumination
- Noise
- Vibration
- Other Physical Environmental Factors

34.0.0 Specialized and Protective Clothing and Equipment
- Specialized and Protective Clothing
- Specialized and Protective Equipment
- Personal Equipment
- Prosthetics

35.0.0 Work Design and Organization
- Work Study, Work Measurement
- Task Analysis
- Skills Analysis
- Job Evaluation

36.0.0 Implementation and Evaluation of Training Procedures
- Industrial Programmed Instruction
- Use of Simulators

37.0.0 Implementation of Selection Procedures
- Intelligence Testing
- Aptitude Testing
- Proficiency Testing
- Interview and Questionnaire Techniques
- Rating Scales

38.0.0 Use of Models and Simulation Techniques

39.0.0 Statistical Techniques and Experimental Design

40.0.0 Use of Computers

TABLE A-2 *List of Applications*[a]

TABLE A-2 (*Continued*)

Building and Construction 79838, 79846, 79996, 80074, 80086, 80148, 80160, 80196, 80262, 80464

Cement and Concrete Industry 80245
Ceramics Industry 79867
Chemical Industry 79851, 79867, 79869, 80074, 80352
Clothing Industry 80391, 80394
Computer Industry 80090, 80360
Conveyor Systems See **Materials Handling**
Cross Cultural Influences see **Nation or Locality**
Cranes see **Materials Handling**
Cycles see **Vehicles**
Cycling see **Sport**

Diving 79966, 80307, 80317
Driving (see also **Vehicles**) 79864, 79873, 79879, 79928, 79962, 79970–79971, 80052, 80139–80141, 80143, 80146, 80152, 80196–80197, 80201, 80210, 80216, 80223, 80228, 80235, 80285, 80287, 80320–80321, 80366, 80374, 80392

Earthmoving Machinery see **Materials Handling**
Education see **Schools**
Electrical and Electronics Industries 79846, 79867, 80092, 80288

Film Industry see **Photographic and Film Industry**
Fire Services 80028, 80260, 80470, 80475
Fishing Industry 80278
Floor Coverings 80204, 80278
Food 80473
Forestry 79992, 80074, 80265, 80275–80277, 80346, 80404
Fork Lift Trucks see **Materials Handling**

Geographical Location see **Nation or Locality**

Hairdressing 80286
Helicopters 80164, 80272
Hospitals see **Medical Services**

Industrial Hygiene 79847, 79850, 79860
Industrial Systems 79851, 79881, 80091, 80243, 80248, 80256, 80380, 80479
Information Retrieval 80163
Inspection 80345–80346, 80388
Iron and Steel Industry see **Metallurgical Industry**

Law 79828, 79837, 79839, 80198, 80205, 80244, 80368

Machine Tool Industry 79846
Maintenance 80231, 80335
Materials Handling 79996, 80030, 80070, 80072–80074, 80080, 80106, 80199–80200, 80213, 80269, 80434, 80440
Mechanical Engineering 80442
Medical Services 79947, 80074, 800159, 80199, 80355, 80359–80360
Metallurgical Industry 79846, 79852, 79869, 80023, 80074, 80193, 80207, 80242, 80292, 80338, 80352, 80396, 80400
Military Services see **Armed Forces**
Mining 79846, 79852, 79951, 80074, 80214, 80316, 80335, 80406

Nation or Locality 79828, 79835, 79837, 79846, 79850, 79852, 79860, 79881, 79921–79925, 79969, 79984, 80008, 80027, 80032, 80038, 80047, 80051–80052, 80112, 80119, 80160, 80179, 80198, 80239, 80241, 80244, 80275, 80297, 80301, 80304, 80325, 80335, 80346, 80350, 80362, 80368, 80392, 80403, 80476
Newspapers 79947, 80224
Nuclear Technology 79844

TABLE A-2 (*Continued*)

Offices 79869, 80087–80088, 80181, 80185–80186, 80249, 80264, 80291, 80455

Petroleum Industry 79846, 80074
Photographic and Film Industry 80215
Physically Handicapped 79849, 79902, 79983, 79999, 80192, 80196, 80199, 80319, 80452
Plastics Industry 79846, 80478
Police Force 80188
Postal Services 80030
Power Stations 79993, 80288
Printing and Publishing Industry 79867, 79947, 80074, 80224–80225, 80357
Proof Reading see **Printing and Publishing Industry**

Quality Control see **Inspection**
Quarrying 79846, 80074

Railways 80156, 80377
Reading 79875–79878, 79882, 79919, 79931, 79934–79937, 79941, 80172, 80271–80272, 80385–80386
Road Safety 79921–79925, 79971, 80143, 80157, 80284, 80364
Roads and Highways 80145, 80152, 80154, 80156, 80158, 80168, 80190
Road Transport Industry 79846, 80074
Rubber Industry 79846, 80478

Safety 79827, 79830, 79836, 79839–79840, 79842–79844, 79850, 79857, 79859, 79862–79863, 79865, 79900, 79946, 79973, 79992, 79996, 80059, 80086, 80148, 80153, 80160, 80198, 80205, 80207–80209, 80211–80213, 80237, 80317–80318, 80321, 80352, 80356, 80358–80359, 80367, 80372, 80390, 80464, 80472–80473, 80479
Sanitation 80003
Schools 79947, 80203, 80222
Ships 79846, 79868, 79947, 79967, 80263, 80278–80279, 80281, 80351, 80460
Shipyards 80267
Shops 80294
Sport 79829, 79832, 79834, 79854–79855, 79885, 79904, 79952, 79975, 79977, 79980–79981, 79989–79990, 79999, 80007, 80012–80014, 80020–80021, 80033, 80036–80037, 80041–80042, 80044, 80047, 80049–80050, 80054–80055, 80057–80058, 80061, 80063, 80065–80068, 80076–80079, 80084, 80093–80094, 80098, 80113, 80117, 80119, 80121, 80123, 80128, 80135, 80137–80138, 80178, 80253, 80259, 80297, 80311, 80314, 80323, 80363, 80398, 80410, 80416, 80418, 80425, 80430–80431, 80435–80436, 80444–80445, 80447
Standard Specifications 80153, 80200, 80222, 80239, 80241, 80264, 80267, 80317, 80325–80326, 80335, 80460, 80462

Textile Industry (see also **Clothing Industry**) 79828, 79846, 79852, 79867, 80244, 80261, 80289, 80293, 80358
Typing 79898, 79914, 80087

Underground Habitation 80316

Vehicle Passengers 80282, 80284, 80463
Vehicles 79880, 79921–79925, 80140, 80147, 80197, 80201–80202, 80210, 80228, 80282, 80329, 80336, 80452
Visual Aids 80153

Welding 79831, 80317
Woodworking 79869, 80346

TABLE A-3 *Sample Abstracts*[a]

18.0.0 DISPLAY-CONTROL DYNAMICS (80194)

80194. **Motor Performance in Relation to Control-Display Gain and Target Width.** L. BUCK. [*Ergonomics*, Jun. 1980, **23**/6, 579–589.]

Five groups of subjects performed a target alignment task using a joystick-oscilloscope system with different control-display gains. Time taken to move to the target depended upon the width of the area into which the joystick had to be placed in order to align the target, while time taken to correct overshoots depended upon that factor and also the width of the target area on the oscilloscope. Movement precision as measured by overshoot rate depended upon target location and not upon target width whether measured on the joystick or oscilloscope. There was evidence of a movement time-overshoot rate trade-off. The results call into question recent views on the significance of control-display gain in the design of the operator-machine interface.

See also 80020

19.0.0 WORKSPACE LAYOUT (80195-80204)

80195. **Ergonomic and Medical Aspects of VDU Workplaces.** E. GRANDJEAN. [*Displays Technology and Application*, Jul. 1980, **2**/2, 76-80.]

VDU's are a new tool in many workplaces. They can produce complaints due to inadequate workplace design, not optimally developed technology, unadapted work organization or psychological attitudes of end-users. Many therapies are available; others must still be studied and developed though therapies are useful only if they are applied.

80196. **Eleventh Congress of the National Association of Physicians for Building and Civil Engineering—Driving of Heavy Vehicles and Personnel Transport in Building and Civil Engineering (XIe Journee d'Etudes du Groupement National des Medecins du Batiment et des Travaux Publics—Conduite de Vehicules Poids Lourds et Transport du Personnel dans le Batiment et les Travaux Publics). (In French.)** [*Revue de Medecine du Travail*, 1979, 7/1, 99 pp.; abstr. in *CIS Abstracts* (*CIS 80-243*).]

Papers presented at the meeting (Reims. 21 Oct. 1978): statutory regulations concerning the operation of trucks and earthmoving equipment (standards applying to contractors' plant, transport of personnel and dangerous goods traffic regulations, rules for drivers); investigation of truck drivers' working conditions, occupational diseases and occupationally induced health damage (obesity, injury to the vertebral column); ergonomics of truck drivers' cabs; driving licence for heavy vehicles and medical fitness; industrial psychology tests to assess aptitude for key posts from the safety viewpoint; problems of personnel transport in the construction industry (medical fitness of drivers, occasional drivers, regulations); rehabilitation of handicapped truck drivers; statistics of accidents involving heavy vehicles; investigation of accidents involving truck drivers in a road making and road repairing firm, and in an earth-moving firm; hazards specific to heavy vehicles according to an analysis of occupational accidents.

80197. **The Ergonomics of Cabins in Lorries—Literature Review (Kuorma-Auton Ohjaamon Ergonomia—Kirjallisuuskatsaus). (In Finnish.)** J. SAARI and M. LAUNIS. [*Institute of Occupational Health (Tyoterveyslaitos), Vantaa, Finland, Katsauksia 25; TTL 21-1979,* 1979, 122 pp.; abstr. in *CIS Abstracts* (*CIS 80-236*).]

The review concentrates on the driver's work posture, visibility, location of controls and other devices, and the driver's seat. Noise, vibration, and microclimate in the cab are also discussed.

80198. **Stairways—Design and Rules for Use (Les Escaliers—Conception et Regles d'Utilisation). (In French.)** R. FOURNIER. [*Travail et Securite,* Nov. 1979, No. 11, 528-545; abstr. in *CIS Abstracts* (*CIS 80-309*).]

Aspects dealt with are: design criteria; number of stairways and required dimensions; covering; nosing of steps; chief accident hazards and prevention measures (loss of adherence, loss of balance by catching of foot, deterioration of stairs by wear and aging, environment, user's behaviour); maintenance. Reference is made to the French regulations.

80199. **Description of the Working Environment of Home Care Personnel Especially Lifting of Persons (Beskrivning av Hemvardspersonalens Arbetsmiljo med Avseende pa Anstrangande Arbetsmoment vid Personforflyttningar). (In Swedish.)** A. ISAKSSON and L. KARLQVIST. [*National Board of Occupational Safety, and Health* (*Arbetarskyd dsstyrelsen*), *Solna, Sweden, Undersokningsrapport No. 1980:2,* 1980, 60 pp.; abstr. in *CIS Abstracts* (*CIS 80-1151*).]

Ergonomic evaluation of arduous tasks performed by home care personnel (lifting of the handicapped). Report conclusions: entrance ways to dwelling premises occupied by the handicapped should be free of obstructions for wheel chairs: lifting equipment should be available in their homes for moving the handicapped; home care personnel should have more training in ergonomics; work in pairs should be encouraged; more men should be recruited for this heavy physical work.

The corresponding European journal is *Ergonomics,* published by Taylor & Francis Ltd. Another European journal, *Applied Ergonomics,* is less technical and features articles that can be directly applied to design problems. It is published by IPC House, Guilford, Surrey, England.

Finally, there are several human factors handbooks that offer concise design information. However, due to the concise format there often are no references in handbooks to give authority to the suggestions and it is difficult for the reader to know which suggestions are supported by considerable data and which are supported primarily by the author's professional experiences. One authoritative handbook is the *Human Engineering Guide to Equipment Design,* H. P. Van Cott and R. G. Kinkade, editors, and published by the U.S. Government Printing Office, 1972. Because each chapter is written and reviewed by specialists this book has become a standard reference work. However, it is over a decade old and no longer represents current practices. A more recent handbook is W. E. Woodson, *Human Factors Design Handbook,* McGraw Hill Co. (New York) published in 1981.

[a] Tables A-1, A-2, and A-3 Courtesy of Ergonomics Abstracts and the Ergonomics Analysis Information Centre at the University of Birmingham, England. The Classification System shown in Table A-1 was devised by J. G. Fox and C. Stapleton.

2 ERROR AND RELIABILITY

Near San Diego, a Pacific Southwest Airlines jet collides with a small Cessna private aircraft. Both planes plummet to the ground killing 237 persons aboard and 7 persons on the ground.

An automatic safety system turns on powerful cooling pumps. The operator turns them off. This led to the infamous Three-Mile Island incident, the worst nuclear disaster involving a commercial power plant in the United States.

You pull an all nighter studying for an important final exam. Finally, at 4 A.M. you collapse into bed forgetting to set your alarm clock. You sleep through the exam and flunk the course.

HUMAN ERROR

All the above are examples of human error. The consequences range from death, to severe property damage, to great mental anguish. Are errors a basic part of human nature that cannot be changed, or is there some way to eliminate or reduce error and the damage errors cause? In this chapter we discuss the nature of human error and what human factors can do to minimize error thereby increasing human productivity and happiness. You will find out how errors are classified, how people detect and correct errors, and how this information can reduce future errors. Then we discuss the relationship between error and coding. Some mathematical aspects of reliability are demonstrated and you will learn to calculate the reliability of simple systems. But this knowledge is more than theoretical. Throughout this chapter you will find real-life examples that illustrate how errors are caused when people and environment interact. Indeed, many human factors analysts believe that minimizing human error is the primary goal of any human factors design. If people never made errors, there would be little need for a science of human factors.

Defining Human Error

We all have a rough idea of what the word "error" means. But communication of technical ideas requires a technical definition. So we will define error as an action that violates some tolerance limits of a system. Systems vary as to the boundaries that define correct action. What may be a serious error in one system may cause no problems at all in a better system. While a sudden right turn in a Chevrolet station wagon moving at as little as 22 miles per hour may cause only a slight problem in vehicle handling, an identical maneuver in a CJ-5 Jeep may turn the vehicle over (Figure 2-1). Errors cannot be defined in a vacuum. Instead, they must be related to the overall system properties, as discussed in the preceding chapter. Note also that the term human error does not imply anything bad about the operator. If a resistor in an electronic circuit had an out-of-tolerance value, you would not call the resistor stupid. Similarly, the occurrence of a human error does not imply that the operator was stupid, or lazy or even careless. As you will see, most errors are due to a flaw in the system rather than a flaw in only the operator.

Human Error Probability

The basic unit of human reliability is the human error probability (HEP). It is the probability of an error happening during some specified task. HEP is defined as the number of errors of a specified type divided by the total number of chances for that error to occur. Expressed algebraically we get:

$$HEP = number\ of\ errors/total\ number\ of\ opportunities\ for\ the\ error$$

This can be illustrated with a simple example. Imagine that a quality-control inspector is examining ball bearings on an assembly line. Her job is to reject any bearings that are not round enough (out-of-round). In one day she inspects 5000 bearings and rejects 400 as defective. However, in this batch there were really 1000 defective bearings. Her HEP is the number of errors—in this case failing to reject 600 bad bearings—divided by the total number of bearings inspected. This is 600/5000 or .12. This job could be better performed (and usually is) by a machine.

Figure 2-1. Vehicle handling tests of a 1980 model CJ-5 Jeep at 22 mph. (Courtesy of the Insurance Institute for Highway Safety.)

Classification of Human Error

There are many ways to organize and classify human errors. No single scheme is perfect and all schemes force the user to make arbitrary decisions in some cases. Nevertheless, some scheme is necessary to organize what is very often an overwhelming set of raw error data. Such organization is the first step toward understanding and correcting the sources of error. Errors can be divided into two classes: intentional and unintentional. An intentional error occurs when an operator deliberately performs an incorrect act because he or she believes that this act is justified. An airplane pilot who descends at a rate faster than that specified in the operating procedures because he or she believes that the procedures are too rigid is committing an intentional error. Similarly, deliberately exceeding some equipment limit, such as loading 110 pounds on a dolly designed to carry only 100 pounds in order to save an extra trip, is an intentional error. Unintentional errors are far more common. The operator did not want an error to happen; it just did. Turning off the headlights on your car while reaching for the windshield wiper control would be an unintentional error.

Errors can also be classified according to their cost. This is not as useful a classification scheme because the same error may have different costs depending on external circumstances. Unintentionally turning off the headlights of your car might have no consequences if your car is stopped at a traffic light but drastic consequences if you are driving along a twisting mountain road. Instead of a scheme ranging from catastrophe (serious damage to property and loss of life) to no cost (you were in the parking lot when you turned off the headlights), it is more useful to have only two cost categories: unrecovered error and recovered error. A *recovered error* has the potential to cause damage but due to luck or to good human factors design that anticipated possible errors, nothing bad actually happened. An *unrecovered error* is one where it was not possible to prevent serious consequences. Recovered errors should not be ignored because they do reveal design inadequacies. Today's recovered error may be tomorrow's disaster.

Human Error Categories Error can be conveniently divided into five categories (Swain & Guttman, 1980).

1. Error of omission.
2. Error of commission.
3. Extraneous act.
4. Sequential error.
5. Time error.

An error of omission occurs when someone skips a part of a task. Thus, a pilot who attempts a landing without first checking that the landing gear is down and locked is guilty of an error of omission. An error of commission occurs when someone performs a task incorrectly. Thus, a carpenter who drives a nail into his thumb has achieved an error of commission. While the remaining three error categories are technically errors of commission, they are important enough to justify having their own classification. An extraneous act is a task that should not have been performed because it diverts attention from the person-machine system thereby creating the potential for damage. A student who misses an important point in class because she was reading the campus newspaper has committed an extraneous act. A sequential error occurs when a task is performed out of sequence. Someone who ignites the kindling in the fireplace before opening the fire damper has committed a sequential error. A time error occurs when a person performs a task too early, too late, or not within the time allowed. A driver who runs a red traffic light has committed a time error and so has one who returns to find his or her parked car ticketed because the parking meter has expired.

A SLICE OF LIFE

Error in the Skies

Midair collisions are tragic and dramatic events that receive intense media coverage. In this box we relate the details of one well-known collision in the skies over San Diego. However, collision is an unusual and relatively rare event. Statistically, as an airplane passenger, it is far more likely that you would be done in by what the aircraft industry calls a "controlled flight into terrain." This jargon means that the aircraft was literally flown into the ground by the pilot with no dramatic failure of the aircraft such as an engine falling off. Therefore, we will also give you the details of one such controlled flight into terrain.

The San Diego Collision

The following discussion is based upon a report in *Human Factors* (Wiener, 1980). Consulting Figure 2-2 will help you to understand the collision better.

A small one-engine private plane—a Cessna 172—took off from Montgomery Field with Lindbergh Field as its destination. The pilot was practicing landing approaches. It then flew northeast and was cleared by air-traffic control to climb as long as altitude was less than 3500 feet. As Figure 2-2 shows at time 0859:50 (8 hours, 59 minutes and 50 seconds) just two minutes before the collision, the Cessna was following instructions.

Pacific Southwest Airlines flight 182, a Boeing 727 jet, was approaching Lindbergh Field from the north intending to land. It reported sighting the airport at 0857 from an altitude of 9500 feet. The air-traffic controller then gave the flight permission for a visual approach

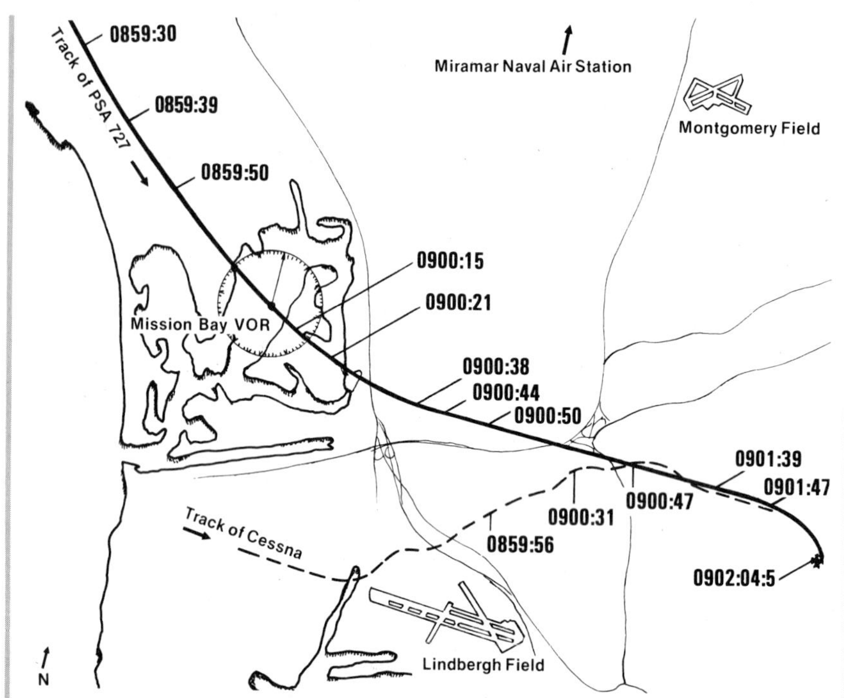

Figure 2-2. A map of the skies near San Diego. (Reprinted from the AOPA *Pilot,* March 1979 © Aircraft Owners and Pilots Association, 1979.)

to Runway 27. A visual approach means that the aircraft can descend at will without getting further permission from air-traffic control. It is the pilot's responsibility to "see and avoid" other aircraft. (Under instrument flight rules the air-traffic controller is responsible for keeping a safe distance between aircraft.)

The approach controller warned PSA 182 about the Cessna at 0859:39. Below is a partial transcription from the cockpit voice recorder for the last 2 minutes before the collision (APC is approach controller):

0859:30 APC TO 182	Traffic is 12 o'clock, one mile, northbound.
0859:35 182 TO APC	We're looking.
0859:39 APC TO 182	Additional traffic's, ah, 12 o'clock, three miles, field, northeast bound, a Cessna 172 climbing VFR out of 1400.
0859:50 182 TO APC	OK, we've got that other 12.
0859:57 APC TO CESSNA	—issues clearance—.
0900:15 APC TO 182 PSA	182, traffic's at 12 o'clock, three miles, out of 1700.
0900:22 182 TO APC	Traffic in sight.

At this point the jet was behind the Cessna and gaining. The controller believed that 182 had the Cessna in sight. It is the responsibility of overtaking aircraft to keep slower aircraft in sight. If the pilot loses sight as he overtakes, he must immediately contact the controller. Here is the remaining conversation with PSA 182 (SAN is the Lindbergh Field control tower):

0900:38 SAN TO 182 PSA 182	Lindbergh Tower, ah, traffic 12 o'clock one mile, a Cessna.
0900:44 182 TO SAN	OK, we had it there a minute ago.
0900:42 182 TO SAN	I think he's passed off to our right.
0900:50 COCKPIT	Is that the one (we're) looking for? Yeah, but I don't see him now.
0900:52 COCKPIT	He was right here a minute ago.
0901:11 COCKPIT	Are we clear of that Cessna?
0901:11 COCKPIT	I guess.
0901:11 COCKPIT	I hope.
0901:21 COCKPIT	Oh, yeah, before we turned downwind I saw him about 1 o'clock, probably beneath us now.
0901:38 COCKPIT	There's one underneath.
0901:47 COCKPIT	–sound of impact–

In an ironic afternote the approach controller responded to an automatic conflict alert that sounded 19 seconds before the collision. At the moment of the crash he notified the Cessna that the jet had the Cessna in sight. Even if the approach controller doubted flight 182's statement that the Cessna was in sight, he had no way of contacting the jet since it had tuned its radio away from the approach controller's frequency in order to contact the Lindbergh tower.

As Wiener (1980) explains, the immediate reaction to this tragedy was to consider limiting the airspace of private pilots, a prime example of blaming the victim. However, the political situation became quite complex and you must consult the original Wiener article for details. It is clear that there is room for human factors improvements, especially in the areas of cockpit visibility, cockpit collision warning systems, and communication. While the current collision avoidance systems and procedures are very good–if airplanes collided every day you would not see reports on the evening news–they cannot attain the goal of zero collisions. No system is likely ever to be perfect and it remains to be seen how much the federal government will spend to improve air safety, which is already good. Indeed, an overzealous government might dictate a cockpit collision avoidance system that reduces air safety by overloading the pilot with too much information. In the words of Wiener: "The very last place in the world that cockpit equipment should be specified is the U.S. Congress."

Controlled Flight into Terrain

The following story of Eastern Airlines flight 401 is drawn from Danaher (1980). Flight 401 was a L-1011 widebody jet approaching Miami International Airport for a night landing. However, it was diverted because the instrument panel did not show that the nose landing gear was properly locked in the down position. The pilot turned on the autopilot, setting it to cruise at 2000 feet. This reduced the cockpit work load so that the crew could devote their attention to checking of the apparent landing gear malfunction. The preoccupied crew failed to notice that the autopilot had been accidentally switched off, sending the plane into a gradual decline.

Miami approach control, although not technically responsible for flight 401 once it had left the area, noticed that 401 was down to an altitude of 400 feet. The controller radioed: "Eastern," ah, 401 how are things comin' along out there?" The crew replied immediately that they were making progress and intended to return for a landing. Less than 30 seconds later flight 401 hit the ground. The impact destroyed the plane and killed 99 people out of the 176 aboard.

It is ironic that the automated equipment functioned perfectly yet did not prevent this accident. The crew relied on the autopilot and failed to notice it had been switched off. The air traffic controller knew that flight 401 was at a dangerously low altitude, yet then-current procedures did not require him to notify the flight of its altitude, since it is the pilot's responsibility to keep track of altitude. (The controller thought his altitude reading might have been wrong since it was possible for the equipment to display incorrect information for a very short period of time between updates.) There is now an automatic minimum altitude warning that alerts controllers to possibly dangerous altitude deviations by planes under their control. The rules have also been changed so that the controller will now notify the pilot. However, maintaining proper altitude is still the responsibility of the pilot.

The major lesson of this accident is that human error cannot always be prevented merely by adding more automatic machinery. In fact, unthinking reliance on automatic systems like the autopilot can at times contribute to accidents instead of eliminating them.

Error Detection and Correction

People very often seem to know immediately if they have made an error. This subjective feeling of having goofed can occur in a wide variety of situations ranging from letting a tart remark "slip out," to typing an incorrect letter. Although recognizing a subjective feeling is a first step toward understanding how errors are detected, we need turn to the laboratory for a richer explanation.

One explanation of error detection states that people more or less automatically monitor their behavior. This checking on self-behavior will be explained more fully in the chapters on feedback (Chapter 12) and information processing (Chapter 6); for now, it is enough to understand that people pause to analyze the correctness of responses they have just made or emitted. In order to test this common explanation of error detection, Rabbitt and Rodgers (1977; 1979) asked people to press buttons in response to visual digits that were displayed. This procedure is called a choice-reaction time task since the experimenter records both the speed (reaction time) and accuracy of responding. The experimenters were particularly interested on the trial immediately following an error—that is, pressing the wrong response button. They called this next response the $E + 1$ response to indicate that it immediately followed an error (E).

One important result obtained by Rabbitt and Rodgers was that error did not occur randomly. Instead, an error was more probable as an $E + 1$ response than after a correct response. People tended on $E + 1$ trials to make the response they should have made on the preceding trial. Let's say that on some trial the digit 7 appeared but the person made some incorrect response other than 7. Then on the following trial the digit 8 appeared. Quite often the person would respond 7—what would have been correct for the preceding trial—even though 8 was now correct.

The picture gets more complicated when we take $E + 1$ reaction times into account. In general $E + 1$ responses are slower, except when they are correct responses because the same signal was repeated. Since the order of signals was random, on some proportion of trials, for example, the signal 7 will be followed by itself. When this happens on an $E + 1$ trial, making the response "7" is correct and also is faster.

The simplest monitoring explanation states that once someone makes an error, she "switches off" while analyzing this error. This predicts that all $E + 1$ responses should have similar delays. But such is not the case. The amount of delay depends on the relationship between the $E + 1$ response and the preceding signal. After an error is made, there is a tendency to make an error-correcting response—that is, the response that should have been made. This tendency conflicts with the signaled response on the $E + 1$ trial and slows down reaction time. However, if by chance the error-correcting response agrees with the new response because the stimulus was repeated, there is no conflict. The reaction time is actually faster because the tendency to make the correct response was already present because of the preceding error.

There is still some debate about the details of this explanation (Welford, 1979). It may be a while before there is universal agreement about how people detect and correct errors (Laming, 1979; Shaffer, 1976). But the concept of *conflict* among potential responses is a theme that will recur often in information processing research and it has important design implications for human factors that we will discuss later.

Error Prevention

Although it is true that understanding how people detect and correct errors is important, from a human factors viewpoint it would be far better to prevent errors. After all, an error that never occurs never needs to be detected and corrected. Preventing errors is a major responsibility of human factors analysts and much of what follows in this text could be considered error prevention.

In order to prevent errors, we must first be able to predict them. There are many techniques that help predict error. The simplest are based on historical data. If we observe workers in an industrial plant, we can discover their error rates. These data can be used to predict future errors in similar industrial settings. More complex techniques are based on mathematical models such as information theory, branching-tree analysis, and so on. The rest of this chapter introduces just enough information theory so that we can discuss error-correcting codes. After that we go through some of the mathematics needed to calculate human reliability: the opposite of error. While it is seldom economically feasible to make any system 100% reliable, there are some straightforward techniques that can increase reliability without drastic cost increases.

ERROR, CODES, AND INFORMATION THEORY

The problems caused by error in systems are well known to engineers. Indeed, there is an established body of knowledge (Hamming, 1980) devoted to "automatic" means of recognizing and correcting errors in complex systems such as computers. In order to understand how such error-detecting and error-correcting codes work, we must first define the nature of codes in general, which leads naturally to an introductory discussion of information theory. Information theory is a body of mathematics that permits us to make quantitative statements about sets of known events. While it aids any discussion of coding theory (Hamming, 1980) it also has many interesting applications to human behavior (see Chapter 5 and Attneave, 1959; Garner, 1962).

Codes

Some of you can remember back to your childhood days when you saved boxtops and sent away for a secret decoder ring that enabled you to exchange "secret" messages with your friends. One common secret code is based on displacing letters of the alphabet a fixed amount. So, for example, letter "A" becomes "E," "B" becomes "F," "Z" becomes "D," and so forth. Another kind of code depends on geometric arrangements of letters such as the standard QWERTY typewriter keyboard illustrated in Chapter 11. Here we can code messages by moving letters an arbitrary amount to the right. This kind of code also wraps around itself, like the alphabet code above, so that instead of falling off the right-hand end of a line (or using a nonalphabetic character) you start the same line over again. If our code required moving a single letter to the right, for example, the letter "P" at the extreme right end of the keyboard would be replaced by "Q" at the beginning of the same line. Try to decode the following message based on moving once to the right: (If you don't have a typewriter keyboard handy, consult Figure 11-4)

JRAQ OZ YTSQQRF OM YJOD NPPL
(Hint: The first word is HELP)

Codes and secret messages have an interesting history (Kahn, 1967) and we hope that deciphering the informal message above has reminded you about codes in general. However, we still need a formal and precise definition of a code. We require two things to establish a code. Each is by itself necessary but neither is sufficient. You have probably already figured out the first necessary element: an *alphabet* or set of *symbols*. While this set of symbols is the literal alphabet in the examples above, actually any symbol will do. Computers use binary codes based upon the symbols 0 and 1. Morse code uses an alphabet of dots, dashes, and blank spaces. The second necessary element is more difficult to guess. It is some fixed relationship among the symbols. In the secret decoding ring example above, the relationship is to add 5. The relationship can get to be quite complex just so long as it remains fixed and does not change on successive instances or trials. To summarize, a *code* is defined as an alphabet of symbols plus a system of fixed constraints among the symbols.

Some codes are easier than others for people to use. While computers are content to mumble strings like "001101010110011010" people are more likely to fumble them. While it is very hard for you to remember a long binary string, using a code will dramatically improve your performance (Miller, 1956). Table 2-1 shows the octal code based on grouping binary digits into sets of three. Once you have learned this table, your memory span for binary digits can be very much improved by encoding a binary string into octal, memorizing the octal digits, and then transforming them back to binary as required. In fact, this effect is so reliable it is

TABLE 2-1 *Octal Codes for Triples of Binary Digits*

Octal	Binary
0	000
1	001
2	010
3	011
4	100
5	101
6	110
7	111

often used to impress students in introductory psychology courses as well as acquaintances at parties. Octal codes are often represented as a number enclosed in parentheses with the subscript 8: $(31)_8$. A normal everyday decimal number could be shown in the same way: $(25)_{10}$. Since these two numbers are mathematically equal, $(25)_{10} = (31)_8$ this shows that Christmas is Halloween: Dec. 25 = Oct. 31. (This reminder is brought to you by CRIME: Committee to Refute Inessential Mathematical Effluvium.)

The Communications Model

Codes imply messages and messages imply communication. Lest you think we are about to begin a heavy discussion of the philosophical bases of language, deep structure, and psycholinguistics in general, we hasten to point out that we are interested only in the technical sense of the word communication. The psychological interpretation of messages and the meaning of messages will not be considered here; this type of problem is illustrated by Figure 2-3. In this chapter, communication refers only to the successful transmission of a set of symbols from one place—the source of the message—to another—the destination of the message.

Figure 2-3. Although a message can be correctly received at the destination, this is no guarantee that it will be understood. The meaning of a message goes beyond technical aspects covered by the communication model. (By permission of Johnny Hart and Field Enterprises, Inc.)

The basic communication model consists of five serial stages.

1. Source.
2. Encoder.
3. Channel.
4. Decoder.
5. Destination.

This can be illustrated by using a telephone system as an example of a communication system. The *source* is you speaking into the telephone, or more precisely, the moving air molecules energized by your vocal tract. The *encoder* is the small microphone inside the telephone mouthpiece. It changes or *tranduces* the motion of air molecules into an electrical signal. The *channel* is everything between your telephone mouthpiece and the earpiece of the telephone used by the person receiving your message: this includes cables, relays, switches, microwave towers, satellites suspended above the earth, etc. The *decoder* is the tiny loudspeaker inside the telephone that tranduces electrical energy into mechanical motion of air molecules. The *destination* is the person listening to you, or more precisely, the ear of this person. Since we are concerned only with the technical aspect of communication, we ignore for now any connection between the ear and the brain and the behavior of the person. If the message received at the ear is identical to that created by the source, our communication system is working perfectly even if the person receiving the message cannot understand it—it is in the Serbo-Croatian language—or fails to act in accordance with the message.

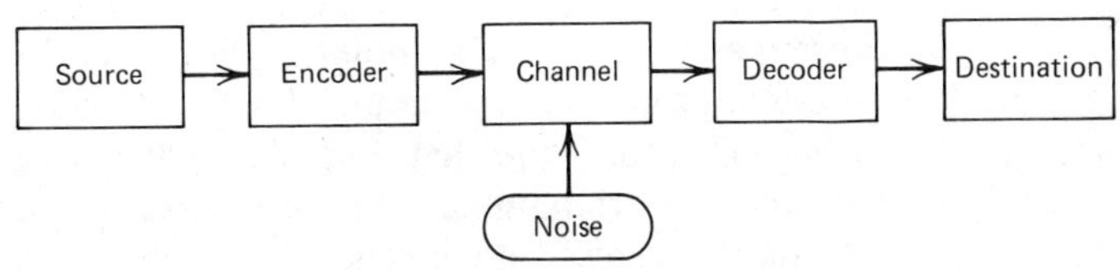

Figure 2-4. The communication model. Information flows from left to right. Note that noise is injected into the channel. The five stages are described in the text.

However, no real communication system is perfect. Messages received at the destination may not be identical to those sent from the source. The culprit or source of error in the model is *noise* injected into the channel (Figure 2-4). For engineers, noise is BAD and entire careers are dedicated to removing, or at least minimized its effects. Technically, noise is information present at the destination but absent at the source. (Be patient, we'll soon define information mathematically.) It is something extra added by the system.

While noise is a nuisance for engineers, psychologists who view the human as an information-processing system, actually spend their careers studying noise. The human is not at all a passive system that receives inputs and dutifully produces outputs that are direct and invariant functions of these inputs. This passive view of the person as a sort of complicated telephone-switching network can no longer be maintained (see Chapter 6). Instead, psychologists study the kinds of active information transformations that people impose on inputs. Since this means that a person's outputs are usually different from the inputs, technically the human has added noise.

We will give one example of how people generate internal noise. The stimulus or input signal is a list of nouns belonging to specific categories such as birds, furniture, etc. The person is required to memorize the list as it is presented and then to recall the words. Perfect performance—no noise added—occurs when the list of words generated by the person is identical to the list that was presented; the same words in the same order. Then the words are presented in different randomized orders and again the person must memorize and recall them. When people perform this task they tend to organize the words by categories (Tulving, 1962); this is called subjective organization. Even though the words are presented in random orders, they come out of the human clumped together by categories. The input information has had noise added to it.

Errors and Codes

Noise in the channel causes errors. Since no finite system can be totally without noise, we need some tools that will fix coded messages that have been distorted by noise. First, our tools must let us know that an error has occurred. We cannot fix a distorted message unless we realize that there was an error. This point is illustrated by the story about the congressman who fled the country while being investigated for accepting bribes. He was acquitted and his lawyer sent him a telegram reading, "Justice has triumphed." The congressman returned a telegram to his attorneys stating "File appeal immediately." Once, we know that a coded message contains an error, we can apply other tools to correct the error. The major tool used to discover errors is called an error-detecting code; error-correcting codes are more complicated since they precisely localize and fix errors.

Error-detecting Codes We will start our discussion of error-detecting codes using binary messages composed of a string of 0's and 1's. When we realize that a modern computer performs 3.6 billion operations in an hour the need for very reliable binary messages is at once apparent. There is a well-known simple way to create an error detecting binary code at the cost of adding one extra binary digit to the message.

1. Count the number of 1's contained in the message.
2. If the number is odd, add a 1 at the end of the message.
3. If the number is even, add a 0 at the end of the message.

This extra added binary digit is called an even-parity check. Error is detected by counting the number of 1's contained in the message. This should always be even. If it is odd, there is at least one error in the message. (This is equivalent to working in modulo 2 arithmetic: divide each number by 2 and keep only the remainder. Some computers have an instruction to do this directly.)

There are some disadvantages to simple parity checks. An extra digit must be added to the message and increasing the length of a message hurts the system by increasing its throughput—the time needed for a message to cross the system—or by increasing system cost. This extra digit is very costly if messages are short. At first, you might think we could get around this by having long messages. Who cares if an extra digit is thrown in after every 5000 symbols? Alas, there are more problems with simple parity checks that compel us to avoid long messages. What if the message has two errors? The parity check will come out OK and the errors will have gone undetected. (In fact, any even number of errors will go undetected. Any odd number of errors will be detected.) You can tell intuitively that as a message gets longer, the chances for two errors increase. (If your intuition is lacking, you can work through the mathematics in the appendix at the end of the chapter.)

Information Theory

Information as a technical term is simply a more precise form of the everyday meaning of the term. If someone tells you that the pope is Catholic, you have not gained any information, either in the everyday sense or in the technical sense. Since you already know that the pope is Catholic, this statement has not reduced your uncertainty about the world at all. But if you are about to flip a coin and someone could tell you if it would come up heads or tails you might be able to think of several ways to use this information. Your uncertainty about the world has been reduced by knowing whether to guess heads or tails. The crucial distinction between technical and everyday usage become obvious when you ask how much your uncertainty has been reduced or how much information have you gained. This question cannot be answered without a formal mathematical definition of information.

Information theory is a mathematical tool that helps us quantify the uncertainty of the world. This is a very necessary function in a world where even weather reports are probabilistic rather than absolute. The unit of information is the *bit*. This is the amount of information present in the toss of a fair coin—that is, the coin has two distinct sides only (landing on its edge doesn't count), and each side has an equal probability ($\frac{1}{2}$) of occurring. Before we flip the coin there is one bit of *uncertainty* about the outcome, heads or tails. After the coin is flipped and the outcome is observed, we have gained one bit of *information*. Information and uncertainty are mathematically equal and are calculated the same way. However, uncertainty refers to future events while information refers to past events: the occurrence of some event reduces uncertainty and gains information.

In order to calculate amount of information, we must have a known set of possible events and the probability of each of these events. When we have two events with equal probability there is one bit of uncertainty.

When one of the two possible events occur, we gain one bit of information. Do you think there is more information gained by going from two alternatives to one or by reducing 100 alternatives to 99? Most people feel that there is less information gained by reducing 100 alternatives to 99 than by reducing 2 alternatives to 1. Information does not depend directly on the number of alternatives that have been reduced. Instead, information depends on the ratios of alternatives before and after some event. Thus, reducing 100 alternatives to 50 yields the same amount of information—one bit—as reducing two alternatives to one, provided, of course, that in both cases all alternatives are equiprobable. This means that information depends on the logarithm of the number of alternatives. It is customary to use logarithms to the base 2 when calculating amount of information, although other bases are rarely used for special situations that we will not discuss. So, if have N equiprobable alternatives we can write

$$\boxed{H = \log_2 N}$$

where H is the traditional abbreviation for information. (H honors Hartley, the man who popularized the decimal information metric.)

For those readers who are a bit rusty we briefly review the concept of a logarithm. (Practice problems with logs are in the workbook.) A logarithm is simply an exponent. It tells us what power the base must be raised to get the desired result. So, the $\log_2$ of 2 is 1: 2 raised to the first power is 2. How much information is present if four equiprobable alternatives are reduced to one? The $\log_2$ of 4 is 2: $2^2 = 4$. So there are 2 bits of information gained. Similarly, selecting one square from a 64-square checkerboard yields 6 bits of information: $2^6 = 64$.

What happens when the world is not willing to provide sets of equiprobable events? We can still calculate H, but now we must resort to a slightly more complex algebraic formula. We now demonstrate that H is really a weighted average. Let h_i represent the amount of information associated with a single event i. Technically, h is called the *surprisal* of an event defined as

$$h_i = \log \frac{1}{p_i}$$

where p_i is the probability of the ith event. (From now on we leave off the subscript 2 on logarithms to avoid cluttering up the equations unnecessarily; in this chapter log always means to the base 2 unless otherwise stated.) Taking an example suggested by Attneave (1959), imagine that a coin has been bent so that it comes up heads 90% of the time and tails 10%. The surprisal of heads is log $1/.9 = 0.15$ bits. Similarly, the surprisal for tails is log $1/0.1 = 3.32$ bits. This makes intuitive sense. If heads comes up you have not gained much information since this happens 90% of the time. What is the total average information H associated with this bent coin?

You might think we should take the mean of 0.15 bits and 3.32 bits, which would give an average H of 1.735 bits. This is wrong. It turns out that H is a maximum when events are equiprobable. A fair coin has only 1 bit of uncertainty. Therefore, a bent coin must have less than 1 bit. It cannot have 1.735 bits. We have erred by not weighting each surprisal by its probability of occurrence. It is incorrect to simply average the two surprisals when one event (heads) occurs 9 times more often than the other event (tails). Using a weighted average we get

$$H = .9\ (0.15) + .1\ (3.32) = 0.47 \text{ bits}$$

This is less than 1 bit and is the correct answer.

Algebraically this weighted sum can be represented as

$$H = \sum^{i} p_i h_i$$

Equation 2.1

But $h_i = \log 1/p_i$. So we can substitute and rewrite

$$H = \sum^{i} p_i \log \frac{1}{p_i}$$

Equation 2.2

A basic rule for logarithms is that $\log 1/a = -\log a$. Using this we can now write

$$\boxed{H = -\sum^{i} p_i \log p_i}$$

Equation 2.3

Information is closely related to coding. Selecting one square out of a checkerboard requires six yes-no decisions if questions are optimally phrased. Imagine that you have to perform this task. You could locate any designated square by asking successive questions that divided the remaining number of squares left in half. Thus, your first question would be "Is the square in the left half of the checkerboard?" After six such questions you would have the selected square located (see Figure 2-5). Thus $H = 6$ bits in this case—represents the minimum number of binary digits into which an event based on equiprobable alternatives could be coded. Similarly, $H = 0.47$ bits for the bent coin means that 100 coin tosses could be perfectly described by a code of 47 binary digits if an optimal coding scheme was used.

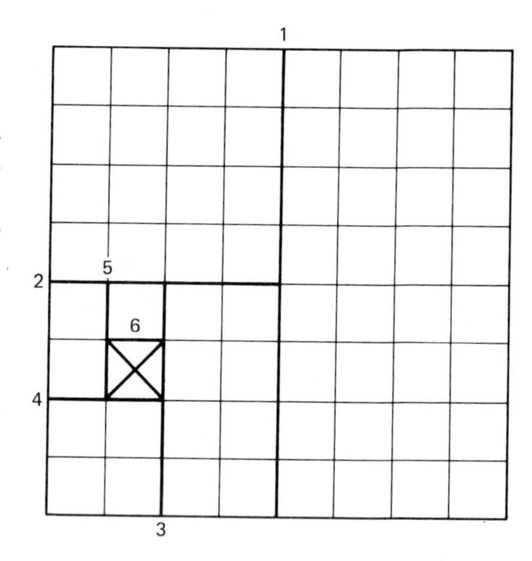

Figure 2-5. How to locate a particular square inside a checkerboard with only six questions. The secret is to ask questions that divide the board in half. Thus, line 1 is drawn and you ask, "Is the X to the left of line 1?" Then the remaining space is divided in half again by line 2. This process continues until line 6, which divides the two remaining squares in half. You might try this with the X in a different square.

Alas, information theory does not tell us how to create an optimal code; it tells only how many binary digits would be needed. It is easy to see that coding 100 tosses of our bent coin by using 100 binary digits, one for each toss, is far from the optimal code. It is harder to figure out better codes but we will give you some hints. We can break up our 100 tosses in smaller groups, say, 7 tosses to a group. We then count the number of tails in each group and use this as a basis for forming a code. You can try this as an exercise or see Attneave (1959) for one solution to this coding problem. Don't be disappointed by this abrupt ending; we return to information theory in Chapter 6.

BEFORE AND AFTER

Error and Paperwork

Modern technology runs on paperwork. Industry and government are flooded with forms that have to be completed to document compliance with regulations, keep track of an item's history, and so forth. A CBS television special on the space shuttle had one worker remark that the paperwork for the shuttle craft would probably weigh as much as the shuttle itself!

Any form that you must fill out, whether it is a government tax form or an industrial inventory form, was designed by a person. If the paperwork designer fails to consider human factors principles, then the users of the forms will make errors. In this box we show you a form that has been poorly designed and an improved version of the same form. How many things can you find wrong in the Inventory Form on page 47?

Inventory Form
Embargo Foundation Garment Corporation

Stock No.	Quantity	Description
M35764822340	100	MEN'S MESH UNDERSHIRT SIZE 40
M35764820340	37	MEN'S V-NECK UNDERSHIRT SIZE 40
M35764821340	41	MEN'S COTTON UNDERSHIRT SIZE 40
W25973881108	43	WOMEN'S FULL-CUT PANTIES SIZE 8
W25973881008	50	WOMEN'S BIKINI PANTIES SIZE 8
W25973881308	11	WOMEN'S MATERNITY PANTIES SIZE 8
W25973883138B	10	WOMEN'S MATERNITY BRA SIZE 38B
M35764823040	5	MEN'S SILK JOCKEY SHORT SIZE 40
W25973882138C	9	WOMEN'S MATERNITY BRA SIZE 38C
W25973882238U	73	WOMEN'S DEEP PLUNGE ANTRON BRA FITS ALL
M35764823140	110	MEN'S COTTON JOCKEY SHORTS SIZE 40
M35764823540	69	MEN'S EMBROIDERED BOXER SHORTS SIZE 40
M35764823240	41	MEN'S POLYESTER JOCKEY SHORTS SIZE 40
M35764823440	71	MEN'S COTTON BOXER SHORTS SIZE 40
W25973882538C	12	WOMEN'S SEAMLESS CUPS POLYESTER BRA SIZE 38C
W25973882438B	51	WOMEN'S PERMAPRESS MINIMIZING BRA SIZE 38B
M35764823340	62	MEN'S NYLON BIKINI SHORTS SIZE 40
W25973882638F	4	WOMEN'S DEEP PLUNGE SHIMMERING SATIN BRA QUEEN SIZE

This form is hard to read and guarantees errors. We can improve it in several ways. First, it is very difficult to read lengthy messages that are in capital letters. Second, the quantities are left adjusted (the digits on the far left line up), which is common on certain computer systems but hard for people to read. Third, the stock numbers are too long for people to remember; we can improve memory by grouping digits. Fourth, the space between the description and the other columns is too big. Fifth, the items should have spaces after every four or five items to help the reader keep track of his or her place. Sixth, the items are arranged in a haphazard manner; they should be ordered according to some sensible criterion that the user can understand. Below is an improved version of the same form.

Inventory Form
Embargo Foundation Garment Corporation

Quantity	Stock No.	Description
100	M-3576-482-23-40	Mesh undershirt
37	M-3576-482-03-40	V-neck undershirt
41	M-3576-482-13-40	Cotton undershirt
5	M-3576-482-30-40	Silk jockey shorts
110	M-3576-482-31-40	Cotton jockey shorts
41	M-3576-482-32-40	Polyester jockey shorts
62	M-3576-482-33-40	Nylon bikini shorts
71	M-3576-482-34-40	Cotton boxer shorts
69	M-3576-482-35-40	Embroidered boxer shorts
50	W-2597-388-10-08	Bikini panties
43	W-2597-388-11-08	Full-cut panties
11	W-2597-388-13-08	Maternity panties
10	W-2597-388-21-38B	Maternity bra
9	W-2597-388-21-38C	Maternity bra
73	W-2597-388-22-38U	Deep plunge Antron bra
51	W-2597-388-24-38B	Permapress minimizing bra
12	W-2597-388-25-38C	Seamless cup polyester bra
4	W-2597-388-26-38F	Deep plunge shimmering satin bra

Note. First letter of stock number indicates men or women. Last two or three characters indicate size: U = fits all, F = Queen size.

It is easy to see which form you would prefer. The improved form eliminates needless repetition of size and sex, fixes the human factors defects described above, and in general will result in fewer reading errors.

RELIABILITY

Error and reliability are opposite sides of the same coin. We define reliability as one minus the probability of error. So, if a system component exhibits an error probability of .05, this component has a reliability of .95. An error-free system has a reliability of 1.0.

There are reasonable mathematical techniques for obtaining accurate reliability estimates for machines and we mention some of them later. But most of this section is devoted to the particular problems that emerge when we try to estimate the reliability of human operators in complex tasks (Drury & Fox, 1975). This is the special province of the human factors analyst.

Human reliability analysis is one part of a series of steps that together form the basis for Person-Machine System Analysis. We list 10 steps in the complete Analysis (Swain & Guttmann, 1980).

1. Describe the system goals and functions.
2. Describe situational characteristics.
3. Describe personnel characteristics.
4. Describe tasks and jobs of the personnel.
5. Analyze tasks and jobs to detect situations where errors are likely to happen.
6. Estimate the probability for each potential error.
7. Estimate the probability that each error will not be corrected.
8. Determine the consequences of each uncorrected error.
9. Devise changes to increase system reliability.
10. Go back to steps 1 through 9 and evaluate these suggested changes.

Steps 1 through 4 have been covered in the preceding chapter. Now we focus on steps 5 through 8.

Analyze Tasks

To perform step 5, the human factors analyst must have a keen appreciation of the mental and physical capabilities of the human operator. Chapters 3 through 6 of this text supply the necessary information. Once this information is digested, the analyst searches for conditions where the tasks imposed on the human are likely to exceed a person's capabilities. Although this sounds simple to do, we stress that no human factors analyst is likely to detect *all* the possible types of errors an operator can create by doing something wrong, or neglecting to do something, etc. But many of the errors that actually occur in industry can be anticipated. In fact, an alert observer with even minimal human factors training can find obvious design flaws when examining tasks and procedures with a critical eye. We list a few representative goofs (e.g., Figure 2-6) that have actually occurred to show you that real improvements can result from anticipating relatively simple design flaws:

- An operator must monitor small gauges 8 feet above her head.
- An operator must open two valves in rapid succession. Controls for these valves are 12 feet apart.
- An operator must remember 80 four- digit numbers that change daily.
- An operator must quickly select one of 24 identical controls that are labeled only with identifying numbers.
- On successive days operators must work in two different control rooms that are mirror images of each other.

Figure 2-6. This is the kind of industrial setting that creates errors. It shows poor human factors design. (From "Human Factors Review of Nuclear Power Plant Control Room Design," Summary Report, November 1976, Electric Power Research Institute.)

This list could be expanded for many pages; indeed, we will discuss more "horrible examples" throughout this text. (If you see no problem with any of these examples, go back to page 49 after you have read Chapters 3 through 6.) But our point should be clear: there are many industrial situations that require people to perform tasks that exceed human capabilities. These excessive demands increase the chances of human error and decrease reliability.

Quantification of Human Error

The basic unit of human reliability is the Human Error Probability (HEP) defined earlier in this chapter. By combining human error probabilities in assorted ways, the analyst can calculate a total reliability figure for human performance in some specific task or job. This desired total reliability figure is not a simple sum or direct multiplication of the HEPs associated with task components because HEPs are seldom independent. The analyst must take the relationship among task components and their HEPs into account.

Where do the basic HEPs come from? How does one know what the HEP is for a task component like reading a three-digit number from a meter two feet away under a given level of illumination, dial size, and legibility, etc.? In an ideal world, this HEP would be estimated from an experiment where personnel who are representative of actual employees perform the exact task for which the HEP estimate is required. In practice, this never happens. It is far too expensive, both in terms of time and money, to perform the extremely large sets of experiments required to estimate even the most important HEPs, let alone all or most of them. Occasionally, the analyst can find a published experiment that is close to what is required. But modifying the results of a "close" experiment to apply a different situation requires great skill.

Unless you are an accomplished human factors analyst, it is far wiser to obtain HEP estimates from published tables. (The creation of such tables is a complex task beyond the scope of this text; just be glad there are professionals who obtain these important data for us.) We will show you how to use and interpret these tables now.

While there are several schemes for quantifying human reliability, we discuss only one in detail. This model is called THERP by its originators (Swain, 1963) and our presentation is based on a recent formulation of this well-known model (Swain & Guttmann, 1980). The basic tool used by THERP is the probability tree diagram (Figure 2-7). This tree represents the possible pathways a sequence of correct and incorrect operations might take. We start at the top of the tree and work our way down.

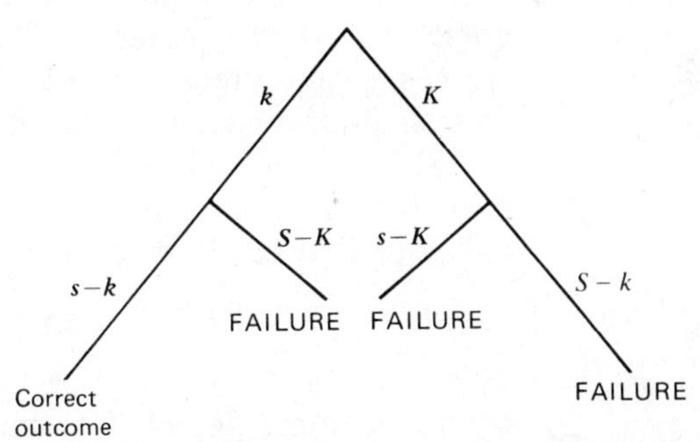

Figure 2-7. Probability tree diagram for starting your car. The probability of selecting the correct ignition key is k. The probability of stepping on the accelerator to get the car started right away is s. Note that lower-case letters correspond to successful actions, while capital letters correspond to failures.

In order to simplify our discussion of THERP, we will use a trivial task as an example: starting your car in the morning. In order to successfully accomplish this task you must perform two steps. First, you must locate the correct key and then you must insert this key into the ignition slot and turn it while stepping on the accelerator. (A more detailed analysis would consider three or even four steps for this total process.) The tree separates into two branches:

k = the probability of successfully locating the correct key
K = the probability of selecting the wrong key

We first travel down the left-hand branch of the tree. Given that you have picked the proper key, two outcomes are possible. First, the car may start, which is, of course, the desired outcome. Second, you may give the car too much or too little gas, so that it fails to start immediately. These outcomes correspond to the lower left hand-branches:

$s|k$ = the probability of a successful one-trial start given that you have the proper key
$S|k$ = the probability that the car fails to start, again given that you have the proper key

(The vertical bar is read "given" and refers to a conditional probability – that is, the probability of some event given that some other event has already happened. If you need review on the topic of conditional probabilities see Chapter 14 or just about any elementary probability textbook.)

The probability of a correct outcome C, in this case, getting your car started immediately, is the product of the two probabilities in the tree:

$$Pr(C) = k(s|k)$$

Any other outcome is failure (F), and the probability can be calculated by considering all other branches in the tree. The remaining left-hand branch shows the probability that the car fails to start even though you are using the correct key, $S|k$. The right-hand branches give the probabilities of starting the car with the wrong key. In this example these probabilities will be low, but it is not utterly impossible to start a car with something other than the manufacturer's ignition key. Thus, the probability of failure can be calculated as

$$Pr(F) = k(S|k) + K(s|K) + K(S|K)$$

This also equals $|1 - k(s|k)|$ because the sum of the probabilities of a correct outcome plus a failure must equal 1.0.

It is extremely important to remember that once we get past the top branch of the tree, all the probabilities that are entered are conditional probabilities. These conditional probabilities already take into account actions that have occurred higher up in the tree. If the analysis incorrectly assumes that branches are independent of one another, there will be a gross *underestimate* of the probability of failure. Finally, note that although we have used binary branches for simplicity, you can have as many branches as are needed at each tree node.

The outputs of the THERP model are estimates of correct or failure probabilities for human behaviors (tasks). Table 2-2 shows some typical HEPS that you might find if you looked in a handbook of reliability. Some HEPS can be quite high, especially in stressful situations. For ex-

TABLE 2-2 *Estimated Probabilities of Error in Various Tasks*

Select wrong control in a group of identical controls having identification labels but not other aids	.003 (.001 to .01)
Turn control in wrong direction in a high-stress situation when design violates strong population stereotype	.5 (.1 to .9)
Operate valves in sequence ($>$10 times)	.01 (.001 to .05)
Operate valves in sequence ($<$10 times)	.003 (.0008 to .01)
Failure to recognize an incorrect status when checking an item in front of your nose	.01 (.005 to .05)

These HEPS are from Swain and Guttmann (1980). The ranges in parentheses indicate 5th and 95th percentile values.

ample, let's discuss the task in Table 2-2 with an HEP of .5. Here is a situation that would fit. You are being pursued by someone in a dark parking lot. To escape you must quickly start your new car. Your old car had reverse gear on the right end of the shifter while your new car has it on the left end. To back out of the parking space, you must throw the shift lever to the left. Your chance of shifting in the right direction the first time you try is only 50%.

It is important to realize that these values are baseline HEPS that describe operations performed independently. Entering them directly in a probability tree diagram would often lead to an underestimate of failure probability. These unconditional probabilities must be modified to obtain the conditional probabilities that take into account various dependencies among task components. For example, an alerting tone that occurred because of an operator error might make the operator less likely to err on her next task. Or, an error may lead to great stress, further increasing the probability of failure on successive tasks. While the techniques for taking dependencies into account are well established (Swain & Guttmann, 1980, Chapter 7) they fall beyond the scope of this text. So any values you calculate will be optimistic since you don't know how to take dependency into account. Nevertheless, even these optimistic calculations will show you how fallible humans can be when designs are not optimal.

Not all human factors experts agree that the best approach to calculating the reliability of person-machine systems involves HEPs such as found in Table 2-2. Adams (1982) has argued that it will be quite difficult to synthesize reliability of human and machine subsystems. Basic error data for machines are usually in the form of number of hours between failures and percentage of a sample of machines that fail in a given time. Human data do not measure reliability in precisely these terms so that combining estimates of human and machine subsystem reliabilities can be awkward. While Adams (1982) has suggested Monte Carlo modeling techniques as a preferred solution, he admits there is presently a shortage of suitable data. While all human factors experts agree that total system reliability cannot be predicted without taking into account the reliability of both human and machine subsystems, these experts disagree as to how this goal may best be achieved.

Redundancy

Redundancy refers to the repetition of information or actions. If systems were perfect, there would be no need for redundancy and in fact any redundancy would be a complete waste of system resources. In practice, imperfect systems depend heavily on redundancy to reduce error and increase reliability. The amount of redundant information can be calculated using information theory. Recall the bent coin example from earlier in this chapter. We stated that with optimal coding a message of 47 binary digits could perfectly describe a series of 100 coin tosses. Therefore, a message of 100 binary digits would be redundant. The amount of redundancy (R) can be calculated from the formula:

$$R = 1 - \frac{H}{H_{\max}}$$

H is the amount of information calculated from some set of events. $H_{\max}$ is the amount of information that would be present if the events all occurred with equal probability. For the bent coin example, $H_{\max} = \log 2 = 1.0$. Therefore, the redundancy can be calculated as $1 - .47/1.0 = .53$. Thus, using a message of 100 digits where 47 would do, results in a 53% redundancy.

A method frequently used to add redundancy to a system requires tasks to be duplicated and/or checked. Thus, one operator can perform a task like opening a valve and a second operator can check that the valve was indeed opened. This, of course, is the same as doing the task twice so you might ask why not just save the second operator's salary and have the first operator check his own work. The answer to this important question takes us back to our discussion of dependency. When two operators perform the same task, the checks are independent of each other. This is not at all true when the same operator checks her own work. A person is far more likely to make the same mistake twice than is the probability that an independent check will fail to reveal an error. It is safer to have independent checks. This is why airplanes have pilots and copilots.

Redundancy can be added by having people check on people, having

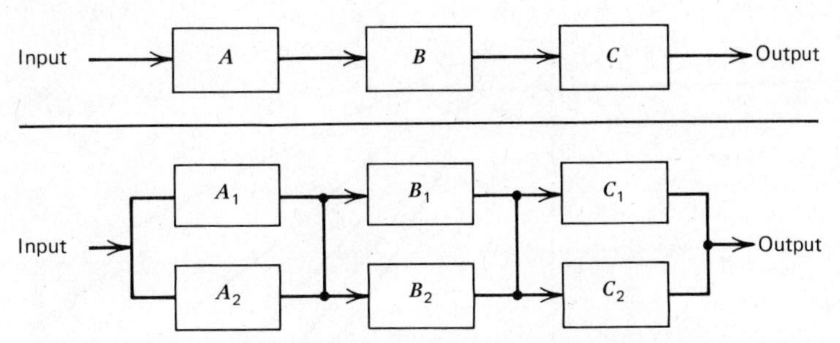

Figure 2-8. Serial (top) and parallel (bottom) arrangements of components. If components are identical, the parallel arrangement is more reliable.

machines check on machines, having machines check on people, and by having people check on machines. When a task is performed twice by independent components of the system—people or machines—we have a *parallel* arrangement of components. In order to calculate the advantage of parallel components we must first examine a system with no parallel components. In such a system each component is performed only once and the next component is performed sequentially until the task is complete. This is called a *series* arrangement of components since information flows directly from one component to the next component in the series.

In a simple series arrangement (Figure 2-8) it is easy to calculate reliability if each task component is independent. (Remember, this is not always true and we must revert to our probability tree analysis with conditional probabilities if task components are dependent.) The reliability of the whole series system is just the product of all the individual component reliabilities.

Reliability of serial system $= R_1 \cdot R_2 \cdot R_3 \ldots \cdot R_i$ where there are i tasks components. This analysis also assumes that a failure in a single task component wipes out the whole system. (If this is false because a successful outcome does not require perfect operation of every component, we must again go back to the probability tree analysis used by THERP.) You can see that as the number of components (i) in a serial system increases, the probability of system failure gets larger. (If you can't see this, try the Workbook project.)

We can greatly improve system reliability by adding redundancy in a parallel arrangement where one (or more) components back up the original component. For the simplified parallel arrangement where all components have the same reliability (r) and the same number of components in parallel for each function (c), the reliability can be calculated as

$$\text{Reliability of parallel system} = [1 - (1 - r)^c]^i$$

where i is the number of task components. By adding enough backups we

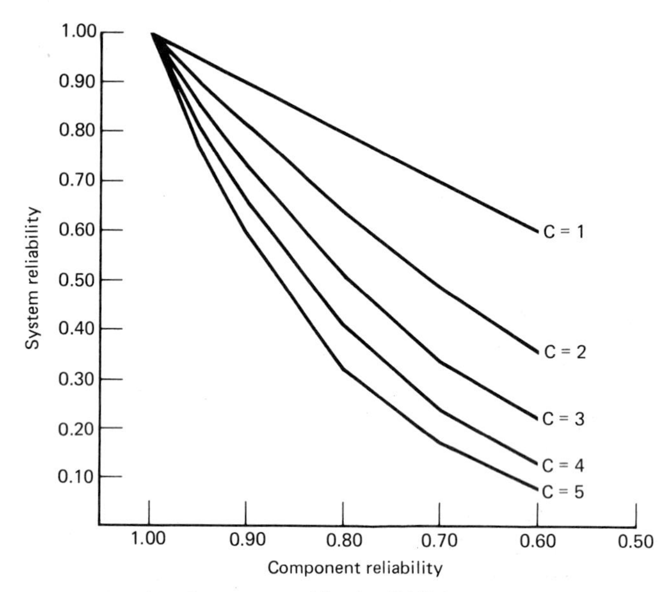

C = Number of components. All unit reliabilities are assumed to be equal.

Figure 2-9. Reliability of a serial system as a function of the number of task components (i) and the reliability (R) of each component.

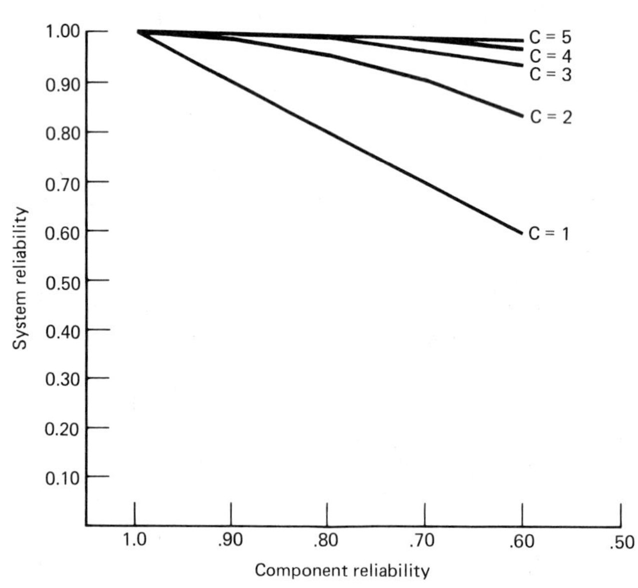

C = Number of components. All unit reliabilities are assumed to be equal.

Figure 2-10. Reliability of a parallel system as a function of the reliability of each component (r), and the number of components in parallel (c). There is only one task component ($i = 1$). Thus, the curves labeled C = 1 in figures 2-9 and 2-10 are identical representing a single "box", e.g., A in figure 2-8. In this figure, additional boxes (e.g., A_1, A_2, . . .) are added in parallel.

can take components with low reliability and still create a system that is highly reliable (see Figures 2-9 and 2-10).

SUMMARY

Minimizing human error is a major goal of human factors. An error is defined as an action that violates some tolerance limits of a system. The human error probability (HEP) is a basic unit of human reliability. There are several methods for classifying errors.

A code is an alphabet of symbols plus a system of fixed constraints among the symbols. Communication is the successful transmission of a code from a source to a destination. Noise makes communication more difficult, leading to potential errors. Special types of codes are used to detect and to correct errors. Information theory is a mathematical tool that can be used to obtain quantitative estimates of error. Information and coding are closely related concepts. But while information theory can tell us if a particular code is optimal, it leaves the creation of such optimal coding schemes to our imagination.

Human reliability analysis attempts to provide quantitative estimates of human error for various industrial tasks. Although the need for such estimates is acknowledged by all, this is becoming a controversial enterprise since some human factors specialists are skeptical about meeting the mathematical and logical assumptions that are required to perform meaningful calculations. Other specialists believe that these estimates have sufficient accuracy to be useful. System reliability can be improved by adding redundancy.

APPENDIX Error-detecting Codes

Error-detecting models (Hamming, 1980) usually assume two things about errors:

1. There is an equal probability (called p) of an error in every position.
2. Errors are independent—that is, an error in one position does not influence errors elsewhere.

Given these two assumptions and a binary message n symbols long, the probability of a correct message is $(1-p)^n$. The probability of a single error is $np\ (1-p)^{n-1}$. The probability of k errors is the kth term in the binomial expansion

$$1=[(1-p)+p]^n=(1-p)^n+np(1-p)^{n-1}+\frac{n(n-1)p^2(1-p)^{n-2}}{2}+\cdots+p^n$$

Let $C(n,k)$ represent the combination of n things taken k at a time. Now, we need the probability of an even number of errors. This is obtained by taking the mean of the next two binomial expansions:

$$1 = [(1-p) + p]^n = \sum_{k=0}^{n} C(n,k)p^k(1-p)^{n-k}$$

$$[(1-p) - p]^n = \sum_{k=0}^{n} (-1)^k C(n,k)p^k(1-p)^{n-k}$$

$$\frac{1 + (1-2p)^n}{2} = \sum_{m=0}^{[n/2]} C(n,2m)p^{2m}(1-p)^{n-2m}$$

Equation 2.1A

where the bracket over the summation sign is interpreted as "the greatest integer in." Equation 2.1A is the probability of an even number of errors. However, this includes the probability of zero errors as the first term in the series of Equation 2.1A. Thus, our final equation starts with $m = 1$, the second term of the series, and is then expanded. It is usually worthwhile to compute only the first few terms of Equation 2.2A:

Equation 2.2A The probability of an undetected error in an even-checking parity code.

$$\sum_{m=1}^{[n/2]} C(n,2m)p^{2m}(1-p)^{n-2m}$$

With Equation 2.2A you can answer all kinds of questions about undetected errors based on the length of the message (n) and the probability of an error (p). (A representative sample of some of these questions resides in the accompanying Workbook.)

The assumption that errors are independent may be adequate for certain types of machines but does not work well for errors of people. For example, in dialing a telephone number, say 423-5728, you might dial 423-5782 interchanging the last two digits. This kind of error can be detected by a *weighted* code. A commonly used alphabet has 37 symbols: 26 letters, the digits 0 to 9 and a blank space. A weighted code multiplies the last symbol in the message by 1, and next to last by 2 and so on. We then take the weighted sum of the symbols in a message, divide it up by 37 and take the remainder. (This is mod 37 arithmetic.) Finally, we add a check symbol at the end of the message so that the weighted sum of all digits (including the new check digit) is 0 mod 37 (has no remainder after being divided by 37). This kind of coding will prevent the interchange error described above as well as many other kinds of errors people make. Weighted coding schemes can be used on credit cards to prevent forgeries. With a 37-symbol alphabet, the probability that a randomly created credit card identification will satisfy the 0 mod 37 check is 1/37. Thus, if you try to make up a credit card number when telephoning in some charge, and the card uses this kind of coding, there is only about a 3%

chance that your forgery will work. A weighted-coding scheme is used by book publishers all over the world. The ISBN (International Standard Book Number) found on all books is based on mod 11 arithmetic. Here is an ISBN code that has sentimental meaning for one of the authors:

0-470-45674-4

Let us check if this is a legitimate number. First, we ignore the hyphens. Then we multiply the last number, which is the check digit, by 1 to get 4. To this we add twice the next to last number (8) to get 12. Continuing on we have this sum:

$$4 + 8 + 21 + 24 + 25 + 24 + 0 + 56 + 36 = 198$$

Dividing this by 11 we get 18 with no remainder. Thus, the author's first book has a legitimate ISBN number. You can try this out on any book of your choice. Since this is mod 11 arithmetic the symbol X is needed to represent a check digit of 10. This is why some ISBN numbers end with an X instead of a digit.

Error-correcting Codes The codes discussed so far have a common flaw. What do we do if an error is detected? The code does not tell us how to find the error. The simplest solution is to transmit the message over again. Repetition always increases reliability.

However, it is possible to devise clever codes that do not require repetition to correct errors. These require more than one parity check. By assigning a value of 1 to a failure of a parity check and 0 to a success it is possible to locate the error by adding the values of the parity checks. The scheme that does this is called a Hamming code. There are many other kinds of error-correcting codes but their technical details fall outside the scope of this text. If you are interested read Hamming (1980). The important point for you to remember is that different codes are better at detecting different kinds of errors.

PART TWO

HUMAN CAPABILITIES

3 HEARING AND SIGNAL DETECTION THEORY

Thump on the desk with your hand. The thump sound you just made was brief and had a moderately low pitch. Snap your fingers (if you can). That clicklike sound was somewhat higher in pitch and even shorter in duration. Such sounds actually involve very complex patterns of air vibration. Measuring or specifying such sound wave patterns can be a difficult job, even with the help of computers and sound analyzers. This chapter discusses the basic aspects of sound and hearing. In addition, the chapter describes how sound wave patterns are specified mathematically, and the important related concepts of system distortion and nonlinearity. These topics form the foundation for the design and evaluation of the systems covered in Chapter 8 (Auditory and Tactile Displays), Chapter 9 (Speech Communication), and Chapter 16 (Noise). These concepts are also important in areas outside of sound and hearing. For example, the techniques of signal analysis are employed to describe sharpness of vision (see Chapter 4) and human control behavior (see Chapter 12). The last third of this chapter is devoted to a discussion of signal detection theory. Signal detection theory is important to the human factors specialist in at least two ways. First, it helps us to understand the functions and limitations of the auditory and visual senses. Second, it is probably the most powerful tool for measuring and predicting human performance in a variety of different tasks. The human factors specialist can often find performance data for one task but needs to predict what performance will be in a different task that involves the same general physical variables. Detection theory provides techniques for making those predictions.

SOUND WAVES AND HEARING

The air around us is a collection of billions of air molecules, constantly moving and exerting pressure on neighboring air molecules and on the molecules of other substances in their paths. When there is a sound or other disturbance in the air, there is a temporary increase and decrease in the density of these molecules, and hence an increase and decrease in air pressure around the disturbance. The air molecules in the region of the disturbance pass on these effects to their neighbors, and the pressure change moves outward through the air as a sound wave.

Sound waves are usually generated by the vibration of a physical object. Figure 3-1 illustrates the vibration of a tuning fork. The tuning fork is a carefully designed device that, when struck, vibrates in a special way known as simple harmonic motion. If we could measure the intensity of the pressure near the fork, we would find that it changes rapidly in time as the fork vibrates. This regular pattern of pressure change is a sine wave, and has the mathematical form:

$$p(t) = A \sin 2\pi ft \tag{3-1}$$

where A is defined as the *amplitude* of the pressure waveform and f is the *frequency* of the fork's vibration, measured in cycles per second, and defined as hertz (one cycle per second equals one hertz, abbreviated Hz). If you remember your trigonometry, you recall that the sine function has a maximum value of 1.0 when the angle is 90° (or 180°) since sin (90°) = 1.0. So the maximum pressure that $p(t)$ can reach is equal to A, when the angle is 90° or $\pi/2$ radians. (Please switch your mind and your calculator over from degrees to radians, 1 radian = 180°/π.) Since sound travels in air at about 1100 feet per second, the pressure you measure (at a given moment) would also depend on the distance from the fork. Note that,

$$\lambda = c/f \tag{3-2}$$

where λ (lambda) is the *wavelength* of sound, and c is the speed of sound. So at a frequency of fork vibration of 1100 Hz, λ would equal 1100/1100 = 1 foot. For that frequency, the pressure peaks would be 1 foot apart in the space around the fork.

Suppose that we had a second tuning fork (in another area) vibrating at the same frequency as the first one. The difference between the sound pressure waves generated by these two devices would depend on the precise moment each was first struck; this timing difference is known as the *phase* difference between the waves and is characterized by Θ in the equations for the sounds generated:

$$p_1(t) = A \sin(2\pi ft) \tag{3-3}$$
$$p_2(t) = A \sin(2\pi ft + \Theta) \tag{3-4}$$

where $p_1(t)$ is the pressure waveform of the fork struck first, and $p_2(t)$ is the waveform of the fork struck second (see Figure 3-2). We have completely specified a sinusoidal waveform if we know its amplitude, frequency, and phase.

The pressure changes in air caused by a vibrating object are actually increases and decreases from the normal atmospheric pressure (about 1000 mbar or 10^6 μbar or 10^6 dynes/cm², at sea level). Since the pressure

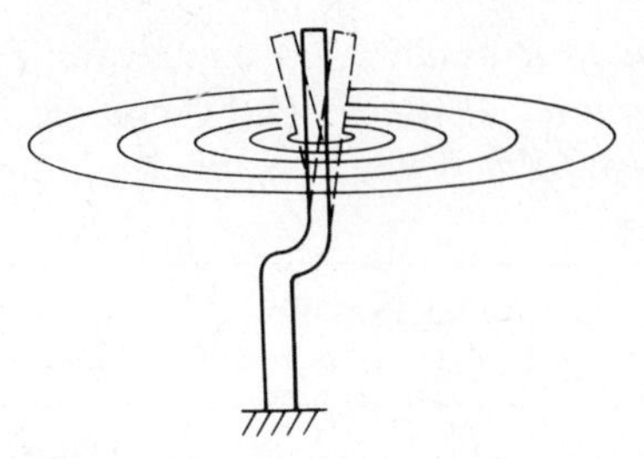

Figure 3-1. Vibration of a tuning fork.

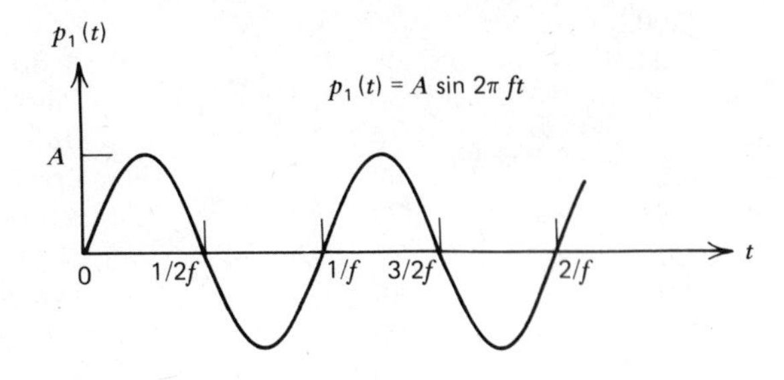

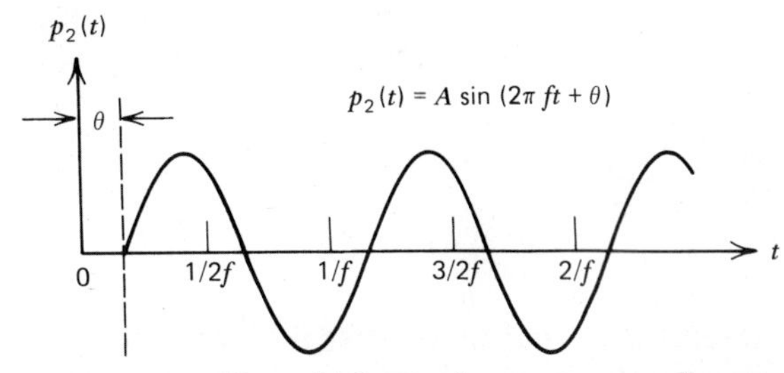

Figure 3-2. Sinusoidal sound pressure waveforms.

waveform $p(t)$ is varying in time, we need to describe its magnitude with a special kind of average measure, the root-mean-square (rms) value. (If the positive and negative pressure changes from the atmosphere pressure were simply averaged, the negative values would cancel out the positive values. So we've got to first square each pressure, and then take the average. The square root of that average value is the rms value.)

Sound Pressure Level

Our hearing is sensitive to sound pressure variations that span an extremely wide range. Under good conditions, we can detect sound waves whose pressure variations are only about 0.0002 μbar (0.0002 dynes/cm²). At the high end, sound pressures of about 200 μbar begin to be painful as stimuli to our ears. This is a range of pressures of $200/0.0002 = 10^6$. Needless to say, a concise scale for describing this wide span of sound pressures would be a great convenience. Such a scale is the decibel scale, abbreviated, dB, based on a logarithmic measure of the ratio be-

TABLE 3-1 *Sound Pressures and Levels for Typical Environment Conditions. (From F. A. White,* Our Acoustic Environment, *p 39, copyright 1975 by John Wiley & Sons, Inc., and reprinted by permission.)*

Sound Pressure (μbar)	Sound Pressure Level[b]	Environmental Condition
0.0002	0	Threshold of hearing
0.00063	10	Rustle of leaves
0.002	20	Broadcasting studio
0.0063	30	Bedroom at night
0.02	40	Library
0.063	50	Quiet office
0.2	60	Conversational speech (at 1 m)
0.63	70	Average radio
1.0[a]	74	Light traffic noise
2.0	80	Typical factory
6.3	90	Subway train
20	100	Symphony orchestra (fortissimo)
63	110	Rock band
200	120	Aircraft takeoff
2000	140	Threshold of pain

[a] 1 μb = 1 dyne/cm^2 = 74 dB is a common reference pressure for instrument calibration. It is also a reference pressure for measurements in underwater acoustics.
[b] Reference is 2×10^{-4} μbar.

tween two sound pressures: the pressure of interest, p, and a reference pressure, p_r. Sound pressure *level* in decibels is defined as:

$$L_p = 20 \log_{10} \left(\frac{p}{p_r}\right) \tag{3-5}$$

where p and p_r are rms measures of sound pressure and p_r is a reference pressure of 0.0002 μbar.

What is the sound pressure level, L_p, of an rms pressure of 0.4 μbar?

$$L_p = 20 \log \frac{0.4}{0.0002} = 20 \log(2000) = (20)(3.3) = 66 \text{ dB}$$

Table 3-1 summarizes sound pressure levels and sound pressures for some typical environmental conditions. Techniques for measuring sound level and assessing the effect on people are discussed in Chapter 16.

Structure of the Ear

Figure 3-3 summarizes the major structural components of the auditory system. Sound is directed into the auditory canal via the pinna, resulting in vibration of the tympanic membrane, the eardrum. Vibrations of the

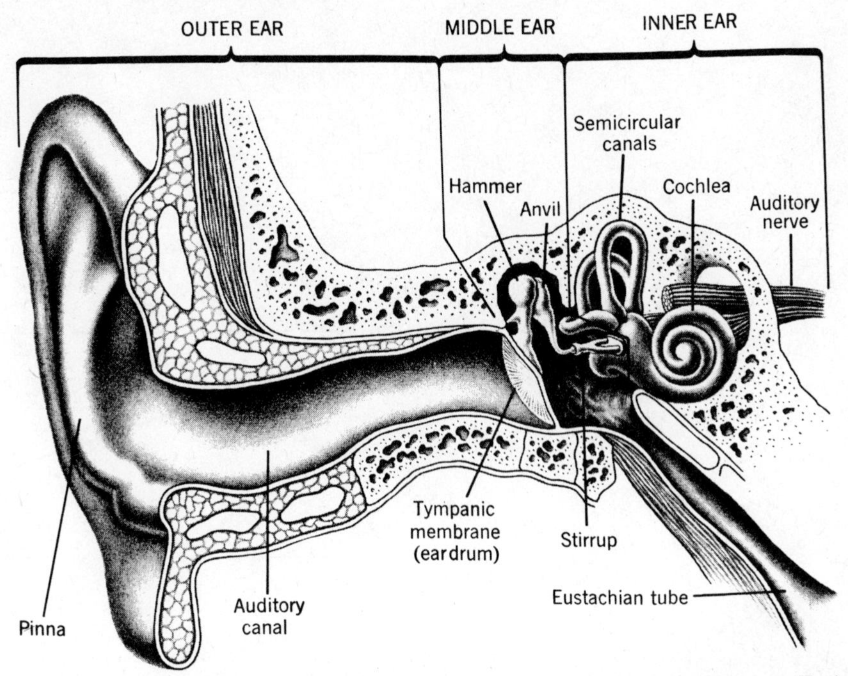

Figure 3-3. Major structural components of the auditory system. (From F. A. White, *Our Acoustic Environment,* p. 106. © Copyright 1975 by John Wiley & Sons, Inc. and reproduced by permission.)

eardrum are conveyed to the inner ear via a chain of bones called the *ossicles,* (the hammer, anvil, and stirrup). The levers and linkages of the ossicles provide an efficient way for the eardrum movement to produce movement of the fluid in the cochlea. The cochlea of the inner ear contains the receptor mechanism for our hearing (see Figure 3-4*a*). The cochlea resembles a coiled tube of about $2\frac{5}{8}$ turns. Within this tube are two passages separated by the basilar membrane, which vibrates when sound causes movement of the fluid in the cochlea. Along the membrane are rows of sensory hair cells that are extremely sensitive to mechanical movement. Movement of the hair cells initiates a complex process that causes electrical signals to be transmitted by the auditory nerve to the brain. If you could observe the membrane during sound stimulation, you would see a moving pattern known as a traveling wave as shown in Figure 3-4*b*. The traveling wave pattern for a low-frequency sound has a maximum displacement farther from the stapes than does the maximum displacement for a high-frequency tone. Somehow the auditory nervous system decodes the displacement patterns that different sounds produce along the basilar membrane.

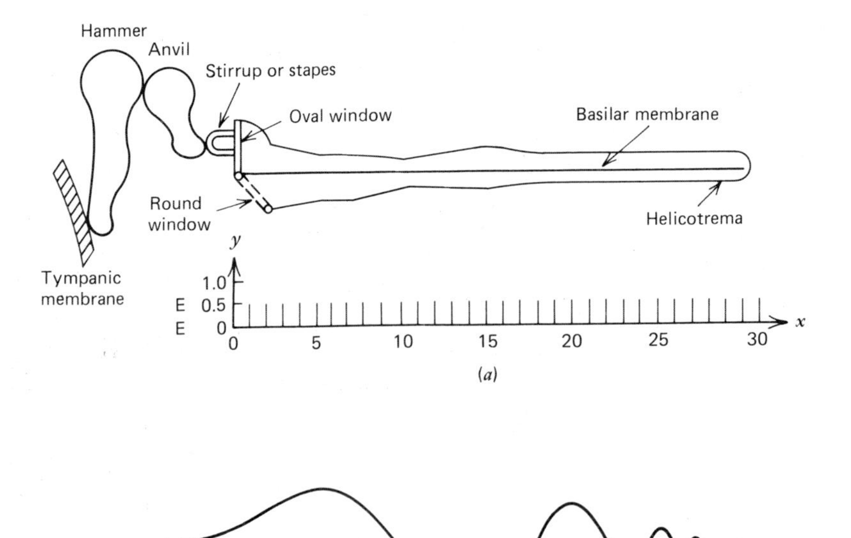

Figure 3-4. (*a*) Schematic diagram of cochlea (uncoiled). (*b*) Traveling wave on basilar membrane. (From F. A. White, *Our Acoustic Environment,* p. 110. Copyright 1975 by John Wiley & Sons, Inc., and reproduced by permission.)

Specification of Complex Sounds

The sound that we hear when we stimulate our ears with a sine wave is a *pure tone*. Usually we hear pure tones only in the laboratory when we generate them with special oscillators or tuning forks. Most sound waves have a more complex shape than the sine wave. This is because most physical objects, when struck, vibrate in a more complicated way than the tuning fork we described. A physical object may vibrate simultaneously in several different simple ways, and we can imagine that there will be a sinusoidal sound waveform generated by each of these different modes of vibration. The resulting sound waveform will be the *sum* of the separate sinusoids associated with each of these modes of vibration. This physical description has an important mathematical interpretation owed to the French mathematician J. B. J. Fourier. Fourier showed that any periodic waveform can be represented as the sum of a number of harmonically related sine waves. In the example illustrated in Figure 3-5, the complex sound wave is composed of two sine waves of frequencies 200 Hz and 400 Hz. Figure 3-6 shows the sine waves that make up a repetitive square wave of 1000 Hz. The lowest frequency present, the fundamental, is related to the repetition rate of the complex

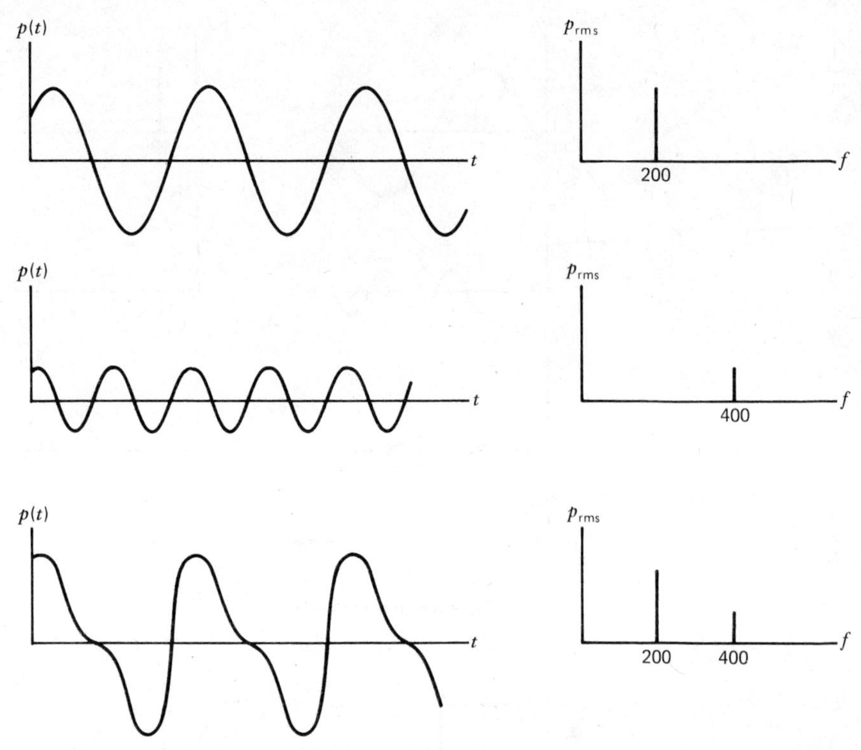

Figure 3-5. Complex waveforms made up by summing two sinusoids of 200 Hz and 400 Hz.

waveform. Figure 3-5 and 3-6 also illustrate a way of portraying a sound as a plot of rms amplitude versus frequency, rather than amplitude versus time. These plots in the frequency domain illustrate the signal's *spectrum*. Complex sounds can have discrete or line spectra, as in Figure 3-7*a*, as well as continuous spectra as shown in Figure 3-7*b*. Continuous spectra are composed of very large numbers of frequency components; many types of noise sounds and nonperiodic sounds include continuous spectral regions. The spectra and effects of different environmental and workplace noise sources are described in Chapter 16.

Suppose that you are given a periodic sound and you need to know the amplitude and phase of the individual sinusoidal components that make it up. Appendix A to this chapter contains the equations needed to compute these values. Instead of computing these values you could feed the sound into a frequency analyzing device. Another technique is simply to listen to the sound. The idea that our ears can perform a crude Fourier analysis of sound is known as Ohm's acoustic law (Green, 1976). To a limited extent, we can separately hear out the individual component

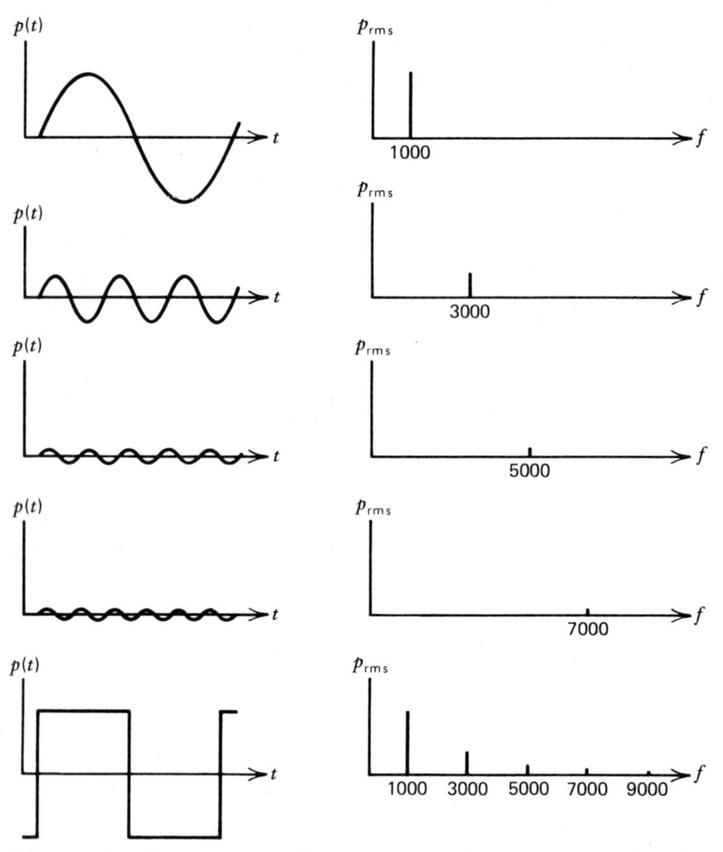

Figure 3-6. Complex waveforms made up by summing many sinusoids of odd multiples of 1000 Hz.

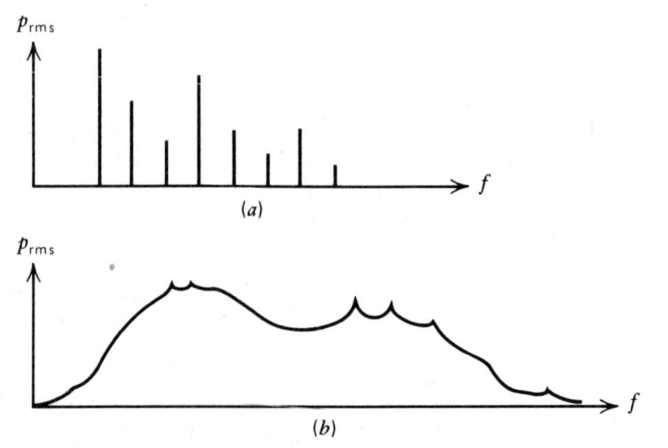

Figure 3-7. (*a*) Sounds composed of discrete spectrum. (*b*) Continuous spectrum.

frequencies of sounds that stimulate our ears. If only one frequency component in a complex sound stimulus is changed, we can usually detect that change.

Sensitivity and Discriminability

Our auditory system is sensitive to a range of sound frequencies from about 20 to 20,000 hertz. Highest sensitivity is to frequencies around 1000 to 3000 Hz. The bottom curve of Figure 3-8 illustrates how our sensitivity, measured as the smallest level required for reporting the tone, varies with frequency. Minimum sound pressure at the best frequency is below 0 dB, or 0.0002 μbar; this pressure corresponds to movements of the eardrum that are smaller than the size of individual air molecules.

We can discriminate a *change* in frequency as small as 2 Hz, but this discriminable change, Δf, depends on the initial frequency, f,

$$\frac{\Delta f}{f} = k \qquad (3\text{-}6)$$

where Δf is the amount of change in the initial frequency, f, and k is a constant. For frequencies above 500 Hz, $k \approx 0.005$. Below 500 Hz, Δf is approximately 4 Hz. Differences in the *level* of a pure tone of approximately 0.5 dB can be discriminated over a wide range of different levels and tone frequencies.

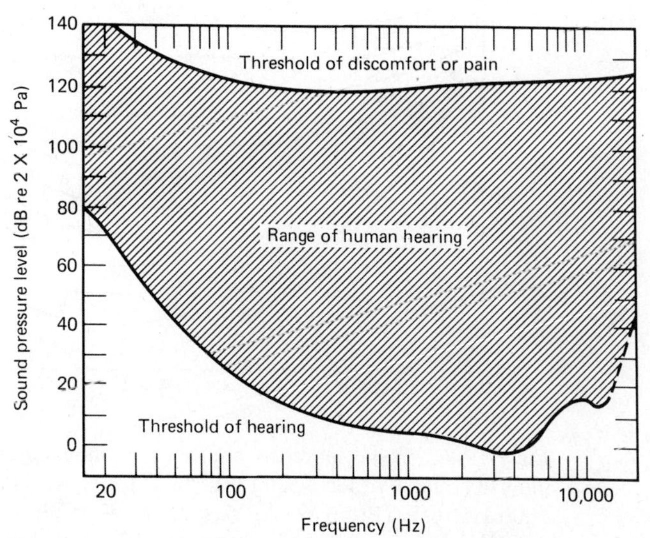

Figure 3-8. The range of human hearing. (From F. A. White, *Our Acoustic Environment*, p. 114. Copyright 1975 by John Wiley & Sons, Inc., and reproduced by permission.)

Loudness

The loudness of a sound stimulus is related to the amplitude of the sound. Suppose that you stimulate a person's ear with a 40-dB, 1000-Hz pure tone, and then ask the person to increase the level until the tone seems to be twice as loud. You'd find out that an increase in level of about 10 dB was required. This relationship between loudness and sound level can be expressed as a scale of loudness, where loudness in *sones, S,* is:

$$S = 10^{0.03(L_p - 40)} \tag{3-7}$$

where L_p is the sound level of the 1000-Hz tone under consideration. A 40-dB, 1000-Hz tone will have a loudness of 1 sone; a 50-dB, 1000-Hz tone will have a loudness of 2 sones.

Because our sensitivity to sound depends on frequency, a 40-dB, 4000-Hz tone or 300-Hz tone will not appear as loud as a 40-dB, 1000-Hz tone. This can be seen in Figure 3-9, which shows the level of tones of different frequency judged to be equally loud as a 1000-Hz comparison tone. The 40-dB curve shows the level (on the ordinate) of different frequency tones that will be judged equally loud as a 40-dB, 1000-Hz tone. For example a 60-Hz tone would have to be almost 60 dB to match that 1000-Hz tone. The unit of loudness level is the *phon*. The loudness level in phons for a sound is the sound level (dB) of a 1000 Hz-tone

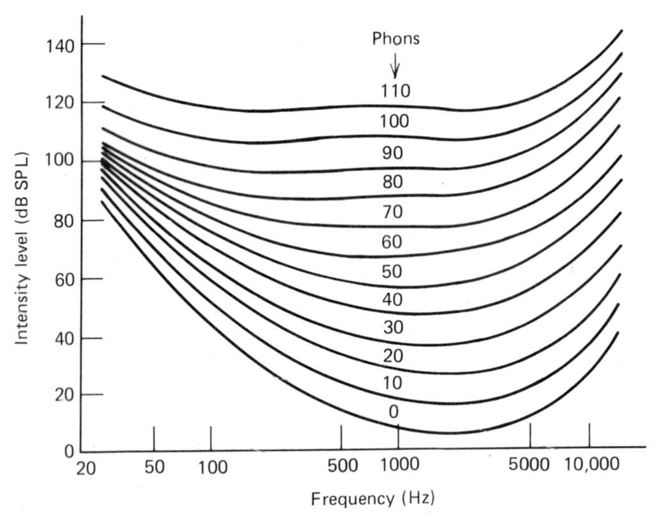

Figure 3-9. Equal loudness contours. From each curve one may determine the intensity and frequency of tones that are equal in loudness to the referenced, 1000 Hz tone. (From W. A. Yost & D. W. Nielson, *Fundamentals of Hearing: An Introduction,* p. 167. Copyright 1977 by Holt, Rinehart & Winston, and H. Fletcher & W. A. Munson, Loudness, its definition, measurement, and calculation, *Journal of the Acoustical Society of America,* 1933, *5,* p. 91, and reproduced by permission.)

judged to be equally loud. The phon measure allows computation of the effective loudness of different frequency tones (or combinations of tones) in terms of the level of a 1000-Hz tone of equivalent loudness. The advantage of the phon measure is that it is expressed in familiar decibel units (see Chapter 16 for a discussion of other loudness level and annoyance measures).

One important aspect of the loudness relationship described in equation 3-7 is that the exponent value of 0.03 changes under conditions of noise interference or hearing loss. Under noise interference conditions, the perceived loudness of a sound can increase very rapidly as the level of the sound is increased. An increase in the sound level of just a few decibels can result in a manyfold increase in loudness. This phenomenon is known as loudness recruitment, and is frequently observed in older persons who have a partial hearing loss. It accounts for why elderly people are sometimes very uncomfortable in the presence of an increased level of sound, such as at a concert. In such environments, the sound loudness may appear much greater than for a younger person with normal hearing.

Pitch

An obvious quality of a sound is its pitch. Pitch usually corresponds to the rate of repetition of the sound or to the frequency region where most of the energy is concentrated. There are important exceptions to this rule, some of practical significance to the human factors specialist. (See the box on the Missing Fundamental). Scaling procedures similar to the sone scale of loudness judgment have been applied to pitch. A standard 1000-Hz, 40-dB tone is assigned an arbitrary pitch of 1000 mels, and other tones are adjusted in frequency until they appear to be twice the pitch of the standard. The resulting relationship is shown in Figure 3-10*a*. You can see that doubling the frequency of the signal does not result in a doubling of the perceived pitch. In musical pitch, however, the scale of pitch depends on the octave frequency relationship, which is a doubling of frequency (see Figure 3-10*b*). For example, the musical note A (above middle C) is at 440 Hz. The A note in the next octave is at 880 Hz. Within each octave are 12 notes equally spaced in frequency (in an even-tempered scale). Some people have the ability to either identify or generate (by humming or playing an instrument) any given musical note. This is called having "perfect pitch." Most errors made by such people in identifying tones are either plus or minus one musical note (semitone) or plus or minus one octave (Lockhead & Byrd, 1981). If you were assessing the performance of an auditory display system that coded information by pitch (see Chapter 8), it would be important to determine the background of the people being tested. The participation of people with musical backgrounds or ability could result in very large variations in the performance of the system.

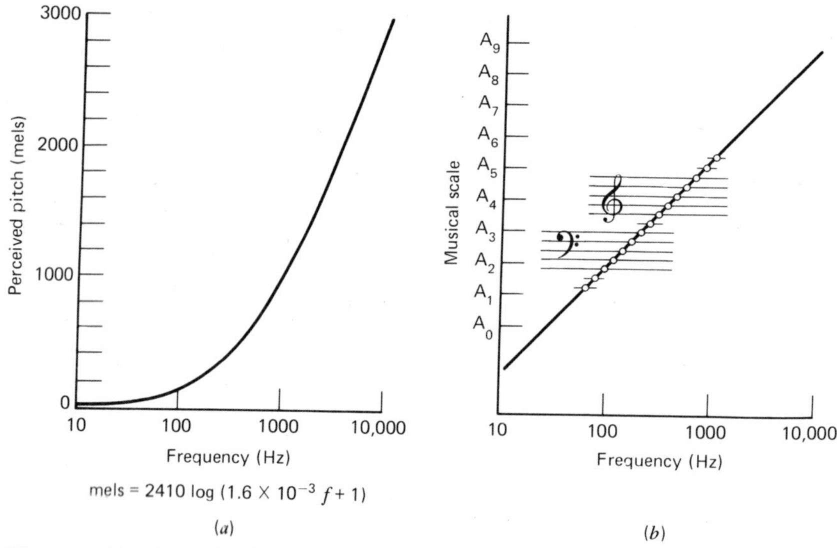

Figure 3-10. Perceived pitch as a function of frequency (*a*), and the musical scale (*b*) (From P. H. Lindsay & D. A. Norman, *Human Information Processing, An Introduction to Psychology*, p. 163. Copyright 1977 by Academic Press, Inc., and reproduced by permission.)

A SLICE OF LIFE

The Missing Fundamental

How can your pocket size transistor radio, with a $1\frac{3}{4}$ inch speaker, generate the low-frequency sounds that you hear when it plays your favorite country-western sound? It can't. That trusty little radio will generate middle- and high-frequency tones, but it just can't hack the lows. (Half the wavelength of a 275-Hz tone is about 24 inches; forget it.) But, you insist, you do hear those lows? Yes, you do—but just the same, they aren't really there. Perceived pitch is a complicated function of the frequency components that are present in a sound. Sometimes we hear a pitch even when there is no sound energy present at the frequency that "normally" corresponds to that pitch. The small transistor radio example is known as the "missing fundamental" effect. Certain sound stimuli that contain harmonically related frequency components, such as at 1200, 1400, and 1600 Hz, will have a pitch that corresponds to a much lower frequency tone at 200 Hz. The 200-Hz tone is the "missing" fundamental shared by the 1200-, 1400-, and 1600-Hz components. There are many experiments dealing with the origin of this effect but the precise mechanisms are still not well understood. Two experiments shown in Figure 3-11 are interesting: First, if you add low-frequency noise to the sound you won't

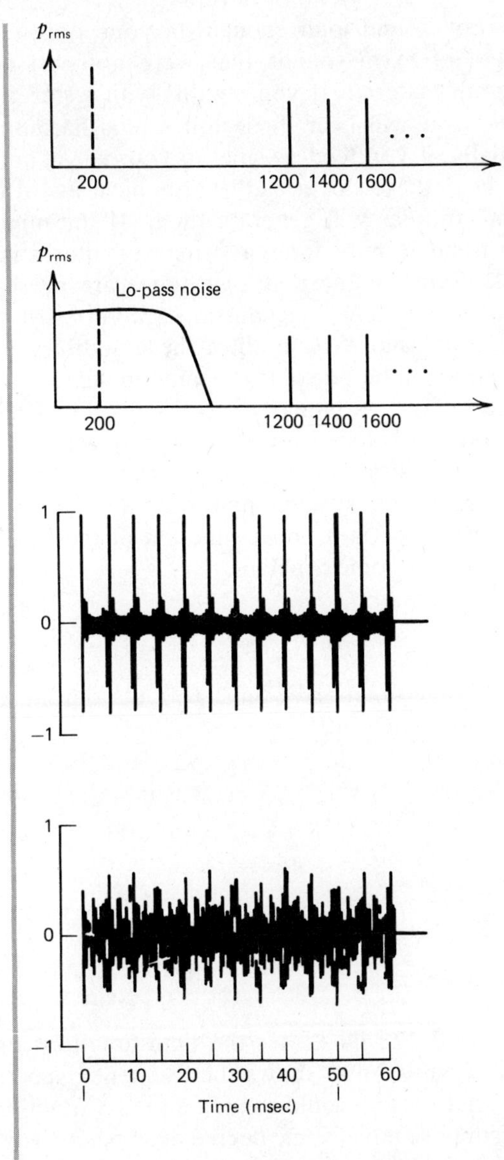

Figure 3-11. Adding noise to the set of high frequency tones (*a* and *b*) does not mask the pitch at 200 Hz. In (*d*) the (12) tones are added at random phases, in (*c*), the components are phase locked. These two conditions produced the same pitch. (Parts c. and d. from R. D. Patterson, The effects of relative phase and the number of components on residue pitch, *Journal of the Acoustical Society of America,* 1973, *53,* 1566, and reproduced by permission.)

wash out the 200-Hz pitch. So the pitch does not seem to arise out of stimulation of the low-frequency sensitive part of the basilar membrane. Second, if you manipulate the starting phase of the components you can arrive at a sound envelope that either repeats itself at 200 Hz (has a lot of *periodicity*) or looks like noise (has no *periodicity*), but this manipulation has very little effect on the perceived pitch! (Patterson, 1973). So neither the place of stimulation on the basilar membrane nor the periodicity of the sound envelope can completely explain the missing fundamental effect.

Distortion and Nonlinearity

Play any sound loud enough on your stereo system, and you'll hear a variety of extra sounds that were not part of the originally recorded program material. If you try this with a stereo test record that has pure tones, you will hear those tones plus harmonics of the original tones. Actually, if you feed a signal to your ear at an intense enough level, you will hear these "overload" harmonics, even though the input signal is clean–the ear will generate them. If the input signal is more complex than a single pure tone, a variety of interesting combination tones will result. Some of these strange tones are present even at relatively low sound levels. The mechanism of nonlinearity that we describe in this section will not explain all of these auditory phenomena. However, the concepts will be relevant to many applications remote from the auditory situation such as human control behavior (Chapter 12). If trigonometry fills you with intense anxiety, you may want to jump ahead to the section Binaural Hearing.

Consider the system input-output relation shown in Figure 3-12*a*. The output, y, is a linear function of the input, x. Any input is simply multiplied by some constant, C_1.

So

$$y = C_1 x$$

If x is some function of time, $x(t)$, the output will simply be equal to C_1 times that function.

That is

$$y(t) = C_1 x(t)$$

So if

$$x(t) = A \sin 2\pi f_1 t$$

then

$$y(t) = C_1 A \sin 2\pi f_1 t \tag{3-8}$$

Figure 3-12*a* illustrates the time functions on the x and y axes of the graph. Figure 3-12*b* shows the frequency spectrum of the input and output signals. What would happen if the x,y function were not quite a straight line; suppose it had some degree of curvilinearity?

Let

$$y = C_1 x + C_2 x^2$$

where C_2 is another constant. What would happen if the sine wave signal were fed into this system? Now,

$$y(t) = C_1 A \sin 2\pi f_1 t + C_2 (A \sin 2\pi f_1 t)^2 \tag{3-9}$$

We can consider the two terms on the right side of equation 3-9 separately. The first term is the same output as in the previous system,

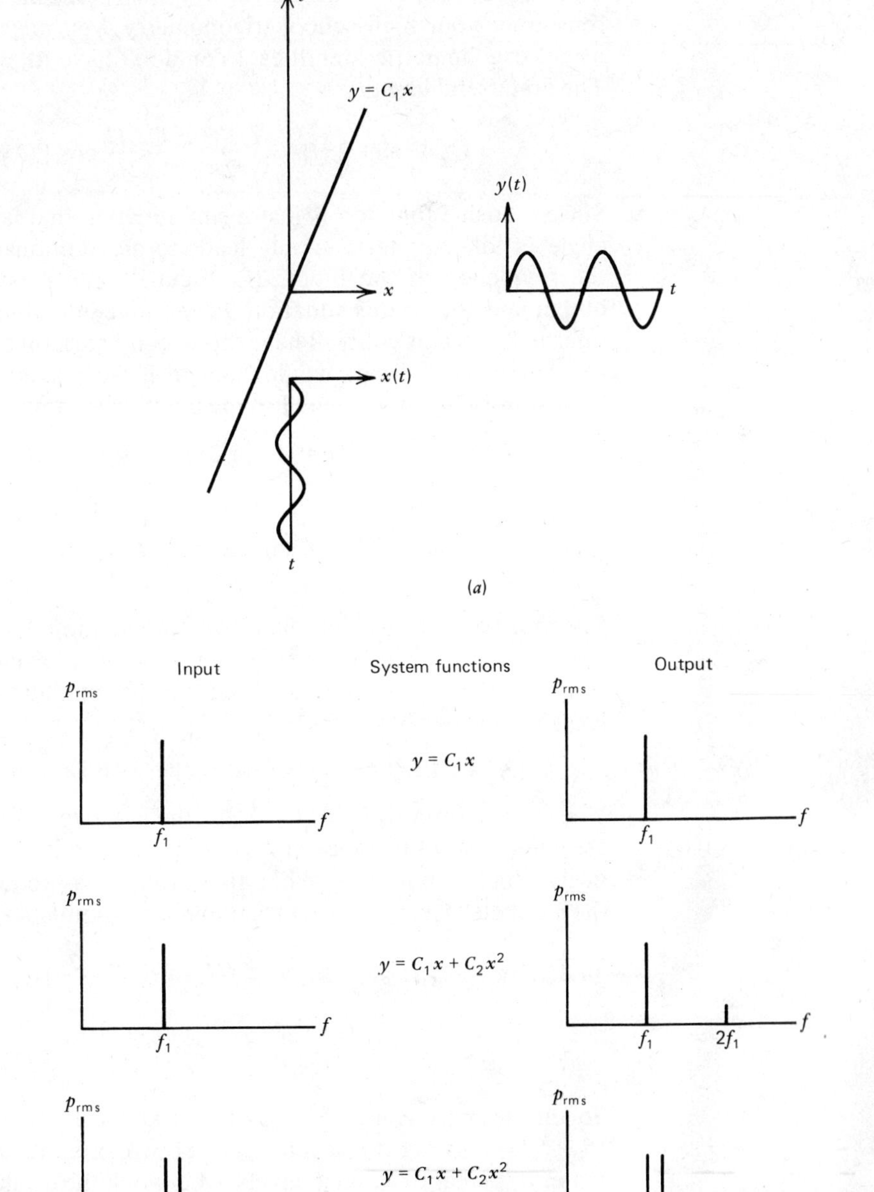

Figure 3-12. Input and Output from a linear system (*a*). See text for a derivation of the output components in (*b*).

$C_1 A \sin 2\pi f_1 t$. The second term yields $C_2 A^2 \sin^2 2\pi f_1 t$. If you could remember your high school trigonometry, you might remember various useful trigonometric identities. (You don't have to, we will supply them.) The first useful identity is

$$C_2 A^2 \sin^2 2\pi f_1 t = \frac{C_2 A^2}{2} - \frac{C_2 A^2}{2} \cos [(2\pi)(2f_1 t)] \tag{3-10}$$

Since a cosine function is just a sine function that is delayed by a phase angle of 90°, this term simply leads to an additional pure tone at *twice* the frequency of the input, $2f_1$. Figure 3-12*b* illustrates the input and output spectra for this situation. Play a tone into your ear at a sufficiently intense level, and you will hear this second harmonic at twice the original frequency. Now what would happen if we played two tones into this slightly curvilinear system? Let the input, $x(t)$, now be equal to

$$x(t) = A \sin 2\pi f_1 t + B \sin 2\pi f_2 t \tag{3-11}$$

Then,

$$y(t) = C_1 (A \sin 2\pi f_1 t + B \sin 2\pi f_2 t) + C_2(A \sin 2\pi f_1 t + B \sin 2\pi f_2 t)^2 \tag{3-12}$$

The first term on the right-hand side of equation 3-12 will lead to sinusoids at f_1 and f_2, just as in the first case we considered. However, the second term on the right-hand side of the equation is more interesting. Expanding it, we have

$$C_2[A^2 \sin^2 (2\pi f_1 t) + B^2 \sin^2 (2\pi f_2 t) + 2AB \sin (2\pi f_1 t) \sin (2\pi f_2 t)]$$

You can see from equation 3-10 that the first two terms inside the brackets are going to lead to tones at $2f_1$ and $2f_2$, respectively, in the same way as in the single input case. What about the cross-product (the last) term in the brackets? Using another trigonometric identity we have

$$2ABC_2 \sin (2\pi f_1 t) \sin (2\pi f_2 t) = \frac{ABC_2}{2} \cos [(2\pi)(f_2 - f_1)t] - \frac{ABC_2}{2} \cos [(2\pi)(f_1 + f_2)t] \tag{3-13}$$

So this term is going to yield tones at the frequencies: $(f_2 - f_1)$ and $(f_1 + f_2)$. Consider the specific case of two tones at 1000 Hz and 1250 Hz fed into the ear at intense levels: you would hear pitch corresponding to tones at 1000, 1250, 2000, 2500, 250, and 2250 Hz! In fact, you would hear other combination tones in addition to these, suggesting that the equation describing the nonlinearity of the ear includes higher-order terms other than x^2; probably at least some x^3 as well. What if the sound levels were not very intense? Well, some of these strange tones also occur at moderate sound levels. A particularly prominent one is known as the cubic difference tone, because it may arise out of nonlinearity of the

C_3x^3 sort. This tone occurs at a frequency equal to $2f_1 - f_2$. In the preceding example, the cubic different tone would yield a pitch at 2000 − 1250 = 750 Hz. The precise nature and origin of nonlinearities in the auditory system remains the subject of intense physiological and psychological experimentation.

Binaural Hearing

We can think of several reasons for having two ears, rather than only one. Having two ears preserves our bilateral symmetry (joke #1), gives us a spare (joke #2), allows us to localize the source of sound stimuli (no joke). Some engineers would point out that having two ears, like having two (or more) antennas, should enable us to process information much more effectively than if we had only one ear. That is certainly the case in hearing. Under the right conditions, we can detect signals in noise that are 15 dB fainter than detectable with one ear alone!

We can also process speech messages in noise with similar advantages. A way to improve a pilot's understanding of speech messages in noise was developed by Licklider (1948) using binaural principles. Suppose that a pilot, wearing headphones, is immersed in an intense noise environment. Each ear gets essentially the same noise input. Licklider found that reversing the wires leading the speech input to one of the pilot's headphones greatly improved speech intelligibility. You can demonstrate the binaural advantage to yourself: go to a crowded cocktail party and try to listen to different conversations around you. Now try it again with one hand tightly clasped over one ear. To understand how these great gains in signal detection and recognition can be accomplished, let's first consider the question of how sound sources are localized with our ears. (Chapter 6 discusses some further consequences of auditory attention and time-sharing).

Sound Localization Figure 3-13 is a top view of a person's head while being stimulated by a sound source in the azimuth plane (the plane containing your ears and parallel to the ground). You can see that the sound wave travels along a shorter path to reach the right ear than the left. The sound wave will therefore reach the right ear before the left, and there will be an *interaural time difference* in the arrival time of the sound at the two ears. This time difference for a typical human head, could be as great as 660 μsec for a sound located directly off one of the ears (at 90° to your nose). We can detect interaural time differences as small as 10 μsec (yes, that's microseconds)!

The interaural time difference provides information about the direction of low-frequency sound sources. Suppose that a signal of 2500 Hz was positioned in such a way that it arrived at the right ear 0.4 msec before the left ear. Since this interaural time difference corresponds exactly to one period of that tone, there would be *no* difference between the waveforms at the two ears, and hence no way for the auditory system to know

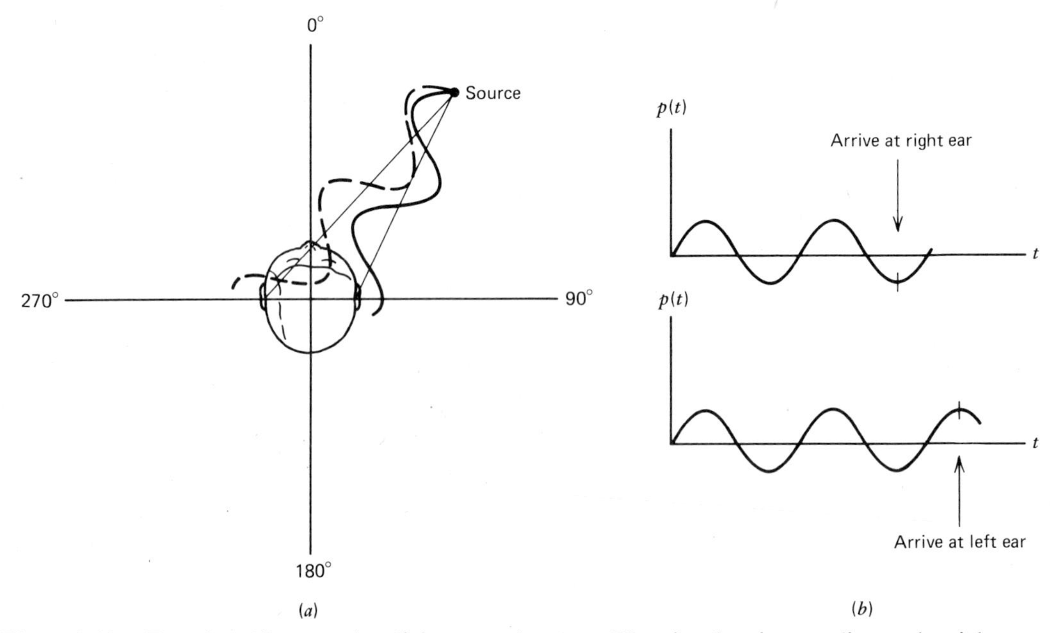

Figure 3-13. Top view of person localizing sound source. The signal arrives earlier at the right ear.

that the right ear signal had in fact arrived one whole cycle earlier. This potential ambiguity in the time cue exists for any signal with a period shorter than about 660 μsec, a frequency of 1500 Hz. Fortunately there is another important cue for the localization of a sound source—the *interaural intensity difference*—that does not have this problem. The sound intensity at the closer ear will be greater than at the farther ear. This occurs primarily because the head produces an appreciable sound "shadow" that attenuates the sound reaching the far ear. This attenuation can be as much as 20 dB for a high-frequency tone of 6000 Hz directly off one of the ears (at 90° to your nose). This shadowing effect is negligible at frequencies below several hundred hertz, so the intensity difference cue is mainly useful for higher frequency components of the stimulus. We can detect interaural intensity differences of about 0.7 dB. What happens if we attempt to localize signals of middle-range frequencies, where it would seem that both these intensity and time cues would be relatively poor? An experiment by Mills (1972) tested how small a separation an observer can detect between the location of two sound sources. As expected, performance was best for either very low or for moderately high frequencies and was worst for middle and very high frequencies.

There are some additional points to consider in sound localization by human observers in real-life situations. First, sound signals are rarely pure tones, and usually contain a rich mixture of low- and high-frequency components. Second, the position of the observer's head is usually not fixed, and head movement can greatly increase the amount of available information about the direction of a sound. Third, there are multiple echoes and sound reverberations present in a normal acoustical environment. These result from the reflection of sound from room and object surfaces and can make sound localization extremely difficult.

Our auditory system seems to have a very effective way of dealing with these potentially confusing echoes and reverberations. Consider a real sound occurring in an echoic environment; the best information about the sound location will be in the time and intensity cues in the *first* sound wave that strikes the ears. Apparently the auditory system processes these first-arriving cues and then suppresses the cues that are in the later-arriving, echo waveforms. This phenomenon is known as the law of the first wavefront or the precedence effect (Gardner, 1968; Wallach, Newman, & Rosenzweig 1949). You can demonstrate the suppression of echo information to yourself in a number of ways. One way is to record some ordinary sounds in several different types of rooms, a small office, a classroom, and a large, unfinished basement. Play back the recorded sounds and you'll hear no obvious differences in the reverberation quality of the sounds. However, if you examine the recorded sounds on an oscilloscope—or play the tapes *backward*—the differences in the actual durations of the echoes in the different rooms will become quite obvious.

Another demonstration of this effect will be familiar to the stereo enthusiast. Put a tape on your stereo system and switch the system to the monaural position, so that exactly the same sound comes from the left and right speakers. If you sit right on a line midway between the two speakers you will be aware of the sound coming from both of them. However, if you move slightly off this midline—to the left or the right—suddenly only the near speaker will appear to be on! Depending on where you are in the space in front of the speakers, the near speaker will appear to be the source of all of the sound; somehow the location cues that come from the far speaker are suppressed because they arrive a few milliseconds after the first cues. Turning off the far speaker does result in a noticeable change in the sound; the suppression seems to apply mainly to the accessibility of location cues in the delayed sound. Stereo engineers know that the time delays that induce this effect can be compensated for by raising the level of the delayed channel an appropriate number of decibels. If you increase the level of the farther channel you can restore the apparent sound location to the middle of the room. This is known to psychoacousticians as a time-intensity trade. A good source of information about stereo recording techniques is Eargle (1980).

Binaural Signal Detection Do interaural time and intensity cues allow us to detect signals that otherwise we could not hope to detect? The answer is yes: this advantage in detection is called the *Binaural Masking Level Difference* (BMLD). In order to control the signal waveform at

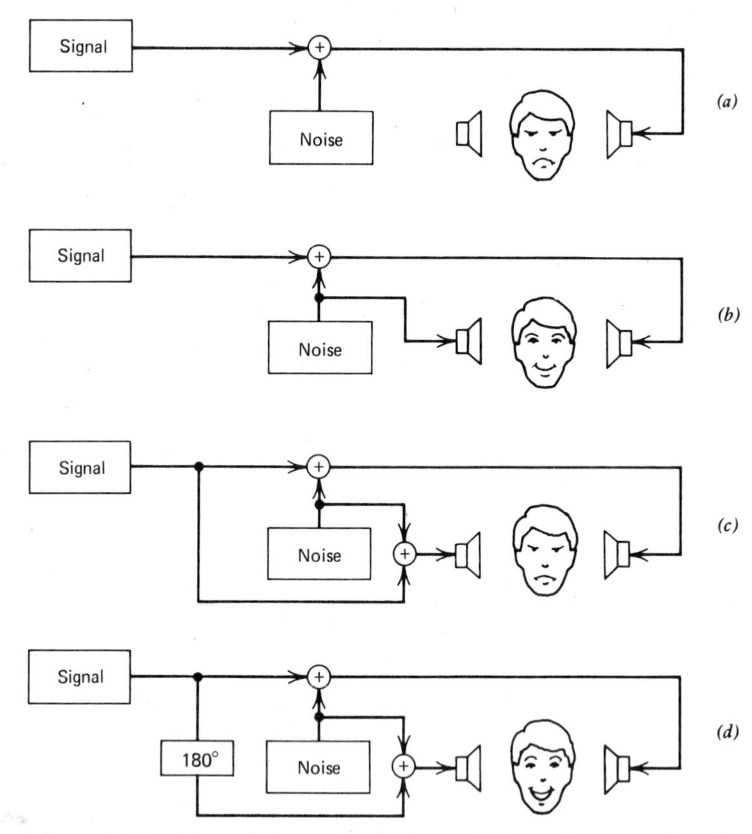

Figure 3-14. Four different BMLD conditions. See text for description of performance in these conditions.

each ear, binaural detection experiments are normally performed with earphone presentation. Consider four basic binaural detection situations in Figure 3-14. In (*a*), the observer is presented with a signal (say, a 500-Hz tone) plus masking noise to the left ear. Suppose that you adjust the signal level until it is just inaudible. Now feed the same noise into the right ear, (*b*) so that you have twice the noise into the system. The signal is now audible, in fact it would take almost 9 dB more noise to make the signal again inaudible in this situation, hence the term "binaural masking level difference" of 9 dB. Now suppose that having fed noise to the right ear, we were to feed the signal *also* to the right ear, as shown in (*c*). Now the signal again becomes inaudible. We could make it audible again by reversing the leads feeding the signal (or the noise) to one of the headphones. This reversal of the signal waveform is equivalent to a shift in phase of 180°, and is shown in (*d*). The result is that the signal is more detectable in (*d*) than in any of the other conditions—the BMLD is as high as 15 dB. The actual size of the binaural advantage depends on the extent of signal or noise differences between the ears; shifts of less than 180° in phase would result in smaller BMLDs. The maximum BMLD is obtained for signals of a few hundred hertz presented in broadband noise (see Chapter 16 for a more detailed discussion of signal detectability in noise). So the mechanism involved in the BMLD must be largely dependent on the time difference rather than the intensity cue.

SIGNAL DETECTION THEORY

Approximately 25 years ago some graduate students from psychology and electrical engineering at the University of Michigan found themselves sharing offices. The engineers were concerned with problems of detecting weak signals in noise; the psychologists were trying to describe the behavior of humans detecting visual and auditory stimuli. They soon realized that the problems they were working on had many common features. Out of their interaction came theories and techniques concerned with the mathematical and the human detection of signals. The theory of signal detectability applied to human signal detection has had a major impact on many areas such as experimental psychology. It has been applied to the analysis of information-retrieval systems, to medical diagnosis, to inspection and quality control in industry (see Chapter 14), to the criminal justice process, to athletics, and to the operation of emergency medical systems (Hutchinson, 1981).

There are several aspects of signal detection theory significant to the human factors specialist. First, it can help us to understand the mechanisms of vision and hearing (and other senses) and the processes of signal detection, discrimination, and recognition. Second, it can provide us with ways to measure and analyze human performance across a variety of different tasks. Third, signal detection theory provides a description of how theoretically perfect or "ideal" detection and discrimination systems behave. Specifying a theoretically "ideal" detection system can tell us (*a*) what the best possible performance can be in a task and (*b*) how performance should depend on the important task variables.

The basic idea of signal detection theory is illustrated in Figure 3-15. Some input, (x), is fed into the first stage of the detection system and un-

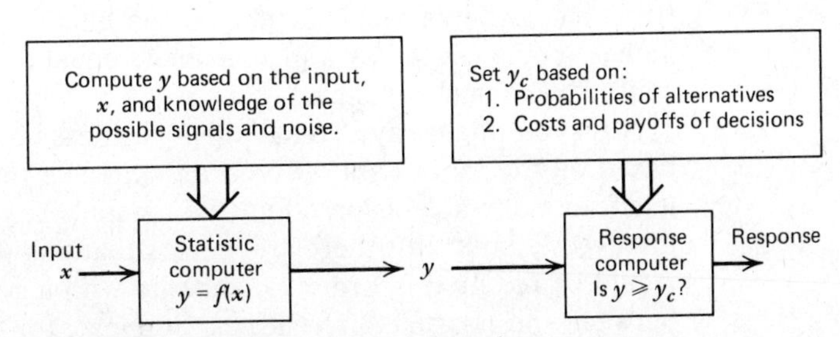

Figure 3-15. The basic signal detection system.

dergoes various computations based in part on data stored in the system's memory. The computed statistic, y, is then compared with a criterion value, y_c and a response is based on that comparison. If $y < y_c$, one response is made; if $y \geq y_c$, the other response is made.

What is the nature of the computation in the first stage? Let's consider the following problem. A person has just walked by your room. You are to report the sex of the person, based on some data we will supply. Let's call the data, x, and for the moment it will be the unknown person's height. By the way, if you *report* male and it *was* a male, you *get* 10¢; if you *say* female and it *was* female, you *get* 10¢; if you are *wrong*—either way—you *lose* 10¢. Ready? The person's height was 5 feet 10 inches. What is your response? Most people would probably say male, right? All but Ingrid, that is. Ingrid lives on the female floor of a dormitory where men are not permitted. There is practically no chance that a male will walk by Ingrid's door; tell her the person's height was 6 feet 4 inches and she'll report "female" everytime.

You would probably agree, even before considering the height data, that some assumptions ought to be made about the relative likelihood (the odds) of having a male or female in the hallway. These assumptions are called *prior* probabilities (or à priori) because they are prior to getting any data about the particular event. Let's keep things simple and assume that they are equal, that is .50 for male and .50 for female. Now, back to the data input: how about 5 feet 3 inches? Probably you'd respond, "Female." How about an observation of 5 feet 6 inches? If you are rational, there will be some value of height below which you will respond "female" and above which you will respond "male"; this is y_c, the criterion value for the response. You may have run into this type of decision task in a class on statistics; that is the mathematical area that was adapted to the general detection situation.

These ideas are illustrated in the diagrams of Figure 3-16. The Normal curves represent our knowledge of the distribution of height for males and females in the population of people expected to be found strolling by your room. We have picked reasonable mean and standard deviation values for these distributions. Given our assumptions about the prior odds and the costs and payoffs of the possible outcomes, it seems reasonable to set the criterion value, y_c, at the intersection of the two distributions. At this value of height, the likelihood of the data ($x = 66.3$ inches) being caused by a male event is equal to the likelihood that it was caused by a female event.

Now you might say, "What's this business about computing a statistic based on the input data; all you did was use the raw input (the height) itself, to make a decision about the person's sex. No computations were involved." True. But there are many situations where some computation would be required in order to come up with a good statistic for deciding between the possible alternatives. Suppose, for example, that instead of

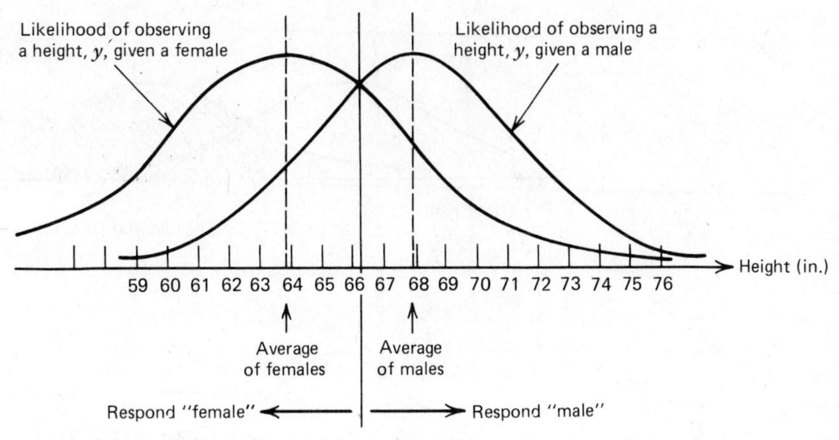

Figure 3-16. The two hypothesis distributions in the sex decision task.

simply giving you the height of the person, we also gave you their weight, foot length, and waist size. Ready? The data are height 5 feet 6 inches, weight 140 pounds, foot 11 inches and waist 33 inches. ("No fair", someone cries, "those data are 'multidimensional.'") Presumably you could come up with a statistic that incorporates all of the data in order to make the decision. One possibility is the Kantowitz Index, KI. Based on his vast experience in decision making, and 2 minutes of thought, one of the authors came up with the KI formula:

$$\text{KI} = \frac{\text{weight} + \text{waist size} + (10 \times \text{height}) + (10 \times \text{foot length})}{16} \tag{3-14}$$

When the computation stage gets the input data it will simply plug them all into the KI equation. A single number, the KI value, will then be passed on to the decision stage. In this case, the KI was $(140 + 33 + 56 + 110)/16 = 21.19$, so we'd probably respond male. Perhaps KI's larger than say, 19, would lead to a male response and smaller ones would lead to a female response.

How well would a person do using this method of discriminating the sex of the unknown person? We could find out by standing in the hallway in question and discretely recording the sex, weight, height, and foot and waist size, of everyone that comes by ("Pardon me. Could you give me some basic information about yourself?"). We can then take this data and construct the frequency distributions shown in Figure 3-17*a*. We can calculate how well a KI-type observer would do by placing a criterion point on these distributions as shown in the figure, and by counting the occurrence of the various outcomes. We know that whenever the KI of some observation yields a value of 19 or better, the observer will say "male." The probability of this happening is related to the number of males possessing KI's greater than 19. The other correct and error proba-

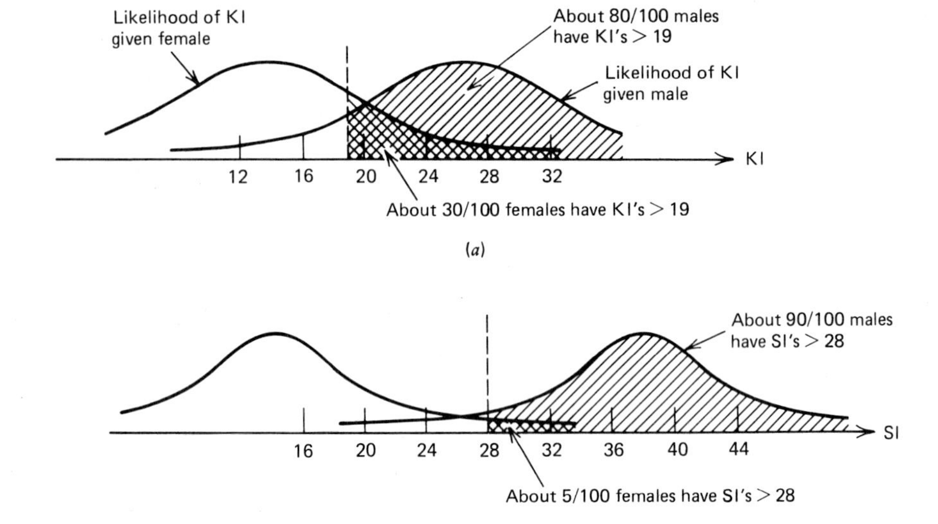

Figure 3-17. Distributions of males and females on the KI and SI measures.

bilities are as shown on the distributions in Figure 3-17*a*. You can see that using this index and criterion value, the number of females that are called males, out of 100 people walking by, will be about 30.

No doubt you can do better than that. If we had accurate knowledge of the population distributions of (and interrelationships among) weight, height, foot length, and waist size as a function of sex, we ought to be able to use that information for the construction of a better statistic. Such a statistic might be the Sorkin Index, SI, which is a different function of these physical parameters, and which incorporates more of our knowledge of the distribution of these parameters over the male and female populations to be expected in the hallway. The SI (a fictitious index) might be defined as follows.

$$\text{SI} = \frac{\text{weight} \times \text{foot length} \times \text{height} + 10 \text{ (waist size)}}{200}$$

Now in the same fashion as before, we could determine how well the SI statistic would do in discriminating one sex from the other. Figure 3-17*b* illustrates the resulting distributions; notice that these new distributions are less spread out than the KI distributions, and are also farther apart. It is clear that if we set some criterion value on the SI statistic, and computed the various probabilities of correct and incorrect, the correct probabilities would be greater and the error probabilities smaller, than when using the KI statistic.

We can simplify this whole business of computing correct and error probabilities and obtaining good statistics by making some very general

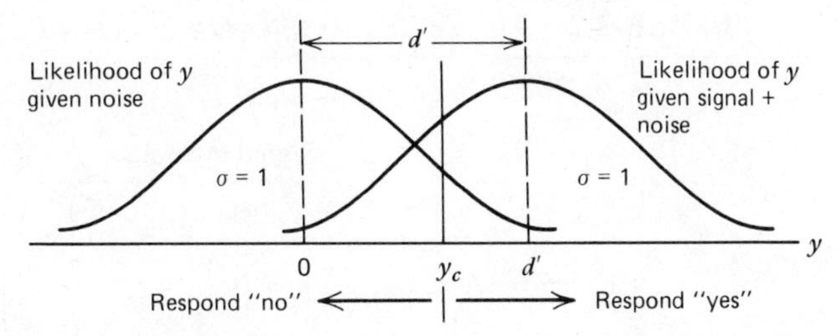

Figure 3-18. Hypothesis distributions on y normalized to means of zero and d', and standard deviation of 1.

assumptions about the distributions in question. If we assume that the distributions of y are Normal and have the same shape, the underlying scale can be transformed in such a way that the standard deviations are equal to 1 and the left-hand distribution has a mean of zero. (In statistical terminology we've just converted to z scores.) Figure 3-18 illustrates these new standardized distributions. We are assuming that the first stage of our detection system transforms the raw, possibly multidimensional input, $(x_1, x_2, \ldots x_n)$, into a single statistic, y, which has the two distributions shown given the two possible types of events. The separation between the means of these standardized distributions is a measure of how well our system can discriminate between the two events, and is defined as d'. We have seen that a good statistical computer will lead to a large value of d', and a poor one will lead to a small value.

Usually we can't directly observe the statistical distributions for the purpose of computing the correct and error probabilities. There's no hallway to directly observe the inputs. In fact it is generally the case in studying human performance that you are given the probabilities of being correct and incorrect, and you want to say something about the distributions of the decision statistic, specifically the value of d'. You may also want to know y_c, or the related measure β (β is the ratio of the heights of the two distributions at y_c). The next section describes the common terms for these correct and error probabilities and the Receiver Operating Characteristic.

Receiver Operating Characteristic

There are four possible outcomes of a detection task, as shown in Table 3-2. The frequency of the four events can be determined from knowledge of the number of *hits* (the observer responds "yes" and is correct) and the number of *false* alarms (the observer responds "yes" and is incorrect). So this pair of values completely specifies the observer's performance. This pair of values can be plotted as one point on a figure known as the Receiver Operating Characteristic, shown in Figure 3-19. In order to compute a pair of hit and false alarms values we need to know the d' and the particular criterion value, y_c. Appendix B gives a procedure for computing the hit and false alarm values.

TABLE 3-2 *The Four Possible Outcomes of a Detection Task*

		CONDITION	
		Signal + Noise	Noise
OBSERVER RESPONSE	"Yes"	𝍸 𝍸 𝍸 𝍸 𝍸 ||| 28 Hits	𝍸 𝍸 |||| 14 False Alarms
	"No"	|||| 4 Missed Signals	𝍸 𝍸 𝍸 𝍸 𝍸 𝍸 𝍸 𝍸 𝍸 𝍸 |||| 54 Correct Rejections
		32 Signal + noise trials	68 Noise trials

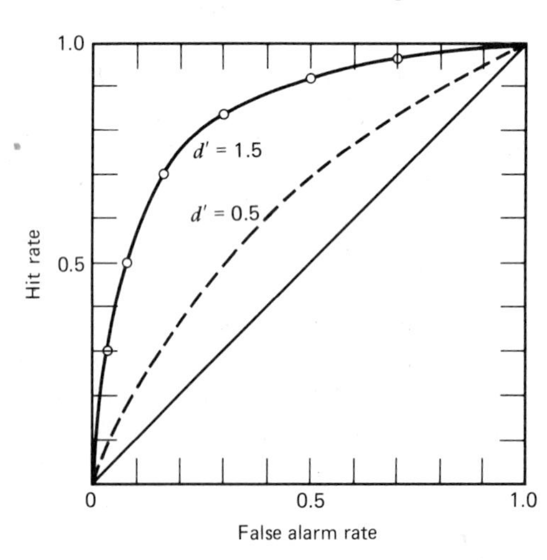

Figure 3-19. Receiver operating characteristic (ROC).

TABLE 3-3 *Hit and False Alarm Rates for* d' = *1.5*

y_c	−.5	0	.5	1.0	1.5	2.0
Hit rate	.977	.933	.841	.692	.5	.308
False alarm rate	.692	.5	.308	.159	.067	.023

The points plotted on the ROC curve were computed for a $d' = 1.5$ and six different values of y_c (see Table 3-3). If these calculations were repeated with $d' = .5$, the result would be points falling on the dashed curve of Figure 3-19. This kind of plot is an extremely useful way to summarize the behavior of an observer in a detection task. The ROC curve shows the hit and false alarm rates possible for a fixed d' and different response criteria that the observer may employ. If each of two observers (or one observer in two different conditions) behaves according to the theory, you could compare the sensitivity, d', of the observers (or the conditions) by comparing the ROC curves that their data fall on. The nearer the points are to the (0,1) corner, the higher the sensitivity; the nearer to (0,0), the more conservative the criterion, the nearer to (1,1), the more liberal the criterion. In some applications, you will know performance (hit and false alarm rates) and will want to calculate the magnitude of d'. Appendix B also provides a procedure for this calculation.

Factors Affecting Signal Detectability

How does the statistical process we have described relate to real biological or electronic detection systems? Suppose that you are trying to detect a faint sound signal in noise. Let's assume that the signal is a brief pure tone and the noise is broadband noise, with all frequencies equally likely to be present. How would you build a gadget to accomplish this task? Let us first use a microphone to convert the acoustic input into an electrical signal. Next, let's feed the signal from the microphone through a narrow band electronic filter. A narrow band filter is a device that will pass energy only at a few frequencies. Obviously we ought to pick the filter so that the frequencies that are passed through include the signal that we wish to detect, but exclude as much of the noise as possible.

We want to measure how much energy has come through the filter during the time that the signal might have been present at the microphone. One way to do this is to take the electrical output of the filter, square it, and then sum up (integrate) the squared output for some time period. (This can be accomplished with some diodes, transistors, capacitors, etc.) What we want to get out of this gadget is a *single* number that corresponds to the energy of the filter's output. That number is a statistic just like the hypothetical detection statistics we discussed earlier. It will have a distribution of different values on trials when only noise was fed into the filter—and another distribution of values on trials when signal plus noise was fed into the filter. That is, something just like the curves of Figure 3-18. If the values of all of the circuitry were chosen carefully, those distributions would be characterized by a high d'. The more accurate the specification of the signal to be detected, the character of the noise, and the possible time of occurrence of the signal, the better we could design that system for detecting the signal, and the higher would be its d'. You can see that if we are fuzzy about the timing of the signal or its frequency characteristics, our resulting d' will be degraded, because the statistic will have to include noise energy that is not relevant to the presence of the signal.

So the system's or observer's knowledge and specification of the signal and noise are vital to performance—whatever kind of system you are considering. This knowledge allows the system to be designed or adjusted so that a good statistic can be computed. If you know what (and when) the signal is that you are supposed to detect (or discriminate or identify) you will perform much better at the task. It is also possible to show, using detection theory, that *uncertainty* about the nature of the signals (or noise) will mainly impair performance when the task is difficult—that is, when the signal-to-noise ratio is low. Consider the following situation. You are playing soccer with a group of friends when Frank shouts that he has lost his good pen. Everyone starts to hunt for the dropped pen in the grass of the playing field. The pen is green. Why is Frank the most likely person to find the pen? Would that outcome be as likely if the pen were red?

An ideal detection system, by definition, uses all of the information available to compute an optimal statistic for the purpose of detecting a particular signal or class of signals, and introduces no new noise in the process. The d' measure can be computed in many specific situations such as the detection of a sinusoidal signal in white noise. In that case, the sensitivity of a theoretical ideal observer is given by:

$$d' = \sqrt{\frac{2E}{N_0}} \tag{3-15}$$

where E is the signal energy and N_0 is the noise power per cycle (see Chapter 16). Equation 3-15 tells you how detectability depends (for an ideal observer) on the physical parameters of the task. This also tells you which physical parameters of the task do *not* affect the behavior of the ideal observer. So detection theory can give you a head start toward figuring out what factors should and should not be important to human performance.

Predicting Performance in Different Tasks

Detection theory has been extensively applied to a variety of tasks other than the simple detection and discrimination of acoustical signals. These applications includes the detection of abnormalities in X rays, the epidemiology of disease, and industrial flaw detection (Hutchinson, 1981; see Chapter 14). The theory has been extended to some types of tasks that are quite different from the single-interval yes/no detection task that we have discussed in this chapter. It's a common occurrence in human factors design that you have performance data in one particular situation, but you need to predict what performance will be in a different, but related task. Suppose that in a simple detection task (in which the observer must detect the signal's presence or absence during a defined time interval), the obtained d' is equal to 1.7. You want to predict performance in a *two*-interval task in which the observer must report which of two time intervals contained that signal (the signal appears randomly in one of the two intervals). If we assume that there are no interactions between the two time intervals (including interference, memory factors, or shared noise in the intervals), then d' in the two-interval task will equal $\sqrt{2}$ times the simple detection d' [e.g., $(\sqrt{2})(1.7) = 2.4$ (Tanner & Sorkin, 1972)]. Another general situation is where a single observation interval contains either signal S_1 plus noise or signal S_2 plus noise. Performance in discriminating S_1 from S_2 is given by

$$d'_{1,2} = \sqrt{(d'_1)^2 + (d'_2)^2 - 2\rho_{12} d'_1 \, d'_2} \qquad (3\text{-}16)$$

where d'_1 and d'_2 are the detectabilities (in noise) of the two signals separately, and ρ_{12} (rho) is the correlation between the signals.

There are a variety of more complicated tasks and task variations. The observer may be asked to respond whether two intervals contained the *same* or *different* signals, or which one of three presented signals was different from the other two. The latter (oddity or triangle) task is used frequently in the evaluation of food products. Another type of task requires the observer to respond whether one or none of a prescribed set of signals has occurred, and if so, which signal it was. This task has been termed "recognition" (Green & Birdsall, 1978). Predicting performance in these complex tasks from basic detection data is sometimes rather involved, in terms of the specific assumptions that must be made about the human observer and the mathematical computations required. Some convenient tables for computing d' (and percent correct) in multialternative forced-choice tasks can be found in Elliot (1964) and Hacker and Ratcliff (1979), for many special variations of the three-alternative forced-choice task (including the oddity task) in Frijters, Kooistra, and Vereijken (1980), and for a variety of discrimination tasks including same-different and ABX designs in Kaplan, Macmillan, and Creelman (1978). General equations for computing d' in different tasks and related applications appear in those references and in Sorkin (1962), Tanner and Sorkin (1972), Macmillan, Kaplan and Creelman (1977), Sorkin and Pohlmann (1973), Green and Birdsall, (1978), Swets and Birdsall (1978), Swets, Green, Getty, and Swets (1978), Frijters (1979), and Smith (1982).

SLICE OF LIFE

Setting the Response Criterion

You may wonder if there is a good way to set the criterion value of the statistic that determines which response will be made. The answer is yes; the criterion value ought to depend on the prior odds and the costs and payoffs of the possible outcomes. In many applications, we are not overly concerned with this part of the system. If an observer is too conservative or too liberal in setting their response criterion, (they are up or down too much on the ROC), it is usually a simple matter to move their operating point: "You're responding 'signal' too often; be more conservative!" This may be the main function of the loudly complaining basketball coach whose purpose is not to change the referee's call but rather to shift the referee's criterion for the next few minutes of play.

There are some applications, however, where major concern is with the operator's criterion rather than their sensitivity. An example is in tasks involving monitoring or vigilance performance. In such tasks an operator must report the presence of signals that occur infrequently over a long time period. Some investigators have reported that the operator's criterion shifts downward during the course of the monitoring session (Craig, 1980). The resulting drop in hits and increase in missed signals may be a significant component of the performance decrement in vigilance tasks. Recent studies suggest some caution in the use of unmodified detection theory measures of performance in these situations. (Craig, 1979; Long & Waag, 1981). However, the measures employed should take into account both sensitivity and criterion effects (Craig, 1981).

The important thing to realize from detection theory is that shifting the criterion in either direction brings a penalty as well as a gain. You can't shift to a more conservative criterion—for fewer false alarms—without also suffering a reduction in the number of hits. As the false alarm rate decreases, *so will the hit rate.* Figure 3-20 illustrates this problem nicely. Increasing the amount of stench agent increases the hit rate but it also increases the false alarm rate. In detection theory, there's no such thing as a free lunch.

SUMMARY

A sound is specified by the intensity and phase of the different frequencies that make up its spectrum. Our auditory systems are extremely sensitive to small changes in sound intensity and frequency. The loudness and pitch of sounds depend on the intensities and frequency relationships among a sound's spectral components. For some sounds, we hear pitch even when no energy is physically present at that frequency. Listening with two ears can allow us to detect and recognize signals that are 15 dB fainter than with one ear alone. Understanding how sound is specified and perceived is important for the design of auditory codes, displays, and effective acoustic and electroacoustic systems such as for speech communication and speech input/output, and for predicting the effects of noise on people. The concepts of spectral analysis and nonlinearity are also useful in assessing visual acuity, the effectiveness of optical systems and in describing human control behavior.

Signal detection theory was originally applied to human auditory detection and discrimination, but has since been used to describe industrial inspection, medical diagnosis, vigilance behavior, information-retrieval systems, and in a variety of other applications. Detection theory

Gas lines need deodorant, not fix, jittery callers told

By JUDY HORAK
Staff Writer

Indiana Gas Co. switchboards were relatively quiet today, after a night when customers in Lafayette, Lebanon and Frankfort kept them lighted up with complaint calls about gas odors.

The company received more than 1,000 calls from customers in the three communities, according to Division Manager George Bowman. And while callers were worried they had gas leaks, the problem actually was "over-odorization" of a gas line south of Lebanon, he explained.

Natural gas is odorless, so utilities add a chemical smell so customers will be aware of possible gas leaks. Indiana Gas adds "Spot Leak 10-09" in a one-part-per-million concentration, Bowman said.

"We were doing some work at the Lebanon (gas) purchase point, south of Lebanon, where we add the odorant to the natural gas," Bowman said. "The first calls came into the Lebanon office at about 5 p.m., and they started at about 7 in Frankfort." Lafayette customers began smelling the stuff between 9 and 9:30 p.m., he said.

He said he doesn't expect the problem to continue north along the gas lines, since the excess odor ingredient "is out of the system now."

Calls from Lebanon customers stopped around midnight, Bowman said, "as the odorant worked its way through the lines."

Callers were assured that the problem probably wasn't a gas leak in their homes. But many asked how the odor could get inside their homes if gas supposedly wasn't leaking.

"Most people have trouble understanding that," Bowman said. "My wife asked me the same question. The odorant in its proper quantities is dissipated in the flames in pilot lights" on ovens, furnaces, water heaters and other gas appliances.

"But if it's in a higher concentration, a certain amount will pass through the pilot lights, escaping into the home. The natural gas that's present still is burned by the pilots," he explained.

Some customers said this morning they still could smell gas near the outside gas meters. Bowman said that could be related to one type of meter, which normally vents some gas. However, he asked any customers who still "smell gas" today to call the utility.

Indiana Gas called in extra switchboard operators Tuesday night, and even tapped the services of General Telephone Co. Indiana Gas phone lines were overloaded, giving callers recorded messages, so Gen Tel operators were alerted to tell callers the problem wasn't considered dangerous.

The Gen Tel operators took names, phone numbers and addresses, and later in the night, Indiana Gas called customers back, offering to check homes for possible leaks. Servicemen visited homes all night long, and still were visiting some this morning.

"We don't want to discourage people from calling us in this type of situation," Bowman said. "That's why we have the servicemen out checking for leaks." In Lafayette, most of the calls came from the south part of town, which is closest to the gas trunk lines. However, customers in Vinton Woods and on Union St. also were affected.

Figure 3-20. Gas deodorant problem. (From *Lafayette Journal and Courier, 60,* September 12, 1979.)

is concerned with the observer's sensitivity in the task and also with the response criterion, the observer's tendency to choose one response alternative over another. Both these aspects will influence the obtained task performance. Important factors that influence sensitivity include such physical parameters as the energy of the signals, the noise intensity, and the relationships among the signals and the noise. Important observer factors include observer knowledge or uncertainty about the signal and noise characteristics and the time of occurrence of the signals. A useful technique for summarizing performance in detection or discrimination tasks is the Receiver Operating Characteristic. This plot allows an approximate evaluation of a detection system's sensitivity and criterion, as well as the comparison of performance of different systems and conditions. An important aspect of detection theory is the ability to predict performance across different kinds of tasks. Equations and tables are available for this purpose.

APPENDIX A

Computing the Fourier Coefficients

Given the periodic function, $f(x)$, defined in the interval, $-\pi$ to $+\pi$

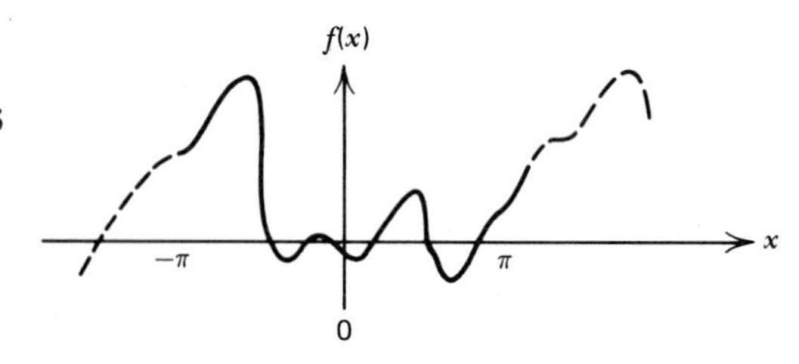

Let

$$f(x) = \frac{a_0}{2} + \sum_{n=1}^{\infty} (a_n \cos nx + b_n \sin nx)$$

or let

$$f(x) = \frac{a_0}{2} + \sum_{n=1}^{\infty} c_n \sin(nx + \Theta_n)$$

where a_n and b_n are computed from the definite integrals.

$$a_n = \frac{1}{\pi} \int_{-\pi}^{\pi} f(x) \cos nx \, dx, \; b_n = \frac{1}{\pi} \int_{-\pi}^{\pi} f(x) \sin nx \, dx$$

and

$$c_n = \sqrt{a_n^2 + b_n^2} \qquad \Theta_n = \text{Arctan}\left(\frac{b_n}{a_n}\right)$$

For example, consider the sawtooth waveform

$$-\pi < x < \pi, f(x) = -\frac{ax}{\pi}$$

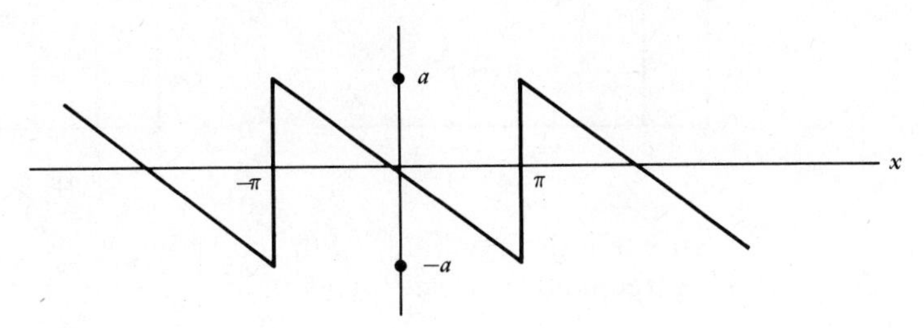

Then

$$a_n = \frac{1}{\pi}\int_{-\pi}^{\pi}\left(-\frac{ax}{\pi}\right)\cos nx\,dx = -\frac{a}{\pi^2}\int_{-\pi}^{\pi} x\cos nx\,dx =$$

$$-\frac{a}{\pi^2}\left(\frac{\cos nx}{n^2} + \frac{x\sin nx}{n}\right)\Bigg|_{-\pi}^{\pi}$$

$$= \frac{a}{\pi^2}\left[\frac{\cos(n\pi)}{n^2} + \frac{\pi\sin(n\pi)}{n} - \frac{\cos(-n\pi)}{n^2} + \frac{\pi\sin(-n\pi)}{n}\right]$$

Note that $\cos(x) = \cos(-x)$, and for $n = 1,2,3,\ldots$; $\sin(n\pi) = 0$. So the expression within the brackets equals zero, and all $a_n = 0$.

$$b_n = \frac{1}{\pi}\int_{-\pi}^{\pi}\left(-\frac{ax}{\pi}\right)\sin nx\,dx = -\frac{a}{\pi^2}\int_{-\pi}^{\pi} x\sin nx\,dx =$$

$$-\frac{a}{\pi^2}\left(\frac{\sin nx}{n^2} - \frac{x\cos nx}{n}\right)\Bigg|_{-\pi}^{\pi}$$

$$= -\frac{a}{\pi^2}\left[\frac{\sin(n\pi)}{n^2} - \frac{\pi\cos(n\pi)}{n} - \frac{\sin(-n\pi)}{n^2} - \frac{\pi\cos(-n\pi)}{n}\right]$$

Again, $\cos(x) - \cos(-x) = 2\cos(x)$, and for $n = 1,2,3,\ldots$; $\sin(n\pi) = 0$. For $n = 1,3,5,\ldots$; $\cos(n\pi) = -1$ and for $n = 2,4,6,\ldots$; $\cos(n\pi) = +1$

Since

$$c_n = \sqrt{a_n^2 + b_n^2} \qquad \text{and} \qquad a_n = 0,\ c_n = |b_n|$$

and

$$\Theta = \text{Arctan}\left(\frac{b_n}{a_n}\right) = \text{Arctan}\left(\frac{b_n}{0}\right) = 90° \text{ or } \frac{\pi}{2} \text{ rad}$$

for all n. So

$$f(x) = \frac{2a}{\pi}\sin\left(x + \frac{\pi}{2}\right) + \frac{2a}{2\pi}\sin\left(2x + \frac{\pi}{2}\right) + \frac{2a}{3\pi}\sin\left(3x + \frac{\pi}{2}\right) + \ldots$$

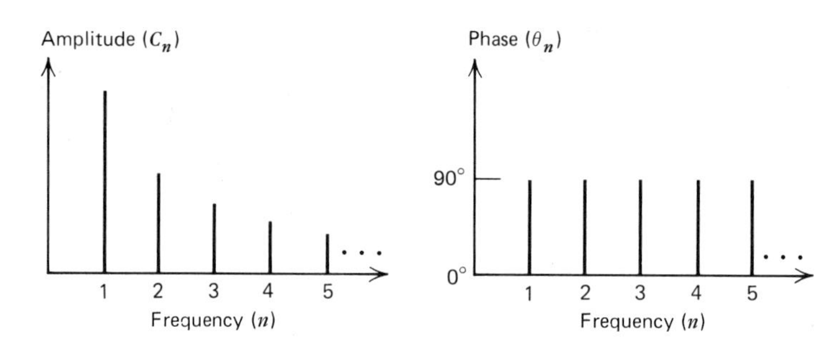

Note that the amplitude of the components are reduced by half each time the frequency doubles (octave).

APPENDIX B

Computation of Detection Theory Parameters

Computation of the HIT and FALSE ALARM rates given d' and y_c

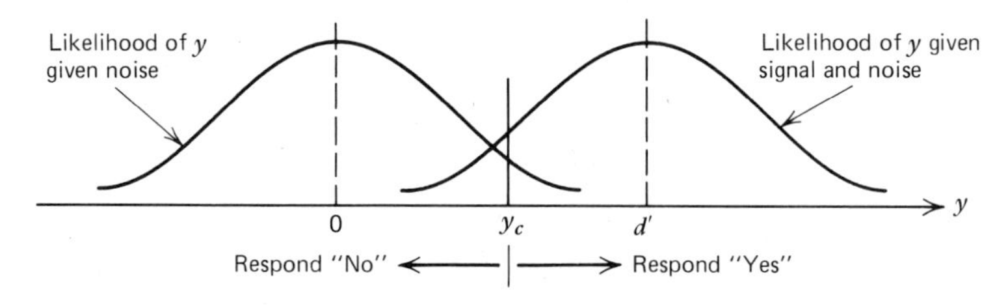

1. Compute the false alarm rate: P(yes/noise) if $d' = 1.5$ and $y_c = 1.0$.

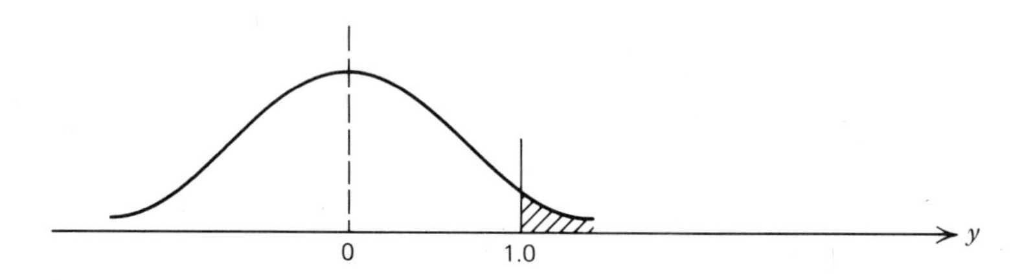

(a) Notice that the total area to the right of $y = 0$ is equal to .5.
(b) Then the false alarm rate (the shaded area), is

$$.5 - \text{area (from } y = 0 \text{ to } y = 1.0)$$

(c) From the Table 3-4 values

$$\text{Area (from } y = 0 \text{ to } y = 1.0) = .341$$

(d) Therefore false alarm rate $= P(\text{yes/noise}) = .5 - .341 = .159$.

2. Compute the hit rate: P(yes/signal + noise) if $d' = 1.5$ and $y_c = 1.0$.

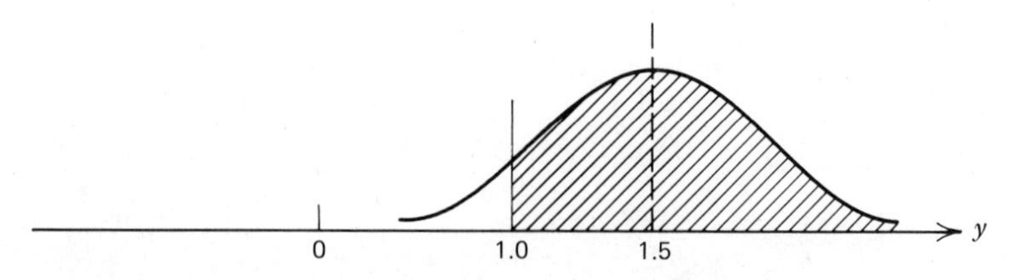

(a) The area to the right of $y = 1.5$ is equal to .5.
(b) The area between $y = 1.0$ and $y = 1.5$ can be determined from the tabled values. Notice that for this distribution (with a mean of 1.5), the area (from $y = 1.0$ to $y = 1.5$) is equal to the area (from $y = 0$ to .5) for a distribution with a mean of zero. So that

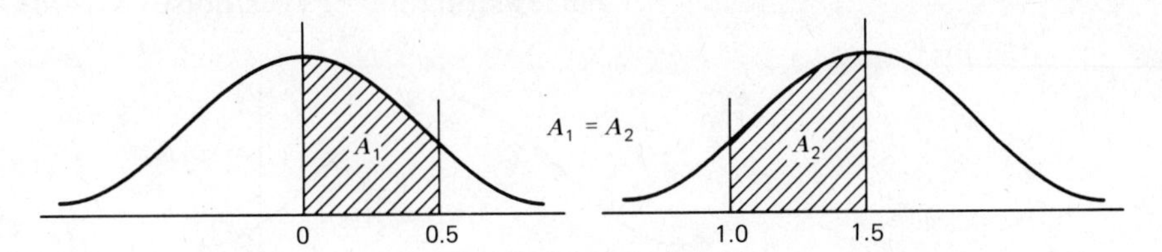

we can get this area from the tabled value at $t = .5$, area $= .192$.
(c) Then the hit rate $= P(\text{yes/signal and noise}) = .192 + .5 = .692$

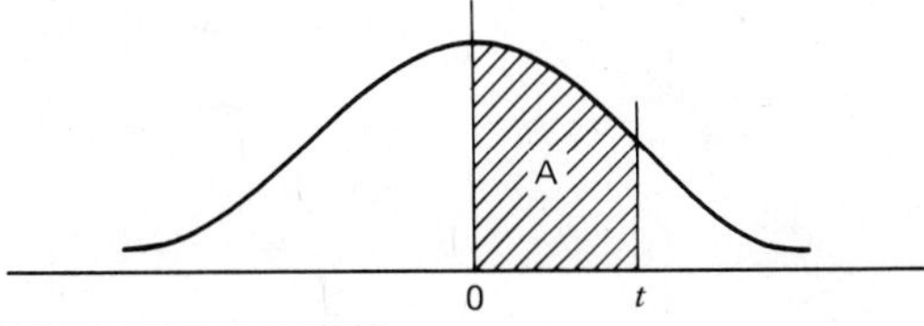

TABLE 3-4 *Tabled Values of the Normal Curve*

t	*Area* (from 0 to t)
0	0
0.1	.039
0.2	.079
0.3	.118
0.4	.155
0.5	.192
0.6	.226
0.7	.258
0.8	.288
0.9	.316
1.0	.341
1.1	.364
1.2	.385
1.3	.403
1.4	.419
1.5	.433
1.6	.445
1.7	.455
1.8	.464
1.9	.471
2.0	.477
2.1	.482
2.2	.486
2.3	.489
2.4	.492
2.5	.494

Computation of* d′ *from the HIT and FALSE ALARM Rates

1. Usually you will have obtained some performance data, in the form of hit and false alarm rates, and need to compute the measure of performance, d'. Suppose you are given these data.
   (a) Hit rate $= P(\text{yes/signal} + \text{noise}) = 28/32 = .875$
   False alarm rate $= P(\text{yes/noise}) = 14/68 = .21$

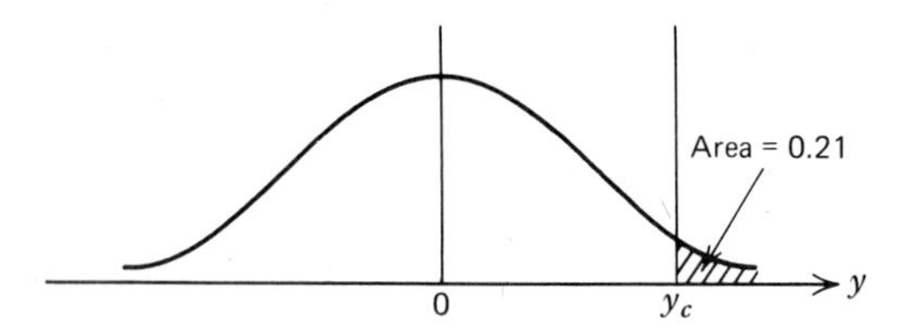

   (b) First draw the noise distribution. Then the area (from $y = 0$ to $y = y_c$) $= .5 - .21 = .29$. Look up the Table 3-4 value of t for an area of .29, $t = .8 = y_c$.

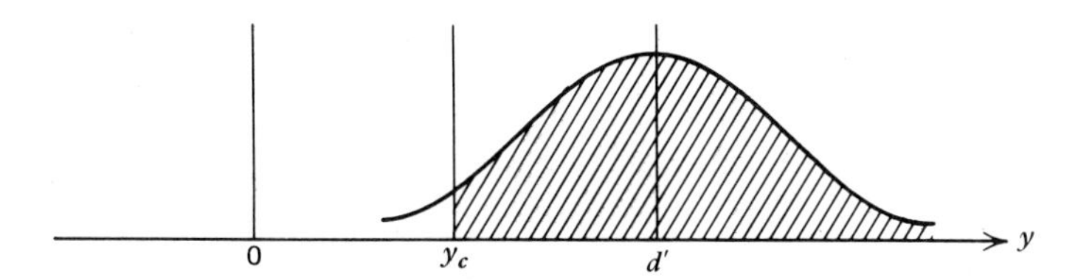

   (c) Now draw the signal + noise distribution. Since the shaded area $= .875$, the area (from $y = y_c$ to $y = d'$) $= .875 - .5 = .375$. Note that this area is equivalent to that from a distribution centered at zero (and reversed),

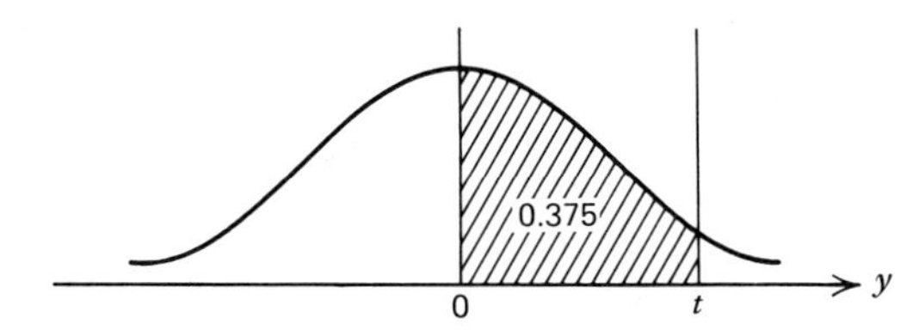

   where $.375 =$ area (from $y = 0$ to $y = t$).
   From Table 3-4, $t \approx 1.15$.

   (d) Putting these values together, we can draw the distributions and calculate that $d' = .8 + 1.15 = 1.95$.

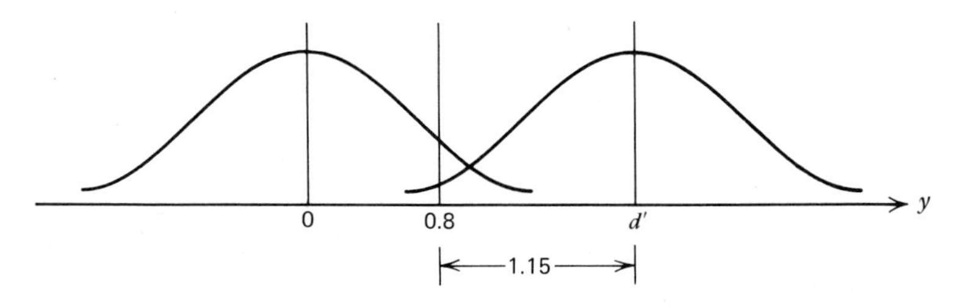

4 VISION

Enter the gunfighter-turned-marshal onto the dusty main street. Tall and lean, his reputation is soft words, steel nerves, and deadly aim. His eyes are dark slits set in an impassive face. He looks around slowly for the outlaw he has warned to leave town. With a sudden, smooth movement he draws and fires. A figure tumbles slowly from a position of ambush on the roof of the general store. A rifle follows the figure to the dust. Could successful gunfighters actually *see* better than ordinary folk?

Your company's annual dance is being held in one of their huge aircraft hangars. The place has been decorated and is jammed with a throng of employees and their guests, two bands, and a platoon of food and drink caterers. You were looking forward to a great evening until you noticed with dismay that the colors of your clothes clash terribly and also make your skin look like you were just released from the hospital. You've worn this outfit before without noticing the problem. Are you imagining things?

You've been driving on the expressway in a hazy drizzle for several hours. There has been no other traffic, and nothing much to look at in the way of scenery—even if the limited visibility allowed you to see any of it. Suddenly a small object seems to fly out from the right side of the road toward you. You quickly realize that it is another car pulling rapidly onto the highway. What caused the delay in your detection and recognition of the other car?

These vignettes illustrate aspects of vision that are related to the performance of quite different types of tasks. The gunfighter example raises the question of visual *acuity:* How well can we detect or discriminate fine detail in visual stimuli, and how does this ability affect performance in different tasks? How does performance depend on the physical properties of stimuli like television images? The party vignette concerns our perception of color and the limitations on our ability to make accurate color judgments. Does the use of color help or hinder performance in different tasks? The driving story actually involves some complex aspects of our perception of distance and size, and the dependence of those judgments on the mechanism that controls the focusing of our eyes. These phenomena are discussed in this chapter, along with the basic processes that underlie them.

THE VISUAL STIMULUS

The physical stimulus for vision is a type of radiant electromagnetic energy. This energy may be produced by heating materials to high temperatures, or by bombarding certain materials with electrons. The energy emitted by such sources is in the form of discrete particles called quanta. A quantum of light energy is called a photon. Light energy is also describable as an electromagnetic wave phenomenon. If you could accelerate an electrically charged particle up and down a short wire, you would generate a changing electric and magnetic field in the space around the wire. This electric and magnetic disturbance would propagate outward from the wire at the speed of light, about 3×10^8 meters per second. The spatial distance between the peaks in that disturbance would be given by equation 1:

$$\lambda = \frac{c}{f} \tag{4-1}$$

where λ is the wavelength, c is the velocity of light, and f is the frequency of oscillation of the disturbance. The range of electromagnetic radiation that is visible to the human eye is relatively narrow. Our eyes are sensitive to very short radiation wavelengths, from about 350 to 750 nm, with maximum sensitivity around 550 nm. A nm (nanometer) is equal to 10^{-9} m.

Most sources of visual energy generate more than one wavelength at a time. Figure 4-1 shows the *spectrum,* the distribution of light energy over wavelength, resulting from the heating of a metal to incandescence. Figure 4-2 shows the spectrum of a mercury vapor lamp; this is called a *line* spectrum and is composed of energy at discrete wavelengths. The tendency for an object to absorb, reflect, or refract (pass) light may depend on the wavelengths of the light hitting the object. For example a red glass filter will pass red (long) wavelengths and tend to absorb energy at other wavelengths. Our visual system is also more sensitive to certain wavelengths of light than to others. This differential sensitivity depends somewhat on the particular state of our eyes – whether we have just been in very bright or very dim lighting conditions. The relative sensitivity of our eyes to light of different wavelengths forms the basis of the commonly used measures of illumination (Chapter 17 has a more detailed discussion of light measures and illumination).

We could specify the total amount of light energy coming off of a source

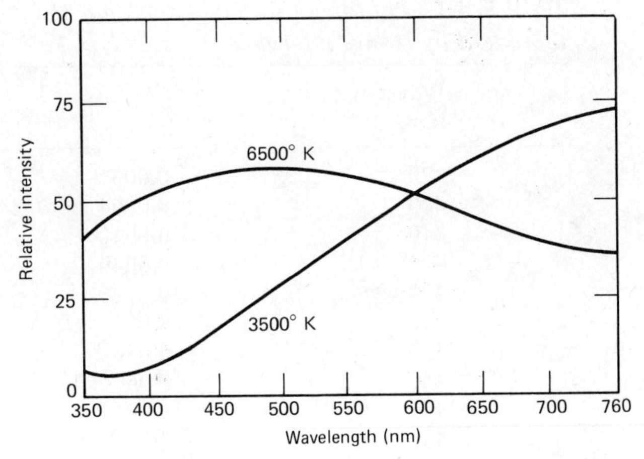

Figure 4-1. Spectra of a tungsten lamp heated to 3500°K and 6500°K. (From *Sight and Mind: An Introduction to Visual Perception* by Lloyd Kaufman. Copyright © 1974 Oxford University Press, Inc. Reprinted by permission.)

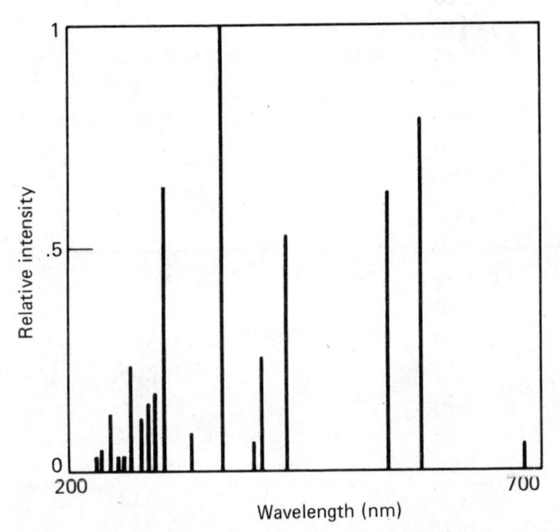

Figure 4-2. Line spectrum of a mercury vapor lamp. (From *Sight and Mind: An Introduction to Visual Perception* by Lloyd Kaufman. Copyright © 1974 Oxford University Press, Inc. Reprinted by permission.)

in terms of the number of photons or watt-seconds of energy generated, but this specification would ignore the fact that the eye was more sensitive to some wavelengths then to others. The accepted measures of light are called photometric measures; these measures include a factor (explained shortly) that takes into account the sensitivity of the human eye. Table 4-1 lists these factors for the human eye under two general kinds of conditions. The photopic function, V_λ, is obtained under normal daylight levels of illumination prior to and during the experiment; the scotopic

TABLE 4-1 *Photopic (V'_λ) and scotopic (V'_λ) luminosity factors tabulated at 10-nm intervals*[a]

Wavelength (nm)	V_λ	V'_λ
390	0.0001	0.00221
400	0.0004	0.00929
410	0.0012	0.03484
420	0.0040	0.0966
430	0.0116	0.1998
440	0.023	0.3281
450	0.038	0.455
460	0.060	0.567
470	0.091	0.676
480	0.139	0.793
490	0.208	0.904
500	0.323	0.982
510	0.503	0.997
520	0.710	0.935
530	0.862	0.811
540	0.954	0.650
550	0.995	0.481
555	1.000	0.402
560	0.995	0.3288
570	0.952	0.2076
580	0.870	0.1212
590	0.757	0.0655
600	0.631	0.0315
610	0.503	0.01593
620	0.381	0.00737
630	0.265	0.003335
640	0.175	0.001497
650	0.107	0.000677
660	0.061	0.000313
670	0.032	0.000148
680	0.017	0.000071
690	0.0082	0.000035
700	0.0041	0.000018

[a] From *Sight and Mind: An Introduction to Visual Perception* by Lloyd Kaufman. Copyright © 1974 Oxford University Press, Inc. Reprinted by permission.

function, V_λ, is obtained after the eye has been in the dark for about 40 minutes before the experiment (there are also some differences in the actual experimental procedure). These values are the agreed-on values used for the purpose of defining photometric measures of light; there is no actual standard human eyeball frozen in nitrogen and stored in the basement of an observatory in Paris.

How are these functions applied? One measures the amount of energy at each wavelength (or within narrow bands of wavelength), multiplies each value by the factor given in Table 4-1, and then adds up the resultant numbers. The formula is given by

$$F = \sum_{\lambda} P_{\lambda} V_{\lambda} \qquad (4\text{-}2)$$

where F is the photometric measure of intensity and P_{λ} is the physical measure in watts or other units (called the radiometric measure). This equation simply *weights* the physical measure at each wavelength by the assumed sensitivity of the average human eye. For example, suppose that we had two light sources A and B. Source A emits 10 W at 420 nm and 20 W at 510 nm. Source B emits 15 W at 555 nm. Which source has the higher photometric intensity? From Table 4-1, $F_A = (10)(0.0040) + (20)(0.503) = 0.04 + 10.06 = 10.1$ and $F_B = (15)(1.000) = 15$. So source B is photometrically more intense even though it emits 15 W less energy than A. At very low light levels, the V'_{λ} function should be used instead of the V_{λ} function.

STRUCTURE OF THE VISUAL SYSTEM

The basic structure of the eye is shown in Figure 4-3. Light enters the eye by first passing through the transparent protective covering called the cornea. The curved surface of the cornea produces most of the focusing of the image that falls on the retina. Next, the light passes through a fluid, the aqueous humor, and then through the pupil, an opening in the iris which can change in diameter.

The pupil controls the amount of light that can enter the eye; opening in dim light, and closing in bright light. However the amount of light control that is exerted by the pupil mechanism is very small compared to the range of sensitivity of the eye. Our visual system is sensitive to light over a range of light intensities of about 10^{13}, or 13 log units of intensity (a log unit is one power of 10). But the pupillary mechanism can produce only about a 16-fold change (less than 2 log units). Furthermore, the speed of the pupil's response to a sudden increase in light level is rather slow, so it is not very useful as a protective mechanism. On the other hand, photographers will recognize that control of a camera's lens opening or aperture is a way of controlling the camera's depth of field – the range of distance over which objects in the camera's view will be in sharp focus. The smaller the lens opening, the greater the depth of field. The pupil mechanism probably functions to maintain a balance between maximizing the depth of field and the amount of light entering our eyes (Haber & Hershenson, 1973). Pupil size also seems to depend on our attentional and emotional state (Goldwater, 1972).

After passing through the pupil, the light passes through the lens. Focusing of the image on the retina is accomplished by changing the refractive power of the lens. This process, called *accommodation,* is

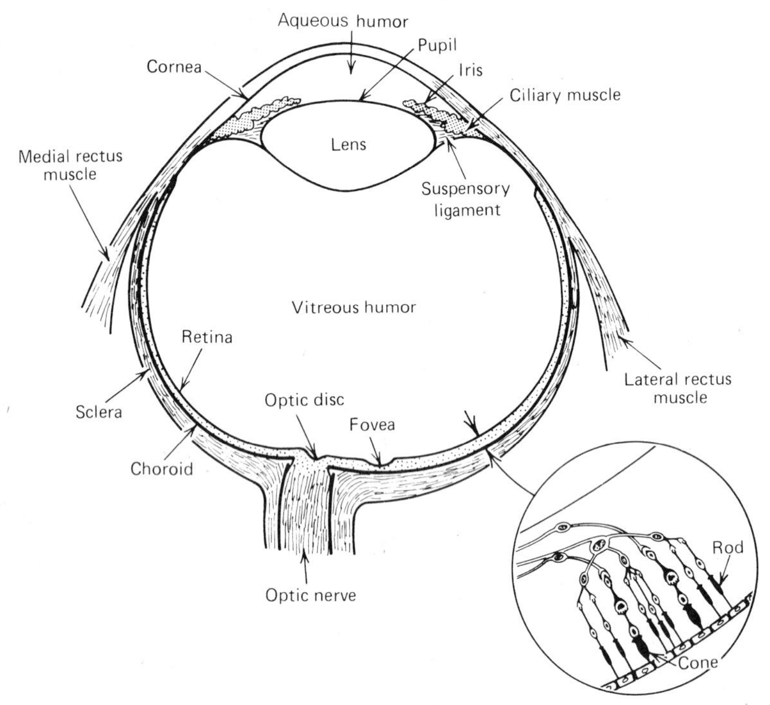

Figure 4-3. Structure of the human eye and retina.

accomplished by increasing or decreasing the lens thickness. The process is relatively slow, taking almost 400 msec for a complete change from near vision to far vision. Light then travels through the fluid called the vitreous humor and then on to the light-sensitive region, the retina.

The retina is a complicated structure taking up almost 200 degrees of the inner surface of the eyeball (see detail of Figure 4-3). The photoreceptor cells of the eye are at the back of the retina, toward the brain and away from the light. Within these cells is a light-sensitive chemical. When a molecule of this photochemical absorbs a photon of light, it undergoes a molecular change that eventually leads to the generation of neural activity in the optic nerve.

The two general types of photoreceptor cells in the retina are indentified as rods and cones because of their characteristic shapes. Figure 4-4 shows how the rods and cones are distributed throughout the retina. The rods are mainly located in the periphery of the eye, the cones in the center. The density of cells is greatest in the center area, called the fovea. In the fovea there is more of a direct one-to-one connection of receptor cells to the fibers of the optic nerve. This central region is optimally located for the parts of the optical image that our eye is directly fixating, and it has the best capacity for resolving fine details of the image. It is also the region responsible for our sensitivity to color; however we will see that

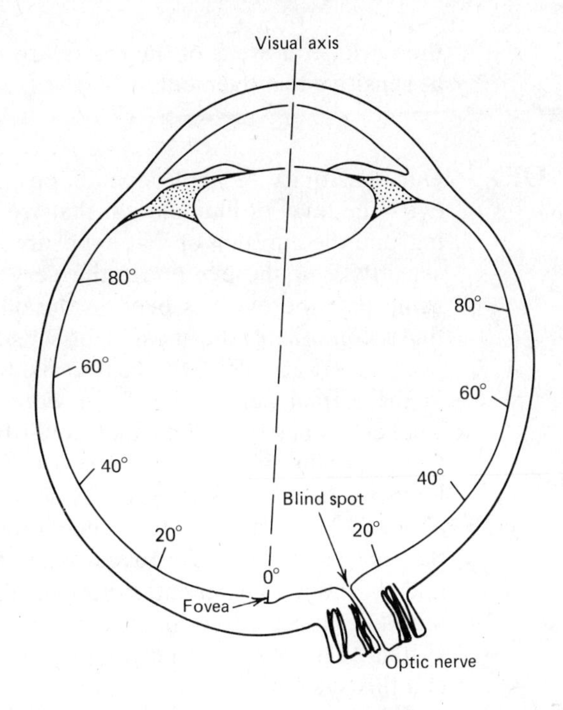

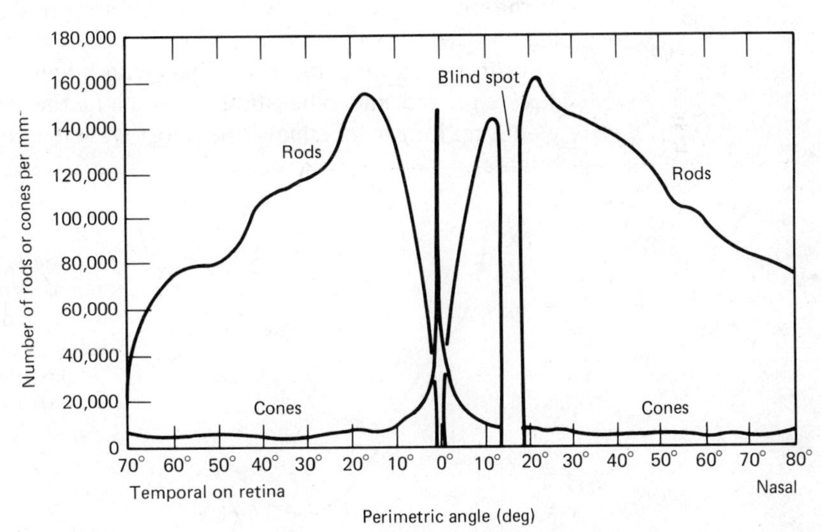

Figure 4-4. Top view of the eye, and the distribution of rods and cones across the retina. (From *Vision and the Eye,* 1967 by M. H. Pirenne, Chapman & Hall Ltd., London, reproduced by permission.)

the peripheral areas of the retina are more sensitive to light and at least as sensitive to movement.

SENSITIVITY OF THE EYE

Our sensitivity to light depends on the prior state of adaptation of our eye—the level of illumination that we have been exposed to before testing, and the duration of that exposure. For the moment, let's consider the sensitivity of the eye under the very best conditions of adaptation: assume that the eye has been in the dark for at least 40 minutes. Under these conditions, the most light-sensitive region of the retina will be a point about 20° off of the central fixation point, in the dense area of rods in the retinal periphery. If we were to test the visual system for the smallest intensity of light that could be seen under these conditions we would obtain the lower curve shown in Figure 4-5. This is the threshold intensity function for the rod system. The most sensitive wavelength is about 510 nm. If we were to repeat the experiment using a light stimulus that only stimulated the fovea of the retina, the cone system, we would find that much higher intensities of light were required to reach threshold. The upper curve in Figure 4-5 is the threshold curve for the cone system. The best wavelength for this system is shifted over to 555 nm, and this system is relatively more sensitive to longer wavelengths. Lights of different wavelengths will have different relative brightnesses depending on whether we are using the rod or cone systems. For example, if you were to match long and short wavelength lights for brightness during the day (e.g., red and blue stimuli), at night the brightnesses of the lights would no longer be equal—the short wavelength light would now appear brighter.

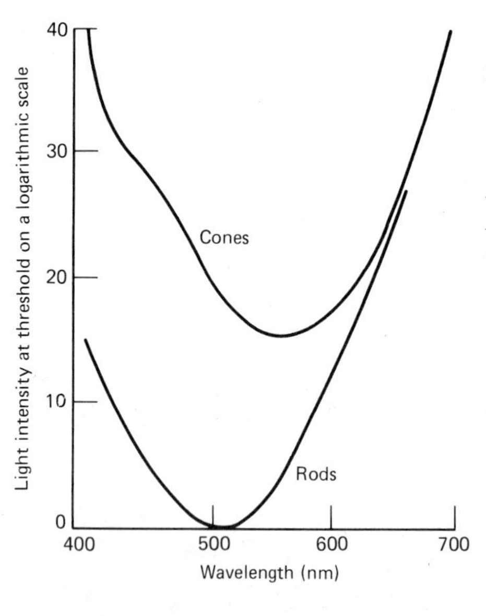

Figure 4-5. Threshold intensity functions for the rod and cone systems. (From S. Hecht & Y. Hsia, Dark adaptation following light adaptation to red and white lights, *Journal of the Optical Society of America,* 1945, *35,* 262, and reproduced by permission.)

Dark Adaptation

Our eyes are sensitive to light over almost a 10^{13} span of intensity, but at any given moment we can accurately discriminate among different light intensities over a much smaller range of about 2 log units. Within this smaller range of intensities we can make very accurate intensity discriminations of less than 1% (Goldstein, 1980). This 2 log unit range of discriminable intensities is called the *operating range* of the system. Its location within the larger 10^{13} span of sensitivity depends on the state of adaptation of the eye. If you expose the eye to a suddenly much dimmer or much more intense light, it will be unable to respond accurately to small intensity changes. When the average level of illumination is changed, the visual system adapts to the new level, and the operating range shifts to a new position. Putting a person in the dark for 40 minutes, pushes this range to its lowest and most sensitive position. The curves of Figure 4-6 illustrate the time course of this adaptation as a function of the amount of time the person is in the dark after initial stimulation to an intense source of light. You can see that the threshold decreases (sensitivity increases) as a function of the time in the dark. There is an initial rapid threshold decrease of about one log unit that flattens out after about 10 minutes; this corresponds to the recovery of the cone system in the fovea of the eye, and doesn't net us much of the ultimate change in sensitivity that we will gain with full dark adaptation. Meanwhile, the rod system has been slowly increasing in sensitivity until after 30 or more minutes an additional decrease of several more log units of threshold has occurred.

If you stimulate a dark-adapted eye with some bright light, the threshold will increase in less than 0.1 sec; in fact it will initially overshoot its ultimate value and then decrease slightly to a level that corresponds to its

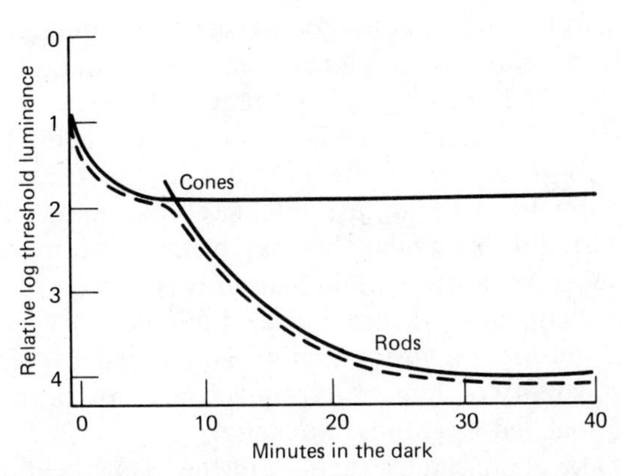

Figure 4-6. Decreases in threshold of the rod and cone systems as a function of time in the dark; the dotted curve is a composite of the rod and cone functions. (From *Sight and Mind: An Introduction to Visual Perception* by Lloyd Kaufman. Copyright © 1974 Oxford University Press, Inc. Reprinted by permission.)

new state of adaptation. The return to dark adaptation depends on the intensity of the light-adaptating stimulus. If you expose a dark adapted eye to dim lights such as in a dimly illuminated room and then return the subject to the dark, full dark adaptation may take only a few minutes. For example, the headlights of an oncoming car at night cause only a short-term impairment of our night vision. Actually, even this temporary loss of adaptation could be critical. That is why in some situations such as on board ship, crew members about to go on watch may stay in areas illuminated only by dim red light, in order to preserve a state of full dark adaptation.

It is important to realize that the usefulness of preserving full dark adaptation depends on the functions to be performed by the personnel. A study by Carr (1967) showed that dark adapted pattern vision is no different under low red or low (nominal) white ambient illumination (0.02 fc). Carr was interested in whether low white illumination and color-coded displays would improve performance in the control room and attack center of a nuclear submarine. He found that the use of nominal white illumination and colored (illuminated) indicator push buttons resulted in significant decreases in the average reaction time needed to respond to indicator signals. In other situations it might be desirable to preserve full dark adaptation, for example, if you need to maintain the ability to detect a very weak light stimulus. Remember that under such conditions the target should be viewed at an angle to your normal direction of gaze, so that it falls on the rod portion of the retina.

TARGET DETECTION AND IDENTIFICATION

When the level of illumination is sufficient for stimulation of the foveal (cone) system, target detectability is best in the direction of gaze, and decreases as a function of distance from the fixation point (target eccentricity). In a simulation of a visual sonar detection task, Vallerie and Link (1968) showed how these variables interact. Their observers had to report the presence of different brightness targets that could occur randomly at different locations in the visual field. The major factors affecting performance were the contrast of the target to the background, the "normalized brightness intensity" expressed in log $(\Delta I/I)$, and the eccentricity or angular distance of the target from the fixation point. (The background screen illumination was approximately 9.12 fL; targets were approximately square and 23.4 minutes of visual angle on a side.) Their results are shown in Figure 4-7. These curves allow estimation of the probability of detection from specification of the target's eccentricity and normalized brightness intensity.

The center and periphery of the visual field have different sensitivities for moving and stationary targets. Rogers (1972) obtained thresholds for low-contrast striped bars that were either stationary or moving, presented at different angles of eccentricity. Thresholds for moving images

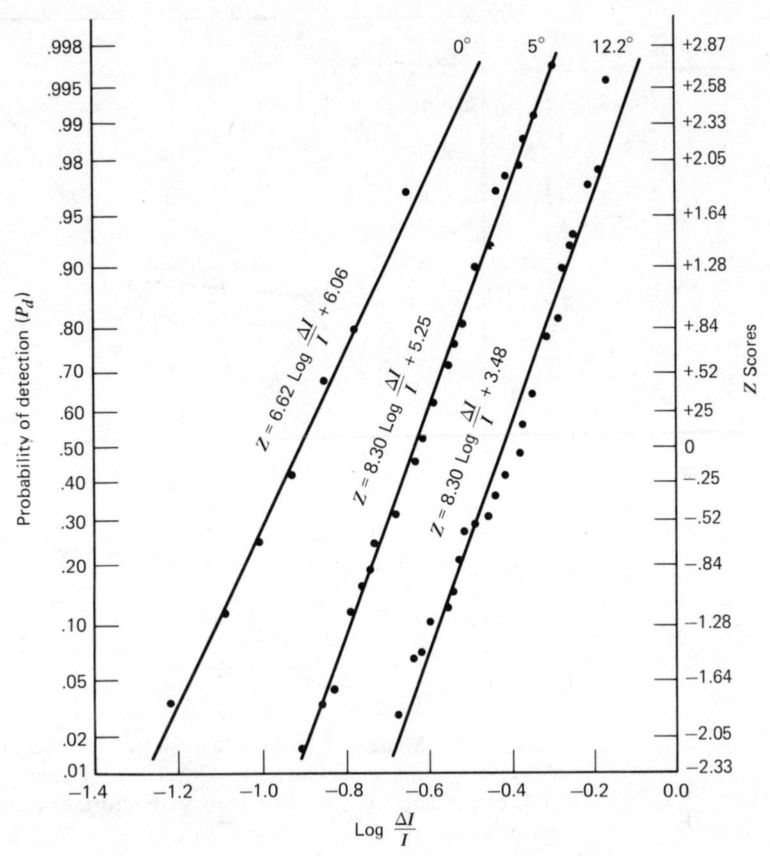

Figure 4-7. Probability of detection as a function of log ΔI/I for three eccentricity angles. (From L. L. Vallerie & J. M. Link, *Human Factors*, 1968, *10*, 409. Copyright 1968 by The Human Factors Society, Inc., and reproduced by permission.)

were found to be lower than for stationary ones, and no differences between center and periphery were evident. The lower curve in Figure 4-8 illustrates this result. However, thresholds for stationary images (upper curve) show a marked elevation for stimuli of greater eccentricity, as in the Vallerie and Link study. So the periphery is a potentially important channel of information, if movement or change can be employed as the signal code. In a later section, we see that obtaining information from peripheral inputs is important in the performance of visual search tasks.

Visual Acuity

The term *visual acuity* refers to the accuracy or sharpness of our pattern vision. There are many ways to measure visual acuity, including the familiar vision tests you have probably had in an opthalmologist's or

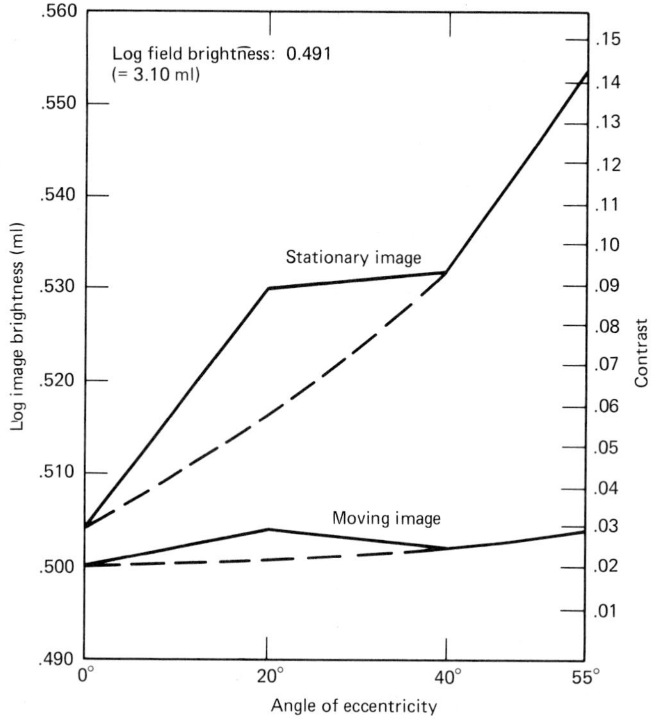

Figure 4-8. Measured contrast thresholds for stationary and moving images as a function of angle eccentricity. (From J. G. Rogers, *Human Factors,* 1972, *14,* 204. Copyright 1972 by The Human Factors Society, Inc., and reproduced by permission.)

optometrist's office. In those tests, you were shown an eye chart consisting of different size characters, and were asked to identify characters or their orientation. These tests measure *recognition* acuity, which is usually stated as a ratio such as 20/100. The ratio describes how well you can recognize the patterns compared to a normal observer. The 20/100 measure says that you can see as well at 20 feet, as a normal observer can see at 100 feet. Other measures of spatial acuity include tests for the detection of a single line (called *detection acuity*) and tests for how well closely spaced lines can be seen as separate (called *resolution acuity*).

It is convenient to describe the limits of acuity on these various tests by stating the smallest visual angle that you require to perform them. Figure 4-9 illustrates what is meant by the term visual angle. It allows us to describe the size of the image on the retina without stating the actual target size and target distance. The tangent of the visual angle is the ratio of the target size to target distance. The limit of acuity in the detection situation is about 2 sec or 0.01 mrad of visual angle; minimal acuity in

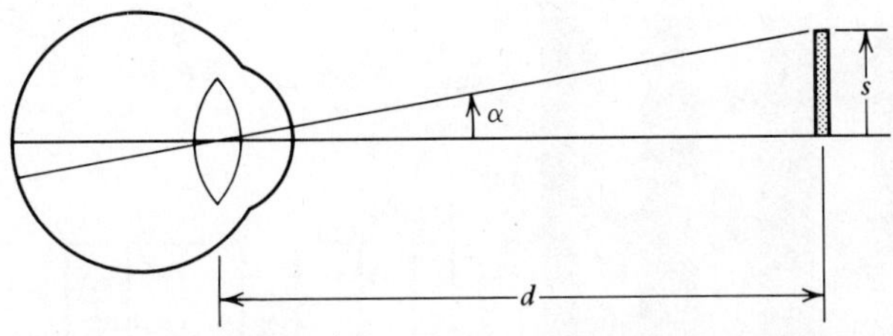

Figure 4-9. The relationship between the visual angle (α) and the object size (s) and distance (d); $\tan \alpha = s/d$.

resolution or recognition is about 30 sec of visual angle. The recognition acuity of the normal observer in the 20/20 measure is defined as 1 minute of visual angle (0.29 mrad). As the level of illumination and the target-to-background contrast is decreased, the minimum visual angle increases (see Chapter 17, the sections on illumination).

We mentioned earlier that the mechanism of accommodation varies the curvature of the lens to bring images into sharp focus on the retina. As an object is moved closer to our eyes, the process of accommodation increases the curvature of the lens to keep the image in focus. Farsighted persons, *hyperopes,* have eyes that are too short for the focusing ability of the lens. For these people the image of the target would be sharply formed at a point just behind the retinal surface, and the resulting image on the retina is blurred. If the target is moved farther from the eye, the image eventually comes into sharp focus. The opposite problem is having an eye in which the distance from the cornea to the retina is too large for the lens; the image is formed in front of the retina. People with this problem have nearsightedness, which is termed *myopia.* Moving the target closer to the eye results in a sharp image. As we age, the lenses stiffen, and we cannot change their curvatures as much as when we were young. That is why many people in their forties and fifties can no longer focus on objects at close distances, and require reading glasses or bifocals to increase the refractive power of their cornea-lens system. This condition is termed *presbyopia.*

Spatial Frequency Analysis An important kind of stimulus that can be used to assess a person's visual acuity is the square-wave grating shown in Figure 4-10*a*. In testing for the limit of your acuity with this pattern, we can decrease the spacing of the grating until you no longer can tell that the grating was present (it might look uniformly grey) or we can decrease the contrast between the light and dark bars, until they appear uniformly grey. After this testing we can tell how your ability to see the bars depends on the bar spacing and the bar contrast. But using this information to predict your perception of stimuli that were not square gratings would

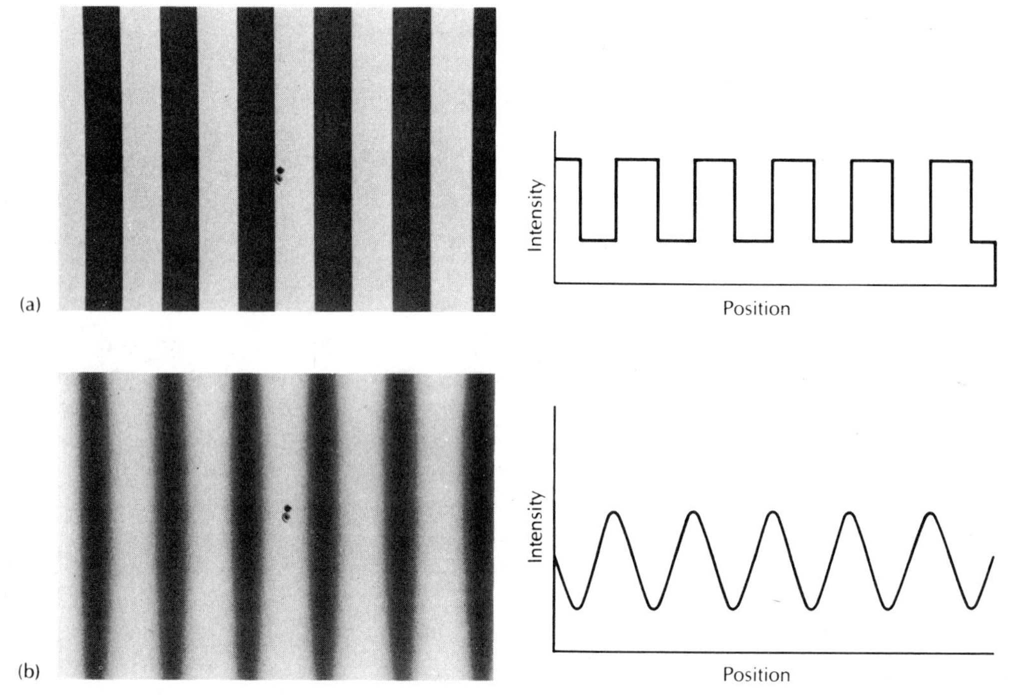

Figure 4-10. Square-wave (*a*) and sine-wave (*b*) grating patterns and related intensity versus position plots. (From T. N. Cornsweet, *Visual Perception,* p. 277. Copyright 1970 by Academic Press, Inc., and reproduced by permission.)

be difficult. A stimulus that is more efficient for this purpose is the sinusoidal grating illustrated in Figure 4-10*b*. The frequency of the grating is the number of fuzzy bars (or cycles) per centimeter across the image, about 1 cycle per centimeter in the figure. The curves in Figure 4-10*a* and *b* illustrate how intensity varies with distance for the square wave and sinusoidal grating patterns. The *amplitude* of the grating is defined as half of the intensity difference between the peaks and valleys of intensity; the average intensity of the wave is shown as the dotted line. Sinusoidal gratings are important because any visual stimulus can be analyzed as a sum of individual (horizontal and vertical) sinusoidal gratings (see the discussion of complex sounds in Chapter 3.) If we could specify our sensitivity to those frequencies, we could predict how well any particular visual stimulus could be seen. This approach is used to specify the effects of different imaging systems such as photographic, computer, or television systems on a picture. These systems pass the spatial frequencies present in the original scene according to a function called the Modulation Transfer Function (MTF).

The results of an experiment that measures how our spatial acuity

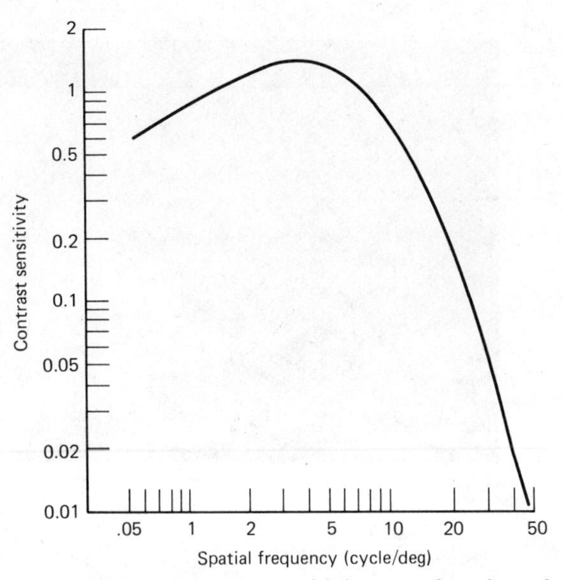

Figure 4-11. Contrast sensitivity as a function of spatial frequency for normal human eyes. (From A. Fiorentini & L. Maffei, *Vision Research,* 1976, *16,* 438, and reproduced by permission.)

depends on the frequency of a sinusoidal grating is shown in Figure 4-11. The figure shows how our sensitivity to the grating's contrast depends on the grating frequency (measured in cycles per degree of visual angle). Sensitivity is greatest at about 4 or 5 cycles per degree and falls off at high and low frequencies. This function allows predictions to be made about the perception of a wide variety of different kinds of visual stimuli. The high-frequency drop-off in sensitivity that is shown in the graph is no surprise; as parts of a stimulus get closer together we know that our limits of resolution acuity will be reached, and we will tend to see the lines as blurred together. But it is surprising to see that our low-frequency spatial sensitivity is also limited.

What that low-frequency effect means is that we are less sensitive to light patterns that change slowly across the visual field. Such patterns will appear to have less contrast than patterns that change around 4 cycles per degree. There is another surprising effect described by this decrease in our sensitivity to low spatial frequencies. Certain patterns such as that shown in Figure 4-12 appear to have a band of lighter intensity along the left-hand area, and a darker region along the right-hand area of the transition region. In fact, these lighter and darker bands do not exist in the physical light distributions of Figure 4-12. These band effects are called Mach bands after the nineteenth-century physicist Ernst Mach. They come about because of the reduced low-frequency response of the visual system.

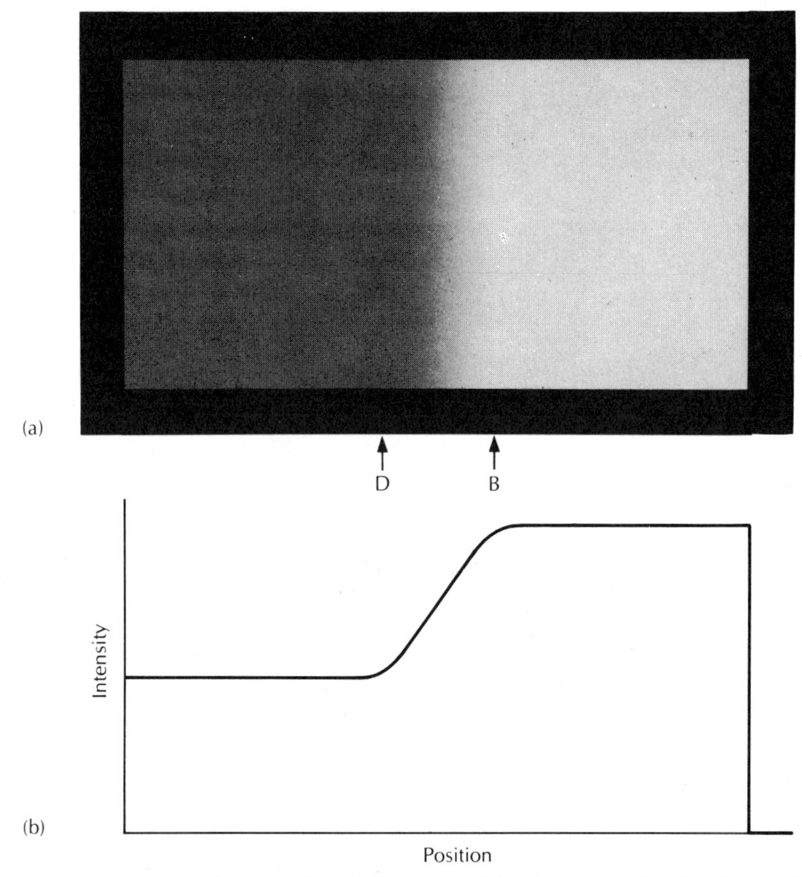

Figure 4-12. (*a*) A mach band pattern and its intensity distribution. A bright band appears at B and a dark one at D. (From T. N. Cornsweet, *Visual Perception,* p. 313. Copyright 1970 by Academic Press, Inc., and reproduced by permission.)

SLICE OF LIFE

Nearsighted People Can't See Well Even with Corrective Lenses

An important aspect of the spatial frequency approach was revealed by a study of severely myopic high school and university students by Fiorentini and Maffei (1976). Since it is mainly the vision for distant objects that is affected for nearsighted people, we might expect that the high-frequency end of the spatial sensitivity function is lower than normal, and the mid- and low-frequency region is the same as normal. In other words, we might think that only the clarity of images of objects that are far away are affected. Fiorentini and Maffei found a quite different result: their myopic students had impaired sensitivity over the *entire* spatial spectrum. (Their students had eyeglass corrections from about −6 to −18 diopters; if your eyeglass prescription is handy, you can check if your vision is in this range.) Figure 4-13 illustrates their results. The lower curves for the myopic students, are depressed from the normal curve throughout the range of spatial frequency. Fiorentini and Maffei concluded that myopic subjects have a permanent visual deficiency relative to normal sighted subjects. This deficiency extends over the whole spatial frequency range and is not simply a reduction of visual acuity at high frequency. Note that this deficiency exists even after their vision has been fully corrected with lenses!

Practically, this suggests that a myopic person sees all patterned stimuli with less effective contrast between the light and dark areas than does a person with normal vision, with a resulting need for a better illuminated and a better arranged reading-working environment. This may be one reason why the military does not select people with cor-

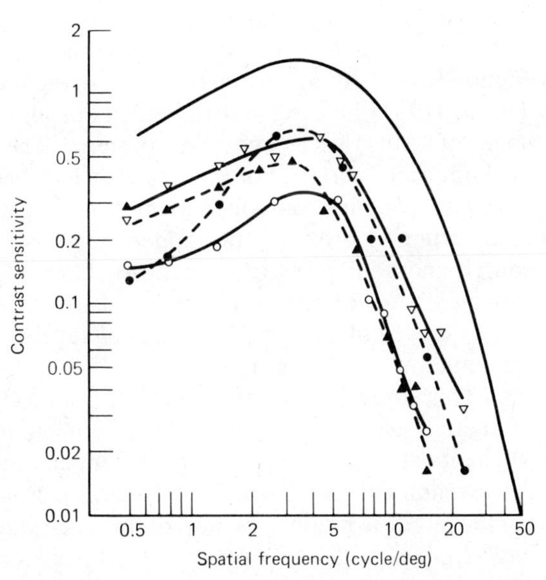

Figure 4-13. Contrast sensitivity functions of four myopic eyes. The upper curve is the average contrast sensitivity function of normal eyes. (From A. Fiorentini & L. Maffei, *Vision Research*, 1976, *16*, 438, and reproduced by permission.)

rected vision for critical jobs involving keen vision, such as fighter pilots.

This idea is supported by a laboratory simulation of a military target acquisition task performed by Grossman and Whitehurst (1979). They assessed the target acquisition performance of 32 subjects under different conditions of target distance (simulated), target type, target masking, target to background contrast, and subject familiarity with the terrain. The investigators also measured the visual acuity of the subjects, which ranged from 20/20 to 20/50. The major task variables were found to be target distance and target masking. Both response time and percent correct detection were highly dependent on the subject's visual acuity; acuity interacted with target distance and masking to increase the performance decrement due to acuity under the more difficult conditions. They concluded that visual acuity should be an important personnel selection criterion for target acquisition tasks. Perhaps the successful gun fighters of the Old West were the survivors of such a selection process.

Image Quality

Specifying how the extraction of information from an image depends on objective measures of image quality is an important practical problem. The spatial frequency approach is especially useful when pictures have been sampled and digitized for processing or storage by a computer (see Figure 9-16 and the discussion of speech technology in Chapter 9). Turpin (1981) investigated the effects of blur and noise on the performance of Air Force photointerpreters. The stimulus materials were medium-scale aerial photographs that had been digitized and then subjected to combinations of blurring (via a spatial frequency filtering and reconstruction process) and added noise (via scanning a random noise source and adding it to the picture). Turpin found major effects on performance due to the noise manipulation and small but demonstrable effects due to blur, at the blur filtering levels and characteristics he used. He found no significant interaction between the blur and noise manipulation. This study represents an important step in relating the characteristics of digitally derived imagery to human image interpretation performance.

The most common type of computer generated display systems employ a television-imaging system in which a picture is generated via a raster composed of a number of horizontal scan lines. A standard TV system has 525 lines in the picture, with 485 lines normally displayed in practice. Higher-resolution systems may have many more scan lines per picture. Erickson (1978) summarized a large amount of data on target acquisition performance as a function of operator, target, and electronic imaging system characteristics. Erickson provided tables and an equation for estimating the television line requirements necessary for certain detection and recognition situations. This calculation assumes a quality of TV imagery at least as good as a standard closed-circuit TV system, and ignores possible (and likely) degrading effects of operator fatigue, training, stress, vibration, target background, and atmospheric attenuation. The system geometry is approximated by equation 4-3:

$$l = \frac{57.29hN}{R(FOV)} \tag{4-3}$$

where l is the number of lines across a target of (projected) size h, N is the number of lines per picture in the system, R is the distance (range) to the target, and FOV is the systems field of view in degrees. Table 4-2 provides the needed scan lines for desired performance with different targets.

Suppose that we wish to detect a raft at a sea with a down-looking TV system. Assume that you have 525-line system (485 usable), a 20° field of view, a raft 2.5 m on a side, and a raft-to-sea contrast of more than 18%. How high an altitude would result in very good performance? From column 2 of Table 4-2 we see that 3 lines are required. Then from equation 4-3

$$R = \frac{(57.29)(2.5)(485)}{(3)(20)} = 1158 \text{ m}$$

TABLE 4-2 *Scan Line and Resolution Line Requirements for Different Tasks*[a]

Task	Scan Lines Required	Performance Level, Percent Correct	Black and White Lines per Target Height
Detection			
Small, isolated targets			
18% inherent contrast	3	100	2
7% inherent contrast	5	100	3
Construction equipment	9	—	6
Three vehicles in moving scene	20	85	14
Recognition (given a detection)			
Ships	10	80	7
Vehicles	10	80	7
Buildings	10	100	7
Bridges	10	100	7
Aircraft	12	80	8
Three vehicles in moving scene	12	85	8
"Large targets"	20	—	14

[a] From R. A. Erickson, *Human Factors,* 1978, *20,* Tables 1 & 3. Copyright 1978 by The Human Factors Society, Inc., and reproduced by permission.

An alternative measure of TV image quality is the number of lines that can be resolved when the system views a test pattern placed at the target location. This measure includes possible factors that might degrade the televised signal, such as noise and spatial filtering. The fourth column of Table 4-2 provides this measure of image quality.

Image Legibility

From the preceding discussion you would expect that the legibility of displayed letters or symbols should depend on the target to background contrast and the character size measured in degrees of visual angle. In most practical display media, such as printed material, signs, etc., the contrast is sufficiently good that it can be ignored as a significant design factor (see Chapter 7 for a discussion of display contrast). That leaves the question of letter size as the most important practical constraint on display design. Clearly, the smaller the letters possible, the more that can be fit in a given display format. This question has been addressed by Smith (1979) in a comparison of published recommendations of letter size with data from a field study of a large number of printed displays. Smith found that mean letter heights ranged from about 0.0019 rad (7 minutes) at the limit of legibility, with over 90% legibility at 0.003 rad and virtually 100% at 0.007 rad. Although these values are consistent with published standards, special viewing conditions and unusual letter shapes may require higher values.

Eye Movements

Our eyes are constantly moving in many different ways. One type of movement is the involuntary eye movement that occurs when we try to hold our gaze on a target. The very small movements that occur are physiological tremors and amount to very fast, extremely small (less than a minute of arc) movement. The eye may slowly drift off of fixation, and when the drift reaches some threshold amount, an abrupt movement will be made to bring the target back into fixation on the fovea. These abrupt eye movements are called saccadic eye movements or saccades. The drift movements are usually about 3 minutes or so of arc per second, and the corrective saccades might be from 5 to 20 or more minutes of arc. If you suddenly move an object that is in an observer's visual field, the eye will move to get the object's image back onto the fovea. The interesting thing about saccadic eye movements is that they are very rapid, and once initiated they take only a brief amount of time. A 40° movement can take only 100 msec. During each fixation the eye might be at rest for 300 msec to a second or more. Some time is required to initiate a saccadic movement, up to 200 msec or more depending on the circumstances.

We can get a great deal of information from only a single eye fixation of a scene (Biederman, 1981), but generally we need to move our eyes around the scene with several fixations in order to bring different areas onto the fovea. Even when a target is within foveal vision during a single fixation, objects in the scene background can have a deleterious effect on target detection and reaction time (Klatsky, Teitelbaum, Mezzanotte,

& Biederman, 1981). For example, jumbling the scene or putting objects in unlikely or impossible positions increases errors and processing time.

In normal viewing, we might make three or more fixations of a picture each second, with less than 100 msec spent in movement time, and 900 msec in fixation time. The fixations tend to be on the highly informative regions of the picture (Mackworth & Morandi, 1967). Another type of eye movement is the pursuit movement, in which the eye attempts to maintain a stable retinal projection of the object as the object moves. Chapter 7 discusses the subject of eye fixations as they apply to the design of visual display systems.

There is a kind of internal gating of the visual system during eye movements. Normally, one might expect that during an eye movement, we would notice a blurring of our vision as the image moves across the retina. This is not the case; apparently the image that is perceived while we are fixated on a portion of the visual field persists during the subsequent eye movement. Campbell and Wurtz (1978) demonstrated this in an experiment in which they triggered a light flash to illuminate a darkened room during different portions of an observer's eye movements and fixations. When the room was illuminated only during the eye movement, the observer reported a smearing of the image. When the room was illuminated during the movement and also during a portion of the subsequent fixation, the scene was reported to be clear and not smeared. Somehow the sharp image formed during the fixation dominates the perception of the scene. Try to observe your eyes in a mirror as you change the position of your eye fixation. Can you see your eyes move? The persistence of the fixated image is an important aspect of our ability to process visual information. Information is available from a persisting image for as long as 300 msec after termination of the stimulus. This storage mechanism is called *iconic storage,* and is discussed further in Chapter 6.

Visual Search

If the stimuli are simple, fairly large, and well illuminated, we can get information from a single fixated field that may be as large as 40°. However, if fine resolution is required the useful field may be very much smaller. Information at the edges of the field may be used to determine the position of the eyes during the next fixation. This idea is supported by experiments by Williams (1966) on the use of peripheral information to select the position of fixation. Williams asked observers to find a specific numbered form in a large visual field (about 40°) consisting of numbered forms of different shapes and colors as shown in Figure 4-14. On some trials the observer was given information about the target form in addition to its number, such as its relative size, shape, or color. The position and direction of eye movements and the search time was recorded. Giving the observer information about the color of the target stimulus reduced search times by nearly 80%. The use of size information was much less evident, and shape information had an even smaller effect. Combining cues had no more effect than the stronger cue alone. The

Figure 4-14. Stimulus display used by Williams (1966) in his study of visual search. (From *Perception & Psychophysics,* 1966, *1,* 316. Copyright 1966 by Psychonomic Press, and reproduced by permission.)

experiments demonstrate that visual information in the peripheral parts of the display field can influence subsequent eye fixations, and that some types of information are much more useful than others for this purpose. A number of studies have examined more complex aspects of visual search such as the dependence of performance on the size of the search field, the number and complexity of the targets, and the nature of the field background. Some recent studies are described in Teichner and Mocharnuk (1979), Robinson (1979), Drury and Clement, (1978) in a series of meeting papers edited by Fisher, Monty, and Senders (1981), and in Carter (1982). The sections on visual inspection and fault diagnosis in Chapter 14 have additional discussion of visual search tasks and models of the search process.

Perception and Movement

Another aspect of target detection and identification, is our ability to detect the movement of objects or parts of the retinal image. How do we perceive what has moved, the object in the retinal field, or our body or eyes? One source of information is derived from a comparison of the brain signals to the eye and head muscles and the output signals from the visual system (Gregory, 1973). The other source of information is contained in the relationships among the parts of the retinal image. Gibson (1968) has pointed out that certain types of transformations of the elements in the retinal field signal movement of the body or eye, while others

TABLE 4-3 *Movement Conditions and the Resulting Perception*[a]

Condition	Resulting Perception
1. Occlusion-disocclusion of background by leading and trailing edges of object.	1. An object moving through the world.
2. Transformation of an entire ambient light array (motion perspective).	2. Locomotion of the observer through the world.
3. Transformation of an entire ambient light array, *and* occlusion-disocclusion.	3. An object moving through the world, *and* locomotion.
4. Shift or rotation of the borders of the field of view, corresponding to the nose and/or eyebrows.	4. Head turning and rotation relative to the body.
5. Displacement or "sweep" of retina behind virtual image of ambient optic array. Shift in gaze line or changing sampling of the optic array.	5. Normal eye-movements relative to the head.
6. Visible motion of an extremity.	6. Motor behavior of the self.

[a] From Table 6-1 in *Perception: An Applied Approach,* by William Schiff. Copyright © 1980 by Houghton Mifflin Company and adapted from J. J. Gibson, *Psychological Review,* 1968, *75,* 335-346.

signal movement of the object. For example if only one element in a background of motionless elements moves, that is probably object rather than body movement. Table 4-3 summarizes some of these transformations and the resulting perceptions of movement.

Stroboscopic Movement Suppose that you view the two lights of Figure 4-15; light A is flashed on and off, and after a brief time, light B is flashed. You would report that the light appeared to move from A to B. At some time interval between the flashing of the two lights (between 30 and 200 msec), there is some movement of the light seen – about 60 msec is highly effective. This kind of motion is called stroboscopic motion and is associated with the apparent movement produced when there is a stim-

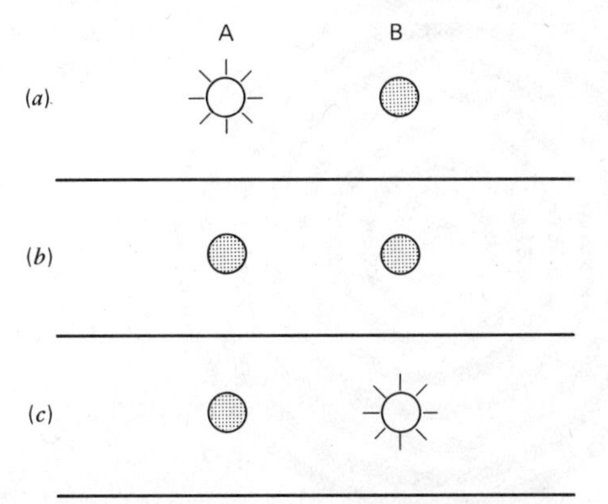

Figure 4-15. Two lights A and B are flashed in sequence. The light appears to move from A to B.

ulation of different points on the retina at different times. The film and television media involve the same phenomena; our eyes are presented with a succession of still pictures which appear to be filled with moving objects. The time between the frames of a motion picture or between frames of a television picture are at the low end of the optimal time separation for the simple light flash movement demonstration.

Stroboscopic lamps are used to "stop" the movement of moving or rotating objects. The brief, periodic flash of the strobe light illuminates a moving target as a series of brief, fixed images. The succession of these stopped images is perceived just as is the succession of stopped images of movie film or television. Suppose that we flash our strobe light 100 times per second at a wheel rotating at 90 revolutions per second. The strobe flash will "catch" the wheel position slightly earlier on every cycle of the wheel's rotation. In the sequence of images of the wheel, the angular position of the wheel's spokes will be decreasing at 10 revolutions per second, and the wheel will appear to be rotating backward. There is sometimes a similar effect for rotating images in television or in films. The perception of realistic movement in film and television depends on more than just the presentation of a succession of changed images. Our experience and expectations about the figure also influence our interpretation of the seen movement (Kolers, 1972).

Movement Aftereffects Sometime after driving for several hours, try the following experiment. Park the car and set the parking brake, then open your door and look down at the ground. You may see the ground moving forward as the car appears to be rolling backward. This illusion (also known as the "waterfall illusion" because if you stare at one place on a waterfall and then look away, you get it very strongly) is related to the aftereffects of the continued movement of images across part of the retina. The effect is very marked if we look at a rotating stimulus like the spiral of Figure 4-16, while it is rotating (copy it and place it on

Figure 4-16. The spiral aftereffect. Copy the figure and stare at it while it is spinning.

your stereo's turntable). After staring at the spiral for a few moments, look away from it at another target. That target will appear to be shrinking or moving away from you, or expanding or moving closer to you – depending on the direction of rotation of the spiral during your fixation of it. These effects reflect the action of motion sensitive circuits in your visual nervous system. Somehow the intense simulation causes those systems to be fatigued or saturated, and subsequent stimulation by nonmoving targets is perceived as movement in the opposite direction.

COLOR

The color of objects and displays is a potentially important channel for conveying information about our environment and the person-machine systems we use. In some types of tasks, color coding of displays can provide a significant improvement in performance; in others no advantage or a disadvantage can result. We saw that color can be very useful in a visual search task. In a major review of color coding research, Christ (1975) analyzed the effectiveness of color coding in a variety of visual target identification and search tasks. If the color of a target is unique to that target and known in advance by the observer, color will aid both identification and search performance. Colors used as features of targets can be identified more accurately than some other characteristics such as size, brightness, and shape, but are at a disadvantage compared to alphanumeric symbols. As the density of targets in the display increases, the relative advantage or disadvantage of color increases. When colors are added to an achromatic display situation, the accuracy of identifying achromatic features of the targets may decrease; that is, there may be an interference from the color code even when it is irrelevant for target identification. Similar results hold for search tasks; color coding can greatly reduce search time when color is a relevant and known code but can also increase time by interfering with the use of achromatic target features.

The use of color as a redundant code, such as in photographs or computer or television displays, may or may not be an advantage. Christ found mainly advantages for color in such situations. However in a recent study comparing colored and black-and-white television displays in a bomber flight simulator, Kellogg, Kennedy, and Woodruff (1981) found no significant differences in performance between colored and black-and-white displays. Under normal viewing conditions, visual acuity is not different whether tested with red, green, or black-white sinusoidal gratings (Nelson & Halberg, 1979).

Because of the importance of the color channel, the human factors specialist should have some understanding of how color is physically specified, and of the principal aspects of color's effect on people. In the next sections we consider some of the technical aspects of color, including color mixing and contrast phenomena, and color blindness.

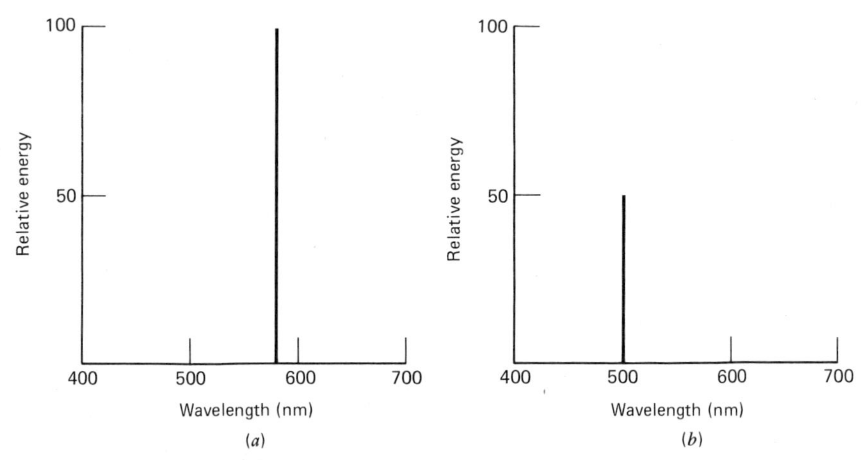

Figure 4-17. Monochromatic stimuli, yellow (580 nm) at (*a*) and green (500 nm) at (*b*).

Color Specification

If you look at light that has been dispersed by a prism you will see a rainbow of colors ranging from the blues of the short wavelengths through the greens of the mid wavelengths to the reds of the long wavelengths. These rainbow colors are called *spectral* colors. The action of a prism is to spread out the light of different wavelengths so that we separately see light of very narrow wavelengths, rather than the complex wavelength mixture of ordinary white light. Spectral colors are composed of light of only a narrow band of wavelengths; we can use a single line spectrum, termed monochromatic stimuli, to represent them (see Figure 4-17). Mixing different monochromatic stimuli together produces a lot of different colors, some of which appear quite similar to the rainbow colors. Most light mixtures that we encounter appear as either "washed out" spectral colors (such as faded blues, greens, or reds), or as nonspectral colors such as pink, brown, and purple.

Three perceptual aspects of a light stimulus are related to its spectrum: *saturation, hue,* and *brightness*. The first aspect, *hue,* refers to the light's color. Hue is related to the color name that we associate with the largest wavelength components in the light's spectrum; if the long wavelengths were the dominant components, we would say the stimulus were red. If you were asked to arrange a set of chips of different hues according to their similarities of color, you might put the yellow near the oranges, the oranges near to the reds, greens near the blues, and so on. What would result is the color circle shown as part of the color solid of Figure 4-18. Around the circumference of the color circle would be the various colors of the rainbow, as well as a few intense nonspectral colors such as the purples. Inside the circle would be various colors that have a spectral color name, but are somewhat weaker in their subjective hue. The perceptual term for this change in the purity of a color is called saturation. Imagine a red light with a good deal of white light mixed in—the result

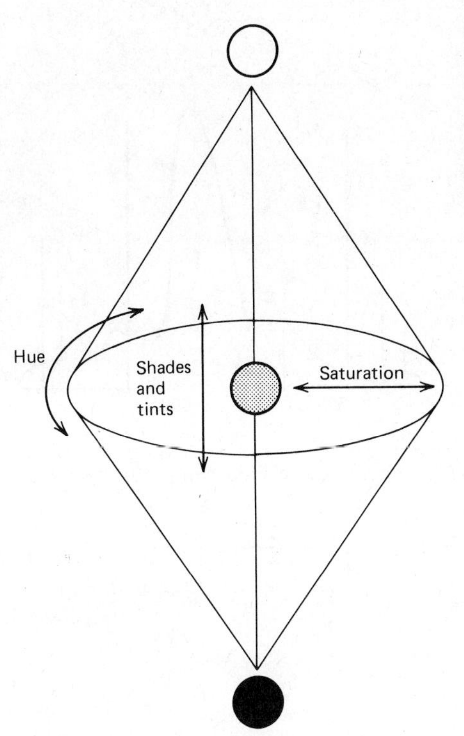

Figure 4-18. The color solid. (From *Sight and Mind: An Introduction to Visual Perception* by Lloyd Kaufman. Copyright © 1974 Oxford University Press, Inc. Reprinted by permission.)

would be a less saturated red. The radius dimension of the color solid represents the degree of saturation of the color. The third aspect of a light stimulus is the brightness, which is related to the photometric intensity of the stimulus. The vertical dimension of the color solid represents this dimension. The color solid is actually a very crude way to represent color perception; other systems for cataloging the appearance of colors are described in Hurvich (1981).

Color Mixing

It is possible to describe a wide variety of color phenomena with a relatively simple set of experiments. Suppose that we wanted to be able to construct any light stimulus in the laboratory, in order to exactly match any real light stimulus that someone happened to come up with, such as the complex mixture of Figure 4-19. A lab gadget we can use to test for the success of our lab-generated stimulus is the split field device shown in Figure 4-20. We will agree that our lab-generated stimulus matches the real-world stimulus if the observer cannot distinguish the line dividing the two halves of the field. It turns out that we can get a pretty good match to any light stimulus by simply mixing light from three suitably chosen monochromatic light projectors. That is, by adjusting the intensity and wavelength of a three-component mixture, we can match the saturation, hue, and brightness of most any stimulus that you can come up with.

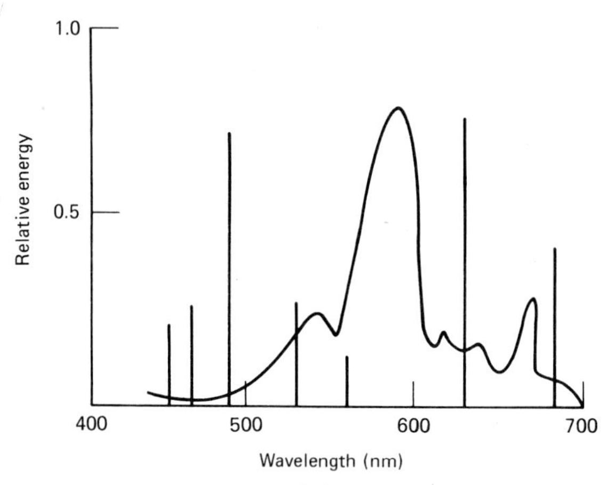

Figure 4-19. Spectrum of light sample to be matched.

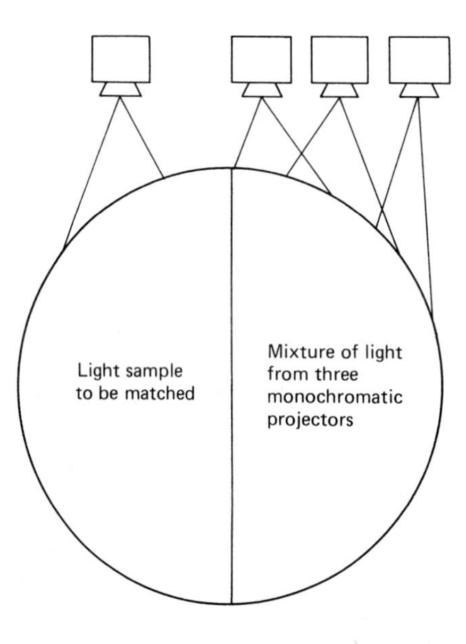

Figure 4-20. Split field device for comparing appearance of two different light mixtures.

Notice that we are describing how different colored *lights* can be mixed to produce any desired color. This is called *additive* color mixing. Mixtures of paints or pigments rather than lights are termed *subtractive* color mixtures and are a bit more complicated to describe. The color of a surface depends on the spectrum of light reflected by that surface (e.g., the wavelengths that the surface does *not* absorb). When two pigments are mixed, the resulting color is determined by the light wavelengths that *neither* absorbs. A mixture of red, blue, and green light thus yields a white light, while a mixture of red, blue, and green paint appears black.

We can make the colored light-matching task even harder for ourselves by fixing the wavelengths of the three projectors that we are using in the mix; only the intensities of the three will be adjusted to match the complex stimulus! We call these three wavelengths, primaries. Three good ones to use are a red of 650 nm, a green of 530 nm, and a blue of 460 nm. Strangely enough, we can still match a large variety of stimuli. In fact, the behavior of the mixing process is describable by some relatively simple mathematical laws. The crucial law is that color mixtures are additive: if stimuli that match are added to other stimuli that match, the resulting new mixtures will match. To specify any colored light stimulus we need only determine the intensity settings of the three primary projectors. This process can be simplified even further by realizing that hue and saturation only depend on the *relative* intensities of the three primaries. Knowing the percent red and percent green, for example, would allow full specification of the light's color. The appendix to this chapter describes how to determine these values for any light stimulus.

There are a variety of other important color phenomena in addition to color mixing. You know that if you stare at a red patch of color and then look at a grey or white surface, you will see a green colored afterimage, the complementary color. (Complementary colors are colors which, when added to the original color, produce a colorless result.) The same thing happens for blues and yellows. There are also complementary color effects that occur simultaneously rather than successively in time. If you look at a grey circle surrounded by red it will appear greener in tint than the same grey surrounded by a white background. Adjacent colored areas induce complementary hue effects.

The influence of a color in one part of the visual field on the apparent color of another part, is related to the interesting phenomenon of color constancy. Color constancy refers to the fact that under ordinary circumstances, our perception of the hue of a surface does not change when the spectrum of the light illuminating that surface is changed. That is why the apparent color of most objects and clothing remains relatively constant even though we view them under different lighting conditions. Apparently our visual color system is designed for making good judgments of the wavelength reflective properties of surfaces, rather than simply the spectrum of the light reaching our eyes from a surface. The color of a part of the visual field depends on the distribution of wavelength over the rest of the visual field. Sometimes color constancy fails, and a color match of makeup or clothes that we thought looked very good, suddenly looks awful under new lighting conditions. In the example given at the beginning of this chapter, the novel spectrum of the hangar lighting system was responsible for the strange colors observed.

Color Blindness

The rules of color mixing that we have described fail for a group of people with deficiencies in their color vision. You probably know at least one man with an obvious color deficiency—perhaps he occasionally wears clothes that, to you, are obviously mismatched in hue. The most common type of color deficiency is called dichromatism, because a person with this color defect can match any color with a mixture of light from just *two* primaries, rather than the three that a person with normal, trichromatic vision requires. The most common forms of dichromatism are termed deuteronopia and protanopia. In both these forms of the deficiency, the person will confuse red and green hues. The protanope also suffers an apparent lack of visual sensitivity to long wavelength light. A very rare form of dichromatism is termed tritanopia, involving a confusion of blues and yellows and a loss in sensitivity at short wavelengths.

A less serious type of color abnormality is classified as anomolous trichromatism, in which there is color weakness rather than an apparent failure of one component of the color system. The person with a deuteranomolous deficiency will confuse red and green hues and will need more green in a red-green mixture than normal, for a match to a yellow. A person with a protanomolous deficiency will also confuse reds and greens but will need more red in a mix to get a yellow. About 2% of the male population (of European ancestry) exhibits dichromacy and about another 6% is color deficient in some way along the red-green dimension; the percentage of color deficiency among women is supposed to be much smaller (less than 1%) (Hurvich, 1981).

Under minimal viewing conditions observers with normal color vision can identify as many as 15 different colors without errors (Feallock, Southard, Kobayashi, & Howell, 1966). Color blind observers (deuteranopes) can identify far fewer colors, about 8, and have particular difficulty with the reds and yellow-greens. Thus color codes should be limited to 8 or 15 standards, depending on the population and viewing conditions. Common color codes like traffic lamps should have their spectra optimized for the discrimination of red and green by color blind persons, by adding appropriate wavelength components to the lights.

SIZE AND DISTANCE PERCEPTION

We do an amazingly good job of estimating our physical position in the space around us. A major part of this positioning task is accomplished by the use of information processed by our visual system. In fact, if we intentionally foul up the correspondence between visual and nonvisual sources of position information, the visual source usually dominates the other even if it is wrong (Rock, 1975). How do we obtain distance and size information from the visual system? Basically, the visual data come from the two curved images that the world projects onto our two retinal surfaces. These images are constantly changing as our bodies, heads, and eyes move, and the relationships between the images and these movements provides a great deal of information. In addition, there are two other potential sources of information available. One is from the process of accommodation; the focusing operation of the lenses provides a source of information about the distance of objects. Because the eyeballs rotate inward and outward in order to track objects at different distances from the head, the convergence angle of the eyes provides another potential source of distance data.

Depth Perception

If you wanted to get a very realistic sense of the depth of a photographically reproduced scene, you would view it with a stereoscope or stereoscopic projection system. Every few years film makers release some films produced with the stereoscopic movie system. In this system two overlapping images are projected on a screen and viewers must wear special polarizing eyeglasses in order to see the separated images, as shown in Figure 4-21. The system is based on the principle of the stereoscope, invented by Charles Wheatstone, the nineteenth-century physicist. Normally our eyes are presented with two slightly different or disparate views of the visual world. The visual nervous system is sensitive to this disparity information. If we are normal observers we see these views as a single fused image in full depth.

Stereoscopic pattern projection can improve the recognition of targets in visual noise. Giarretto (1968) found pattern discrimination performance was much better with stereoscopic presentation, and the improvement increased with increased levels of noise. The detection process is similar in principle to that obtained in binaural (auditory) signal detection (see Chapter 3). Mountford and Somberg (1981) performed a study of two types of stereographic display systems in an airborne fire control situation. They presented three-dimensional displays of dynamic com-

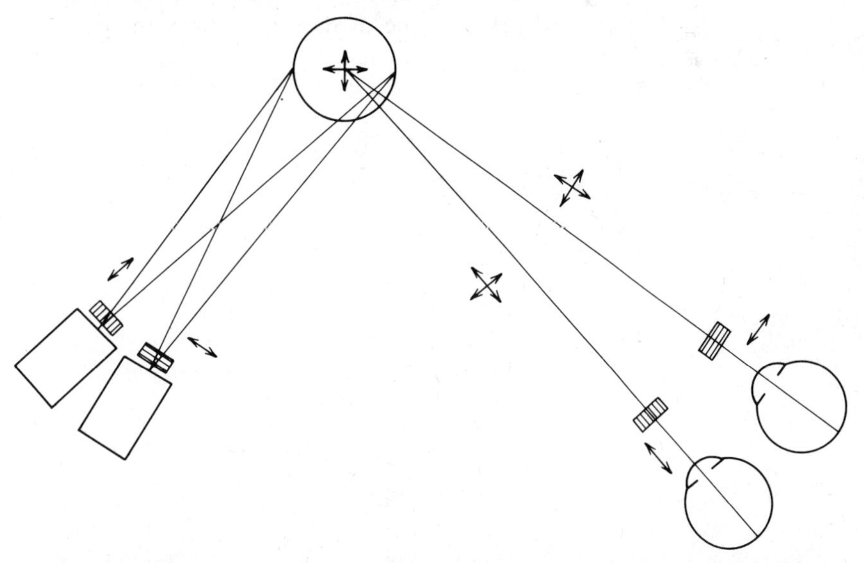

Figure 4-21. Stereoscopic film projection system using polarized filters in front of the projectors and the viewers eyes. Each eye sees only the images projected by the appropriate projector. (From *Sight and Mind: An Introduction to Visual Perception* by Lloyd Kaufman. Copyright © 1974 Oxford University Press, Inc. Reprinted by permission.)

bat situations using a system with simple line graphics. In spite of some problems with one of these systems they reported strongly favorable user response to the 3-D presentation mode in conveying time and trajectory information.

Monocular Distance and Size Cues

Quite a bit of distance information is available from a single retinal image. For example, the partial overlap of part of one object's contour by the contour of another, closer object is called *interposition.* Figure 4-22 illustrates this cue, and also the major cue of *linear perspective.* If the size of an object is fixed (a common situation in nature), then you can see that the visual angle will be inversely proportional to the distance from the object. This is the essential aspect of linear perspective and texture perspective: as the distance to objects in the environment increases, their image size decreases. Other consequences of perspective include the increased density of the detail of surfaces as distance increases, and the fact that lines and contours converge as the angular separation of their component parts decreases with increasing distance to the object. These cues are employed in visual display systems to provide information about target size and distance.

Because we normally have the use of a succession of two dimensional images, rather than a single one, we also have additional cues to distance that are a function of the *changes* that occur in the visual field as a consequence of movement. There is a perspective associated with the differential speed of movement of parts of the retinal image as a function of the distance of the object from the observer, called *movement paral-*

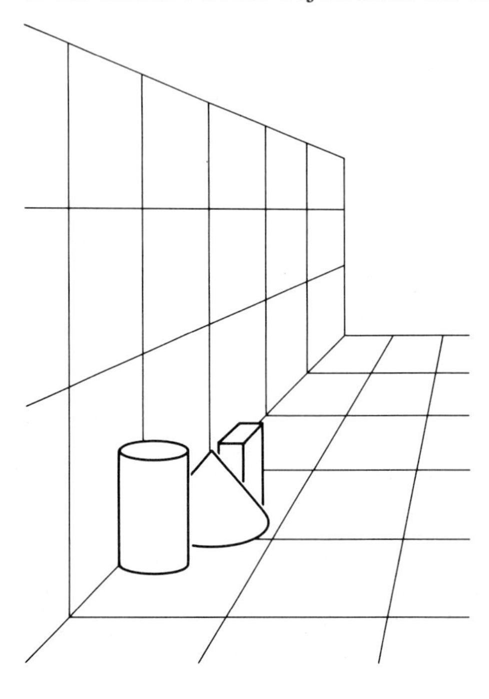

Figure 4-22. Cues of interposition and perspective.

lax. Imagine looking out your car window at a farm scene while you are driving; the telephone poles fly by much faster than the cows in the scene or the farm houses in the far distance.

Suppose that you asked an observer to estimate the distance or the size of an object; how would various cues interact to determine the response? It turns out that the apparent size of an object is intimately related to the distance that it appears to be from the observer, a principle known as Emmert's law. The statement of the law is as follows: apparent size is proportional to apparent distance,

$$s = kd \tag{4-4}$$

where s is apparent size, d is apparent distance, and k is a constant of proportionality. You could make rather accurate judgments of the size of an unfamiliar object as long as you had available a full range of normal distance cues. Thus, as the object was moved farther away from you and its retinal image size decreased, your estimates of its size would, following equation 4-4, be constant and accurate.

The judgment of size and distance depends on the eye being focused accurately on the stimulus object. If the visual cues necessary for proper focus are reduced or removed (see the following Box), our judgment of the object's size and distance will probably be incorrect. Our ability to judge the size of physical objects as constant is termed size constancy. Generally, our estimates of the size of objects in our view is extremely accurate. But sometimes, particularly when visual information is degraded or unusual, our estimates of size and distance get fouled up. The visual illusion known as the Ponzo illusion is explainable in these terms. The strong perspective cues given by the converging lines in Figure 4-23 seem to dominate the apparent depth of the scene and overcome the other cues that tell us that these two lines are on a flat surface. The perspective cues make the upper object appear to be farther away, hence larger.

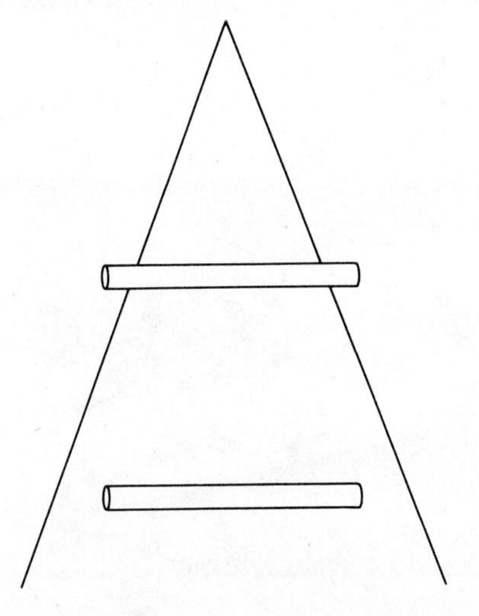

Figure 4-23. The Ponzo illusion.

A SLICE OF LIFE

Can Visual Illusions Cause Accidents?

When a pilot makes a nighttime landing approach over water toward a brightly lit city, a so-called "black hole approach," there may be certain illusory aspects of the scene (see Figure 4-24), depending on the accommodative state of the pilot's eyes. The runway may suddenly seem to be larger, closer, and lower in the visual field than when seen under clear daylight conditions. This misperception of the approach scene has apparently led to misjudgments of the aircraft position relative to the runway. In some instances, commercial airliners have landed in the water short of the airport (Roscoe, 1979). Another general kind of situation involves the opposite effect: when pilots make landings with imaging flight display systems (projected at unity magnification) or with flight periscopes or simulated contact visual systems, the imaged runway seems to appear smaller, farther away, and higher in the visual field than when viewed directly (Roscoe, 1979, Hull, Gill & Roscoe, 1982). Landings with these systems tend to involve faster and longer approaches, and harder touchdowns. What accounts for these errors in size and distance perception? These phenomena are related to the size-distance illusions that we described in connection with departures from Emmert's law. Both apparent size and apparent distance depend on the accuracy of the eyes focus, which in turn is sensitive to the presence and extent of texture gradients and other perspective cues in the visual field.

The eyes accommodative mechanism has a preferred resting setting called the "dark focus," which varies from person to person but is usually equivalent to a point of focus less than 3 feet away (Leibowitz & Owens, 1975, 1977). Suppose that the visual texture gradient leading out to an object is obscured is some way. The resulting focus will

Figure 4-24. Night Landing over water.

be a compromise between the focal stimulus value of the object itself and the pull of the dark focus. This misaccommodation will result in erroneous estimates of the object's size and distance. Objects closer to or farther away from the person than this resting distance will be perceived respectively as larger or smaller than normal, will be poorly focused, and errors in performance may result. When you stare out a car's windshield at night, your eyes focus (in diopters) halfway between your dark focus and optical infinity, or at a distance of about 2m on average (Owens & Leibowitz, 1976). On a rainy, dark night, it is quite possible that your eyes will become accommodated to an even nearer distance, a point quite unsuited to the accurate detection of targets at a distance from your vehicle (Mathews, Angus, & Pearce, 1978). This phenomenon, called "night myopia," can also occur if you are driving through a fog or snowstorm, in which case it would be labeled "empty field myopia." This situation also can occur when objects are viewed via microscopes or other imaging display systems where the clarity of the image viewed does not depend on the eye's focus. The result, termed "instrument myopia," is a less than optimal sharpness and a misperception of the size of the target and its distance (Hennessy, 1975). The other side of the coin is when the object is actually closer than the resting accommodation distance of the observer, resulting in a situationally induced hyperopia.

In the night landing situation, there may be rapid changes in the accommodative setting as the aircraft approaches the landing area. The precise nature of the relationship between pilot judgment and visual accommodation distance is undergoing study (Roscoe, 1979). Some techniques to minimize the effects of these factors include the following.

1. Wear special lenses (as recommended by United Airlines).
2. Select personnel with appropriate resting accommodation levels.
3. Use lead-in light markers (buoys) for airports approached over water.
4. Use special display/instrument designs that incorporate distance focus aids.

SUMMARY

The performance of a person working at a visual target detection or identification task depends on a number of factors having to do with the person's visual system and the characteristics of the visual task. Factors relating to the person's visual system include their state of adaptation, their spatial acuity, and their direction of gaze. Factors relating to the task include the target's intensity, contrast, size, location, movement, and color, the nature of the background, and the characteristics of the optical or electrooptical system employed in the task. These factors interact to determine the speed and accuracy of human performance that is obtainable. In addition, a number of spatial, color, movement, and size-distance phenomena exist that can significantly affect performance in what otherwise would seem a straightforward visual situation. An understanding of these person and system factors and their interactions can enable the human factors specialist to predict system performance in a wide variety of visual task environments.

APPENDIX

Computing Color Coordinates for a Light Stimulus

Any colored light mixture can be matched by a suitable mixture of light from three monochromatic light projectors, termed primaries. Knowing the relative amounts of each primary (the percent red, percent green, percent blue) yields sufficient information to specify the hue and saturation of the unknown mixture. This kind of specification is the basis for Figure 4-25, which is a plot of the International Commission on Illum-

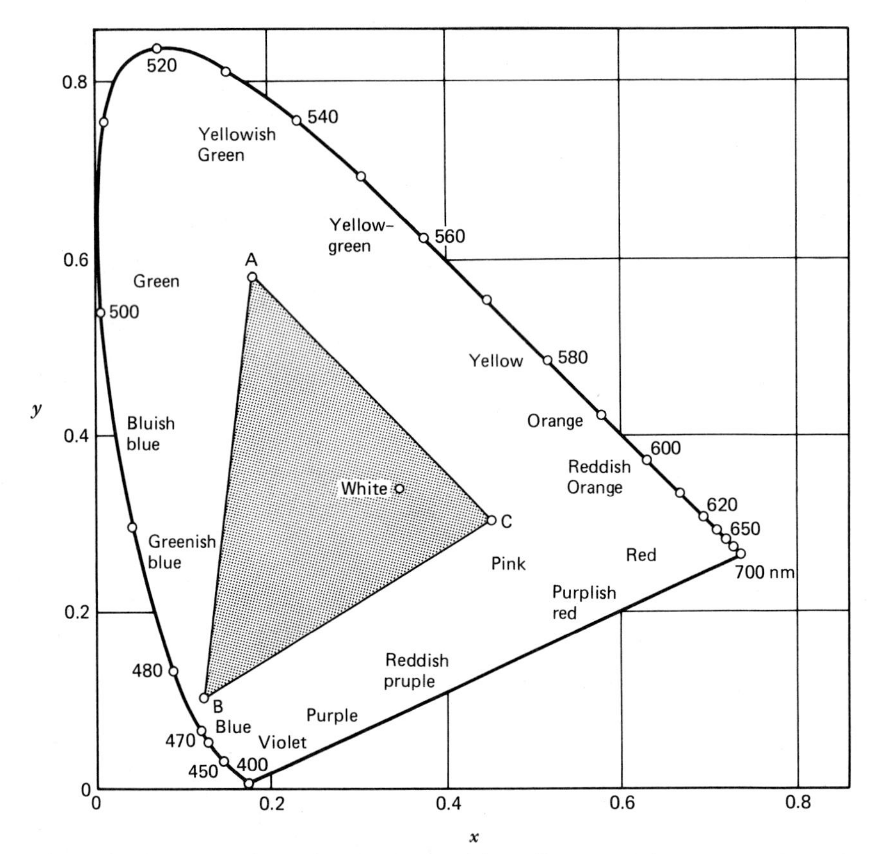

Figure 4-25. Two-dimensional chromaticity diagram. The range of colors possible by mixing primaries A, B, and C is bounded by the triangular area connecting them.

ination (CIE) chromaticity coordinates. The ordinate is the relative fraction of the "green" CIE (y) primary and the abscissa is the relative fraction of the "red" CIE (x) primary. (These primaries are actually mathematically defined primaries rather than realizable colors; they are used for convenience in computation.) Mixtures of two colors in this space fall on the line connecting them; mixtures of three colors fall in the area bounded by the colors, as shown in Figure 4-25. If you were generating colors via some printing or photographic process that used A, B, and C as primaries, you could not get colors outside of that triangular area.

Notice an interesting thing about the diagram. The spectral colors exist as points in this space, along the curved line around the space. We've said that we can match any complicated mixture of wavelengths with three primaries; that principle also allows us to match any monochromatic stimulus. In fact, the amounts of the three CIE primaries that are needed to match any monochromatic stimulus are extremely important in color technology. These values are given in Table 4-4. This table allows us to match any new colored stimulus without going to the trouble of actually doing a matching experiment. Suppose that we need to match the spectral stimulus of Figure 4-17*a*, 100 units at 580 nm. From Table 4-4 we would need (100) (0.9163) = 91.63 units of the x primary,

TABLE 4-4 *CIE 1931*[a] *Tristimulus Values* $\bar{x}$, $\bar{y}$, *and* $\bar{z}$

Wavelength, (nm)	Color-Matching Functions			Wavelength, (nm)	Color-Matching Functions		
	$\bar{x}_\lambda$	$\bar{y}_\lambda$	$\bar{z}_\lambda$		$\bar{x}_\lambda$	$\bar{y}_\lambda$	$\bar{z}_\lambda$
400	0.0143	0.0004	0.0679	560	0.5945	0.9950	0.0039
410	0.0435	0.0012	0.2074	570	0.7621	0.9520	0.0021
420	0.1344	0.0040	0.6456	580	0.9163	0.8700	0.0017
430	0.2839	0.0116	1.3856	590	1.0263	0.7570	0.0011
440	0.3483	0.0230	1.7471	600	1.0622	0.6310	0.0008
450	0.3362	0.0380	1.7721	610	1.0026	0.5030	0.0003
460	0.2908	0.0600	1.6692	620	0.8544	0.3810	0.0002
470	0.1954	0.0910	1.2876	630	0.6424	0.2650	0.0000
480	0.0956	0.1390	0.8130	640	0.4479	0.1750	0.0000
490	0.0320	0.2080	0.4652	650	0.2835	0.1070	0.0000
500	0.0049	0.3230	0.2720	660	0.1649	0.0610	0.0000
510	0.0093	0.5030	0.1582	670	0.0874	0.0320	0.0000
520	0.0633	0.7100	0.0782	680	0.0468	0.0170	0.0000
530	0.1655	0.8620	0.0422	690	0.0227	0.0082	0.0000
540	0.2904	0.9540	0.0203	700	0.0114	0.0041	0.0000
550	0.4334	0.9950	0.0087				

[a] From *Commission International de l'Éclairage,* National Physical Laboratory, Teddington, 1932, pp 25, 26.

TABLE 4-5 *Primary Values Needed for Match to the Stimuli of Figure 4-17a and b and Their Combination*

	x	y	z
Stimulus 4-17*a*			
100 units at 580 nm	91.63	87.00	0.17
Stimulus 4-17*b*			
50 units at 500 nm	0.25	16.15	11.60
Stimulus 4-17*a* plus *b*	91.88	103.15	11.77

(100) (0.87) = 87 units of the y primary, and (100) (0.0017) = 0.17 units of the z primary. Now make the same computations for a match to the stimulus in Figure 4-17*b*, 50 units at 500 nm. This would require (50) (0.0049) = 0.25 of x, (50) (0.3230) = 16.15 of y, and (50) (0.2720) = 11.6 of z.

Now, suppose you need to construct a match for a new stimulus that is composed of the *mixture* of the stimuli in Figure 4-17*a* and *b*, how much of the primaries would you need? All you do is *add* the results of your *separate* calculations for each wavelength of the complex mixture; the result is how much of each primary will match the new stimulus. Table 4-5 summarizes this calculation. In the same way, *any* complex stimulus can be constructed from its spectrum and the tristimulus values of Table 4-4.

For very complex stimuli, this calculation is a bit tedious; can it be automated? We just did the following mathematical operation:

$$\text{amount of } x \text{ primary} = \sum_{\lambda} x_\lambda P_\lambda \tag{4-5}$$

$$\text{amount of } y \text{ primary} = \sum_{\lambda} y_\lambda P_\lambda \tag{4-6}$$

$$\text{amount of } z \text{ primary} = \sum_{\lambda} z_\lambda P_\lambda \tag{4-7}$$

where P_λ is a measure of the intensity of the stimulus at each wavelength. How could these computations be performed automatically? Suppose that you could find a light meter that reacted to light energy (e.g., by producing a voltage) in a way proportional to that of Table 4-4 for the x function. That is, it reacted readily to wavelengths around 610 nm and less to others. Point the meter at a small patch of light in a real colored scene in front of you. The meter would produce a voltage proportional to the computation of equation 4-5; it would tell us how much of the x primary is needed in a mixture to match the unknown patch in our scene. If we *simultaneously* did the same thing for y- and z-sensitive devices, we would have all of the information needed to reproduce that patch of color with our three primary projectors. That is essentially the process performed in a color television receiver, or with three layers of pigment as in a color film slide, or with closely spaced colored dots in color printing.

5 PSYCHOMOTOR SKILL

Try this simple exercise: Close your eyes and extend one arm out in front of you with your index finger pointing away from your body. Now with your eyes still closed, touch the end of your nose with the tip of your finger. Most people can do this easily. This ability is an example of psychomotor skill. You have a marvelous ability to control and plan movements. Anyone who has observed a gymnast, or concert pianist, or a skilled dancer is well aware of how finely movement can be controlled. But we tend to take this ability for granted in ourselves. When was the last time you gave any thought at all to the mental and physical skills needed to pick up a pencil or throw a switch up to turn on a light? We hope that touching your nose has reminded you that some thought and skill is required to control your own movements. In this chapter we discuss the theoretical basis for planning, modifying and executing motor movements. Until recently, psychologists did not have experimental techniques for studying the fine details of movement control. Fortunately, new theoretical advances have made this an exciting research area for scientists and there are even new journals—*Journal of Motor Behavior* and *Human Movement Studies,* for example—that specialize in this field as well as many articles in more general journals. In order to appreciate some of these new techniques we have to learn more about information theory. So this chapter continues the discussion of information theory started in Chapter 2. If you did not read that part of Chapter 2 very carefully, now would be a good time to review it. This chapter assumes that you understand elementary information theory as presented in that chapter.

We start our discussion of psychomotor skills with the topic of reaction time. This topic is more than 100 years old in experimental psychology and we will see how new techniques have built on older foundations. Then we go through the mathematics of bivariate information theory. If you understand simple information theory, you will have no trouble with the bivariate extension. From this we return to reaction time and then go on to movement control. This chapter ends with a discussion of motor programs that are hypothetical computer programs that run inside your head. Along the way, we touch on such practical topics as automatic robot arms and typing.

REACTION TIME

"Time is money." How often have you heard that familiar phrase? Modern society stresses the importance of time and human factors is no exception. The time it takes a pilot to read a crucial dial on the cockpit instrument panel can determine life or death. So can the time it takes a radar operator to determine that two blips on her screen are about to merge, indicating a collision between an oil tanker and a freighter in the English Channel (Figure 5-1). As we will see in later chapters, human factors specialists go to great lengths to help the human operator respond more rapidly to system demands. But before you can really understand the applied aspects of reaction time, we must first develop the theoretical base that helps us to explain why people are slow in some situations yet fast—sometimes too fast—in others.

The ABC of Reaction Time

Let's go back to eighteenth-century England and the age of kings to discover how the study of reaction time got its start. We are observing the nightly grind at the Royal Observatory and you are there. A clock ticks every second. Kinnebrook, the new assistant observer, is on duty at the telescope nervously peering at the sky. His attention is focused on a hairline inside the telescope eyepiece. His job is to record to the nearest tenth of a second precisely when Venus crosses the hairline. He must do this by listening to the ticking of the clock. Kinnebrook duly notes the crossing time in the Observatory log. However, his boss has recorded a shorter time. They discuss the matter:

BOSS: Kinnebrook, your time is too long. Can't you get it right, you dummy.

KINNEBROOK: I'm observing as fast as I can.

BOSS: If I can get it right, why can't you? Are you slow because you use all your energy chasing after the milkmaid?

KINNEBROOK: Keep Griselda out of this. Maybe my times are correct and your times are too fast.

BOSS: You're fired! Never again darken the Royal Observatory door. And take Griselda with you!

This would have been the end of this tale but for another astronomer named Bessel who thought that perhaps poor Kinnebrook wasn't such a dummy after all. Bessel suspected that all persons might observe the

Figure 5-1. Collision of the Italian tanker *Vera Berlingieri* and the French freighter *Emmanuel Delmas* as they burn off the coast of Rome, June 26, 1979. (United Press International.)

crossing with slightly different reaction times. And when astronomers began to systematically compare each others crossing times, they did indeed find consistent differences. Different people recorded crossing times differently. This effect was named the "personal equation" because it was systematic—that is, astronomer A tended to always be faster than astronomer B, etc. One can imagine the annual convention of the Royal Astronomical Society with drunken astronomers trying to determine who had the biggest personal equation.

The personal equation would have remained only an astronomical oddity were it not for the Dutch physiologist Donders. Donders realized that he could use the personal equation to calibrate the time needed for different mental operations. Donders established three different types of reaction time experiments that even today are still called Donders A, B, and C reactions.

In the A reaction (Figure 5-2), there is only one stimulus and only one response. This is often called a simple reaction. A light goes on and the observer must respond by pressing a control button. Simple reactions occur frequently in sports. The start of a 100-yard dash or the center jumps in basketball are examples of simple reaction tasks. The time between the onset of the stimulus—for example, the starting gun—and the response is called simple reaction time. Donders believed that simple

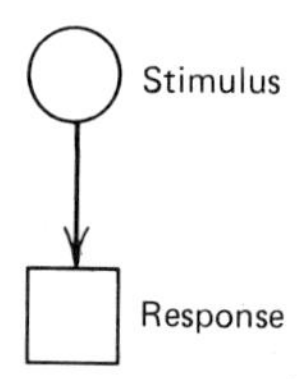

Figure 5-2. The simple reaction time task. A light (represented as a circle) goes on and a response (represented as a square) is made.

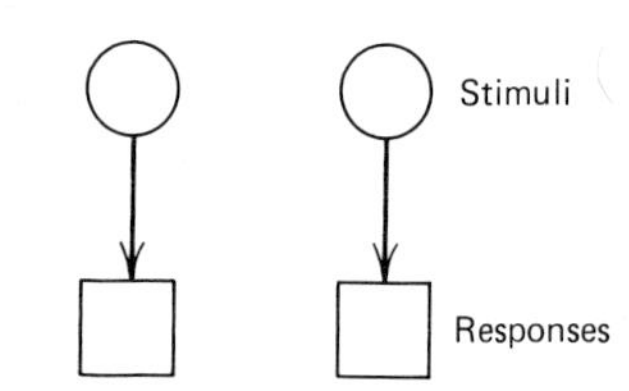

Figure 5-3. The choice reaction time task. Two (or more) lights are mapped to responses. Each stimulus has its own unique response.

reaction time provided a baseline for mental events. Simple reaction time took into account the time required for basic factors like the conduction of impulses through nerves. Once this baseline was evaluated it is possible to estimate the times required for more complex mental operations. Donders used B and C reactions to explore more complicated situations. In the Donders B reaction there is more than one stimulus and each stimulus has its own unique response (Figure 5-3). This is often called a choice reaction situation since it requires identifying one of several possible stimuli and choosing the correct response. When your car is at a traffic light you must perform a choice reaction. If the light is red you step on your brake pedal. If it is green you respond by stepping on your accelerator pedal.

The Donders C reaction has several stimuli but only a single response. This means that no response should be made to one or more stimuli (Figure 5-4). Unless the correct stimulus occurs, the operator must do nothing. You perform a C reaction whenever you patronize a takeout restaurant. Until your number is called, you must not respond to other stimuli. The C reaction requires identification of stimuli but no response selection since only one response is allowed. Information theory tells us that there is no information present with only a single alternative. There is no response information in a C reaction, although there is stimulus information.

To review, the Donders A reaction, also called a simple reaction, measures baseline time for nerve conduction. The Donders B reaction, also called a choice reaction, requires the mental operations of stimulus identification and response selection. The Donders C reaction, also called a C reaction, requires stimulus identification but no response

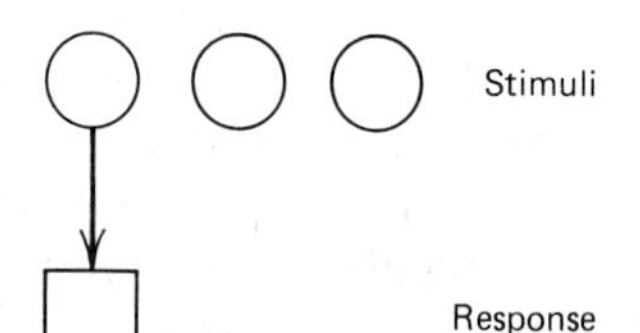

Figure 5-4. The Donders C reaction. There are several stimuli but only one of them is mapped to a response.

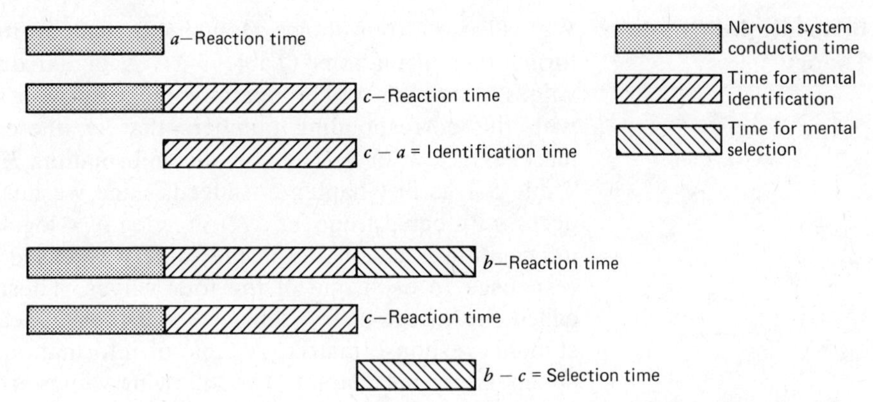

Figure 5-5. An example of Donders' subtractive logic. Subtracting *a*-reaction time from *c*-reaction time (top panel) gives an estimate of mental identification time. Subtracting *c*-reaction time from *b*-reaction time gives an estimate of mental selection (bottom panel). The three different shadings show the proportion of reaction time alloted to each mental process.

selection. We are now ready to appreciate the clever part of Donders' idea. By subtracting appropriate pairs of reaction times, we can obtain estimates of the time required for the mental operations of stimulus identification and response selection. If we subtract A reaction time from C reaction time, the result is an estimate of stimulus identification time. If we subtract C reaction time from B reaction time, we get an estimate of response selection time (Figure 5-5). This subtractive logic makes a strong prediction. Since the B reaction requires both stimulus identification and response selection, it should take longer than either the C reaction or the A reaction. Similarly, A reaction time should be the fastest and C reaction time should fall in between. Many experiments have obtained this ordering of A, B, and C reaction times. This is strong support for Donders subtractive method.

INFORMATION THEORY AND REACTION TIME

In Chapter 2 you learned about elementary information theory and how to calculate information for one set of events. This is called univariate information theory because only one set of events is used. However, the real power of information theory emerges when *two* sets of events are considered. One set is made up of stimuli that usually represent states of a machine, for example, display lamps that tell the operator if valves are open or closed. The other set is made up of responses that are available to the operator, for example throwing switches. This is called bivariate information theory and an important part of bivariate information theory deals with the relationship between stimulus events and response events. Now we explain the elements of bivariate information theory and show how it helps us to describe human behavior in reaction-time tasks.

Bivariate Information Theory

We will use a four-choice Donders B reaction to illustrate bivariate information calculations (Table 5-1). A digital display can take on the values 1 through 4. The operator must throw a switch to open a valve with the corresponding number—that is, there are four valves each marked 1, 2, 3, or 4. The stimulus information, $H(S)$, is calculated from Table 5-1 as in Chapter 2. Indeed, since we notice that all four stimuli occur with equal frequency $H(S) = \log N = \log 4 = 2$ bits. The response information is calculated in the same way based on the total number of responses to each one of the four valves. These totals are sometimes called the marginal frequencies since they appear in the margin of the stimulus-response matrix. A table of information values for $p \log 1/p$ is appended to this chapter. The following values are taken from that table: for $p = .25$, $p \log 1/p = 0.5$; for $p = 0.3$, $p \log 1/p = .5211$; for $p = 0.2$, $p \log 1/p = 0.4644$. Response information, $H(R)$, equals $.5 + 0.5 + 0.5211 + 0.4644 = 1.9855$ bits. As a partial check, note that this result is less than 2 bits the maximum amount of information possible when events are equiprobable. Since all responses did not have the same totals, $H(R)$ should be less than log 4.

Although we have calculated numbers for $H(S)$ and $H(R)$ we lack a number that describes a relationship between them. This is transmitted information, $T(S{:}R)$. If we represent stimulus and response information as circles in a Venn diagram (Figure 5-6), transmitted information is the intersection of the two circles. It is information present in both the stimulus set and the response set. In terms of the communication model discussed in Chapter 2, transmitted information is that portion of the message that was successfully carried through the channel to the receiver. Transmitted information can never exceed the *smaller* of $H(S)$ or $H(R)$.

In order to calculate $T(S{:}R)$, we must first calculate the joint information $H(S,R)$. This is the information inside the stimulus-response matrix (Table 5-1). (For those of you who are familiar with Venn diagrams or set theory, it is the union of the stimulus and response sets.) Again we take the following values from the appendix: for $p = 0.2$, $p \log 1/p = 0.4644$; for $p = 0.05$, $p \log 1/p = 0.2161$; for $p = 0.15$, $p \log 1/p = 0.4105$; for $p = 1$, $p \log 1/p = 0.3322$. We calculate the joint information $H(S,R)$ by adding the $p \log 1/p$ values for each of the cells inside the stimulus-response matrix: $.20 \log 1/0.20 + 0.05 \log 1/0.05 + 0 \log 1/0 + \ldots .\ 0.15 \log 1/0.15$.

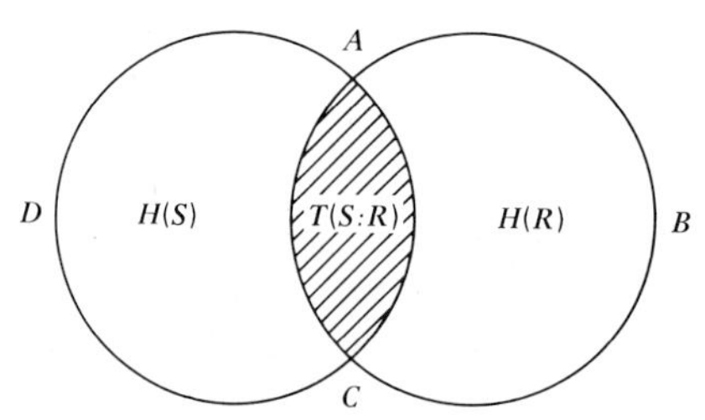

Figure 5-6. The relationship among informational quantities. Transmitted information, the shaded area, is the intersection of $H(S)$ and $H(R)$. The joint information $H(S,R)$ is the union represented by the outline obtained as follows: Start at point A, then go to B, then to C, then to D and back to A.

TABLE 5-1 *S-R Matrix Used to Calculate Transmitted Information (see text for explanation)*

	Responses				
Stimuli	Valve 1	Valve 2	Valve 3	Valve 4	Σ
1	20	5	0	0	25
2	5	15	5	0	25
3	0	5	15	5	25
4	0	0	10	15	25
Σ	25	25	30	20	100

$H(S) = 2$ bits.
$H(R) = 1.9855$ bits.
$H(S,R) = 3.1086$ bits.

All the 0 log 1/0 terms drop out since 0 times anything is 0. The remaining sum equals $0.4644 + 0.2161 + 0.2161 + 0.4105 + 0.2161 + 0.2161 + 0.4105 + 0.2161 + 0.3322 + 0.4105 = 3.1086$ bits. Now that we know $H(S)$, $H(R)$, and $H(S,R)$ there is simple formula for T(S:R).

$$T(S:R) = H(S) + H(R) - H(S,R).$$

Therefore, using our previous calculations $T(S:R) = 2.0 + 1.9855 - 3.1086 = 0.8769$ bits. This is not very good performance as can be seen by comparing actual transmitted information to the maximum possible transmitted information in this situation (Table 5-1). Transmitted information cannot exceed the smaller of $H(S)$ or $H(R)$, which in this case is 1.9855 bits. We can express actual transmitted information as a percentage of the maximum transmitted information. In this case that is 0.8769/1.9855 or 44%. It is clear that our control panel should be redesigned to increase amount of transmitted information.

Reaction Time Meets Information Theory

We know from everyday experience that as situations become more complex it takes more time to reach a decision. It takes less time to order dinner from a fast-food franchise where a short menu is displayed on a bulletin board than from a fancy French restaurant where the wine list is 14 pages long. However, this general principle provides only rough guidance to the human factors specialist who must predict precisely how long it will take a person to react and decide. Transmitted information allows us to quantify the time needed for decision and reaction. The relationship between reaction time (RT) and transmitted information is called Hick's law after the psychologist who discovered it (Hick, 1952). Hick varied the number of equiprobable alternatives in a choice-reaction time situation and found that the rate of gain of information was constant. Mathematically, this is a simple equation:

$$RT = a + bT(S{:}R)$$

There are three ways to vary amount of stimulus information. First, we can change the number of equiprobable alternatives. Second, we can hold the number of alternatives constant and vary the probabilities of these alternatives. Third, we can vary the sequential dependencies between successive events. In the first two methods each event has the same chance of being followed by any other event. The third method changes this so that some events are more probable given that other events have just occurred. For example, in written English the letter "q" is always followed by "u." Since Hick only varied information the first way, there was at least the possibility that reaction time depended on number of alternatives instead of on calculated information. However, another experimenter soon did a study that varied information in all three ways. Hyman (1953) also found reaction time to be a linear function of information, regardless of how stimulus information was varied. Thus, during the 1960s psychologists were convinced that Hick's law was the basic relationship between sets of events and reaction time. But a new experimental paradigm developed in the late 1960s (Forrin, Kumler, & Morin, 1966) suggested that reaction time might be a linear function of the number of alternatives and not the amount of information—that is, there was no need to take the log of N. This new technique is somewhat more complex than choice reaction time, since it involves a list of items. These items are usually single digits. A person is first required to memorize a small set of digits. Then a single digit is presented. If it matches one of the digits in the previous list, the correct response is "yes." If it is a new digit, the correct response is "no." So, for example, you might have to memorize the list "3,7,8." Then the test digit is presented. If it was "7" you should respond "yes." If it was "5," you should respond "no." The point of interest is how long it takes someone to respond "yes" or "no."

Sternberg (1967, 1969) made psychologists aware of some important theoretical conclusions about mental processing that could be drawn from these kinds of data. The empirical results are easy to explain: reaction time is a linear function of the number of digits in the list that was memorized. The theoretical implications are more complicated.

One important question is whether processing is parallel or serial. As you search through the memorized list to check for possible matches do you proceed one digit at a time—serial processing—or are you able to compare the single test digit to the entire memorized list simultaneously—parallel processing? Simple parallel and serial models make different predictions for the reaction time function. If you process items in parallel, increasing the size of the list that was memorized should have no effect since all the digits are compared simultaneously anyway. But a serial processor should take longer as more study digits are added to the list since each extra digit must be compared separately. Sternberg concluded that search was serial since reaction time increased with list size. While later workers have questioned this interpretation (Townsend, 1974), the empirical finding of linear increases (rather than logarithmic increases predicted by Hick's law) has not been disputed. This then creates a problem for the human factors analyst: Which relationship – linear or logarithmic – is correct? A heroic attempt to resolve this issue was published by Briggs (1974) who fitted both linear and logarithmic functions to seven years worth of published reaction time research. After examining 145 studies, Briggs concluded that 62% of the data was fitted better by logarithmic functions and 38% better by linear functions. However the linear fits came about mostly when stimuli were digits rather than other kinds of stimuli. Thus, from a practical point of view one should use Hick's law unless stimuli are digits.

There is a serious difficulty with the above rule of thumb. There are many differences between the choice reaction time task and the Sternberg memory search task besides using digits as stimuli. For example, choice reaction time maps a unique response to each stimulus – that is, there are as many responses as there are stimuli. But the memory search task has only two responses – yes and no – regardless of how many stimuli are used. This is called a many-to-one mapping of stimuli onto responses versus the one-to-one mapping used in a choice reaction task. Human factors specialists could not do their jobs without rules of thumb. But such convenient rules are not substitutes for theoretical understanding of mental processes. Until we obtain a strong theoretical model of reaction time that explains why linear fits are best some of the time and logarithmic fits best at other times, we are stuck with our empirical rules of thumb. Nevertheless, it would be a mistake to be content with rules of thumb at the expense of theory that will eventually replace crude guidelines. There is considerable interest in explaining the determinants of reaction time (e.g., Kornblum, 1973; Teichner & Krebs, 1974; Grice, Nullmeyer, & Spiker, 1982) and such models may eventually replace the rules of thumb now used in human factors.

Important Determinants of Reaction Time

There are many aspects of stimuli, responses and the relationship between stimuli and responses that can alter reaction time. The physical characteristics of stimuli such as size, shape, color, intensity, discriminability are all of concern in practical situations. Similarly, the control – switch, lever, knob, wheel, voice key, etc. – plays a role in establishing speed of reaction (Chapter 10). This kind of precise information is best obtained from handbooks and manufacturer's catalogs. Here we focus on more global issues and discuss only two of the important factors that control reaction time: stimulus-response compatibility and amount of practice.

Figure 5-7. An improved shower control offers separate adjustments for water temperature and volume. (Courtesy of Moen, a Division of Stanadyne.)

Stimulus-Response Compatibility Think back to the last time you took a shower. In order to get the water set to your taste, you probably had to readjust hot and cold faucets, alternately freezing and scalding yourself, until finally you achieved the desired water temperature. You have been the victim of poor human factors design—in this case, failure to consider control-display compatibility. You had to go through a back-and-forth (or more technically, a recursive) adjustment because the controls were not set up to directly allow you to manipulate temperature. Newer shower controls (Figure 5-7) allow the user to rotate a dial to set temperature and to pull the dial to set water volume. This is a much better design because the dial directly controls the desired parameter.

Control-display relationships (explored more fully in later chapters) are one instance of the more general concept of stimulus-response compatibility. Compatibility is a learned relationship. In the United States you throw a switch up to turn on a light. Thus, up is the compatible response. But in England the reverse is true: a switch thrown up will extinguish a light. So one way to determine compatibility is to take a vote and simply ask people what response should go with some particular stimulus. This majority opinion is called a population stereotype. The most compatible relationship is the one that most people say is the most compatible. This definition may strike you as being somewhat circular. But this does not make it any less useful for the human factors specialist. However, there have been attempts to refine the concept based on laboratory experiments to which we now turn.

The classic demonstration of stimulus-response compatibility effects was published in 1953 (Fitts & Seeger, 1953). Compatibility was changed by altering spatial relationships (Figure 5-8). The geometry of stimulus and response arrays can be manipulated in ways limited only by the designer's ingenuity. Fitts and Seeger found that reaction time was fastest and error rates lowest when the geometry of stimulus and response arrays corresponded directly. Thus, stimulus-response compatibility does not depend only on the type of stimulus array or only the type of response array. Instead, it depends on the relationship between the two arrays. Even though this laboratory experiment is far more precise than simply taking a vote to determine population stereotypes, the logical definition of compatibility is still circular. The arrangement with the fastest reaction time and lowest error rate is the most compatible. There is no independent way to measure compatibility without actually doing the experiment. The

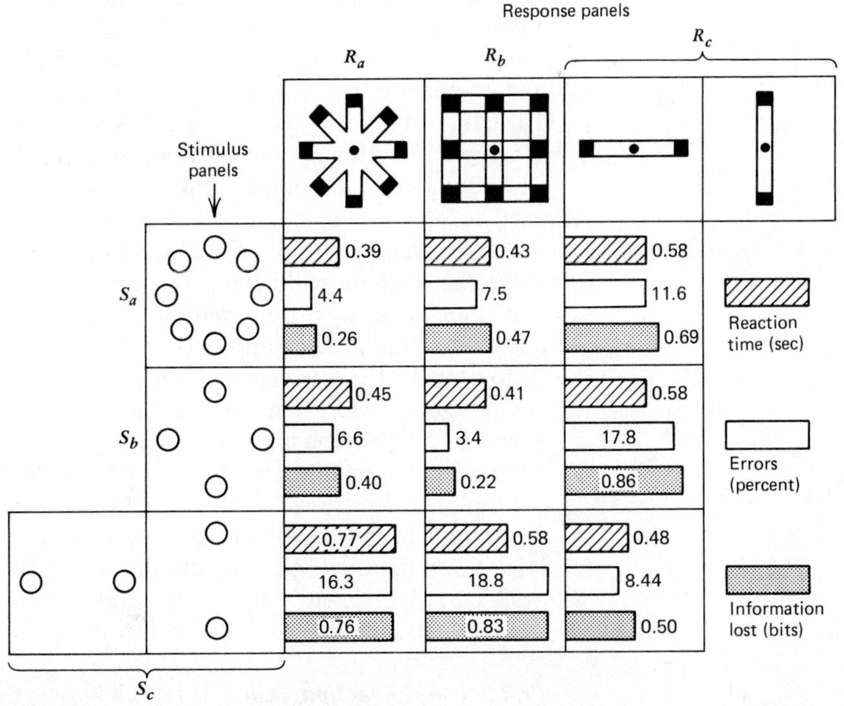

Figure 5-8. Spatial arrangements used to test stimulus-response compatibility. (From Fitts & Seeger, 1953 with permission of APA.)

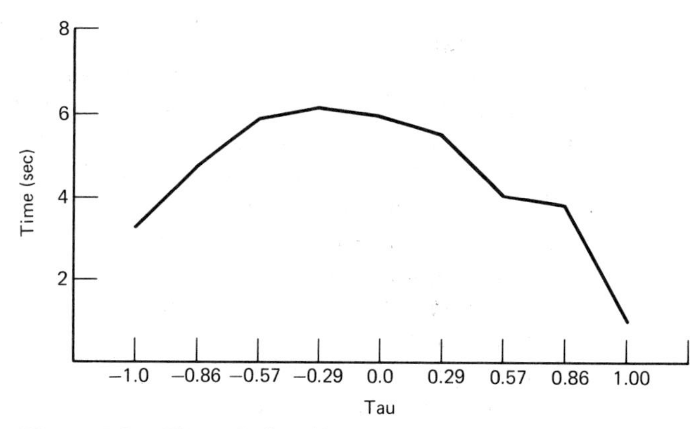

Figure 5-9. The relationship between reaction time and compatibility, where compatibility is measured by tau, a correlation coefficient. (From Morin & Grant, 1955 with permission of APA.)

most compatible mapping is the fastest and the fastest is the most compatible. A later experiment (Morin & Grant, 1955) provided a clever way around this difficulty. Their arrays were horizontal rows of stimulus lights and response keys that could be numbered from one to ten from left to right. Before doing the experiment, they calculated the correlation (using Kendall's tau – a nonparametric index) between lights and keys. A correlation of 1.0 meant that light 1 was controlled by key 1, light 2 by key 2, and so on. A correlation of −1.0 mean that light 1 was controlled by key 10, light 2 by key 9, etc. Figure 5-9 shows their results. This inverted U-shaped function agrees with our intuitive feelings about compatibility and population stereotypes. But because the correlation coefficient was used to measure compatibility a priori, this experiment is not circular. Although this approach to compatibility is elegant, it is limited to only those cases where numbers can be meaningfully applied to both stimulus and response arrays. For example, how would you assign numbers for a linear stimulus array mapped to a circular response array? The human factors analyst must often be content with using population stereotypes and circular logic. While more recent research on compatibility (see Kantowitz, 1982 for a brief review) offers the promise of help for the human factors analyst, so far this is only on the horizon rather than inside the tool box. For now the human factors specialist must be content with the statement that stimulus-response compatibility alters the slope of the Hick's law function. As compatibility increases, the slope decreases. We shall consider specific examples of stimulus-response compatibility in later chapters when the topic of control-display relationships is discussed.

Practice and Reaction Time It is well known that practice has a dramatic effect on reaction time. Not only does practice decrease reaction time in general, it also changes the slope of the Hick's law function. (See Welford, 1976, Chapter 4 for a review of experimental results.) Effects of stimulus-response compatibility in particular can be diminished through large amounts of practice. After several days of practice, the speed and accuracy of incompatible stimulus-response mappings approaches that of unpracticed compatible mappings. Theoretical explanations of this results are complex with little agreement about specific details of the process. However, it is generally agreed that practice reduces the amount of information processing required by an internal translation stage that maps stimulus representations onto response commands. Exactly how this is accomplished is unknown at present, although a reduction of memory load is probably involved, perhaps by the substitution of rules for the rote memory of specific stimulus-response response pairings (Duncan, 1977). Teichner and Krebs (1974) plotted reaction time as a function of practice (number of trials) for 59 published experiments. The following equation summarizes their results

$$RT = K \log_{10} NT + a$$

where NT is number of trials. With digits as stimuli and key presses as responses they found that both the slope and intercept of this practice equation changed according to the number of stimulus-response alternatives. For example, with only two alternatives, $K = -.099$ and $a = .725$. With eight alternatives, $K = -.217$ and $a = 1.54$. Thus, the effect of practice is more pronounced as number of alternatives increases. Furthermore, since these curves eventually converge, after very large anounts of practice there is hardly any effect of number of alternatives. This means that the slope of the Hick's law function becomes quite flat with operators who are extremely well practiced.

Speed-Accuracy Trade-off

Reaction time is a widely used dependent variable in human factors research. At first glance, there seems to be little problem in measuring and defining reaction time in the laboratory or in some applied setting. Indeed reaction time is often thought of as the minimum time required to respond to some stimulus or perform some operation. While it is well-known that the same person may not exhibit exactly the same reaction time from trial to trial, it would be incorrect to attribute this only to human variability and so to ignore this variation. Recent research has demonstrated systematic differences in reaction times that are controlled by specific task characteristics.

Think of trying to type a term paper that is due in one hour. You are typing as fast as your stubby little fingers will go. But alas, your typing is not very accurate. You could type better by slowing down your pace. This is one example of speed-accuracy trade-off. The same task can be performed either rapidly with a higher error rate, very slowly with a much lower error rate, or anywhere in between. In short, people have the ability to trade speed for accuracy. This decision about whether or not to stress speed or accuracy in some task is quite similar to selecting a decision criterion in signal detection theory (Chapter 3). And just as beta in signal detection theory can be manipulated by instructions and payoffs, so can the decision about speed versus accuracy.

The concept of an operating characteristic that has proved so useful in signal detection—where you recall, it is called a receiver operating characteristic—is equally valuable in reaction time situations. A speed-accuracy operating characteristic (Pew, 1969) plots speed on the horizontal axis against accuracy on the vertical axis (Figure 5-10). Note that the function flattens out as maximum accuracy is approached. The usual instructions given to a worker call for working as fast as possible without making any errors. This is the point labeled optimum in Figure 5-10. This point represents a theoretical perfection that is seldom, if ever, obtained. Instead, most workers will operate to the right of the desired optimum. Since the speed-accuracy operating characteristic is flat to the right of the optimum point, even experienced workers cannot determine how close they are to the optimum point without risking errors. If a job description stresses avoidance of error, a prudent worker will keep well to the right of the optimum point. The cost of such caution, however, is a substantial increase in reaction time. This cost may exceed the benefits gained by zero errors. It may be more effective to reach a system performance of zero error by adding redundant elements to the system (Chapter 2) and allowing each element to have a small error rate. For example, a machinist who must work to fine tolerances may take a long time to produce a finished part. If it is possible to economically correct small errors, more in-tolerance parts may be produced in a given time by having two machinists check each other's work. (Of course, if an error destroys a part then a slow work rate must be accepted.) In general, people do not perform well when they must adopt a decision criterion of zero errors. A more realistic system design should allow for human error rates of at least 2% for relatively simple tasks.

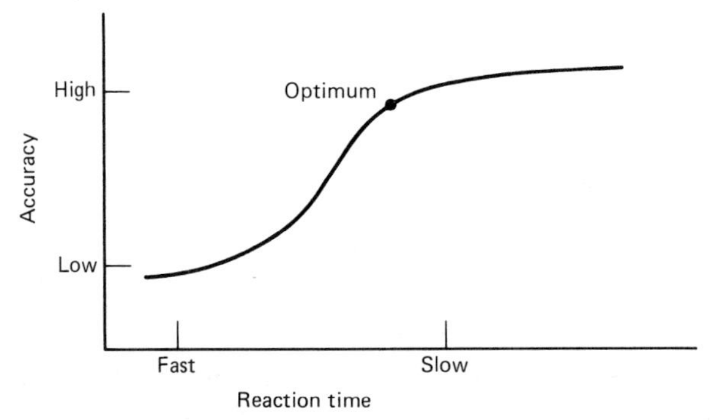

Figure 5-10. A speed-accuracy operating characteristic.

While the practical importance of the speed-accuracy operating characteristic is obvious, there is a subtle theoretical issue also at stake here (Pachella, 1974). Experiments cannot be directly compared unless there is some assurance that participants are operating at the same point of the operating characteristic. Otherwise particular results may be due to a speed or an accuracy stress rather than other independent variables. The best way to avoid this problem is to systematically manipulate payoffs and so trace out entire operating characteristics as is often done in signal detection research. But this is costly and not always practical. The minimum that is required is a careful reporting of error rates in all reaction time experiments. Reaction time is not a simple univariate dependent variable. It and error rate are opposite sides of the same coin. It is impossible to understand reaction time data without knowledge of errors.

MOTOR CONTROL

The study of control of motor movements predates human factors and can be traced back to the nineteenth century (Woodworth, 1899). Woodworth divided a voluntary positioning movement, as it was called then, to distinguish it from a conditioned reflex movement, into two distinct phases. The first phase was the initial lift-off and was called initial impulse. The second phase he called current control because feedback information was used to modify and control the current movement as it progressed. This two-phase distinction is still very much with us today, although as we will see, the terminology has been altered to keep in step with modern computer jargon. The combined efforts of researchers in physical education and psychology have given the area of motor control new vitality and the outpouring of published research reports threatens to swamp workers with new advances (e.g., Stelmach, 1978; Stelmach & Requin, 1980). Because the field is changing so rapidly it makes little sense to present the finer details of theoretical controversy. Thus, we focus on accepted basic results realizing that the explanations for these results will probably change in the next few years. The most basic relationship in motor control relates movement time to physical parameters that define the movement; it is called Fitts' law.

A SLICE OF LIFE

Robot Arms in Industry

For most people the word "robot" conjures up images of clanking metal monstrosities bent on taking over the world in a late-night television rerun or perhaps the more amiable robots R2D2 and C3PO in Star Wars. In reality, industrial robots are merely sophisticated tools that are logical progressions of automation. The earliest robots were

R2D2 and C3PO in *Star Wars*. (UPI.)

numerically controlled tools that did jobs like drilling holes in metal fabrication. These tools were designed for a single type of job, and while the pattern of holes to be drilled could be changed by altering the numeric program, the tool was still restricted to one type of industrial task. Modern robots are far more flexible and the same robot can be found in many different factories doing quite different jobs.

Once they are programmed, industrial robots require no human supervision, except for routine maintenance. This distinguishes them from a related class of tools used by humans to manipulate objects at a distance in dangerous environments. The human-operated arms are called teleoperators. A robot arm is more sophisticated since it must function on the production line without a guiding hand.

We all tend to take motor control for granted. Few people have difficulty in picking up a small part and placing it accurately. But getting a machine to accomplish this seemingly trivial task is far from simple. The tool must use assorted sensors to detect objects and then correctly grasp, rotate and place these objects (Bejczy, 1980). Manipulator control is quite difficult. The typical robot arm has about six joints that must move together in an organized fashion. Complex trigonometric transformations must relate the joint positions to the work-space coordinates. Even worse, the motion dynamics of the arm are not linear and depend on the states of the joints. Thus, the control program must continuously and rapidly evaluate complicated second-order differential equations. You may not know calculus but your arm does!

Robot arms work properly in industry because the tasks they perform are fixed and repetitive. The work-space coordinates do not change and no adaptive behavior is required from the robot arm. When task conditions change, the robot arm can get jammed. However, new improvements in robot intelligence are on the horizon. It is easy for you to select a small part from a bin and pick it up. This is hard for a robot when the parts are jumbled up. But progress is being made by having the arm hold up the part in front of its television camera. If the wrong part was selected, the robot puts it back and tries again. However, this is a slow process and must be speeded up before it can be used on the production line. Additional sensors will be used to determine how much force the robot is exerting on an object it holds. Ultimately, it may be possible for the same robot to pick up an egg without crushing it and to lift a heavy metal casting without dropping it.

One advantage of robots that is often mentioned is their ability to serve in hostile environments that threaten human health and safety. However, it is important for human factors specialists to realize that no amount of sociological benefit will induce a factory to install equipment unless this equipment shows a profit or government laws and regulations leave no other option. Robots do offer the promise of improved profits and productivity because they can work continuously without need of coffee breaks and eight hours of sleep each night. Robots do not necessarily work any faster than humans and often are

Unimate® robot arms in industry.

set to match a human pace because few factories are entirely automated. But in harsh environments, robots also suffer (Engelberger, 1977). While humans are constructed of flame-resistant components, robot arms need oil, and a tiny leak in a high-temperature environment can be disastrous. People are immune to noise in electrical lines but robots can do great damage as a result of noise spikes. People do not usually emit sparks but a robot that is spray painting can start a fire if a spark occurs in this volatile atmosphere. Dust and dirt can penetrate almost any opening and play havoc with gears and cogs. For a robot to be economical it must have a high mean time between failures despite these environmental hazards. Furthermore, its downtime for normal maintenance must be low. The president of Unimation, a major manufacturer of robot arms, has concluded that to be successful in the marketplace robots must withstand industrial environments, have downtimes of only 2%, go 10,000 hours between major overhauls and provide a return on investment of 25% (Engelberger, 1977). The human still offers tough competition to the robot and it will be quite some time before people are replaced entirely by robots on the production line.

A Unimate® robot arm in a hostile environment.

Fitts' Law

Reduced to its essence, motor control allows us to move from here to there. Many industrial assembly tasks require such precise movements as picking up a nut and threading it onto a bolt, placing a cotter pin through a tiny hole, or in the integrated-circuit manufacturing industry placing a

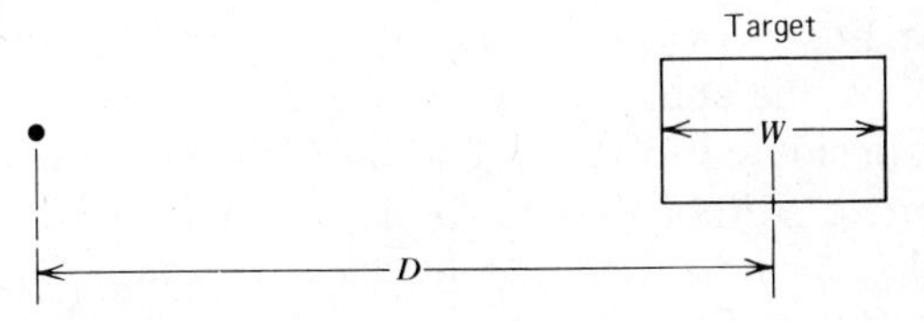

Figure 5-11. The two parameters that determine the index of difficulty for movement. D is the distance from the starting point to the center of the target. W is the width of the target.

tiny mask on a silicon chip observed through a microscope. Fitts' law lets us predict the time needed to accomplish these kinds of accurate movements. Figure 5-11 shows the two movement parameters of concern: distance to a target and target width. As distance (D) increases it takes longer to move to a target. Similarly, as target width (W) becomes narrower it also takes longer to land on target—it would take you longer to pick up a dime you found lying in the street than to pick up a half dollar in the same location. (At present inflation rates you probably wouldn't even bother to pick up a penny.) Mathematically, this can be best expressed with our old friend—information theory. (You didn't really think you could escape from this chapter with only one dose of H.) Fitts defined an index of difficulty (ID) as $\log(2D/W)$. The 2 in the numerator lets us avoid taking the log of 1 to the base 2 if $D = W$. This ID then is a simple number measured in bits that takes both movement distance and target width into account. Fitts' law states that it is the ratio of distance to width that controls movement time (MT):

$$MT = a + bID$$

As you might suspect, Fitts' law has been replicated extensively and there is no doubt that it provides an excellent fit to data. But until recently it existed primarily as a laboratory result that had little impact in industry. This changed when two industrial engineers needed to predict movement times for assembly work performed under a microscope (Langolf & Hancock, 1975). Traditional work measurement systems used in industrial engineering—called predetermined time systems—were useless in microscope work because they underpredicted times by up to 50% and were inconsistent. Langolf and Hancock had operators manipulate tiny bronze disks as small as .031 inch with a tweezers. The operator had to grasp the circle with the tweezers, move and position it within the boundary of a target, tap two more targets, return and grasp the circle again, etc. This simulated basic motion sequences required in industry. For all these types of motion, Fitts' ID was able to accurately predict movement time. Thus, a basic laboratory finding from experimental psychology was better at predicting industrial motions than the standard predetermined time systems developed with no theoretical base solely for use in industry. We hope this dramatic example explains why this human factors text that is primarily concerned with real-life application contains four chapters devoted to a theoretical understanding of human capabilities. Now we continue with some of the theory behind Fitts' law.

Explanations of Fitts' Law

Communication Model The original explanation proposed by Fitts was based on the communication model and information theory. The human was considered to be generating information whenever a movement was executed. Although this kind of explanation is no longer considered novel (and indeed not mathematically rigorous, Kvålseth, 1981), at that time (Fitts, 1954) it was a considerable advance. The idea that movements created information was a minor revolution. Now, it may not be apparent to you how movement to the kind of target specified in Figure 5-11 generates information. After all, you know that generation of information requires a set of events and a target looks like only one event, not enough to produce information. We now offer some informal observations to give you the flavor of how information is created by movement. What happens when you miss a target and land short? You have selected an incorrect target. Now imagine a whole row of alternative targets stretching from the starting point, to the nominal target, and beyond. Movement to the correct target requires that it be selected from this array of alternate targets. How many alternate targets are there? As distance to the target increases, more alternate targets can be squeezed in. Thus, more information is gained. Similarly, as target width decreases, again more targets can be squeezed in thereby increasing information. So information should be directly proportional to distance and inversely proportional to target width. This is why D is in the numerator of Fitts' ID and W is in the denominator. Movement generates information because it requires selecting some particular target from a set of possible targets.

This explanation of Fitts' law suggests a particular interpretation of the slope of the line relating movement time and ID. The unit of the slope is time/bit. The reciprocal of the slope is bits/sec, an index of channel capacity. So the slope measures channel capacity for the movement-generating system. The original task used by Fitts required people to rapidly and continuously tap back and forth between two target plates. Later, Fitts and Peterson (1964) required discrete movement: the subject made only a single movement from a home position to the target. While Fitts' law holds for both continuous and discrete tapping tasks, the slopes differ. The discrete task has a lower slope and therefore a higher channel capacity (Figure 5-12). The continuous task is more difficult because it requires continuous monitoring of position or else the subject will drift off target. In the discrete task, this monitoring occurs after the target has been hit and is not added into movement time. Thus, the extra feedback processing required in the continuous task is responsible for the lowered channel capacity.

Feedback Model This class of models assumes that ongoing movement is monitored by comparing the relative positions of the target and the stylus, pointer, or limb in motion. Feedback information about current position relative to the goal controls corrective movements as the target is approached (Crossman & Goodeve 1963 cited by Keele, 1968). This correction is based not on deviation from some preselected path—as would, for example, corrections of an airplane pilot being told she was 50 feet above the glidepath—but on the relative positions of goal and limb. When this is put mathematically (in a first-order difference equation), movement time is predicted to be a linear function of *ID*. Since the feedback model and the communications model make exactly the same predictions for movement time, one cannot tell them apart by only examining Fitts' law. Indeed, for many years the feedback model was considered to be as correct as Fitts' original explanation and its mathematical derivation was more elegant. But newer research has conclusively rejected this explanation by examining the fine-grained details of movement trajectories as we discuss as follows.

Nonlinear Models In terms of control theory (see chapter 12) the feedback model based on a difference equation is a linear model. (If you don't know exactly what a linear model implies, don't worry; we'll get to this in Chapter 12.) Linear models make a strong prediction when position or velocity of a movement is plotted over time. Early research on aiming movements lacked the technical resources to monitor the actual position of a limb and merely noted the time a movement started and ended. More recent research (Langolf, Chaffin, & Foulke, 1976) has carefully examined how movements are swept out over time and their results show that any linear model, including the feedback model described above, must be rejected. This is sad because linear models have the great virtue of mathematical simplicity. But it appears that the human is not a linear system. However, all is not lost since there are many nonlinear control

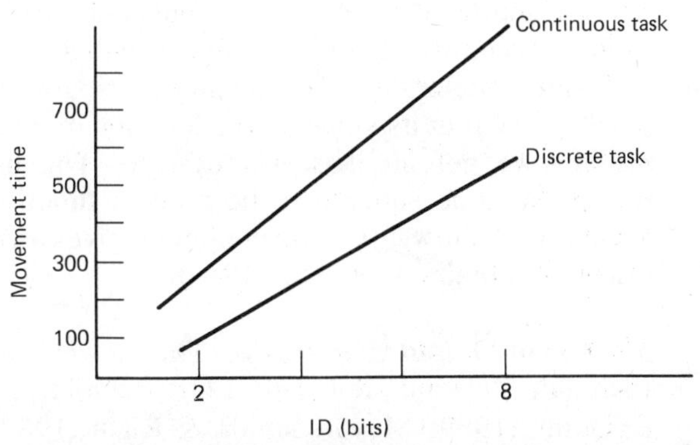

Figure 5-12. Movement time as a function of *ID* for continuous and discrete tapping tasks. (Adapted from Fitts & Peterson, 1964.)

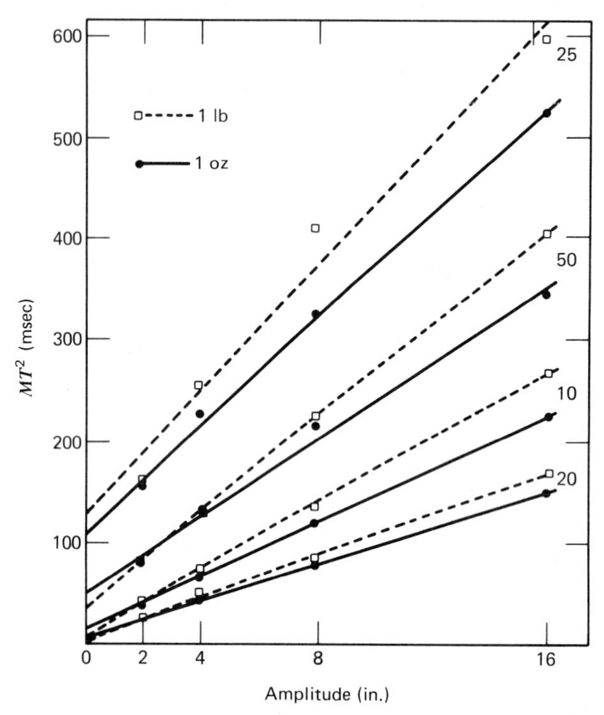

Figure 5-13. Movement time squared as a function of distance (amplitude). Target width ranges from 0.25 to 2.0 inches. Dashed lines indicate a heavy (1 lb) stylus and solid lines a light (1 oz) stylus. The predictions of the bang-bang model fit these data quite well. (From Kantowitz & Knight, 1978.)

models that are well known to control engineers. One of them called a bang-bang model—sometimes also called an on-off or maximum effort controller—has been applied to Fitts' original data (Kantowitz & Herman, 1967; Kantowitz, 1978). In this application the nonlinear bang-bang controller accelerates with maximum capability until halfway to the target and then decelerates with maximum capability over the remaining distance. If we plot its velocity as a function of time—this is called a velocity profile—we get an isosceles triangle. The bang-bang model predicts movement time squared to be a linear function of movement distance. Figure 5-13 shows that this prediction gives a very accurate fit to the data originally obtained by Fitts (1954).

Movement-Output Variability The newest explanation of Fitts' law (Schmidt, Zelaznik, & Frank, 1978; Schmidt, Zelaznik, Hawkins, Frank, & Quinn, 1979; Meyer, Smith, & Right, 1982) combines two opposite effects of output variability. Instead of measuring movement time as did earlier studies, this approach demands that people produce a specified movement time and measures the error associated with production of movement time. This error is directly proportional to the distance moved. However, the force impulse that generates movement has two opposite components: impulse amplitude and impulse duration. The impulse amplitude is inversely proportional to movement time squared while the impulse duration is directly proportional to movement time. Combining these two factors leaves a net relationship of the impulse being proportional to the reciprocal of movement time. Thus, the error of movement is proportional to distance divided by movement time.

At first this may sound very much like older explanations of Fitts' law. However, there is a basic difference. The feedback model states that the number of corrective movements is what controls movement time. The output variability model states that movement time is limited by the person's ability to produce accurate forces for definite periods of time. Indeed, a more general version of the output variability model (Kvålseth, 1981) based on information theory offers some advantages over the earlier formulations of Fitts' law. It can be applied to one- and two-dimensional movements, does not collapse when target width is equal to zero, and can also handle speed-accuracy trade-off. In applied settings the human factors specialist will let the situation determine which model should be used. If physical parameters of a target are known the traditional explanations will suffice. If movement time is known or specified, then the output-variability model can predict movement error.

Motor Programs

Implicit in all explanations of Fitts' law and motor control is the distinction between open- and closed-loop behavior. Closed-loop behavior is guided by feedback and corresponds to what Woodworth called current control in 1899. Open loop or ballistic motion corresponds to Woodworth's initial impulse. Modern theorists have borrowed the term "motor program" to explain behavior that is open loop.

The strongest arguments for the existence of motor programs, defined as a set of muscle commands that are executed sequentially without interruption (Keele & Summers, 1976), are by default. How else, the argument goes, could a skilled pianist, athlete, etc., perform so rapidly if it were necessary to monitor feedback after each movement? A motor program is more than a specific list of commands. Instead, it is held to be a general prototype for a class of movements (Klapp, 1977). With the proper parameters, the same motor program can generate many different movements. While some researchers are skeptical about the ultimate utility and correctness of motor programs (Adams, 1976), a concept that has led to many debates about the minimum time needed to process feedback information, for the moment motor programs are "in." However, it is too soon to state how this theoretical concept can help the human factors specialist.

FOCUS ON RESEARCH

Typewriting and Motor Programs

The study of movements in typewriting has a long history (Coover, 1923; Lahy, 1924). These early workers discovered that letter pairs were typed faster when each letter was typed with a different hand. But the modern typewriter keyboard is not designed to take advantage of this fact. Indeed, the keyboard layout of the modern typewriter—called the QWERTY design since these letters occur in the top row—is far from optimal and some better designs are discussed in Chapter 11. Here, however, we focus not on keyboard design per se but instead deal with the theoretical concepts that describe how people control their fingers while typing.

As you might expect, current explanations of typing rely heavily on the idea of a motor program. Since the motor program is a general concept, it requires lots of detail before a clear model of a specific task such as typing can emerge. While this research has become popular of late (Shaffer, 1975, 1978; Sternberg, Monsell, Knoll, & Wright, 1978; Ostry, 1980) it is still too early for agreement about the model for typing. So you should view this box not as an ultimate explanation of typing but instead as a description of how scientists go about trying to reach an explanation.

An English psychologist has been in the forefront of this research effort and we will focus on one of his studies as typical of the kind of research now being conducted. Shaffer (1976) has proposed a general model that explains how performance—the movement of specific motor effector units like fingers—is derived from mental intentions. The psychologist, of course, cannot observe intention and is forced to measure only performance. Theory is required to bridge this gap and experiments are required to test the theory. When Shaffer's model is applied to the task of typing (Shaffer, 1978) there are two important components. First, intention must generate the sequence of response elements, in this case movements that depress the typewriter keys. Second, a rhythm must be generated to time the sequence of response elements so that the letters are typed efficiently as fast as possible. The motor program must therefore handle two kinds of representations: spelling sequences of letters and the response commands that control keyboard responses.

Shaffer (1978) was especially interested in the timing of motor programs. He suggested several mechanisms a typist might use to keep an efficient rhythm. He called these the Random Walk process, the Feedback process and the Metronome (or clock) process. A random walk process occurs if the typist generates rhythm by relying on memory to recall how fast the last key was pressed. A feedback process occurs if the typist has an ideal pace in mind and continually adjusts the rate of typing to equal this ideal pace. A metronome process occurs when the typist has an internal clock that generates regular pulses to control when keys should be hit. All of these processes are hypothetical and cannot be directly observed. So Shaffer studied the reaction times of typing responses in order to make inferences that would hopefully reject one or more of these alternatives, thereby narrowing down the field. Shaffer used a statistical calculation called autocorrelation to decide which, if any, of the three processes was correct. An autocorrelation is just an ordinary correlation coefficient applied to a series of data, in this case the reaction times to strike typewriter keys. So Shaffer calculated the correlation between the time to strike the first key with the time for the second, then the correlation between the second key and the third, etc. (This is technically termed autocorrelation with a lag of one.) Without going into the mathematical proofs we merely state that the three process models make different predictions about the size of the autocorrelation. The Random Walk process predicts a positive value, the Feedback process a negative value, and the Metronome process zero autocorrelation. Results showed that the autocorrelation was zero for typing prose, thus ruling out Random Walk and Feedback processes. Therefore, it seems reasonable to suppose that typing is controlled by a kind of internal clock that synchronizes response commands.

Unfortunately, this internal clock does not keep a perfect beat, or if it does, the typist deliberately ignores it at least some of the time, much as a pianist might deliberately slow down or speed up parts of a performance to achieve some desired expressive effect (Shaffer, 1980). Thus, the story is more complicated. The motor program concept originally came from experiments, like those on Fitts' law, where a relatively simple movement was required. Real-world tasks like typing are more complex. A motor program that cannot be modified as it runs has difficulty explaining complex tasks. This leaves the scientist two choices: either make the motor program more complicated or find a better explanation altogether. So far, the researchers have opted for the first choice (e.g., Ostry, 1980). If the motor program is expanded to have several levels or hierarchies, it can explain the finding that the time between striking typewriter keys—called interresponse time—is not constant (Ostry, 1980). A perfect metronome would have equal interresponse times for successive letters. But results show that interresponse time is at first short, then jumps up and finally returns to the original level. The price of expanding the complexity of the motor program is a decreased ability to make very precise predictions.

As a motor program gets more powerful it can do more and more. Science progresses by rejecting flawed theories. If a theory can never be rejected because it is so complicated that it can predict anything, the theory is useless. Human factors specialists require very precise predictions before being convinced that theory is useful. Most of the time, they settle for imprecise prediction based on rule of thumb, rather than theory. Somehow we expect a theory to be more accurate than a rule of thumb and so demand greater precision from theoretical predictions. It is clear that the motor program concept has not advanced to the point where it will be routinely used by human factors specialists concerned with typing and data entry devices. But the next few years may result in motor-program theories that are at least as good as present rules of thumb described in Chapter 11. Or, the motor program concept may turn out to be a dead end and only a current fad. This is the challenge and the frustration of basic research.

SUMMARY

The scientific study of psychomotor skills goes back over 100 years. Donders tried to estimate the time required for mental operations by subtraction. He devised A, B, and C reaction tasks that are still used today. However, application of information theory to the study of reaction time greatly expanded our ability to predict reaction times with different numbers of stimulus-response pairs. The linear relationship between information and reaction time is known as Hick's law.

Stimulus-response compatibility can have a large influence on reaction time. Highly compatible mappings, such as pointing to a stimulus, produce faster reaction times than less compatible mappings. It is important for the human factors specialist to ensure that displays and controls are compatible. Practice decreases reaction time and can eventually minimize the influence of low stimulus-response compatibility. Well-practiced operators have Hick's law functions with lower slope than do unpracticed operators. Operators can trade off speed for accuracy. Demanding faster performance will increase error rates.

The study of motor control has important implications for performance in many industrial assembly tasks. Fitt's law states that movement time is a linear function of the information generated by a movement. It can be used to predict industrial motions such as found in assembly work under a microscope. Several theoretical models have been put forward to explain Fitt's law. It is probably too early for the most recent of these models to be applied directly to the practical problems that are of greatest interest to the human factors specialist. Nevertheless, some familiarity with newer models will help the specialist to avoid pitfalls in applying older models that are better known.

APPENDIX

Values of $\log_2 n$ and $p \log_2 1/p$

n or p	$\log_2 n$	$p \log_2 1/p$	n or p	$\log_2 n$	$p \log_2 1/p$
1	0.000	.0664	51	5.672	.4954
2	1.000	.1129	52	5.700	.4906
3	1.585	.1518	53	5.728	.4854
4	2.000	.1858	54	5.755	.4800
5	2.322	.2161	55	5.781	.4744
6	2.585	.2435	56	5.807	.4684
7	2.807	.2686	57	5.833	.4623
8	3.000	.2915	58	5.858	.4558
9	3.170	.3127	59	5.883	.4491
10	3.322	.3322	60	5.907	.4422
11	3.459	.3503	61	5.931	.4350
12	3.585	.3671	62	5.954	.4276
13	3.700	.3826	63	5.977	.4199
14	3.807	.3971	64	6.000	.4121
15	3.907	.4105	65	6.022	.4040
16	4.000	.4230	66	6.044	.3957
17	4.087	.4346	67	6.066	.3871
18	4.170	.4453	68	6.087	.3784
19	4.248	.4552	69	6.109	.3694
20	4.322	.4644	70	6.129	.3602
21	4.392	.4728	71	6.150	.3508
22	4.459	.4806	72	6.170	.3412
23	4.524	.4877	73	6.190	.3314
24	4.585	.4941	74	6.209	.3215
25	4.644	.5000	75	6.229	.3113
26	4.700	.5053	76	6.248	.3009
27	4.755	.5100	77	6.267	.2903
28	4.807	.5142	78	6.285	.2796
29	4.858	.5179	79	6.304	.2687
30	4.907	.5211	80	6.322	.2575
31	4.954	.5238	81	6.340	.2462
32	5.000	.5260	82	6.358	.2348
33	5.044	.5278	83	6.375	.2231
34	5.087	.5292	84	6.392	.2113
35	5.129	.5301	85	6.409	.1993
36	5.170	.5306	86	6.426	.1871
37	5.209	.5307	87	6.443	.1748
38	5.248	.5304	88	6.459	.1623
39	5.285	.5298	89	6.476	.1496
40	5.322	.5288	90	6.492	.1368
41	5.358	.5274	91	6.508	.1238
42	5.392	.5256	92	6.524	.1107
43	5.426	.5236	93	6.539	.0974
44	5.459	.5211	94	6.555	.0839
45	5.492	.5184	95	6.570	.0703
46	5.524	.5153	96	6.585	.0565
47	5.555	.5120	97	6.600	.0426
48	5.585	.5083	98	6.615	.0286
49	5.615	.5043	99	6.629	.0140
50	5.644	.5000	100	6.644	.0000

6 HUMAN INFORMATION PROCESSING

In contemporary society the term "information processing" conjures up images of pocket calculators, integrated-circuit chips, and a microprocessor in every living room. Our society has indeed been dramatically altered by machines that process information (Figure 6-1). A recent book (Wise, Chen, & Yokely, 1980) projects the effects of microcomputers to the year 2000 and forecasts powerful impacts. In this surge of technology we tend to forget that humans have been powerful information processors since well before the advent of the computer revolution (Figure 6-2).

Figure 6-1. This dime-sized chip, the BELLMAN™-32A microprocessor, contains almost 150,000 transistors and offers processing power comparable to the minicomputers of the early 1980s. (Courtesy of Bell Laboratories.)

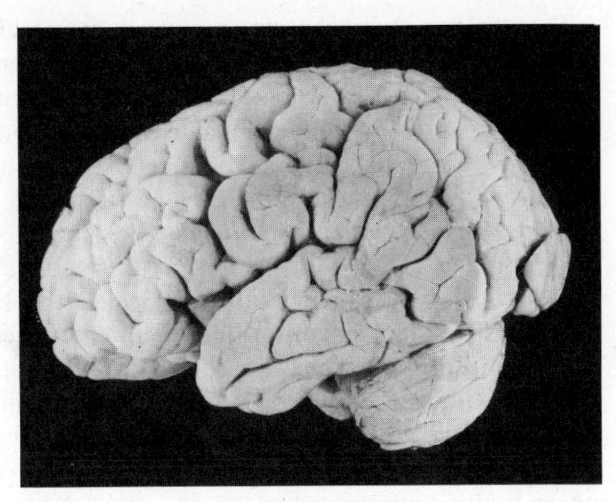

Figure 6-2. This processor is larger than a dime but still fits neatly inside your head. It contains from 10 to 100 billion neurons and has a proven history of reliable operation. (Courtesy of Neuropathology, New York State Psychiatric Institute, State of New York Department of Mental Hygiene.)

In this chapter we examine the human's ability to process information. The information-processing approach differs from more traditional views of the human that have been formulated by experimental psychologists. It rejects the old idea that the human is a passive recipient of information or stimuli. Instead, people are regarded as active, dynamic systems capable of great flexibility in the ways they choose to handle and transform incoming information. While entire books have been published about human information processing (e.g., Kantowitz, 1974) we limit our discussion to only two important areas: memory and attention. No adaptive system can function without these capabilities. So we review current theoretical models and selected empirical findings that will help you to understand how humans attend and remember information. In order to demonstrate the practical importance of these models for the human factors specialist, we conclude with a detailed discussion of mental work load, an important area of human factors research that is closely related to attention.

THE INFORMATION-PROCESSING APPROACH

There are several crucial assumptions shared by researchers who adopt the information-processing approach to the study of human behavior. The most important assumption is that behavior is determined by the internal flow of information within a person. Since this information flow is internal and invisible, special techniques and methodologies are used to allow inferences about this postulated information flow. We discuss some of these techniques later in the chapter. But differences in methods should not be permitted to hide the basic goal of all information-processing research, which is to map internal information pathways.

The information-processing approach uses techniques that are in many ways similar to those used by engineers designing large systems. The human is regarded as a complex system and experimental psychologists try to discover what happens inside the "black box." However, engineers have a considerable advantage since they can insert probes and meters within their black boxes; the psychologists cannot. Although some researchers use psychophysiological measures like brain waves to help peer into the black box, at present this technique cannot yet plot the hypothetical information pathways inside the human. Thus, the effort to understand internal information flow proceeds primarily by testing alternate representations based upon different arrangements of subsystems with different properties. It is not sufficient to create a model that will duplicate the behavior of humans, although this is of course a necessary requirement for any information-processing model. A female singer and a tape recording made with the proper brand of tape might both be able to shatter a slender crystal goblet, but no one would claim that this duplication of behavior proves that the singer and the tape recorder produce auditory signals by the same internal processes. Thus, the information-processing theorist must duplicate not only behavior but also the internal patterns of information flow before an acceptable explanation of human thought and action can be found.

Information-processing models differ in the number and arrangement of subsystems. Many possible arrangements are reasonable so that each theorist must try to show how their model is superior to other competing models. There is seldom complete agreement about which model is best and this can confuse the human factors specialist who wishes to learn only a little about information-processing models. While the models discussed later in this chapter are popular, generally accepted, and based on substantial data, it would be rash to claim they are permanent models. Even extremely good models are eventually replaced by newer theories, or even by older theories that are reborn due to new data or new techniques. (See Kantowitz & Roediger, 1978, Chapter 1, for a discussion of models and theories in experimental psychology.) It takes a while for this kind of theoretical progress to get communicated from laboratory to field applications (Kantowitz, 1981). But as models improve so does our ability to use these models to solve human factors problems. Therefore, prudent human factors specialists take time to update their knowledge of basic research models that can help them find good directions for solving applied problems.

The typical information-processing model represents the human cognitive system as a series of boxes connected by an assortment of arrows. The boxes represent subsystems that perform different functions and processes that route information to and from the various boxes. Each box represents a generalized kind of information transformation that goes on inside your head. As the models become more refined, the level of detail represented by a box becomes finer. A box that represents a relatively fine level of detail is often called a stage of information processing (Sternberg, 1975; Taylor, 1976) or an isolable subsystem (Posner, 1978). The precise definition of a stage is mathematically sophisticated (Townsend, 1974) but we will not be far off if we think of a stage as corresponding to a single transformation of information. In general, the output of a stage will not match its input. For example, one common model of memory assumes that printed words received through the eyes get recoded into a format that is related to how the words sound when read aloud. This transformation occurs even though people were not asked to pronounce the words. So a visual input has been transformed into an auditory (i.e., acoustic or phonological) output. This kind of transformation is common in machines. A computer transforms punched holes in cards into electrical impulses. A telephone transforms electrical signals into air vibrations. So it is not surprising that the human information processor is capable of these kinds of internal information transformations.

Different arrangements of stages are required to model the flexibility of the human information processor (Figure 6-3). The simplest arrangement occurs when several stages are linked in a straight line (or in cascade) with the output of one next becoming the input of the succeeding stage. This is called serial processing because no stage can perform its own transformation of information until it receives the output of the preceding stage in the chain. This, of course, will not happen until that stage has

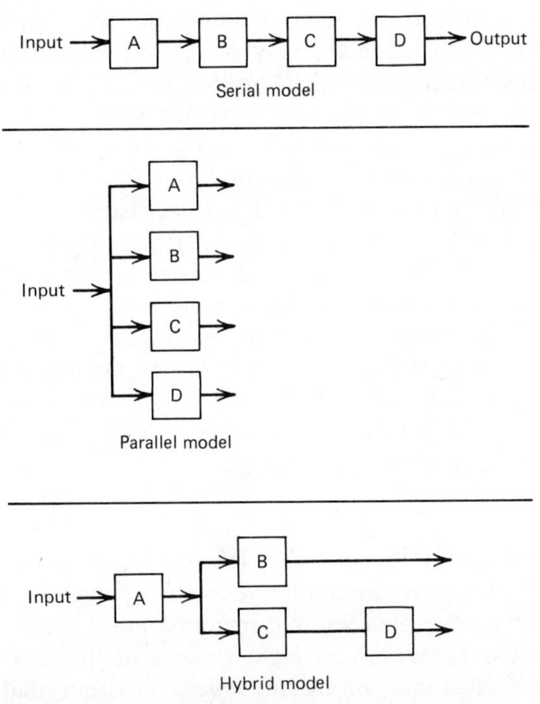

Figure 6-3. Serial, parallel, and hybrid models.

received information from its preceding stage. So the stages are like a row of dominoes. Tipping the first domino over starts a reaction but this is propagated slowly down the line of dominoes. Similarly, serial processing models require each stage to wait its turn before producing an output.

If a stage need not wait for other stages to finish, the arrangement is called parallel processing. In parallel processing several stages can access the same output simultaneously. Each parallel stage can do its own thing without having to wait for other parallel stages to complete their processing. An arrangement with both serial and parallel components is called hybrid processing. Hybrid processors are often more powerful than serial or parallel processors but this extra power is gained by making the model more difficult to understand and analyze. Since many people find serial models easier to understand, most information-processing models are serial. But our crystal ball predicts that parallel and hybrid models will become dominant soon.

Although we now have an excellent scheme for classifying the structure of a model into three categories – serial, parallel, and hybrid – you may be surprised to find that structure alone cannot determine the predictions that a model will generate. We must also know the "price" each stage demands for performing its transformation of information. This is called resource allocation or capacity. Capacity is a hypothetical construct that controls how efficiently a stage operates. In some models, it is assumed that each stage has adequate capacity to do its own thing, regardless of how many other stages are operating and how complex these operations might be (Sternberg, 1969). Other models assume that capacity is limited so that stages must compete for available processing resources (Kantowitz & Knight, 1976). In these models, a stage cannot always operate as efficiently as if it were the only stage in the system. Other stages may divert the necessary capacity. By cleverly selecting assumptions about capacity it is possible to make serial systems mimic parallel systems and vice versa (Townsend, 1974). Thus, in order to generate predictions for a model, we must specify both the structure of the model and its capacity assumptions. The best models of human information processing specify (1) the number and configuration of internal processing stages, (2) the capacity requirements of individual stages, and (3) total availability of capacity and rules that govern distribution of capacity to individual stages.

MEMORY

The ability to remember information is an essential and vital ingredient of human information processing. Even the simplest adaptive system cannot function without memory, and people are complex adaptive systems. Since human memory is a limited resource, the human factors specialist must be careful to design systems that do not overload memory.

This is especially important in emergency situations and whenever the system is approaching peak loads, for it is under these circumstances that a human memory failure is most likely and most devastating. The task of designing systems that do not overload memory is complicated because there is more than one kind of memory subsystem used by the human information processor. Memory for operating procedures differs from memory for recent events like the onset of a warning signal. In this section, we discuss three memory subsystems called sensory memory, working or short-term memory, and long-term memory with an eye toward the practical implications of how each subsystem contributes toward memory failures. We conclude with a box that will help you to improve your own memory.

Sensory Storage

The sensory storage system holds information from the eyes (iconic storage) and ears (echoic storage). This information is stored for a brief time—less than one second—after which it must enter working memory or be lost.

The classic demonstration of iconic storage (Sperling, 1960) has people look at an array of letters: for example, three rows of letters presented for less than 100 msec. When observers are asked to report as many letters as they can—this is called a whole report—they can correctly name only four or five even if many more had been displayed. But people do much better when they are asked to report only a single row of letters; this is called a partial report. Even though they were not told which row to report until after the stimulus was removed, people were able to report almost an entire line of three to six letters correctly. These two sets of results from whole and partial reports present an interesting dilemma. If someone could correctly report an entire row using the partial report procedure, why can they not report all the rows using a whole report procedure? Sperling's answer was that the representation of the letters in iconic storage faded away so rapidly that the time taken to report one row was long enough so that the observer forgot all the other rows. That is, the observer knew almost all the letters in all rows when the report was started. But by the time one row had been reported, all the remaining information was lost. In order to test his idea, Sperling deliberately introduced a delay between the time the letters were turned off and the partial report was started. Even though people had to report only a single row, the delay caused forgetting. When the delay was increased to one second, partial reports were no better than whole reports without a delay. Thus, it is the delay that was responsible for forgetting in the whole report procedure.

Another way to test the delay characteristics of iconic memory used dot patterns like those in Figure 6-4. Each dot pattern alone looks like a random collection of points. But when the two patterns are superimposed, they create a three-letter word. By imposing a delay between the two

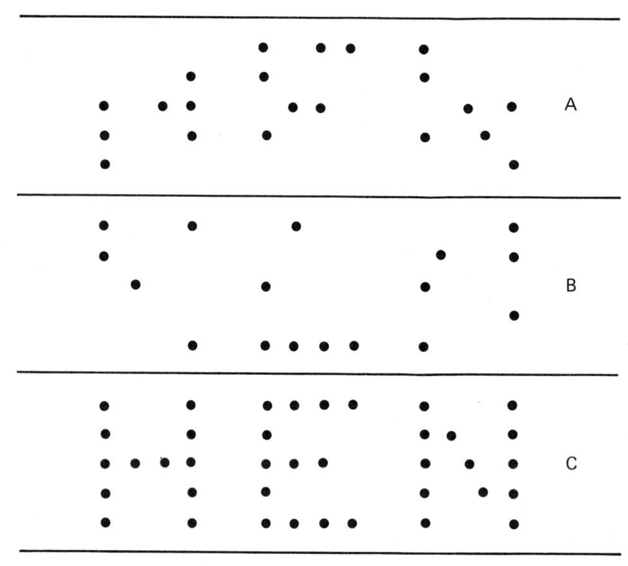

Figure 6-4. When patterns A and B are superimposed they spell HEN, pattern C.

patterns, we can study iconic storage (Eriksen & Collins, 1968). If the delay is less than 100 msec, the word can be read. As the delay is increased beyond 100 msec, it becomes difficult or impossible to read the word. This means that the information in the first dot pattern remains in iconic storage for only 100 msec, the same duration as found by Sperling.

Similar results have been obtained for echoic storage of material presented through the ears. Psychologists can create a "three-eared man" by presenting letters to the left ear only, the right ear only, and to both ears with equal intensity. Items presented to both ears sound as if they occur in the middle of your head. So sounds can be presented to left, right and "middle" ears, analogous to the three rows of visual stimuli used by Sperling. Partial report is superior to whole report, although the advantage is smaller than that found for iconic storage. Furthermore, the decay rate is slower taking more than one second (Darwin, Turvey, & Crowder, 1972; Rostron, 1974). Exactly how much longer is being debated, with estimates of two minutes or even longer being offered (Klatzky, 1980, p. 43).

Working Memory

Sensory memory can hold information for only a brief period. Therefore, other memory subsystems are required for longer storage. Information from sensory storage must be transferred to working memory, or as it is sometimes called, short-term memory. Working memory can be controlled to a much greater degree than sensory storage. While you cannot recycle information through sensory storage, there are a variety of control mechanisms you can apply to manipulate information in working memory. The most common of these control processes is called *rehearsal.* Rehearsal may be as simple as repeating a telephone number over and over to yourself. Rehearsal maintains information in working memory. If rehearsal is stopped, information can be lost from working memory.

The importance of rehearsal was demonstrated in a classic experiment that tried to eliminate rehearsal (Peterson & Peterson, 1959). Such elimination should lead to forgetting. People were told to remember a three-letter item such as PAJ. However, as soon as the item was presented they had to count backward by threes. This counting backward made it very difficult to rehearse the items. After a period, called the retention interval, the person stopped counting and tried to recall the three letters. You may think that remembering only three letters is a trivial task and that no forgetting should occur. So did most psychologists. But the Petersons found considerable forgetting (Figure 6-5) using their new technique. Later research (Wickens, 1972) discovered that interference among different three-letter items was an important source of this forgetting. Trying to recall later items is hampered by exposure to other items early in the experiment. This interference is called proactive inhibition. We can temporarily eliminate proactive interference by changing the kinds of items, for example, from letters to numbers (Figure 6-6). After the kind of item is shifted, there is a temporary improvement in recall.

The Peterson-Peterson technique for studying working memory was used in a recent study of communication between pilots and air traffic

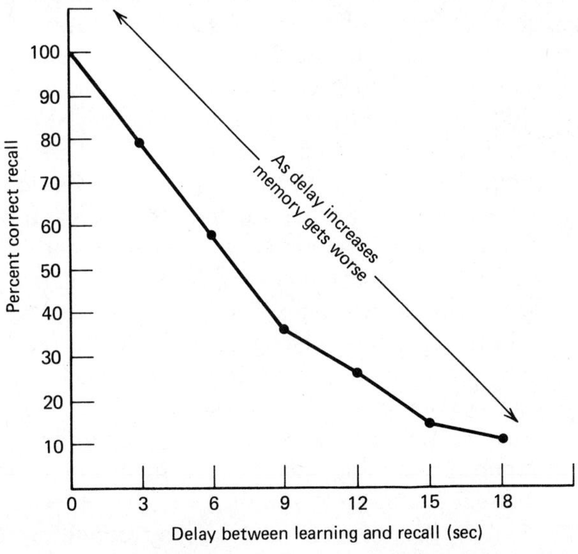

Figure 6-5. Counting backwards by threes decreases correct recall as retention interval increases.

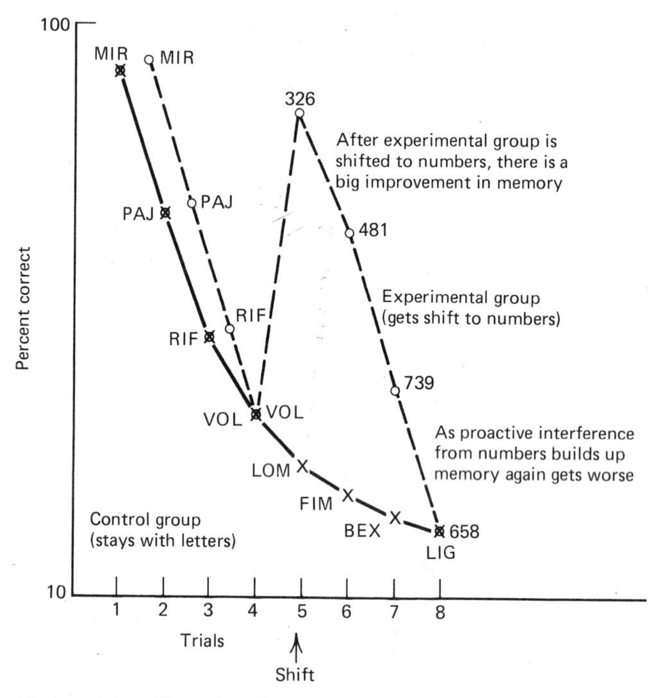

Figure 6-6. Changing from letters to numbers yields a temporary release from proactive interference.

controllers (Loftus, Dark, & Williams, 1979). Messages from controllers to pilots often contain numeric information (Table 6-1). Pilots must change their radio frequencies as they move from one zone of control to another, must change their transponder setting to enable the controller to identify the blip produced on a radar scope, and must fly at designated altitudes. All these actions must be based on working memory of communications received from the ground. The pilot is not always able to act immediately on receipt of the message. She may have to first put down a banana that was lunch, or finish a conversation with a passenger, or locate a map. These kinds of intervening activities are similar to counting backward in the laboratory since they can be expected to prevent rehearsal of the message.

Loftus and his colleagues examined three kinds of memory information: place (name of air control center), frequency (radio setting), and transponder code. Using a laboratory setting, they created sets of messages containing these three kinds of information and tested recall. An important variable was memory load. In the low memory load condition, messages consisted of either a transponder code or a place name plus a radio frequency. In the high-load condition, messages contained both transponder codes and place/frequency information. Thus, a typical high-load message might be "Contact Seattle Center on one-two-eight-point-

TABLE 6-1 *Messages used by air traffic controllers (Adapted from Loftus et al, 1979)*

Message	Interpretation
Contact on 132.4	Change radio frequency to 132.4 MHz
Descend to 6500	Decrease altitude to 6500 feet
Squawk 4256	Set transponder code to 4256
Wind 090 at 15	Wind direction is 90 degrees, speed 15 knots
Fly heading 175	Change direction to a heading of 175 degrees

nine, Squawk 4273." (Incidentally, the word "squawk" is used for transponders because a transponder parrots back a code and parrots squawk. Just a little humor in the sky.) Results clearly showed memory to be worse under the high-load condition. Furthermore, as delay increased the rate of forgetting was higher under high memory load. The practical implications are obvious. Controllers should limit their instructions to pilots to one at a time.

This study can be criticized because the subjects were college students rather than licensed pilots. Therefore, how do we know that the findings can be applied to real pilots? This type of criticism of laboratory work with naive populations is often raised in human factors. The best reply is that all people tend to process information in similar ways. Indeed, there would be little value in formulating any theories of human behavior if people behaved randomly or inconsistently. While it is true that there are individual differences, say person A can remember more items than person B, both A and B will be affected the same way by other independent variables such as delay. So even though A may have a better memory, this better memory will still remember less at longer retention intervals. A second point raised by Loftus is that it is precisely those pilots with less experience who have the greatest difficulty in remembering and acting on messages from the ground. Therefore, the results should probably generalize to newer pilots, the group that needs help the most. Of course, it is always possible that pilot training alters something basic about human information processing in general and memory in particular. While there are no immediate grounds to suspect this, it would be prudent to repeat the study with pilots and even better to do this in the air. But practical limitations often prevent optimal experimental conditions. It would be difficult and expensive to replicate this study in the air. In fact, the small likelihood of obtaining different results with licensed pilots gives this replication a high cost-benefit ratio and most experimenters would be content that results apply to pilots as well as to college students.

It is clear that working memory has a limited capacity, otherwise there would have been no forgetting in the experiments described above. The human factors specialist would like to know exactly how many items can be stored in working memory before information will be lost. Unfortunately, experimental psychologists have been unable to agree on this magic number. But a range of from five to nine items encompasses most reasonable estimates. Part of the problem is that more information can be jammed into a working memory slot by clever encoding schemes and this makes it hard to determine precisely how many slots there are. For example, a three-letter nonsense syllable (PAJ) might be expected to fill up three slots: one for each letter. But if the syllable is pronounced as one sound, perhaps only one slot is needed. Or if a syllable such as AFL is recognized as an abbreviation for American Football League, it may also require only one slot. This recording of stimulus information has been called chunking (Miller, 1956) and is similar to some coding schemes (called bit packing) used to get more information into computer memory slots. Thus, we cannot evaluate the storage of information in working memory without considering the encoding of this information. Indeed, some memory theories view encoding as the basic memory process (Craik & Lockhart, 1972) and a great deal of effort has gone into studying memory by comparing encoding and retrieval cues (see Kantowitz & Roediger, 1980 for a review). Thus, the question of exactly how many slots exist is regarded as an oversimplification and we doubt that the human factors specialist seeking a simple answer (e.g., working memory has 5.6 slots) will ever find it. Indeed, there is some doubt that working memory is truly an all-purpose subsystem that is capable of carrying the heavy load placed on it by memory theorists (Crowder, 1982).

It is known, however, that the more slots that are filled, the less working space is available for other calculations and problem solving (Baddeley & Hitch, 1974). Maintaining information in working memory cuts down our ability to perform other mental work at the same time. So the human factors specialist should always try to minimize the amount of information that must be kept in working memory. We will return to the problem of limited resources and attention later in this chapter.

Long-term Memory

It takes effort or capacity to maintain information in working memory. But once information has been transferred to long-term memory from working memory, it is there forever. No rehearsal is needed to maintain items in long-term memory. However, merely because information is available in long-term memory does not guarantee that it is accessible. The major problem associated with long-term memory is retrieval—finding the information inside long-term memory. While retrieval from working memory is relatively easy—indeed, many models of working memory assume that retrieval is automatic—this is not the case for long-term memory. There is so much information contained in long-term memory that locating it is not always possible.

Most models assume that information in long-term memory is coded according to its meaning. But once that common point is left, there is

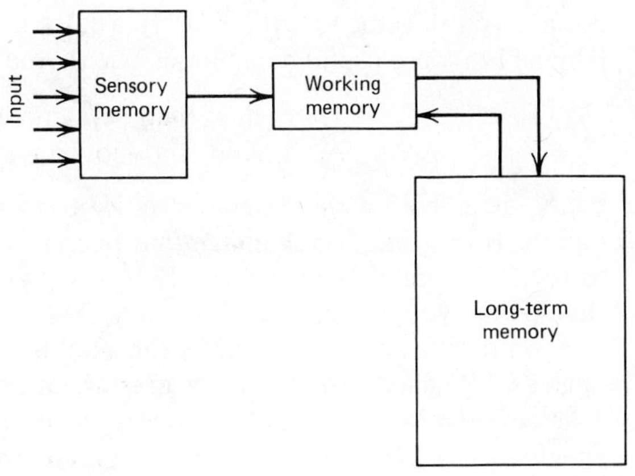

Figure 6-7. The traditional "three-box" model of human memory.

considerable debate as to how information is located and retrieved from long-term memory and indeed there is some question about how many long-term memory subsystems exist (see Klatzky, 1980 for a review). Space limitations prevent elaboration of these details.

The important points for the human factors specialist to remember about memory are summarized in Figure 6-7. Working memory is at the center of the human memory system. Information cannot be placed or retrieved from long-term memory without passing through working memory. Efficient use of the human as a system component cannot be accomplished without efficient use of human working memory.

OUTSIDE THE LABORATORY

Improving Memory with Mnemonics

The typical human factors specialist is a very practical person with a deep concern for improving the efficiency of systems involving people. While abstract theoretical discussion of memory can be useful by pointing out human limitations that need be factored into the design process, the human factors specialist is even more impressed by concrete suggestions that will improve human memory. Experimental psychologist have studied one class of techniques, called mnemonics, to discover why and how they improve memory.

A mnemonic device is any formal scheme to aid memory, most often of a list of items. This could be a shopping list or an industrial checklist. A mnemonic device your probably have heard about is tying a string around your finger to remind you to do something or other. This is a weak mnemonic because although it tells you to do something, it does not aid you in remembering exactly what it is you are supposed to do. Rhyme is a more useful mnemonic. The rhyme

> Thirty days hath September,
> April, June, and November . . .

really does help you remember how many days there are in each month. It has one disadvantage that becomes obvious when you need to recall the number of days in February. You must recite the entire rhyme until you finally reach February. This is inefficient.

A more efficient mnemonic is the method of loci. Loci is Latin for "places." Ancient Greek orators used the method of loci by making up a list of places, for example, rooms inside a building. The speaker imagined himself moving within the building from room to room in a fixed order. Each room held an image of some point the speaker wished to make. So, for example, if the speaker wanted to end his oration with the phrase "Carthage must be destroyed!" he would place a mental picture of Carthage reduced to rubble in the last room of the building. The method of loci not only helped the orator remember all his debating points, but also ensured that they emerged in correct order since the orator took a fixed mental pathway through the rooms.

Let's see how the method of loci can help you remember your class schedule for Monday. Your first class is human factors, then music, then digital circuit design. After these morning classes you are supposed to meet Mary for lunch. Then back to chemical engineering laboratory and scientific Russian. For our loci we will take the pathway you use to enter your dorm room: (1) the driveway leading up to the dorm, (2) the lobby, (3) the elevator, (4) the hallway outside your room, (5) your clothes closet inside your room where you hang your coat, and (6) your bed. This list of places and their order is easy to remember since you go through this sequence by walking up the driveway and eventually hanging up your coat and collapsing on your bed. (Of course, if you had more classes to remember you would need a larger list of loci.) Now, to remember your first class form a mental image of a giant gear with a person inside of it blocking your driveway. Successive mental images link each class with a place on your list. Thus you could then imagine a tuba in the lobby, and a pile of integrated circuits tumbling out of the elevator when the door opens. For your lunch appointment you could imagine Mary inside a giant hot-dog bun in the hallway outside your room. For your afternoon classes you could imagine a beaker heated by a Bunsen burner spewing out a cloud of fumes inside your closet, and the entire Bolshoi ballet troop cavorting on your bed. Figure 6-8 is an artist's rendition of these images. To remember your Monday classes, all you have to do is to mentally walk into your dorm room being careful to look for each image as you pass by. This method really works quite well. Laboratory experiments with college students found that people using the method of loci did from two to seven times better than a control group memorizing the same material but without any special instructions on how to remember (Bower, 1972).

The method of loci is effective because it supplies strong retrieval cues that help you find information in memory. Memorizing a list without any special techniques is like dumping a few books on the floor of

Figure 6-8. Mental images utilizing the method of loci.

a library that is so overcrowded that books are falling off the shelves. It is hard to find the right books when you need them later. The method of loci provides an organized set of shelves or pigeonholes where memory items can be neatly stored. The loci tell you what shelves to look at to find memory information. There are several other effective mnemonic techniques that are based on retrieval cues. For a description of some that have been validated by research see Higbee (1977).

ATTENTION

As a technical term, attention has two opposite meanings. First, it refers to our ability to ignore extraneous events and to focus in on one class of important events. We call this selective or focused attention. Your ability to maintain a conversation in the midst of a noisy cocktail party where you must focus on one person and ignore the din around you is one example of selective attention. Second, it refers to our ability to do more than one thing at the same time. We call this divided attention or timesharing. Your ability to watch television and do homework simultaneously is an example of divided attention. Since the same theoretical models are often called on to explain both focused and divided attention, it is important to distinguish between these two kinds of attention in order to avoid excessive confusion. (A limited amount of confusion should not disturb you excessively but an excessive amount will limit your comprehension.)

Selective Attention

The impetus for current interest in selective attention can be traced back to a problem faced by air traffic controllers in the early 1950s. At that time, controllers received messages from pilots over loudspeakers in the control tower. (Now headphones are used except at very small airports.) Hearing the voices of many pilots from the same loudspeaker was a task that controllers found difficult. So Broadbent (1958) recommended multiple loudspeakers so that each pilot's voice could be mapped to a unique spatial location. This change aided the controller and resulted in better performance. It was easier to pay attention and to select the proper messages with several loudspeakers.

Since it is difficult to obtain precise scientific data at cocktail parties and in control towers, a laboratory version of these real-world situations was rapidly developed to study selective attention. This is called the dichotic listening task and is illustrated in Figure 6-9. Separate messages are presented to each ear. The listener is required to repeat the message from only one ear. This repetition is called *shadowing* a message. It guarantees that the listener is paying attention to the message. Recall of

Figure 6-9. This figure shows correct performance when a person is required to shadow the right ear in a dichotic listening task. A common error would have the person saying "mice eat cheese."

the other message—the one that was not shadowed—is usually quite poor.

Results from the dichotic listening task—as well as many other tasks we do not mention (see Broadbent, 1958)—helped Broadbent devise his famous limited-capacity channel model (Figure 6-10). This model has been the guiding light for much information-processing research for over two decades and its importance cannot be overstated. While the model is showing signs of age (Broadbent, 1971) so that many of its details require alteration, the basic conception of the human represented in Figure 6-10 has proved to be accurate. Even more importantly, the idea of representing the human by a model that specifies internal information flow is now so completely accepted that it is hard to realize that in the 1950s it was a true scientific revolution. We will only discuss highlights of the model; see Broadbent (1971) for details. Information enters the senses and then goes directly to short-term storage. (Sensory storage was not understood in 1958. Broadbent's model was responsible for the recent explosion of research on short-term memory that has occurred since 1958.) A selective filter limits the information that can leave short-term memory. This filter was supported by the dichotic listening results already mentioned. The filter could be set or tuned to accept messages from some particular channel, say, the left ear or perhaps high-pitched sounds of a female voice as opposed to the lower pitch of a male voice. This information would get through the filter and other information would be rejected and lost. So the poor recall of unshadowed information in dichotic listening was explained by the inability of the unshadowed message to pass through the selective filter. The filter protected the limited-capacity channel from being overloaded. (More on this part of the model soon.) Because the filter preceded the channel, this class of model is known as an early-selection model. Selectivity precedes detailed analysis of incoming information in early-selection models.

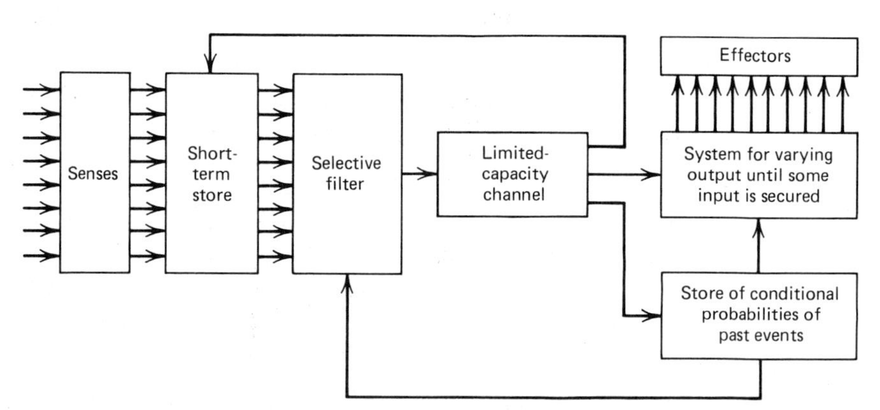

Figure 6-10. The original early selection, limited-capacity channel model. (From Broadbent, 1958.)

The original filter mechanism soon came into disrepute as new data revealed inconsistencies. For example, information in the unattended ear could be processed when it was high-priority speech: the listener's name in the unattended ear would occasionally be noticed even though the filter should have removed it. Similarly, grammatical constraints as in a sentence that started in one ear, switched to the other, and then returned to the original shadowed ear, would be recalled entirely even though the filter should have rejected parts of it. Thus, other theorists proposed that the filter merely attenuated or weakened information rather than eliminating it entirely (Treisman, 1969; Treisman & Gelade, 1980). Other theorists rejected the entire class of early-selection models and have argued that all incoming data makes contact with long-term memory (Keele, 1973; Norman, 1968). Broadbent himself modified the filter mechanism extensively (Broadbent, 1971) to allow selectivity at early and late stages of information processing. So there is less than complete agreement about how selectivity is imposed by various filters. This issue is less important now because newer models of attention do not rely as heavily on explicit filter mechanisms (Duncan, 1980; Kahneman, 1973; Kantowitz & Knight, 1976; Navon & Gopher, 1979). These models emphasize divided attention, to which we now turn.

Divided Attention

The basic tenet of the limited-capacity model is that the human can transmit information only at a finite rate. (Information theory was discussed in the preceding chapter.) People can only do tasks that require a small number of bits of information to be transmitted each second. If the task requires too many bits per second, people will make errors thereby reducing the amount of information transmitted each second. This basic principle is a truism in human factors, contained in virtually all handbooks and texts. While the limited-capacity-channel model is still a reasonable first approximation of human capabilities in most tasks, we will see that recent findings show the bottleneck that is represented by the channel to be far more elusive than previously thought. Ironically, years of research effort aimed at localizing the bottleneck in information processing have decreased the accuracy of the basic construct.

The limited-capacity model states that tasks can be successfully combined as long as the total information rate of the channel is not exceeded. So, if the channel can transmit 12 bits per second and a task requires only 9 bits per second there is 3 bits per second of spare or excess capacity available. If we add a second task that requires up to 3 bits per second, there will be no performance deficit and the two tasks can proceed together. But if we add a task requiring more than 3 bits per second, then the channel will be overloaded and performance on at least one of the two simultaneous tasks will suffer. Thus, we can discover how much capacity the channel has only by imposing an overload on the human information processor. This technique is widely used in engineering where, for example, metal alloys are placed in hydraulic presses and subject to increasing pressures until the metal fails. Overloading a system often can provide valuable insights into how the system operates under normal loads. Of course, with humans a gentler method of imposing this overload is required and hydraulic presses are prohibited. Such overload can be generated in two ways. First, we can keep the amount of information constant but speed up the rate with stimuli occurring more and more rapidly. As the time between successive stimuli decreases, overload is increased. Second, we can maintain the time parameters or speed of a task but can add information either directly to the primary task itself or by imposing additional task demands to be performed simultaneously. Both these techniques have been used extensively to study divided attention.

Psychological Refractory Period The major advantage of laboratory research is control over the experimental environment. While life outside the lab is complex and messy. the controlled laboratory environment allows the scientist to simplify and refine the experimental situation so that only, or at least primarily, factors of direct interest influence experimental outcomes. Overloading the human by increasing the rate of incoming stimuli requires a minimum of two successive stimuli. While a real-world environment consisting of only two stimuli would be rare, this is the optimal number of stimuli for laboratory study. Any additional stimuli only increase complexity without expanding the potential for gaining knowledge and understanding. In the psychological refractory period task, two stimuli are presented in close temporal succession, usually less than 500 msec apart. The time between the two stimuli is called the interstimulus interval (ISI) and as it gets shorter the stimuli occur closer and closer together. An ISI = 0 means that the two stimuli occur at the same time.

The earliest study of the psychological refractory period (Telford, 1931) found that reaction time to the second signal was considerably delayed. It was as if the brain turned off after receiving the first signal. Thus, the phenomenon was called the psychological refractory period by analogy to the refractory period of a single neuron that will not respond to a second input when it is stimulated by signals that arrive close together. While this explanation is no longer believed (see Kantowitz, 1974a), the descriptive name has remained.

As the ISI is decreased, the human is more and more overloaded. This can be measured by recording reaction time to the two stimuli (Figure 6-11). When the two stimuli are far apart—at large ISIs—reaction time is fast. As the ISI gets smaller, reaction time increases. Reaction time is often interpreted as an index of load or capacity (but see Broadbent, 1965 for possible dangers associated with this line of reasoning). Thus, at short ISIs the human is badly overloaded. The limited-channel model explains the delay in reaction time to the second stimulus as being due to

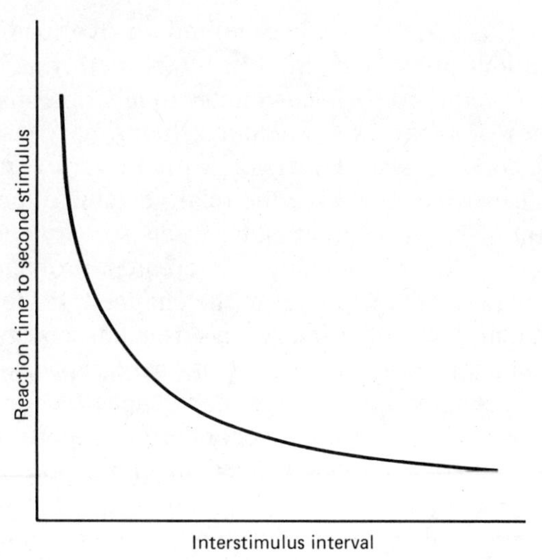

Figure 6-11. The psychological refractory period effect.

processing of the first stimulus. Until the first stimulus clears the channel, the second stimulus cannot be processed with normal efficiency. The overlap between first and second stimuli is greatest at shorter ISIs, so reaction time to the second stimulus is most delayed when the two stimuli occur close together.

However, later research discovered that there is also a delay to reaction time for the first stimulus (Herman & Kantowitz, 1970). This finding cannot easily be explained by the limited-capacity channel model without making complicated assumptions about interruption of processing. Since the channel is empty at the time the first stimulus occurs, there should be no reaction time delay. But data show that reaction time to the first stimulus also decreases as ISI increases, although first reaction times asymptote before second reaction times (Kantowitz, 1974b). Newer models are superior to the limited-capacity channel model because they can explain reaction times to both first and second stimuli, whereas the limited-channel model can only handle reaction times to second stimuli in the psychological refractory period task. A review of models and results in the psychological refractory period and related paradigms can be found in Kantowitz (1974a).

Timesharing While results from dichotic listening and psychological refractory period tasks originally provided much of the support for the original limited-capacity channel model (Broadbent, 1958, 1971), we have seen that later research in these areas caused the model to be modified and/or questioned. But it was results from timesharing studies that drove the final nail into the camel's back, to mix a metaphor. The timesharing task is yet another method for overloading the harried human information processor. Simultaneous performance of two independent tasks is required. It is customary to label the main task—that is, the task of interest to the experimenter—the primary task and the remaining task the secondary task. However, without very explicit instructions such as payoff matrices that give the relative value of performance on each component task, this distinction between primary and secondary tasks can be arbitrary; this distinction has created problems in measuring mental workload (mentioned later in the chapter). If the capacity demanded by a particular level of primary task remains constant, performance on the secondary task is an indicant of spare capacity. So as the primary task level is manipulated to use more capacity, for example by filling more slots of working memory, secondary task performance should get worse. This technique is widely used in human factors where degraded secondary-task performance is regarded as an index of mental workload. But before discussing workload, we first must understand how the limited-capacity model makes predictions for the timesharing task because this model underlies much of the human factors research on mental workload.

To aid our explanation we will assume some arbitrary capacity demand for our primary task. This primary task has two levels: 6 bits per second and 8 bits per second. The human channel is assumed to be limited to 10 bits/sec. Therefore, either version of the primary task can be performed alone without decrement. We now add a secondary task that requires 4 bits/sec. If we assume that people first allocate their limited capacity to the primary task and then use what is left over for the secondary task, we can generate predictions for the limited-capacity model. (This assumption is not always correct and the rules for allocation of attention or capacity are not well specified. See Kahneman, 1973, for an introduction to this issue.) The easy version of the primary task can be successfully combined with the secondary task since both tasks together require 10 bits/sec and this is the limit for our hypothetical human channel. But when the more difficult primary task is combined with the secondary task a total of 12 bits per second is required to perform both tasks. This exceeds available channel capacity so that performance will decline. If the primary task is protected by allocating capacity to it first, then secondary-task performance will decrease while primary task performance remains constant. When we put these predictions on the same graph (Figure 6-12) we see that the basic prediction of the limited-capacity channel model is an interaction between task difficulty (6 or 8 bits/sec) and whether the primary task is performed alone or with the secondary task. For the easy primary task, we get a horizontal line while the line for the hard primary task slopes down. (Any departure from parallel lines indicates an interaction.) Although we will not work through the logic, the same interaction is also predicted for experiments where the secondary task has two levels and only dual-task performance is plotted.

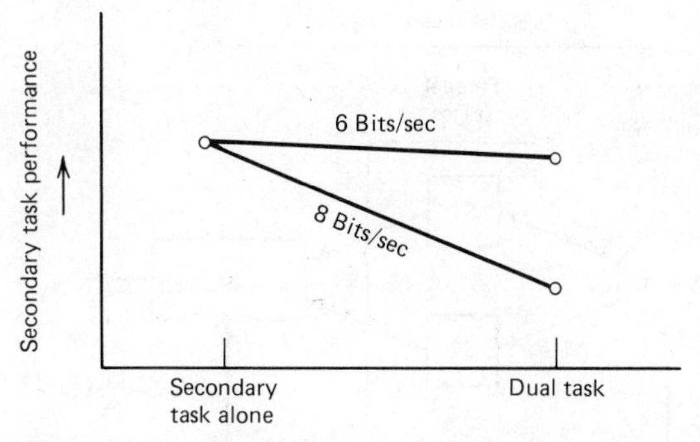

Figure 6-12. Predictions for the limited-capacity channel model when two tasks are combined. Primary-task performance is not shown but is assumed to remain constant.

Finally, it is the interaction that is important rather than only the horizontal line in Figure 6-12. As long as the line for the difficult task falls more rapidly than the line for the easy task, the limited capacity channel is supported even when the line for the easy task is not quite horizontal.

Early experiments supported this prediction of an interaction. For example, Herman (1965) combined an auditory judgment task with a tracking task and obtained the basic interaction of Figure 6-12, although other parts of his results suggested minor modifications to the Broadbent (1958) model. But later efforts found combinations of tasks where the predicted interaction failed to occur. For example, Allport, Antonis, and Reynolds (1972) combined auditory shadowing with sightreading of music. These two tasks could be combined with little loss of efficiency in either, contrary to predictions of the model. This was a major blow and indeed Allport titled his article "A Disproof of the Single-channel Hypothesis." Allport suggested that the human has several independent channels or processors that operate in parallel as an explanation of his results. This is a dramatically different suggestion and not all researchers accept the idea of multichannel operation. Many other new models have been created to handle this failure of the limited capacity channel. (See Lane, 1981, for a review.) We will discuss one hybrid model created to solve this problem, but it is far too early to state which model will become the accepted successor to Broadbent's explanation of attention.

A hybrid model falls in between the limited-capacity channel and the multiprocessor models. It has some stages that operate in parallel and others that are in serial (Figure 6-13). This model was used to explain findings in a timesharing experiment that combined Fitts' law tapping (see Chapter 5) and digit naming (Kantowitz & Knight, 1974b, 1976). Results of these experiments showed the predicted interaction when single-task performance was contrasted with dual-task performance. But within dual-task performance, the size of the decrement did not increase as the component tasks became more difficult or more complex. This latter

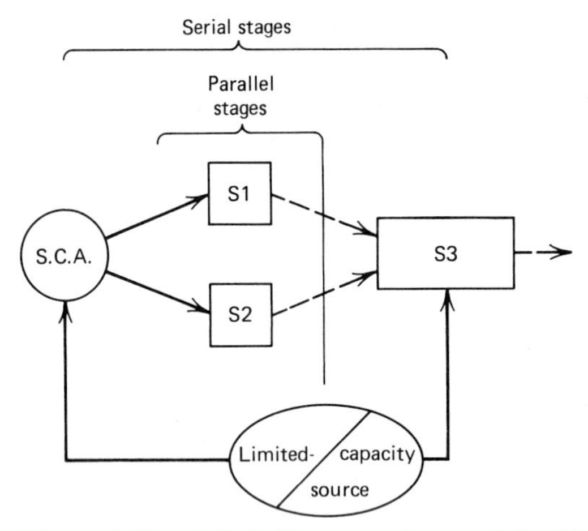

Figure 6-13. A hybrid processing model with serial and parallel stages. The parallel stages represent early or perceptual information processing. They are controlled by a static capacity allocator (S.C.A.) that divides capacity between them. (From Kantowitz & Knight, 1976.)

finding contradicts the limited-capacity channel whereas the former finding rules out independent multiprocessors. The hybrid model retains a limited-capacity source but places some contraints on how this capacity can be allocated. Without going into technical details, we merely state that the parallel stages account for the dual-task results of additivity (no interaction), while the serial stage (S3) that represents response processing accounts for the interaction between single- and dual-task conditions.

The original limited-capacity channel model was a milestone in human information processing research. It has been greatly modified (Broadbent, 1982) and even rejected by more recent findings (Kinsbourne, 1981). But no model is eternal and it is a pleasant fate for a model to be swallowed up in the rush of research it generated. By any standard, the limited-capacity channel was a resounding success. In its earliest form, it placed the bottleneck of information processing, the filter mechanism, at an early stage. However, this bottleneck has tended to retreat into the organism as more research is accomplished. For example, the hybrid model as well as other recent models (see Kantowitz, 1974a and Lane, 1981 for reviews) allows information to enter in parallel with no bottleneck. The bottleneck occurs only when responses must be emitted. This type of model has strong implications for human factors design. It suggests that people are very good at taking in information but slow down when many responses must be made. So control panels should be arranged to minimize the number and frequency of required control activations because limiting responses will help the operator more than limiting perceptual display information. Note that this design suggestion is more specific than that implied by the limited-capacity channel (Broadbent, 1971) in its revised form, which would limit both input and output loads far more equally. So as models improve, so do design suggestions.

The only certainty in human information processing research is that models change rather rapidly. The human factors specialist must keep abreast of the latest findings by at least reading review articles (e.g., Lane, 1981) as they are published. Communication between basic and applied researchers is at present a weak link in the chain of knowledge (Kantowitz, 1981) that can be strengthened only by deliberate efforts.

Mental Work Load

As our society becomes more and more technological (see Chapter 1), more and more workers are engaged in mental work rather than physical work. Of course, even such physical work as digging ditches requires some mental effort to guide the shovel but most would agree that the mental requirements of most modern jobs exceed the physical requirements. An office worker or executive seated at a video display terminal (see Chapter 11) does some physical work pressing keys but this part of the job is less taxing than the cognitive effort that determines which keys should be pressed. Physical workload and energy expenditure can be measured fairly accurately by such means as oxygen consumption and heat generated by the body. Since more and more work is based primarily on mental, rather than physical, work, considerable effort has gone into establishing methods for measuring mental workload. The ultimate goal would be a single number that describes or predicts how much mental work load is associated with any given task. Unfortunately, it is not likely that this goal will ever be accomplished (Johanssen, Moray, Pew, Rasmussen, Sanders, & Wickens, 1979). Mental work load is too complex a concept to be summarized by a single number.

A recent NATO conference on mental work load (Moray, 1979) concluded that there is no single definition of it. Many operational definitions were offered depending on the background of the individual researcher. Good arguments can be made for relating mental work load to information processing and attention (experimental psychologists), time available to perform a task (system engineers), control engineering (electrical and industrial engineers), and stress and arousal (physiological psychologists and physiologists). Behavioral measures of mental work load have been classified into 14 tasks or methods divided, as was Caesar's Gaul, into three categories (Williges & Wierwille, 1979): subjective opinions, spare mental capacity, and primary task.

Subjective opinions can be collected either by rating scales or by questionnaires and/or interviews. A rating scale provides a psychometric technique for ordering opinions in a mathematically consistent manner whereas an interview or questionnaire obtains data that are not quantified. The most popular rating scale is the Cooper-Harper scale used by test pilots to quantify how well aircraft handle (Cooper & Harper, 1969). More recent efforts at subjective assessment of mental work load (Eggemeier, 1981) have used a mathematical procedure called conjoint measurement to obtain a work load scale (Reid, Shingledecker, & Eggemeier, 1981). While this work is at an early stage of development so that not all the psychometric assumptions have been validated empirically, it does appear to be promising. Multidimensional scaling is another approach that has been used (Derrick, 1981). Perhaps this concentration of effort will soon produce a reliable and valid subjective scale of mental work load (Moray, 1982).

Spare mental capacity is the largest category. It is based on the assumption that attention is limited and that the human channel has an upper bound (Rolfe, 1971). Measures of spare capacity have been derived from information theory (Senders, 1970) and assorted computer simulations. But by far the most data have been collected using the secondary task paradigm described earlier in this chapter (Ogden, Levine, & Eisner, 1979). Since human factors specialists have only recently become aware of the limitations of the limited channel model (Sanders, 1979) that imply there is no universal secondary task (Pew, 1979), much of these secondary-task results are difficult to interpret. While a few researchers have tried to use theoretical models to guide selection of secondary tasks (Wickens, 1979; 1980), most have used arbitrary combinations of primary and secondary tasks that prevent generalization of results beyond the particular combination of tasks selected. Since it is not practical to perform experiments with all possible combinations of the wide variety of tasks used to assess spare mental capacity, it is clear that progress in this area will depend more on new theoretical developments than on acquisition of more data based upon the incorrect limited- or single-channel model.

The primary-task method is based upon the assumption that increasing mental work load will cause a decrement in primary-task performance. Indeed, one researcher goes so far as to state that if the mission was successfully completed, there was no overload (Albanese, 1977)! This approach may be successful if work load is indeed quite high. But analysis of a primary task that is associated with low to medium mental work loads is often too insensitive to reveal effects of additional tasks or procedural alterations.

FOCUS ON RESEARCH

Can Brain Waves Measure Mental Work Load?

Since physiological measures of physical work load have proved so successful, it is only natural that researchers try to obtain physiological measures of mental workload (Wierwille, 1979). One important physiological measure that is related to attention is the cortical evoked potential. This is recorded by placing electrodes on the scalp or neck.

There are strong potential advantages for using brain waves as a measure of mental work load. First, unlike the secondary-task technique where primary-task performance can be changed by inserting the secondary task, recording brain waves does not alter the primary task. Second, no assumptions about the correctness of the limited-capacity channel model are required to interpret brain waves.

One recent laboratory study used a simulated air-traffic control task to see if brain waves were related to workload (Isreal, Wickens, Chesney, & Donchin, 1980). Subjects monitored a display that containing moving squares and triangles that represented aircraft. Work load was manipulated by varying the number of aircraft on the screen. There were two kinds of display changes that had to be detected. Course changes referred to a square leaving its old straight-line trajectory. Flash detection referred to a change in the brightness of a square. Triangles merely added noise to the display and were supposed to be ignored. In dual-task conditions, a second auditory reaction-time task was added. Results showed that auditory reaction time increased as work load increased, as would be expected. The more crucial question was: What happened to the brain waves?

These results are shown in Figure 6-14. These graphs show a particular portion of the evoked potential, called the P300 component, for all eight subjects (S1 to S8) in the experiment. The P300 waves look different for the three work load conditions (eight display elements, four display elements, and only auditory task) and this was confirmed by statistical analyses. In general, P300 effects agreed with secondary-task reaction time effects. Furthermore, recording the P300 wave did not disrupt primary-task performance. So this study supports the use of brain waves to measure mental work load.

However, not all studies reach this conclusion. Brain waves of this type are elicited by appropriate stimulus events, in this case the flashing of the squares. It is not always easy to find an appropriate stimulus that fits in with the task, especially outside the laboratory. Another study that examined evoked potential in a working environment was less successful (Wastell, Brown, & Copeman, 1981). Brain waves were recorded from telephone switchboard operators who were carrying out their normal duties. The ring tone, which indicates that the party's phone is ringing, that operators hear in their headsets was used as the stimulus event that triggered the evoked potential, just as the flashing square was used in the air-control study. This is a normal stimulus that is already part of the job and hence should not be expected to create any attentional artifacts. Two different types of switchboard were studied: the old kind where operators plugged in cords and the newer cordless model. Although there was a tendency for the brain wave N1-P2 component to be higher on the cord board, this effect was

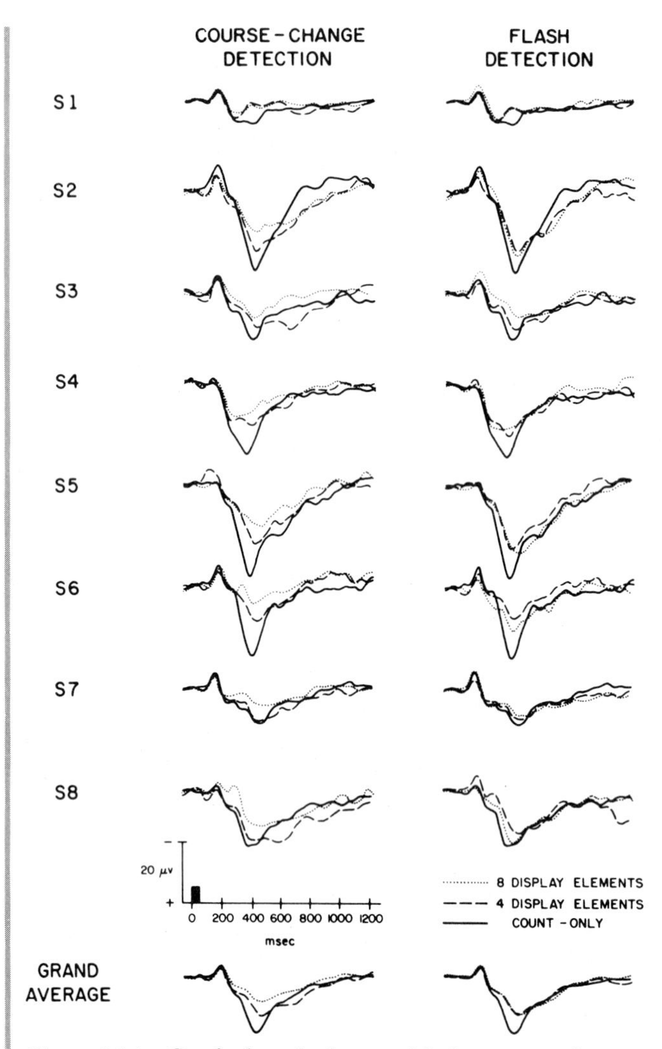

Figure 6-14. Cortical evoked potentials for course-change and flash detection tasks. (From Israel et al., 1980. Copyright 1980 by The Human Factors Society, Inc. and reproduced by permission.)

not statistically reliable. So this study failed to demonstrate an attentional difference between the two types of telephone switchboard.

There are many apparently slight differences in techniques used to record and score brain waves that may explain why some studies find reliable effects and others do not (Isreal, Wickens, Chesney, & Donchin, 1980). While some physiological techniques such as evoked potentials appear to offer promise, a review of these methods concluded that they are not yet sufficient to measure work load by themselves and should be combined with behavioral measures (Wierwille,

1979). Perhaps when physiological researchers agree on exactly which brain wave component should be recorded to reflect attention, and exactly how it should be scored, then more uniform results can be expected. But lacking identical procedures and a theory that relates brain waves to behavior we cannot yet use evoked potentials to define mental work load.

SUMMARY

Human behavior is determined by the internal flow of information. This flow of information cannot be observed directly and so must be inferred. Many competing models try to map internal information pathways and there is seldom universal agreement as to which model is best. Indeed, even when a majority of researchers agree on the virtues of some particular model, this agreement is transient and dissolves when newer models are presented. The rapid progress (or at least rapid change) in "approved" information-processing models makes it difficult for the human factors specialist to keep abreast of current findings.

Information-processing models are based on serial, parallel, and hybrid arrangements of internal stages. However, structure by itself does not completely specify performance. We must also know (or assume) how the internal system allocates resources or capacity.

The dominant model of human memory identifies three distinct storage subsystems. Sensory storage holds information from the eyes and ears for a brief time. Working (or short-term) memory is used to maintain information for longer periods. Rehearsal maintains information in short-term memory. If rehearsal is interrupted, information can be lost. No rehearsal is required to maintain information in long-term memory. There is so much information stored in long-term memory that retrieving items is not always possible. Techniques aimed at improving recall from long-term memory use mnemonics that help provide retrieval cues.

Selective attention refers tò your ability to focus on source of information embedded in a set of several sources. Early models of attention attributed selectivity to a perceptual filter, but current models contain selective mechanisms that operate after perceptual processes have been completed. Divided attention refers to your ability to perform more than one task simultaneously. The limited-capacity model states that tasks can be combined successfully only if the total information rate of the channel is not exceeded. While this model is still important, if only as an approximation that can be used by the human factors specialist, new results have demonstrated that it is incorrect in many details. A variety of replacement models have been suggested, but it is doubtful that any of them will have the powerful influence once held by the limited-channel model. These new models have helped identify methodological limitations in human factors studies that attempt to measure mental work load.

PART THREE

HUMAN-MACHINE INTERFACES

7 VISUAL DISPLAYS

J. R. BUCK

Program in Industrial and Management Engineering
The University of Iowa
Iowa City, Iowa 52242

"Hi! I'm Jim. Who are you?" appears on the pad of paper. Immediately below it reads, "Nice to meet you Jim. I'm Marie."

The pilot read the altimeter. Turning to the copilot she said, "At 1290 meters we have more than enough altitude to clear these mountains." Just then grey rocks penetrated the craft.

In both situations above visual displays were used. The notes on the pad of paper and the dial in the aircraft cockpit are both simple examples of visual displays. While the situations and consequences of these examples differ greatly, both of the above examples illustrate that visual displays are used in communication between two people or between the aircraft and the pilot. In fact, the purpose of displays is to communicate. Since both Jim and Marie had extreme hearing problems, they elected to write notes as visual displays. In flying, there are no altitude signs to show how high the aircraft is and clouds often obscure the ground or nearby mountains. Therefore, instruments are used to communicate important information to the pilot. It was a tragic human error when the two pointers on the altimeter were misread as 1290 meters (m) instead of 912 m; 88 m short of the mountain top. Visual displays can miscommunicate if they are not designed for the people who read them. Ask Jim and Marie.

COMMUNICATIONS AND DISPLAYS

You communicate with other people and with machines for purposes of entertainment, learning, and performing a job. Letters, books, television, and electronic devices (e.g., games and calculators) are typical examples of such communications. When thinking about the communication process you recognize that there is both an information sender and a receiver in a message transfer. Following this transfer the receiver processes this information based on knowledge, objectives, and the situation at hand. When this processing is completed, then the receiver usually changes roles and becomes a sender of information back to the original source. That single information-exchange loop may complete the communication or additional exchanges may occur. A single exchange loop is illustrated in Figure 7-1 as an extension to the information theory concept shown earlier. The purpose of abstracting this loop of the communication process (Figure 7-1) is to consider various needs of the process whether it is people-to-people, machine-to-people, or machine-to-machine communications and the role of displays in this process.

Displays are devices, no matter how simple, which are used by the information sender to communicate with a human receiver. This page of this book is a display. Those devices used by human information senders to machine receivers are called "controls" (a sometimes elusive attempt to infer human supremacy). Control devices, such as electric light switches, and their design philosophies are discussed in Chapter 10. The principal distinction between displays and controls is in the capabilities of the senders and receivers of the information to be communicated. Most control devices are also displays because they typically provide information to people on the condition of the control such as whether the switch is on or not.

The following discussion first starts out on the theme of what considerations need to be made in obtaining effective communications with or without visual displays. A time-tested axiom in human factors is that an analysis of communications needs to be made before starting on the design of the system and this discussion provides some basis for this analysis. Next the discussions focus on various forms of visual displays; how they differ and how they work. Finally the focus is on human performance in various criteria with various forms of visual displays. Necessary fea-

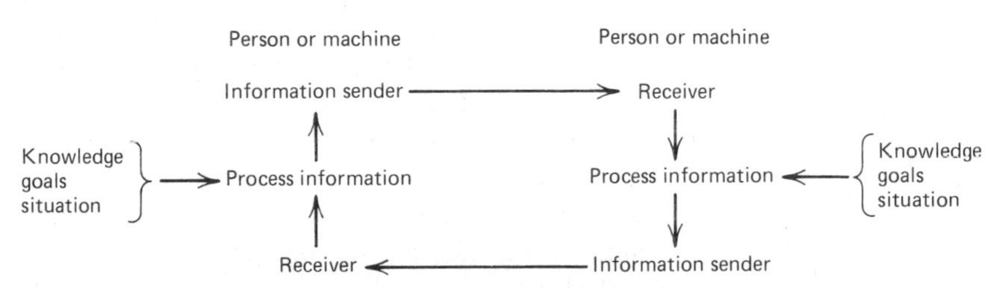

Figure 7-1. Schematic of communications process.

tures dealing with visibility, distinguishability, and interpretability are covered.

Effective Communications

In order to provide a basis for viewing issues on the proper selection of visual displays in human factors design, this and the next few subsections are addressed to important features of effectively communicating with people. Since the role of displays is to communicate, displays need to be chosen to meet necessary conditions of communication, provide the types of information needed, be relevant to the situation, and fit the people who need or want the communication.

A first requirement of communications is compatibility to the senses of the receiver. Visual messages, as on this page, audio messages, the touch of the hand, body language, or a kick in the shin under the table are different forms of sensory message forms. A 0- to 5-microvolt (μV) electric signal is not within the range of the human senses and therefore an inappropriate medium for human communications. The second necessary condition for communications is language compatibility. Any message sent that is not in a language known by the received due to experience or training will not be received. The dots and dashes of Morse code are meaningless to a person who can not first translate the signals to letters and the letters into meaningful words. Even after the language compatibility is assured, a third necessary condition for human communication is intelligence compatibility where the words are translated into meaningful and relevant thoughts by the receiver. Going beyond the necessary conditions of communications to achieve highly effective human communications involves such issues as providing the right type of information at the right time, meeting relevance needs, and matching the form of message to the specific group of human receivers. Information that does not meet these issues is redundant, clutter, noise, or misleading.

Types of Information

As a student of human behavior, you ought to notice what people talk about under different situations so that you can design the natural information forms needed by people to do different jobs. In the current situation I attempted to transmit *instructional* information, that is, telling someone how to do something. Two other closely related types of information are *command* and *advisory* information. Commands are direct orders of required actions or constraints. Advisory information is either recommendations of actions or notice given for warnings or caution so that someone can prepare and plan for an action. Other types of information transmitted to people are *answers* to direct *inquiries*. Some inquiries relate to the identification of objects. Other inquiries are about the status of variables such as the outside air temperature, the time of day, or the speed of the car. The expected specificity of precision of the answer to the inquiry is, however, highly situation dependent. Another general type of information is related to conditions in the past or future. These types of information may be respectively noted as *historical* and *predictive*. All of these types of information occur in communications between people as well as communications between machines and people.

Relevancy with Receiver's Goals, Model, and Situation

Effective communications contain information which is relevant to the receiver's goals. If I told you that the Washington Monument in the District of Columbia (United States) is 555 feet 5.5 inches tall (169.3 m), then you would likely say, "So what!" That is, unless that is the answer to the next question on the examination you are writing. Relevance depends on current roles, goals, and situations.

Another feature important to communications is that the information sent to a human operator is compatible with the operator's viewpoint of the system, commonly referred to as the *operator's model*. An auto driver who views the engine to be analogous to a wound-up rubber band does not need to know the oil pressure. Race drivers, on the other hand, view the engine's operations in a highly technical fashion and often demand to know the status of many other operating details including the oil pressure. This race driver example also illustrates that the operator's model is shaped by roles and goals as well as training and experience.

The *situation* effect should also be emphasized. Flying has often been described as hours of boredom punctuated with moments of stark terror. Communications during the boring period usually consist of system monitoring and flight planning, perhaps intermixed with fantasies. Information desired during the system monitoring is largely on checking to see if everything is operating satisfactorily and making qualitative inquiries of status. In those moments of stark terror the operator's situation changes abruptly and so does the type and precision of information needed. It is at these times when the consequences of inadequate speed and accuracy are so severe. Another important point about situation effects on communications is that changing situations can alter the operator's model. Changing situations are particularly noticed in flying when there are orientational changes to the pilot.

Populations of People Receivers

As with all communications, the receiver's viewpoint is most important because of language and relevancy issues. In different situations there are different populations of people who would receive the information. Displays in elementary schools serve young children whereas highway signs serve adults. Although both those circumstances have somewhat specialized audiences, there are wider variations between different people in each audience than information needed in a control pulpit of a steel-rolling operation or inside of an army tank. Identifying the user population is essential to the selection and design of visual displays.

Beyond the obvious language issue in communications, there are subtle but important considerations in effective information transfer. One of these issues is known as *population stereotype behavior,* which refers to long-term habits and well-ingrained knowledge that a particular population of people have. For example, most people from Europe and the Americas are accustomed to reading words and symbols horizontally from left to right. Other cultures in Asia and the Middle East have different long-term reading habits. Sequential information presented in one fashion is subject to misinterpretation by people of a different population stereotype behavior. Even in situations where the persons receiving the communications are well trained on a particular form of communications, emergencies arise that cause people to revert to old habit patterns. Another subtle point is that every population develops colloquial expressions beyond the basic language that can alter the intended meaning of a message dramatically. A foreign automobile maker, exporting autos to the United States, who named a particular car model "Lemon" would be apt to find sales disappointingly low due to unintended communications.

The preceding discussions on effective communications show a variety of important issues that pertain to any communication. There is, however, an important difference between people-to-people and machine-to-people communications. Perhaps the most important difference is that machines are rarely adaptive to the human operator's informational needs and so it is critically important that the designers address the communications issues before selecting and locating forms of visual displays. Therefore a communications analysis needs to be made along with a task analysis so that provisions are made in the design to account for the issues of relevant and timely machine-to-people communications via visual displays.

Essentially a task and communication analysis is made by listing the activities that an operator performs as a sequence of tasks. Tasks may be described as finely as appears feasible. If these tasks are to be performed at specific locations, then the locations are also designated and the environment factors at these locations should be noted. Goals associated with each task are denoted. Any precedence between tasks is shown. The communications part of the analysis is achieved by defining the information requirements desired to achieve the task goals and subgoals including the precision requirements and the type of information. If there are time requirements on task performance, then that time data should also be noted. The population of people who are to perform the task with their variabilities in abilities should be further noted.

VISUAL DISPLAYS: DISTINCTIONS, SPACE, AND TYPES

Visual displays, as the name implies, are displays that require primarily the visual sense. Because of features of the visual process, as discussed in Chapter 4, there are specific advantageous characteristics of visual displays and constraints on them as well. Vision is directionally controlled by the eye movement and so display location is very important because visual displays that are not looked at, aren't seen. While that feature of vision is constraining, it should be noted that vision is the most developed of the human senses. Three physical dimensions can be viewed simultaneously; two easily with very high resolution (i.e., ability to perceive high amounts of detail). These features of vision makes this sense best for communicating *location* information and very good for compacting highly precise information detail. Since vision is a well-learned sense in growing up and for use in everyday life, it is easy to create (1) illusions of the third physical dimension, (2) pictorials that are easily recognized, (3) relationship diagrams, and (4) information of a historical or predictive nature; all of which can be readily interpreted by most people with little or no instruction. Visual displays are extensively used for many of these reasons.

Types of Visual Displays

The world around us is filled with many types of visual displays. In fact, there is such variety that a taxonomy of types is virtually impossible. Distinguishing names are given to various visual displays to signify differences within or between various forms of visual dimensions; displays that use other sensory modes of communication (e.g., auditory or tactile) are discussed elsewhere. One distinguishing dimension is whether the display is *static* or that it is *dynamic* due to some changing feature or features. Dynamic displays are further characterized by how the feature changes, *continuously* or *discretely*. Since continuously changing displays typically change analogously to some variable of the system, these displays are called *analogue* displays, such as a clock changes analogously with the passage of time.

Many displays are distinguished by the form of information to be communicated, as the earlier discussions in this chapter expounded. This distinguishing dimension of all displays is usually the most important in human factors design. Typical types of information include status, warnings, predictions, history, commands, advice, and instructions.

Another important dimension of display differences is the mode of visual display coding. Some of the variety of codings used in displays are alphanumeric symbols (single symbols, words, or abbreviations), pictures or pictorials of various levels of abstraction, or purely abstract codes such as shapes, colors, lines, or figures. This psychological dimensions varies from the conventional letter symbols of a language to the highly faithful pictures of objects by various degrees of abstraction as well as by sensory and perceptual characteristics such as color or perspective.

A further dimension for distinguishing display types is by the technology of the displays that translate the machine language of information or the message from another person to the receiver of that message. Visual display technology varies from simple signs to mechanical-chemical-electronic devices of varying sophistication. For example, thermometers are relatively simple mechanical-chemical devices that visually display the temperature variable. Warning lights are simple electromechanical devices. However, cathod ray tubes (CRTs), plasma tubes, and liquid crystal displays are technologically more sophisticated display devices. One of the purposes in distinguishing the technology dimension of a visual display is to denote the capabilities, or limitations, in what can be communicated in a given display or to denote the requirements of a given display technology on the system. For example, signs without mechanical features can not change with time but they are inexpensive and require little system support (e.g., electric power). On the other hand, CRTs have many capabilities in the type of coding, dynamics, and the types of information but they require considerably more physical space and system support.

Various types of visual displays are classified by these and other dimensions such as the shape of the display or orientation of the display device, the format of displayed symbols, or movement patterns of dynamic displays. Distinguishing names are given to distinguish these features, or to state someone's opinion as to the usefulness, accuracy, or quality of the device.

Display Dimensions and Space

Although sculpture is a true three-dimensional physical display, most visual displays present one or two *physical* dimensions for presenting information. A third continuous dimension of physical representation may be noted through binocular, auto-stereo-scopic or holographic devices (Mountford & Somberg, 1981), by perspective illusion or time-frame compressions where a series of pictures are quickly presented (Roscoe, 1981). These physical dimensions usually denote distance in space, time, a condition of a variable (e.g., temperature or pressure), or time derivatives (e.g., rate of change or acceleration). A single physical dimension contains a metric of the represented space (e.g., distance in meters, time in minutes, pressure in kilograms per square meter or speed in kilometers per hour). Within a single physical dimension there is a *minimum discriminable unit* of measure that is deemed to be the required maximum precision for an operator's task. There is also a *range* of values that this physical display dimension must show. When these units of measurement are uniformly scaled, then the ratio of the range to the minimum unit specifies the dimensional *space* of the scaled variable being presented. This space is analogous to the alphabet of a symbol. A thermometer with the range from −40 to 67°C, which can be visually read to the nearest 0.5°C, has a space of 214 units (i.e., $[67 + 40] \div \frac{1}{2}$). A two-dimensional display may consist of the above temperature space as a vertical axis and a horizontal time space of 120 minutes. The space of this two-dimensional display is the product of the two spaces or 25,680 temperature-time units. The physical space in analogue displays that is needed to communicate that information to a human depends on the visual acuity of the observer (see Chapter 4).

Symbolic verbal displays provide a set of symbols, such as alphanumeric symbols, where the minimum discriminable unit is each symbol. The range is the collection of different symbols or the combination of different symbol combinations. If there are M different symbols and N character locations in a display, then there are M^N possible symbol permutations in the display space. While there is a minimum amount of physical display space required to have each of the full set of symbols accurately recognized and distinguished from adjacent symbols, the potential information space of symbolic displays is typically far greater than for the analogue displays. However, symbolic verbal displays depend on the the receiver's knowledge of the symbols and their combinational meanings for effective communication.

Pictorial visual displays provide a figure on a display that physically corresponds to the space and object information being presented. With corresponding physical figures, pictorial displays do not require as much knowledge as do symbolic displays, but they are not as compact either. Accordingly, the physical space of analogue, symbolic, and pictorial forms of visual displays needed to communicate information of specified precision depends on the type of display chosen.

Physical Characteristics of Visual Displays

Hardware used to make visual displays takes on many forms. Light from the displays to the eyes of the observer is the principal feature of visual displays. Either this light is reflected from the display faces or it is projected by the display (e.g., a warning light or light-emitting devices such as LEDs or CRTs). Generally the light-projected displays have light-emitting elements that are simply turned on or off to create the displayed symbols and figures against some background. Since the physics of reflected light is different from projected light, so is vision, and these differences should be reflected in the design of these devices and their selection for particular applications.

Regardless of the way the display light reaches the eye, visual displays differ extensively in physical size, symbols used, directions and elements of movement, the scales used, colors employed for symbols and background, and many other features. For example, a one-dimensional analogue display with black symbols on a white background can:

1. Be shaped linearly in either a vertical or horizontal orientation, circular, semicircular, or some similar general form.
2. Have a fixed pointer and moving scale or a moving pointer and fixed scale.
3. Show a full or partial scale.
4. Present a wide variety of pointer and symbol designs.
5. Be scaled incrementally in units, logarithmically, or some other fashion with increases in any direction of the scale.

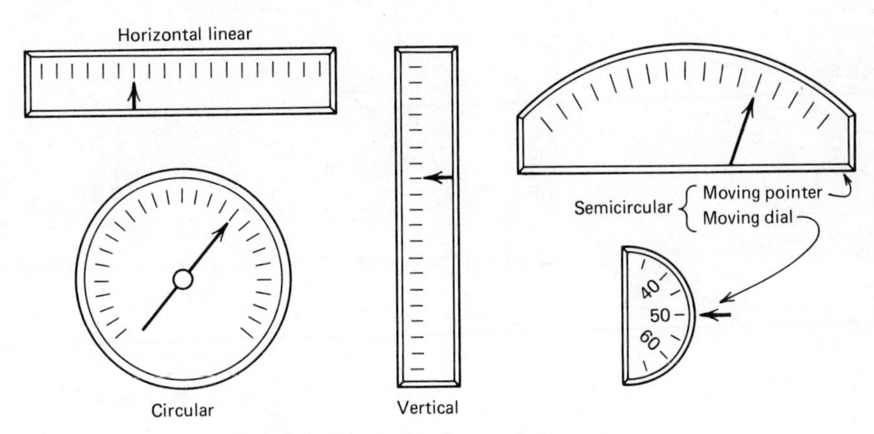

Figure 7-2. Examples of analogue display variations.

Figure 7-2 illustrates many of these various design options for this simple common form of analogue display. When you consider the other forms of physical features of visual displays mentioned above, the potential of an almost endless variety can appear almost overwhelming. Fortunately results from a lot of research make the problem of display design and selection less of a problem than it might otherwise appear.

Cathode ray tubes (CRTs) and other forms of electronic displays have some special characteristics that deserve particular mention because of their high use in computer and control systems. In the CRT an electron gun at the rear of the tube can emit directional impulses across the tube face. The most typical design plan for moving the gun is through first a horizontal scan at the top of the tube, then a small downward movement (a raster), another horizontal scan, followed by another downward movement, and so forth until the entire face of the tube is covered. It then starts all over again. When the gun is in a particular orientation, an impulse may or may not be fired. If an impulse is fired, it will strike a phosphor dot on the inside face of the tube and the phosphor will start to glow in a color and intensity that depends on the phosphor and the impulse. Phosphors glow for a time period and begin to fade. In order to make the emitted glow appear to be constant, the electron gun must return to that orientation and fire another impulse for regeneration before the phosphor fade is perceptible. With slower regeneration rates of the electronic gun and with phosphors that fade faster, there is a greater pulsation of the emitted light as illustrated in Figure 7-3*a* to *c*. Another important feature of CRTs is the size of the phosphor dots. Larger dots cause lumpiness in the light emitted as Figure 7-3*d* to *f* shows. Finer phosphor dots yield more raster lines for greater resolution in the picture quality, but this requires more signal control for the electronic gun as a price for this added quality. Current television CRTs have over 500 raster lines whereas higher-resolution television will have about twice as many lines.

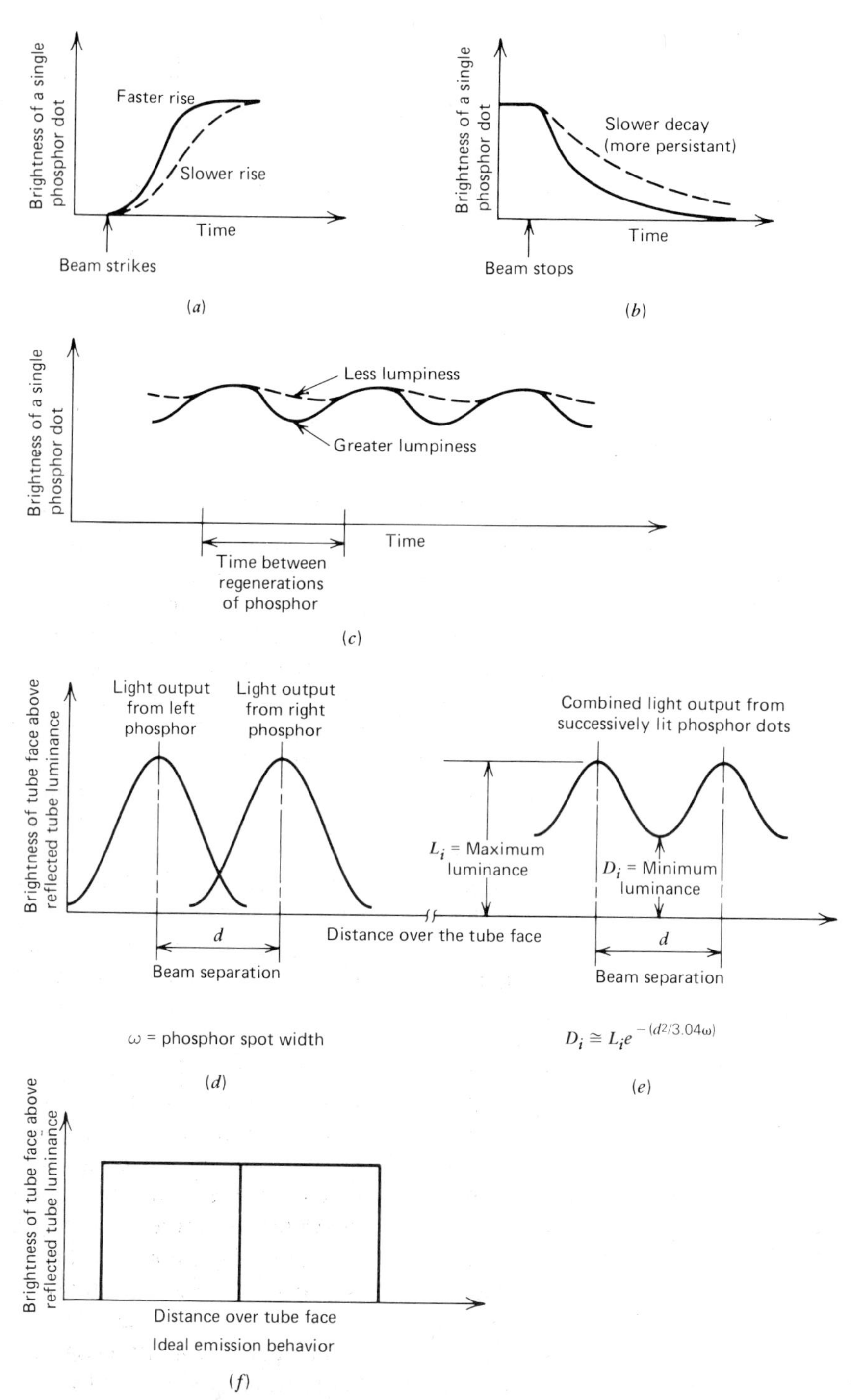

Figure 7-3. Cathode ray tube display characteristics. (From Gourd, 1968.)

Also different mixtures of phosphors emit different wavelengths of light to give different colors of emitted light. Colored CRTs have clusters of three types of phosphor dots where the emitted light color depends on the impulse strength on each dot in the cluster. A different signal control is employed in the electron gun for each color. The resolution quality of color CRTs is essentially the same as monochromatic CRTs with the same number of raster lines.

Human Performance and Visual Display Selection

Displays serve to communicate information and information is needed for decision making and performing the proper control action in a timely manner. Various features of human performance that contribute to the timely execution of correct actions in actual use are important human performance criteria. Speed and accuracy are obvious important criteria. Other criteria include (1) speed in learning, (2) comfort, (3) absence of fatigue in long-term use, (4) small variation effects due to individual differences in the population, and (5) performance stability under degraded environments and stress. The relative importance of these different performance criteria is highly dependent on the specific system under design. Fast-acting systems, such as aircraft, necessitate both high speeds and accuracies. Slower-acting systems, such as large ships and nuclear reactors, require very high accuracies, but small differences in the speed of communications are not highly critical. The good news is that visual displays have been highly studied so that much is known about speed and accuracy aspects of human performance. Moreover, visual displays that are usually read faster are typically read more accurately, reducing the need for making tradeoffs. The bad news is that little is known yet about the other criteria.

Three necessary steps to effective communication with visual displays were shown earlier to consist of the following properties: (1) *visibility* in that all critical elements of the displays are seen, (2) *distinguishability* exists between all parts and symbols of a visual display, and (3) *interpretability* of all display variations into appropriate actions. In other words, if one cannot see the display well enough, confusion will result in the meaning of the displayed information. If you get confused, you cannot consistently select an appropriate response. Fortunately many of the visibility and distinguishability features of visual displays are intuitively correct or follow from knowledge of the vision process (Chapter 4). Also there has been a great deal of experimentation performed to verify the points of common sense and to extend to features of visual displays that are not all that intuitively obvious. Some selected highlights of these issues are discussed below. Numerous other issues are covered in a variety of handbooks and textbooks on this subject, including (1) Van Cott and Kincade (1972), (2) Woodson and Conover (1966), (3) McCormick and Sanders (1982), (4) Murrell (1965), (5) Carroll (1976), (6) Engel and Granda (1975), and (7) Ostberg (1974).

Display Visibility Considerations

Many common-sense points about visual displays are evident from even superficial knowledge of the vision process and some fundamental thinking about display reading. First, a display that is not located within the field of normal vision will not be seen unless the operator is trained to specifically scan that display on a periodic basis or some alarm is used to direct the operator's attention to it. Since the normal range of vision is approximately $\pm$ 30° vertically and $\pm$ 80° horizontally about the typical line of sight, a display outside of that field stands a good chance of not being seen without training or alarming. Of course, operators do not have a fixed visual point of view all day, but the general principle here is that most people tend to have a typical focus direction that approximately specifies the field of view (e.g., the driver of a car) and that is the best place to locate visual displays. Some other principles on display location are discussed later.

Assuming now that a person is looking at a visual display, the question posed is, "How can we assure and improve visibility of that display?" One common-sense principle for display visibility is that displays should show clear correspondence and that elements do not block the visibility of other display elements. Pointers and numerals on analogue displays should align with the display scale so that visual interpolation is not required and that the pointers neither obscure the numerals nor the scales as illustrated in Figure 7-4*a* for both preferred and poor cases of visibility. Another common-sense principle pertaining to analogue displays is that the scale of a display itself describes order and range that should be clearly visible to the operator. Display scales that are not fully exposed do not show the scale range, and scales that are not marked orderly with sufficient symbols impose a need for the operator to visually scan disperse symbols, count markers, and/or to interpolate the indicated scale value. Figure 7-4*b* illustrates these cases. A third common-sense point in the proper design of analogue displays is that the smallest scale division required for the task must be visible to the operator. In order to achieve this visibility, the smallest scale width which must be recognized by the operator must be of a sufficient visual angle to meet visual acuity (Chapter 4) requirements. Results from dial reading studies by Murrell (1965), see Figure 7-5, show that both dial reading speed and accuracy deteriorate rapidly when the visual angle is much less than 5 minutes of visual arc. That is, when the minimum display scale distance is divided by the distance between the eyes and the display, then this angle tangent should be at least 0.0015. Visual angles of three or more times this tangent are typically recommended as a safety factor.

Regardless of the type of visual display used, display visibility is improved with greater contrast between the figures/symbols and their background (Cornog & Rose, 1967; Gould, 1968). It follows that any visual glare will reduce this contrast and the display visibility. An associated point of common sense is that black figures/symbols on a white background or the reverse polarity provide high contrast for *reflective* dis-

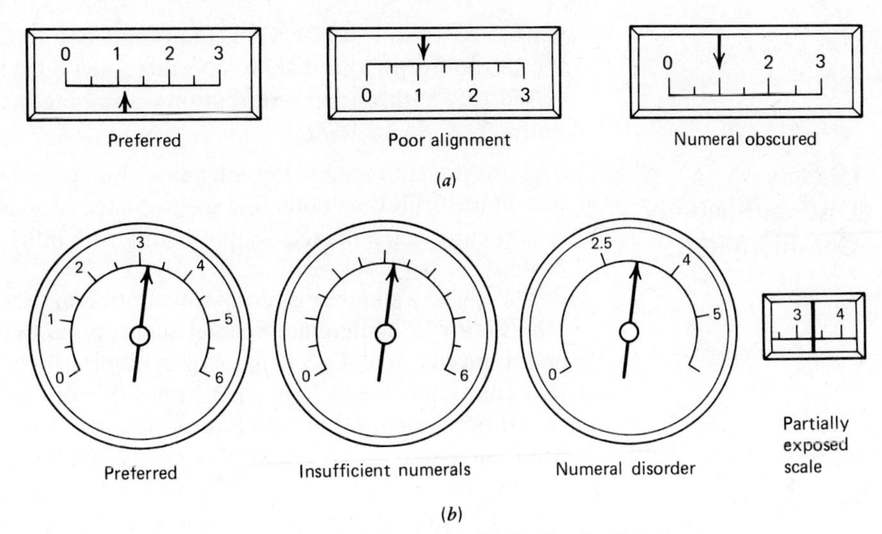

Figure 7-4. (*a*) Pointers in dial displays. (*b*) Scales in dial displays.

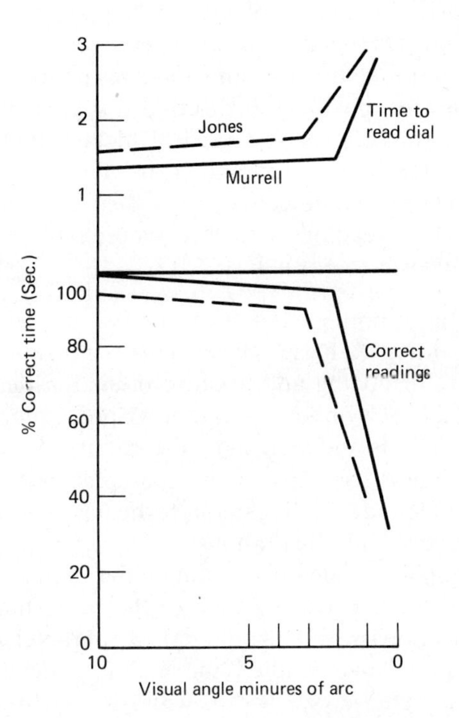

Figure 7-5. Display reading performance as a function of the visual angle.

play and such background/figure combinations are usually recommended; and more so for older operators (Olson & Bernstein, 1979; Sivak, Olson, & Pastalan, 1981). However, color is often desired for coding purposes and trade-offs between the addition or benefits of the color-coding effect and the loss of contrast should be considered (Carter, 1979;

Carter & Cahill, 1979). An additional common-sense consideration is the durability of the display contrast, particularly if the displays are exposed to weather (e.g., international shipping label signs, Konz, Chawla, Sathaye, & Shah, 1972).

Display Distinguishability Considerations

The problem of recognizing a display, often called display distinguishability, is more difficult than detecting the presence of a visual signal. Empirical test results can often enhance display distinguishability and lead to better display design.

There are a number of common confusion factors with visual displays. One factor is inadequate spacial separations between and within figures and symbols or between lines of symbols. In reflective visual display, a half character-height line separation typically suffices to keep visual line separations clear. Line separation, length, and other features of reading print on paper have been extensively studied through eye motion behavior. Fewer visual fixations and fewer reverse-direction eye motions without a loss in comprehension are indicators of improved reading variables (Tinker, 1958, 1963). When the display elements are not all in the same plane, parallax problems (i.e., alignment changes due to the angle of viewing) can result and so spacial separation must be increased to assure that this source of confusion does not result due to modest changes in the viewing position. Projective displays present a problem due to light-scattering effects and vertical flicker. Because of these problems, full character-height separations are typically recommended. Another desired feature in projected visual displays when there is considerable time spent in reading is to have screen tilt and brightness under the user's control (Lee & Buck, 1975).

Symbol Confusion

Symbols are another source of confusion due to a loss of distinguishability (Cornog & Rose, 1967). This confusion results because symbols are a form of coding and because of similarities between different symbols or lack of symbol resolution. Alphanumerical characters are well-learned symbols and as long as the combinations of these characters are familiar, then there is usually no coding problem. There may be a detectability problem depending upon the design. *Old English* and *old German* script are obvious illustrations of alpha characters that are difficult to distinguish, not only due to unfamiliarity but because of complexities and letter similarities. Because of the importance of accurate and fast communications with alphanumerical characteristics in wartime, military organizations have studied the legibility design problem extensively. Character style, stroke width, width-to-height ratios, and polarities are some of the more investigated variables that have been studied. For most applications bold vertical and black alphanumerical characters with width-to-height ratios between 3:5 and 1:1 and stroke widths equal to about one-sixth the height have been found to be generally superior for legibility in reflected displays. The uppercase NAMEL (Navy Aeronautical Medical Equipment Laboratory) style is an example (Figure

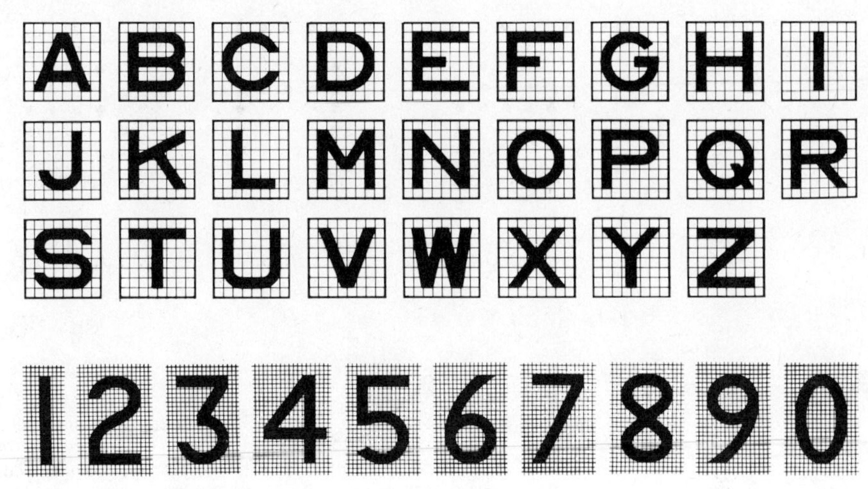

Figure 7-6. The NAMEL symbol alphabet.

7-6) that illustrates many of these properties. However, it would take a quite noticeable variation in the character variables to cause a substantial decrease in distinguishability. With projected displays, the irradiation of light spreads over the darker regions and so a white symbol on a black background (negative polarity) should have a narrower stroke width (around one-ninth the height) for better symbol legibility. Also, negative polarity symbols tend to give better legibility under low ambient light levels (Kelley, 1968), such as an airplane cockpit at night, but positive polarity tends to be better under normal and high light levels. There have been a number of other alphanumerical fonts purposed, besides the NAMEL system, including the Berger and Mackworth systems, which were developed in the United Kingdom, the unusual Lansdell system that was developed in Canada for high variations in light levels and viewing angles, the U. S. Air Force-Navy AND system, and the COURTNEY (Rowland & Cornog, 1958) and LEROY (Shurtleff, 1967) for CRTs and computer systems. Except for the Lansdell system, these others all tend to be modest variations of bold styles.

If these other character properties are maintained, then an alphanumeric symbol legibility (distinguishability) is determined by the character height and the viewing distance. The combined effect of these factors is the viewing angle that subtends the character height. Currently recommended heights yield 10 to 24 minutes of visual arc (Military Standard Specification 1472B). Numerous other studies have supporting results. For example, under good viewing conditions Smith (1979) shows that 90% of the characters presented are correctly recognized when heights are at the lower limit and 100% legibility is assured when heights are at the upper limit.

Some of the forms of electronic displays and electronically actuated printing have created resolution problems in alphanumeric character

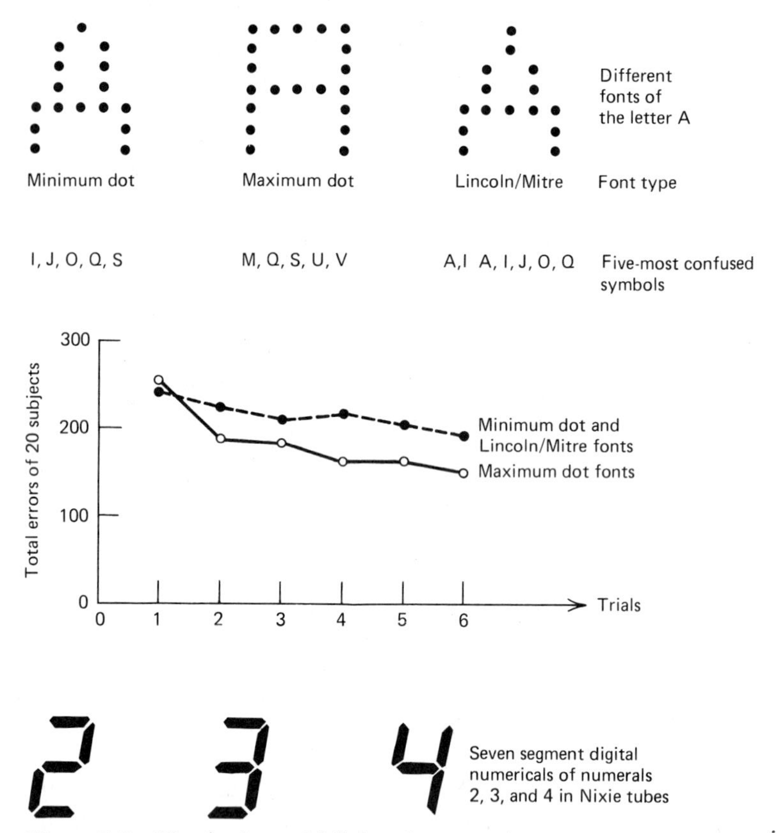

Figure 7-7. Matrix-dot and Nixie tube examples.

distinguishability. The 5 × 7 dot matrix and Nixie tubes are particularly noted cases in point. In a 5 × 7 matrix of dots there are 35 possible dots from which to describe all of the alphanumeric characters and so a variety of character fonts are possible (Figure 7-7*a*). Tests have shown that the maximum dot font provides fewer reading errors than the others (Maddox, 1979), as Figure 7-7b illustrates, and that recognition accuracy was also affected by the character size and dot luminance at viewing distances beyond about 1.5 meters (Snyder & Taylor, 1979). Nixie tubes have seven segments for numerical character displays (Figure 7-7*c*) where recognition accuracy also depends on the selected fonts (Van Nes & Bouma, 1980). These devices both appear to have resolution deficiencies. CRT displays also use a dot matrix to generate alphanumeric characters where a minimum of 10 raster lines for 15 minutes of visual arc are recommended for adequate resolution (Gould, 1968). At a 4:5 width-to-height ratio, the CRT display would have at least 80 dots for resolution, over twice that of the 5 × 7 dot matrix. Greater resolution adds more redundancy to the displayed symbols so that reading accuracy is less affected by missing symbol elements, extraneous noise, or glare spots.

TESTING SYMBOL DISTINGUISHABILITY

The increased use of symbols and pictorial displays in current times for distinguishing various control functions is a source of possible confusion. To test out how well people can correctly identify the appropriate coding, it is often necessary to familiarize or train subjects on the symbol representations from the coded meaning. Testing can then be performed by presenting the symbol and requesting the subject to identify the meaning by depressing an appropriate button. For every presented symbol there is a subject's response to an identified meaning. If the correct meanings of the symbols are listed as rows of a matrix and the subject's identified meanings as columns in that matrix listed in the same order as the rows, then every correct identification is listed on the major diagonal of the matrix (i.e., upper left to lower right-hand corner). Errors are contained in cells off of the major diagonal. Figure 7-8*a* shows an adaptation of data (Green & Pew, 1978) where the response percentage of presentations for a typical subject are shown. Other symbols are indicated in the final column. This diagram is known as a confusibility matrix. Figure 7-8*b* shows the same data in changed order of the rows and columns to help in-

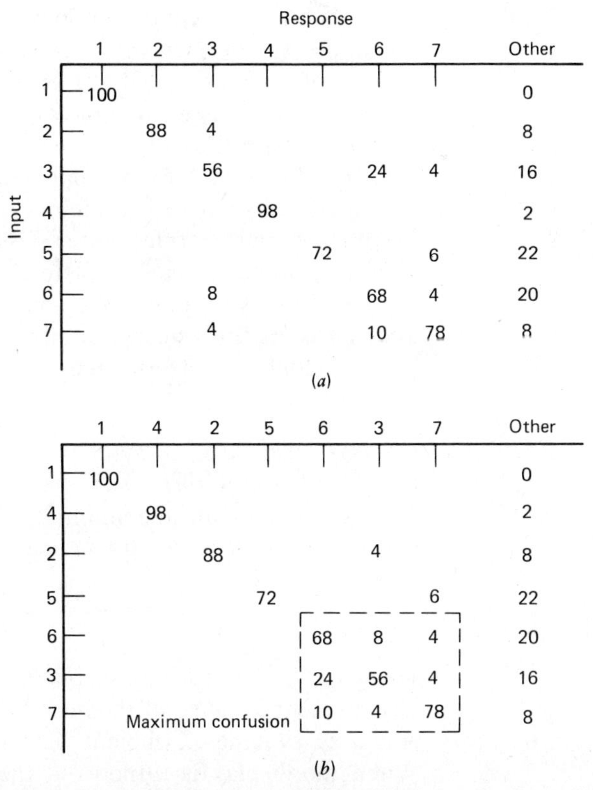

(*a*) Response

Input	1	2	3	4	5	6	7	Other
1	100							0
2		88	4					8
3			56			24	4	16
4				98				2
5					72		6	22
6			8			68	4	20
7			4			10	78	8

(*b*)

Input	1	4	2	5	6	3	7	Other
1	100							0
4		98						2
2			88			4		8
5				72			6	22
6					68	8	4	20
3					24	56	4	16
7					10	4	78	8

Figure 7-8. Distinguishability-confusibility matrices on data adapted from Green and Pew, 1978. (*a*) Original confusibility matrix. (*b*) Revised confusibility matrix.

tensify the collections of symbols and pictorials where there is greatest confusion. In this case the meanings and symbol confusions were greatest between for meanings 3, 6, and 7. With this knowledge, the designer can concentrate on further improving symbol distinguishability where confusion is the greatest. After improvements are made and verified, then a retest under degraded environmental conditions will show the stability of this character set. Also information analysis (Chapter 5) provides another basis for the analysis of a confusion matrix.

Also note that response times could be measured in an identification task for symbols and pictorials since stronger associations tend to have a shorter response time. However, response times in such tasks are highly affected by learning, particularly with more abstract symbols. Also the response times are strongly affected by the number of meanings in the alphabet of symbols or pictorials so comparisons should not be made between different sizes of alphabets.

Since symbol distinguishability is usually a function of learning, it is typically important to estimate the ultimate performance of very well-learned display reading in order to compare two forms of displays. Another feature of importance is the rate of learning. For example, Figure 7-7b shows the learning effects of various dot fonts in terms of decreasing subject errors over six trial sessions. A useful technique for this evaluation when differences in performance are not so clear is a fitted learning curve to the data. Learning curves are prescribed functional forms that show performance changes over sequential trials where parameters of the function denote ultimate performance and learning rates (Buck, Tanchoco, & Sweet, 1976). These parameters can be fit to empirical data for alternative display types. Then each alternative's parameters can be used to make direct comparisons for a design choice. Also a different learning curve can be used for each performance criteria.

Contrast

Contrast also helps in symbol detectability. In projected light displays (e.g., CRTs) in a *dark room* the internal display contrast is determined from the maximum and minimum luminance levels; respectively as L and D. In this case the luminance contrast C is described as follows.

$$C = \frac{L - D}{L + D}$$

However, projected light displays are rarely used in a completely dark environment because of the need for other tasks such as writing notes or actuating switches. Ambient light in the room reflects off the display. When the display is turned off, then the luminance from the reflected ambient light of K millilamberts changes the contrast by adding to both the maximum and minimum internally projected luminance so that the luminance contrast is as follows.

$$C = \frac{L + K - (D + K)}{L + K + D + K} = \frac{L - D}{L + D + 2K}$$

It is easy to recognize that contrast C decreases as the reflected ambient luminance K gets larger. Also, character detectability decreases with C (Howell & Kraft, 1959); very appreciative decreases were found with small characters (height $<$ 16 minutes visual arc) when $C < 90\%$ and for larger characters when $C < 86\%$. As a result, these authors recommend a luminance contrast of 94% for small characters and 88% for large ones. Typically rooms with CRT displays have a dimming control for light levels and shields are often installed around the CRT face to further improve the display contrast.

Motion Reference within Visual Displays

A source of movement reference within a visual display occurs when there are multiple pointers or multiple lines. "When the big hand is on 12 and the little is on . . ." is a familiar expression we heard as children learning to tell time from the clock on the wall or a disparaging slur to some adult whose intelligence is questioned. Other visual displays besides clocks are shown with multiple pointers. Figure 7-9*a* shows an altimeter where the short, heavy pointer indicates altitude in units of 10,000 feet (or 1000 m) and the long, slender pointer indicates altitude in units of 1000 feet (or 100 m). However, it is easy to reverse this pointer reference and think that the aircraft is at about 42,500 feet (4250 m) instead of nearly 24,500 feet (2450 m). The command display with two points, as shown in Figure 7-9*b*, indicates the speed wanted by the commander and the current speed but which is which? These examples describe the reference confusion that results with multiple pointer displays and the reason why human factors people recommend avoiding multiple pointer displays. The altimeter information, for example, could be separated into two adjacent displays to avoid this confusion as Figure 7-9*c* illustrates. Fitts (1951) found out that two large sources of errors by pilots in reading visual displays was misreading multipointer dials and misinterpreting the direction of dial movement.

Multiple-line displays also create confusion when the graphed lines cross. A circular strip chart is shown in Figure 7-10*a* where the air humidity and temperature are simultaneously shown by two lines and confusion effects can be seen when line crossovers occur. In Figure 7-10*b* this same type of chart is shown with separate line codes used to reduce this confusion. Alternative colors could be used for line coding as well. Figure 7-10*c* shows a horizontal multiple-line display denoting the trend of two system variables over time. In contrast, Figure 7-10*d* shows two horizontal single-line displays in the same orientations. While the later case of separating the lines on two displays eliminates the line confusion problem, it would be natural to expect that operators could read the two point values of these displays faster and more accurately. However, Shultz (1961) found that there was very little difference between the speed and accuracy of these two types of graphs for the point reading

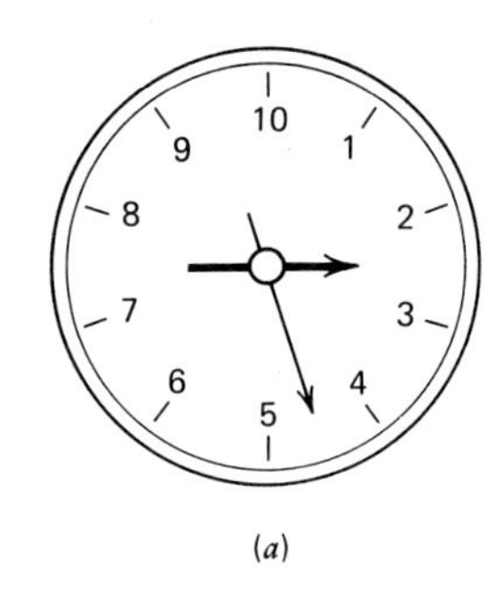

(*a*)

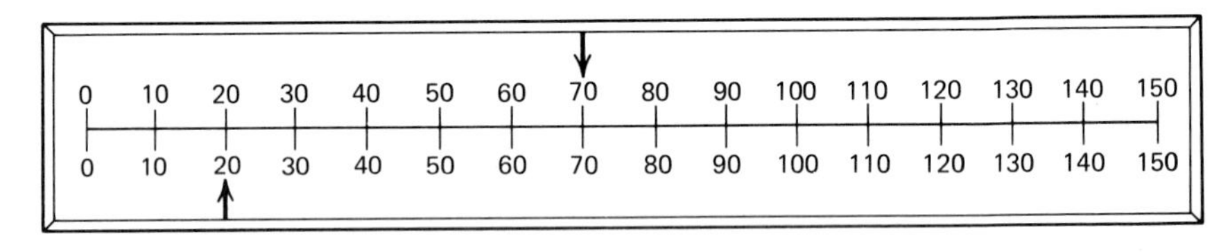

(*b*)

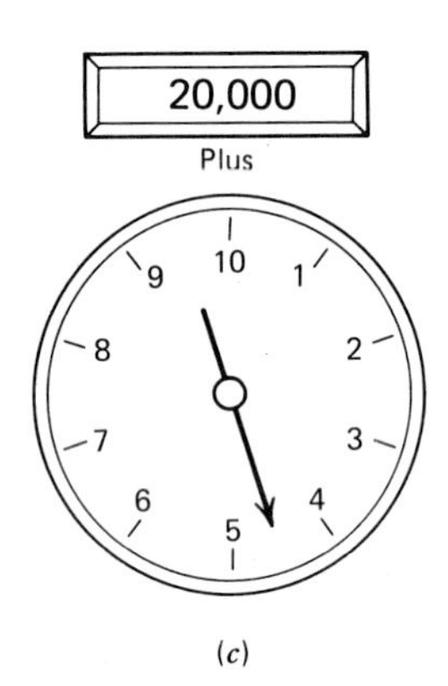

(*c*)

Figure 7-9. (*a*) Multiple pointer altimeter. (*b*) Multiple pointer command display. (*c*) Altimeter dial plus counter.

task. On the other hand, when people needed to compare the trend effect between two lines, then the response speed was much better with the multiline display, particularly when the lines were color coded. No differences were found between these two display forms in reading or comparing accuracies. Although it is important to reduce reference confusion on visual displays, you must be alert to the operator's task so that you don't achieve the opposite result. Incidentally, Shultz (1961*b*) also compared line graphs with bar graphs (both vertical and horizontal bars) and found both speed and accuracy of distinguishing trends to be superior with the line graphs both without and with missing data points (Figure 7-11); errors were highly correlated with response times. Even labeling names on graphs can seriously affect the locating time (Noyes, 1980).

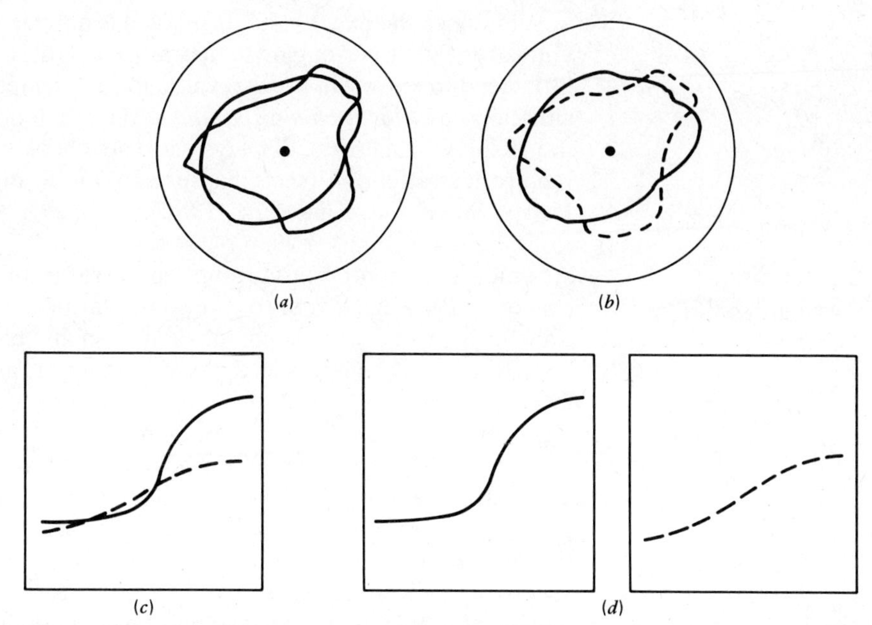

Figure 7-10. (*a*) Temperature-humidity recording on a circular chart. (*b*) The same temperature-humidity recording on a circular chart but with coded humidity. (*c*) Multiple-line display (see Shultz, 1961). (*d*) Two single-line displays.

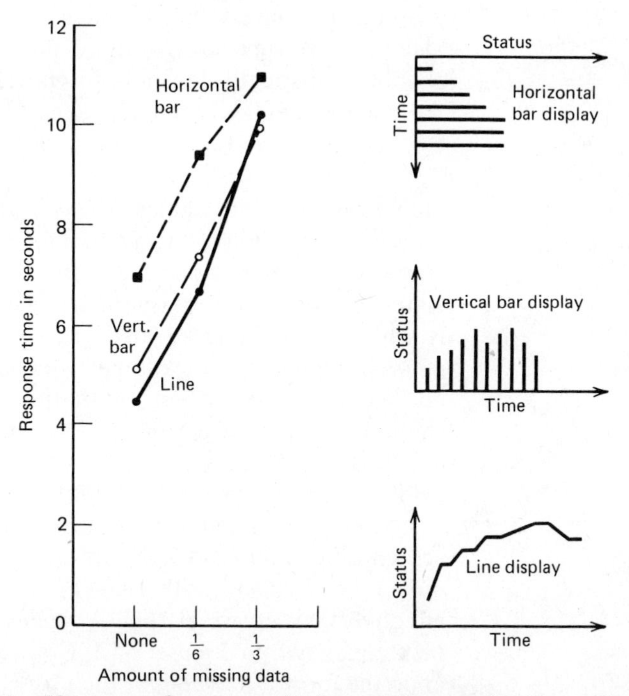

Figure 7-11. Human performance with bar and line graphs. (From Shultz, 1961.)

Sometimes the exact value displayed is not needed. For example, while you care if your car engine temperature is Hot, Cold, or Normal there is little need to know the precise value for engine temperature. This kind of situation calls for a *check reading* rather than a quantitative reading of the display. Analogue displays can be check read more quickly and require less attention (see Chapter 6) than do digital displays (Hanson, Payne, Shively, & Kantowitz, 1982).

Demanding Distinguishability

At times it is important to prompt an operator, to draw attention to some part of a visual display, or to give some warning. This form of information transfer is frequently accomplished visually by activating a warning light (e.g., a low oil-pressure warning lamp on an automobile dash), by increasing the brightness of a projected light, underlining a word on a CRT, light flashing, and by combining these and other techniques. In these cases, the fundamental purpose is to draw the operator's attention to a particular display or display location for immediate concern and possibly action. Flashing lights are highly demanding of attention; particularly at about 3 Hz. Sometimes 6 Hz and 9 Hz flashing is used to respectively indicate lower levels of priority. However, flashing frequencies much above 9 Hz approach the point of critical flicker fusion and tend to appear more as a steady light. Another device that is used for *attention demandingness* is light brightness. Light brightness for attention-demanding warning lights can be as much as 50 to 75 times the background brightness but brightness that is much above that can impair the operator's vision. Lower levels of warning and visual forms of operator prompting should be distinctly lower in brightness. It should be kept in mind that the use of attention demanding features for some specific tasks are distractions to the performance of other tasks.

Code Distinguishability

Numerous different codes are used to aid human operators. A commonly seen code in factories is color-coded pipes where pipes carrying high-pressure air are painted a different color than those carrying chilled water. This coding tells operators and maintenance personnel which line is which. Codes can be made of shapes, size, line weight or dash frequency, symbols, or location as well as color. Combinations of shape and color can be used when multidimensional coding is required; that is, shape may be used to code a function and color to code danger. However, not all modes of coding can be compatibly used together (e.g., symbols and shape). Of all the above modes of coding, results from numerous experiments (e.g., Smith & Thomas, 1964) show that visual search time is 50 to 70% faster and more accurate using color code over most other forms of code. Hitt (1961) shows that other codes can be more effective than color, at times, depending on the operator's mediational task, as shown in Figure 7-12. Colored lights can be effectively used to code the form of warning or the prompting. In these cases, red and green

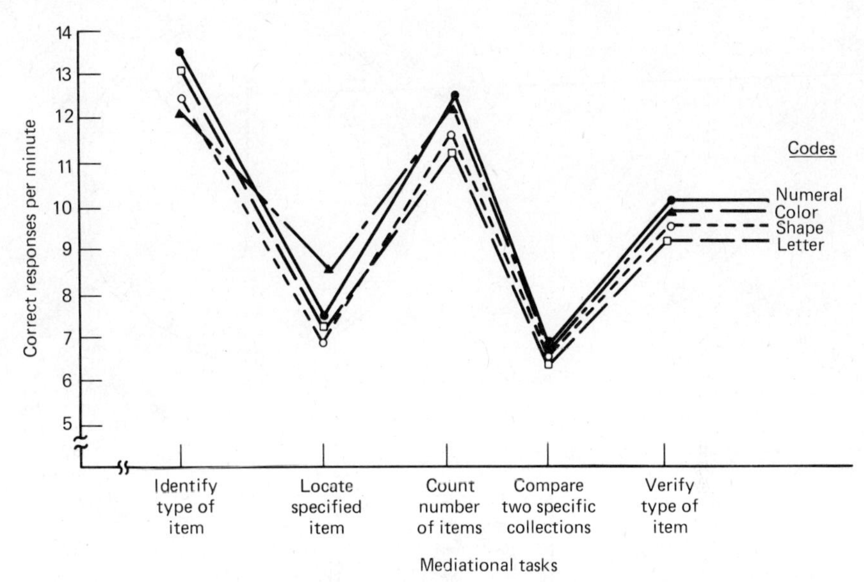

Figure 7-12. Accuracy-speed in performing various mediational tasks using alternative codes. (Adapted from Hitt, 1961.)

are excellent colors; yellow is good but is easily confused with white, and blue or purple tend to be far poorer. In general coding with colored luminants, colors that are farther apart on the CIE diagram (see Chapter 4) are more distinguishable (Chapanis & Halsey, 1955). Shape coding of labels can typically aid in attention demandingness if the shapes have sharper angles (Cochran, Riley, & Douglas, 1981).

Derivative Distinguishability

Rate-of-change or derivative information may be communicated through a visual display directly or inferred by observing the display over time. Acceleration is derivative information from speed. As you watch the speedometer of the car decrease while coasting along a level road, you can estimate the amount of braking or additional power needed before making your turn. Some visual displays show derivative information much more clearly than others. Figure 7-13 shows two forms of dials with the dashed line showing the changed position of the pointer after a few minutes of time. Also a digital counter is shown on this figure with the dashed display appearing a few minutes later and a strip chart is shown. Of these four displays, the digital counter gives the very precise direct information quickly, but derivative information is most difficult to infer because such inference must be made by abstract calculation. Also rapid rates of change with digital counters are very difficult to read and trying to do so can induce nausea. Either of the two dials provide a reasonable means for visually estimating a rate of change. The shaft center of rotation on the circular dial provides an anchor point for more

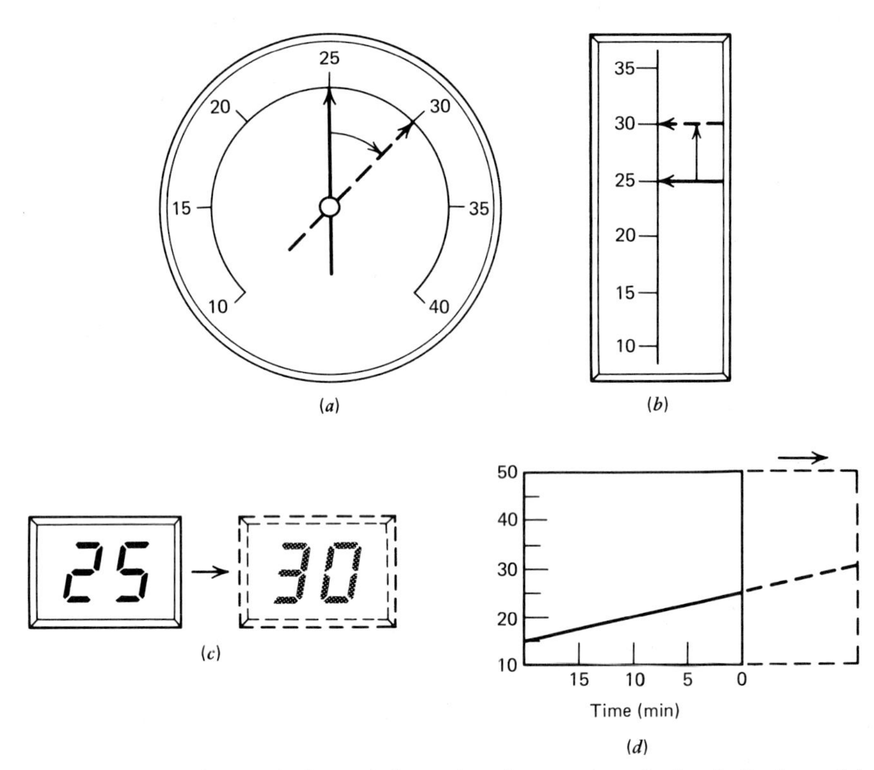

Figure 7-13. Inferred derivate information from types of visual displays. (*a*) Circular dial with pointer change of angle. (*b*) Linear dial with pointer change in displacement. (*c*) Digital counter with change in numerals. (*d*) Strip-chart recorder with trend line change.

easily estimating the angular pointer movement. Therefore, derivative inference is usually a bit faster and more accurate for circular dials than for vertical dials (Murrell, 1965). Finally, the strip chart recorder provides a direct means for estimating derivative information through the slope of the graph. In very slow acting systems (e.g., large ships or nuclear reactors), it may be very difficult to infer derivative information without a display similar to the strip chart recorder.

Display Interpretability Considerations

Even when display information is easily distinguished from other information, such information may not be interpreted accurately. A display labeled "Gute-faktor" with a reading of 95% may be clearly distinguished from one that said 87% but this potential information would be meaningless to a person who does not understand at least some German. Just because a display may be written in the language of the operator, it does not mean that the information will be quickly and accurately understood. Information format design is an important part of that communication (Engel & Granda, 1975). Until display information is unambiguously and effectively integrated into action, then there is some question about the interpretability (Lees, 1974).

Signs and labels are often poorly designed for meaning. Many examples of needless words and ambiguous meaning are present in everyday life. The key to effective communications in planning signs or instructions is to use the language simply and clearly or use symbols or pictorials such as international road signs. Even when you are convinced that you have done that, empirical testing is still recommended. Assembly directions, which accompany products that the buyer must assemble, are particularly prone to misinterpretations. Many frustrated parents have endured such holidays. Pictorials accompanying written instructions are extremely helpful to naive persons in assembling products (Konz & Dickey, 1967). Empirical testing of alternative instructional forms can be used to identify ambiguous steps that can be corrected before marketing products.

Display Compatibility

Two principal concerns of display compatibility are that display information is not in discordance over time or from different displays and that this information is compatible with the control actions. In the first case of *stimulus-stimulus compatibility* it is obvious that contradictory information creates operator confusion. Imagine a pilot who sees two altimeters with grossly different results. It is just as bad to have one display showing adequate cooling while another display shows that the nuclear core is heating up. Therefore logic testing of display information, as a first step, and physical simulation testing is highly recommended as a means of detecting and correcting for unintended incompatibilities of this type. The second form of *stimulus-response compatibility* is not so obvious. As mentioned earlier, controls for operator-to-machine communication are in part displays that inform the operator which control is associated with which displayed variable. This communication is provided by the location, movement, and orientation relative to the displayed variable. Many industrial process control panels contain trim control devices near the corresponding display devices. Figure 7-14 illustrates different degrees of display-control compatibilities. In general, better display-control compatibility results when controls are nearer the desired display movement, in the same plane as the display, and the movements of the control are the same as the desired display movement. This form of compatibility is discussed further in Chapter 10.

Motion and Control Interpretability

Visual displays are supposed to communicate information correctly regarding the status of location, position, and orientation, as well as changes in those features over time. When an airplane goes into a banking maneuver, the pilot needs to know the degree of bank and when recovering he or she needs to know when the craft is level again. Communicating

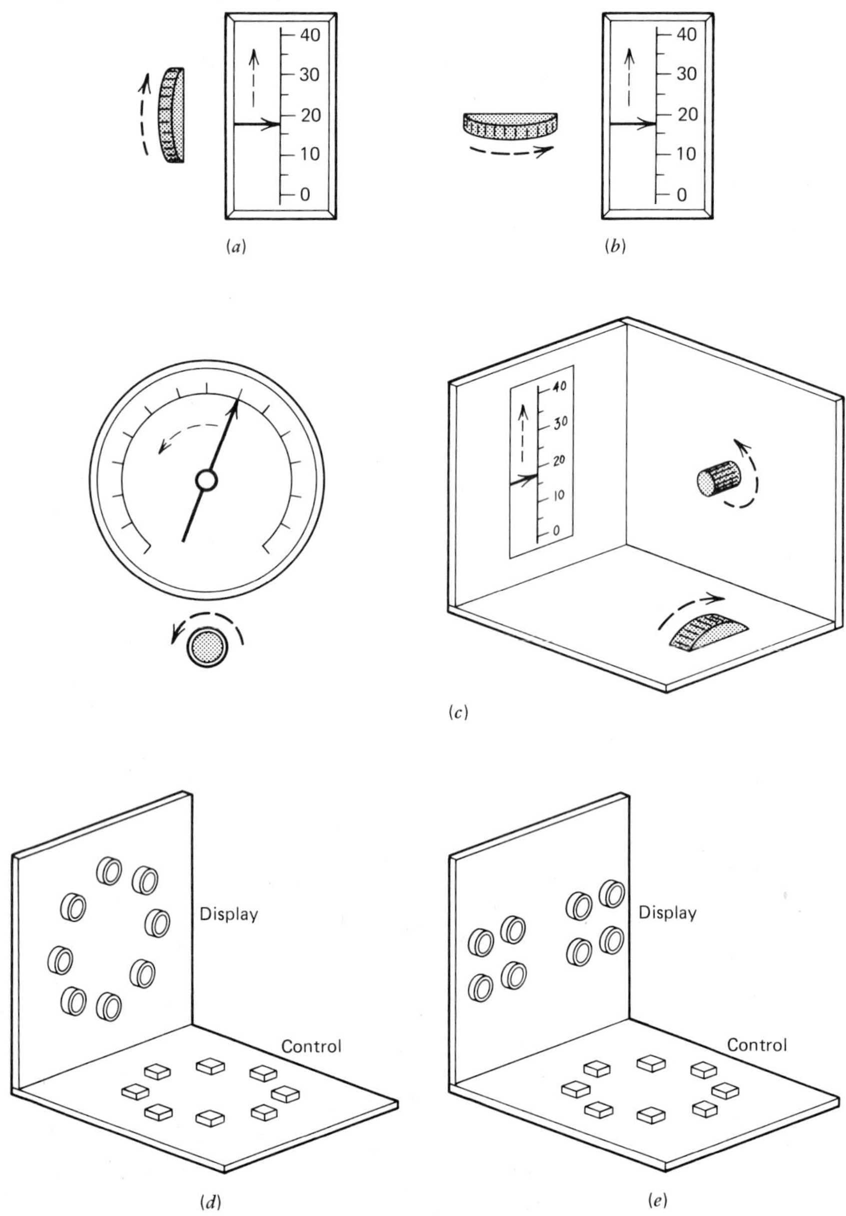

Figure 7-14. Illustrations of display-control compatibilities. (*a*) Compatible location and movement. (*b*) Compatible location but incompatible movement. (*c*) Incompatible locations and movement. (*d*) Compatible orientation of controls to displays but direct association is slightly unclear. (*e*) Incompatible orientation and obscure association.

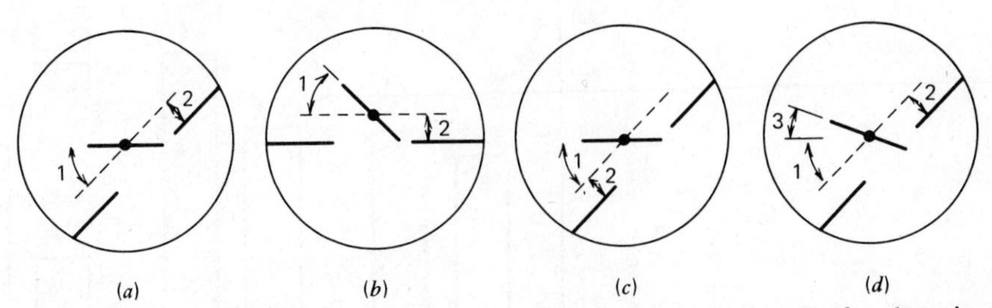

Figure 7-15. Alternative versions of artificial horizon indicators for denoting aircraft bank and pitch. (*a*) Moving horizon. (*b*) Moving airplane. (*c*) Kinalog. (*d*) Frequency separated. (1) Bank angle. (2) Pitch angle. (3) Aileron position. (adapted from Roscoe, Johnson, and Willigres 1980)

this information to a pilot through a visual display is complicated by the fact that the pilot's orientation model is affected by the craft position and past maneuver. Early versions of an aircraft display that gave this bank information were referred to as an artificial horizon and they consisted of a circle with a moving line that remained parallel with a flat horizon. An aircraft symbol was frequently shown in the middle of the display (Figure 7-15*a*). This display was made to appear as the true horizon would to a pilot looking through the windshield and so it is said to have an inside-to-outside viewpoint. Besides the bank angle, the horizon line would move above or below the aircraft symbol to show the pitch angle. Some people complained about this display saying that the reference of a fixed aircraft and a moving earth was backward. An an alternative visual display of this information, these people proposed an outside-to-inside viewpoint where the earth was the fixed reference with a moving aircraft symbol as shown in Figure 7-15*b*. Advocates of the outside-to-inside viewpoint supported their argument by stating that most other aircraft displays had that viewpoint. Opponents of the outside-to-inside viewpoint said that display was opposite to what the pilot saw when the horizon was visible. To counter these arguments, Fogel (1959) proposed a third alternative, illustrated in Figure 7-15*c*, which he called a kinalog display. Because both the aircraft symbol and the horizon line moved, when the joystick was changed, the aircraft symbol moved rapidly to show the control change and then the aircraft symbol slowly leveled as the horizon line moved gradually to the bank angle. When a control movement action is indicated directly on a display, it is known as display quickening and the kinalog display provides this feature. The essential concept of the kinalog display was to separate information on the quick (high frequency) effect of the joystick control action or quickening from the slower (low frequency) response of the aircraft. A further development of this display by Johnson and Roscoe (1971) showed the high-frequency aileron position to indicate the rate of roll, as shown in Figure 7-15*d*, and they called it a frequency-separated display. Tests

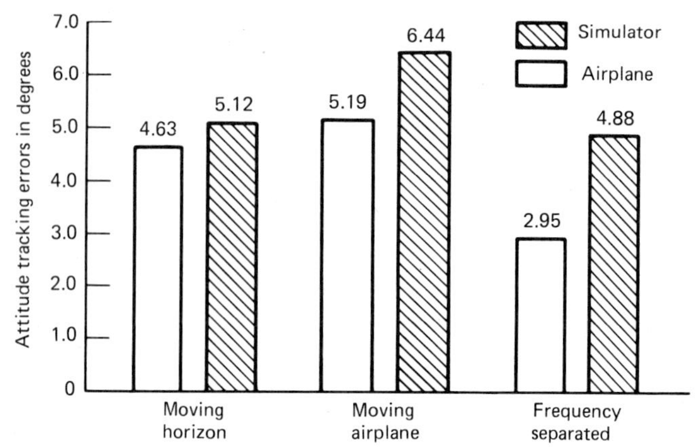

Figure 7-16. Human control accuracy with different types of artificial horizons.

performed by Beringer, Willeges, and Roscoe (1975) demonstrated that aircraft-handling performance of average experienced pilots with the frequency-separated display was equal or superior to that with the traditional forms of artificial horizon displays. Figure 7-16 shows the control error differences of these displays for both simulator experiments and actual flight tests. Both the kinalog and the frequency-separated display designs show the expected *direction* of abrupt control actions separately from the effects of that control action on the craft motion. This separation appears to have reduced the confusion of viewpoint differences but, as importantly, the added information of control action action aids the pilot in short-term planning.

Another case of motion interpretability with visual displays is the controversy between *pursuit* and *compensatory* displays, illustrated in Figure 7-17. Pursuit displays show a separate cursor and target where the object is to move the cursor onto the target. Differences between the horizontal and vertical distances between the cursor and target denote current errors needed for control action. In the compensatory display, the current control errors are shown directly but not the relative locations of the cursor or target. Thus the pursuit display communicates information on locations and changes in locations over time, leaving the operator to interpolate errors visually. It would therefore seem that the direct command information in compensatory displays would provide better human performance in controlling the motion of the cursor. That conjecture tends to be true *if the target is stationary*. However, target motion patterns are relatively easy to anticipate in pursuit displays but these patterns are considerably more difficult to interpret with compensatory displays. Pursuit displays are preferred modes of communications with changing targets because the movement pattern information allows operators to plan ahead.

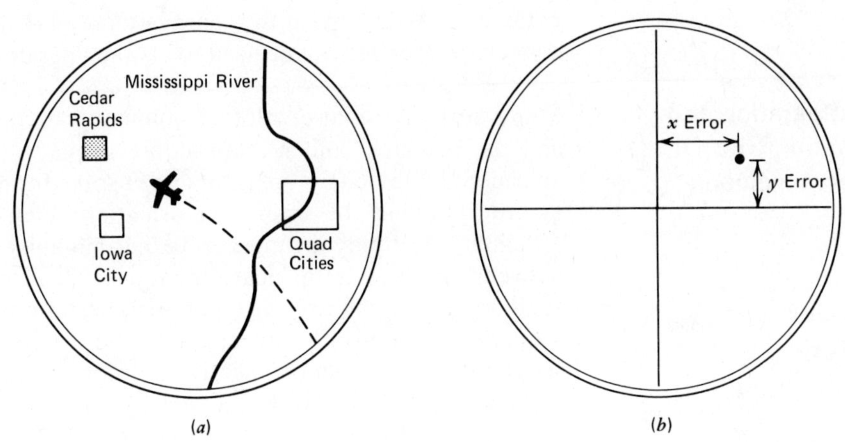

Figure 7-17. (*a*) Two-dimensional pursuit display. (*b*) Two-dimensional compensatory display.

INFORMATION INTEGRATION AND AIDING

The purpose of visual displays is to communicate with people with respect to their knowledge, goals, tasks, and situation. In many cases the objective is to match the display to the person viewing it as an aid in performing such mediational tasks as identifying, comparing, counting, verifying, or locating. In a study several years ago Hitt (1961) found that among various forms of CRT coding (pictorials, geometric shapes, alphanumerics, and colors), performance on all of these forms of mediational tasks could be described as combinations of the first and last. He also found that numerical codes provided the greatest accuracy per unit of time except for the locating process where color coding was better. Color coding was found to be a close second in all other processes except identifying where letter and shape codes were modestly better. More recent studies on color coding show that search time in locating an object is less with color coding using a few colors, but as more colors are added the search time takes longer (Carter & Cahill, 1979). It is important to note that display codes need to be made to be compatible with the tasks that people perform with the display information.

Another operator task that can be aided with a proper selection and placement of visual displays is *monitoring*. Operators must frequently scan displays of system parameters to see that the parameters are all within norms. If these parameters all have rather constant norms during steady-state operations, then the displays that show these parameter values can all be aligned for normal values showing commonly at 12 o'clock or 9 o'clock positions for locating deviate parameters quickly. When there is a particularly large number of parameters, periodic computer checking is frequently done where a warning is flashed to the operator to show when a parameter is out of bounds and to denote which parameter. With only a few parameters, marked and/or color (e.g., red danger zone) codes can be used to aid the operator's monitoring task.

Integrating and Aiding with Display Arrangements

Many systems require a lot of communications between the operators and the machine, and so alternative display arrangements need to be considered. Obviously operators need some form of display organization in order to find the desired information accurately and quickly. One principle for display arrangements that minimizes the time spent looking between displays is to locate them according to their *sequence of use*. If all operators who use the equipment perform tasks in exactly the same sequence and a display appears once and only once in the sequence, then sequential arrangement is clearly an optimum arrangement. Even if there were some small deviations between operators or between repetitions of an operator, this principle could be argued effectively. Fitts, Jones, and Milton (1950) did a study where the eye movements of pilots were recorded during aircraft instrument landings. In order to determine sequences in the visual scanning, an analysis was made of the pairwise sequential glances between displays and Figure 7-18 shows the study results. These pairwise glances are represented as links with the percentage of occurrences shown. By reducing the distance between instrument pairs with the strongest links (i.e., greatest percentage), the average pilot can spend less time in acquiring needed information during the critical landing period (see the box on Link Analysis). Also shorter eye and head movements speed up the acquisition of visual display information (Robinson, 1979). However, note that some pilots may spend

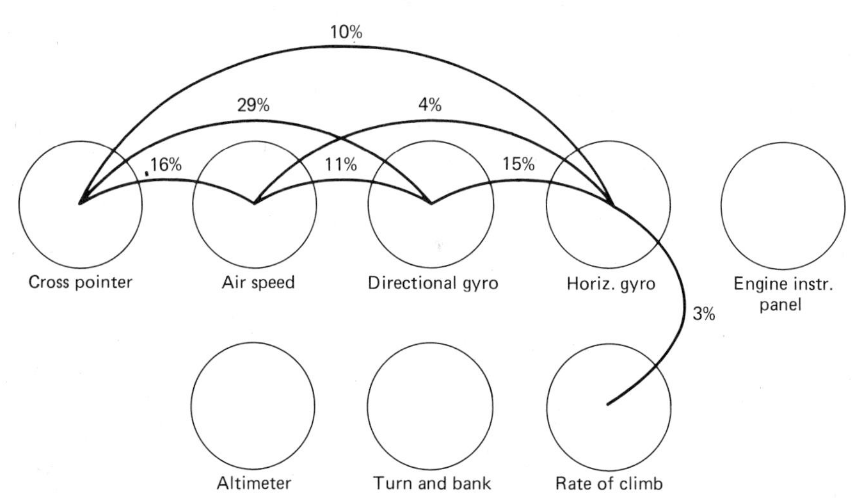

Figure 7-18. Link chart of eye motions between aircraft instruments during instrument landing approach (links with less than 3% omitted).

more time than others in acquiring information and that other periods of flight (e.g., takeoffs) may have a distinctly different sequence. Although the sequence-of-use principle does not solve all or even most problems in display arrangements, it does provide a good start in a number of cases.

If there is no dominant sequence of tasks but the operator tends to take care of those tasks that deal with some function (e.g., adjust the speed of the motors) and then move on to another function (e.g., check the oil pressure of the engines), then it makes sense to arrange the displays by *function groupings*. In this way, the within-function tasks can be handled more quickly. Whenever the functions are handled sequentially, then this and the previous principle both pertain.

Another display arrangement principle is to locate the *most important* visual displays in the most easily seen location and the less important displays progressively more peripherally. Generally designers denote displays that are more critical to safety or mission achievement as measures of importance. When these criteria are not particularly relevant, then the *more frequently used* displays are viewed as more important or else this frequency criterion is factored in with subjective trade-offs. Locations that are most easily seen are those nearest the center of typical viewing such as on the panels below and above the windshield of a vehicle. An example of the use of this principle was provided by Cole, Milton, and McIntosh (1954), where they measured the average duration of visual fixations, the frequency of fixations, and the proportion of time spent on seven important cockpit displays during climbing maneuvers on a constant heading. Each of these measures of visual behavior by pilots indicate the importance pilots attribute to the information displayed. A more recent technology, called Heads-Up Displays (HUDs), puts the most important display information right on the windshield or on helmet visors.

These various design principles are rarely purely descriptive of the jobs that people perform in different phases of job activity (Christensen & Mills, 1967). Consequently, all of these principles need to be fused together and tested dynamically in mock-ups particularly when the space or real estate available for displays is highly limited (Gravely & Hitchcock, 1980). More recently, there have been efforts made to aid designers with mathematical techniques in this display-locating task (Pulat & Ayoub, 1979; Topmiller & Aume, 1978). Before performing the task of locating the physical displays within the available real estate, the designer needs to address the question of what information can be combined effectively within a single display in order to reduce the number of displays and improve the operator's attention resources for faster and more accurate information detection and integration (Josefowitz, North, & Trimble, 1980). It is of particular note that the principles for locating displays are also applicable to locating information within displays.

LINK ANALYSIS

Link analysis is a technique performed to assist in the redesign of display, other equipment, or people arrangements. A link is an arc or connection between a pair of elements (i.e., people or machines) that denotes a specified relationship. Common relationships used in human factors are (1) the relative frequency of an operator going from one element to another, (2) communication frequencies, or (3) relative importance. The *link value* is a numeric quantity associated with that relationship.

When the elements of the system under study are shown in their relative locations in space such as on a plan drawing, link lines can be drawn between the elemental pairs and the link values may be added to summarize relevant information. Conventionally, machines are shown by their plan shapes, drawn to scale, and people are symbolized by circles. A link drawing made in this fashion denotes people and machine locations along with their respective links and link values. If the links denote the frequency of an operator who moves from machine to machine as indicated in Figure 7-19, then a link drawing with the highest-frequency links at the shortest distances would indicate the least movement effort and travel distance needed by the

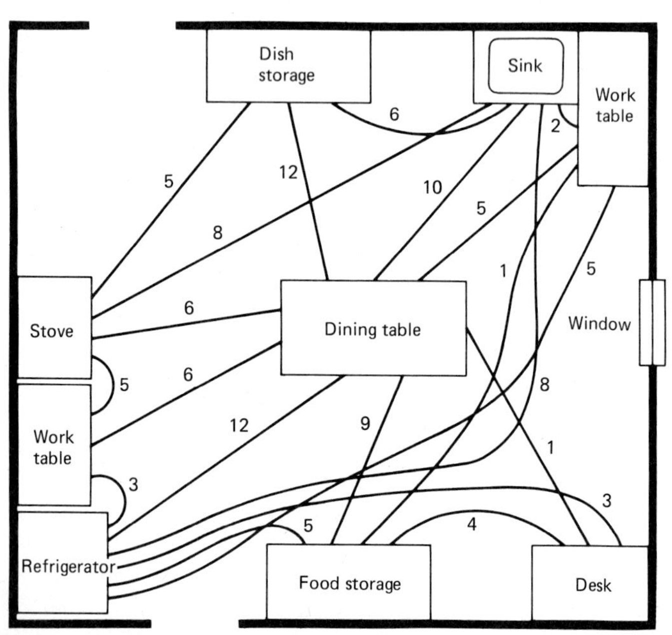

Figure 7-19. A home kitchen example of a link chart where average travel frequencies are shown per meal.

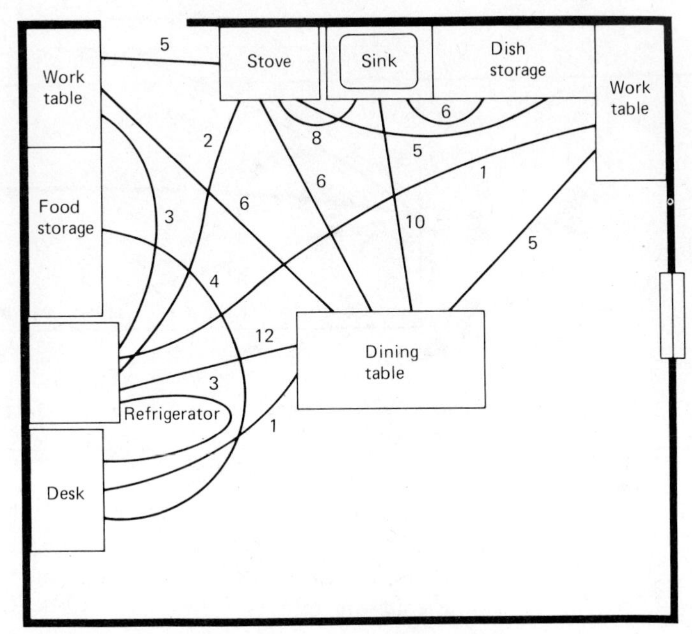

Figure 7-20. Some layout improvement in the home kitchen.

operator. Figure 7-20 shows improvement in the layout of the kitchen example. Links with longer distances and higher frequencies are good candidates for location changes with those that have shorter distances and lower frequencies. Alternative layouts of equipment can be considered by rearranging their locations and redrawing the links in order to visually evaluate the improvements and to discover further improvements. Chapanis (1959), shows examples of link analysis for improving the layout of the combat intelligence center for the operators and their equipment on board a cruiser.

Historical and Predictive Displays

Historical displays denote a chronology of past conditions of a variable over time, typically on a last-in, first-out basis. A stripchart recorder is an example. Far back in the window is a long history of the variable. The position of the pointer is the current status of the variable, the slope of the recording is the speed or first derivative, and the recording curvature denotes the acceleration or second derivative. Direct inferences of these forms of derivative information in conjunction with the current variable itself are information integration features that make historical displays very useful display forms, particularly with slow-acting, high-inertial systems (e.g., large ships and many industrial processes). The historical recordings also unburden operator memory requirements. However, the

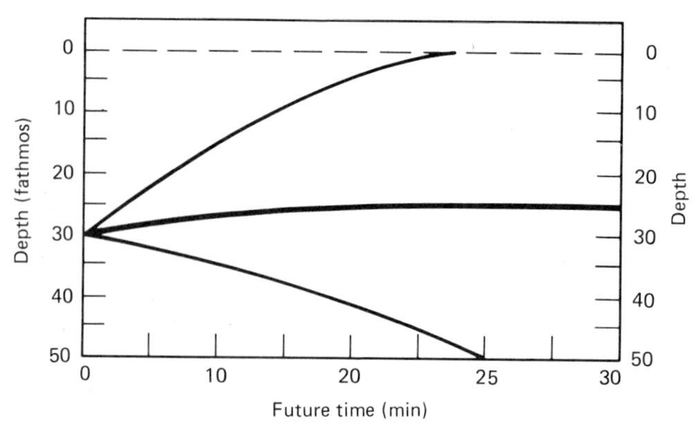

Figure 7-21. Projection and boundary form of predictor depth display for submarine control operations.

most appealing feature of historical displays is that they provide a basis for operator adaptation, prediction, and planning. With high-inertial systems, control actions must take place well in advance of the desired time that the effects are needed. Historical displays allow for planning these control actions and anticipating the timing of these actions.

While historical displays allow the operator to predict the future by extrapolating from the past, *prediction displays* provide the extrapolation directly. Predictor displays have the intelligence to make these predictions of system dynamics based on current status of variables and position of the controls as well as the recent history. One form of predictions display is the projection and boundary type, which denotes the condition of the variable in question t minutes in the future, *given that no further control change is made,* and the maximum boundaries of that variable at extremes in control changes. Figure 7-21 illustrates projection and boundary form of predictive depth for a submarine where the upper and lower boundaries form a maneuvering envelope. Predictor displays of this type show an operator what can be done, leaving the operator free to decide; whereas command displays unburden operators from the mediational tasks of estimating system dynamics, the space of possible control actions, or deciding on a specific action, but they do not show the effects of alternative control actions. However, studies made with these forms of predictor displays (Kelly, 1968; Palmer, Jago, & O'Connor, 1980) in contrast to other forms of visual displays have indicated that safer, more accurate, and more economical control can be obtained.

Information Control and Integration

Increasing complexities of industrial and power generation processes have occurred during the past few decades along with a recognition of the importance of greater information availabilities. Control rooms for these processes got larger and larger in order to accommodate all the visual displays needed. This situation started to cause problems for the control room operators due to the vast amount of displayed information, much of which was only needed occasionally. As an alternative to a vast array of visual displays that are dedicated to a specific type of information, a more recent idea was that one or more CRTs should be used where the operator can call up the specific types of information desired. A further embellishment of this idea was to provide *zooming* capability where the control operators could focus on a specific part of the system and get higher and higher detailed information on that part or zoom back on the overall system without the detail clutter. One of the first information display and control systems to enjoy these capabilities was made by Foxborough Corporation for industrial process control. More recently Westinghouse Corporation has expanded on this idea by devising integrated forms of overall system monitoring and alternative ways to control the focus and zoom in controlling nuclear reactor power generation.

The recent development of computers coupled with CRTs have allowed combining integrated information more effectively and they provide the user with information flow control. Figure 7-22 shows the cockpit of a contemporary commercial aircraft where much of the engine management display information (i.e., engine data, caution indicators, and flight engineering data) have been combined and presented in four

Figure 7-22. Cockpit in a Boeing 757 aircraft. (Courtesy Collins Division, Rockwell International.)

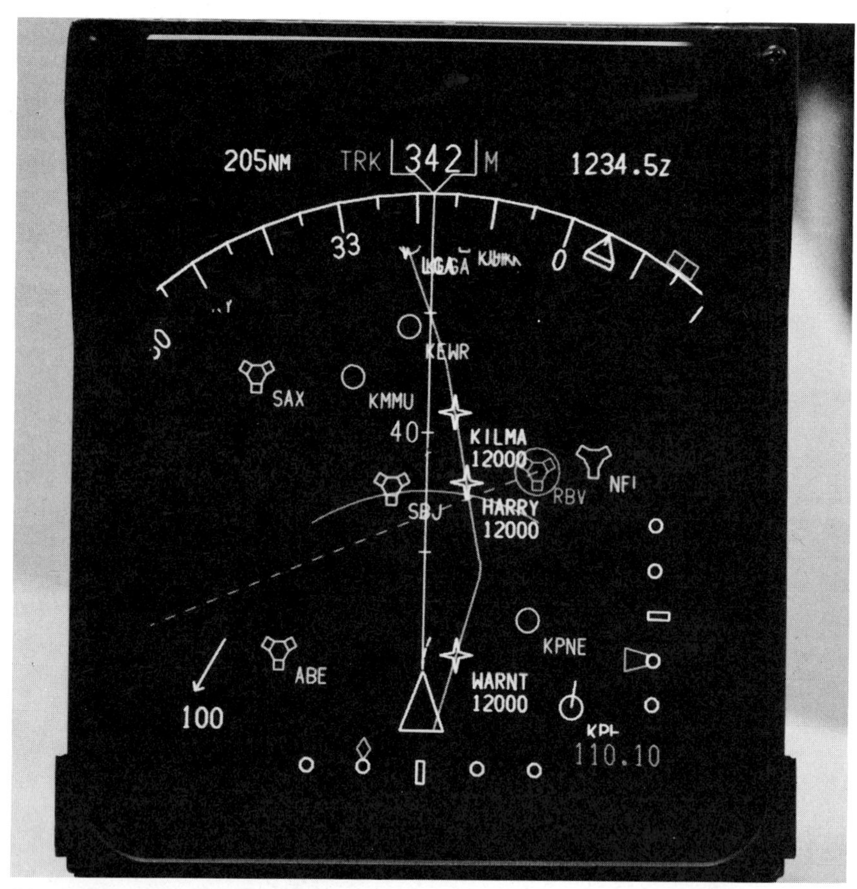

Figure 7-23. Navigation display in the Boeing 757 aircraft. (Courtesy Collins Division, Rockwell International.)

CRTs between the captain and first officer. Besides some traditional instrumentation, there is also a separate navigational CRT display and a primary flight CRT display (e.g., roll, pitch, horizon, and glideslope deviation) provided in front of each officer. The navigation display, illustrated in Figure 7-23, combines enhanced radar information along with identified ground locations, air speeds, and headings. Weather information can also be added through pilot control.

SUMMARY

Visual displays associated with various machines are part of the communication process. The design, in terms of display selection and location, can substantively improve or deter communications so that system performance is altered toward the desired objectives or toward disaster. For these reasons, task and communication analysis must precede the design process. Knowledge of the user population must also be available at the onset of design. With this background set forth, the foregoing discussions show many basic notions of various visual displays and contemporary display technologies available to meet the operators' display needs and to fit the operators' abilities to sense, discriminate, and interpret display information for effective communication. However, the effectiveness depends on moment-to-moment compatibilities with the operator's tasks and on long term system performance. Marie and Jim both agree!

8 AUDITORY AND TACTILE DISPLAYS

I think fire engines, ambulances, and speeding police cars are magnetically attracted to my car. I have had several near misses (or near "hits") with various types of speeding emergency vehicles at intersections. If your own car is near mine when you hear a siren—look out! I'm heading for the side of the road. I don't care where or how far away that siren is; get me out of there!

AUDITORY DISPLAYS

This little confession illustrates some of the problems with one common type of auditory display: the warning signal used by emergency vehicles. What are the limits to the information that we can obtain from auditory displays in different applications? How does the human factors specialist decide when to recommend use of the auditory channel rather than the visual? What factors are important in setting the information rate of an auditory channel or its manner of coding data? These are some of the topics considered in this chapter.

With the exception of speech, most auditory channel applications will be limited to much lower information rates than are possible for the visual channel. Medium-to-high information rate applications of the auditory channel (roughly 3 bits per second or higher) may be needed in situations where the visual system is not available (i.e., where hearing is employed in an aid for the blind) or where vision is otherwise precluded by environmental or task demands. Medium-to-high rate applications may be indicated in situations where environmental conditions can impair vision, such as anoxia or g-force stresses in high-altitude flight.

Most uses of the auditory channel will probably involve low-to-middle information rates (less than about 4 bits per second) such as in auditory warning or alerting signals. When the visual channel is heavily loaded with data, as in an aircraft cockpit or is dedicated to certain displays by virtue of head position and eye direction, the only way to provide additional data may be via the auditory channel. The auditory input channel is to some extent omnidirectional, so that important information can be communicated even in the case of primary operator attention to the visual channel. The operator may have to constantly move his or her head and eyes, making a fixed visual display unusable as an alarm device. Finally some low-information rate applications may, by the very nature of the low rate aspect of the task, make it desirable to have an auditory display channel in addition to a visual channel carrying the same information. In detection situations involving the long-term monitoring of displays for the occurrence of infrequent signals, an auxiliary auditory channel may provide a significant improvement in the system's effectiveness. Remember that economic criteria, rather than ultimate performance, may dictate the choice of the auditory channel. An example is the use of a portable radar system for small boats. This radar employs an acoustic display system because of cost, weight, and power advantages over a visual display system (see Figure 8-1). These advantages override the

Figure 8-1. "Whistler" portable radar system for small craft. (Copyright Controlonics Corp., Westford, Mass.)

fact that much better person-system performance is obtainable with a conventional visual display and rotating antenna system.

Another important aspect of the auditory channel is the nature of the code used to communicate information over the channel. Signals can be essentially uncoded, in that they are "raw" versions of the actual data conveyed, such as the engine sounds employed by a mechanic for diagnostic purposes. Other examples of uncoded acoustic signals are the normal sounds of a power plant. Leaking relief valves, noise in steam lines, and the clicking sound of electro-mechanical control devices all provide characteristic acoustic signatures that may be monitored in parallel with information from the primary visual data channels. Such auditory information may be an effective way to give a "feel" for a plant's normal operation. The operator is thus provided with continuous status information about the plant without diverting attention from other activities (Eckert & Woods, 1981). Some applications of the uncoded auditory channel may take advantage of the human's ability to learn to discriminate auditory signals from noise, particularly when the designer does not know the signal and noise characteristics in advance. The human may be the best system for performing auditory detection or recognition tasks in an uncertain environment that requires a high degree of flexibility.

Coding signals for the auditory channel that are derived from nonacoustic sources can involve varying degrees of complexity. Obviously one can manipulate the spectral, temporal, and intensive aspects of a sound in many different ways. The spectral character of warning signals is usually dependent on considerations having to do with their masking, alerting, or localizing characteristics in the expected environment. Codes where the frequency of a tone is varied may be used in medium rate applications such as auditory tracking tasks and direction indicating displays. More complex tonal codes might be used in situations where the speech channel was otherwise dedicated, where channel security was desired, or where the environmental noise possessed speech characteristics and hence would subject a speech signal to masking. The speech code is usually the appropriate code to use in high-information rate applications, since rates as high as 250 words per minute are possible, as compared to rates of 30 words per minute, for example, with highly trained observers using the Morse code. The use of nonspeech codes in high rate applications usually means an extensive training requirement and some loss of system flexibility.

Low-Information Rate Applications

Alarm Signals According to Fidell (1978), the optimization of audible warning signals has been a continual problem for human factors specialists, and in spite of many efforts to standardize such signals as emergency evacuation signals, fire alarms, and other military and nonmilitary warning signals, there has not been much success in this area. Fidell points out

TABLE 8-1 *Test Signals Heard by Drivers (from Fidell, 1978)*[a]

Signal	Description
1	Federal PA200 with CP 100 high power transducer in "high-low" mode at 0° (on axis)
2	Dunbar-Nunn Unitrol 800 with Atlas HPR 370 transducer in "yelp" mode at 0° (on axis)
3	Carson SA 410 with 890R transducer in "wail" mode, at 0° (on axis)
4	B and M S8 electromechanical siren 5 sec on and 5 sec off, at 0° (on axis)
5	Frequency modulated (3 Hz) sinusoid within one-third octave band centered at 1600 Hz
6	Signal 5 switched smoothly on and off at intervals of 333 msec

[a] From *Human Factors,* 1978, *20,* 22. Copyright 1978 by The Human Factors Society, Inc., and reproduced by permission.

that the proponents of such criteria as novelty, meaningfulness, and distinctiveness (among others) have now generated a huge number of different types of warning signals including elephant cries and women's screams!

In an attempt to simplify the design problem, Fidell advanced the hypothesis that the effectiveness of warning signals, in terms of the level needed for detection and rapid response, is simply related to the detectability of the signals measured in a detection theory way (see Chapter 3). Suppose that you have a set of signals of quite different spectra and time distribution, but you know that these signals are *equally detectable,* based on detection tests in the laboratory with trained observers performing at reasonably high d' values ($d' \approx 4$). Now suppose you increase the level of those signals relative to the background environmental noise by 10 dB, so that the d' would increase (if you could measure it) to about 40 or more. The modified signals would now be effective as warning signals in that they would be distinctly audible and clearly recognizable. Furthermore, they would also have approximately equal merit in eliciting rapid and correct responses under emergency conditions. This is an extremely useful result because it means that the human factors specialist can design a warning signal for a unique acoustic environment, and using known psychoacoustic detection data, can set the desired levels of that warning signal. Fidell's study tested this hypothesis with several different vehicle warning signals in a multitask driving simulation, so we can have some confidence in the generality of his result (see Table 8-1). The American National Standards Institute's (1979) recommendation for immediate evacuation signals in industrial installations is consistent with this principle. They suggest using a unique signal (relative to the acoustical environment), such as an amplitude modulated tone. Table 8-2 describes the ANSI recommendation.

TABLE 8-2 *ANSI (1979) Recommendations for Immediate Evacuation Signals*

1. The fundamental should be at or below 1000 Hz and the modulation rate should be less than 5 Hz.
2. The signal level should be at least 10 dB above the maximum overall typical ambient noise and at least 75 dB everywhere evacuation is considered essential.
3. If levels greater than 115 dB are required, consideration should be given to the use of visual alerting signals.

You can probably think of many good reasons for *not* choosing warning signals that are unique to specific environments. For example, The Committee on Hearing and Bioacoustics and Biomechanics (CHABA) of the National Research Council was asked by the National Fire Prevention Association to recommend an acoustic signal usable as a standard national fire-alarm signal (Swets, Green, Fay, Kryter, Nixon, Riney, Schultz, Tanner, & Whitcomb, 1975). The recommendations of the CHABA committee are interesting because in addition to recommending a specific solution to the standard fire alarm question, they faced a classical human factors design problem: they were asked to design a signal that was specific to fire alarm situations, and would be universally recognized by all people and easily distinguishable from other alarms. They were also asked to consider the problem of adapting existing fire alarm systems to the proposed signal.

The problem is difficult because there are large variations in the spectral and temporal character of the sound in different sorts of buildings such as schools, hospitals, offices, theaters, factories, and apartment houses. A specific signal that was easily detectable in one of these environments might be difficult or impossible to detect in another. Furthermore, the existing types of signaling systems in these environments have a variety of different power spectra and modulation characteristics; a complete changeover would be impractical. The Committee's solution to the problem was to recommend a standard *temporal profile* for the fire alarm signal, rather than any special signaling spectrum or signaling device. The suggested profile is shown in Figure 8-2.

The temporal code suggested is a repeating sequence of two short signals followed by a long, then followed by a time-out. The nominal duration of each short period could be $\frac{1}{2}$ sec. (Suggested values are 0.4 to 0.6 sec for an on period, and 0.3 to 0.6 for an off. The fluctuation in on level should be less than 2 dB, the fall during the off period should be at least 10 dB within 0.1 sec, the on transition should also be within 0.1 sec.)

The committee further recommended that the signal should be clearly audible and distinguishable from potential background interference. The level of the signal should exceed the prevailing equivalent sound level

Figure 8-2. Temporal code for national standard fire signal.

(L_{eq}, see Chapter 16), by 15 dB or exceed by 5 dB any peak sound level having a duration greater than 30 sec, whichever is greater. (If levels greater than 130 dB are required, further consultation with health authorities would be accomplished to reduce the possible risk of hearing damage.)

There are many advantages in choosing a temporal code rather than a specific spectral characteristic for the alarm signal, in addition to the mentioned reasons of minimizing environmental masking and being able to use existing devices. Tactile or visual signals could be used to warn deaf lodgers of apartments or hotels. Emergency vehicles such as fire engines and police cars could direct the standard temporal signal toward the threatened building to warn occupants and to alert the people in the immediate area of the burning building. To eliminate the possible confusion with other warning signals in use, particularly the confusion with the signals from emergency vehicles, the Committee also recommended that all *moving* emergency vehicles, including fire engines, employ a single standard alarm signal: the high-low or "continental" alarm signal.

A SLICE OF LIFE

Where Is that #!%*#!! Fire Engine?

At the beginning of this chapter we raised the problem of emergency vehicle warning sirens (see Figure 8-3). The design of such sirens involves a variety of difficult human factors design problems. The emergency vehicle siren is subject to some rather important limitations as an effective auditory display of vehicle distance and direction. For one thing, the windows in your own vehicle are probably closed in the summer or winter, greatly reducing your ability to quickly acquire this information and take appropriate action. You may also have your radio or tape system on, adding a significant masking factor to the siren detection and discrimination problem. As part of a study of the ambulance siren problem, T. Caelli and D. Porter (1980) interviewed 12 experienced ambulance drivers about the adequacy of siren function on the road. Most of them reported that car drivers would not respond to their siren until the ambulance was within 100 m of the rear of the car, or 200 m approaching from the front. Most felt that the ambulance siren was very poorly heard or localized by car drivers under normal driving conditions.

Figure 8-3. Beware of speeding emergency vehicles! (Joel Gordon.)

Caelli and Porter tried to study the siren problem experimentally in order to quantify the factors involved in a driver's problem of estimating the distance, location, and movement of an emergency vehicle. Figure 8-4 illustrates their test conditions. They tested the detection and localization of an ambulance in up to 16 different conditions including four locations and three angles of movement (at 40 km/h) with respect to the fixed car. The observer in the car listened with the car engine off under two conditions: driver window open, or all windows closed and radio on (about 60 dB level). After a 2-sec burst of the siren, the observers had to indicate the direction of the siren, its distance from them, and if the ambulance was stopped or moving. The authors evaluated one conventional alternating ("hee-haw") siren and an electronic siren unit (PI-300, International Sound and Signals) that had three siren types: (1) 12 cycles/minute wail, (2) 180 cycles/minute yelp, and (3) electronic hee-haw at 35 cycles/minute. These sirens all produce sound levels of at least 120 dB at the front of the emergency vehicle.

In our Chapter 3 discussion of sound localization, we pointed out that we are good at localizing sound sources that are directly in front of our face, compared to directly off one ear or the other. However, when a sound source is close to the plane that cuts our head

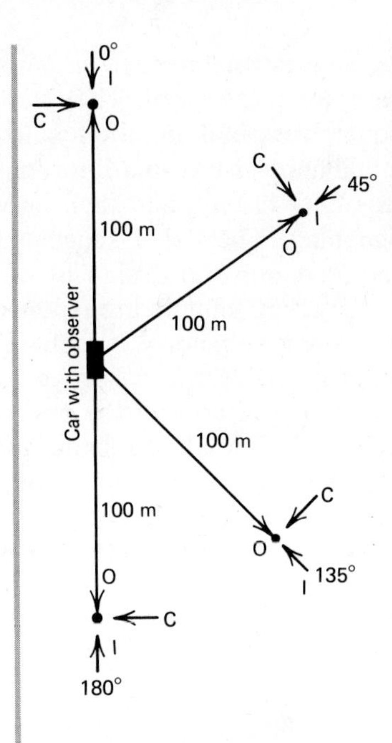

Figure 8-4. Experimental configuration of car and ambulances showing ambulance position and direction. (From T. Caelli & D. Porter, On difficulties in localizing ambulance sirens, *Human Factors,* 1980, *22,* 720. Copyright 1980 by The Human Factors Society, Inc., and reproduced by permission.)

in half vertically (front to back), we tend to make reversal errors in the front-back direction of the source. Caelli and Porter found that many localization errors were due to reversal errors in the front to back position of the sound. Localization errors also depended on the cycle rate of the siren employed, with the largest errors occurring with the slowly varying siren type (electronic wail of about 12 cycles/minute). Localization errors were also greatest where one window was open, probably due to a tunnel effect induced by the single opening in the car. Closing all of the windows reduced the error rate but also made it much more difficult to detect the sound of the siren in the first place, as well as increasing the error in estimating the distance of the emergency vehicle.

Generally, *all* distances were overestimated by the observers (up to a factor of two times), and this overestimate increased as the position of the ambulance moved toward the rear of the observer. Amazingly, the observers were very certain of their localization and distance judgments, even though they were often wrong by more than 20°, were reversed in direction, or were only at half the distance they believed to be from the emergency vehicle. Worst localization was when the ambulance was approaching the observer. A very dangerous situation is when an ambulance is approaching an intersection, because drivers at different positions and orientations in the intersection will get different and hard to interpret localization information;

Caelli and Porter suggest that in such situations flashing visual signals are probably superior to the siren. They concluded that the auditory system is inadequate for obtaining the localizing information required in the typical ambulance siren situation. Ambulance drivers ought to sound their sirens only briefly, and then only when they can see the car they are signalling. They also suggested that a warning device could be incorporated into the design of cars that would signal the presence of an emergency vehicle in response to triggering by a radio signal in the emergency vehicle. We might add to this notion a *localizing* auditory display inside the passenger compartment of the car: eight sound sources placed around the perimeter of the car's interior that would selectively respond and signal the direction and approximate distance of a speeding emergency vehicle. At the present time we would anticipate that such a device would have a high cost-to-benefit ratio; therefore you should not expect to see this system in next year's new car models.

Vigilance Tasks There is a large body of research on the changes that occur in human signal detection performance during very long periods of monitoring, particularly when the occurrence of signals may be infrequent and random over time (see Buckner & McGrath, 1963). Performance in such *vigilance* task situations may be significantly improved by the addition of an auditory display channel. This advantage may be gained even when the signals involved are more discriminable on the visual display channel. A study by Colquhoun (1975) evaluated this advantage in a simulation of a sonar monitoring task. Observers had to monitor visual only, auditory only, or combined displays for very long periods of time (up to 10 hours each day, for 12 days). Colquhoun reported that signal detection rates were consistently higher under the auditory display than the visual display condition, and higher yet with both auditory and visual displays together. The auditory displays seem to net an advantage in vigilance situations because of their superior attention-gaining capacity. Evaluation of the mechanisms operating in situations involving simultaneous visual and auditory displays is difficult because of the possible interactions between sensory events in each channel, the possibility of shared internal noise sources, and differences in observer response criteria between the channels. However, it appears that one can obtain a practical improvement in performance in the vigilance task by employing both an auditory and a visual display and having them monitored concurrently. This strategy maximizes the detection of signals and ensures that the visual displays will be available for the immediate processing of localization and identification information from the targets (Colquhoun, 1975).

Medium Information Rate Applications

Probably 99% or more of the applications of the auditory display channel involve use of multiple tone and/or voice announcement signals. The great use of the auditory channel in aircraft can be seen in Tables 8-3 and 8-4, which summarize currently used cockpit auditory signals. A Federal Aviation Administration (FAA, 1977a, b) series of reports pointed out the following facts.

1. There is no consistent utilization philosophy for auditory alerting signals.
2. There is only a partial standard for the acoustic codes employed.
3. The number of signals employed is increasing; newer widebody aircraft have 14 to 17 alerting signals (including some voice signals).

The potential for pilot confusion currently exists. A number of specific design guidelines for the design of cockpit warnings are discussed in these FAA reports; the interested reader is urged to employ them as a guide.

A NASA report by Cooper (1977) suggested that the number of auditory signals should be limited to four or five even though some current aircraft now have 30 or 40! The report pointed out problems with using very loud warning signals and suggested that voice warning systems are highly desirable as a potential means of coding a variety of warnings. The problem of loud and aversive auditory signals was also addressed in a study by Patterson and Milroy (1979) of warning signals on the Boeing 727 and BAC-111. They concluded that takeoff and undercarriage warnings were far too loud (over 100 dB) given the background noise level and would probably disrupt communication. They suggested that a reduction of 15 to 20 dB would not decrease the effectiveness of the signals. Another study by the same authors (Patterson & Milroy, 1980) studied the learning and retention of aircraft warning signals. They found that although large sets of auditory warnings can be learned and remembered, it takes considerable training time and requires regular retraining. They recommended the use of six or fewer warnings with distinctive spectral and temporal characteristics.

Similar problems with auditory warning signals exist in the alarm and annunciator-warning systems used in nuclear power plant control rooms (Kantowitz, 1977; Seminara, Gonzalez, & Parsons, 1977). In a review of nuclear power plant control room design, they found that a great deal of operator time was used for responding to auditory alarms, including determining the cause for the alarm, correcting the cause, clearing the alarm, and resetting the associated display panels. They pointed up a number of problems including the poor use of the auditory channel for conveying information about the urgency or cause of the condition, and inadequate use of signal intensity. In some cases signals were masked by noise, in others they were so intense as to be aversive and disrupting. A major

TABLE 8-3 *Summary of Currently Used Cockpit Aural Alerts (FAA, 1977a)*

Alert Condition \ Airplane	Type of Aural Alert Applied								
	707/727	727	737	747	DC-8	DC-9	DC-10	L-1011	BAC-111
Altitude alert	"C" chord A	Tone A, B "C" chord A	"C" chord A	"C" chord A	Horn	Horn**	"C" chord B	"C" chord B	•
APU fire	None	Bell A	Bell A	Bell A	None	None	None	Bell E	Bell
Attitude displays disagree	None	None	None	Tone B	None	None	Wailer (provisional)	None	•
Autopilot disengage	None	Wailer A (some acft)	None	Wailer A	None	Click	Wailer B	Wailer C	•
Call on interphone	Chime G	Chime G	Chime G	None	Chime	Chime	Chime	Chime N	•
Close proximity to ground, gear up	Warbler & voice	Warbler & voice	Warbler & voice	Warbler & voice	Warbler & voice	Warbler & voice	Warbler & voice	Warbler & voice	Warbler & voice
Cockpit call from flight attendants	Chime	Chime	Chime	Chime	Chime B Chime C	Chime B Chime C	Chime B Chime C	Chime H	•
Cockpit call from ground crew	None	None	None	Chime E	Chime	Chime	Chime	Chime P	•
Cockpit call to flight attendants	Chime	Chime	Chime	Chime D Chime L	Chime	Chime	Chime	Chime M	•
Decision height	Tone C	Tone C	Tone C	"C" chord A	None	Tone	None	Tone H	•
500-foot terrain warning	None	Tone	None	Tone	None	Tone	None	Tone	•
Emergency evacuation	Tone B	Tone B	None	Chime F Tone B	None	Horn	Horn	Tone F	•
Engine fire	Bell A	Bell A	Bell A	Bell A	Bell B	Bell B	Bell C	Bell E	Bell
Excessive airspeed	Bell D	Clacker A	Clacker A	Clacker A	Clacker D	Clacker D	Clacker C	Clacker F	Bell
Excessive sink rate	Warbler & voice	Warbler & voice	Warbler & voice	Warbler & voice	Warbler & voice	Warbler & voice	Warbler & voice	Warbler & voice	Warbler & voice

* Characteristics unkown.
** Not delivered but available.

TABLE 8-3 *(continued)*

Alert Condition \ Airplane	707/727	727	737	747	DC-8	DC-9	DC-10	L-1011	BAC-111
	Type of Aural Alert Applied								
Excessive terrain closure rate	Warbler & voice	Warbler & voice	Warbler & voice	Warbler & voice	Warbler & voice	Warbler & voice	Warbler & voice	Warbler & voice	Warbler & voice
Flap load relieve inoperative	None	None	None	None	None	None	None	Buzzer B	•
Galley overheat	None	None	None	None	None	None	None	Tone E	•
Inadvertent "duck under" GS	Voice	Voice	Voice	Voice	Voice	Voice	Voice	Voice	Voice
Low cabin pressure	Horn E	Horn E	Horn E	Horn F	Horn L	Horn L	Horn H	Horn S	•
Negative climb after takeoff	Warbler & voice	Warbler & voice	Warbler & voice	Warbler & voice	Warbler & voice	Warbler & voice	Warbler & voice	Warbler & voice	Warbler & voice
SELCAL	Chime J	Chime J Chime H	Chime H	Chime K	Chime None	Chime None	Bell None	Chime M None	• •
Instrument comparator alert	Clacker B	Clacker B	Clacker B	None	None	None	None	None	•
Smoke in cargo area	Bell A	None	Bell A	Bell A	None	None	None	Tone E	•
Smoke in lower galley	None	None	None	None	Horn J (on 60 models)	Horn K	Horn G	None	•
Stabilizer in motion	Clacker	Clacker	Clacker	Clacker					
Unsafe ground condition	None	Horn A	None	None	None	None	None	None	•
Unsafe in-flight condition	None	Horn E	None	None	None	None	None	None	•
Unsafe landing condition	Horn A	Horn A	Horn A	Horn B	Horn C	Horn C	Horn D	Horn R	•
Unsafe takeoff condition	Horn E	Horn E	Horn E	Horn F	Horn L	Horn L	Horn H	Horn S	Horn S
Wheelwell overheat or fire	Bell A	Bell A	Bell A	Bell A	None	None	None	Bell E	•

• Characteristics Unknown.

TABLE 8-4 *Aural Alert Characteristics (FAA, 1977a)*

Aural Alert	Frequency, Hz	Loudness, dB	Description
Horn A	200 to 443	90 ± 5	Continuous
Horn B	220 to 280	93 ± 5	Continuous
Horn C	635	89	Continuous
Horn D	602 and 657	85 ± 5	Continuous
Horn E	200 to 443	90 ± 5	Same as horn A, except interrupted
Horn F	220 to 280	93 ± 5	Same as horn B, except interrupted at 3 Hz
Horn G	116 and 259		Continuous
Horn H	602 and 657		Same as horn D, except interrupted at 1 Hz
Horn J	140	91	On 0.5 seconds, off 0.8 seconds in variable sized groups, 2 seconds between groups
Horn K	60	85	On for 0.5 seconds, off for 1 second
Horn L	635	84 to 98	Interrupted at 0.6 Hz
Horn M	625	95	Interrupted at 0.6 Hz
Horn N	325 and 390	94	Two tones alternating at 0.25 Hz
Horn P			"Ooga" horn
Horn R	300	86	Continuous
Horn S	300	90	333 ms period with a 50% duty cycle
Tone A	1000		Continuous
Tone B	2800 ± 300	90 ± 5	Beeper tone, pulsating at 1.5 to 5.0 Hz
Tone C	800	Increasing	Tone that increases in volume over a 3-second period
Tone D	400	Increasing	New system for McDonnell Douglas airplanes, application uncertain
Tone E	1.4 k to 2.0 K	90	Alternating tone
Tone F	3 k	77	333 ms period with a 50% duty cycle
Tone G	700 to 1.7 k	90	Pulsating tone
Tone H	1.0 K		Pulsating tone
"C" chord A	461 to 563 567 to 704 691 to 845	95 ± 5	Intermittent
"C" chord B	512, 640, 768	90	Sound duration 2 seconds
Buzzer A	300, 600, and 900	90 ± 5	
Buzzer B	90	81	2 seconds
Warbler & voice	400 to 800	85 to 96	Three "whoops" per second, followed by voice saying "pull up." Some of the airplanes indicated do not have this system and some have the warbler without voice
Wailer A	130 ± 20 to 200 ± 30	93 ± 3	2 to 4 Hz of variation between longer and higher frequencies—minimum variation 49 Hz—mod 4.76 Hz
Wailer B	640		
Wailer C	130 to 200	88	
Bell A	600 to 10,000	93 ± 5	Continuous
Bell B	750	87	Continuous, striker frequency, 1.3 Hz similar to telephone
Bell C	640 and 648		Continuous, two tones alternating, striker frequency, 12.5 Hz
Bell D	600 to 10,000	95 ± 5	Same as bell A, except interrupted

TABLE 8-4 *(continued)*

Aural Alert	Frequency, Hz	Loudness, dB	Description
Bell E	100		"Gong" type bell – electrically activated
Clacker A	1000 to 2400	86	Modulated at 5 to 10 Hz
Clacker B			Repetition frequency, 1 Hz
Clacker C	512		Repetition frequency, 4.76 Hz, sounds like clucking of a chicken
Clacker D	Two tones, clicks	84 to 96	Repetition frequency, 9 Hz
Clacker E	335	87	Similar to a square wave, modulated with very distinctive clicks at 10 Hz
Clacker F	2500	86	Two burst in a 20-ms interval repeated at a 340-ms rate
Chime A	620	87	Repeating, 1.5 second repetition rate
Chime B	750	76 to 84	Single stroke gong-like sound, when mechanics call, interrupted at 0.85 Hz
Chime C	4700	76	Single stroke gong-like sound
Chime D	727 to 947	95 ± 5	"High chime", single stroke gong-like sound
Chime E	477 to 497	95 ± 5	"Low chime", single stroke gong-like sound
Chime F	727 to 947 and 477 to 497	95 ± 5	High-low chime combination of chimes: D and E repeated at a rate of 3± 1 Hz
Chime G	588	95 ± 5	"High chime", single stroke gong-like sound
Chime H	588 and 488	95 ± 5	High-low chime not repeated
Chime J	588 and 488	95 ± 5	Same as chime H except fast repeat
Chime K	588 and 488	95 ± 5	Same as chime H except it does two cycles and stops
Chime L	727 to 947 and 477 to 497	95 ± 5	Same as chime F except it does two cycles and stops
Chime M	587	85	Single chime in most configurations
Chime N	587/487	85	Single high-low chime
Chime P	487	85	Low chime not repeated
Click			Actual sound of disconnect of the autopilot lever

problem in the nuclear application that is not shared by the aircraft application is the occurrence of nuisance or false alarms.

Operators at all plants complained about the high number of nuisance alarms. The reasons for their occurrence varied. At one plant there was a blank, supposedly nonfunctional, annunciator window that would occasionally alarm. The maintenance and operational people had been unable to determine its cause, but an acknowledgement, silence, and reset were required on each occasion. In many cases alarm set-points were known by operators to be too sensitive to normal transients. As a consequence slight deviations or transients, thought of as normal, would set the alarm off even though no further operational action was required. Maintenance or calibration operations often caused recurring alarms that were a nuisance. The net result of the many false alarms is a "cry-wolf" syndrome which leads to lack of faith in the system and a casual attitude towards the constant presence of certain alarms. On many occasions operators were observed to casually silence and acknowledge alarms without further concern or surveillance of plant status; the alarms had become "old friends" (Seminara, Gonzalez, & Parsons, 1977).

Speech Message Signals The need for greater alarm specificity in both the nuclear and aircraft applications suggests that more and more use will be made of alarm and warning signals incorporating brief speech messages. The brief voice message may be a very effective way to communicate critical information to an operator or an aircraft crew member, without interfering with a concurrent visual task, without dependence on the direction of eye and head position, and without the training required for a novel acoustic code. Experiments on female-voice cockpit warning systems for Air Force and Navy aircraft (cited by Simpson & Williams, 1980), have indicated a response time advantage for using a voice warning system in conjunction with a visual warning display, as compared with the visual display alone. Cooper (1977) has pointed out that the modular design of aircraft electronic equipment may be a problem in integrating cockpit voice warning systems so that there is some priority control of the messages—for example, so that an emergency situation doesn't result in several voices shouting at you all at once.

FOCUS ON RESEARCH

Extra Words Help, Extra Tones Don't Help

There is some controversy about the best way to present voice warning information to the crew member or operator. Several human factors guides (including the FAA 1977a, b guides cited) suggest use of an alerting tone just prior to the voice warning signal. Simpson and Williams (1980) point out that in cases where the characteristics of the warning voice are sufficiently distinctive (and intelligible in the given environment), such a prior alerting signal is probably not necessary. In cases where the warning voice were female, it would probably be so different from other voices in the cockpit that its spectral characteristics alone would fill the alerting function. The probability of female pilots, crew members, and air controllers removes this distinctive aspect of the female voice. Synthetic speech (see Chapter 9) offers one technique for constructing voice warnings that combine aspects of high distinctiveness and intelligibility for warning and alerting signals. Simpson and Williams assessed the use of prior alerting tones in synthesized voice warning systems as well as the effects of adding certain extraneous words to the brief warning messages. Their experiments employed a general purpose aircraft simulator at NASA-Ames Research Center that was programmed with the flight

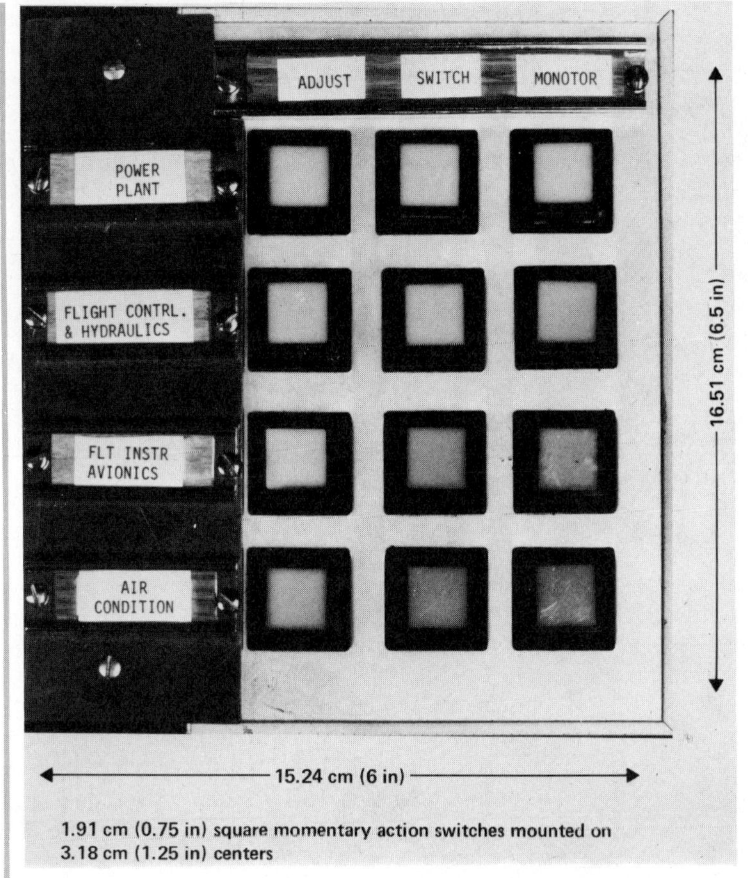

Figure 8-5. "Immediate action" response box employed in Simpson & Williams (1980) study. The authors' note that nobody noticed the misspelling of monitor. (From Human Factors 1980, *22,* 322. Copyright 1980 by The Human Factors Society, Inc. and reproduced by permission).

dynamics of a Boeing 727. Warnings were generated by a VOTRAX VS-6 speech synthesizer to randomly issue one of six warnings. The warnings required the subject to rapidly push one of 12 response buttons. The response box is shown in Figure 8-5. The possible warnings and their correct responses are shown in Table 8-5. When a warning was followed by an incorrect response, the warning message was terminated and the response was scored as an error. The subjects were qualified airline pilots who were given instruction in the simulator and the warning system; all the pilots were responding accurately and rapidly to all the warning signals by the end of the initial training period. The Simpson and Williams study included a number of interesting design features that we will not discuss here; the crucial aspects of their study involved two manipulations of the warning stimuli: First,

TABLE 8-5 *Correct Immediate Action "Row" by "Column" Button Responses (Simpson & Williams, 1980)*[a]

Warning	Immediate Action	Row	Column
Landing gear not down	Gear handle DOWN	FLIGHT CONTRL. & HYDRAULICS	SWITCH
Tank boost pumps out	Crossfeed OPEN	POWER PLANT	SWITCH
Cabin pressure dropping	Man/auto selector MANUAL	AIR CONDITION	SWITCH
Collision traffic __ o'clock	AVOID	FLIGHT CONTRL. & HYDRAULICS	ADJUST
Flight instruments disagree	Select reliable system	FLT INSTR AVIONICS	SWITCH
Engine fuel filter bypass	Engine MONITOR	POWER PLANT	MONITOR

[a] From *Human Factors,* 1980, *22,* 324. Copyright by The Human Factors Society, Inc., and reproduced by permission.

warning messages were either presented in a stripped-down key word format, or with an extra word that provided additional semantic context for the message. Table 8-6 illustrates these two message formats along with the fact that the semantic context messages averaged about $\frac{1}{2}$ sec longer in duration than the key word format message. The second aspect of their study tested the effect of an alerting tone of $\frac{1}{2}$ sec duration followed by an off period of $\frac{1}{2}$ sec, just prior to each message. The time sequences of these conditions and the resulting average response times are graphically illustrated in Figure 8-6. You can see that there is *no* advantage in using an alerting tone in these situations and neither is there a *disadvantage* in using the longer semantic context message. Bear in mind that these tasks were embedded in a busy flying task and

TABLE 8-6 *Warning Message Formats and Durations in Seconds (Simpson & Williams, 1980).*[a]

Semantic Context	Duration (s)	Keyword	Duration (s)
Landing gear not down	1.78	Gear not down	1.25
Tank boost pumps out	1.67	Boost pumps out	1.32
Cabin pressure dropping	1.77	Pressure dropping	1.31
Collision traffic one o'clock	1.88	Traffic one o'clock	1.36
Engine fuel filter bypass	1.87	Fuel filter bypass	1.40
Flight instruments disagree	1.88	Instruments disagree	1.48
Mean message duration (s)	1.80		1.35
Standard deviation	0.080		0.070

[a] From *Human Factors,* 1980, *22,* 322. Copyright by The Human Factors Society, Inc., and reproduced by permission.

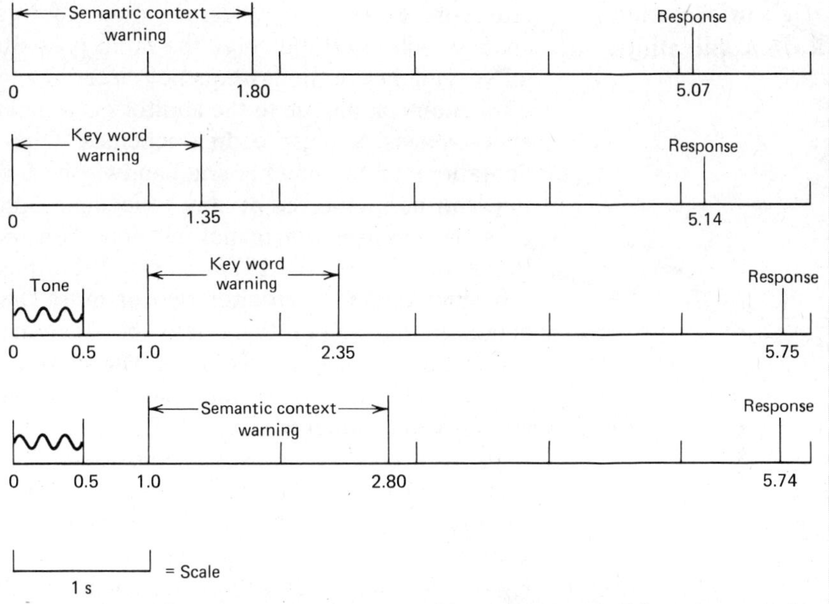

Figure 8-6. Mean times for four warning message conditions in study by Simpson & Williams (1980). (From *Human Factors,* 1980, *22,* 326. Copyright 1980 by The Human Factors Society, Inc. and reproduced by permission.)

that there was probably a distinct change in the voice quality of the warning message compared with other ongoing cockpit speech and communications. Even if the alerting tone could be effectively shortened (including the necessary off period) to about $\frac{1}{2}$ sec, there would still be a net loss in reaction time to the message-plus-tone warning. The caution to be noted, however, is that the speech warning message must be qualitatively different from other voices heard in the operating environment. As other components of the system become computerized, the probability of increased usage of synthesized speech increases, and the synthesized speech warning could lose its distinctiveness. In that case, a prior tonal warning signal might be necessary.

The other result of the Simpson and Williams study is that adding about 300 msec of speech, the extraneous word, to the speech warning actually reduced average reaction time to the message, rather than increased it. The authors concluded that additional linguistic redundancy in the message reduces the overall attentional capacity required for speech comprehension. The average processing time per word is thus reduced because some words or syllables become more predictable within the overall message context. This is an important factor to be considered by the designer of voice warning systems, because more information can be effectively transmitted with only a small increase in message size and little or no increase in response time.

High-Information Rate Applications

Attempts to convey moderate rates of information via the auditory channel will usually involve the rapid presentation of sequences of stimuli varying in duration, frequency, intensity, or spatial aspect. A number of phenomena unique to the auditory channel may function to reduce the effectiveness of these coding schemes. One set of limitations arises in limitations on the number and bandwidth of different frequency channels that can be monitored at about the same time. Experiments on the processing of information in multiple detection and recognition tasks (Sorkin, Pohlmann, & Woods, 1976; Swets, 1982) have indicated that observers can simultaneously monitor two or more frequency channels in simple detection tasks, but there may be decrements in performance when multiple responses are required. There is an extensive literature on the problem of attention to multiple input signals. (See the discussion of attention in Chapter 6).

There are other important constraints on our ability to employ the auditory channel for the processing of sequentially arriving information. Bregman (1981) and his colleagues have demonstrated that auditory sequences are perceptually organized in certain stereotyped ways, depending on the most likely source and location of the sound. Suppose for example, that a sequence of alternating high- and low-frequency tones were employed to encode data. Bregman's studies have demonstrated that these high- and low-frequency sequences will be separately perceived as high- and low-frequency "streams" and that it will be much more difficult to obtain timing and order information *between* tones from different streams than from tones *within* each stream. Other constraints on our perception of tonal sequences are discussed by Kubovy (1981), Jones, Kidd, and Wetzel (1981), Deutsch and Feroe (1981), and Sorkin, Boggs, and Brady (1982).

A second important set of constraints applies to an observer's ability to extract information from complex, word-length tonal sequences. There are a number of interesting experiments in this area by Watson and his colleagues (Watson, 1976; Watson & Kelly, 1981) and by Warren (1974). Apparently there are large differences in an observer's ability to resolve certain temporal and spectral portions of sound sequences. These differences depend, in part, on an observer's uncertainty or familiarity with these sounds. We have a limited capacity for storing the fine-grain information present in a complex word-length tone pattern, even though we can resolve the components of that pattern in a simple detection or discrimination (minimal uncertainty) situation. Adult observers have a tendency to allocate their memory and attention to certain frequency and time elements in a tonal pattern; late arriving, high-frequency components are usually resolved better than early arriving, low-frequency components. A specific pattern or tone sequence can serve as a "signature" that somehow guides selective attention and memory mechanisms toward specific components of the sequence that would otherwise not be resolved. These limitations and listening tendencies almost certainly are involved in our ordinary processing of speech; their specification will be an important requirement for the design of efficient nonspeech codes for the auditory channel.

TACTILE DISPLAYS

You've just dropped in on the auto body shop where your recently damaged car is being repaired. A worker is checking the surface of some holes that have been filled in the car's fender. He is using a technique that you have not seen before. He is moving a thin piece of paper across the fender surface, rather than touching the surface directly with his fingers. Can he detect the texture of the surface better this way?

You are checking the proposed design of the controls in a new, complex system. The designer wishes to incorporate some form of tactile pressure or vibratory feedback in the main control lever of the system. You are asked whether this will improve the operator's ability to control the system.

These two examples of tactile displays only brush the surface of the possible situations involving tactile stimulation. We seem to take the skin sense somewhat for granted as a way of gaining information about our environment, considering it a poor cousin to the dominant senses of vision and hearing. Yet the tactile channel can be an extremely important part of the total sensory input to the human component of a system. A simple example is in the use of different control knobs in the cockpit of an aircraft or in an underwater repair system. Coding of knob shape can be a critical factor in reducing the chance of operator error under high-information load or high-stress conditions. At the other extreme of application of the tactile sense, are the high-information rate displays that employ the tactile channel as a means for overcoming deficiencies in a person's vision or hearing. We will discuss systems that have been developed for communicating visual information via the skin as a partial replacement for inadequate or lost vision, and some systems proposed for communicating speech information to the deaf. Between these two extremes are a wide range of practical applications of the tactile channel.

The Tactile Sense

Tactile stimulation can be accomplished via a number of different methods that present mechanical, thermal, chemical, or electrical energy to the skin. The perceptions that we report, given these different types of energy input, can take the form of pressure, vibration, pain, itch, warmth, or cold. Mediating this variety of tactile sensations are a number of different receptor structures in the skin, including free nerve endings and encapsulated nerve endings. There seems to be no precise correspondence between the particular receptor and the type of sensation elicited. One important type of receptor is the Pacinian corpuscle, which is sensitive to mechanical indentation of the skin.

The Pacinian corpuscles are relatively deep in the skin and are also in muscle tendons and joints. There are a number of other types of structures nearer to the skin surface that are sensitive to pain, pressure, cold, and warmth. The Pacinian corpuscle appears to be primarily responsive to rapid changes in the mechanical stimulation of the skin. According to Lowenstein and Skalak (1966), the Pacinian corpuscle functions as a mechanical system for shaping the response of its associated neural fiber. That is, the Pacinian corpuscle limits the type of mechanical stimulation of the skin above it that gets passed on to its encapsulated nerve fiber. The system responds primarily to rapid applications or removals of skin pressure, rather than to steady pressure of different amounts. The situation is similar to that of the temporal and spatial sensitivity functions described in Chapter 4; the Pacinian system is primarily responsive to mid and high frequencies of temporal stimulation, and has very poor low-frequency sensitivity. This system is thus a major component in our perception of rapid vibration of the skin.

Our tactile sense has been categorized in two distinct ways. The first type of touch sensation has been referred to as *cutaneous sensitivity* or *passive touch* to distinguish between it and the second, more complex function called *active touch*. In the passive touch situation, the skin is stimulated, and the observer's responses are recorded. In the active touch situation, the observer actively explores a surface or object with his or her fingers or hand. These two different modes of operation of the skin sense result in quite different perceptions and have different properties.

Sensitivity of the Cutaneous System

There is a large variation in the range of the sensitivity of our skin to physical energy, depending on where on our body the energy is applied. Our tongue and lips are especially sensitive to mechanical pressure, but our feet and back are not. Figures 8-7 and 8-8 illustrate the results of a study by Weinstein (1968) on the distribution of skin threshold sensitivities to pressure for male and for female subjects. The lowest thresholds are in the face area, with the fingers and upper body (for the male) following in sensitivity. Overall sensitivity for females is apparently better than for males. Another important measure is the ability to distinguish a stimulus composed of two separated pressure points from a single pressure stimulation (pressure stimulation is via application of a force-calibrated nylon filament). Figure 8-9 illustrates these two-point discrimination thresholds for males. Again, lowest thresholds are in the facial and finger regions, but now the fingers and hand are distinctly superior to any other part of the body in this task. Weinstein also assessed how well a subject could localize the position of a point stimulus relative to a reference stimulus. These point localization thresholds are shown in Figure 8-10. The minimum energies needed to produce a touch sensation can be as small as a few hundreds of an erg. [An erg is about 10^{-7} watt-seconds (W-sec).]

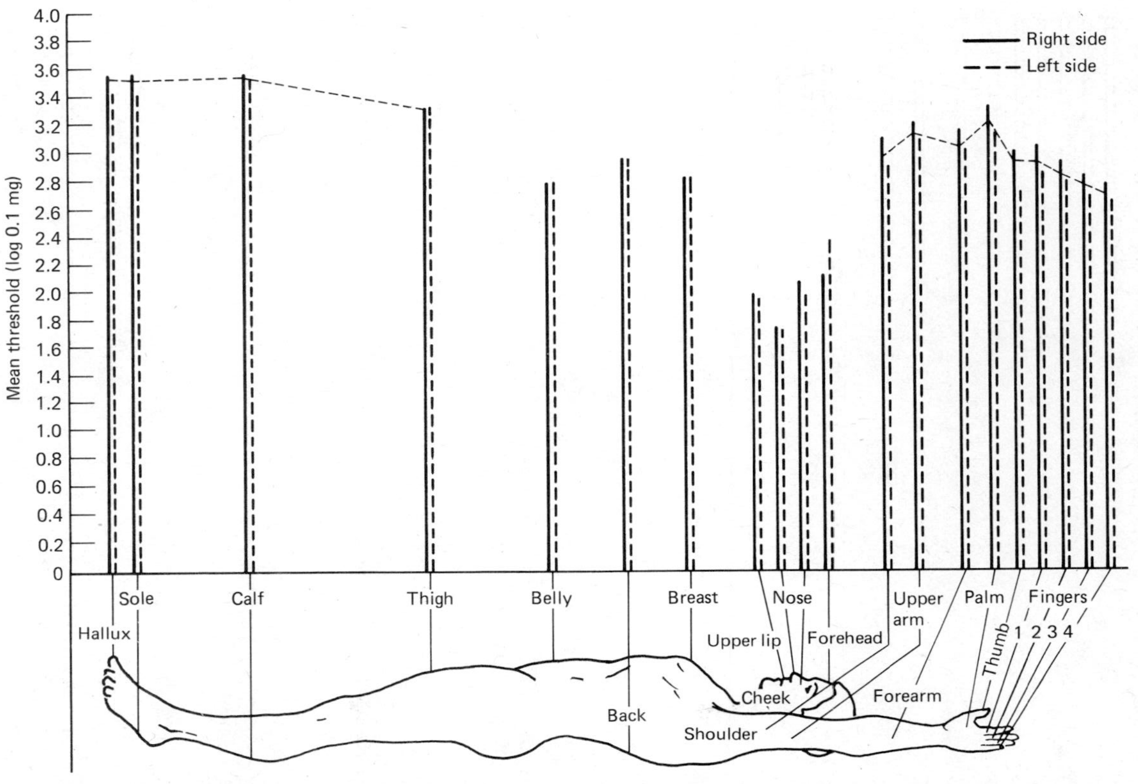

Figure 8-7. Pressure sensitivity thresholds for males. (From S. Weinstein, Intensive and extensive aspects of tactile sensitivity as a function of body part, sex, and laterality, in D. R. Kenshalo (Ed.), *The Skin Senses,* 1968, pp. 195-218. Courtesy of Charles C. Thomas, Publisher, Springfield, Illinois.)

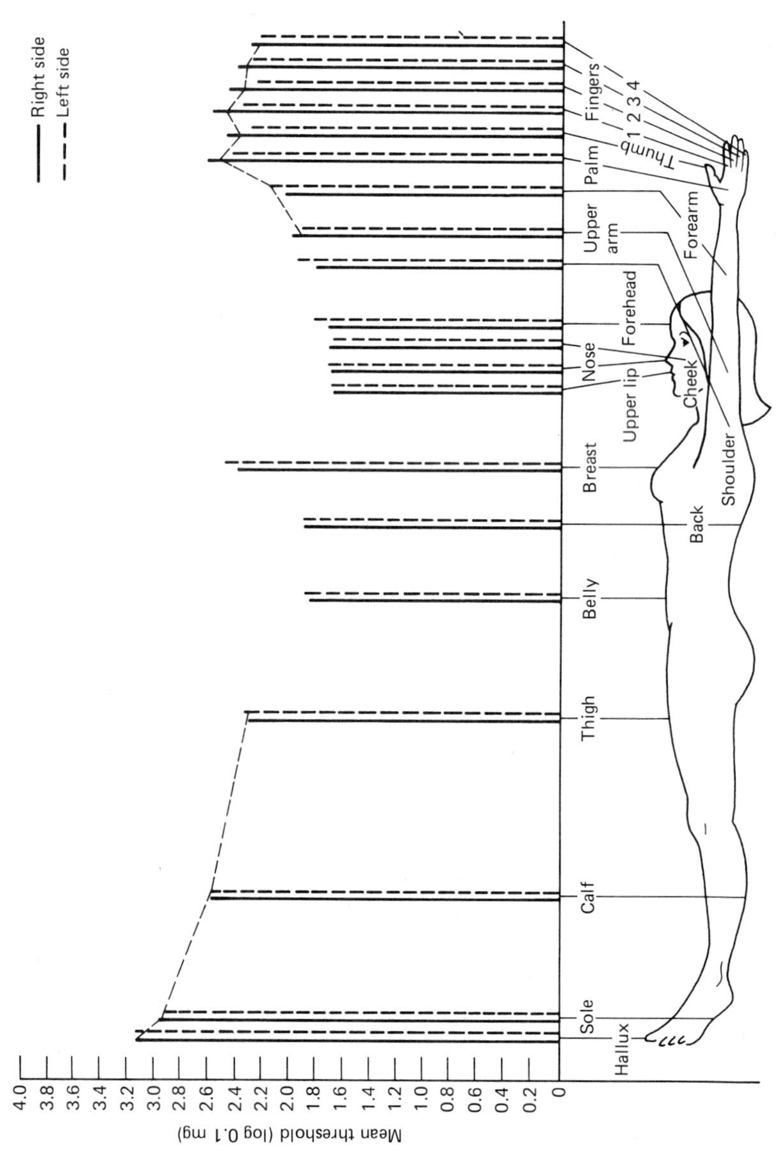

Figure 8-8. Pressure sensitivity thresholds for females. (From S. Weinstein, Intensive and extensive aspects of tactile sensitivity as a function of body part, sex, and laterality, in D. R. Kenshalo (Ed.), *The Skin Senses,* 1968, pp. 195-218. Courtesy of Charles C. Thomas, Publisher, Springfield, Illinois.)

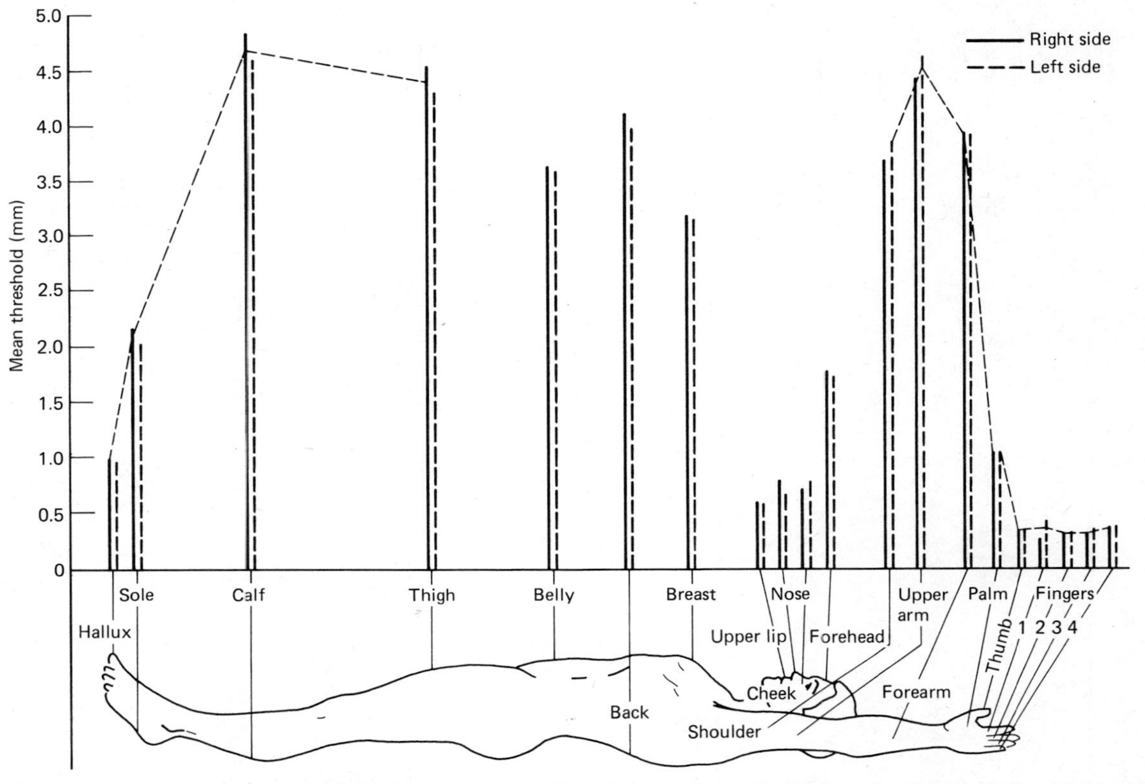

Figure 8-9. Two-point discrimination thresholds for males. (From S. Weinstein, Intensive and extensive aspects of tactile sensitivity as a function of body part, sex, and laterality, in D. R. Kenshalo (Ed.), *The Skin Senses,* 1968, pp. 195-218. Courtesy of Charles C. Thomas, Publisher, Springfield, Illinois.)

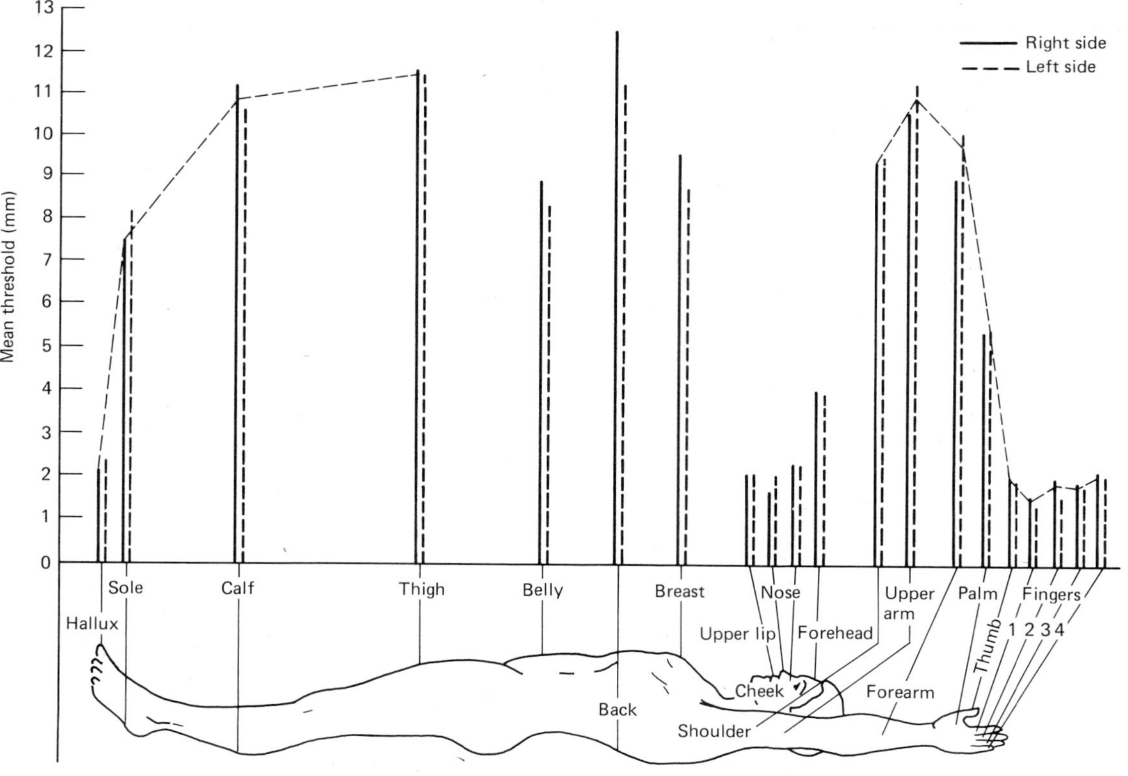

Figure 8-10. Point localization thresholds for males. (From S. Weinstein, Intensive and extensive aspects of tactile sensitivity as a function of body part, sex, and laterality, in D. R. Kenshalo (Ed.), *The Skin Senses*, 1968, pp. 195-218. Courtesy of Charles C. Thomas, Publisher, Springfield, Illinois.)

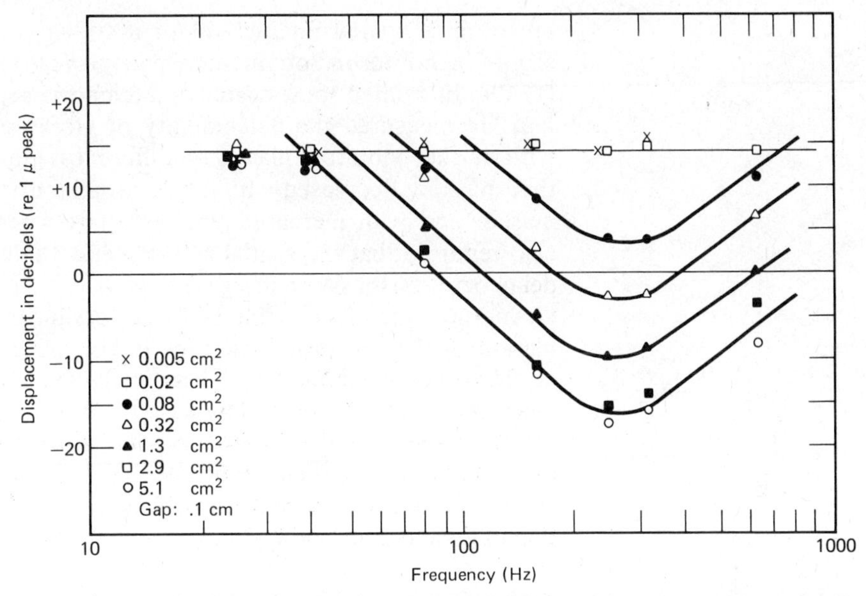

Figure 8-11. Vibrotactile thresholds as a function of frequency. (From R. T. Verrillo, Effects of contactor area on the vibrotactile threshold, *Journal of the Acoustical Society of America,* 1966, *35,* 1965, and reproduced by permission.)

Sensitivity to changing patterns of pressure such as produced by a vibrating surface extends over a wide range of vibratory rates, perhaps from 10 Hz to several thousand hertz, with the threshold amplitude of the vibration pattern being as small as a few hundredths of a millimeter. Figure 8-11, from experiments by Verrillo (1966) shows how the vibrotactile threshold increases at frequencies below and above 200 Hz, for stimulated areas larger than 0.02 cm². The Pacinian corpuscle is probably the major subsystem responsible for our sensitivity to vibrotactile stimulation. Sensitivity to pressure and alternating patterns also depends on such factors as the temperature of the skin and its adaptive state with respect to prior stimulation. We seem to adapt quickly to low-pressure, large-area stimulation. Are you normally aware of the pressure produced by the action of your clothing against your skin?

In Chapter 4, we discussed some surprising interactions that occur between the light and dark areas of a visual target, specifically, the dark and light bands (Mach bands) that occur at the borders where a light and dark area meet. We described these phenomena in terms of a lowered sensitivity of the visual system to low spatial frequencies. Similar effects also occur in the stimulation of adjacent areas of the skin. There are areas of the skin where a stimulus will act to reduce or inhibit the sensation of a simultaneously presented stimulus that is a few centimeters away. Beyond this separation, no interaction will occur. The presence of spatial interactions or masking is of significance in the design of high-information rate tactile displays (see later discussion). Gilson (1969a, b) studied some spatial and temporal aspects of cutaneous masking. He measured the detectability of vibration on the upper left thigh with masker vibrators placed on different regions of the body. He found that masking decreased with greater spatial distance between signal and masker and with increased time delay between signal and masker. He also reported that the spatial masking effect interacted with the temporal delay between masker and signal onset.

Another type of skin stimulation is possible using direct electrical stimulation of the receptor structures in the skin, termed electrocutaneous or electrotactile stimulation. The threshold sensitivities for electrocutaneous stimulation can be as low as 10^{-7} W-sec. There are some practical problems associated with electrotactile stimulation such as a very small dynamic range between the absolute threshold and the level where pain is experienced, and a large amount of variability depending on the location of the electrical contact point and the nature of the contact.

Active Touch In the preceding discussion we described some aspects of our sensitivity to mechanical and electrical stimuli presented passively to our skin. But most of the time we employ our sense of touch in quite a different way than this passive mode. Generally we touch objects as part of an active process to determine their characteristics, and the resulting perceptions are quite different from those that result when someone stimulates parts of our skin, such as in a two-point touch threshold determination. Active touch also involves the stimulation of our internal sensory systems that signal the position and stress of the tendons and joints, and their interaction with the tactile receptor systems (as well as interaction with the visual and acoustic senses). Gibson (1962) has shown that we can identify shapes much more accurately if we can actively feel the shape surfaces rather than having the same surfaces pushed onto our skin in a passive mode. Actually, we virtually never employ our skin surfaces in a passive mode. Think of the terms normally used to describe the tactile sense: we press, rub, and squeeze surfaces to acquire information about object and surface properties. In all these cases, our mode of information acquisition is clearly in the active rather than the passive mode.

An interesting aspect of our touch system's sensitivity to the orientation of undulations on an object's surface has been described by Gordon and Cooper (1975) and in related studies by Lederman (1978). Gordon and Cooper demonstrated that the orientation of a raised region in a surface can be detected more accurately if a person moves a thin piece of paper across the surface under their fingers rather than their bare fingers alone. (See the brief example at the beginning of this section). Lederman (1978) evaluated some hypotheses about the cause of this phenomena. She showed that the perceived roughness of a surface is heightened when an intermediate sheet of paper is used between the fingers and the sur-

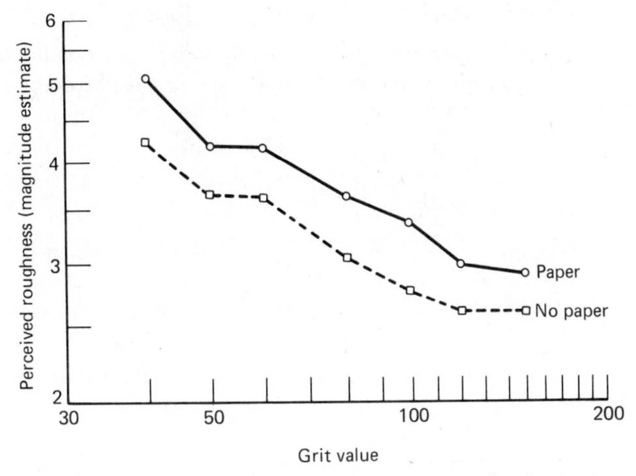

Figure 8-12. Perceived roughness (magnitude estimates) as a function of sandpaper grit in paper and no-paper conditions. (From S. J. Lederman, "Improving one's touch" . . . and more, *Perception & Psychophysics,* 1978, *24,* 155. Copyright 1978 by the Psychonomic Society, Inc., and reproduced by permission.)

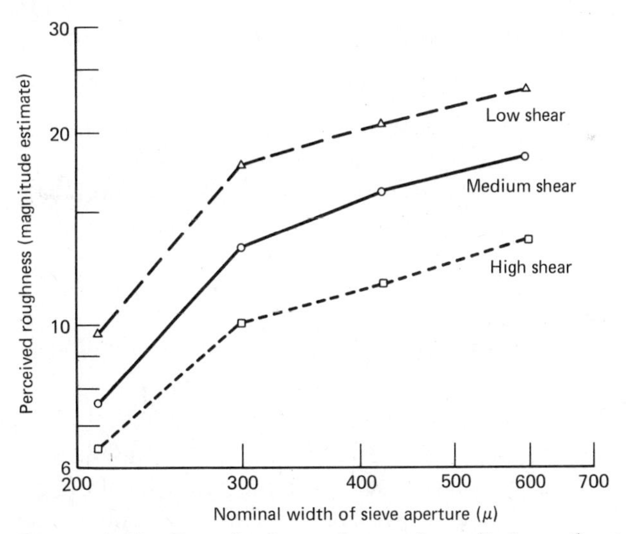

Figure 8-13. Perceived roughness (magnitude estimates) as a function of nominal width of seive aperture and shear. (From S. J. Lederman, "Improving one's touch" . . . and more, *Perception & Psychophysics,* 1978, *24,* 158. Copyright 1978 by the Psychonomic Society, Inc., and reproduced by permission.)

face, and that this effect depends on the magnitude of lateral (shear) forces acting on the skin (See Figure 8-12 and 8-13). Under bare finger touch, shear forces are produced by finger pressure on the surface. These shear forces act to mask the signals that convey information about surface roughness and orientation. The phenomena is relevant to the question of whether thin hand coverings may, in some circumstances, enhance the perception of information available in the tactile channel. Such thin coverings might be usable in aids such as tactile maps for the visually impaired.

Tactile Displays and Controls

Specific design guidelines for shape or texture coded controls, knobs, and dials may be found in appropriate human factors handbooks such as Van Cott and Kinkade (1972). It is possible to design sets of shape- or texture-coded controls in which confusions between any pair of controls within the set are minimal. Essentially one performs an information transmission experiment where the confusions among the controls are assessed (see Chapter 2 and 6). Controls that contribute to confusions can be eliminated from the set used to code the system functions. In some situations, the adequacy of the tactile control code may be critically important. An example is in situations where there is limited vision due to the wearing of equipment such as diving or radiation protection equipment. Carter's study (1978) investigated the speed and accuracy of the performance of Navy divers in adjusting a set of knobs of the type that might be used in underwater applications. Carter found that divers wearing gloves in different temperature and pressure conditions generally performed better with much larger-diameter knobs than are normally available (see Figure 10-6). They recommended tactile markers on the knob surface to enhance setting accuracy. Wearing protective equipment can seriously degrade the tactile feedback available in these situations. Interactions between a degraded visual channel and the tactile channel also may be a significant problem for the human factors specialist to consider. The box on tactile-vision interaction highlights this point.

FOCUS ON RESEARCH

Interactions between Degraded Sight and Touch Information

A study by Banks and Goehring (1979) clearly illustrates the interdependence between our use of the visual and tactile senses in performing tasks. The purpose of the study was to assess performance and physiological stress connected with different diving equipment and underwater working conditions. A major concern was to study possible interactions between the effects of wearing gloves and working in murky waters where visibility is very poor. As visual conditions deteriorate, more diver reliance on his or her tactile sensory channels is required. If, in addition, it is necessary to wear gloves that limit the tactile channel, a significant and deleterious effect on performance may result.

In the Banks and Goehring study, divers had to perform a standard underwater task incorporating functions similar to those performed by military salvage divers, such as assembly and disassembly of hatches, valves, shackles, bolts, and nuts. Figure 8-14 shows how the mean

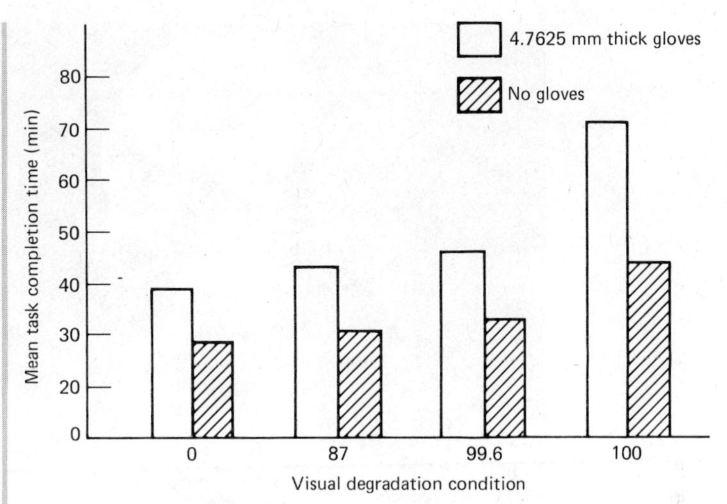

Figure 8-14. Relationship between diver performance and luminance degradation for both gloves and no gloves tactile conditions. (From W. Banks & G. S. Goehring. The effects of degraded visual and tactile information on diver work performance, *Human Factors,* 1979, *21,* 413. Copyright 1979 by The Human Factors Society, Inc., and reproduced by permission.)

time to complete the task depended on the illumination condition and whether or not the diver had to wear 4.7625 mm thick, three-fingered gloves. Performance was degraded by a factor of 2 between the 0 and 100% vision conditions. Wearing gloves increased the time by at least 50% in all conditions. There was a clear interaction between the tactile and visual degradation conditions. The pattern of errors made followed roughly the same pattern. The error rate increased by a factor of 8 between the 0 and 100% degraded visual condition for the no-gloves case and by a factor of 13 for the gloved case. It is clear that both work time and errors are increased by degrading the visual and tactile channels and that interactions between these channels can lead to significant performance effects. Additional interactions may also result from the fact that decreased visibility and the requirement for thicker gloves frequently go along with more dangerous and stressful underwater environments. Curley and Bachrach (1981) investigated the ability of divers to make weight discriminations and identify objects while working inside of the 1 atm armored diving suit shown in Figure 8-15. This suit's upper limbs are equipped with hand-activated mechanical manipulators. The ability of divers to discriminate weight (at $> 80\%$) was approximately 0.7 kg compared to 0.2 kg under normal, unsuited conditions. In underwater conditions of reduced visibility, the divers depended completely on tactile cues to identify known objects and tools. These studies suggest that concern for improved underwater visibility and provisions for adequate tactile feedback should be important considerations in the design of underwater work systems.

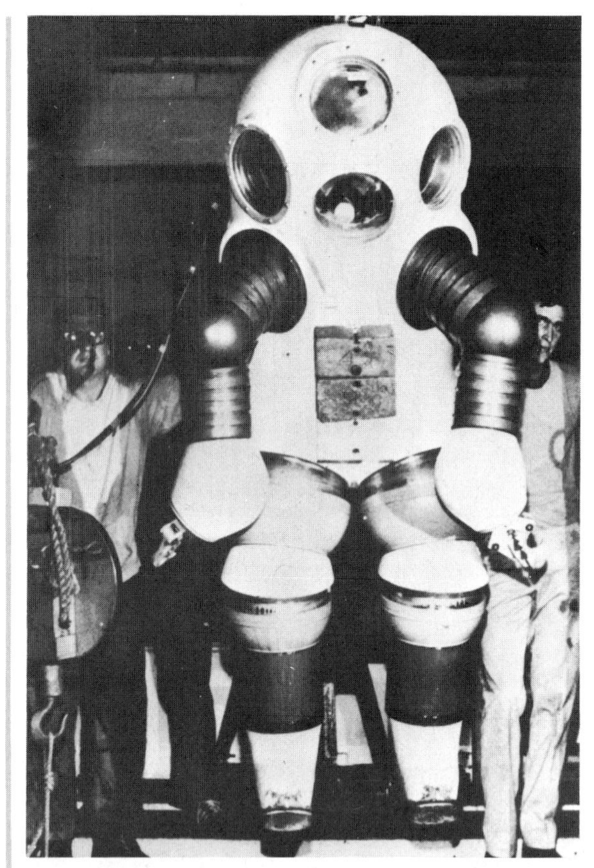

Figure 8-15. One atmosphere diving suit. (Courtesy M. D. Curley, Naval Medical Research Institute.)

Tracking Displays

Experiments have been performed on tactile displays as the primary display or as an auxiliary display in a target tracking task. These displays have employed vibrating stimulators (composed of arrays of vibrators), a protruding displacement display, and a linear airjet display. Possible applications of the tactile channel in tracking tasks are in high visual and auditory work load situations such as the control of aircraft, and in remote-controlled systems where visual feedback is poor or occurs too late for action. An example suggested by Hirsch (1974) is in the control of fine machining operations where the operator must be constantly aware of the rate of metal removed by the machining operation. If feedback could be supplied to the operator about the tool pressure on the workpiece, the machining or milling operation could be more precisely monitored. In a similar manner, a ground controller of a remotely controlled vehicle could be provided with change-rate information not otherwise available (unless the controller were actually in the vehicle) and this information would not compete for attention with the input from the visual modality. In the diving suit applications already discussed, the provision of augmented tactile feedback from the mechanical manipulator systems would fall in this category.

Triggs, Levison, and Sanneman (1974) reviewed a number of different types of vibrotactile and electrocutaneous tracking displays including some simulating aircraft control situations. An interesting feature of the tactile tracking application is that you can make direct comparisons of performance such as rms error, between tracking with tactile display systems, and tracking with visual display systems. Remember that in conveying tracking error or rate data via a tactile display it is often necessary to time sample and quantize the data so that they can be presented (see Chapter 9) because of the inherently poorer resolution of most tactile display systems. For example, in one of the situations investigated by Triggs et al. (1974) the output control was a two-axis, force-sensitive hand control and the input was either a visual display of aircraft pitch and roll or one of several electrotactile and vibrotactile display configurations. (Pitch is aircraft rotation around an axis connecting the wings; roll is rotation around an axis running from the nose to the tail.) One such display was an X-Y display of 13 tactile stimulators (tactors), 7 in a line crossing on a common tactor. In the polarized display condition, the mode of presentation involved the sequential activation of the tactors, sweeping from the center tactor outward, alternately horizontally and vertically. The display thus swept these dimensions at a fixed rate. A circular tactor array consisting of a center tactor plus two concentric circles of tactors was also evaluated. Again, the sequential display mode involved sweeping outward from the center tactor depending on the desired data to be displayed. So the tactile information is inherently limited by the scan rate (e.g., "sampling rate" as in Chapter 9) and the number of tactors and their range of amplitude (e.g., "quantization level" as in Chapter 9). Triggs et al. found little difference between equivalent electrocutaneous and vibratory two-dimensional tracking displays. The form of information coding had a significant effect, with the polarized activation of tactors superior to conditions in which the tactile signal simply began on one side of the tactor array and swept to the other side. Performance was superior on a standard visual display showing continuous data, but this can be attributed to the quantization and sampling nature of the tactile channel, since performance on "equivalent" quantized and coded visual displays was essentially the same as on the tactile system.

Hirsch (1974) studied a single-axis tracking task simulating the attitude control of an aircraft with and without rate information supplied via two vibrators (on the thumb and index fingernails). He found improved performance on a visual plus tactile system over the visual-only system. Providing error rate information via the auxiliary tactile channel apparently is an effective way to add "quickening" information (see Chapter 12) to

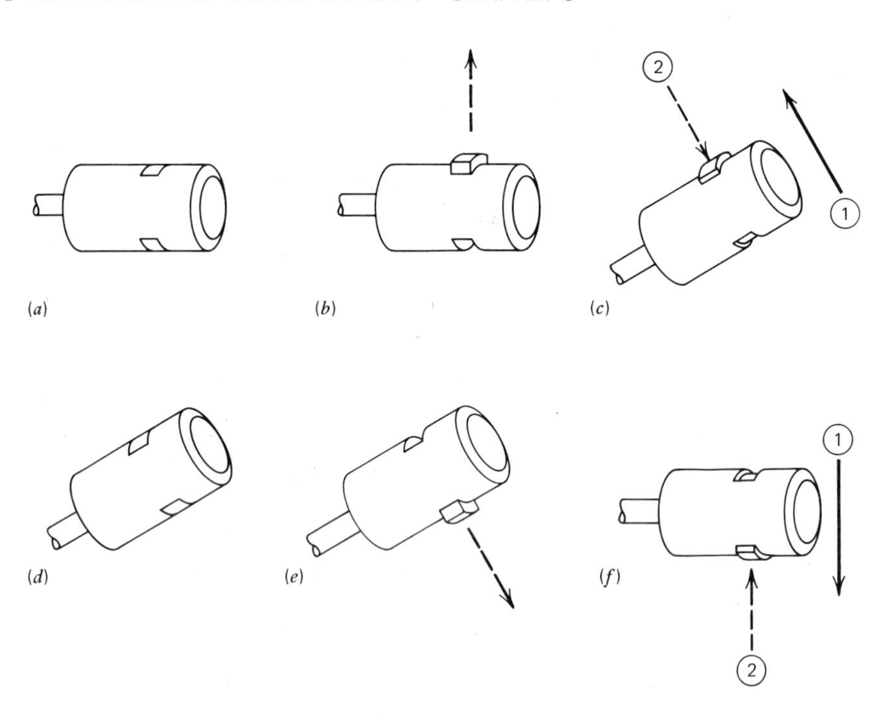

Figure 8-16. Control-display relationships for tactual display studied by Jagacinski, Miller & Gilson (1979). The displayed error is indicated in (b) and (e). The control movement shown as ① results in a decrease in the displayed error, ②. (From *Human Factors,* 1979, *21,* 80. Copyright 1979 by *The Human Factors Society,* Inc., and reproduced by permission.)

a person-machine control system without overburdening the primary visual channel. The use of tactual displays as a practical supplement to control tasks with high visual loading has been validated in studies by Gilson and his associates (Jagacinski, Miller, & Gilson, 1979). One form of such a display is a variable height slide embedded in a control handle, as shown in Figure 8-16. The slide is controlled by a servomechanism, which causes it to protrude either fore or aft to a degree corresponding to the error signal displayed. The operator can move the control stick in the direction in which the slide protrusion is reduced, thereby eliminating the error in the desired control position. Jagacinski et al. (1979) compared this technique to a visual display in a single-axis tracking task. They found that performance in the tactile controller task followed essentially the same pattern of dependence on the type of error signal information available, but was superior for the visual display over the tactile display. This difference could have been due to a variety of factors including the fact that neither the visual nor tactile systems were optimized for presentation of the displayed information. An aspect of performance in the tactile tracking task that was also reported by Triggs et al. (1974) was the nonlinear mode of subject behavior in the tactile condition, compared to the visual tracking situation. Subjects in the tactile task seemed to be operating in a "bang-bang control" or "pulsed mode," rather than in a continuous or proportional-controller mode. That is, their movements involved repeated and large displacements of the control rather than small, apparently continuous tracking movements. This nonlinearity may have been produced by time constants inherent in the electromagnetic slide control in the Jagacinski et al. study or display quantization and sampling in the Triggs study. Whether it is an inherent characteristic of the tactile mode is an important question for further experimental study. It may be that use of wider-bandwidth tactile displays will eliminate this potential problem. One such possibility for a wide-band display is reported in an experiment by Loomis and Collins (1978). In their study, a water jet is directed against a thin watertight membrane that is between the water jet and the skin. This stimulation technique allows the essentially frictionless travel of a point stimulus over the skin. Their study showed an exquisite sensitivity of the cutaneous sense to rapid and small changes in the position of the point stimulus. Use of a control system in which such a tactile display were employed could conceivably produce tracking performance more nearly matching the typical proportional controller behavior observed in visual tracking systems.

Stick Shaker Many aircraft are required to have a kind of tactile display built into the pilot and copilot's control column. A motor driven shaker device is usually mounted on the copilot's control column and mechanically linked to the pilot's control column. A stall sensor on the aircraft's wing triggers the device to shake the control columns with an obvious mechanical (and audible) vibration when a stall is imminent. (A stall occurs when there is insufficient aerodynamic lift to prevent the aircraft from falling.) Appropriate pilot action is to push the control stick forward to increase airspeed. False alarms are a rare occurrence except under some conditions when the aircraft has been waiting on the ground for extended periods in heavy, freezing rain, and the stall sensor false triggers during the subsequent takeoff. Since that condition is specific and can be anticipated, the pilot can then ignore the stick shaker signal without any deficit in his or her ongoing performance. The shaker system is interesting because it is the only tactile display system known by the authors to be used in commercial aircraft.

High-Information Rate Displays

The most dramatic types of tactile display system are those developed for the high-information rate demands imposed by the need to substitute the tactile information channel for that of vision or hearing. A variety of such systems have been developed; we review some of these systems and discuss some of the problems involved in employing the tactile modality as the main channel in a high-information rate system.

Tactile Communication of Speech Suppose that you had the misfortune to lack a functioning sense of hearing; to assist you we supplied a device that electronically amplified the sound in your environment and supplied that sound through a device that vibrated against your skin in synchrony with the sound. Such simple, direct coding of sound was tried more then 50 years ago. It is no surprise that such a device is not very useful for conveying speech information. In Chapter 9 we point out that most of the important information carrying speech frequencies are between several hundred and about 2500 Hz. Yet the skin's sensitivity to vibratory frequency drops off rapidly above 1000 Hz. In addition, the tactile channel has much poorer temporal resolving power than the ear, perhaps 50 or 100 times slower. Some methods have been tried that intentionally slow the speech rate, but these do not appear promising for the perception of continuous speech in real time.

Another technique involves coding of the speech frequency spectrum, prior to presentation to the skin sense. For example, one approach (Kiedel, 1974) placed a fluid stimulated membrane against the skin of a subject's arm. The displacement pattern of the membrane conveyed information to the subject about the spectral properties of the signal, in a manner similar to the displacement pattern of the basilar membrane of the ear. Reasonable communication of spectral information with this device required the slowing of the speech input to a fraction of real time. Another technique has been to analyze the speech spectrum with a bank of tuned filters, and present the output of these filters as a coded vibratory signal to the skin. The frequency channels used in these devices are similar to those used in voice vocoders (see Chapter 9), but instead of reconstructing the speech sound, the outputs cause vibration of a vibrator located in different positions on the skin. Many versions of this technique have been attempted, but success has been limited and extensive training is necessary (cf. Pickett, 1980; Rothenberg, Verrillo, Zaborian, Brachman, & Bolanowski, 1977).

An alternative approach to attempting full communication of the speech message to the skin, is to reduce the load on the tactile channel in one or more ways. One way is to use the tactile channel only to convey some of the information not available in the normal communication situation, rather than overload the channel with all of it. For example, some features of speech are difficult to pick up by lipreading. If these features could be communicated over an *auxiliary* tactile channel, a considerable improvement in communication ability might result (Scott & De Filippo, 1976). A second approach is to reduce the channel load and learning requirement by utilizing a limited and more natural tactile coding of the speech information. One device, developed at the Central Institute for the Deaf, employs three stimulators producing an electrotactile, a vibrotactile, and a spread-versus-compact skin sensation. Their experiments demonstrate that people with normal hearing and language experience can complement lipreading successfully with such a device (cited by Pickett, 1980). One possibility for using a natural tactile channel for speech communication already exists. It is known as the Tadoma method and is used by the deaf-blind. The subject puts his or her hands on the speaker's face so that they can feel the vibrations of the larynx with the lower fingers, the opening and closing of the jaw, and some of the airflow changes at the mouth. This method encodes information directly related to the speech-generating movements of the speaker. One possibility is the development of an electronic substitute for the Tadoma method in which it would not be necessary for the subject to place his or her hand on the speaker's face. Ultimately, it may be possible to construct an electronic speech recognition system that could recognize a large vocabulary of words or sound segments spoken by varied speakers and present the properly coded output to the subject via the tactile channel (Kirman, 1973, 1974). Some of the technical problems involved in the front-end design of such a system are discussed in Chapter 9.

Conveying Visual Information via the Tactile Channel Many display systems have been developed to study the feasibility of using the tactile channel for the transmission of visual information. One commercially available system, the Optacon (Optical-to-Tactile-Converter), manufactured by Telesensory Systems, is composed of a photosensitive array (of 6 col by 24 rows) that drives a like array of vibrating pins measuring 1.1 by 2.7 cm. as shown in Figures 8-17 and 8-18. The user places his or her fingertips on the vibrating array while the camera array is passed over the pattern of light and dark to be "read," such as a single letter or pattern. The coding of light-sensitive array to vibrator array is direct; an O-shaped pattern on paper in front of the light sensor produces an O-shaped vibrotactile pattern on the fingertip. Moving the sensor causes the vibrotactile array to move in a corresponding manner, and this movement may in fact improve the acquisition of information from the array. Generally it takes quite a bit of training to produce high reading rates from this system; from 10 to 12 words per minute after 9 days of training to 30 to 50 words after extensive practice. Craig (1977) has reported on two (sighted) observers who demonstrated remarkable ability to read with the Optacon after less than 20 hours of experience with the device. His observers were able to read at rates of 70 to 100 words per minute after this relatively short period. These observers were much better than other people at discriminating and recognizing vibrotactile patterns, but were not also superior at tasks dependent on visual or general cognitive abilities. So another human factor in the design of

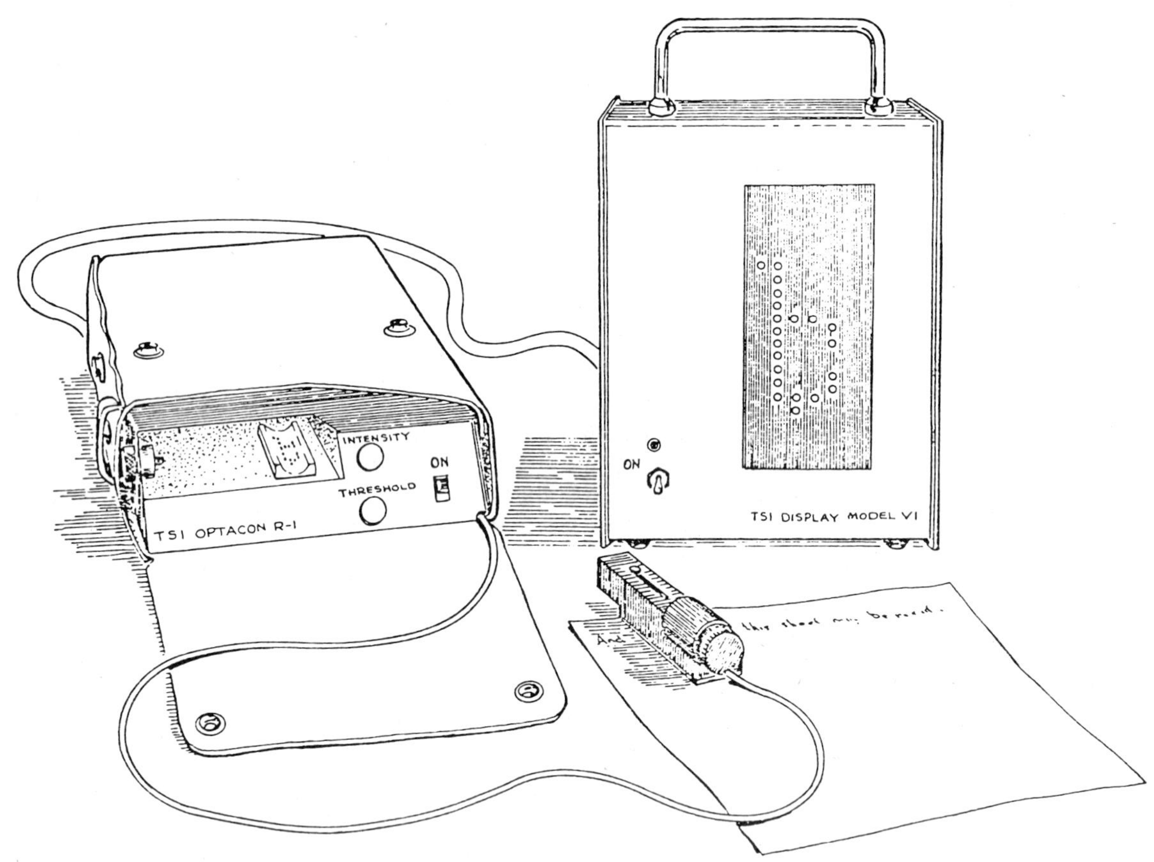

Figure 8-17. Optacon camera and visual and tactual displays (Courtesy Telesensory Systems, Inc.)

tactile displays is the existence of large individual differences in the ease and training time with which different persons may use these devices.

Another general type of system that has been studied for the substitution of tactile information for visual information is the Tactile-Vision Substitution System (TVSS) shown in Figure 8-19. White, Saunders, Scadden, Bach-Y-Rita, and Collins (1970) reported on experiments with this system for converting an optical image into a tactile display of the pictorial information presented to a television camera. The video image is converted and transmitted to an array of 20 by 20 vibrators on the skin of the subject's back. After some practice with this system, subjects can discriminate between certain types of figures and objects, and can identify which one of a small group of objects has been placed in front of the camera. However, they reported considerable difficulty in tasks where the internal details of the pattern had to be identified, rather than the outside contour. Whether refinements to this system will be able to yield more accurate and faster processing of visual information is a question for future research.

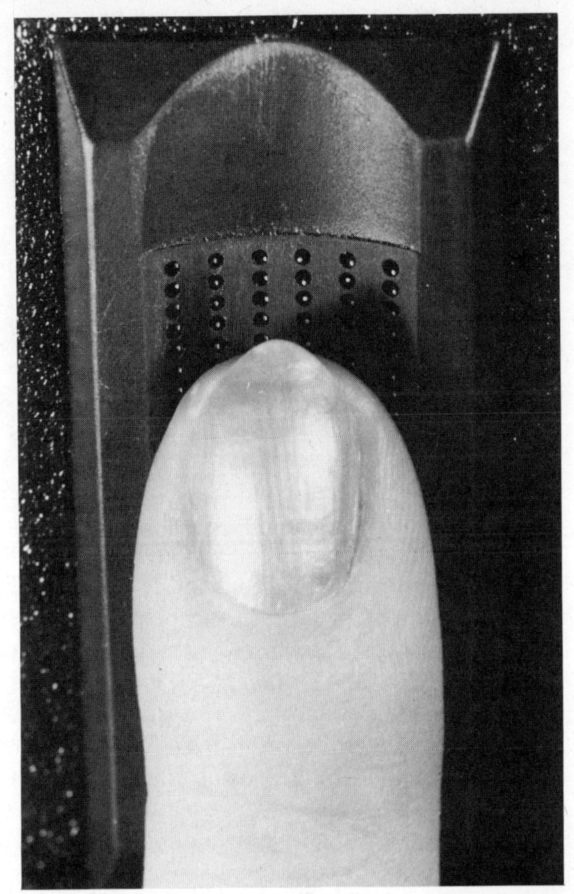

Figure 8-18. Optacon tactual array. (Courtesy James C. Craig.)

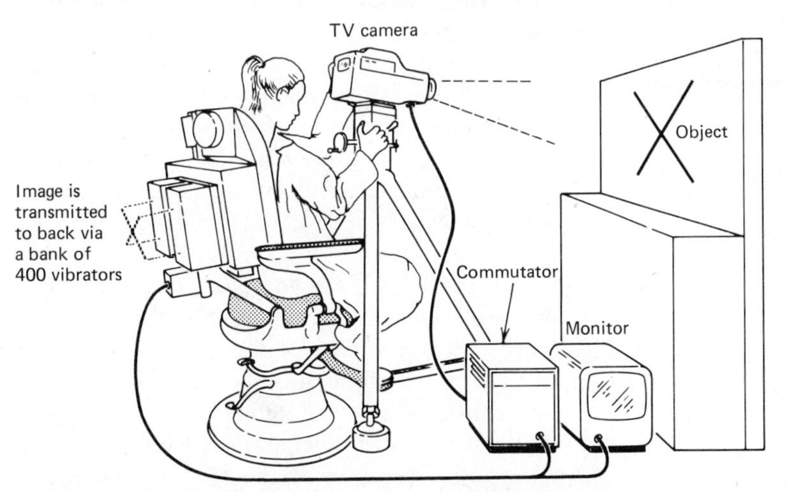

Figure 8-19. Schematic representation of the tactile system, TVSS. (From B. W. White, F. A. Saunders, L. Scadden, P. Bach-Y-Rita, & C. C. Collins, Seeing with the skin, *Perception & Psychophysics,* 1970, *7,* 23. Copyright 1970 by Psychonomic Journals, Inc., and reproduced by permission.)

Coding Considerations The problems with the TVSS system and the characteristics of behavior with the Optacon have led some investigators to question the notion that conveying (or attempting to convey) a direct pictorial representation of the visual image is the best way to code the signal for the tactile channel. This controversy has been called the pictorial versus the coded approach (Geldard, 1974). To some extent a "naturally" coded signal, such as in the pictorial approach, may avoid some of the extensive training and familiarization required by devices that employ novel and unusual coding schemes. However the ultimate usefulness of a system in terms of information and error rates may depend on using a coding plan that is optimized for the tactile channel, and hence is quite different from normal visual-pictorial coding. That a simple, static, pictorial code may not be optimal for tactile communication can be seen from a variety of experiments which compare various presentation and coding modes. Craig (1974, 1980) and Sherrick (1982) have addressed the questions of type of coding, for example, static, scanned or slit-scan (where the display is swept across the matrix of tactors, rather than temporarily frozen for a frame). In one study Craig (1974) found no large differences in observer's performance when the pictorial code was disturbed by scrambling the connection between the camera and the tactor array in a TVSS type system. In a study comparing static and scanned modes using the Optacon, Craig (1980) found that although prior studies have generally found a superiority of a moving tactile display (probably attributable to greater elaboration of spatial contour), the advantages or disadvantages of a scanning mode depend on a number of factors such as forward and backward masking. Thus the relative advantage of a static versus scanned model will depend on the fraction and speed of the scanning, the area of skin being stimulated, the amplitudes of the stimulation, and the ensemble of targets to be discriminated. A static display mode might be best in situations requiring a minimal amount of temporal masking, such as those requiring very brief display times. High levels of intensity will enhance the recognition of brief duration patterns and also increase the effectiveness of those patterns as maskers of adjacent patterns.

Movement-related Phenomena The tactile display designer ought to be aware of several movement-related phenomena that can result in these display systems. One of these is the apparent motion that results from the successive stimulation of different sites on the skin. This can induce a sensation of a strong "gouging" effect on the skin (Sherrick & Rogers, 1966). Apparent movement depends on the time between the successive stimuli, the duration of the stimuli, and the distance between the skin locations. Related to this effect is sensory *saltation* (Geldard, 1975). If a few tactile stimulators are activated in sequence, they may yield this strange percept: a series of taps that appear to "hop" along the skin between the locations of the vibrators on the skin. A third phenomena is known as the phantom of lateralization (von Bekesy, 1967) and involves the perception of a single tactile sensation located between two sites of tactile stimulation. The position of this phantom stimulus can be shifted by controlling the onset times and intensities of the real tactile stimuli. These phenomena will be either problems or useful aspects of future tactile display devices.

Craig and Sherrick (1982) point up an important aspect of the current state of our knowledge about high rate tactile displays: the display should be tested with the kinds of stimuli (e.g., letters or visual objects) that will be used and the kinds of subjects who will be using it. The area of high information rate tactile displays is one in which we may anticipate rapid developments over the next decade.

SUMMARY

When the visual channel is heavily loaded, the auditory or tactile channel may be an appropriate way to communicate information to the human. The auditory channel is particularly well suited to serve a warning or alerting function. It is heavily used for this purpose in some aircraft cockpit applications. Problems of existing auditory alarm signals include masking and interference by other signals, confusability of signals, and problems with signal localization. An important area of application is the use of brief speech messages for alerting or warning signals. The use of the tactile channel may be particularly important under conditions of degraded visual input, as in underwater work environments. Some applications of the tactile channel include displays for tracking and control tasks, and systems for encoding visual information such as the Optacon. One question common to the design of both auditory and tactile systems is whether the code should be natural or artificial. A "natural" code preserves the spatial and temporal relationships that exist among the elements of the signal in the original (usually visual) channel. An "artificial" code employs a coding scheme that is optimized for information transfer in the new channel. The requirement for extensive user training is an essential aspect of this question. A second general problem is the existence of special sensory phenomena that limit the rate of information transfer. In the auditory sense these include sequential phenomena such as auditory "streaming." In the tactile sense, these include temporal masking, spatial masking, and apparent movement phenomena. Developments in technologies such as speech processing and image processing will probably result in new systems employing the tactile and auditory channels.

9 SPEECH COMMUNICATION

Your boss has given you the following problem, and wants an answer by 5 P.M. Crews consisting of three or four workers in protective clothing will work in a very noisy environment for brief periods of time. They will need to communicate with each other using a small set of commands and responses. Given the level and frequency characteristics of the noise in that environment, will speech communication be so poor as to warrant the use of expensive radio (or other) communications gear?

You are riding in a small airplane that your pilot friend has chartered. Your friend has just requested landing permission from the airport at your destination. The radio responds with a stream of incomprehensible chatter. You look questioningly at your friend. "Oh," she says, "The tower just said to get into the landing pattern for runway 12 behind the Beechcraft, and to watch the 10 Knot crosswind." Have your ears suddenly gone bad?

The supervisor of your engineering group has just entered your office with a cassette tape of the output of an inexpensive new speech synthesizer. She wants to use this synthesizer in the new toy doll that you are designing. The doll will have a fairly large speaking vocabulary. After carefully telling you what the doll will say, she plays the tape. "See?" she says, "Sounds perfectly good to me." You have some misgivings about committing yourself to this synthesizer without further tests. Should you resist a quick decision this time?

You are trying to speak to a group of factory workers over the noise of several metal working machines. You find yourself shouting and repeating important words. You begin to wonder if this strategy will result in improved communication of your message. Will it?

These examples are meant to illustrate the range of speech-related problems and phenomena that you may encounter. The human factors aspects of speech communication have changed dramatically in the past few years. The widespread use of microprocessors has brought about an entirely new world of speech technology, including the computer generation of speech and the computer recognition of speech and speakers. This technology has already spread to many consumer, industrial, and military applications. With it has come a variety of human factors problems that cannot be solved by the application of existing tables and formulas. Many of the human factors problems of this new technology have yet to be clearly defined; two examples that we consider later in this chapter are in the areas of speech input to computer and the intelligibility of synthetic speech.

Probably the best tool that a human factors specialist can have in this area is a good understanding of the basic principles of speech production and perception. We try to achieve that goal in the first section of this chapter. In the second part of the chapter we consider the problem of assessing the characteristics of a communication channel. In the last section of the chapter we consider the new area of speech technology, with discussion of the principles of waveform digitization, speech bandwidth compression, and speech synthesis.

SPEECH PRODUCTION AND PERCEPTION

One strategy for studying the speech signal is based on speech production: How do we use the different parts of our vocal system (our chest, tongue, vocal cords, lips, etc.) to encode the information that we want to convey in speech? That is, what particular movements and positions of these organs correspond to the consonant and vowel sounds that we intend to generate? A completely different strategy is to study the acoustic properties of the transmitted speech signal. How can we decode a complex speech signal to tell what speech message (e.g. consonant, vowel, etc.) was transmitted? This approach studies the perceptual side of the speech system. You probably realize that if speech scientists could completely answer either of these general questions, our job would be very easy—and they might be out of work! In fact, both of these approaches have been valuable in acquiring information about the speech system, but a good many questions still remain.

You can appreciate how complex the problem is by considering that even though a specific place for the tongue usually accompanies a particular vowel sound, the exact location of the tongue depends on what sounds were made immediately before and after the one that we are studying. Furthermore, when we make that vowel sound in rapidly spoken speech, we may never actually reach that tongue position. That is, the "normal" tongue position that we associate with producing a particular vowel, might be only a target position for the tongue, rather than an absolute requirement. Trying to pin down the speech sounds from the perceptual side is at least as difficult. In a given period of spoken speech, you don't know precisely where the sound for one syllable ends and another begins. Furthermore, the exact acoustic properties of a given syllable may be different each time you repeat them, or when they appear as parts of different words! So we face large problems of complexity and variability in describing speech stimuli.

Phonemes

To the nonspeech scientist, the basic units of speech are the vowel and consonant sounds. Speech scientists have defined a smaller unit: the phoneme. The phoneme is defined as the smallest speech sound that can change the meaning of a word. Phonemes form syllables and words; in order to define or produce a word, we need only to generate a specific sequence of phonemes. Unfortunately, the phoneme is really more of a theoretical definition than it is a precise description of the spoken segments of sound that form our speech alphabet. Most phonemes can't be characterized by a specific sound that would allow us to perfectly identify it, if we could hear it all by itself. Table 9-1 illustrates the phonemes of English and the phonetic symbols used to indicate them.

We can gain some understanding of speech production by examining the way phonemes are produced by our speech mechanism. Figure 9-1 shows the major parts of our vocal tract. Figure 9-2 is a schematic diagram of this system, showing the energy source for pushing air through it, and the important air cavities. A vowel is generated by vibration of the vocal cords in the larynx by air pressure from the lungs. The vibratory motion of the vocal cords is something like your lip vibration when you play a trumpet or trombone. The vocal folds vibrate at 80 to 400 vibra-

TABLE 9-1 *The Phonemes of English and Their Phonetic Symbols (Clark and Clark, 1977)*[a]

Consonants				Vowels		Diphthongs	
p	pill	θ	thigh	ɩ	beet	ay	bite
b	bill	∂	thy	I	bit	æw	about
m	mill	š	shallow	e	bait	əy	boy
t	till	ž	measure	ε	bet		
d	dill	č	chip	æ	bat		
n	nil	ǰ	gyp	u	boot		
k	kill	l	lip	ʊ	put		
g	gill	r	rip	ʌ	but		
ŋ	sing	y	yet	o	boat		
f	fill	w	wet	ɔ	bought		
v	vat	ʍ	whet	a	pot		
s	sip	h	hat	ə	sofa		
z	zip			*i*	marry		

[a] From *Psychology and Language: An Introduction to Psycholinguistics,* copyright 1977 by Harcourt Brace Jovanovich, Inc., and reproduced by permission.

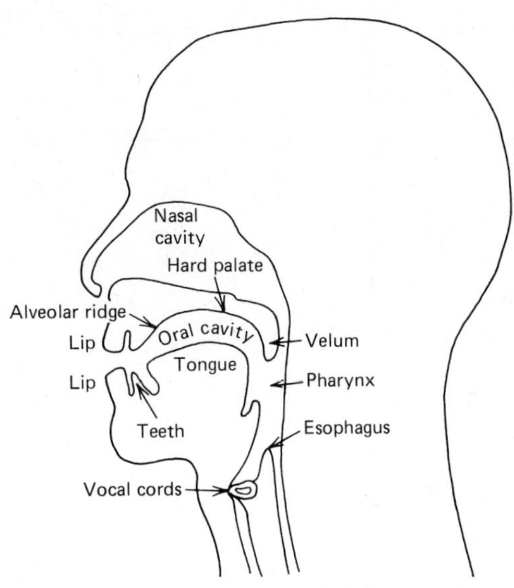

Figure 9-1. Important parts of the speech production system. (From *Psychology and Language: An Introduction to Psycholinguistics* by H. H. Clark & E. V. Clark, p. 181. Copyright 1977 by Harcourt Brace Jovanovich, Inc., and reproduced by permission.)

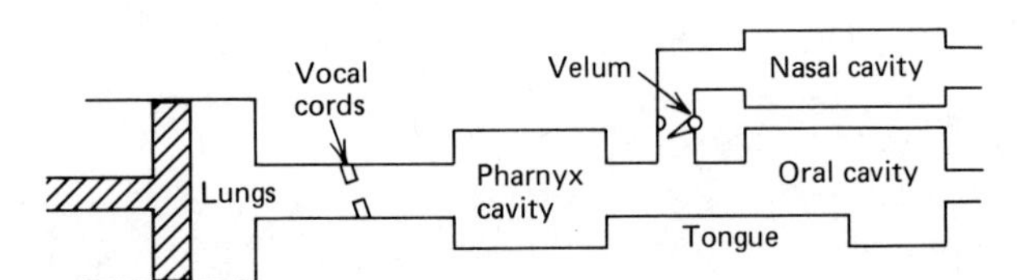

Figure 9-2. Schematic diagram of the speech production system.

tions per second. These sound vibrations are then acted on by the various vocal cavities such as the pharynx, the oral cavity, and the nasal cavity, and by their respective openings. When you move the slide on your trombone you change the dimensions of the internal cavity of the instrument—the same thing happens when you place your tongue in different positions inside your oral cavity. As we change the size and shape of these cavities and openings, we accentuate or minimize different frequencies that were present in the original vibrations of the vocal cords. For the vowel sounds, the most important control of these structures is in the tongue and lips. Try to make the various vowel sounds of Table 9-1 and notice how the position of your tongue changes for each sound.

The generation of the consonant sounds differs in several respects from the vowels. When we generate consonant sounds, we constrict part of the mouth to stop or restrict the passage of air. The major differences among the consonants have to do with where and how in the oral cavity

TABLE 9-2 *Place and Manner of Articulation of the English Consonants (after Clark and Clark, 1977)*[a]

Manner of Articulation (Unvoiced / Voiced)	Place of Articulation: **Bilabial** The lips are together.	**Labiodental** The bottom lip is against upper front teeth.	**Dental** The tongue is against the teeth.	**Alveolar** The tongue is against the alveolar ridge.	**Palatal** The tongue is against the hard palate.	**Velar** The tongue is against the velum.	**Glottal** The glottis is constricted.
STOPS Complete closure and release of pressure	p / b			t / d		k / g	
FRICATIVES Constriction that produces turbulence		f / v	θ / ð	s / z	š / ž		h /
AFFRICATES Complete closure followed by constriction					č / ǰ		
NASALS Velum is lowered to allow air through nasal cavity	/ m			/ n		/ η	
LATERALS Shape tongue to enable opening at sides				/ l			
SEMI-VOWELS Shape tongue to enable opening in middle	/ w			/ r	/ y		

[a] From *Psychology and Language: An Introduction to Psycholinguistics*, copyright 1977 by Harcourt Brace Jovanovich, Inc., and reproduced by permission.

we make that constriction, and whether there is also a vibration of the vocal cords. Table 9-2 summarizes these aspects of consonant generation.

These dimensions of phoneme production can provide a way to classify any speech segment into which phoneme was transmitted; you only have to specify where the element fell on the place, manner, and voicing dimensions. However, we have pointed out that the production of such a sound segment depends on what sounds precede and follow it, so the classification task is actually more difficult. Furthermore, for natural-sounding speech, there are also aspects of speech production that extend in time over more than one sound segment. Stress and intonation pattern are examples. Stress has to do with the different emphasis given to syllables in a spoken sentence, and usually involves a change in the intensity, pitch, or duration of the vowel part of a syllable. Intonation generally involves a change in the pitch of the voice, as in the rising intonation at the end of a question. The speed of transmitting phonemes in spoken speech is normally about 12 phonemes per second. We can understand speech at rates as high as 30 or more phonemes per second (Liberman, Cooper, Shankweiler, & Studdert-Kennedy, 1967).

Speech Spectrograms

We can examine the acoustic nature of the speech signal, and try to determine which phoneme was generated from the acoustic information, rather than from the production information. A useful tool for this purpose is the speech spectrogram, shown in Figure 9-3. The horizontal axis

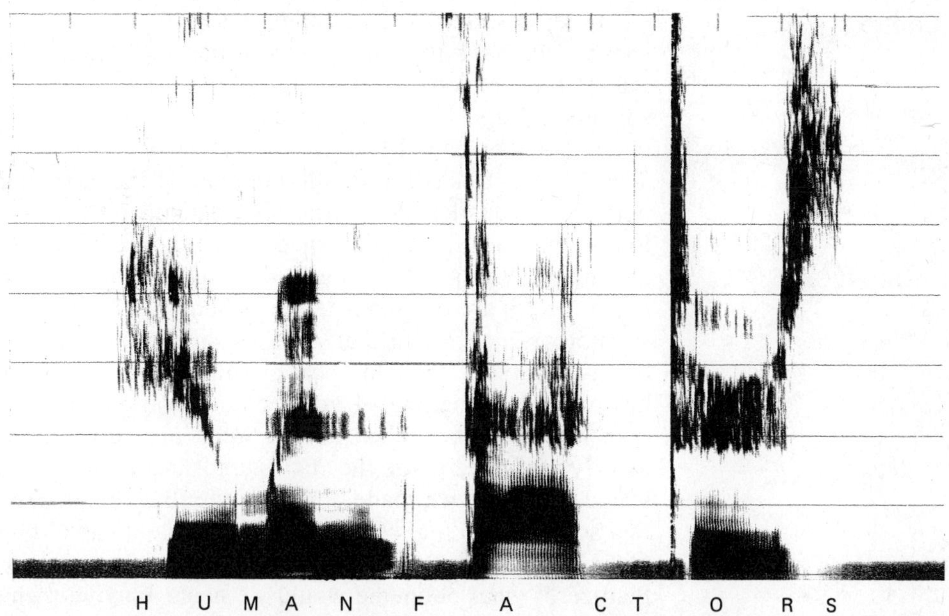

Figure 9-3. Speech spectrogram of "Human Factors," female speaker. (Courtesy of Robert Sorkin.)

of this plot is the time from the beginning to the end of the utterance, and the vertical dimension is frequency from 0 to 8000 Hz. The darkness of the lines indicates the intensity of the sound at each frequency.

The second word spoken in the spectrogram of Figure 9-3 is "factors." You can see certain frequencies are emphasized during the vowel segments of the sound. The enhanced frequencies are at approximately 800, 2200, 3000, and 4200 Hz. These bands of frequencies are known as formants, and correspond to the mechanical resonances of the cavities of the vocal system. When we change the position of the tongue, for example, we change the size of the oral cavity and pharnyx, and thus the frequencies at which these cavities enhance the sound frequencies generated by the vocal cords. The first formant is defined as the lowest resonant frequency, and the second formant is the next-highest resonant frequency.

The situation is somewhat different for consonants. For consonants, there are quite rapid changes in the formants, and sometimes the energy spread is very broad and abrupt. In some cases the sound source is very much like noise. The consonants of English probably carry most of the information conveyed in a word. The problem is that there is no simple one-to-one relationship between the consonant phonemes and the spectrographic patterns associated with those phonemes (Liberman et al., 1967).

Effects of Phoneme Context

The spectrographic patterns for many of the consonant sounds are identical; in order for us to determine which consonant phonemes were spoken, we have to look at cues in the vowel sounds that immediately follow the consonant. For example, to discriminate among the consonants /b/, /g/, or /d/, we would need to know the change in the frequency of the second formant of the vowel phoneme that came after it. The important cues for identifying most of the consonant phonemes are not fixed acoustical properties, but depend on the speech sounds that follow them in time. Some speech sounds are relatively independent of the immediate speech context; the vowels and the consonants /s/, /z/, /š/, /ž/, /č/, /ǰ/ are examples (Cole & Scott, 1974).

Another speech cue that depends on the immediate context is voice-onset-time (vot), the time between the release of air pressure from the lips and the beginning of vocal cord vibration, as in the syllables /ba/, /pa/, /da/, and /ta/. This time between pressure release and voicing is usually about zero for the /ba/ sound and about 60 msec for the /pa/ sound. Suppose we made a tape of the /ba/ sound. We could make that /ba/ sound into a /pa/ sound by adding a silent gap of 60 msec at the right place in the tape. But suppose that gap were less than 60 msec but more than zero; what phoneme would we hear? Playing with such an artificial stimulus illustrates an important speech phenomenon known as *categorical perception.* Suppose you listened to a list of such sounds that has its vot vary continuously from 0 to 60 msec. You would be impressed by the fact that this list sounds like the sequence: ba, ba, ba, ba, ba, pa, pa, pa, pa. There is a *sudden* transition in the phoneme that you hear, even though you know that the actual physical cue varies slowly from one extreme to the other. Apparently our speech perception system forces us to put these inputs into discrete categories, and this process depends on our analysis of the immediate speech context as well as the specific speech sound.

The dependence of our speech perception on the stimulus context extends quite a bit deeper than the preceding or following phoneme sounds. The characteristics of a speaker's voice affects our perception of individual phoneme sounds, as does the grammar of the spoken sentences, and the listener's expectations and familiarity with the spoken material. In later sections of this chapter, we will see that all of these contextual factors are critical in determining the intelligibility of a speech channel.

A SLICE OF LIFE

The Voiceprint Controversy

Can you use the spectrographic pattern of a person's speech to identify that person reliably? The answer to this question is at the heart of the "voiceprint controversy." A strong proponent of the voiceprint, L. G. Kersta (1962), has argued that the answer is yes. Kersta coined the term *voiceprint* to indicate an analogy between the use of the voice spectrogram and the fingerprint for identification purposes. Voice spectrograms have been used in a variety of legal applications, and by some police departments. But there is a great deal of controversy over the scientific basis of its use in identification, and about the errors that can arise from its misuse. In 1970, at the request of the Acoustical Society of America (ASA), a group of acoustical experts (Bolt, Cooper, David, Denes, Pickett, & Stevens, 1970) reviewed the matter. The ASA group considered a variety of questions related to the use of speech spectrograms for identification, including the assumed analogy to the fingerprint. We have reproduced part of their comparison of the fingerprint and the voiceprint in Table 9-3. The table summarizes many of the important differences between the two techniques. In terms of accuracy, dependence on environmental factors, permanence, and susceptibility to the effects of emotion and deception, voiceprints probably have more in common with handwriting samples than with fingerprints (Hawley, 1977). The Acoustical Society group was very concerned that the use of voiceprints could result in high rates of false identification depending on the particular type of identification task, the training of the examiner, and other factors. They concluded that there was no scientific basis for accepting the voiceprint as a reliable means of identification (Bolt et al., 1970).

TABLE 9-3 *Comparison of Fingerprints and Voiceprints*[a]

Finger Ridge Patterns	Voice Patterns
1. Patterns are inherent in anatomy, not changeable in kind (i.e., they cannot be changed from one pattern to another). Parts of pattern, large or small, can only be obliterated.	**1.** Patterns are only partially dependent on anatomy and are changed by the articulatory movements needed to realize the language code.
2. Details of patterns: (a) Are permanent. (b) Are not affected by growth (aging merely changes the size or the print grain). (c) Are not affected by habits (calluses merely change the print grain).	**2.** Details of patterns: (a) Are just as variable as the overall patterns. (b) Are affected by growth. (c) Are affected by habits (learning new dialects and voice qualities).
3. Pattern similarity depends entirely on underlying anatomical structure.	**3.** Pattern similarity depends primarily on acquired movement patterns used to produce language code and only partially on anatomical structure.
4. Patterns result from a direct transfer from the skin of the finger to the surface touched by it.	**4.** Patterns result from an analysis of voice sounds that are related only indirectly to the vocal anatomy of the speaker. Moreover, the transmission channel from speaker to spectrograph is vulnerable to acoustical and electrical distortions.

[a] From Bolt, R. H., Cooper, F. S., David, E. E. Jr., Denes, J. B., Pickett, J. M. & Stevens, K. N. *Journal of the Acoustical Society of America,* 1970, *47,* 607.

The question of identification accuracy was addressed in a large study by Tosi, Oyer, Lashbrook, Pedrey, Nicol, and Nash, in 1972. This study employed a population of 250 speakers and 29 examiners selected and trained for the study. The variables that they investigated included the number of words and samples, types of recording condition, word context, intraspeaker variability (whether the utterance was repeated during the same or a subsequent recording session), and the type of experimental trial. The types of experimental trial included two main aspects: how many speakers were to be compared with the unknown voice, and whether or not the examiner knew that the unknown voice sample was included in the known set of comparison speakers.

In an actual police situation, an error could lead to the selection of an innocent person or the rejection of the guilty person from prosecution. In the Tosi study, overall rates of false identification were about 4.3% and false eliminations about 4.6%. An important question is whether the real-world situation, including the use of auditory comparisons of the voices, different and larger populations of speakers, professional examiners, possibly disguised voices, female voices, and the use of additional clues, would result in lower or higher error rates than this study. Tosi and his colleagues felt that lower rates would result. The Acoustical Society group (Bolt, Cooper, David, Denes, Pickett, & Stevens, 1974) analyzed the Tosi study and reconsidered their own earlier judgments. They pointed out that the panels of examiners used in the Tosi study occasionally exhibited considerable disagreement as to what constituted a match when given the same voiceprint matching task. The average error rates tend to mask this variability. Bolt et al. (1974) estimated that under real-world conditions, the actual error rates would increase substantially, rather than decrease. Their position of scientific caution remained essentially unchanged from their earlier 1970 analysis. What about yourself—would you want to risk your freedom on that 4.3% error rate?

CHANNEL INTELLIGIBILITY AND QUALITY

In this section we will discuss the evaluation of speech communication channels such as telephone and radio communication. Evaluating a speech channel depends on what aspects of communication are considered to be important. The intelligibility of the speech may be a major concern. You may need to specify how sensitive the channel will be to disruption by noise, or how rapidly messages may be transmitted over the channel. In some cases where speech is easily understandable, the sound "quality" may be of major interest: Will the channel's users perceive the speech as being of acceptable quality? Solving problems of intelligibility or channel quality will usually involve some compromises in either the training or the instrumentation for a system. So an early human factors analysis of the channel can be important in arriving at an effective system design.

Articulation and Intelligibility

Roughly speaking, the terms intelligibility and articulation refer to the same general question: How good is speech transmission over the channel? Articulation testing usually means a check on the recognition of syllables or phonemes, while intelligibility testing usually refers to the comprehension of words, sentences, or the total message. Some of the many articulation and intelligibility tests that have been devised are illustrated in Table 9-4. The stimuli in these tests include nonsense syllables, words, and sentences. In some cases the correct answer can be

TABLE 9-4 *Types of Articulation and Intelligibility Tests*[a]

Stimulus Material		Stimulus (Response)
Nonsense syllables		monz,nihf,nan,zeef,....
Phonetically balanced (PB) monosyllabic words (the occurrence of different phonemes is approximately the same as in spoken English)		smile,strife,pest,end, heap,...
Bisyllabic spondaic words (equal stress on both syllables)		again,farmer,football,...
Modified Rhyme Test		rang(rang,fang,gang,hang,bank,sank) bark(mark,bark,park,dark,lark,hark)
Triword Test		badge, bayed,mat(batch,base,bat, bash,bathe,base,bayed,bays,beige, mat,fat,that,rat,vat)...
Interrogative sentences		What do you saw wood with? (saw) What letter comes after B? (C)
Five key word sentences		*Deal* the *cards from* the *top,* you *bully* *Jerk* the *cord,* and *out tumbles* the *gold.*
SPIN test (Speech in babble noise)		
	High predictability:	The watchdog gave a warning growl. (growl)
	Low predictability:	The old man discussed the dive. (dive)

[a] After Egan (1948), House, Williams, Hecker, and Kryter (1965), Kalikaw, Stevens, and Elliot (1977), and Sergeant, Atkinson, and Lacroix (1979).

predicted with some accuracy from the preceding speech context, as in the high-predictability item in the SPIN test. If you suspect that the type of stimuli and the predictability of the response have a large effect on intelligibility, you are right! We will see that the intelligibility measured by these different techniques is consistent with what we have learned from information theory and detection theory in Chapters 2 and 3.

Message Predictability and Familiarity

In Chapter 2 we discussed the influence of message set size and message probability on the computation of the information metric. The more messages that may be transmitted, the higher the average amount of information for a message. The highest average amount of information per message is the case when all the messages are equally probable. The information measure has a very real interpretation; it is directly related to the cost of successfully communicating messages over the channel. Suppose that you wanted to send a telegraph message. Years ago you could send certain types of prewritten telegraph messages, such as wedding or birthday greetings, for very low cost. You simply picked out a previously composed standard message, let's say number 347: CONGRATULATIONS ON YOUR RECENT MARRIAGE. BEST WISHES AND GOOD LUCK. MANY HAPPY RETURNS OF THIS DAY. SORRY WE COULDN'T BE THERE TO SHARE IT WITH YOU. Ordinarily, you had to pay for a message by the number of words that you used. With the standard message, however, you simply paid a small, flat fee. Why the cost reduction? To send message 347, the operator had only to transmit the three-digit code number 347 a sufficient number of times until it was received correctly with some degree of confidence. A lot easier than sending the actual 26 words with the possible repeats and corrections that would be required under noisy conditions! Of course, if your message was incorrectly received as 247—CONDOLENCES. WE SHARE YOUR LOSS . . .—this economy could prove to be rather expensive!

We also considered the effects of other important channel characteristics in the discussion of signal detectability in Chapter 3. One is rather obvious; the higher the signal-to-noise ratio, the better will be detection, discrimination, and recognition performance. A second aspect has to do with the particular conditions of reception, such as whether the listener uses both ears, and whether the signals sent to the ears are identical or different. We will consider these two aspects later in this section. A third important aspect is the listener's knowledge or uncertainty about the messages to be received. In Chapter 3 we discussed the large effect on detection that resulted from changes in the observer's knowledge of the signal (and noise) characteristics. This factor is related to the listener's familiarity and experience with the stimuli to be used in the intelligibility test, and to whether or not the message set is specified to the listener before the message is transmitted. The smaller is the set of possible messages, and the more familiar is the listener with the possible messages, the better will be performance.

You can see from Table 9-4 that the variety of articulation and intelligibility tests pretty much cover the ground of these information and detection theory factors. Figure 9-4 shows how intelligibility increases as a function of the speech signal-to-noise ratio. Notice that the intelligibility is poorer for the nonsense syllables than for the phonetically balanced (PB) word lists. In a phonetically balanced word list, phonemes occur with approximately the same frequency as they occur in English. A study of the effects of message set size and intelligibility was performed by Miller, Heise, and Lichten (1951), who showed that words chosen from specified lists of 2, 4, 8, 32, 256, or 1000 words were identified better, the shorter the size of the list. So the effect of vocabulary size is consistent with our theoretical expectation.

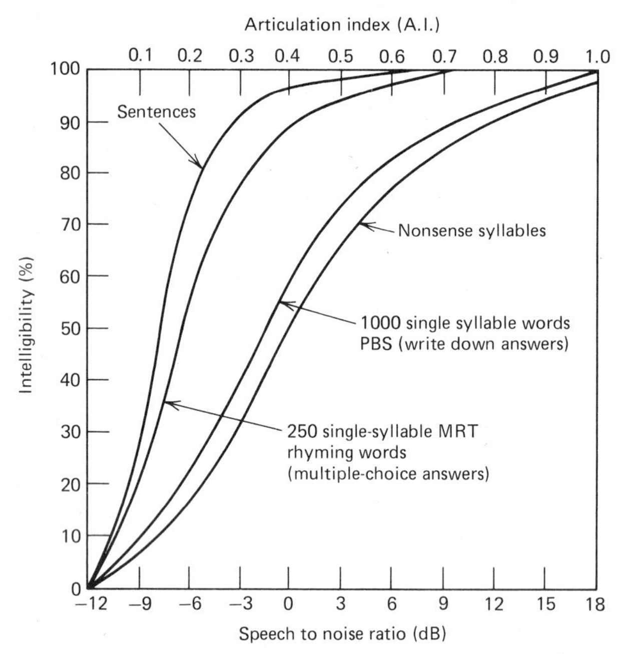

Figure 9-4. Intelligibility of different test materials as a function of speech-to-noise ratio. (From J. C. Webster, Speech Interference Aspects of Noise, p. 197, in D. M. Lipscomb (Ed.), *Noise and Audiology,* copyright 1978 by University Park Press, and reproduced by permission.)

Suppose your message set consists of 16 nonsense syllables and you provide the listener with a list of those syllables and some training in identifying them. You would expect him or her to do a good deal better at the task then if he or she had never seen the list of possible messages. Message sets where the response alternatives are not defined or constrained to the listener will be more difficult and more susceptible to the degrading effects of noise. These factors of listener familiarity and specification of the message set are involved in the use of defined vocabularies in certain communication environments. An example is the international word-spelling alphabet: ALPHA, BRAVO, CHARLIE, etc. Using this vocabulary allows very high intelligibility for alphabetic information under noisy conditions and with speakers and listeners of different nationalities. Similar effects are illustrated by the air traffic controller example at the beginning of this chapter. The vocabulary of words used is actually relatively small. Your pilot friend had only to analyze the message based on that small set of possible ones. Experience with the possible message set would have allowed you to "decipher" the control tower's instructions.

Effects of Sentence Context

The sentence context can provide important information for speech intelligibility, as shown in the superior results with sentence material in Figure 9-4. The size of the message set is effectively reduced, and the predictability of the possible messages is increased by the sentence context. An experiment by Miller, Heise, and Lichten (1951) showed that words presented in five-word sentences were identified more accurately than words presented in isolation under noisy conditions. These effects are shown in finer detail by an experiment by Miller and Isard (1963). They presented people with lists of three different types of sentences:

1. Grammatical sentences:
   Trains carry passengers across the country.
2. Anomolous sentences:
   Trains steal elephants around the highways.
3. Ungrammatical strings of words:
   On trains hire elephants the simplify.

The lists of sentences were designed so that they contained the same words equally often, despite the differences in word sequences. The grammatical sentences conformed to the grammatical rules and meaningfulness of ordinary English, while the anomolous sentences preserved the grammatical rules but not the meaning. As Figure 9-5 shows, accuracy was greatest on the grammatical, a little less on the anomolous, and worst on the ungrammatical sentences. The experiment showed that listeners employ information at higher levels than the segments of phonemes, to help identify which words were transmitted.

There are a variety of speech phenomena that illustrate this point, that information from sentence structure and meaning is used to recognize particular phoneme segments. Under conditions of noise or other degradation of the speech signal, a listener will fill in perceptually what is missing, and "hear" the speech as normal. Warren (1970) produced an interesting demonstration of this with an illusion known as the phonemic restoration effect. He presented subjects with a recorded normal sentence in which a short portion of the sentence had been removed and replaced with the sound of a cough. Most people who hear the recording don't

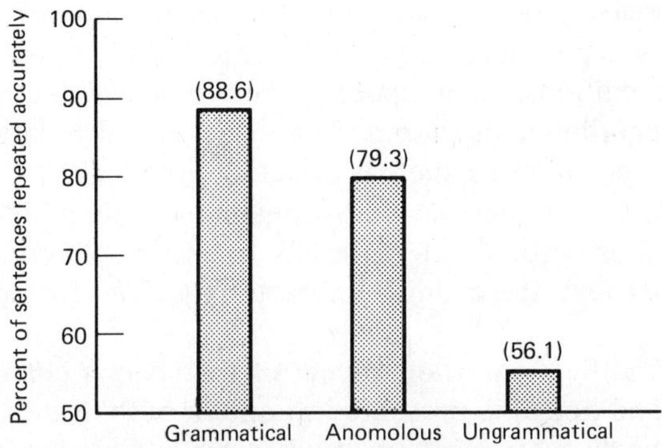

Figure 9-5. Percent of sentences received correctly as a function of the sentence context. (After Miller & Isard, 1963.)

report that there are any sounds missing from the sentence. Those that do usually select the wrong syllable as having been replaced. A tone in place of the cough has a similar effect. Even when you know how the recording is made, you still hear the added sound as being present along with a normal and complete sentence. You're not sure where in the sentence the extra sound occurred. This phenomenon is significant to us because the obliteration of parts of normal speech is a common occurrence in ordinary communication, particularly in noisy situations. We fill in these wiped out segments automatically; add to these disruptions the sloppiness and speed of normal speech and you can see that our perception of speech must involve high levels of information processing. Sometimes when we hear such fouled-up speech, we automatically misperceive a reasonable substitute (from a phonemic, syntactic, and semantic viewpoint). The result can be embarrassing:

> Your classmate says . . . "This teacher's a winner" and you thought she said, "meet you for dinner."

Speech Intensity

An important channel characteristic is the speech signal-to-noise ratio, shown as the horizontal dimension of Figure 9-4. The intensity of speech depends on the speaker and the amount of vocal effort the speaker is expending. Speech levels at different amounts of vocal effort range from peak pressures of 70 dB at a whisper, to 110 dB at a shout. At a given level of vocal effort the range of levels span about 40 dB. The range across different speech sounds is about 30 dB. A system designed to reproduce the extreme range, from the minimum levels of soft speech sounds during a whisper to the peak levels of loud speech sounds during a shout, would have a dynamic range of 80 dB. But it is possible for trained speakers and listeners to communicate effectively over systems with dynamic ranges of 20 dB or less.

Under ordinary conditions, a speaker will automatically increase vocal effort to compensate for an added background noise. As the noise increases, the increase in effort soon becomes insufficient to compensate for the increasing noise (Kryter, 1970). Apparently if you speak in noisy conditions, your speech is more intelligible than it would be if it was recorded in quiet and the noise were added later. However, at some point of vocal effort such as shouting, intelligibility drops and the advantage is lost. You can see that in testing a channel it is a good idea to use previously recorded materials in order to avoid the complex interaction between the channel characteristics and the speaker's way of speaking.

Binaural Listening

Another important aspect of the speech channel is whether the speech (and noise) is presented to one or both of the listener's ears. Recall from the discussion of binaural hearing in Chapter 3 that a 15-dB advantage in detection can result from the binaural presentation of the signal. Similar advantages can be gained from the presentation of speech signals over the two auditory channels. The effect is dependent on the ability of the auditory system to separate the speech and noise signals, using an analysis of the differences in sound phase and intensity at the two ears. This analysis is also involved in the so-called cocktail party effect, which describes our ability to selectively attend to different conversations in a crowded, noisy room. This process is made possible by our having two ears and our being able to localize the direction of different sound sources. Plug up one ear, and our understanding of speech in such conditions would be severely degraded. The processes underlying selective attention have been studied in the laboratory; see the discussion in Chapter 6 for some examples. Briefly, we are able to use information residing in physical cues about the desired and undesired signals to screen out irrelevant signals. The cues in the cocktail party situation involve the spatial location of the speaker—but we can employ other cues such as the frequency range of the speaker. For example, it is much easier to selectively attend to a woman speaker against a babble of men's voices, than to a man's voice in the same background. Signals that are physically similar to the desired input will produce the greatest interference with it, because we are unable to do any filtering on the basis of physical differences between the desired and undesired signal.

Distortion of Speech

Sometimes the communication channel will distort the speech signal in some way. We can tolerate a variety of different types of distortion before intelligibility is severely degraded, but certain types are much worse than others. Some types of deliberate distortion of the speech signal have been used to reduce the normal range of frequencies of speech or to more effectively code the intensity range of speech for radio transmission. Another type of manipulation removes time segments of the speech signal. In one technique the speech is turned on and off at a rapid rate; during the off periods the channel can be shared by another speech message. In addition to distorting the original speech waveform in some way, all of these techniques remove some information from the speech signal. In the last section of this chapter we discuss the general problem of coding the speech signal for efficient transmission, and the related topics of computer synthesis and recognition of speech.

Peak Clipping One technique for modifying the intensity characteristics of the speech signal is to compress the range of intensities that the channel will pass. Peak clipping is the process by which positive and negative peaks of the speech waveform are clipped off; only the remaining waveform is transmitted. The remaining waveform can be amplified so that its new peaks are equal to the old ones. Figure 9-6 illustrates the effects of peak clipping a speech waveform. Suppose that regardless of the moment-to-moment intensity of your speech, your microphone always transmitted either a full positive or a full negative voltage. This is an example of infinite peak clipping, and is approximated by the 20-dB clipping case

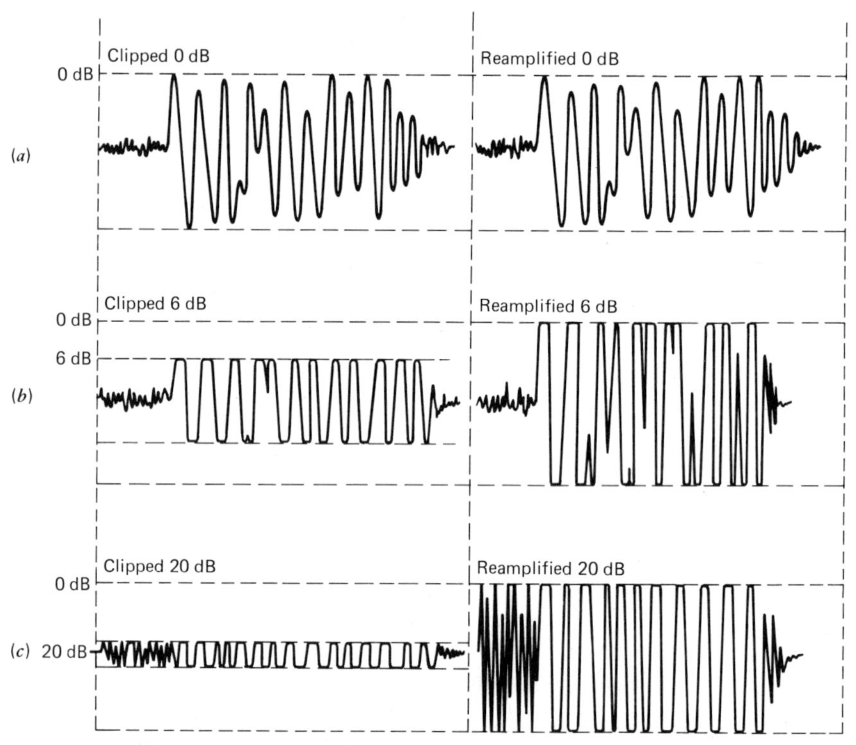

Figure 9-6. Speech samples before and after 6 dB and 20 dB of peak clipping. The right-hand samples are re-amplified until their peak amplitudes are the same as the original sample. (From J. C. R. Licklider, D. Bindra, & I. Pollack, The intelligibility of rectangular speech waves, *American Journal of Psychology,* 1948, *61,* 3. Copyright 1948 by Karl M. Dallenbach, and reproduced by permission of University of Illinois Press.)

shown in Figure 9-6. Peak clipping and then reamplifying the waveform to the previous peak amplitude has the effect of increasing the power of the consonant sounds relative to the vowels. It also increases the average speech power, making the speech more intelligible in certain kinds of noise than unclipped speech.

Filtering the Speech Spectrum The ordinary speech spectrum spans a frequency range from below 100 Hz to over 8000 Hz. In fact, the contribution of different frequencies to the intelligibility of speech varies somewhat over this spectrum. You can determine the importance of different regions of the speech spectrum by feeding the speech signal through a bandpass filter, a device that passes a defined band of frequencies and rejects any frequencies outside that band. You could, for example, reject all the frequencies above 1700 Hz without much effect on intelligibility. Conversely, you could filter out all the frequencies below 1700 Hz, with much the same effect. Different speech sounds will be affected by these different manipulations; the consonants will be hurt much more by rejecting the high frequencies, and vowels will be hurt more by cutting out the lows. These results are related to the general effects of noise on speech. In high noise levels, the vowels will remain intelligible longer as the noise level increases, since more of the speech energy is in the vowels. If we filter out and discard a 3000-Hz band of frequencies centered at 1700 Hz, intelligibility will go to zero. On the other hand, if we only pass those middle frequencies and reject everything else, there is almost no effect on intelligibility. How much of a bandwidth is needed for essentially no decrease in intelligibility? Perfectly satisfactory communication is obtainable with a band of frequencies from 800 to 2500 Hz. Highly trained speakers and listeners can communicate effectively with defined message sets over channels having much narrower bandwidths. Remember that eliminating these "extraneous" speech frequencies may remove nonspeech information such as cues about the speaker's sex, age, or emotional state, and may have drastic effects on the perceived quality of the communication channel.

Articulation Index

If some frequency regions are more important than others to speech intelligibility, then some regions should be more sensitive to interference from noise. If we could specify those frequency regions, we would have a way to predict the disruptiveness of a given spectral distribution of noise. This idea is the basis for the Articulation Index (AI) (Beranek, 1947; French & Steinberg, 1947).

The Articulation Index (AI) is a technique for predicting the effects of noise on speech by considering the speech signal-to-noise ratios in a set of frequency bands spaced across the speech spectrum. Originally, a set of 20 different frequency bands were defined that were assumed to each contribute an equal amount to speech intelligibility. Under ideal circumstances, each band contributes one-twentieth or 0.05 to the maximum AI score of 1.0. Under noisy conditions some of these bands would contribute proportionally less than that 0.05 maximum, depending on the signal to noise ratio within that band. A simpler procedure for computing the AI is outlined in Figure 9-7 and Table 9-5.

The more that the speech spectrum of the speaker in a particular application differs from that of the noise, the larger will be the resulting AI. In Figure 9-4 the AI was plotted across the top of the graph, showing how intelligibility varied as a function of the AI. Figure 9-8 shows how AI varies with the level of a fairly representative noise spectrum, that might be found in a typical "open plan" office layout (Pirn, 1971).

The AI measure seems to be a good predictor of speech intelligibility over a variety of noise conditions and speech manipulations (Kryter, 1962) but is inappropriate for systems where there is much speech processing for compressing the speech band. In those cases, you may have to directly evaluate the intelligibility of the system under consideration. An AI value below 0.3 will usually be unsatisfactory for most ap-

TABLE 9-5 *Computation of the AI*[a]

The ANSI procedure involves comparing the noise levels and the peak speech levels in 15, one-third octave, bands.

1. Plot the one third octave spectrum of the speech signal to be evaluated; this is shown in Figure 9-7 as the idealized male speaker spectrum.
2. Plot the one-third octave noise spectrum; a sample spectrum is plotted in the figure.
3. Determine, for each band, the difference in dB between the peak speech level and the noise level. (If the noise is bigger, assign a zero difference; if the signal is more than 30 dB bigger, assign it a 30 dB difference.)
4. Multiply these dB differences by the appropriate weighting factor given in column 3, and enter these numbers in the fourth column.
5. Add up the entries in the fourth column and you have the AI. (The reason for the 30-dB limitation is because the speech signal is assumed to have a 30-dB dynamic range within each band. The calculation is actually an estimate of the proportion of that 30 dB "used up" by the noise in each band.)

Band	Speech Peaks Minus Noise-dB	Weight	Column 2 × 3
200	4	0.0004	0.0016
250	10	0.0010	0.0100
315	13	0.0010	0.0130
400	24	0.0014	0.0336
500	26	0.0014	0.0364
630	26	0.0020	0.0520
800	24	0.0020	0.0480
1000	21	0.0024	0.0504
1250	18	0.0030	0.0540
1600	18	0.0037	0.0666
2000	15	0.0037	0.0555
2500	15	0.0034	0.0510
3150	6	0.0034	0.0204
4000	8	0.0024	0.0192
5000	12	0.0020	0.0240
		AI = 0.5357	

[a] This material is reproduced with permission from American National Standard, *Methods for the Calculation of the Articulation Index,* ANSI S 3.5-1969, page 17, Figure 6. Copyright by the American National Standards Institute. Copies of this standard may be purchased from the American National Standards Institure at 1430 Broadway, New York, N.Y. 10018.

plications. An AI of 0.5 could be satisfactory for a limited vocabulary situation; an AI of 0.7 would be required for sentence comprehension better than 99%. Automatic techniques for assessing the AI of a given channel have been developed. One of these involves the calculation of a Speech Transmission Index (STI) that allows consideration of the effects

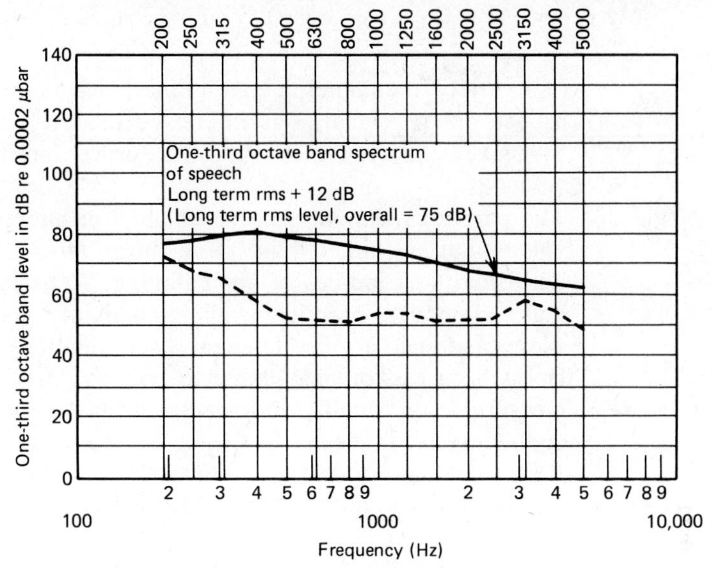

Figure 9-7. Example of the calculation of an AI by the One-Third Octave Band Method. (This material is reproduced with permission from American National Standard, *Methods for the Calculation of the Articulation Index,* ANSI S3.5-1969, p. 17, figure 6. Copyright 1969 by the American National Standards Institute. Copies of this standard may be purchased from the American National Standards Institute at 1430 Broadway, New York, NY 10018.)

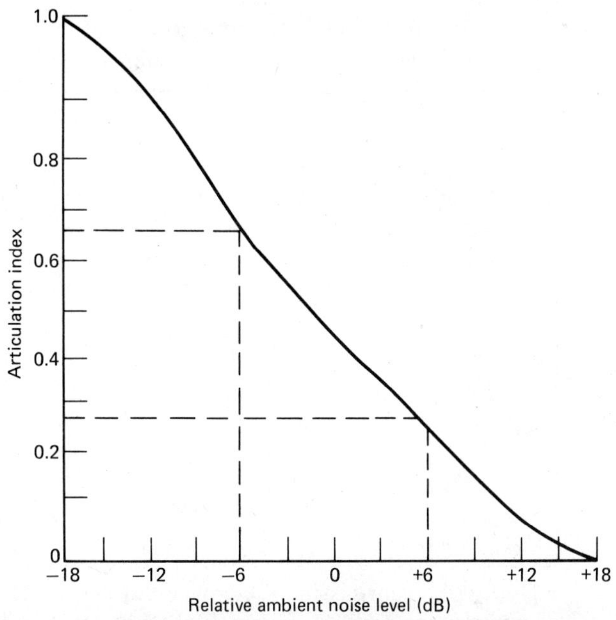

Figure 9-8. AI as a function of background noise. Each 6 dB increment in the ambient-noise level generally results in an AI decrease of 0.2. (From R. Pirn, Acoustic variables in open planning, *Journal of the Acoustical Society of America,* 1971, *49,* 1342, and reproduced by permission.)

of a variety of channel manipulations, such as distortions in the temporal aspects of the signal. The interested reader is referred to an article by Steeneken and Houtgast (1980) for further information about this index.

Channel Quality

The particular communication channel of interest to you may have AI values that are larger than 0.8 or more. How do you evaluate channels with such high indices of intelligibility? Webster (1978) suggests that at high AI values, practically all intelligibility tests will yield such high scores as to be very inefficient at discriminating among different systems. In such cases you may have to employ tests involving reaction time measurements, quality judgments, or tasks involving the presence of competing messages or of a speech babble as the noise.

One technique asks people to compare the quality of speech on a test channel and a standard channel. In an experiment by Munson and Karlin (1962), listeners compared speech on several communication channels to speech on a high-fidelity monaural system. Munson and Karlin measured how much noise had to be added to the high fidelity channel to make it appear equal in quality to the channel under test. The Transmission Preference Level (TPL) measures this noise. Figure 9-9 illustrates this measure for some practical types of speech circuits. Webster (1978) has calculated the AI values that correspond to these values and they are included in the figure. It is evident that in some applications we might need to specify the quality of a given channel at AI's larger than 1.0. The reason for this is not simply the quality as perceived by a prospective user. The speed and ease with which speech can be understood and in-

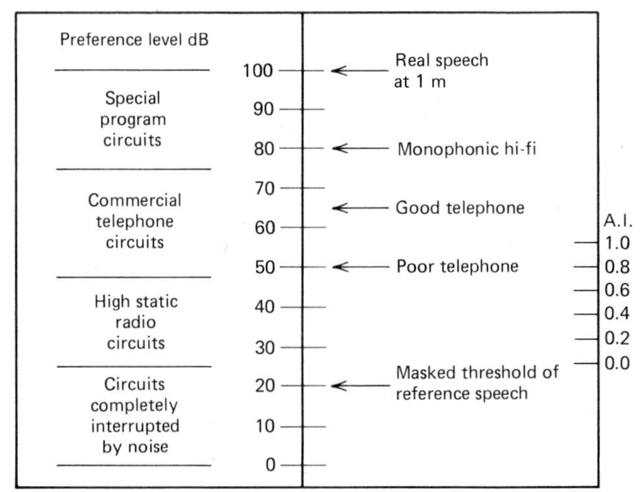

Figure 9-9. Transmission Preference Levels and AI values for different types of communication circuits. (From J. C. Webster, Speech Interference Aspects of Noise, 211, in D. M. Lipscomb (Ed.), *Noise and Audiology,* copyright 1978 by University Park Press, and W. A. Munson & J. E. Karlin, Isopreference method for evaluating speech-transmission circuits, *Journal of the Acoustical Society of America,* 1962, *34,* 773, and reproduced by permission.)

formation processed from the channel may be an important function of the channel characteristics at AI's greater than 0.8. Developments in computer speech may make it even more important to distinguish among systems performing in this range (See the discussion of Speech Synthesis by Rule).

Noise Criterion Curves

Beranek Blazier, and Figwer (1971) developed a rating scheme that has been employed to describe the acceptability of noise in offices, and similar workplaces. Acceptability is assumed to depend on both the effectiveness of speech communication in the noise and on the perceived loudness of the noise. To determine the Noise Criterion (NC) value, you compare the measured background noise spectrum with the curves shown in Figure 9-10. For example, a workplace has a NC value of 30, if none of the octave band levels of the measured noise exceed the plotted values for NC-30 (and at least one exceeds the next smallest NC curve value). A

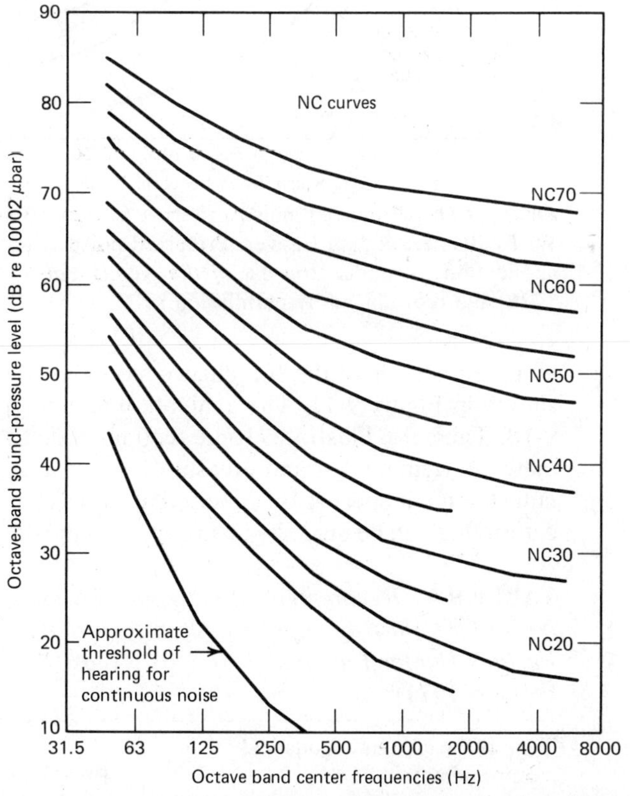

Figure 9-10. Noise criterion (NC) curves for rating rooms and offices (see text). (From L. L. Beranek, W. E. Blazier, & J. J. Figwer, Preferred noise criterion (PNC) curves and their application to rooms, *Journal of the Acoustical Society of America,* 1971, *50,* 1224 and T. J. Schultz, Noise criterion curves for use with USASI preferred frequencies, *Journal of the Acoustical Society of America,* 1968, *43,* 637, and reproduced by permission.)

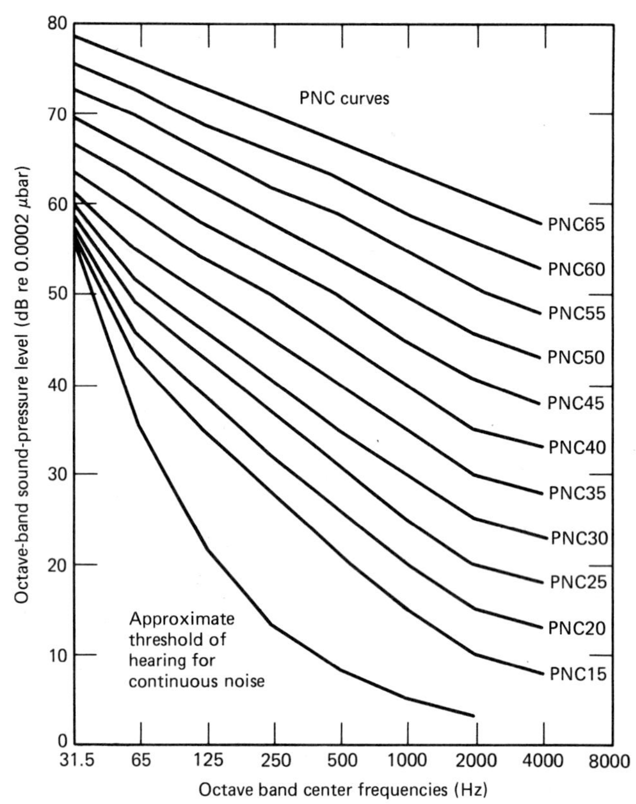

Figure 9-11. Preferred noise criterion curves (PNC). (From L. L. Beranek, W. E. Blazier, & J. J. Figwer, Preferred noise criterion (PNC) curves and their application to rooms, *Journal of the Acoustical Society of America,* 1971, *50,* 1226, and reproduced by permission.)

modified version of the NC curves has been suggested by Beranek and is shown in Figure 9-11. The evaluation technique is the same as in Figure 9-10. Table 9-6 illustrates some recommended PNC curves and approximate (A-weighted) sound pressure levels for a range of work, living, and entertainment spaces. It is likely that the NC or PNC value is the specification that will be made by an architect in the design of a building.

TABLE 9-6 *Recommended Category Classification and Suggested Noise Criteria Range for Steady Background Noise as Heard in Various Indoor Functional Activity Areas* (Beranek, Blazier, & Figwer, 1971)[a]

Type of Space (and acoustical requirements)	PNC Curve	Approximate Sound Level[b] in dB(A)
Concert halls, opera houses, and recital halls (for listening to faint musical sounds)	10 to 20	21 to 30
Broadcast and recording studios (distant microphone pickup used)	10 to 20	21 to 30

TABLE 9-6 *(Continued)*

Type of Space (and acoustical requirements)	PNC Curve	Approximate Sound Level[b] in dB(A)
Large auditoriums, large drama theaters, and churches (for excellent listening conditions)	Not to exceed 20	Not to exceed 30
Broadcast, television, and recording studios (close microphone pickup only)	Not to exceed 25	Not to exceed 34
Small auditoriums, small theaters, small churches, music rehearsal rooms, large meeting and conference rooms (for good listening), or executive offices and conference rooms for 50 people (no amplification)	Not to exceed 35	Not to exceed 42
Bedrooms, sleeping quarters, hospitals, residences, apartments, hotels, motels, etc. (for sleeping, resting, relaxing)	25 to 40	34 to 47
Private or semiprivate offices, small conference rooms, classrooms, libraries, etc. (for good listening conditions)	30 to 40	38 to 47
Living rooms and similar spaces in dwellings (for conversing or listening to radio and TV)	30 to 40	38 to 47
Large offices, reception areas, retail shops and stores, cafeterias, restaurants, etc. (for moderately good listening conditions)	35 to 45	42 to 52
Lobbies, laboratory work spaces, drafting and engineering rooms, general secretarial areas (for fair listening conditions)	40 to 50	47 to 56
Light maintenance shops, office and computer equipment rooms, kitchens, and laundries (for moderately fair listening conditions)	45 to 55	52 to 61
Shops, garages, power-plant control rooms, etc. (for just acceptable speech and telephone communication). Levels above PNC-60 are not recommended for any office or communication situation	50 to 60	56 to 66
For work spaces where speech or telephone communication is not required, but where there must be no risk of hearing damage	60 to 75	66 to 80

[a] From the *Journal of the Acoustical Society of America,* 1971, *50,* 1227.
[b] Position A on standard sound level meter; see Chapter 16 for definition.

CONFLICTING SYSTEM GOALS

Designing Halls for Speech or Music

The design of large halls and auditoriums has long involved a complex merging of art and engineering (see Figure 9-12). Many of the important acoustic criteria for concert halls have been defined and analyzed. Table 9-7 summarizes some of these aspects of concert halls. These characteristics will be dependent on the hall volume, width, ceiling height and slope, shape, cross section, and the presence of absorbing and reflecting materials.

One of the more obvious features of a hall is the reverberation time, T. You may have noticed that some halls seem more "live" than others. The reverberation time, T, is the time it takes for an abruptly terminated sound to decay 60 dB in level. In some halls, the presence of sound-reflecting surfaces can result in moderate levels of reverberant sound for many seconds after the termination of the direct sound from the stage. Figure 9-13 illustrates some typical reverberation times for different types of halls and functions. The reverberation

Figure 9-12. A view toward the stage of the Avery Fisher Hall auditorium during a New York Philharmonic performance in Lincoln Center for the Performing Arts. (Photo by Norman McGrath © Lincoln Center for the Performing Arts.)

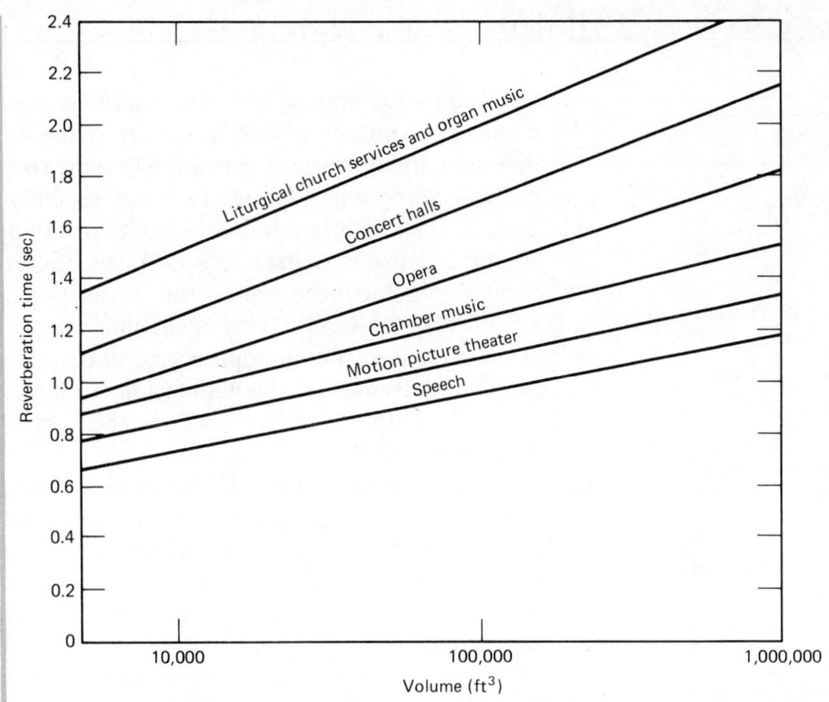

Figure 9-13. Typical reverberation times for various auditoria and functions. (From F. A. White, *Our Acoustic Environment,* p. 440. Copyright 1975 by John Wiley & Sons, Inc., and reproduced by permission.)

TABLE 9-7 *Important Characteristics of Concert Halls* [Modified from Baxa & Seireg (1980); Schroeder, Gottlob, & Siebrasse (1974)]

Characteristic	Definition or Description
1. Definition	Ratio of energy arriving during the first 50 msec to energy in the total time period of sound.
2. Diffusion	Measure of the directional distribution of sound energy arriving at the listener's seat.
3. Echo	Related to the presence of detectable reflected sounds distinct from the exponentially decaying reverberation.
4. Intimacy	The time delay (gap) between the direct sound and the first reflection at the listener's seat.
5. Liveness	The reverberation time, *T*. (Related is the sustenance of bass and treble sound.)
6. Loudness	The ratio of direct to total (reflected) sound.
7. Spatial impression	A measure of the interaural incoherence or binaural dissimilarity at the listener's seat.

characteristics should also be relatively constant for the bass frequencies. Another desirable aspect is the extent of apparent sound diffusion, the observer's sensation that the sound is reaching him or her from all directions (White, 1975). A recently defined aspect of good concert halls is interaural coherence, or the dissimilarity between the sound arriving at a listener's two ears (Schroeder et al., 1974). The more dissimilar these signals, the better you sense the spatial character of the music. Imagine switching your stereo system from the stereophonic to the monophonic position and you'll get the idea of the possible extremes on this dimension. In fact, it is said that wide halls with low ceilings tend to produce the "dreaded monophonic effect" (Schroeder, 1980).

An interesting problem is that many of these criteria for music halls are not desirable for auditoriums that are used for speech. The important criteria for lecture halls include *short* reverberation times (and low ambient noise levels): T values of the order of 0.3 to 1.7 sec are desirable, depending on the room volume. (For a large room, you need some "liveness" or the speaker has to shout to be heard in the back.) Short reverberation times are clearly incompatible with music applications. Another desirable aspect for speech is a highly directional sound field. You want the speaker's speech to appear to be coming from his or her position on the stage, rather than from the ceiling or rear wall. Again, this aspect may be incompatible with the desire for good sound diffusion in music. How can these conflicting design criteria be reconciled in the same hall? Electronic amplification and delay systems have been used to change the effective reverberation time of halls (White, 1975). As knowledge in architectural and psychological acoustics increases, it may be possible to design halls with electronic sound systems that can provide a controllable electroacoustic environment that is suitable for any type of musical or theatrical performance.

SPEECH TECHNOLOGY

Communication engineers have long been interested in compressing the speech channel to as narrow a band of frequencies as possible, while still preserving good intelligibility and speech quality. The reason for this interest is that a given radio or wire channel has a practical limit on the band of frequencies that it can pass. A given channel can carry a limited number of voice or other (video, etc.) messages. Advances in satellite and fiber-optics telecommunication have somewhat reduced the need for more efficient use of bandwidth, but there are still many applications where it is desirable. One such application is in the Mobile Radio Telephone Service, where bandspace is at a premium.

There is an important new reason for learning how to encode the speech signal efficiently. The availability of inexpensive and efficient digital electronics has made it practical to perform various computations on speech signals. The applications of this new technology include the computer processing of speech for encryption or privacy, low data rate communications, the storage and generation of synthetic speech by computer from internal coded information or from written text, and the recognition by computer of spoken material or of particular speakers. Today, we can see widespread use of speech encryption devices, of children's toys that speak (see Figure 9-14), of computer instruction devices that generate spoken instructions to workers, and the spoken entry of limited speech vocabularies to computers. A highly popular add-on device to the home computer is the speech synthesizer. As usual at the beginning of a new technology, it is difficult to give specific directions for the design of systems that are optimal from the human factors standpoint. In this section, we discuss briefly some basic aspects of the new technology and some of the potential problems that the human factors specialist should understand.

Figure 9-14 Speak 'n Spell™ Toy. This toy utters words for children to spell. (Photo by B. Kantowitz, Copyright 1983 TEXT-EYE.)

Waveform Sampling and Quantization

The first operation that must be performed on a waveform, if we are going to get it into a computer, is to convert it from analog to digital form. Figure 9-15 illustrates a typical analog signal (a portion of a vowel sound) that we have digitized for computer processing. Notice that the sampling operation selects a number of particular moments in time to be measured, the rest of the waveform is ignored. In fact, this sampling operation does not discard as much information as you might think. Suppose the signal of interest to us contains frequency components in a band, W, that is 3000 Hz wide. (Perhaps it includes frequencies from 500 to 3500 Hz.) We can sample the signal at a rate of at least $2W$, or 6000 times per second, without losing any information about these different frequency components. The reason for this is that it takes only two pieces of information to specify the amplitude and phase of any particular frequency component; to reconstruct one second of our signal we would need a number of samples equal to two times the number of frequency components. This principle is known as the *sampling theorem.*

What kind of precision is needed to specify the amplitude of each of those $2W$ samples? This question is called the *quantization* problem. We need to divide the amplitude scale (as we did the time dimension) into a number of small but measurable increments. But how small must each of these increments be? In Figure 9-15 we quantized each sample to one of

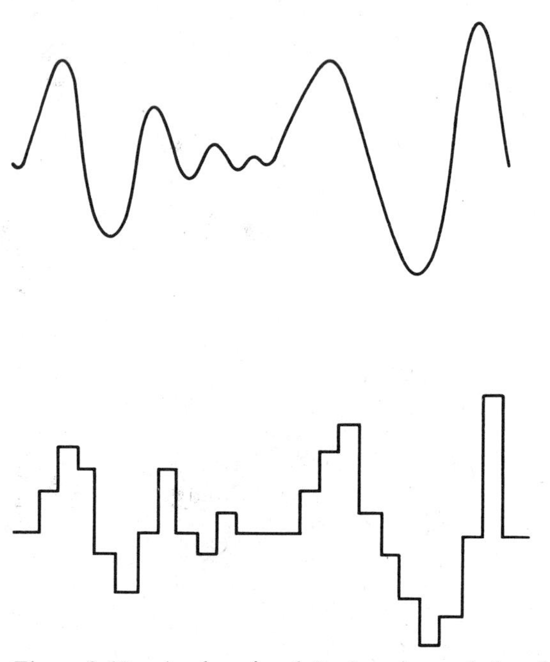

Figure 9-15. Analog signal (top) and sampled and quantized version of analog signal (bottom). Each sample can take one of 16 values (4 bits).

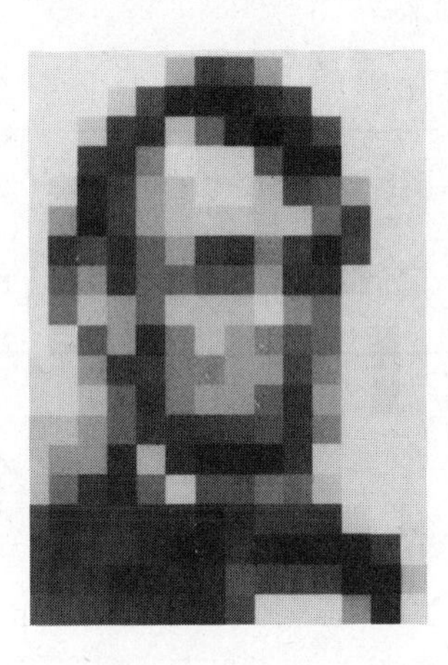

Figure 9-16. Coarsely sampled and quantized picture. Squinting at the picture reveals the subject. (From Harmon & Julesz, 1973.)

16 values (e.g. 4 bits). You can see that the result is a bit crude. A quantization precision of 12 bits would specify each amplitude sample to a precision of one in 4096, or about 0.02%, a value you might think was compulsively accurate for speech purposes, and you would be right. But you can see that both the sampling rate and the precision of quantization will affect how well the final waveform matches the original waveform.

The quantization and sampling operations have another effect, in addition to discarding information that is in the original signal. These operations add noise to the signal. Imagine a sinusoidal signal that is quantized to a precision of 4 bits, at a sampling rate of 20 samples per cycle. The resultant waveform will still look approximately sinusoidal, but it will have various square edges at different points along the signal. If you analyzed what frequency components were present in this new signal, there would be many more frequencies present than in the original sinusoid, which only had one component. Of course that is to be expected whenever we perform some nonlinear operation such as sampling and quantizing. But these additional components would be heard as noise—and they might even interfere with our ability to hear the information in the original signal that we wanted to preserve. One of the nicest examples of this consequence of sampling and quantization is the Lincoln picture made by Harmon and Julesz (1973) and shown in Figure 9-16. The original picture has been spatially sampled so that there are only 266 blocks (sampling). Each block has been given one of 16 grey levels (quantization). The reason that it is hard to see Lincoln is because of the added noise produced by the sampling and quantization operation. However, if you squint at the picture (or hold it far away from your eyes) you can filter out the interfering frequencies (see Chapter 4 for a discussion of spatial frequency analysis). Then Lincoln appears! Precisely the same sorts of phenomena occur in the processing of speech signals. In fact Schroeder, Atal, and Hall (1979) have suggested an interesting way to make the noise components less evident in highly processed speech by adjusting the intensities of the speech frequencies so that they prevent us from hearing the noise frequencies!

Bit-Rate Reduction

Choosing the sampling rate and the precision of quantization will obviously depend on the specific application. For the high-fidelity digital reproduction of music, we might want a sampling rate of 50,000 samples per second and a quantization of 16 bits per sample. Transmitting such signals would take a transfer rate of 50,000 × 16 or 800,000 bits per second. Your typical home computer might have 48,000, 8-bit bytes or 48,000 × 8 = 384,000 bits in its memory. So you could use it to store only 384,000/800,000 or about one-half second of that high-fidelity digitized music. Fortunately, the speech channel can be a lot narrower in bandwidth then the high-fidelity music channel. An 8-bit quantization and a 5000-Hz sampling rate would result in quite acceptable speech. The bit rate would then be: 5000 × 8 = 40,000 bits per second. Your home computer could hold: 384,000/40,000 or almost 10 sec of that speech.

Actually we can hedge that last figure in a number of ways. We can cut down the quantization accuracy; remember that amplitude clipping of the speech waveform does not have a large effect on intelligibility—so we could use much less than the 8 bits (in fact we could compress the range of speech intensity by using a log transformation prior to digital conversion). A more elegant approach is to restrict the intensity range (in bits) of each subsequent sample to some increment or decrement that depends on the intensity of the prior sample value. This approach takes advantage of the fact that there is a correlation between the intensity of successive speech samples. The speech mechanism is a physical system that has mass and inertia, and so there are physical limitations on the sounds that can be made right after other sounds. This fact has led to the development of special pulse code modulation techniques that take advantage of this predictable aspect of speech. A very elegant technique carries the procedure even further; essentially it works by quantizing only those parts of the signal that cannot be predicted from the already coded signal (Atal & Hanauer, 1971). Such systems have produced speech with rates as low as 2400 bits per second. Generally, these techniques have produced quite usable speech at 9600 and 16,000 bits per second, but the voice quality and intelligibility do not meet the standards of good telephone-quality speech (Goldberg, 1979). There are working systems in the field, such as a voice security system, with bit rates lower than 9600 bits per second.

Elegant on-line processing of speech signals may have to await the presence of computers faster than those now available, at least on a practical basis. In such high-speed systems, it will be possible to perform more advanced computation on the speech input; it may be possible to extract information about the particular phoneme sequence, transmit that data, and then reconstruct the speech signal from the phoneme sequence at the destination. We mentioned at the outset of this chapter that phoneme rates in normal speech are normally about 12 per second; if it takes say, 6 bits, to encode each phoneme, you can see that a bit rate of 720 bits per second would suffice to transmit speech. At such a rate, the 48,000-byte memory we mentioned earlier would hold about 9 minutes of speech. By the way, that 48,000-byte memory could hold almost an hour and a half of material spoken from written text (printed text = 75 bits/sec, Flanagan, Coker, Rabiner, Schafer, & Umeda, 1970). So generating phoneme sequences from text would result in even greater advantages, not to mention all of the possible applications, such as computer spoken books for the blind.

Speech Synthesis by Rule

It is possible to generate intelligible speech by specifying the operations that will cause a phoneme sequence to be generated by a computer speech synthesizer. These systems were developed out of the pioneering work of speech scientists and engineers during the last 30 years. Better speech generation involved the use of rules for making the acoustic transitions between phonemes and for modifying the pitch and timing of each speech segment as a function of the phoneme context. The logical next step was to construct a program for generating speech from printed text. Several systems have been developed for accomplishing this function. A system for automatically converting printed wire-wrapping instructions for a worker on a production line was informally tested by Flanagan, Rabiner, Schafer, and Denman (1972). In such a task it is efficient for the worker to keep both hands on the equipment, and not to divert his or her eyes from the work. Instructions were conveyed via computer-synthesized speech from a wire-wrap instruction list coded in a computer card deck. Although the worker mentioned that the synthetic speech did not sound natural, no difficulty was encountered in using it, and no wiring errors were made during the tests.

The MITalk system is a text-to-speech system that automatically produces naturally sounding synthetic speech from unrestricted English text. Pisoni and Hunnicutt (1980) conducted a perceptual evaluation of the MITalk system that included testing phoneme intelligibility, word recognition, and listening comprehension. They found that the quality of the system was quite satisfactory for a wide range of text-to-speech applications.

It's quite likely that the human factors specialist will be asked to make evaluations and decisions about the effectiveness of proposed and existing systems for encoding and generating speech at low bit rates. It is difficult at this point to present any rigid formula, such as the AI, that would allow performance predictions to be made. Two recent approaches to the problem of developing objective measures of channel quality for such systems are proposed by Barnwell (1979) and Mermelstein (1979); you may want to refer to these studies or to the *IEEE Transactions on Acoustics, Speech, and Signal Processing* for further information about this problem. Recently, Pisoni (1982) discussed the evaluation of voice response systems. He found that there can be important perceptual and cognitive limitations when synthetic speech is employed. For example, the processing time required for certain tasks is greater when synthetic rather than natural speech is used. It may be necessary to directly evaluate such systems with suitable articulation or intelligibility tests of the sort that we have described, or with performance tests in the actual proposed application. This type of evaluation, prior to fixing of the system design, may be easier to perform than in the past, since it is likely that the prospective synthesis system will be implemented on a general purpose computer prior to fixing it in an integrated-circuit chip. This possibility makes the argument for early human factors evaluation even more compelling.

Speech recognition

The other side of the speech synthesis coin is the automatic recognition of spoken words by machine. A range of success with different systems that recognize words and sentences has been achieved. Several of the successful speech understanding systems developed under the support of the Advanced Research Projects Agency have been reviewed by Klatt (1977). One aspect of less sophisticated systems that are designed to recognize words spoken in isolation, is the need to "train" the computer system to recognize speech spoken by particular speakers. In order for these systems to create templates for the words in their vocabulary, they must be presented with spoken examples of the words that they will be expected to recognize. This dependence on human operator input can involve some significant human factors problems. This issue was discussed by Carpenter and Lavington (1973) in a study of a real-time speech recognition system that was designed to control a CRT cursor with a five-word spoken command vocabulary. One problem they found was that the human operators changed the way that they spoke the same words during the experiment. The human speakers modified the way the words were pronounced in highly variable ways, that depended on the performance of the system (if it moved the cursor in the correct way or not). These changes in pronunciation usually did not increase the computer's accuracy. Although the human operators appeared to be hunting for the "right" pronunciation, that pronunciation usually did not match any of the speech templates the machine was using. The resulting interaction between the computer system and the speaker resulted in instability in the performance of the system, as well as operator frustration. The Carpenter and Lavington experiment highlights the importance of considering human factors aspects of the design of such systems.

SUMMARY

The human factors specialist will be involved in many systems that have the speech channel as a vital component. The production and perception of speech are highly complex processes; for example, speech perception depends on many factors including the acoustic, syntactic, and semantic context of the input signal. Techniques are available for the evaluation of the intelligibility of speech communication channels. These techniques require specification of the nature of the speech messages (particularly their predictability and familiarity) and the noise background (particularly its level in different frequency bands relative to the speech signal). There are several reasons why the human factors specialist should be aware of new techniques for evaluating speech communication systems. First, measures of intelligibility may not accurately assess the perceived quality of a channel, as judged by prospective users. Second, intelligibility measures may be imprecise measures of performance (such as processing time) in tasks that make high-level cognitive and memory demands on the user. Third, existing measures may be inappropriate under conditions of extensive reduction in bandwidth or bit rate, such as with digitally encoded or processed speech or with synthetic speech. Under these conditions, the potential impact of system or signal characteristics on information processing or user acceptability may be significant.

10 CONTROLS AND TOOLS

Just as displays (Chapters 7 and 8) allow the machine to communicate to the operator, controls allow the operator to communicate to the machine. Thus, the communication concepts discussed in Chapter 7 apply equally here. And just as some displays work better than others, some controls are more suited to the human's psychomotor abilities (see Chapter 5). In this chapter we discuss the types of controls that are available, control features, relationships between controls and displays, and design of tools.

TYPES OF CONTROL

A control is any device that allows a human to transmit information (see Chapter 5) to a person-machine system. An astonishing variety of physical devices have been built to aid the human in transmitting this information. Controls can be mechanical, electromechanical, electronic, optical, magnetic, and piezoelectric to list some of the more common varieties. However, it is more useful for the human factors designer to describe controls according to three dimensions or categories that apply regardless of the physical implementation of the control device. These three categories are (1) discrete versus continuous operation, (2) linear versus rotary operation, and (3) one- versus two-dimensional operation. Some combination of these three categories will describe almost all control devices currently in use.

Discrete Versus Continuous Control

A discrete control can assume only a limited number of finite states. For example, a light switch can be either on or off, up or down. There is a sharp transition from one control state to an adjacent state. A continuous control, ocassionally called an analog control, can assume a large number of states. For example, a dimmer control for a light can continuously alter light output from none to full brightness. There is a smooth and gradual transition from one control state to the next.

TABLE 10-1 *Guidelines for Control Design*

Use a discrete control when:
- There is a limited number of control states (<25).
- Numerical or alphabetical information must be entered.
- On-Off or Yes-No information is required.
- Only small mechanical forces are required or available.

Use a continuous control when:
- There is a large number of control states (>25).
- The operator must exert substantial force.
- Speed of operation is more important than accuracy of control setting.

Continuous controls are used when fine adjustments are necessary and when there are many control states but not enough physical space to provide a discrete control. Discrete controls are used when there are only a small number of control states, say less than 25-30. One advantage of discrete controls is the tactual feedback they provide when operated. For example, a skilled driver can feel the control engage as she shifts from second to third gear, before releasing the clutch pedal. Table 10-1 gives design guidelines for selection of discrete and continuous controls.

How fine an adjustment can be made under optimal conditions with a continuous control? Recall that the human psychomotor system can generate about 10 bits/sec (see Chapter 5). Ten bits equals 1024 equiprobable alternatives. This implies that the upper limit for fine adjustment if only one second is allowed for control movement is on the order of 0.1% accuracy. If less time is available for control movement, accuracy will decrease. If unlimited time is available, accuracy will depend on the bandspread of the control. Bandspread refers to the range of physical control action. A linear slide that can move a distance of 10 cm has half the bandspread of a slide that can move 20 cm. If unlimited adjustment time is available, any desired accuracy can be achieved by having enough bandspread. In most situations bandspread is limited by physical workspace parameters: a slide that is 100 m long is just not practical in a typical workspace. And even if it were, the adjustment time would be considerable.

Linear Versus Rotary Controls

The distinction between these two types of control is obvious. Linear controls such as pushbuttons, toggle switches, and slides move in a straight line. Rotary controls such as a telephone dial, door knob, and circular volume controls on radios and televisions move in an arc. Several examples of linear and rotary controls are shown in Figure 10-1. At present there is no well-established theoretical basis for choosing a linear versus a rotary control. The design decision is based upon physical

One—dimensional control

Two—dimensional control

Discrete control

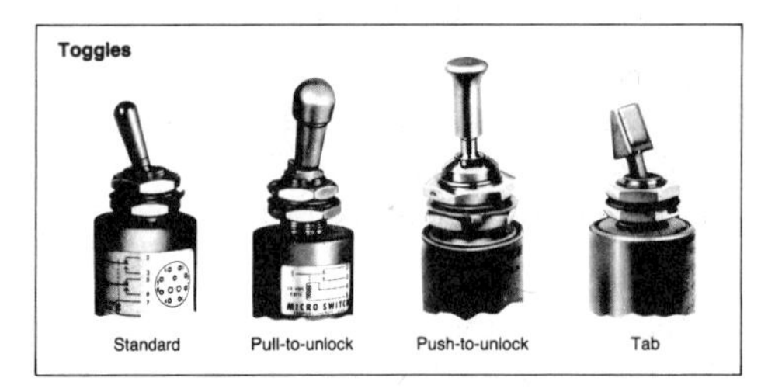

Toggle Switches

Calculator Keyboards

Continuous control

Automobile brake pedal

X—Y Potentiometer Joystick

Figure 10-1 a. Some examples of linear controls. (Top left & right, courtesy of Microswitch, a Honeywell Division; bottom left, David S. Smith; bottom right, courtesy of Measurement Systems, Inc.)

characteristics of each control and upon display-control relationships (to be discussed later in this chapter).

One- Versus Multidimensional Controls

This distinction is also obvious, being based on the geometry of control motion. Two-dimensional controls such as joysticks, keyboards, track-balls, and manual automobile shift levers are used when mechanical or electronic arrangements demand control outputs that require vector rather than scalar format. (A scalar quantity can be specified by a single number; a vector quantity requires two numbers.) Examples of one- and two-dimensional controls are shown in Figure 10-1. A single two-dimensional control is generally preferable to using two one-dimensional controls to communicate the same information. However, one important exception to this rule occurs when the display information is presented

One—dimensional control

Two—dimensional control

Discrete control

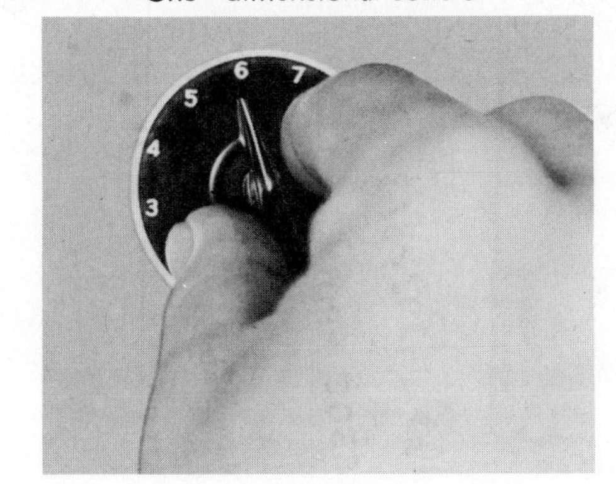

Rotary Switch

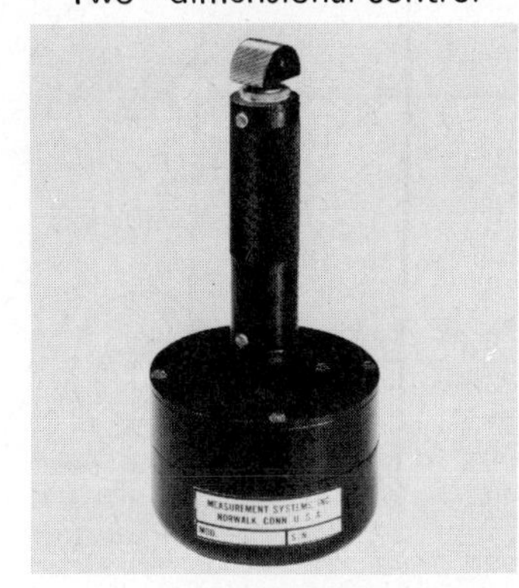

Multi—position toggle switch

Continuous control

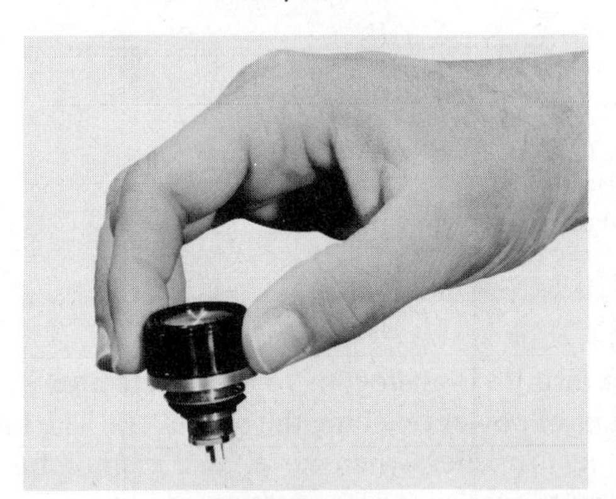

Spring return pot

Track ball

Figure 10-1 b. Some examples of rotary controls. (Top left, courtesy of The Digitran Company; top right & bottom, courtesy of Measurement Systems, Inc.)

as two scalar values. Then control-display relationships may be more important than optimal design of the control considered by itself.

CONTROL FEATURES

Although many different types of controls are available, there are certain features that are common to all. Control knobs must have a shape and a size. Controls all offer some resistance to movement. There must be some label on or adjacent to a control. And the designer must face the issue of protecting a control from accidental activation.

Shape Coding

Some of the earliest human factors research during World War II arose due to inadequate shape coding of controls in military aircraft. Two important controls in propeller-driven airplanes are the throttle and the

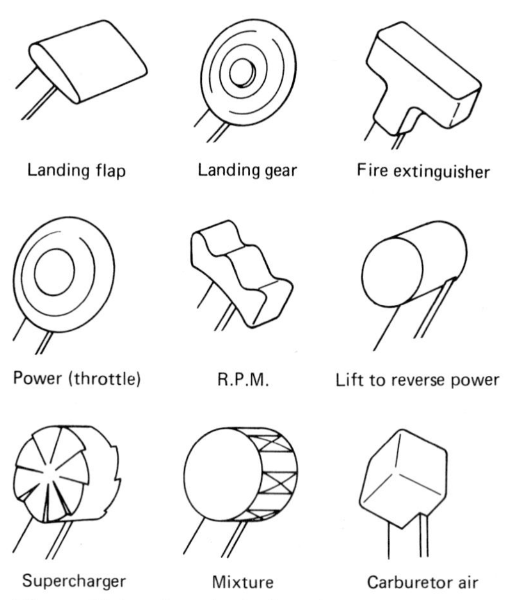

Figure 10-2. Standard aircraft controls using shape coding.

pitch control. The throttle is analogous to the accelerator in your car and controls the amount of power reaching the propellers. The pitch control is used to feather the propeller when an engine cannot function due to damage. Feathering a propeller in a multiengine aircraft minimizes the drag imposed by a defective propeller and allows the aircraft to remain flying on its other engines. In certain aircraft these two controls were located side by side with identical knobs. In the stress of battle, a pilot who wished to feather a damaged engine would occasionally turn off the throttle by mistake. At first, this mistake was classified as pilot error, at least for those pilots who returned to tell the tale. But human factors researchers (Fitts & Jones, 1947) determined the pilots had been "set up" for this error due to the ease of confusing throttle and pitch controls. The problem was solved by changing the shapes of the knobs so that the pilot could distinguish them by touch alone. Now control shapes for military aircraft have been standardized (see Figure 10-2).

Various "alphabets" of standardized shapes that can be distinguished on touch alone are available in handbooks. Researchers obtain these data by starting with a large set of shapes. Blindfolded subjects are then asked to make same-different judgments after touching two shapes. The results of these comparisons can be analyzed using signal detection theory (see Chapter 3) or other psychophysical techniques. Most of the time subjects in these experiments wear gloves. If shapes can be dis-

Figure 10-3. A set of shapes that can be discriminated by touch alone.

criminated when you are wearing gloves, they will cause no problem to the naked hand. However, the reverse is not always true. Figure 10-3 shows one such standardized shape alphabet.

Size Coding

As a coding scheme, varying size is not as effective as varying shape. The human can make absolute judgments—that is, touching only one knob at a time—of at best three practical sizes. The relative sizes of two round knobs needed for size discrimination are shown in Figure 10-4. About a 20% difference in knob size is required. Fortunately, shape and size coding can be combined in the same set of knobs.

Size coding has been used for historic reasons in some industries. For example, a power plant has large switches to control large motors. In the early days of power plant design, a large switch was needed to carry the large electric current. Now, of course, the switch on the control panel does not carry the full current of the motor; it merely operates a relay or other device that in turn controls the motor. Nevertheless, the electric industry has continued to use large switches on control panels (Figure 10-5). These require a great deal of space and make the control room so large that operators have trouble getting to switches. Nevertheless, the large switch does remind the operator that a large expensive motor is

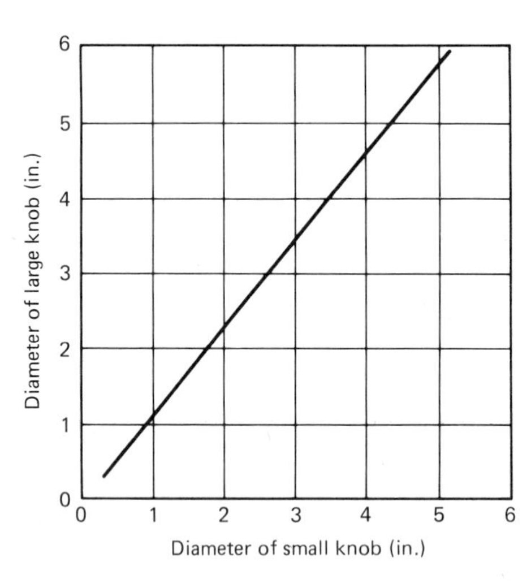

Figure 10-4. Desirable relationship between knob sizes when only touch is used to discriminate between small and large controls. (From Chapanis & Kinkade, 1972.)

Figure 10-5. Control panel for Dresden nuclear power plant. Note size of controls and lack of shape coding. (Copyright B. Kantowitz, TEXT-EYE.)

being controlled so that due caution is necessary. However, this same reminder to the operator could be better established by shape coding small switches. This would greatly decrease the size of the control panels.

The size of a knob is not too important under normal conditions (Jenkins & Connor, 1949). But an interesting study conducted underwater reminds us that handbook data cannot be generalized to all condi-

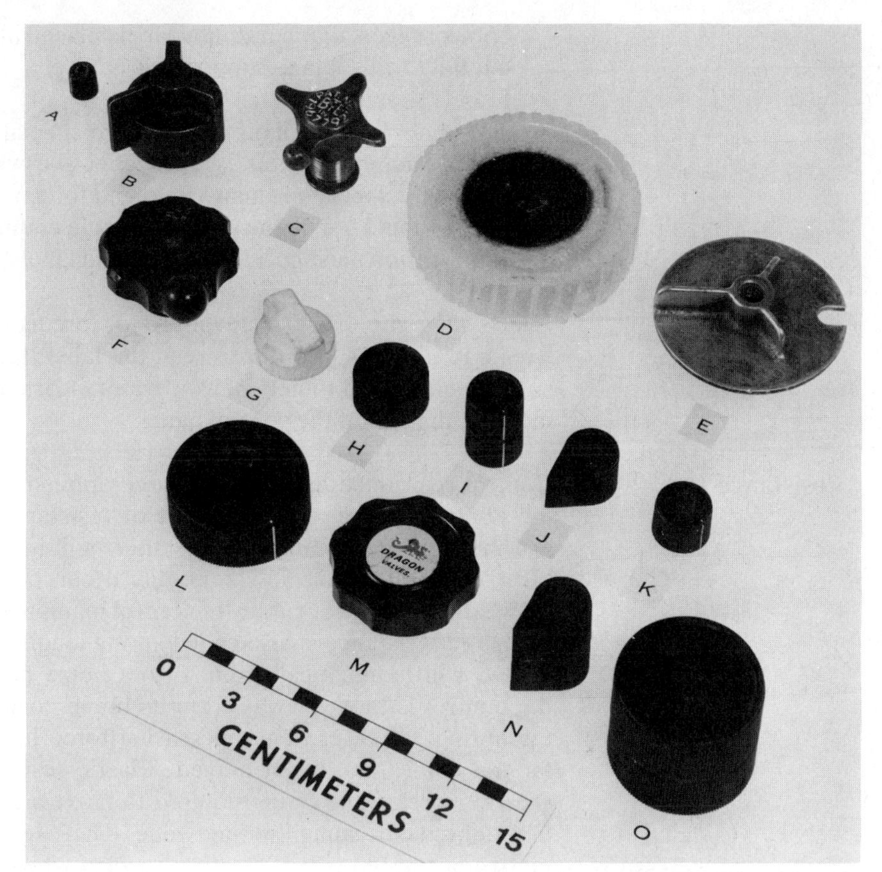

Figure 10-6. Knobs used in the underwater test of shape coding. (From Carter, 1978.)

tions. Fifteen knobs of varying shape and size (Figure 10-6) were tested by six Navy divers wearing gloves (Carter, 1978). The surprising results were that a very large knob with a diameter of 10 cm was easiest to adjust underwater. Shape coding with a fin across the knob (Knob G in Figure 10-6) improved accuracy. The optimal underwater knob is large in diameter, has a height of 2 cm, and a fin placed diametrically across its top.

Labels

Labels are a simple and inexpensive control feature. They have the advantages of low cost, can be applied to all controls, and require little operator training if common symbols and terms are used. Some design guidelines for control labels are as follows (Chapanis & Kinkade, 1972).

- Locate labels systematically and consistently in relation to controls. Do not put some labels above controls and others below. Many hand-

books suggest that horizontal labels placed above controls are optimal, but fail to cite data to support this advice.
- Labels should be brief.
- Avoid abstract symbols that require special training.
- Color coding of labels is most effective when a special color is reserved for special meanings (e.g., red for emergency).
- Color should not be the primary or sole coding method.
- Illumination must be adequate to read labels.

Labels have the major disadvantage of forcing the operator to look away from any displays in order to read the label. Labels are most useful as a supplementary aid to help new operators learn a control panel. They also increase the size of the control panel.

Resistance

Control resistance opposes the force applied by the operator in moving the control. There are four kinds of resistance: elastic, static, viscous, and inertial. The amount of resistance built into a control determines the "feel" of the control, how smoothly it can be moved, and the trade-off between speed and accuracy of control manipulation.

Elastic resistance is also called spring loading. Controls that are spring loaded will automatically return to a home position when released. As the control is moved away from its home position, the force exerted by the spring increases. This increasing force helps the operator feel how far the control has been moved. Elastic resistance can be designed to greatly increase the force required to move a control as the control position approaches some limiting value. The "dead-man" switch on a locomotive is spring loaded to return to zero when released by the operator. This stops the train.

Static friction opposes the movement of an object that is at rest. Once the object starts to move, static friction diminishes dramatically. You may have learned this by trying to push a large, heavy piece of furniture without wheels. Although it takes a lot of force to get it started in motion, you must be careful not to fall over once it starts moving since much less force is required. Static friction makes it more difficult for the operator to precisely adjust a control.

Viscous friction is proportional to the velocity of motion. It tends to damp out irregular motion and resists rapid movements. Hydraulic door closers work by having a piston move through a thick (viscous) fluid. Viscous friction makes it easier for the operator to make smooth control adjustments.

Inertial friction is proportional to the acceleration of motion. It resists rapid changes in velocity. Inertial resistance comes into play primarily with larger controls, such as hand cranks. While it does smooth out motion, it also makes it difficult for the operator to make precise adjustments without some overshooting.

A SLICE OF LIFE

Pilot Error or Design Flaw?

This case history of a twin-engine airplane crash that killed 27 people is based on the report of the human factors team that investigated it after an official conclusion that pilot error was the cause of the accident (Bowen & Mauro, 1980). The plane was approaching runway 14R at Chicago's O'Hare Airport with the captain and the copilot carefully going through landing checks. A third pilot, catching a ride to Chicago, sat in the jump seat behind the crew and also monitored the landing procedure.

Passing over the middle marker—half a mile from touchdown at an altitude of 200 feet—the captain released the autopilot and took control using visual references to land the plane. But on this fatal flight the flight recorder recovered after the crash showed that the plane entered a gentle climb and turned toward the left. Somehow the aircraft controls had been set this way and the climbing turn continued. Air speed fell off. The copilot called this to the captain's attention and after a brief 3-sec delay, the captain applied full power. It was too late. The nose-up position of the aircraft became worse when power was applied. The plane was standing on its tail and stalled. The plane was too low to recover and crashed. The entire episode from the start of the climbing turn to impact took only 35 seconds.

After many years of investigation the official conclusion was that the accident had been caused by pilot error. The widows of the three pilots did not accept this conclusion. They sued the manufacturer of the autopilot. Since they did not believe that the pilots caused the climbing turn, the widows tried to prove that the autopilot was at fault. But the crash destroyed the autopilot. One alternate explanation was that the test button on the autopilot had been pressed. This would cause a climbing turn to the left. Since it was adjacent to the jump seat, where the third pilot was seated, this seemed possible. But how could the third pilot have pressed the test button without being aware of it?

The official investigator had sat in the jump seat and found it very awkward to reach the test button. Thus, this possibility had been rejected early in the official investigation. The human factors consultants built a very accurate mock-up of the cabin. The official investigator was 5 feet 7 inches tall, while the third pilot had been 6 feet tall. The mock-up showed that a person between 5 feet 10 inches and 6 feet 2 inches tall could have easily hit the test button with his shoulder, elbow, or hand. Since the button required only 8 ounces of pressure for activation, the person in the jump seat would be unaware of hitting it. The jury awarded $3.5 millions to the widows. (See Chapter 20 for discussion of the legal aspects of this product liability case.)

This accident was due to the inadvertent activation of a control intended for use only as a maintenance aid. The control had not been protected against activation during flight. A 10-cent ring guard placed around the test button would have saved 27 lives as well as the millions of dollars spent investigating the accident and paying indemnities.

Protection from Accidental Operation

As the preceding box tragically illustrates, the designer must protect controls from inadvertent activation. Chapanis and Kinkade (1972) identified seven methods for protecting controls.

1. *Recessing.* The control is mounted so that it does not protrude above the plane of the control panel.
2. *Location.* Place controls where they are unlikely to be hit. This, of course, is not as secure a method and should not be used where high costs are associated with accidental activation.
3. *Orientation.* Have the direction of control movement on an axis where accidental forces are less likely. This too is not a secure method.
4. *Covering.* Place a guard or cover over the control. This is a very secure method for controls that are not often used. However, operating personnel are likely to disable the guard if they must use the control frequently.
5. *Locking.* A control can be placed in a channel that requires two successive movements in different directions for activation. Thus, for example, if the channel is shaped like a T, the control must be moved vertically and then horizontally to get from the bottom of the channel to the top. However, operators will find this annoying if the control must be used frequently.
6. *Operation Sequencing.* A series of interlocking controls must be activated in the correct order before the desired action takes place. If extreme protection is required, as in launching a nuclear missile, two operators may have to go through the appropriate operation sequencing. A less dramatic example is the Reset key on the Apple II microcomputer. It is placed adjacent to the Return key, which is frequently used. This was a design error since it guaranteed that users would accidentally strike Reset. Later models solved the problem by requiring two keys, Reset and Control, to be pressed simultaneously when the operator desired the reset function.
7. *Resistance.* A stiff control is more immune against accidental activation. If it is accidentally pressed, it is more likely that the operator will be aware of this error.

Several illustrations of correct and incorrect control protection can be found in Meagher (1977). Common errors include guards that only protect part of the control and using toggle switches or knobs with pointers that protrude. Emergency stop switches should be placed on both sides of equipment that has moving parts. During an emergency, the operator may not be able to reach a single stop switch on one side of a machine.

CONTROL DESIGN

There is no universal set of guidelines for all controls. The specific parameters such as operating force and spacing depend very much upon the particular type of control in use. These values can be found in human factors handbooks (see the appendix at the end of Chapter 1). Here we concentrate on two specific types of controls, pushbuttons and foot pedals, that are widely used. Although the design practices for pushbuttons and foot pedals cannot be literally applied to different kinds of controls, our discussion stresses the logic behind specific design solutions. Thus, the discussion represents the kind of logical analysis a human factors designer conducts in order to produce effective controls. Once specific control decisions are made, the designer must locate controls effectively. This general problem is discussed in Chapter 15, Workspace Design; here we only mention a new method for quantifying control accessibility.

Foot Pedals

The pedal designer must deal with two problems: first, how much force must be exerted on the pedal and second, the speed and accuracy with which the pedal can be reached. In order to specify pedal force, the designer must know the seating arrangement and the geometric relationships between the pedal and the operator. Important angles and distances are shown in Figure 10-7. Pedal force can be divided into two categories depending on whether the force is generated by rotating the ankle to move the pedal or by pushing with the entire leg. If large forces are required the entire leg must be used. Then optimum design (Roebuck et al., 1975) requires the pedal to be almost at seat height with the leg being almost straight when the foot is placed on the pedal. There should be a backrest and optimum angles (Figure 10-7): up to 30 for alpha, 150 to 165 for beta (knee angle), and 80 to 90 for gamma (angle between foot and tibia).

If small forces are required, the pedal can be lower. The actual force exerted depends on the angle of the pedal (see Figure 10-8). Alpha should be between 10 and 15, beta between 90 and 150, and gamma between 90 and 120.

The existing empirical research just described allows the designer to specify pedal force and geometry without much difficulty. However, purely empirical attempts at measuring speed and accuracy have been less successful (Roebuck et al., 1975, p. 389) and "do not provide strong guidance." The designer must resort to theoretical knowledge about human psychomotor skill. Speed and accuracy of movement is extremely well predicted by Fitts' law (see Chapter 5) which, as you undoubtedly

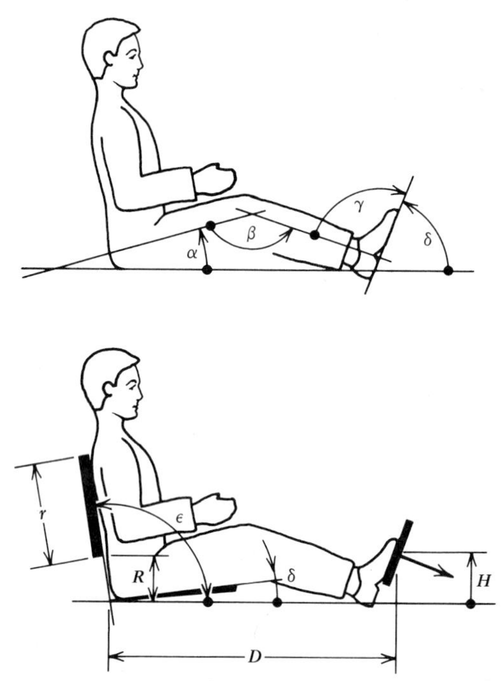

Figure 10-7. Crucial dimensions for pedal design. (From Roebuck et al., 1975.)

remember, states that movement time is a linear function of an index of difficulty. This index of difficulty is the logarithm (to the base 2) of twice the movement amplitude divided by the target width. However, all of the research described in Chapter 5 was based upon hand movements. Fortunately, Fitts' law has also been successfully applied to foot movements (Drury, 1975). Since the pedal will be hit if any part of the sole of the shoe strikes it, the effective target width is nominal target width (W) plus the width of the sole of the shoe. When this sum is used in the index of difficulty, Fitts' law holds with a correlation of .98. So the same theoretical basis for predicting hand movements also works for foot movements. Thus, for two pedals mounted in the same plane, speed and accuracy of movement can be predicted quite accurately with no further empirical efforts. The designer can use the Fitts' law constants provided by Drury (1975).

What happens when two pedals are not in the same plane? Fitts' law has been applied only to coplanar foot movements. Should the accelerator pedal and brake pedal in automotive vehicles be in the same plane or should one pedal be higher than the other? An empirical study (Glass & Suggs, 1977) tested this, as well as a combined brake-accelerator pedal, in order to find the arrangement with fastest foot travel time. The brake pedal vertical position was varied from 6 inches above the accelerator

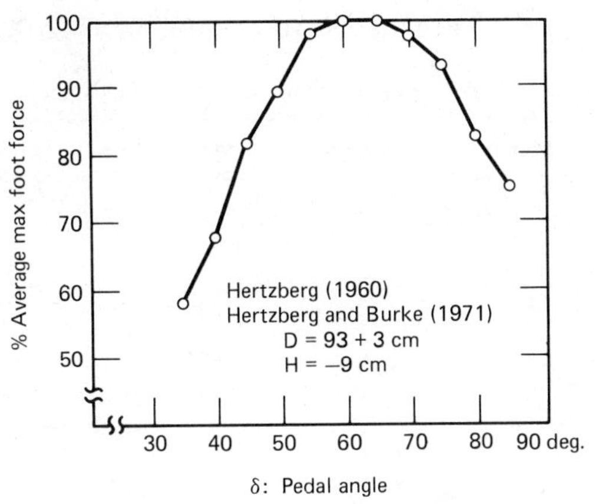

Figure 10-8. Maximum foot force as a function of pedal angle. (From Roebuck et al., 1972.)

pedal to 6 inches below. The lateral distance between the pedals was held constant at 2.5 inches. Best results were obtained with the brake pedal 1 to 2 inches below the accelerator pedal. This saved 28 msec or about 12.5% of standard foot travel time. At a speed of 55 mph this equals a 2.3 foot shorter stopping distance.

Even faster movements were obtained with a combined brake-accelerator pedal. This pedal could move in two directions. (This is expressed technically by stating that the pedal had two degrees of freedom.) Angular rotation about an axis determined the acceleration. Vertical displacement controlled braking. This single pedal had a total reaction time of 270 msec, only 74% of the reference (two pedals) foot travel time. However, the single pedal was given low preference ratings by users, most of whom were drivers already accustomed to a dual-pedal system.

Pushbuttons

The pushbutton is an extremely common discrete control. It is available in three dominant types.

1. Momentary contact – push on, release off.
2. Latching – push on, lock on.
3. Alternate action – push on, push off.

The momentary-contact pushbutton is used where a pulse is required, for example, to make a fine adjustment. A latching pushbutton is used to energize a process or piece of equipment that will stay on for some time. The alternate-action pushbutton is used where it would be tiring or inconvenient for an operator to hold a button down for some time, but where it will soon be necessary to turn off the process or equipment. In this section, we review control design in the specific context of pushbuttons. Our discussion is based in part on guidelines reviewed by Moore (1975).

TABLE 10-2 *Design Parameters for Pushbuttons (from Moore, 1975)*

	Diameter (mm)	Displacement (mm)		Resistance (g)		Control Separation (mm)	
	Min	Min	Max	Min	Max	Min	Preferred
Type of operation							
Fingertip							
One finger—randomly	13	3	6	283	1133	13	50
One finger—sequentially	13	3	6	283	1133	6	13
Different fingers—randomly or sequentially	13	3	6	140	560	6	13
Thumb (or palm)	19	3	38	283	2272	25	150
Applications							
Heavy industrial push-button	19	6	38	283	2272	25	50
Car dashboard switch	13	6	13	283	1133	13	25
Calculating machine keys	13	3 (acceptable)		100	200	3 (acceptable)	
Typewriter	13	0.75	4.75	26	152	6	6

The pushbutton has four operational requirements. Accessibility means that the pushbutton can always be reached by the operator. In some cases, extra STOP buttons must be added to industrial equipment to guarantee accessibility of this function. Ease of use means that displacement, force, and spacing are appropriate for the operator. Freedom from error means that the operator can easily select the correct button and that buttons are protected from accidental activation. Safety means that the button is physically constructed to avoid pinching fingers and insulated to prevent electric shock even if the operator's hands are wet. Position of controls can affect safety. Thus, the START button for a drill press should not be located behind the drill chuck where the operator's clothing can get caught on the drill when reaching for the button.

The pushbutton has seven human factors considerations: physical parameters, coding and labeling, feedback, panel design, panel position, standardization, and stereotypes. Physical parameters include size, shape, separation, displacement, and resistance. Table 10-2 gives recommended parameters for pushbuttons. (Similar tables for other controls can be found in handbooks.) Coding and labeling help the operator identify pushbuttons. Redundant coding with more than one dimension, for example, shape coding plus symbols, facilitates operation. Feedback is information provided to the operator by the pushbutton itself. The most common feedback for buttons is tactile and/or auditory. The operator must be able to determine that the button has been pressed and engaged. Some pushbuttons have lights that give visual feedback but these are far more useful as memory aids at some later time than as immediate indicants of successful pushbutton engagement. Panel design refers to pushbutton layout according to what the buttons do (function), the order they must be operated (sequence), relative importance (priority), amount of use they receive (frequency), and common expectancies (stereotypes). Panel position refers to layout so that the operator can see the pushbuttons and so that they can be reached safely. Standardization means that the same design, coding, and layout should be used within a company and hopefully within an industry. (This latter goal seems unlikely without government intervention. Companies seem to prefer their own unique design and often require equipment manufacturers to customize a standard design even though they have no evidence that their modification is better than the standard design.) Stereotypes means that the control-movement expectancies for the population of workers should be met. In summary, taking these factors into account will ensure that pushbutton design will lead to improved operator performance.

The self-service elevator is a machine where pushbuttons are commonly used. However, the arrangement of these buttons in existing elevators violates many of the principles listed above. A photographic survey of 25 elevator control panels (Smith, 1979) grouped design problems into three categories: standardization, visual noise, and labeling. Each control panel was unique. There were even panels where numbering schemes read from right to left. This lack of standardization imposes a hardship on users, especially in emergency situations where the user must quickly find an emergency control. Little attempt was made to group related control functions together. This resulted in visual noise as for example when the emergency alarm was located next to open-door and close-door buttons while all other emergency controls were on a separate panel on an adjacent wall. Labels were another source of confusion. Many floors were labeled with letters rather than numbers. Which button would you press when faced with these common abbreviations: M, P, SB, L, LL, MEZZ, and G? Emergency pushbuttons were labeled Alarm, Emergency Alarm, Emergency Call, Push to Stop Elevator, to Be Used in Case of Fire, and Full Emergency Stop. No instructions or emergency procedures were posted to inform passengers how to use these various emergency controls. Even the simple pushbutton can be misused by poor design. Suggestions for improved elevator control panels based on empirical user preferences can be found in Smith (1979).

Accessibility Index

All human factors specialists know that important controls must be easily accessible. But until quite recently, there was no quantitative method for comparing alternate control placements. When several control arrangements are possible, how does the designer know which one is best? This question is answered by calculating an accessibility index (Banks & Boone, 1981). The index takes three important factors into account: whether the operator can reach the control conveniently, how often the control is used, and the relative physical position of controls with respect to the operator.

The ability of the operator to reach a control is determined by the reach envelope. (Chapter 15 explains how reach envelopes are derived.) This is a spatial area that can be reached by an operator from a fixed position, usually a seat in front of the control panel. Reach envelopes vary according to the size of the operator. Larger operators have greater reach envelopes. A good accessibility index will penalize controls outside the reach envelope, especially if the controls need be used frequently. Controls that are used frequently should be closer to the operator and the accessibility index should also take this relationship into account.

The formula for the accessibility index proposed by Banks and Boone (1981) is:

$$I = r - \sum^{s} \left[\frac{\sum^{n} \hat{f}}{\sum^{N} f} \right] \div s$$

While it may appear complex at first, actually each term has a simple interpretation:

r is the (Pearson) correlation coefficient between the distance from the operator to the control and the ranked frequency of use of the control. To calculate r, we make two columns of numbers. The first column is the physical distance measured between the control and the operator. The second column is obtained by having operators rank how often each control is used. While the first column will be the same for all operators, in general the second column will not since operators may disagree about which controls are used most often. Nevertheless, such subjective rankings are much used in human factors because obtaining objective data by having an independent observer report how many times each control is used can take a great deal of time, effort, and money. Even though subjective rankings are imperfect, experienced operators are usually consistent in their rankings.

Controls are grouped into two categories: either within or without the reach envelope for each individual operator. f is the rank of each control within the reach envelope and $\hat{f}$ is the rank of each control outside the envelope. The formula shows that these ranks are averaged; n is the number of controls outside the envelope and N is the total number of controls.

Summing over s, the number of operators used to collect the data, yields an average weighted ratio of controls falling outside the reach envelope (the numerator of the fraction in the formula) to the total sum of all weights for all controls (the denominator). Since reach envelope differs according to the size of the operator, it is important to test a wide range of operators who accurately represent the population of operators, from the smallest to the largest.

The accessibility index will range between -2.0 and $+1.0$. The closer the index approaches 1.0, the better the control arrangement. To make this clearer, let us take a hypothetical control arrangement where (1) more frequently used controls are physically closer to the operator, and (2) all controls fall within the reach envelope. Because of (1) the correlation r between frequency of use and distance will be perfect giving a value of $+1.0$. Because of (2) the numerator in the fraction will be 0. Therefore, the index will equal $1.0 - 0 = +1.0$ telling us that control accessibility is optimal. (Additional examples can be found in the Workbook.)

Although the accessibility index was successfully validated (Banks & Boone, 1981), it was never intended as the sole measure of acceptability for control panel design. It was not intended to take into account such important criteria as control spacing, grouping by function, sequencing, etc. Future work will lead to more sophisticated formulas. But the formula above represents an important first step in quantifying what was an ill-defined qualitative concept. As human factors research advances, we can expect more quantitative statements to replace imprecise verbal descriptions of human factors standards and goals.

CONTROL-DISPLAY RELATIONSHIPS

The human sends information to a system by moving a control. The system very often acknowledges receiving this information by altering a display. The new state of the display aids the human in determining if the control movement has achieved the desired outcome. If not, another control movement would be made. Thus, the operator must activate a control and then immediately observe a display; this pair of operations will be repeated until the system reaches the desired state. Therefore, it is essential to design controls and displays so that they function well together. A control or display that works well alone may no longer be optimal when it is part of a control-display system. This section discusses some of the major control-display relationships that must be considered for controls and displays to work well together.

Control-Display Ratio

You have just joined the firm of Stereo Harmony And Frequency Technology (SHAFT), a well-known manufacturer of medium- and high-priced FM tuners. Your first assignment is to put the finishing touches on a new medium-priced tuner. This tuner has a horizontal display with a pointer moving across the spectrum from 88 to 108 MHz. There is room for a 2-inch circular knob to control pointer movement. Your job is to determine how many rotations of the control knob should be required to move the pointer across the display from 88 to 108 MHz.

Since you want to be able to tune in a station as quickly as possible, your first attempt requires only one rotation to sweep over the FM spectrum. The pointer flies rapidly across the display. But before releasing your design you have the good sense to test it on some potential users. To your dismay, none of the test subjects are able to accurately tune in any FM station without taking less than 10 secs in adjustment time. Although they can quickly get to the general area on the tuning display, it takes them quite a while to get locked on to the desired station. Back to the drawing board! (Good thing you tested your design on people other than yourself and your friends.) After some thought, you realize that the control must move many more rotations to allow people to make fine tuning adjustments. So you redesign the tuner so that it requires 24 turns of the knob to move across the spectrum. You have spent hours trying out this new design, but just to be safe, again test using a sample of users. Alas, it still doesn't work properly. While they can tune in any station with ease once they get the pointer to the general vicinity, it takes far too long to travel from one end of the display to the other.

The amount of control movement required to generate display movement is called the control-display ratio. The first design attempt when only a single rotation caused a large display movement is an example of low control-display ratio. A high ratio requires many rotations or large control movements to generate a small display movement. Total time to tune in an FM station is based on two components: travel time and adjustment time. Travel time is time needed for the large motion required to get the pointer in the general vicinity of the desired station. Adjustment time is time needed to fine tune and reach the precise display setting. The relationship between control-display ratio and these two times is shown in Figure 10-9. It is impossible to minimize travel time and adjustment time with the same control-display ratio. Instead, the human factors designer will minimize the sum of these two times by selecting a ratio that is a compromise between optimal travel and adjustment times. In the case of the SHAFT FM tuner, 12 dial rotations would lead to a reasonable control-display ratio. However, you have an even better idea. Instead of one 2-inch control you place two concentric controls on the

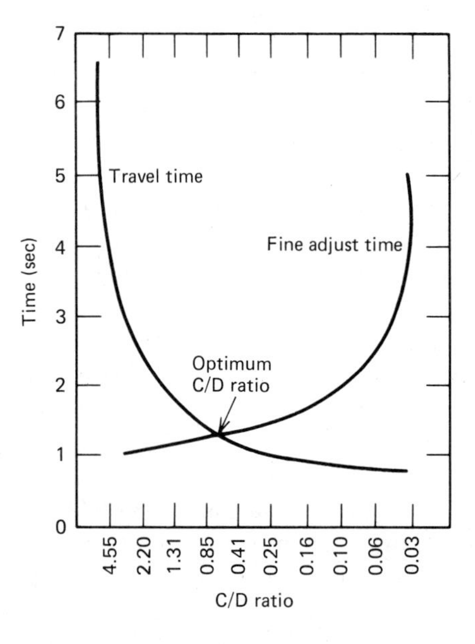

Figure 10-9. Time to adjust a control as a function of control-display ratio. (Adapted from Chapanis & Kinkade, 1972.)

same shaft (or SHAFT). The large knob is used to minimize travel time and has a low control-display ratio. The small inner knob is used for fine tuning and has a high control-display ratio. You determine the correct knob dimensions by looking in a handbook or an article on control design (Bradley, 1967).

More recent research has suggested that the U-shaped function obtained by adding the two curves in Figure 10-9 is not a universal truth for all controls but instead is valid only for knob rotation (Buck, 1980). Since the display width was constant in this older research—that is, the size of the FM horizontal scale was fixed—different control-display ratios were obtained by changing the amount of control movement required. These same ratios could have also been achieved by keeping control movement constant and changing the width of the display scale. Such an experiment was conducted using a joystick control and an oscilloscope display (Buck, 1980). Results were complex. Holding control-display ratio constant did not result in constant movement time. When control-movement amplitude and width was varied, Fitts' law (see Chapter 5) held for three out of four conditions regardless of the oscilloscope target width and amplitude. Movement time was controlled by target width for both control and display, rather than by control-display ratio. These results, although not as straightforward as the human factors designer would prefer, question the importance and utility of control-display ratio as the best predictor of movement time.

Population Stereotypes

Population stereotypes were first discussed in relation to stimulus-response compatibility (see Chapter 5). For example, in the United States the dominant population stereotype for turning on a light is to throw a switch up. Although the concept of compatibility was developed in relation to spatial mappings, it is easily extended to other human factors areas such as verbal labels and pictures (Smith, 1981). Before reading any more text, take the following short quiz. Write your answer in the book. Then compare your answers to those of a sample of 92 male engineers, 80 women who were friends or relatives of these engineers, and 55 human factors specialists (mostly male) tested by Smith (1981). No peeking!

POPULATION STEREOTYPE QUIZ

(from Smith, 1981)

1. Knob turn

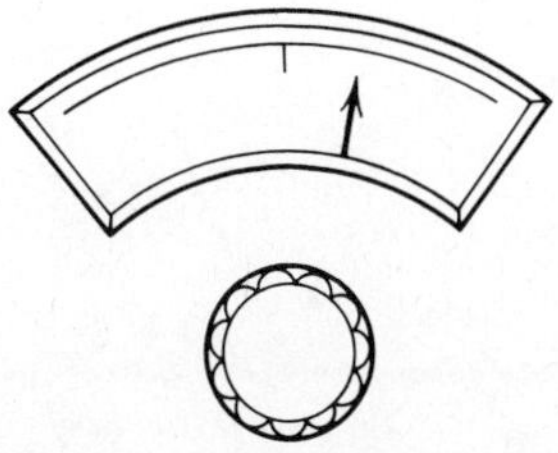

1. To move the arrow-indicator to the center of the display, how would you turn the knob?

   ______ clockwise

   ______ counterclockwise

2. Quadrant labels

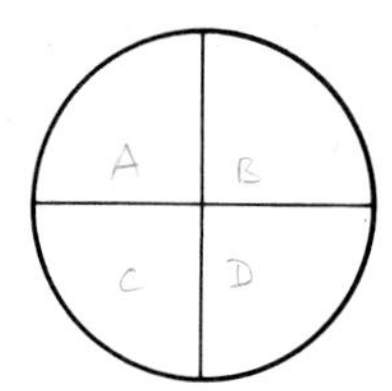

2. In what order would you label the four quadrants of a circle?

Write in the letters A, B, C, and D, assigning one letter to each quadrant.

3. Numbered keys

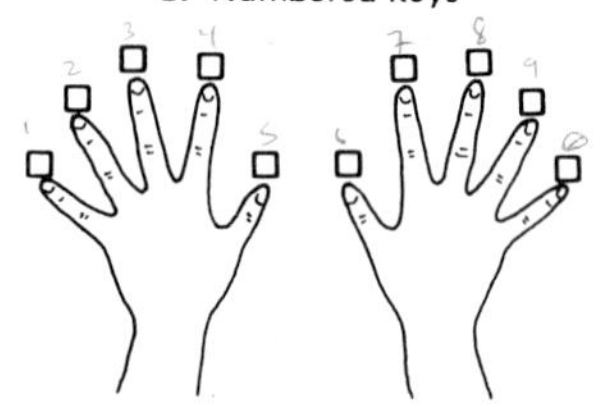

3. A worker is required to duplicate numbers as they appear on a screen, by pressing 10 keys, one for each finger.

Label the diagram to show how you would assign the 10 numerals to the 10 fingers.

4. Cross faucets

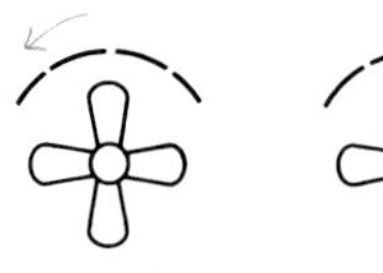

4. Here are two knobs on a bathroom sink, looking down at them.

Put an arrow on each dotted line, to show how you would operate these knobs to turn the water on.

5. *"Pressure High"*

Working with a fire crew, the hoseman yells "Pressure High."

What should be done to the water pressure?

_______ Lower the pressure
_______ Raise the pressure

6. River bank

6. Here is a river, flowing from east to west.

Is the church on the

_______ left bank?
_______ right bank?

7. Highway lanes

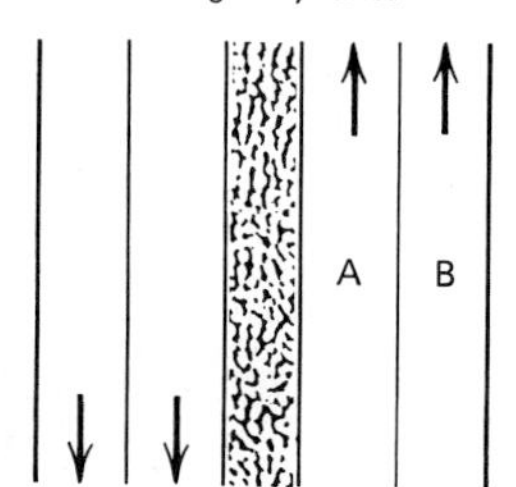

7. On the four-lane divided highway pictured here, which is the outside lane?

_______ A
_______ B

8. Lever control

8. To move the arrow-indicator to the right of the display, how would you move the lever?

_______ push
_______ pull

9. Lever faucets

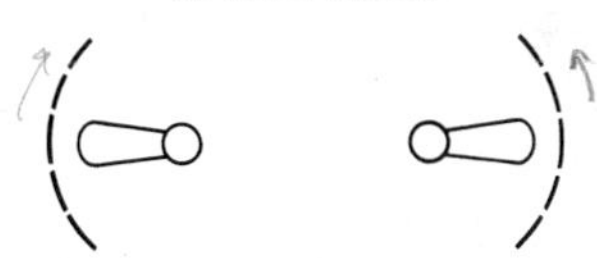

9. Here are two knobs on a bathroom sink, looking down at them. Put an arrow on each dotted line, to show how you would operate these knobs to turn the water on.

10. Digital counter

Knob

10. To increase the number displayed in the window, how would you turn the knob?

_______ clockwise
_______ counterclockwise

Answers

1. Counterclockwise:

Engineers (%)	97%
Women (%)	94%
HF (%)	91%

2.

	Engineers (%)	Women (%)	HFS (%)
Clockwise from upper right	33	26	45
Clockwise from upper left	19	11	5
Counterclockwise from upper right	34	3	5
"Reading" order	14	54	43
Other		6	2

3.

	Engineers (%)	Women (%)	HFS (%)
Ascending left to right	70	70	84
Ascending outwards from thumbs	18	16	5
Other	12	14	11

4.

Left faucet	Right faucet	Engineers (%)	Women (%)	HFS (%)
C	C	17	34	22
C	CC	23	20	13
CC	C	13	26	16
CC	CC	47	20	49

5.

	Engineers (%)	Women (%)	HFS (%)
"Lower the pressure" (Error)	66	48	78
"Raise the pressure" (Command)	34	53	22

6.

	Engineers (%)	Women (%)	HFS (%)
Left bank	18	16	13
Right bank	82	84	80
Other (not answered)			7

7.

	Engineers (%)	Women (%)	HFS (%)
A outside	50	51	20
B outside	50	49	80

8.

	Engineers (%)	Women (%)	HFS (%)
Push	76	59	71
Pull	24	41	25
Other (not answered)			4

Note: Only 70 engineers and 63 women received this question.

9.

Left faucet	Right faucet	Engineers (%)	Women (%)	HFS (%)
Back	Back	25	33	16
Back	Forward	3	3	5
Forward	Back	10	4	7
Forward	Forward	62	60	71

10.

	Engineers (%)	Women (%)	HFS (%)
Clockwise	87	79	95
Counterclockwise	13	21	5

Note. Only 46 engineers and 43 women received this question.

We hope that comparing your answers with those of the three test samples was illuminating. While some of your own personal stereotypes agreed with the majority in the samples, others did not. In cases where population stereotypes offer great consistency, for example, item 1 in the quiz, the human factors designer has no problem. But when the population is not in good agreement about what the stereotype should be, for example, item 5, the designer will be unable to satisfy enough people without considerable ingenuity of design.

A traditional example of the population stereotype problem is called the "four-burner range design" (Acton, 1976; Chapanis & Lindenbaum, 1959; Ray & Ray, 1979; Shinar, 1978). The basic problem is designing the linkage between controls on the front of the range and burners on top of the range (Figure 10-10). There is no dominant population stereotype: the largest group to record a single control-display linkage was 13% for B A C D (Shinar, 1978). Asurvey of 49 models available in appliance stores revealed no standardization and in some cases the same manufacturer had used different display-control linkages on different models (Shinar, 1978). The ingenious solution that solved this problem (Figure 10-11) was created in 1959 (Chapanis & Lindenbaum, 1959) but it is almost impossible to find a manufactured range using this design (Shinar,

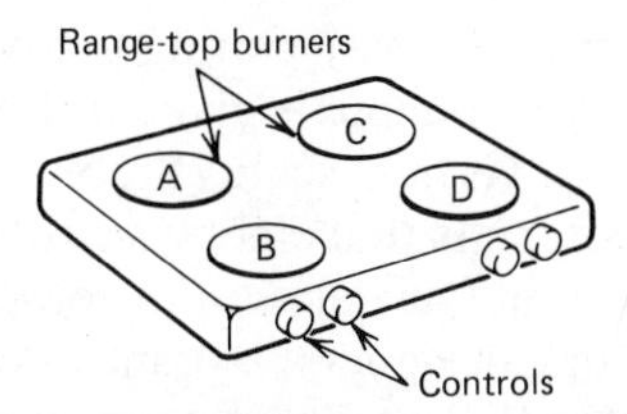

Figure 10-10. Which control do you think should operate each burner?

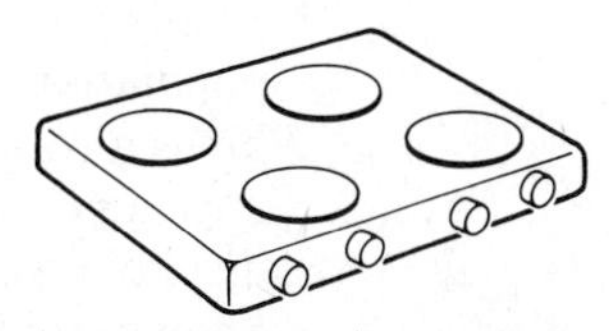

Figure 10-11. A solution to the population stereotype problem. Offset burners line up with unique controls.

1978). Despite the inconvenience and danger associated with mistakenly turning on an incorrect burner—especially with electric ranges that do not offer immediate feedback telling the user which burner is on—manufacturers have ignored one clever design for more than two decades.

A more recent example of S-R incompatibility occurs in the modern helicopter (Hartzell, Dunbar, Beveridge, & Cortilla, 1982). The altitude indicator is located to the right of the pilot's line of sight while the airspeed indicator is on the left. (These locations are based on tradition developed in fixed-wing aircraft—that is, what we have called "historical imperative" in this text.) However, the control for altitude is located at the pilot's left while airspeed is controlled by the stick in the right hand. This design shows a clear disregard for S-R compatibility.

The effects of this design were tested in the laboratory by requiring subjects to control two displays (altitude and airspeed) represented by parallel vertical lines. The task was to move a pointer on the display to a specific target area. (This is the Fitts' law task discussed in Chapter 5.) Controls were either on the wrong side as in real helicopters (contralateral) or on the same side as the display being controlled (ipsilateral). On the average, reaction time for the ipsilateral control-display configuration was 74 msec faster than for the contralateral. Hartzell and his colleagues concluded that this advantage for the S-R compatible arrangement was enough to make a practical difference to a pilot flying a nap-of-the-earth mission calling for very low altitude high-speed flight.

In general, the designer should select the most compatible control-display relationship. For example, if the operator is required to follow (or track, see Chapter 12) a dot moving across a screen, performance will be better if the control is a light pen that only has to be pointed at the dot rather than a lever (Neilson & Neilson, 1980). The most compatible relationship is pointing your finger, and moving a light pen is far more similar to this than moving a lever. Even better would be simply touching your finger to the screen. New technology now makes such a touch-sensitive screen and a preliminary test of this in air traffic control was very successful (Gaertner & Holzhausen, 1980). In fact, manual keyboards can be eliminated entirely by having the computer draw a picture of a keyboard on the screen with the operator pointing to "virtual keys" instead of pressing real keys.

It is possible for a control-display pairing to involve more than a single population stereotype. An example is given in Figure 10-12, which combines a rotary knob with a vertical scale (Brebner & Sandow, 1976). One strong population stereotype is that clockwise rotation leads to increasing display values (see item 1 in Quiz above). Another population stereotype known to human factors designers is called Warrick's principle (see Loveless, 1962). It states that the indicator will move in the same direction as the nearest point on the control knob (Figure 10-12). If the knob is to the right of the display, then Warrick's principle and the clockwise-

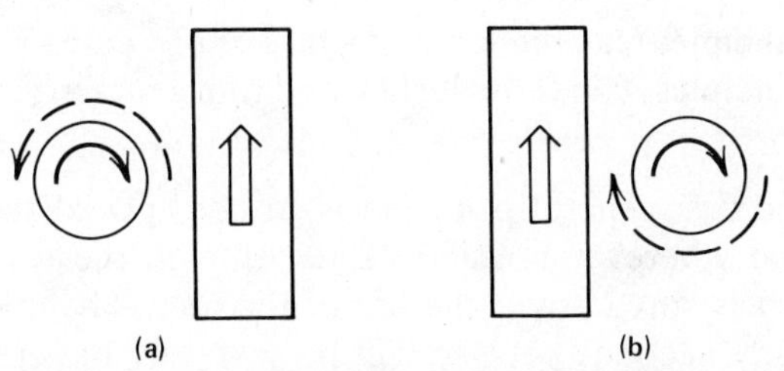

Figure 10-12. The solid arrow shows the clockwise population stereotype for increasing the display indicator. The dotted arrow shows Warrick's principle. They conflict in (*a*) and agree in (*b*).

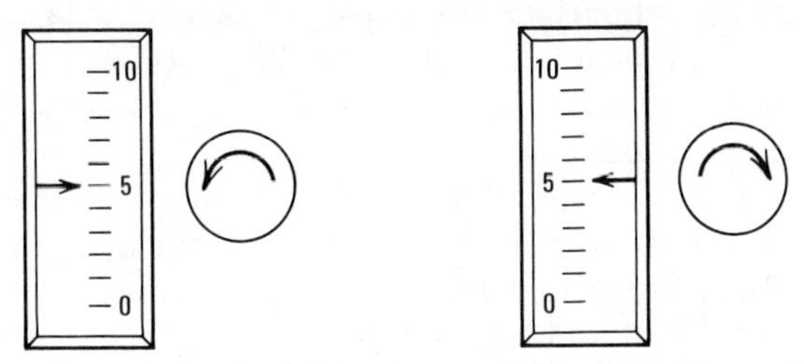

Figure 10-13. Effects of scale side. The arrow shows the direction of control rotation to increase indicator values.

rotation stereotype are in agreement. But if the knob is placed to the left of the scale, Warrick's principle requires counterclockwise rotation and the two population stereotypes conflict. The situation becomes even more complex when a numerical scale is placed alongside the display. A third population stereotype, called scale side (Brebner & Sandow, 1976), must also be taken into account. The scale-side principle states that the indicator will move in the same direction as the part of the control knob located in the same relative position on the knob as the scale is on the display (Figure 10-13). So, if the scale is on the left side of the display, the scale-side principle predicts that operators will base their control movement on the left side of the knob, regardless of which side of the display the knob is on. If this sounds complicated, it is intended to. The point for you to understand is that the designer cannot simply use a single population stereotype. Optimal design requires a configuration where all salient population stereotypes are in agreement. Conflict among stereotypes will degrade control-display relationships.

TOOLS

Name three common tools. When faced with this request, most people name handtools used in industrial operations such as pliers, screwdrivers, hammers, and saws. We tend to take for granted the tools we use everyday such as pens, spatulas and soup ladles. The human factors of handtool design apply to both industrial and consumer tools. Tool design principles can be grouped into three categories (Konz, 1974): general principles, handgrip design, and tool geometry.

General Principles

The first general principle is to use specialized tools that save operator time wherever possible. Even though these tools cost more, this extra cost is spread over the life of the tool. The designer must calculate how many seconds per use will be saved by the specialized tool. Let us take the example of a nurse who uses a tool 10 times a day (Konz, 1974) with a useful tool life of 2 years. The special tool costs $10 more than a general-purpose tool. If the wage cost (salary plus benefits plus taxes) for the nurse is $5 per hour, the labor cost per second is .14 cents. The break-even point for the tool is the capital cost, 20 cents per use, divided by the labor cost, or 20/.14 = 143 seconds. If the special tool will save 1.4 seconds per use, it should be purchased.

A tool that combines two functions is more efficient since the operator does not have to switch tools. This saves the time required to reach, grasp, move and release. Dual-function tools such as a claw hammer that drives and removes nails or a cutting pliers that bends and cuts wire save time. Another kind of dual-function occurs when tools can be used by either hand. Although only 10% of the population is left handed, right-handed people will also benefit by switching hands to reduce fatigue. Fatigue is also reduced by using power tools for jobs where mechanical energy must be supplied by the tool. Motors don't get tired. The human at best (pedaling a bike) is only 23% efficient as a source of mechanical power.

Grips

There are two major types of grip. The power grip is formed when the hand grasps a cylinder by making a fist with the thumb reaching around to lock on the first finger. A handsaw is manipulated with a power grip, Force is applied along the axis of the forearm. A precision grip is used for small tools like pens and knives that must be accurately controlled. The tool is held between the thumb and one or more fingers. A common error is to make the size of the grip related to the size of the tool. Thus, small screwdrivers have tiny handles and big screwdrivers have big handles. However, the size of the hand using the tool often has little to do with the size of the tool, especially since women are now employed in jobs that in the past were limited to men. Hand dimensions don't change when a small tool is put down and a big tool picked up (see Chapter 15 on engineering anthropometry). Separate tools with different size grips are needed for large people (most men and large women) versus small people (most women and small men).

For a power grip, grip diameter should be 1.5 inches (Ayoub & LoPresti, 1971). A larger diameter prevents the thumb from locking over the first finger and a smaller diameter cuts into the hand and fails to provide enough surface area for a nonskid grip. For precision grips, diameters should be at least 0.25 inch (Konz, 1974). Provided a grip gives enough

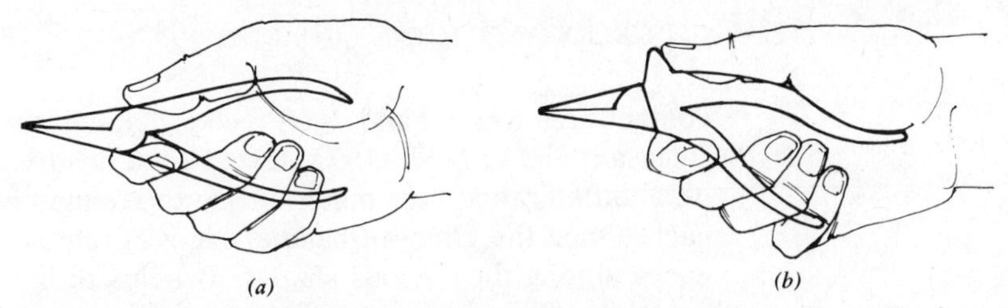

Figure 10-14. Standard (*a*) and improved (*b*) design for a pliers. The improved tool is bent and has a stop for the thumb. (From Roebuck et al., 1972.)

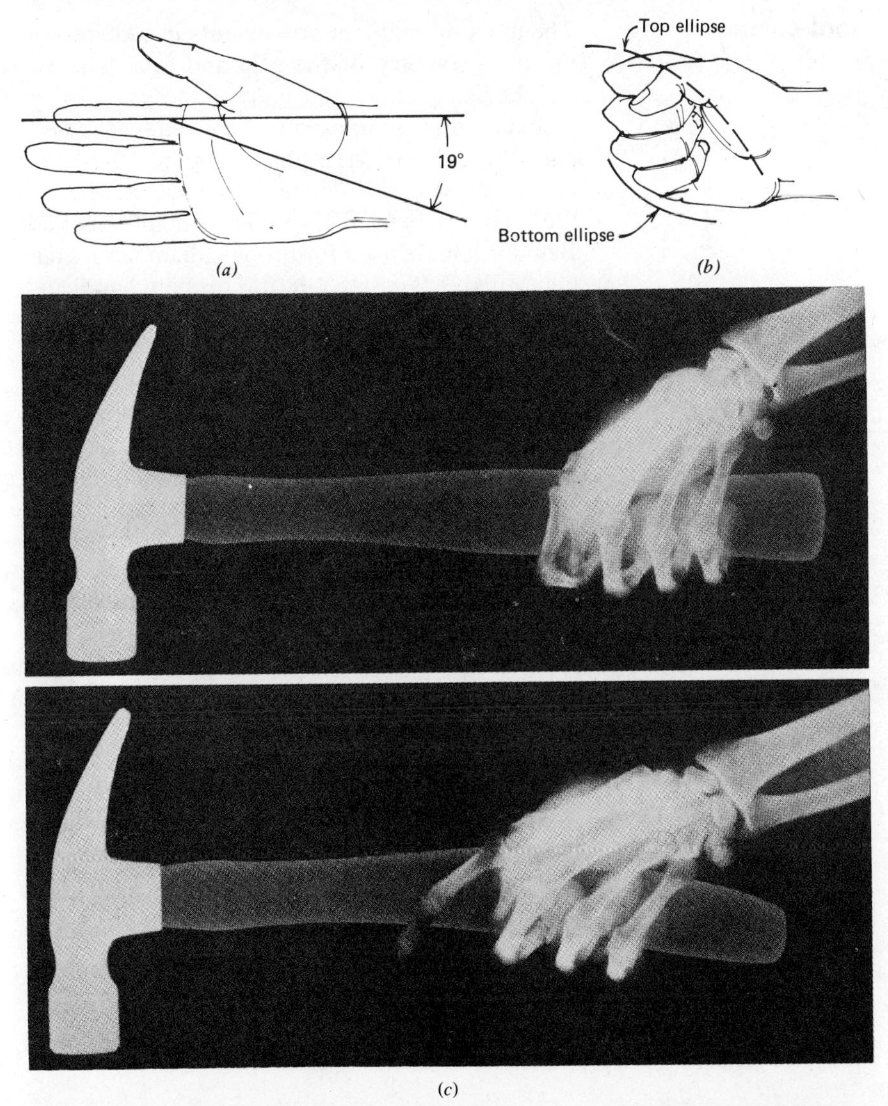

Figure 10-15. (*a*) 19° angle formed by index finger and lifeline under ball of the thumb used as basis for angle of the Bennett handle. (*b*) Top and bottom ellipses used as a basis for shape of the Bennett handle. (Copyright 1980 The Human Factors Society. Used by permission) (*c*) The Bennett design, a 19° double ellipse bent handle was the result. (Photo Courtesy J. Bennett)

contact area to the hand, the precise shape of the handle is unimportant. Pheasant and O'Neill (1975) tested 13 screwdrivers and 4 knurled cylinders to determine how much torque the human hand could achieve with each. Once the effect of handle size was removed, there were no differences among the various shapes. It helps to have the grip surface compressible with a high coefficient of friction. Wood, rubber, and soft plastic work well. Bare metal is a poor choice.

Tool Geometry

The most dramatic improvements in tool design have occurred by changing the geometry of the grip and tool. The basic principle is that tools should bend, instead of making the wrist bend. Bent wrists can lead to tenosynovitis, an impairment caused by the tendons being rubbed as they pass through the wrist bones. Figure 10-14 shows a standard pliers that causes wrist stress versus a redesigned pliers where the tool, instead of the wrist, is bent. This same principle accounts for the success of the Bennett handle used for brooms, hammers, and sports equipment (Emanuel, Mills, & Bennett, 1980). This bent handle uses two elipses to conform to the shape of the closed hand (Figure 10-15).

SUMMARY

Controls allow people to transmit information to machines. Controls can be discrete or continuous, linear or rotary, one- versus multidimensional. Some important control features include shape coding, size coding, resistance, and labels. Controls must be protected from accidental activation.

There is no universal set of guidelines for control design. Examples of design problems and solutions were discussed for foot pedals and pushbuttons. Placement of controls can be evaluated using the accessibility index.

Controls do not exist by themselves. The control-display relationship must be considered by the designer. Population stereotypes must also be taken into account.

Tools ranging from pens to shovels are used by people every day. Specialized tools are worthwhile if they save enough operator time. All tools should have proper grips. The basic principle is that the tool should bend, rather than making the wrist bend.

11 DATA ENTRY

H. E. DUNSMORE
Department of Computer Sciences
Purdue University
West Lafayette, Indiana 47907

INTRODUCTION

In this chapter we consider some of the issues concerning data entry. First, we look at some features common to or that might be used with data entry devices. We begin with a minor controversy—whether modern VDT's (video display terminals) are really superior to the hard-copy (i.e., print on paper) terminals that the VDT's are rapidly making extinct. Then we consider such items as the effect of keyboard and screen height, distance from the operator, viewing angle, illumination, glare, brightness, and flicker. Virtually all input terminals use the common QWERTY keyboard that has been on typewriters since Moses found it in the Red Sea. In this chapter we discuss some experimentation that investigates variations on keyboard design.

We investigate positioning devices that are used to indicate a choice of items or the place on the screen where input or output is to appear next. Such positioning devices include light pens, joysticks, and "mice." We consider the means of making simple selections such as pointing and typing, and look at the use of "menus" for presenting lists of potential selections.

We discuss recent research work that considers the effect on the data entry operator of varying display speeds (i.e., the rate at which information is presented on the screen). We also investigate the effects of speed stress (get the stuff in fast) and load stress (do several things at once).

We next take a look at the typical means of interaction with today's computer systems: input via a keyboard and output via a video screen.

What if the computer user could speak his or her input into a microphone? What if the computer system could speak back the output? Would either (or both) of these alternatives be helpful? Some recent experimentation suggests that they may be. The first section ends by considering graphics devices such as plotters, graphics terminals, and tablets. These tools have led to a revolutionary new means of industrial automation called Computer-Aided Design/Computer-Aided Manufacturing (CAD/CAM).

In the second section we examine text editors that are used for entering and changing documents, programs, and data. We direct our attention to *interactive* editors (i.e., text editors that are used at a terminal in a conversational manner with the computer system), and look at two types of these known as line editors and screen editors. We ask "which is better?" and we look at the languages that are used for conversing with these editors. We report on experiments concerning the commands used in text editors and the "flexibility" of editors (i.e., does the editor allow abbreviations or synonyms for its terms?).

DATA ENTRY DEVICE FEATURES

The typical data entry device for use with most computing systems is the video display terminal (VDT) with an attached keyboard. Some information may be entered via more exotic devices such as light pens, mice, and joysticks (honestly!; more about these shortly).

Video Display Terminals

The original input/output terminals were typewriter-like devices that displayed the conversation between the user and the computer system by printing it on paper. The earliest typewriter devices (see Figure 11-1) fall squarely into this category. But, the standard terminal is rapidly becoming the VDT (see Figure 11-2). They are becoming cheaper to buy (or lease), can display output much faster, require *no* expenditure for paper, and entail less expense for maintenance (since there are fewer moving parts). However, there still is a little controversy as to whether the trade-off is entirely positive. For example, the paper terminal creates a tangible record (which you can go back and look at as needed) of the data entry process. Since VDT's usually can display only 20 or so lines at a time, it is possible that some valuable information disappears from the screen before the user realizes that it should have been reckoned with. Information is lost when it "rolls" off the screen of the simplest VDT's, but more "intelligent terminals" allow for retrieval of previous screens of information. On the other hand, the loss of information at roll-off is still far more common. Do all the other VDT features make up for this inconvenience?

Zacarias, Benham, Dreyer, and Duffy (1981) experimented with an interactive question-answering system that could be used on either a paper or video terminal. The video terminal was about four times faster in displaying information than the paper terminal, but interactions were so

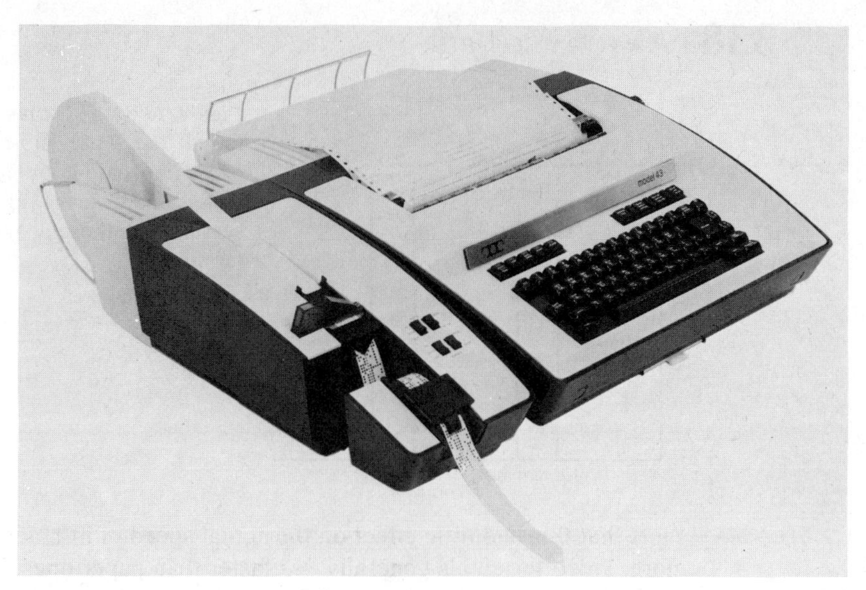

Figure 11-1. This is the Teletype model 43. The earliest interactive data entry devices were Teletypes. (Courtesy Teletype Corporation.)

Figure 11-2. The Video Display Terminal (VDT) has a keyboard and a television-like screen. (Courtesy Hewlett-Packard.)

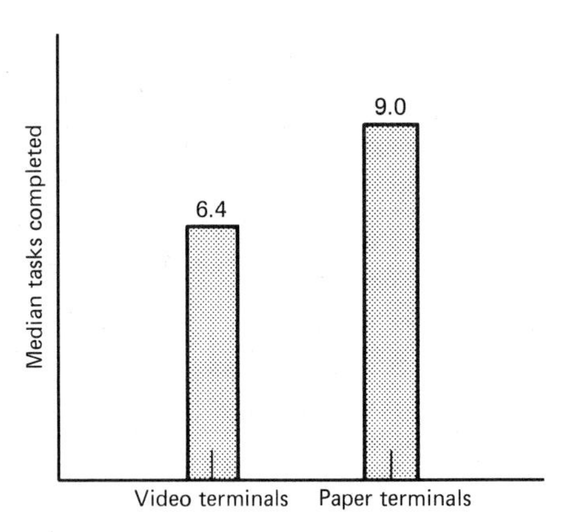

Figure 11-3. Zacarias et al. (1981) found that those subjects who used paper (i.e., hard-copy) terminals completed more tasks in a 30-minute session than those people using video terminals.

short that this had little effect on the actual speed of interaction. Furthermore, video terminals generally *are* faster than paper ones. Each subject was trained briefly in the use of the interactive system, given a warm-up period, and required to perform as many tasks as possible in a 30-minute session. The tasks given to the subjects were sufficiently difficult to ensure that the VDT would roll a lot of information off the screen during the experimental session. Figure 11-3 shows the mean number of tasks completed correctly for each of the two groups of subjects. The subjects who used the paper terminals were more productive than those who used video terminals.

We would not want to conclude from this experiment that VDT's should be scrapped in favor of the old paper dinosaurs. The experiment probably only demonstrates that the ability to retrieve rolled-off information (as a record of previous interactions with the system) would be a worthwhile feature for VDT's as well as paper terminals.

Another concern in using VDT's is the visual fatigue factor. VDT's operate by directing electrons toward a phosphorescent screen. Characters are nothing more than places where the screen is "lit up" by electron bombardment. If the characters are too bright, too dim, too "flickery," too small, too close together, etc., then the productivity (and the sanity) of the person doing data entry can be affected. In fact, in a study of over 100 office workers who regularly worked with VDT's, Dainoff, Happ, and Crane (1981) found that 45% reported symptoms of visual fatigue.

Stammerjohn, Smith, and Cohen (1981) examined a number of factors concerning VDT use including illumination, glare, and contrast. Proper *illumination* is necessary so that the user can read both the screen and any printed materials nearby. Note the trade-off. As illumination in the room gets brighter, stuff on paper can be read more easily but it makes the screen harder to read; as the room gets darker, the characters on the screen look better, but it is hard even to find the paper. *Glare* occurs when reflected light makes it difficult to read characters on the screen. In such a case, the screen can even act like a mirror—displaying a perfect image of the overhead light to the detriment of the information on the screen. At other times, glare just destroys the contrast. *Contrast* is the difference in luminance between characters and their background. Notice that the words in this book are printed in black on white paper. The contrast is great. But, what if the characters were printed in gray and the paper was slightly grayish. It could make reading much more of a chore than it ought to be.

Furthermore, the height and angle of the screen and keyboard are important factors as well. Keyboards that are too far off the work surface lead to the same fatigue that bank employees experience during a robbery when forced to keep their hands up for long periods of time. In addition, the keyboard should be tilted slightly toward the user to allow easier access to the keys at the top of the keyboard.

It is also important for the user to be the right distance from the terminal screen. Obviously, this can make a difference. If you are 20 feet away, it becomes quite difficult to determine what is on the screen (and may require seeing through walls). Also, it is possible to get too close, which not only makes it hard to see the characters but may lead to nose marks on the screen. A distance of about 450 to 500 mm (i.e., about 18 to 20 inches) is recommended by Cakir, Hart, and Stewart (1979). The screen should be approximately at eye level and tilted away from the user only slightly (Cakir et al., 1979). In a survey of VDT users Stammerjohn et al. (1981) found that 95% of them positioned themselves 500 to 700 mm (i.e., 20 to 28 inches) from the screen, and 82% viewed the screen from angles of 10 to 30°.

Keyboard Design

Virtually all keyboard-type data entry devices use the standard QWERTY typewriter layout (see Figure 11-4). This keyboard was designed with typing letters and documents in mind. There may be a better way to arrange the keys for use with a computer system.

To show that keyboard design is a relevant factor in the speed and accuracy of data entry, Conrad (1966) considered only the 10 digits. He created two keyboards like those displayed in Figure 11-5. Obviously, the one on the left, which he called "high compatibility" because of its intuitive appropriateness, must be easier to use than the "low com-

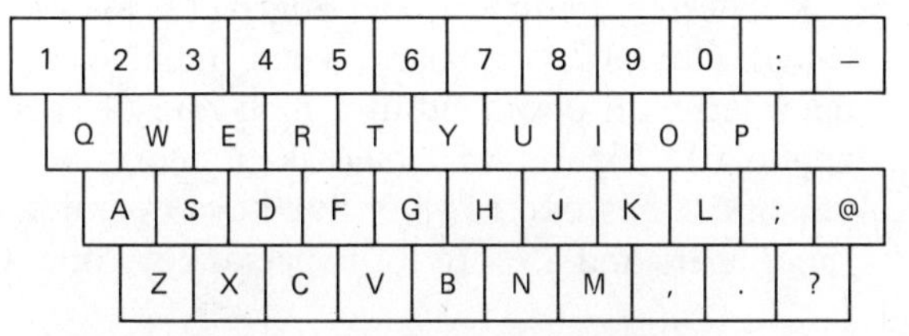

Figure 11-4. The "QWERTY" keyboard.

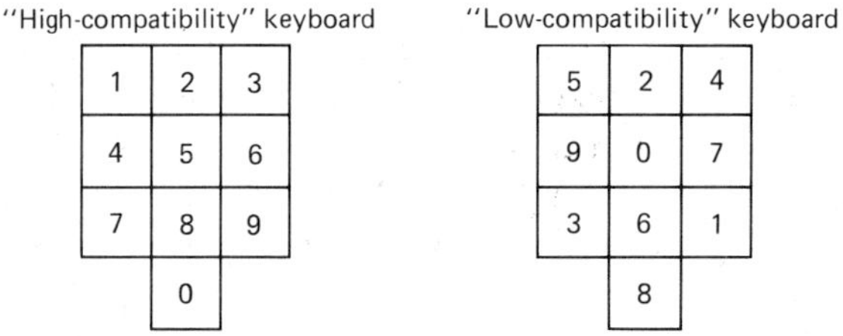

"High-compatibility" keyboard

1	2	3
4	5	6
7	8	9
	0	

"Low-compatibility" keyboard

5	2	4
9	0	7
3	6	1
	8	

Figure 11-5. Conrad (1966) created two keyboards. Notice that the "high-compatibility" version is typical of the key arrangement on most telephones.

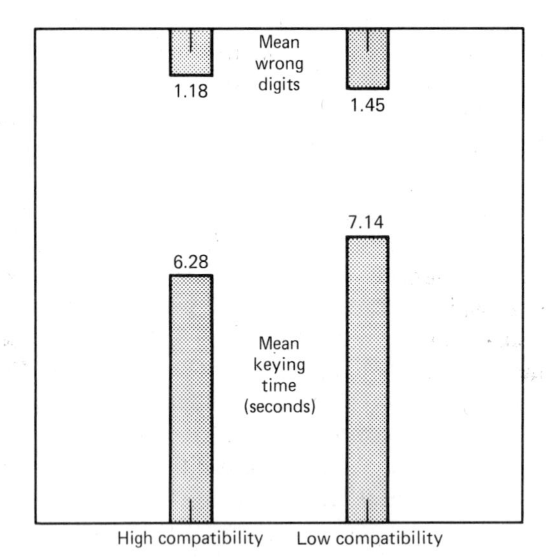

Figure 11-6. Conrad's (1966) subjects were faster and made fewer errors with the "high-compatibility" keyboard.

patibility" one on the right. And such was, indeed, the case. His subjects used both keyboards, viewing an 8-digit number in an aperture for 3 sec and then keying in the number from memory. They repeated this task a number of times for each keyboard. Figure 11-6 shows that the subjects took longer to enter the numbers and keyed in more wrong digits with the "low compatibility" keyboard. These results suggest that keyboard design is related to productivity and that inappropriate configurations may make data entry more difficult than it would otherwise be.

Rochester, Bequaert, and Sharp (1978) designed a miniature keyboard called a "chord" keyboard. Their intent was that this keyboard could be operated with one hand at a high rate of speed—much like a stenotype keyboard. Figure 11-7 shows a chord keyboard for a right-handed person. It has two rows of five keys operated by the index, middle, and ring fingers and a set of four special keys operated by the thumb. Instead

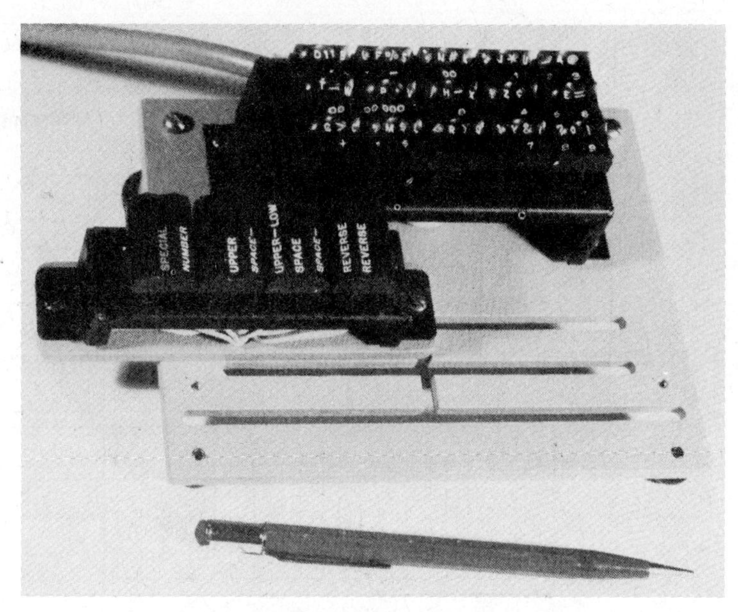

Figure 11-7. This chord keyboard was designed by Rochester et al. (1978). Any number of keys can be pressed simultaneously. The thumb key plate can be moved to the right for left-handed use.

of pressing keys, the user presses "dimples" on the keys (no fooling) that cause one, two, or four keys to depress. Furthermore, the operator may press up to three dimples at once. (This is where the "chord" name came from). Figure 11-8 shows the chord keyboard layout. Certain very popular combinations of characters can be typed with one hand movement: for example "and," "the," "ing," "est," "und," "thi," and "tre."

The designers conducted a simulation involving typical typing rates, typical stenotyping rates, and an analysis of the chord keyboard's similarity to the steno machine. They concluded that trained typists could increase typing speeds by 10% using the chord keyboard. An obvious advantage of this device is its one-handed mode of operation leaving the other hand free to do something else (e.g., use a positioning device like those discussed in the next section).

Positioning Devices

Data entry comes in at least three flavors: entering and manipulating text, programs, and data. In all three cases a critical activity is specifying the "position" at which one wishes to be.

For example, what if you want to change a word four lines back in a document, enter a declaration back at the beginning of a program, or indicate the quantity of an inventory item in a box drawn on the VDT? All of these examples first involve indicating to the system the position

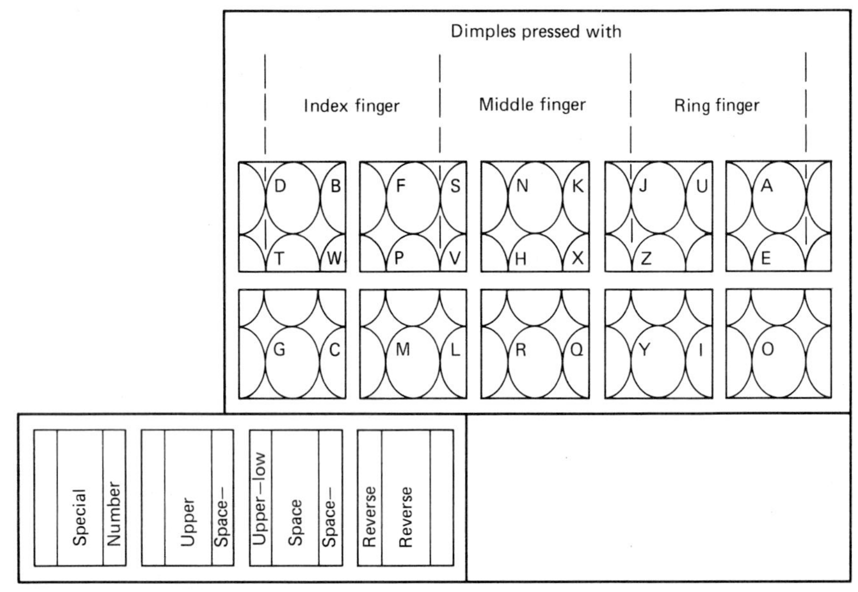

Figure 11-8. This diagram shows the configuration of the chord keyboard. The characters "the" form a chord that can be entered by pressing the dimples that press down both keys in the first, third, and fifth columns.

at which you want to enter data. For many data entry applications, commands are available to specify the desired position. For other applications special keys are available.

But, there are even more exotic ways of indicating position and that is where the "mice and joysticks" come in. Since people generally indicate a position or an object by pointing, input devices that include a means of pointing are a natural means of human communication (Ohlson, 1978). This can make input more pleasing and (possibly) more efficient.

Only slightly exotic (and really fairly common) is the "light pen." The light pen (see Figure 11-9) is simply a ballpoint penlike device that may be pointed at a video screen to indicate a position. The light pen contains a light detection device in its tip. The user moves it where he or she wants it and then either touches it to the screen or pushes a button (or something similar). A detector senses the pen's position and passes that information back to the computer. The detection process is more sophisticated than we are able to discuss here. But, basically it involves timing information. Since the image on a VDT is refreshed several times a second, it is possible to tell what point was being refreshed when the pen sensed light. That is the place to which the pen is pointing.

The light pen provides a direct means of input, while the others discussed below are indirect. That is, for the following devices something is moved (but not up to the glass face of the screen) and this movement

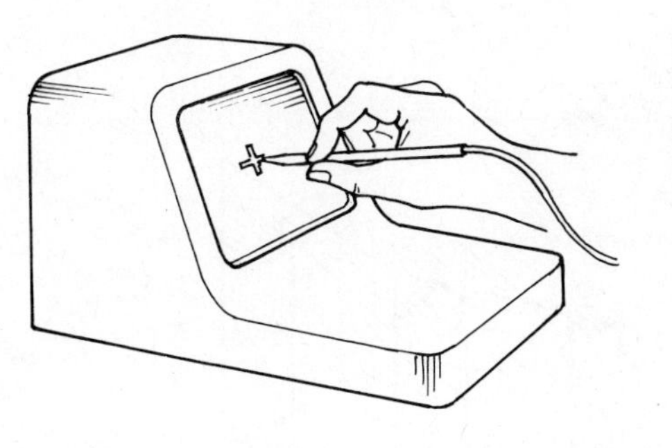

Figure 11-9. The "light pen" is pointed at or touched to a screen to make a choice or to indicate a position.

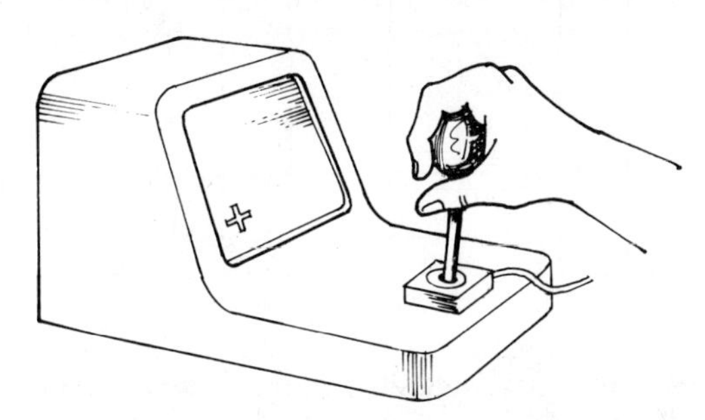

Figure 11-10. The "joystick" is familiar to users of many video games. It is an indirect way to move a point around on the video terminal screen.

causes an indicator on the screen (called a "cursor") to move around. For example, the "joystick" (see Figure 11-10) is simply a lever mounted in a base that detects pressure (some versions even allow a little joystick movement) in any two-dimensional direction. (Some even allow three dimensions.) If you push the joystick forward, the cursor on the screen moves toward the top. If you push it left, the cursor moves left, etc.

Finally, we get to the "mouse" (Newman & Sproull, 1973). The mouse was developed by the Stanford Research Institute and performs like the top of a joystick with two little wheels beneath it. The wheels are mounted at right angles in order to detect motion in both horizontal and vertical directions. As the user rolls the mouse around on a flat surface, the cursor moves correspondingly. When the mouse rolls forward, the cursor moves up. When the mouse rolls left, the cursor moves left. (Whatever happens, however, the cursor never eats cheese.)

In an experiment comparing positioning via light pens, joysticks, and mice (English, Engelbart, & Berman, 1967) the critical measure was the time to move from the keyboard, indicate position with one of the devices, and then return to the keyboard. Figure 11-11 shows that experienced subjects were fastest with the mouse, a little slower with the light pen, and a lot slower with the joystick. On the other hand, inexperienced subjects were fastest with the light pen, a little slower with the mouse, and (like the experienced ones) a lot slower with the joystick. Thus, it appears that the mouse and light pen are both good positioning devices, but that the use of the joystick seems more time consuming. Most users

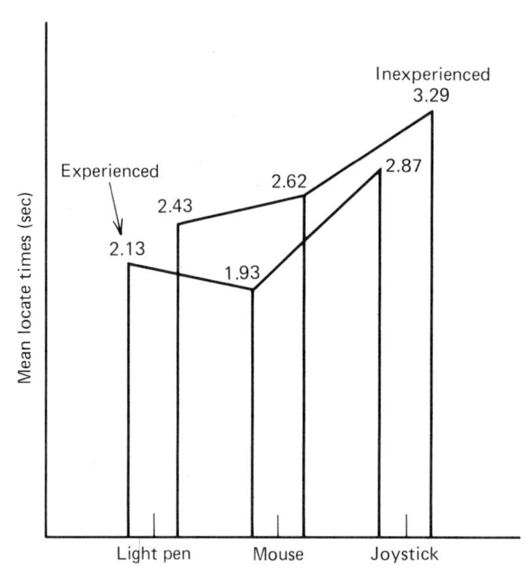

Figure 11-11. English et al. (1967), for both experienced and inexperienced subjects, found little difference in the light pen and mouse. But, all subjects were much slower using the joystick.

of mice and light pens report that they prefer the mouse for long data entry sessions because it avoids the arm fatigue that people usually experience in pointing the light pen for long periods of time.

Simple Selection

One of the simplest means of data entry is just to select something from a list of those presented by the system. Positioning devices can be helpful in this type of activity. Earl and Goff (1965) examined the "point-in" (i.e., something like a light pen) method and the "type-in" (keyboard) method of data entry. Their subjects were seated at a console with a simulated light pen and simulated keyboard data entry device. Subjects were presented with a screen of from 6 to 18 words and three words to search for in that list. All words found in the list were to be marked with the "light pen" and all words not found were to be typed on the "keyboard." Figure 11-12 shows that input was much faster with the "point-in" device. In addition, notice that subjects made fewer errors pointing than typing. "Point-in" errors were such things as failing to point to a desired word in the list or pointing to one that was not desired. "Type-in" errors included typing words more than once, and typing redundant words. Thus, it seems that some sort of device for simple selection may be superior to requiring that data entry always consist of typed-in material.

On the other hand, if data entry is limited to a keyboard it may be possible by some enhancements to use the keyboard efficiently for simple

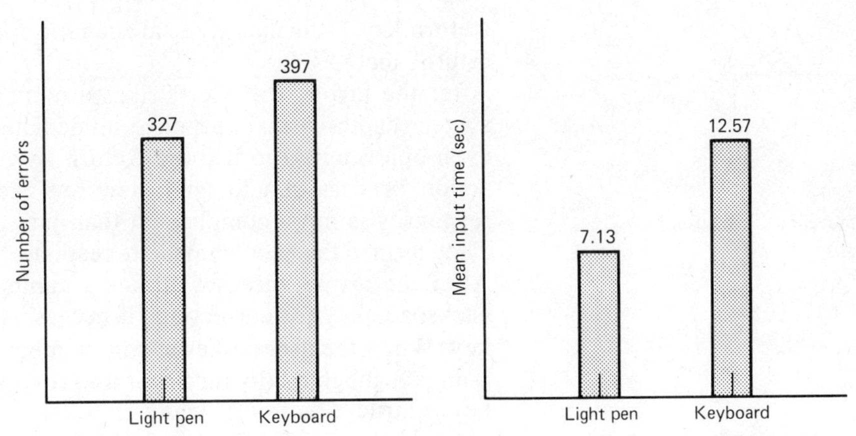

Figure 11-12. Earl and Goff's (1965) subjects used a light pen to input words found in a list and a keyboard to input those not found. The input time results shown are from the two extremes (i.e., all words found and all words not found). The errors are the total from the entire experiment.

selection. Communication with nonprogrammers using interactive systems is typically done via menus. The user is presented several choices from which he or she selects the appropriate one that will either lead to a desired system action or to another menu.

In a menu system, the user is given a list of possible choices. She or he responds by typing in a number, letter, word, or phrase by which the system is able to detect a choice. Some menu systems are more clever than others allowing several synonyms for menu items or allowing an unambiguous substring in place of an entire word or phrase (e.g., STAT instead of STATISTICS).

Backstrom, Cole, and Duffley (1981) conjectured that the simplest of menu response scenarios could be enhanced to increase productivity and satisfaction. Specifically, they investigated three ways of responding to the user's selection. Consider the following example.

What type of investment are you considering?

1. Stocks
2. Mutual Funds
3. Real Estate
4. Money Market

Which? →

Backstrom et al. (1981) constructed a prototype interactive system using menus for user communication in which user responses could be varied as follows.

1. After the arrow, the user strikes a number (in this case 1, 2, 3, or 4), using a backspace key for corrections if necessary, followed by the

Return key. This most typical scenario they termed the "response-return" method.

2. After the arrow, the user strikes a number (1, 2, 3, or 4) and the system captures that response immediately. There is no need (or even opportunity) to use the Return key. In this method the interaction process should go faster, but recovering from erroneous responses is more complicated than just using the backspace key. They termed this the "immediate response" method.
3. After the arrow, the user strikes a number (1, 2, 3, or 4), using a backspace key for corrections if necessary, followed by the Return key. When the user strikes a valid number, the corresponding menu item is highlighted (by putting it into reverse video) until the Return key is struck. Normally white characters appear on the screen on a black background. "Reverse video" simply means that the characters are black on a white background. If the user realizes (aided by the highlighting process) that he or she has not selected the item desired, he or she can use the backspace key, strike another number, and its corresponding menu item will be highlighted. This process can continue indefinitely until the Return key is used. Backstrom et al. (1981) felt that this method should be more satisfying than the "response-return" method and much less error prone than the "immediate response" method. They termed this the "highlight response" method.

Their subjects had little or no computing experience. Each subject was required to perform several equivalent tasks on each of the three systems. They recorded the number of correct transactions completed and the number of errors committed. In addition, at the conclusion they asked each subject to rank the three systems in order of preference. The results appear in Figure 11-13. There was no significant difference in

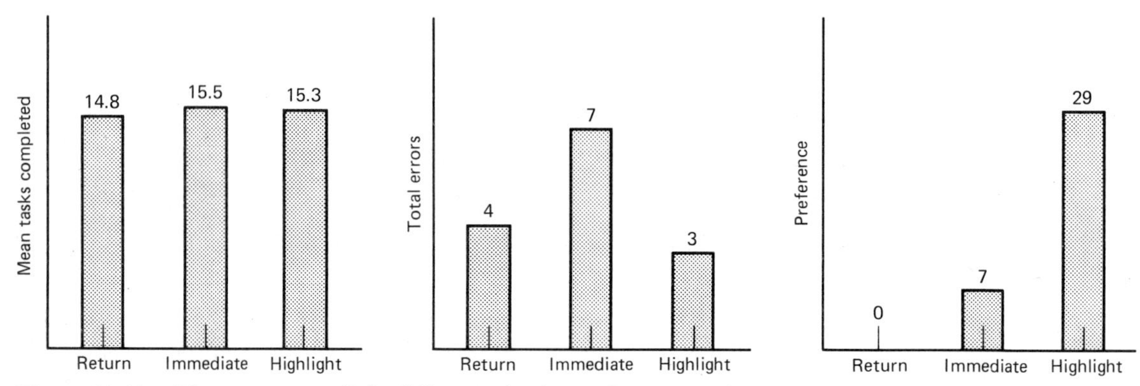

Figure 11-13. There was very little difference in the performance of Backstrom et al.'s (1981) subjects. But, they made the most errors with the "immediate response" system and obviously preferred the "highlight response" system.

productivity, but the mean number of tasks completed under "immediate response" (15.5) was the best and "response-return" (14.8) was the worst. Also, their subjects made the most errors with the "immediate" system and the fewest with the "highlight" system. On the other hand, no one preferred the "response-return" system over the other two. Eighty percent favored the "highlight response" system. Thus, inexperienced users certainly preferred the highlighted system even though there seemed little difference in performance among the three systems.

System Response and Distractions

The rate at which a VDT displays information on the screen and the attendant delays must affect user performance. In particular, if the rate is too slow, then the user wastes a lot of time waiting on the system to respond to display commands or to indicate the effect of other commands. On the other hand, if the display appears too rapidly, then information may disappear from the screen before it has been read by the person using the system. The latter problem is not a serious one really, because in most cases systems can be instructed to present information a "page" (i.e., one screen full) at a time. Given one of the two problems, any user would opt for the latter. Far more maddening is the terminal that plods along slower than the desires of the human user.

Terminal speeds are generally given in terms of "baud rates." A rate of 10 baud is approximately 1 character per second (cps). The early teletypes transmitted information at 110 baud (i.e., about 11 characters per second), which is much slower than humans can read it. In recent years, data transmission speeds have typically been between 300 baud (30 cps) and 1200 baud (120 cps). The latter is about the upper limit (or beyond for some people) of human reading speed. Now, baud rates of 2400 (240 cps) to 9600 (960 cps) are becoming more common. Obviously, at these upper speeds people are not expected to read stuff as it rolls along on the screen; that's impossible. Instead "paging" facilities are required.

Miller (1977) conducted an experiment in which he had some subjects use terminals operating at 1200 baud (120 cps), while other subjects used terminals running at 2400 baud (240 cps). Also, for some subjects the display rate was held constant, while for others it was highly variable with the 1200 or 2400 baud rate representing the *maximum* baud rate. He recorded the time to complete a series of retrieval tasks and the number of keystrokes required by each subject. In addition, a questionnaire was used after the experiment to determine each subject's attitude about the version of the system he had used.

Miller found no difference in performance or attitudes between the two constant baud rates. On the other hand, the variable display factor was very detrimental to both performance and attitude. That is, those subjects for whom the speed of system responses was variable performed more poorly and enjoyed their terminal session much less than those for whom the system response times were more standardized (Figure 11-14).

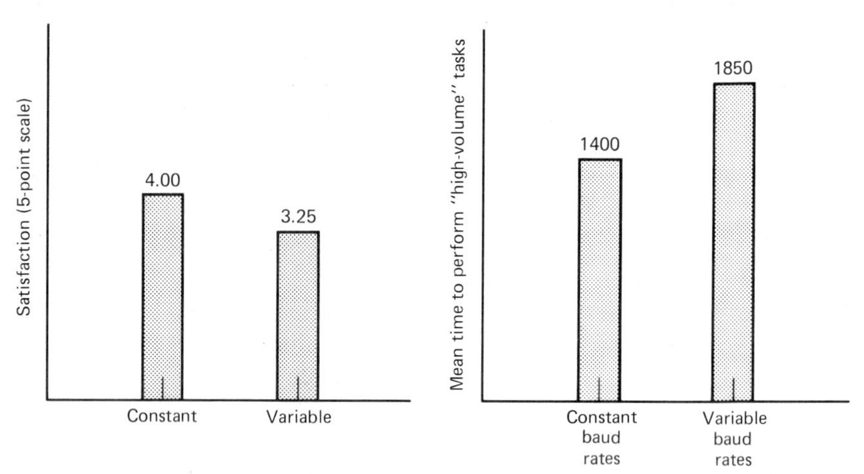

Figure 11-14. Overall, Miller's (1977) subjects were more satisfied with constant speeds. Furthermore, for "high-volume" tasks they were nearly 25% faster with constant speeds.

Thus, for interactive data entry the suggestion from this experiment is that minor speed differences may not be as important as we might expect. However, this does *not* suggest that there would be no disparity between speeds as different as, say, 300 and 2400 baud. Obviously, wide differences in speed must have an effect. On the other hand, it appears that regularity of system response is far more important. When the system performs unpredictably, it may affect the data entry process much more than the mean speed.

Still another factor that may affect the data entry process is the "load stress"—the number of tasks that must be handled simultaneously. Data entry is frequently carried on in a situation where several things are going on at the same time. For example, the clerical employee entering purchase order information may be talking to the person making the order—answering questions about projected delivery date, quality of materials, etc. Or, the person at the nuclear power plant console may be watching several dials, gauges, and screens while simultaneously entering reactor control information that depend on her interpretation of these dials, gauges, and screens.

Goldstein and Dorfman (1978) constructed a visual display of three dials and a subject response panel of three buttons that could be pushed to reverse the direction of the pointer in the respective dials. The task was to keep the pointer in each dial between two target lines. The experimenters varied the speed with which the pointers moved and varied the load stress by requiring subjects to attend to one, two, or all three dials simultaneously. To avoid really confusing the issue, the pointers all moved at a constant speed when subjects were working with two and three dials.

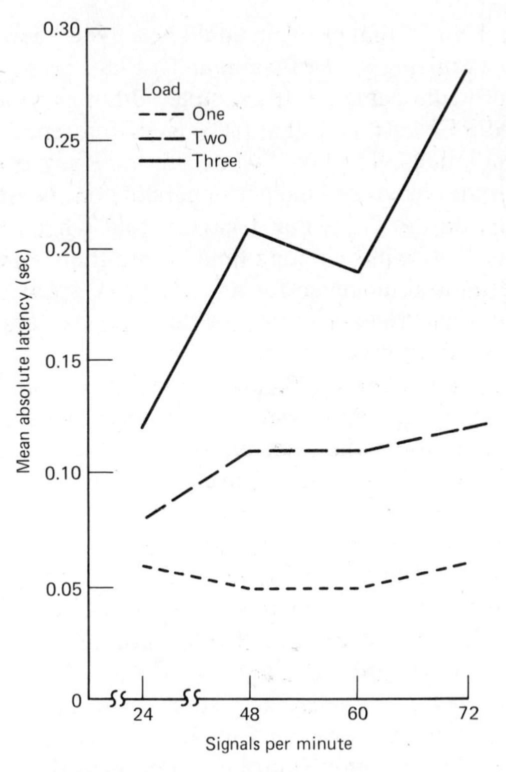

Figure 11-15. Goldstein and Dorfman (1978) used the performance measure "absolute latency" (the difference between where the pointer was stopped and the target line where the subject was trying to stop it). Notice that latency gets worse with two and then three dials, and that pointer movement speed starts to have a marked effect with three dials.

Each subject was required to work with all combinations of speed and load factors. Figure 11-15 shows that load stress definitely affected performance. The subjects performed less well for two or three dials than for one. Increased system speed seemed to affect the subjects negatively only in the three dial case in which performance was only half as good at the fastest speed as at the slowest. The implications of this experiment for data entry are that if the person using the system is subjected to various distractions, other tasks, or even a good deal of system output to digest, then their speed of input to the system may suffer.

What About Spoken Input and Output?

We have assumed throughout this chapter that data entry would be via the "typical" mode of input typed at a keyboard while viewing output from the system as printed material on either a screen or paper. This mode of input/output has become predominant because technology has made it available at a relatively inexpensive cost and because users find it acceptable. But how far can technology take us? Certainly science fiction suggests that spoken input and output are just on the horizon. We would speak into a microphone and receive information from the computer verbally as well.

Let us begin by ignoring the rather insurmountable difficulty that most natural languages are absolutely horrible vehicles for technical interaction. Even if that problem could be solved, there is a tremendous difficulty in recognizing spoken commands. First, no two people pronounce things exactly the same. (For example, although you and I can both type the word REPLACE so that there is no difference, I may say something that sounds like "Re-plays" while you may say something like "Red-plos.") Second, our voice changes depending on the time of day, our health, etc. Thus, on the day when I have a cold what I say may sound like "We-pways." It will be a long time before the technology is reliable to recognize natural language for any arbitrary speaker on any given day. Note again that "recognize" is not the same as "understand." That's a whole other ball of wax.

In spite of these difficulties, there are a number of usable, but limited, speech recognition systems currently available. Many can handle only isolated words (i.e., not phrases or sentences). Some require extensive "training" in which the future user is required to speak every word in the vocabulary (several times!) in a prearranged order. Those speech recognition systems that try to handle phrases or that do not require training are much more error prone, generally allow much smaller vocabularies (50 words or less), and may not work at all for some speakers.

Welch (1980) compared the speed and accuracy of voice input to a typical keyboard and a light pen device. The voice input system was one of the better commercial systems available in the late 1970s. However, subjects were slower and made more errors with it than when entering data with the other (more conventional) devices. There was one bright spot for the voice input device: with it subjects made fewer errors confusing "0" with "O" and "l" with "I"—probably because they could keep their eyes on the text. Subjects using the other devices had to glance back and forth. In a later experiment, Welch added a second task that involved pushing a button while performing data input. Figure 11-16 shows that under these circumstances the speed of voice input was degraded less severely than the speed of either keyboard or light pen input.

Spoken output is not as difficult a problem as spoken input. Anyone who has ever called up the time-of-day telephone number must realize that that poor woman doesn't sit there all day long reciting the time (correct to the second). The message is composed of short recordings of hours, minutes, and seconds. Also, the spoken messages heard in the subway at the Dallas-Fort Worth airport are often claimed to come "from a computer." But, you realize of course that all the computer is doing is selecting the proper prerecorded message to play at each stop. Prerecorded verbal output from computers is completely possible today. Furthermore, synthesized pronunciation of words and phrases is even possible—if you don't mind listening to an accent that sounds like a bored Eastern European. (For more information on speech communication, refer to Chapter 9).

But, is there really some benefit to be gained from verbal input and output? Should there be a crash effort made to upgrade the technology to

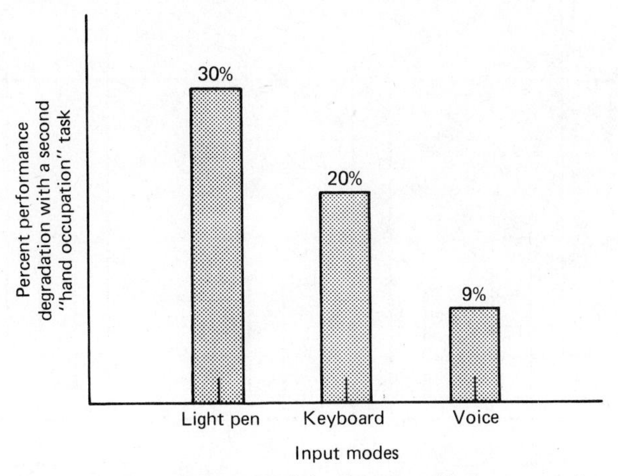

Figure 11-16. When Welch (1980) added a second task to the primary one of data input, he found that the amount by which performance was degraded was much less using voice input than either of the other two conventional modes.

achieve this? The results of some experiments concerning verbal interactions with computer systems provide us with a qualified "yes" answer.

Hammerton (1975) asked subjects to type particular letters and digits on a terminal keyboard (instructions) when these particular characters appeared among a group of presented characters (data). Thus, an instruction might be "Type A, R, Q, 3, 1." The associated data might be "D8 R7 S3 Q1" from which the subject would type "R3Q1." Instructions and data could be presented via a VDT-type screen or via a headphone. Subjects were randomly assigned to one of five treatments.

1. Both data and instructions via the screen.
2. Both data and instructions via the headphone.
3. Data via the screen and instructions via the headphone.
4. Data via the headphone and instructions via the screen.
5. Both data and instructions via both methods.

Figure 11-17 shows the results in terms of number of data sets processed correctly and number of errors. Notice that condition (3) is far superior to the others, and that condition (1) is better than all the rest—among which there appears to be no difference. Thus, we may conclude that displaying the data visually was far superior to its presentation via the headphone. However, with visual data, an audio communication of the instructions led to the best results. The suggestion is that the *separation* of data and instructions is helpful, and that the audio instructions-visual data means is the best. So, we may conclude that there are some circumstances in which audio output would be worth including in the repertoire of a computer system.

Mountford and North (1980) investigated the potential benefit of voice input versus typed input. Their subjects were simulated pilots for

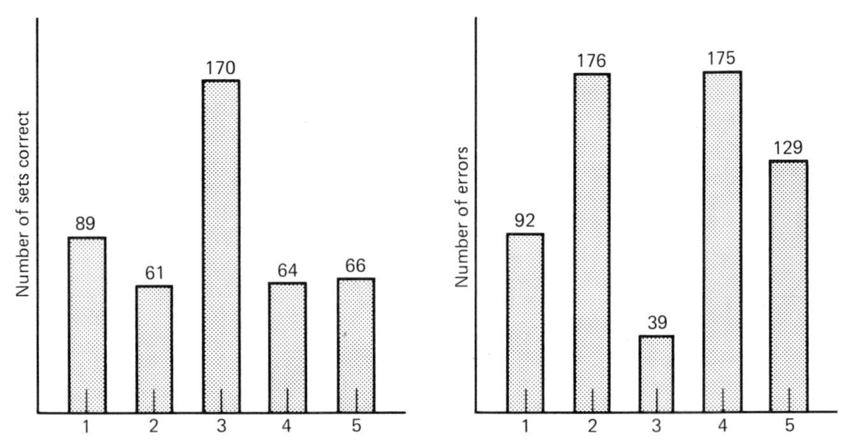

Figure 11-17. Hammerton (1975) found that his subjects in group (iii) performed better (more tasks performed correctly and fewer errors). Treatment (iii) consisted of data via screen and instructions via headphone.

whom one task was to keep their airplane on the proper course by using a joystick with their right hand (all subjects were right handed). As another task, subjects were sometimes asked to select radio channels during the simulated trip. They could choose the radio channel either by keying in the appropriate characters at a keyboard or by speaking it into a microphone. The voice recognition system was a real system in use at the Honeywell Speech Recognition Applications Laboratory in Minneapolis. But it only had to be able to recognize a group of prearranged radio channel identifications. Furthermore, channel information was provided to the "pilots" either on their screen (visually) or through their headset (verbally).

Figure 11-18 indicates that (when performing the tracking task) tracking errors were the least for the tracking task alone and increased only slightly when radio channels were selected by voice data entry. On the other hand, tracking errors increased by more than 400% when radio channel selection was done via the keyboard. Also, Figure 11-18 shows that channel selection scores were higher for voice input than for keyboard input. These results offer evidence that there are some circumstances in which voice input may be superior to keyboard input, particularly when the data entry task is complicated by some other activity (cf. the tracking task).

Toney (1981) examined both typed and spoken input to the computer and visual and spoken output from the computer. Subjects were asked to perform some fault diagnosis tasks with the aid of a computer system. Each was randomly assigned to either (1) type input to the system or (2) speak input into a microphone; and each received information from the system in either (3) visual form (i.e., characters displayed on a VDT) or (4) spoken output via headphones.

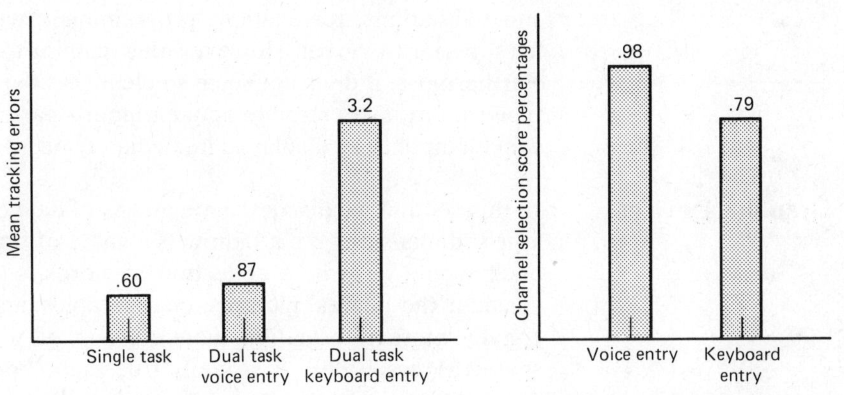

Figure 11-18. Mountford and North's (1980) tracking performance was degraded very little when radio channels were selected by voice entry (right chart). Also (left chart), the subjects made fewer channel selection errors with voice data entry.

Although system components (1) and (3) were easy to implement using available equipment, (2) and (4) had to be simulated by the experimenter. For spoken input, the experimenter listened to what the subject spoke into the microphone and transcribed that into typed input for the system. For spoken output an appropriate message, computer generated from previously digitized human speech (see Chapter 9), was played into the subject's headphones. However, postexperimental interviews indicated that *each* subject was convinced that the system he or she was using was real.

Figure 11-19 shows that the time to solve the fault diagnosis tasks was less for those subjects speaking into a microphone than for those typing input, that the time was less for those receiving verbal rather than visual output, and that those subjects with spoken input and output took less time by far to complete their assigned tasks. The results shown in Figure 11-19 are a composite over all tasks. Toney also found that these differences became even stronger as the level of difficulty of the tasks increased.

We should probably not conclude that spoken input and output are the most desirable modes for every situation. It seems clear that there are

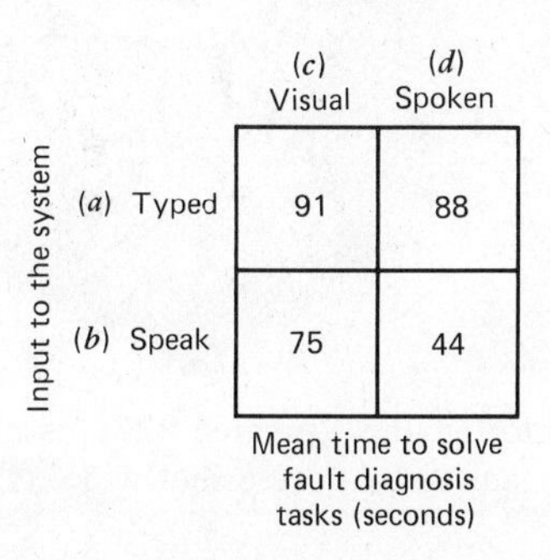

Figure 11-19. Toney (1981) found that spoken input to the computer and spoken output from the computer led to the best performance.

some situations, for example programming, when visual output may be far superior to verbal. However, this remains to be supported empirically. Furthermore, it does not seem so clear that there are any situations when typed input is superior to spoken input—particularly if it is possible for spoken input to be displayed immediately on a VDT.

Graphics Devices

In this section we discuss some means of data entry (and output) that add a new dimension. We all know the value of graphics. Without them this book would only be a collection of words. Flip through the pages and consider the figures, pictures, charts, tables, and diagrams. Graphics can convey more information than a group of words occupying the same space (Myers, 1980). Is it really true that a "picture is worth a thousand words"? Probably so, maybe even more than a thousand in some circumstances. From time immemorial (i.e., about 1945) computer data entry and output had always been restricted to characters. Beginning in the 1970s there has been a tremendous advance in computer graphics devices including video terminals with picture-drawing capabilities and devices that allow us to input hand-drawn figures.

The graphics revolution began with paper plotting devices in the 1960s. These plotters (see Figure 11-20) typically use a ballpoint or felt-tip pen that is moved along the surface of the paper in any direction. Plot sizes can range from about 8 x 10 inches to about 6 x 6 feet. Figure 11-21 shows an example of what a plotter can do. Plots can be generated in colors by using more than one pen during the plotting process.

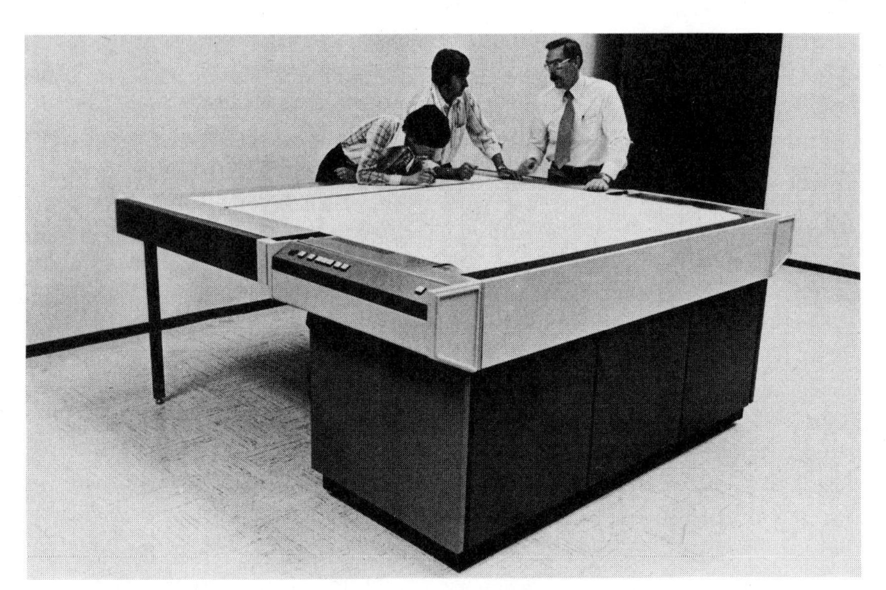

Figure 11-20. This plotter, the Versatec 8272, is one of the largest ever manufactured. It can draw figures that are 6 feet wide. (Courtesy of Versatec.)

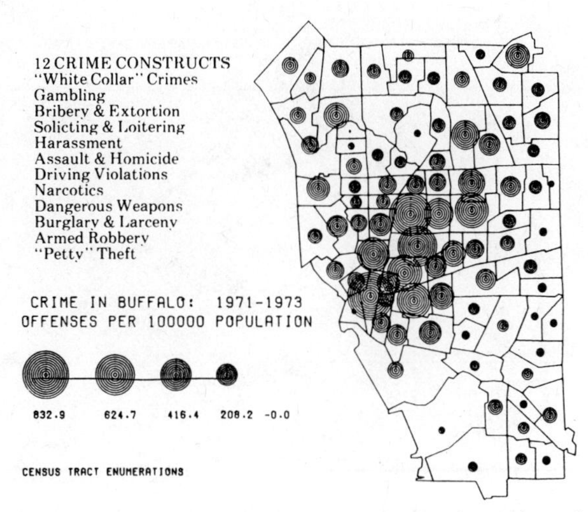

Figure 11-21. This "crime map" showing crime densities in Buffalo between 1971 and 1973 was produced at the Geographic Information Systems Laboratory at the State University of New York at Buffalo. (Courtesy K. Brassel and J. Utano, SUNY Buffalo.)

Graphics video terminals became popular in the 1970s (see Figure 11-22). Currently such graphics terminals allow full-color capability, animation, the appearance of depth, and high-resolution images (i.e., there are a lot more points in the display matrix on the screen than in the typical color television set) (Myers, 1979). With graphics terminals we are able to modify images and to look at hidden surfaces of displayed designs.

Graphics input usually employs a device called a *tablet* (see Figure 11-23). As a stylus moves along the surface of the tablet, its tip receives signals from the grid of lines beneath it (Ohlson, 1978). These signals can be decoded to determine exactly where the stylus is. The location of this point is transmitted to the computing system. As the user sketches a figure on the tablet, the series of points generated can be used to produce this figure on a graphics terminal (and can even be stored for later recall and modification). Modern tablets do not even require a special stylus. These "touch-sensitive" devices generate high-frequency vibration waves. The point of their disruption by a pencil, pen, or even a finger can be detected.

Computer-Aided Design/Computer-Aided Manufacturing (CAD/CAM) is a new kind of industrial revolution being made possible by graphics devices. In the "design" phase engineers sketch directly on graphics tablets and view their results on graphics terminals. Sophisticated computer software gives them the ability to see these images from various vantages (e.g., rear, side, and top views) and to analyze transformations of their designs using computer simulations ("What if we make

Figure 11-22. The video terminal shown here is the Tektronix 4027A color graphics terminal. From a total of 64 available colors, up to 8 colors may appear on the screen simultaneously. (Courtesy of Tektronix.)

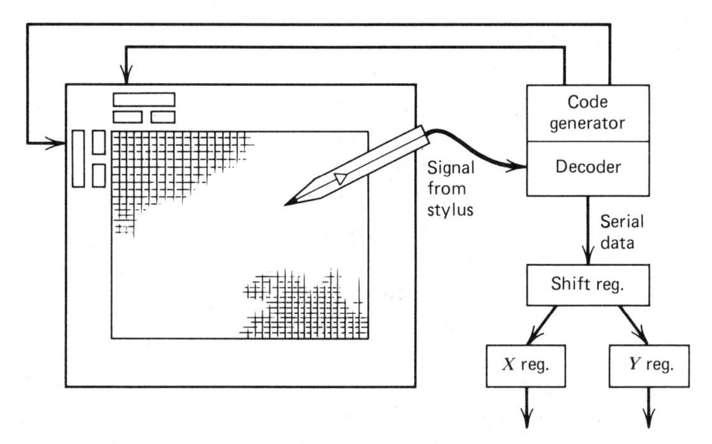

Figure 11-23. The operation of the tablet is described in this diagram. Notice that the signal from the stylus is decoded into x and y components.

Figure 11-24. This engineer is working on the design of an airplane using a CAD/CAM system that includes a graphics terminal and a light pen. (James Sugar/Black Star.)

it three feet longer and two inches wider?"). Figure 11-24 shows an engineer using a CAD/CAM system.

TEXT EDITORS

The first part of this chapter should have convinced you (if you were not already) that there are many ways of using computers other than programming. In fact, in recent years we are beginning to see a number of computer users who are "word processors." They do a surprising amount of productive work without writing any programs at all. They enter text, add to it, revise it, and have it printed later in formatted form. They produce letters, technical reports, business reports, and books. What they do is often quite sophisticated: moving text around and manipulating it in various clever ways. But they do not have to program in order to do their business. These people use text editors and text formatters.

```
.ce
The Cold, Dark Night
.sp+1
.ce
by Elwood Scuggins
.pg
The wind howled out of the west as the sun slipped slowly down into the brown
valley.
Mrs. Robbins scowled as she peered about for the last of her chickens to herd into
the henyard.
She had read
.ul
The Grapes of Wrath
and knew that her lot was not much better.
.pg
There was something she had to do before the man from the bank came back to
padlock the barn.
Back inside the house the embers were still glowing.
She carefully took the picture down from above the fireplace.
It was her only remembrance of Sam.
And even Sam could not help her now.
```

The Cold, Dark Night

by Elwood Scuggins

The wind howled out of the west as the sun slipped slowly down into the brown valley. Mrs. Robbins scowled as she peered about for the last of her chickens to herd into the henyard. She had read <u>The Grapes of Wrath</u> and knew that her lot was not much better.

There was something she had to do before the man from the bank came back to padlock the barn. Back inside the house the embers were still glowing. She carefully took the picture down from above the fireplace. It was her only remembrance of Sam. And even Sam could not help her now.

Figure 11-25. **The lines at the top are from the text file complete with formatting commands. When this text file is run through a text formatter the lines with initial periods are not printed. They are commands to the formatter: ".ce" means to center the next line, ".sp+1" means to skip one line, ".pg" begins a new paragraph, and ".ul" requests that the following line be underlined. Notice that the lines within paragraphs are run together in the printed text even though each sentence begins on a separate line of the file.**

The text editor is a tool that is used by both programmers and non-programmers. The programmer enters programs via a text editor (instead of punching into cards as was prevalent in the old days). But, the most appreciative text editor user is probably the word processor. This person, who may not even be a programmer, can enter vast quantities of text for later printing without much regard for getting spacing, spelling, or even order of text correct initially.

Since the text as entered is saved within the computing system (typically in a disk file) it can be manipulated, altered, and revised as often as necessary. In addition, along with the text can be entered a number of text formatting commands (such as directives to skip lines, to center lines, to line up right margins, etc.). When the user is happy with the proposed document, then the system can be called on (using some sort of text formatter) to produce a printed version. Figure 11-25 shows a sample of some text and interspersed formatting commands and the result of submitting this to a text formatter. This is not only an efficient way of producing final versions, but also has probably already saved quite a few trees due to fewer intermediate versions being printed.

Line Editors

The heart of a good word processing system is a good text editor. The first text editors were those that could be used on typewriter-like terminals. They operated in what is called a "line" mode in which all lines of text as well as directives to the editor to change words, delete lines, or move lines were entered as separate lines. If you have never used a line editor before, look at Figure 11-26. Our line editor doesn't actually exist, but is an example of the way line editors usually work. Notice that there are really two modes in which lines are entered. Some lines (i.e., all those after the LINES command and before the COMMANDS command) are taken by the text editor and stored away. They are the lines of text that are being entered. They can be altered, moved, or deleted. Later they will be printed when the user wants to see the document.

```
LINES
The wind howled out of the west as the sun slipped slowly down into the brown
valley.
Mrs. Robbins scowled as she looked about for the last of her chickens to heard into
the henyard.
COMMANDS
FIND "looked"
→ Mrs. Robbins scowled as she looked about for the last of her
REPLACE "looked" with "peered"
→ Mrs. Robbins scowled as she peered about for the last of her
FIND "heard"
→ chickens to heard into the henyard.
REPLACE "heard" with "herd"
→ chickens to herd into the henyard.
LINES
She had read
The Grapes of Wat
and knew that her lot was not much better.
COMMANDS
FIND "at"
→ The Grapes of Wat
REPLACE "at" with "rath"
→ The Grapes of Wrath
```

Figure 11-26. This sample editing session with our line editor shows the use of the LINES, COMMANDS, FIND, and REPLACE commands. There are many more including commands to delete or move lines around in the file.

On the other hand, some lines (i.e., all those beginning with COMMANDS and including LINES) are not stored anywhere but request the text editor to do something immediately. For example, LINES indicates that all following lines are to be treated as lines of text until the COMMANDS command is given. FIND asks the editor to look through the lines of text that follow until it finds the next occurrence of some characters. The line where the character string is found becomes the "current" line. (It's like putting your finger down on a line in this book and saying that all changes I specify will be made to this line until I move my finger). REPLACE asks the editor to replace the first occurrence of some characters on the current line with some other characters.

There are some things about text editors that users have to get used to. Notice that our FIND always moves forward from the current line looking for some characters. Our editor has another command LOOKBACK, which does the same thing as FIND except it moves backward from the current line. A lot of editors have one command that does both by having the user specify an optional parameter. For example, "FIND +" (or just "FIND") might be the same as our FIND and "FIND –" might be our LOOKBACK. Furthermore, notice that the characters that are found (or replaced with REPLACE command) do not have to be words in the English language sense. We can surely find the word "hopfuly" (ugh!) in order to change it to "hopefully"; but the same thing could be accomplished by the commands:

```
LOOKBACK "hop"
REPLACE "hop" WITH "hope"
REPLACE "ly" WITH "lly"
```

MOVE is used to pick up a group of lines and move them somewhere else (much like you might take a pair of scissors, cut this paragraph out and paste it in the book somewhere else). COPY is very similar to MOVE except it leaves the original lines and just puts a copy of them somewhere else. DELETE is a command to be careful with; it can be used to delete (i.e., remove, scratch, destroy, gone, can't get it back no matter how hard you try) the current line. If DELETE is followed by a number this means to delete that many lines beginning with the current one.

You must (since you are clever enough to be reading this clever book) have thought of a potential problem. You must have said to yourself, "Self, what if I am entering lines of text and I want a line that is simply the word COMMANDS. If I just type in

```
COMMANDS
```

the text editor thinks I want it to stop accepting text lines and start accepting commands. What can I do?" Well, there are several ways of doing it. Try to think of one. We'll give you the answer in a few paragraphs.

But, the real question is: Why should you have to figure out a clever way of doing such a thing? Why doesn't the text editor let you do this easier? And those are good questions. Some text editor designers solve this problem by using a special key sequence or characters that they don't think you'll ever want to use for a real line of text.

Notice also that our text editor has nice mnemonic commands and that command lines even look a little like the way you would describe things to your Aunt Martha. Some text editors aren't as nice as ours. For example,

```
REPLACE "hopfuly" WITH "hopefully"
```

might be done with some editors via something like

```
s.,./hopfuly/hopefully/p
```

Generally, editors that use terms and syntax that are more like English are easier for people to use—especially for nonprogrammers to use. Many editors that have nice mnemonic terms allow experienced users (they might still be nonprogrammers, but experienced nonprogrammers) to enter abbreviations of terms to save time. Thus, our system allows you to say

```
REP "hopfuly" WITH "hopefully"
```

or even

```
REP "hopfuly" "hopefully"
```

Here's "an" answer to the problem posed above. (There is no "the" answer, since there are several ways of doing it.) Enter the lines

```
XOMMANDS
COMMANDS
REPLACE "X" WITH "C"
LINES
```

and then go merrily on your way. Do you see why it works?

Screen Editors

Line editors do have some disadvantages. First, a lot of command lines have to be entered just to move around in the lines of text. Second, to replace text you have to type the bad stuff as well as the good stuff. (Remember, in the last section we found that if we had misspelled "hopefully" we would have to type in the word "hopfuly" in order to say what we wanted to replace. In fact, in our editor we had to type in "hopfuly" twice . . . doubly annoying.)

Recently, developers of text editors have introduced "screen" editors on video display terminals in which a whole "window" (about 20 lines) of text can be seen at once. In addition to the lines of text the screen also contains an indicator (usually a white box, underline character, arrowhead, or some such) called a "cursor" that indicates the current position where text will be entered when it is typed. Changes are made by moving the cursor up, down, left, or right via some special keys (these are sometimes called "function keys") or other positioning devices, and then by simply typing over existing text.

Screen editors make finding and replacing text very easy, but there are some problems associated with them. First, they usually take a lot more system (central processing unit) time than line editors since any of the entire screen contents are eligible for change at any moment (not just the "current" line). Second, instead of typing in commands, frequently commands are entered by special key sequences that sometimes have little mnemonic value.

Until recently (because of the prevalence of hard-copy terminals), line editors have been the typical text editors. However, now with VDT's becoming *the* means of input/output, the use of screen editors is becoming much more common. (Note that you can run a line editor on either a hard-copy terminal or a VDT, but screen editors require a video terminal.) What are the disadvantages of line editors that screen editors rectify? First, recall that the line editor user has to type a lot of commands to make things happen. Furthermore, in order to correct a misspelled word you may have to key in the erroneous spelling as well as the correct. With a screen editor, a lot of things can be accomplished with function keys instead of full-fledged commands (i.e., lines can be deleted, blank lines can be added, lines can be moved, etc.). Also, corrections can be made by simply typing "over" the erroneous stuff. But, screen editors have their faults, too. First, they usually place a greater demand on the system's resources (i.e., they take a lot more CPU time) than a similar line editor. In the experiment discussed below a typical user's session took about 8 CPU sec for a line editor and about 29 CPU sec for a screen editor. Also, those function keys that do things without the use of commands may have little mnemonic value (i.e., it may be hard to remember which special key deletes a line if none is marked "delete a line").

To test the ease of learning and use of a line editor and a screen editor, Adkins, Boss, Driscoll, & Michtom (1980) recruited novices who had never used either type of editor. This provided a presumably unbiased pool of subjects. The subjects were randomly split into two groups—assigned to each of the two editors. Each subject was taught how to make some basic changes (add, change, and delete characters and lines) using either the line or screen editor. The amount of time necessary for completing a tutorial was used as an index of ease of learning. The number of corrections made to a document in a limited period of time was used to assess performance.

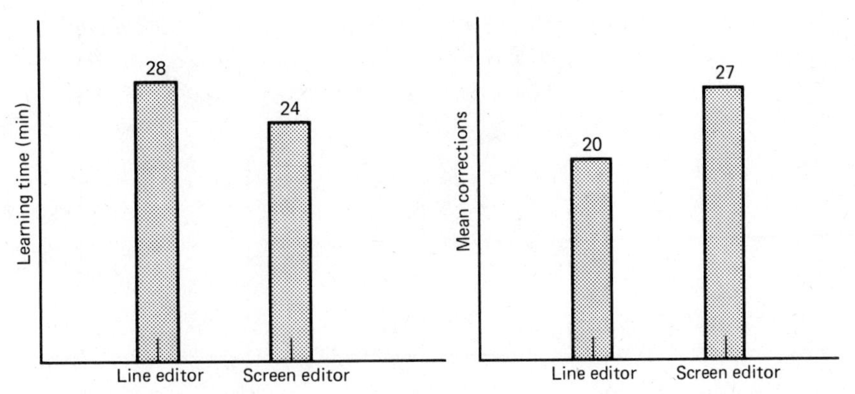

Figure 11-27. Adkins et al. (1980) found that their novice subjects took less time to learn how to use a screen editor than a line editor and that the screen editor users could complete more tasks in a 15-minute editing session.

Figure 11-27 shows that the subjects using the line editor took more time to learn how to use it than those using the screen editor and that the mean number of corrections made with the screen editor was about 35% more than those made with the line editor. Thus, for these novice users the screen editor was superior to the line editor in terms of both ease of learning and in performance.

In a similar experiment (O'Keefe, Brandes, Collins, Crow, Moore, & Notarnicola, 1981) the keyboard for the screen editor did not employ special function keys. Subjects who had never worked with an editor were randomly assigned to either the line or screen editor, trained in its use, and directed to perform some corrections to the text of a computer program. Once again the typical screen editor performance (26.2 changes correctly made) was superior to the line editor performance (18.6 changes).

Editor Command Languages

Earlier in this chapter there appeared the opinion that editors that use English-like terms and syntax are easier for people to use. This is complicated by the fact that the best terms (and command language) for experienced users may not be the best for inexperienced users. Furthermore, when does an "inexperienced" user become an "experienced" one? After 20 hours, one year, or never? This is one of those questions that it really is impossible to answer in general terms. (You know the old saying "All generalizations are false—even this one.") The crossover from "inexperienced" to "experienced" depends on the nature of a person's use as well as the nature of the computer system.

Ledgard, Whiteside, Seymour, & Singer (1980) sought to show that English-like terms and syntax are indeed easier for people to use. In order to do this they employed two text editors—one a standard editor available from a major computer manufacturer that used typical com-

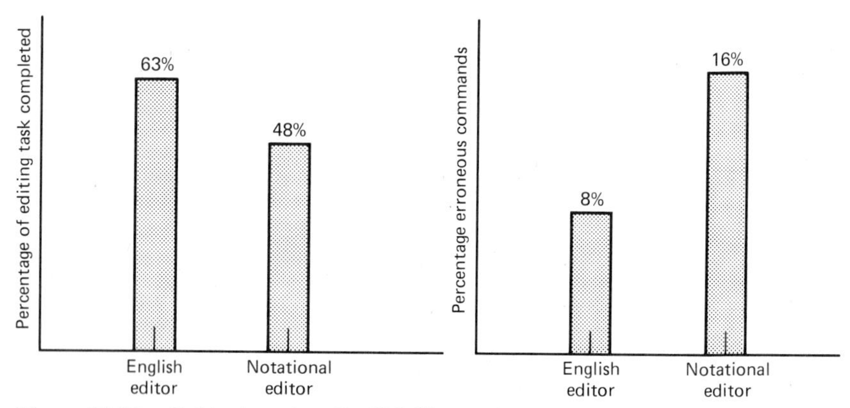

Figure 11-28. Subjects using English-like terms and syntax completed more of their editing tasks and committed fewer errors than subjects using a typical, notational editor.

mands that they called "notational" in nature. Remember the example 's.,./hopfuly/hopefully/p'. This is the type of cryptic, programming-language type of command that they called "notational." They then constructed a new version of the editor with equivalent capability but more "conversational" in nature—like our 'REPLACE "hopfuly" WITH "hopefully" '.

Ledgard et al. (1980) asked subjects to do some text editing using both versions of the editor. Their results are summarized in Figure 11-28. Their subjects completed more tasks and made fewer errors with the "English" editor. Since the two editors differed *only* in their command language, this experiment shows that this factor should be considered in designing editors. Also, the subjects indicated a strong preference (92%) for the "English" editor at the conclusion of the experiment. Furthermore, in a small test after the experiment, Ledgard et al. asked three users with a lot (at least 50 hours) of experience with the "notational" editor to use the "English" one. These users were just as productive with the latter as the former even though it was new to them.

However, as this is being written (it is 1982—20 years from now you readers will probably get a big chuckle out of this) it is not possible to communicate with computers via a true natural language like English. Thus, in designing data entry interfaces of the "English" variety, the suggestion is that the designer choose terms from the language (like REPLACE) that have some intuitive meaning. It is not suggested that every English term must be honored; if you try to use SUBSTITUTE instead of REPLACE, the system should probably say something like "Huh? I don't know what SUBSTITUTE means."

Another feature of editor command languages is how *flexible* they are. For example, some editors might allow us to use REP as an abbreviation for REPLACE if we like or even to call REPLACE something else (like

SUBSTITUTE). We might be able to make up our own commands and give them names. However, most editors are not nearly so flexible.

Scapin (1981) sought to compare the learning and recall of editor commands that differed in context and redundancy. "Context" was either operational (defined without reference to a computer) or functional (defined in terms of what the computer would do as a result of the command). "Redundancy" referred to whether the term itself was used in its definition. Although redundancy is not very desirable for dictionaries (cf., "Orange is the color of oranges"), it can be very useful for defining computing terms ("REPLACE is a command that replaces a word with another").

Two groups of subjects participated in the experiment: some experienced computer users and some with little or no experience. Subjects were asked to learn the definitions of commands for a hypothetical editor. A week later they were given cue definitions and asked to supply the appropriate command. The cue list consisted of five operational and five functional definitions.

Figure 11-29 shows that experienced subjects were better able to recall command words when they had been originally shown redundant definitions. On the other hand, inexperienced subjects were just the opposite. They recalled commands better if they had been shown nonredundant definitions. Scapin explains this by suggesting that experienced subjects

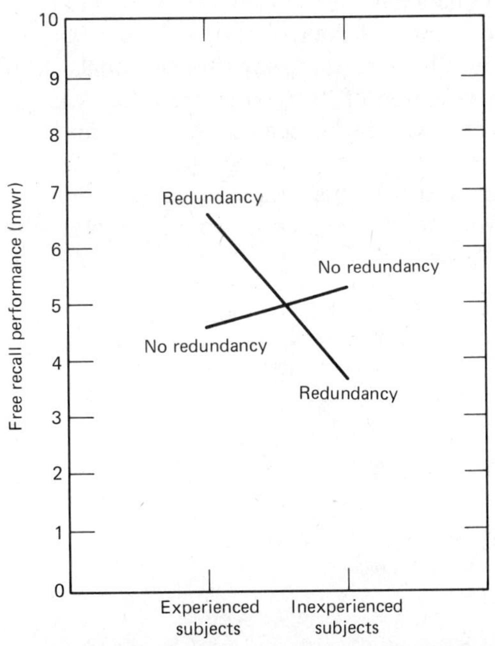

Figure 11-29. Scapin (1981) found that using the English word in the definition of a computing command was helpful for experienced subjects but harmful for inexperienced ones.

probably know several possible command terms for any operation and that identifying the term in the definition may help them to select the correct term from all the competing ones. On the other hand, the additional information provided by the definitions without redundancy may help inexperienced subjects interpret and understand the meaning of the term.

This experiment supports what we have all suspected: that experienced computer users and inexperienced ones have different characteristics. Thus, a command language developed with one set of users in mind may not be very useful for another set of users.

SUMMARY

In this chapter we have considered some of the issues concerning data entry. The standard terminal is rapidly becoming the VDT (Video Display Terminal). A concern with these devices is visual fatigue caused by problems with illumination, glare, and contrast. We discussed research that suggests that the design of keyboards and positioning devices is relevant to the speed and accuracy of data entry. Results indicate that the "mouse" and light pen are both good positioning devices. Furthermore, it has been found that a system that responds with fairly standard delays is superior to one with rapid responses sometimes and long delays at other times. We outlined research results that suggest there may be some promise in pursuing voice input and spoken output.

We concluded with a discussion of text editing (i.e., entering and changing programs and documents). The first text editors were "line editors" that could be used on typewriter terminals. More recently we have seen the introduction of "screen editors" on VDT's in which a whole "window" of text can be seen at once. Some research has suggested that screen editors are superior to line editors when considering both ease of learning and user performance. Furthermore, it appears that the more "English-like" the commands are for a text editor, the easier it is for the editor to be used.

PART FOUR

HUMAN-MACHINE SYSTEM PROPERTIES

12 FEEDBACK AND CONTROL

B. H. KANTOWITZ AND J. R. BUCK

The Question in Point is
"Whither Thou Goest?"

For When Thou Has Arrived,
How Dost Thou Knowest?

(With Due Apology to Poets Everywhere)

In the broadest sense of the word, *control* is really decision making. It entails the choosing of goals, planning to achieve those goals, a recognition of uncertainties and risks, and the initiation of the selected plan. Control means the capturing and releasing of energy as extension of the body and minds of people. In this latter sense of a control system, a human controller operate devices that direct the release of power in such a way as to achieve the intended results. Sails on boats are unrolled to capture the wind power and propel a boat in the direction guided by the rudder. More generally a control system may be viewed as being analogous to Figure 12-1. In a purely manual control system a person serves all the functions in Figure 12-1 except that of the environment. The human body serves as the source of power, the mind as the controller, and the effectors are the physical movements of the body members. With purely automatic systems people set the goals and input information; the rest of the functions are electromechanical. More common mixed manual-automatic systems vary between these two extremes where people serve varying roles in this basic control system. In any of these cases, the junction of control is the device or human action that regulates the amount and type of energy at some point of application such as the opening of a valve or the depressing of the throttle foot pedal of a car.

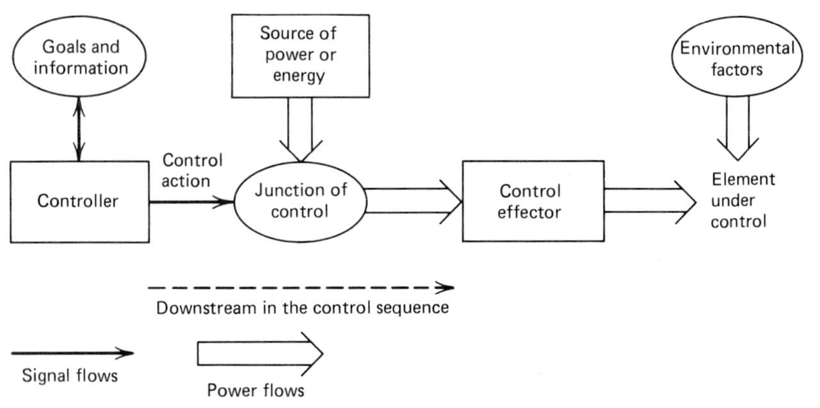

Figure 12-1. A basic control system.

The control effector applies this released energy or power to a mechanism that causes the physical movement desired, such as engaging the transmission of an auto to make the wheels rotate backward. When these two actions are performed, the backward movement of the car is the controlled element subject to such environmental factors as the garage door.

While the basic control system shown in Figure 12-1 is reasonable, it ignores one very important feature known as *feedback*. A person who starts to back out the car from the garage guides the control process by what he or she visually observes. If the rapidly approaching garage door is seen, then an immediate change in control action is likely along with some verbal utterances. Therefore, a repair is needed to the model of the basic control system as shown in Figure 12-2 (and perhaps the garage door as well). The detector shown in this figure senses information about the element being controlled and this information is sent to the controller. If the element under control is progressing satisfactorily with the set

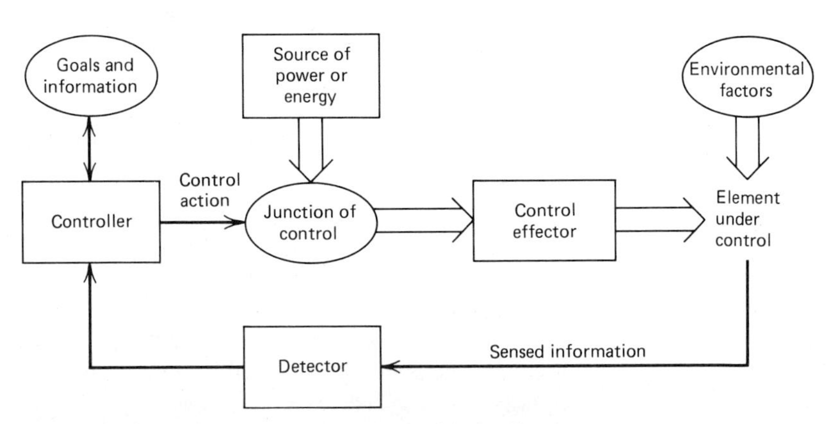

Figure 12-2. A basic control system with feedback.

goals, then no change is needed by the controller. Unsatisfactory progress requires a change in the control action and that is the role of feedback in system control. However, feedback is a concept with more general implications for human factors and that discussion follows next. Later the discussions will return to system control with various forms of feedback.

THE FEEDBACK MODEL

The ability to utilize feedback information is an essential characteristic of both human and machine systems. Negative feedback is defined as a flow of information (or energy) counter to the main flow within a system. Thus, the direction of negative feedback is always opposite that of the main system flow. Positive feedback, as you might expect, flows in the same direction as the main flow. In this chapter when we use the term "feedback" by itself, we mean negative feedback. Negative feedback helps to increase the stability of systems while positive feedback can cause a system to run out of control. Since this definition of feedback is abstract, we will go through some concrete examples to give you a better feeling for the concept. Then we will offer a more formal explanation of feedback.

You have been dating a member of the opposite sex for several weeks and you believe that both of you have grown quite fond of each other. It is time for your relationship to progress toward greater intimacy in your opinion. You try to convince your partner by such romantic devices as presents and verbal utterances like "I love you." You gauge your progress by the feedback information your partner supplies in reaction to your romantic devices. The main flow of information is from you to your intended. Feedback is conveyed by facial expressions, body posture, verbal statements and so forth. This feedback causes you to modify your own behavior. For example, if statements of eternal love fail to produce the desired outcome, you might try something more drastic like threatening suicide or proposing marriage.

Many mechanical systems use feedback to control their operation. One of the earliest of such devices is the flyball governor for a steam engine invented by James Watt in 1788 (Figure 12-3). As the engine shaft rotates faster and faster, centrifugal force forces the two balls to fly apart. The mechanical linkage translates this motion so as to close the valve and decrease the supply of steam. This slows down the engine and causes the shaft to rotate more slowly decreasing centrifugal force and allowing the balls to descend. This opens the valve that supplies more steam that raises the balls that closes the valve that diminishes the steam that slows the shaft that lowers the balls . . .

Feedback systems also control ecology. A decrease in the population of bobcats is soon followed by an increase in the rabbit population. This supplies more food for bobcats and their population increases. More

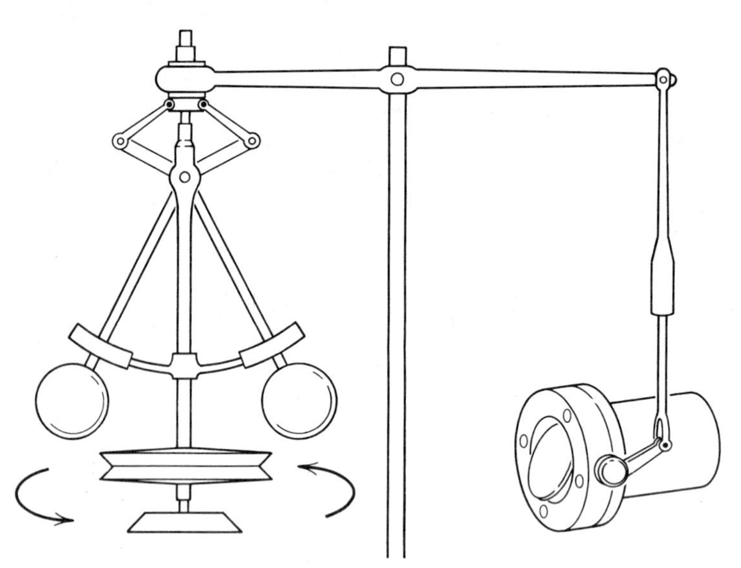

Figure 12-3. Flyball governor for a steam engine.

hungry bobcats are bad news for rabbits, so the rabbit population declines. This feedback process continues indefinitely, regulating the population of prey and predator alike. If either population starts to get too large, feedback helps to stabilize it at a lower level.

Our final example is your home thermostat. You set it to maintain a comfortable room temperature of 68°F. Should the temperature fall below this value, the furnace comes on. This causes room temperature to rise toward the desired value and when this is accomplished the furnace shuts off. The main flow of information and energy is from your furnace to the room. The thermostat provides feedback from the room to the furnace.

Components of a Feedback System

All of the examples above can be represented by a simple diagram (Figure 12-4). We will use the last thermostat example as we work our way through Figure 12-4. The input signal is the desired value of the system output. It acts as the reference against which the output is evaluated. In the case of your home thermostat, the input signal is 68°F. You set this value by turning the thermostat dial to the desired room temperature. The comparator compares the input signal to feedback information. In your thermostat the comparator is a bimetallic strip that

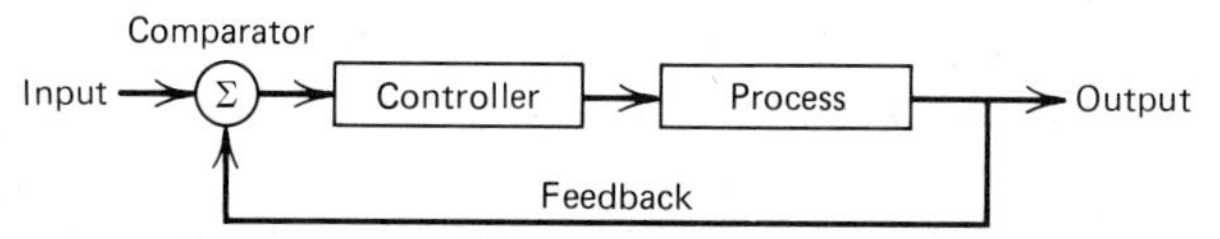

Figure 12-4. Elements of a feedback control system.

bends according to temperature. If the input signal and feedback agree, the output of the comparator is zero. The comparator output tells the controller if any action is required. If the input to the controller is zero, nothing need be changed. However, sometimes the input signal and the feedback do not agree. The comparator takes the algebraic sum of input and feedback. Thus, it calculates a value equal to the input signal minus the feedback signal. When the feedback signal is less than the desired input value because your room is too cold, the comparator output is no longer zero. This tells the controller to activate the process. In this case the process is the heating of air by your furnace so the controller turns on the furnace. The furnace stays on until feedback information reveals that room temperature has increased enough so that the output of the comparator is once again zero. At this point the controller turns off the furnace.

Most feedback systems do not operate with 100% precision. It is unlikely that your own thermostat keeps room temperature exactly at 68°F. Room temperature may drop as low as 65°F before your furnace goes on and may exceed 68°F before your furnace finally shuts off. Comparators often have a small band of values, say 66 to 70°F for a thermostat, where deviations from the desired input are not detected. This is called *dead space*. It prevents the system from continually turning on and off to correct minor variations of system output. Feedback systems do not guarantee that system output will always be precisely equal to the desired input. Feedback systems introduce corrections that tend to bring the system back to its desired state.

A Mathematical Example The qualitative description of a feedback system does not really show you the degree of control that such a closed-loop feedback system is capable of. This can be best understood by comparing the system in Figure 12-4 to a similar system that does not use feedback. Feedback systems are called *closed-loop* systems because you can trace a complete path, or loop, from the comparator through the controller and process, and finally back to the comparator via the feedback path. If the feedback path is eliminated, the system is called open loop. Of course, an open loop system is just a straight line with no loop at all. We use the term "open loop" as a contrast with "closed loop."

Figure 12-5*a* shows an open loop system with a gain of 10. The gain tells us how much the input is multiplied. So if an open loop system has

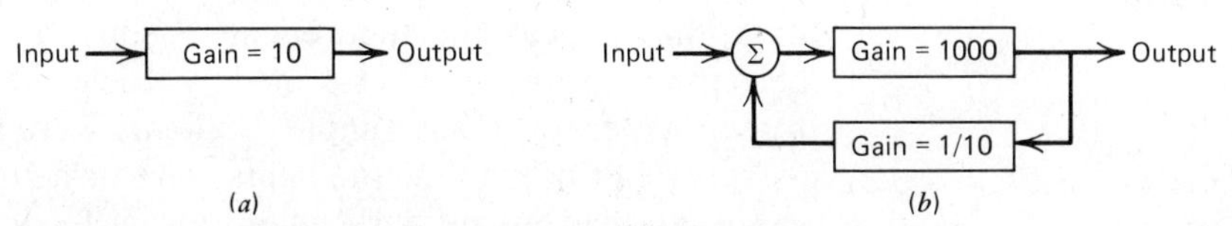

Figure 12-5. Open- (*a*) and closed-loops (*b*) systems.

an input of 1.0, and a gain of 10, the output of the system will equal 10. (Note that we have intentionally omitted units. The system could be amplifying voltage or current or anything else that can be measured.) Figure 12-5*b* shows a closed-loop system with similar characteristics. The gain of the amplifier in this system is 1000. However, because there is negative feedback with a gain of 1/10, the output of the system is not 1000 times the input. It is only 10 times the input. In order to understand this, we must work through some simple algebra to see how the closed loop system functions.

Let I stand for the input and P represent the output. We can write the following equation:

$P = 1000\ (I - P/10)$ since the quantity $I - P/10$ is the signal coming out of the comparator that gets amplified 1000 times. Now we solve this equation for P as a function of I:

$$P = 1000I - 100P$$
$$101P = 1000I$$
$$P = \frac{(1000)I}{101}$$
or $P =$ approximately $10I$

So both systems in Figure 12-5 have an output that is 10 times the input. Now let us assume there is some disturbance in the system—low voltage, etc.—that causes the gain of the amplifier to drop by 10%. What is the new output of both systems with a drop in amplification?

The open loop system is easy to figure out. The output drops by 10%. For the closed loop system we must again use algebra:

$$P = 900(I - P/10)$$

Solving this equation we find that the output is still very close to 10 times the input. The 10% drop in gain has had virtually no effect upon the closed-loop system! Indeed, for the closed-loop system to have a 10% loss of output, the gain must drop from 1000 to 90. This is an impressive demonstration of how negative feedback acts to stabilize system output.

Processing Visual Feedback

People use visual feedback to guide motor movements (see Chapter 5). The amount of time required to process feedback information is important for both practical and theoretical reasons. The classic study aimed at measuring this time was performed by Keele and Posner (1968). The logic of their experiment was simple. Students were paid to move a pointer to a target. Half of the time the lights were turned off so that the movement was performed in the dark where visual feedback information was not possible. The lights were turned off randomly, so that people could not decide in advance whether or not to use visual feedback. Thus, any planning that occurred before the movement was the same in light-on and light-off conditions.

Movement time was controlled by instruction. The students were told to generate intended movement times of 0.15, 0.25, 0.35, and 0.45 sec. The experimenters recorded movement accuracy. The time required for visual processing was estimated as the shortest intended movement time for which accuracy was better with the lights on. For the 0.15-sec movement time the target was missed 68% of the time with lights on and 69% with lights off. For the 0.25-sec movement time results showed 47% misses with lights on and 58% with lights off. Therefore, the time to process visual feedback must be less than 0.25 sec. Since participants in the experiment actually took 0.19 sec during the 0.15 intended movement time condition—that is, they were slightly slower than instructed—Keele and Posner concluded that the time required to process visual feedback information was between 190 and 250 msec.

Since this conclusion agreed with results obtained from an earlier tracking experiment (tracking is discussed later in this chapter), where people had to keep a target centered on an oscilloscope screen while the display was blanked out for short times to prevent feedback processing (Pew, 1966), this estimate has been widely accepted for many years. Indeed, it was the great faith in this estimate that is in part responsible for the present popularity of motor programs (see Chapter 5) as an explanation of control for rapid movements that require less than 250 msec. Nevertheless, at least some researchers believe that this estimate is too high (Adams, 1976). By measuring signals associated with the movement of muscles, which occur before overt movements, evidence has been obtained for processing times as fast as 10 msec in monkeys (Evarts, 1973) and 50 msec for humans (Sears & Davis, 1968). If these much shorter estimates are correct, even the performance of concert pianists could be explained by closed-loop feedback models. This, of course, would remove much of the theoretical need for open-loop motor programs.

PRACTICAL CONTROL SYSTEMS

There are many many control systems in everyday use. In these systems there are numerous roles for human factors, such as aiding human operators or automating some functions that are difficult for people to do. The manner of aiding and the reason for automating is often dependent on the control level of the human operator and the order of control that the operator must exercise. These concepts and their implications for human factors are given in the following paragraphs. These features and implications in practical control systems show a background need for some elementary control theory in performing human factors design.

Control Level

Control systems are often represented as hierarchically organized systems. This means that several levels interact to produce the system output. A large business firm is one example of hierarchical organization. Clerical workers are the lowest level of control and are responsible for the execution of daily details necessary in corporate life. At a higher level are the managers who supervise the clerical workers. Above them are vice-presidents and the chief operating officers. The work of the firm is organized according to hierarchies. Chief executive officers do not type billing statements and clerks do not set overall goals for the business. Each level of the hierarchy does a different kind of job but all levels must communicate and interact for the firm to prosper.

A large ship provides another example of hierarchical control and control levels within this hierarchy. At the most inner level is the immediate rudder positioning control that keeps the ship making a straight movement rather than zigzag movement with each wave and trough. This inner loop control requires frequent correction and near constant attention. At a modestly higher level is the basic change in ship heading. This level of control is made less frequently and the maintenance of this heading control requires one to take periodic observations of the compass rather than using constant attention. A highest level of ship control is exercised by the captain who specifies the required ship locations and target times to be at those locations. In effect, this highest level is really *goal selection*. The navigator must translate these goals into a plan of action, which denotes general course heading and speeds as an intermediate control level action between the captain and the person who calls for a basic change in the course heading. Thus control levels are hierarchical from the outside loop of goal selection and planning through the intermediate loops of planning and adapting to the innermost loop of almost continuous adjustment.

Control Aiding and Automation

Control aiding is achieved by providing the human operator with information or capabilities that facilitate task performance. Automation, on the other hand, is the transfer of a human task to a machine task. Cruise control on an automobile is a form of automation where an electromechanical device holds the car at a near-constant speed and thereby freeing the driver from the constant attention demanded by this task. This form of automation is often called *control augmentation* because it assists humans in performing inner-loop control tasks.

Display augmentation is a form of operator aiding where the operator is informed, advised, instructed, or told what to do. *Command displays* are cases where a person is told what to do and sometimes how to do it. Normally if one always knew precisely how to perform some command, then the procedure of completing the commanded task would be an excellent candidate for automation. Therefore command displays typically tell one what to do but leave room for some judgment in how to do it. For example, the classical engine speed telegraph aboard older ships simply instructed the engine room of the propeller speed desired by the bridge. The rate of change from the current propeller speed to the new speed was up to the engine room personnel subject to the captain's temper. For the most part, however, display augmentation serves to show the system condition relative to typical goals of operators. The problem in display augmentation is the form of the information relative to the control actions needed and the goals set. Compensatory displays (see Chapter 7) show the *error* in the current system status toward the desired whereas pursuit displays show the current status and the goal separately, leaving the operator to infer the error through the status-goal difference. In either of these form of display augmentation, the operator must still decide on the control action to take.

Control Order

Orders of control refer to derivatives of control action with respect to time. Zero-order control is displacement such as moving a lever 5 cm to the right. Let X_t be the position of the control at time t; a first-order description of control. It then follows that a first-order control refers to the velocity of change in the displacement as $\Delta X_t = X_{t+1} - X_t$ for a single unit of time. As this unit of time gets smaller and smaller, then ΔX_t approaches the derivative dX/dt. Second-order control is simply the acceleration or

$$\Delta^2 X_t = \Delta X_{t+1} - \Delta X_t = X_{t+2} - X_{t+1} - (X_{t+1} - X_t) = X_{t+2} - 2X_{t+1} + X_t$$

Here again, as the units of time became smaller, then $\Delta^2 X_t$ approaches d^2X/dt^2.

An important consideration of control order in human factors is in regards to the type of control device the operator is operating. On a car the driver depresses a foot pedal to increase the rate of fuel into the engine. When the car is moving along the highway on level ground and the pedal is depressed further, the speed of the car increases. For this reason, the throttle foot pedal is called an accelerator pedal even though the foot pedal position determines the speed of the car relative to the highway grade. Another foot pedal on a car engages the wheel brakes so that a greater brake pedal displacement causes more friction within the wheels and usually faster stopping. Both of these foot pedals are second-order control devices; one for increasing the speed and the other for decreasing the speed. Acceleration or deceleration is caused by the operator's rate of change in positioning the devices.

A second important human factors consideration about control order is display augmentation. If the primary goal is to be at a particular location at a specified time, then one can establish milestones in accomplishing primary goal as a series of subgoals and subsubgoals. This plan may be carried to any degree of time resolution from the outer-loop primary goal to an instant-to-instant inner-loop command. When a mile-

FOCUS ON RESEARCH

Reducing Order of Control

As the order of control increases, it becomes more difficult for the human operator to control a system. Thus, a basic human factors design philosophy is that the human should be made to function as a zero-order controller when practical (Birmingham & Taylor, 1954). More often, however, a low order of control, such as first-order (rate control) is the optimum choice (Roscoe & Kraus, 1973). In some systems the order of control does not remain constant. Thus, in addition to the problems involved with higher-order manual control, the operator must also detect that control dynamics have changed. While people can adapt to slowly changing system dynamics (Ince & Williges, 1974), such a system places additional burdens on the operator.

The airplane pilot is one example of an operator who could function better if changes in flight dynamics were eliminated and order of control reduced. One experimental system, called the Performance Control System (PCS), has had great success in improving pilot performance (Roscoe & Bergman, 1980). This system reduces the pilots workload by eliminating the need for the pilot to sense and control higher-order loops involving bank angle and vertical speed (see Figure 12-6). Thus, the pilot functions as a zero-order controller of bank angle and vertical speed. This equates to first-order control of heading and second-order control of lateral position, rather than third-order or higher, and first-order control of vertical position (altitude). Experiments performed in a flight simulator showed fewer pilot errors by a factor of ten when PCS was compared to normal flight control (Figure 12-7). Furthermore, performance of a secondary digit-canceling task (see Chapter 6 for discussion of secondary-task methodology) was also improved using PCS. It is quite clear that real benefits are gained when airplane manual control dynamics are redesigned to make the human a low-order controller.

stone is reached too late, an increase in average speed is needed as a control correction. How much change should be made in that foot pedal and how fast? Display augmentation must provide that information. Sometimes the answer to the second part of the question depends on who is paying for the gasoline. In some of the newer, more-expensive cars there is a device that can be set that records the elapsed distance and time to tell the driver the needed speed to reach the preset milestone goals. A more sophisticated device could then monitor the change on the

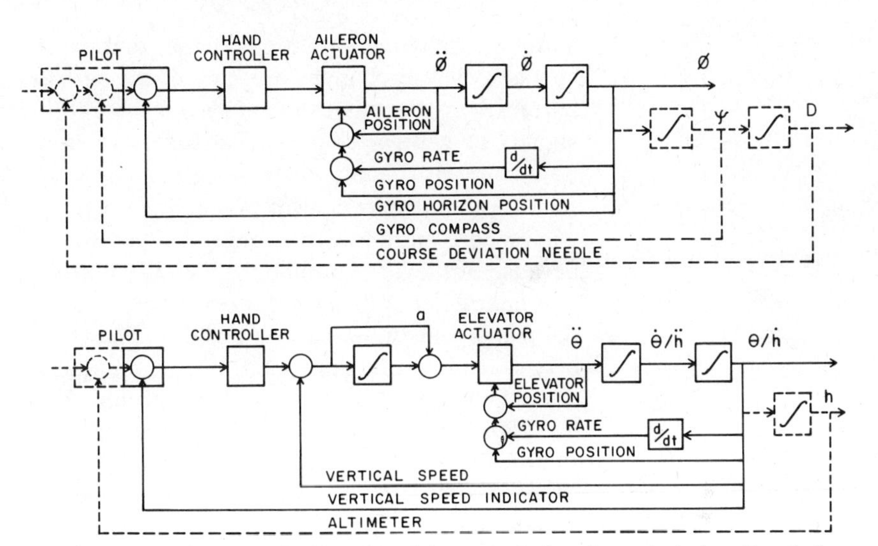

Figure 12-6 PCS model. (Reprinted by permission from Aviation Psychology by Stanley Roscoe © 1980 by The Iowa State University Press, Ames, Iowa 50010.)

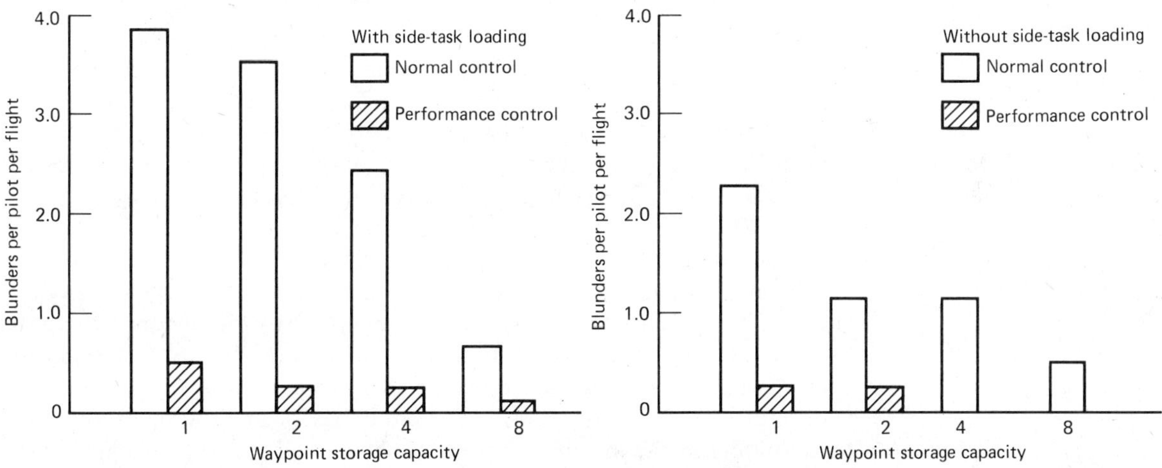

Figure 12-7. Area navigation procedural blunders as a function of computer waypoint storage capacity for normal control and flight performance control, with and without side-task loading. (Reprinted by permission from Aviation Psychology by Stanley Roscoe © 1980 by The Iowa State University Press, Ames, Iowa 50010.)

throttle foot pedal or brake to tell the driver if the control device change is appropriate as a fully or partially augmented display.

Adaptive Control

Perhaps the most important function of a human operator in system control besides goal selection is planning. System control starts with goals and a plan for reaching those goals. However, outside disturbances and unanticipated situations arise that necessitate plan and/or goal changes. These changes are part of the process of adaptation. Another part of this process is changing the perception of how the system responds to control changes. Suppose you are driving north on I-69 and now you must change over to I-80 and head west. That strong side wind on I-69 that required a lot of steering corrections now becomes a headwind where there is now need for a lot of throttle corrections, particularly because of the large number of highway patrol cars in this vicinity. One's *internal model* of how the car responds to control changes must change or the plans/goals must change or both. The human factors question posed here is, "How does one effectively communicate changes needed in the internal model of system control and aid the driver in adapting his or her plans and/or goals?" One thing is sure: any information being presented about the system behavior must fit with the existing internal model of the system because that is the reference for change. An experienced system operator can receive direct information and recognize most of the implications, while a naive operator needs to know the meaning of the information in terms of control implications.

It is sometimes said that the best window to the future is from the past. Unless there is reason to believe that an abrupt change was made that uncouples the past, then projection from history provides a good forecast. *Historical displays* in the forms of strip-chart recorders or similar devices provide such information. That is, these forms of displays indicate, by extrapolation, what will likely happen if nothing is done. With a bit more sophistication, a historical display can be extended to make the future projections directly as a *predictive display* (see Chapter 7).

Adaptation in control requires some feedback information. If a change in system response occurs such as the traction of a car right after the start of a rain, then a common practice is to provide a minor control action as a test signal such as lightly touching the car's brakes. The system response to the test signal provides an idea of changes in the system response due to the observation of system behavior and the operator's internal model of the system. In more sophisticated control systems there are sensors employed to select information that can be used in conjunction with an engineering model of the system that automatically tunes parameters of the model to keep the model up to date. Some of these sophisticated systems then use these engineering models to drive *preview displays* where predictions of system performance are shown and changed by *off-line* control actions. That is, operators can test alternative control actions ahead of the need for them, evaluate the quality of the alternative control actions, select the one desired, and have that action be performed automatically at the proper time. The key to this technology is the engineering model of the system and the incorporation of the human controller into that model (Birmingham & Taylor, 1954).

BEHAVIORAL CONTROL THEORY

The preceding discussions of feedback and control has been mostly qualitative. In order to appreciate the power of feedback and control models more fully, some mathematics are necessary. The following discussion assumes that the reader knows enough about elementary calculus to understand the terms "integral" and "derivative." We first introduce some elementary mathematical treatments of feedback and control, including Laplace transforms. But our introduction to these topics will not be mathematically rigorous. Our goal is only to provide enough intuitive understanding so that you can follow the concepts of control theory. Detailed mathematical proof will be completely omitted. So even if you have only a minimal mathematical background don't give up. You will still be able to grasp the basic ideas. Once these ideas are explained we go on to demonstrate how they are used to improve design of person-machine systems.

Some Mathematical Preliminaries

Imagine a container filled with water (or beer if you prefer) with a tap on the bottom. Once this tap is opened the flow of liquid depends on how much still remains inside the container. The rate of flow will at first be high, and gradually taper off. (The shape of such a mathematical function that plots volume as a function of time is called an exponential decay function.) While this qualitative description is correct, a far more precise description can be made by using mathematics.

Let x represent the volume of liquid in the container. Then the derivative $\frac{dx}{dt}$, the rate of change of volume over time, represents the flow through the tap. Mathematically we can write

$$\frac{dx}{dt} = -Kx$$

where K is a constant. The minus sign in front of K indicates a negative rate of change since the volume of liquid is decreasing.

Figure 12-8 shows this exponential decay function. The slope of the function is greatest at time zero. As liquid pours out, and time increases moving along the function to the right, the slope decreases showing that liquid is pouring out less quickly. In control theory the term "time constant" describes how quickly the system decays, or in this case, how fast the liquid pours out. The dotted line in Figure 12-8 shows what would happen to volume if the rate of decay were constant. The time for this dotted line to reach the X axis is called the system time constant, *tau*. It turns out that tau is always 63% of the total decay distance. So we can also define tau as the time required for the system to decay to 37% (1 − 0.63) of its original value. Later on we examine some examples where the time constant of a system is important. Now we complicate the simple system of Figure 12-8 by pouring water into the top of the container while liquid is still running out of the bottom.

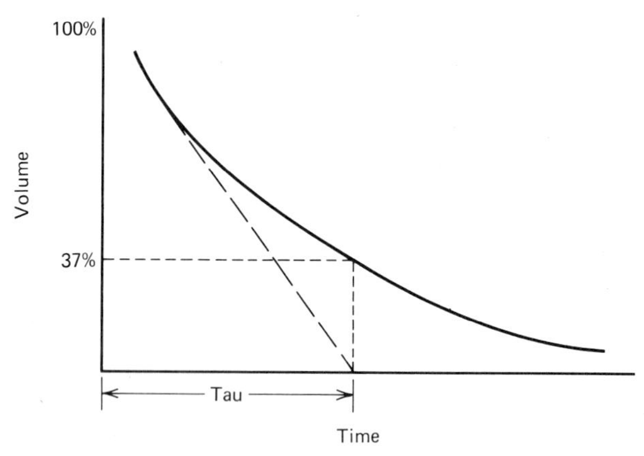

Figure 12-8. An exponential decay function.

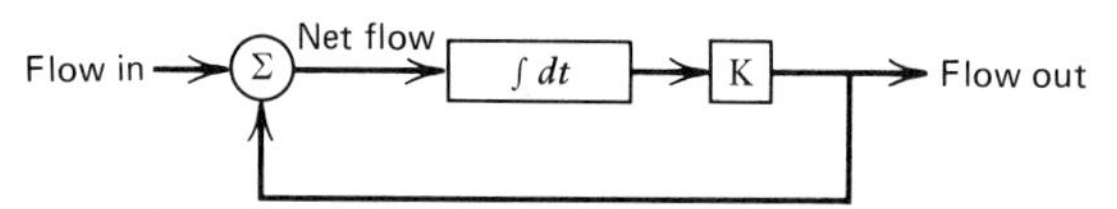

Figure 12-9. A feedback model of the container with liquid flowing into the top and out from the bottom.

This new system can be represented by the feedback loop of Figure 12-9. Since flow is the derivative of volume, volume is the integral of flow. Mathematically, we can write

$$\text{volume} = \int_0^t (\text{top flow in} - \text{bottom flow out})\, dt$$

assuming that the container was empty when we started at time zero. The output of the comparator is net flow, the difference between the flow coming in at the top and that leaving at the bottom. This is integrated to yield volume. Then volume inside the container is multiplied by K to obtain the flow out of the bottom. This gets fed back to the comparator and the loop is completed.

Figure 12-9 allows us to predict the response of the system to various kinds of input functions. For example, if an impulse function occurred at time zero—you dumped a large bucket of water into the container—the volume would increase as a step function and then decay exponentially. For a step input, volume would increase at a decreasing rate until some constant volume was obtained. Having a quantitative description of the system permits precise predictions of how the system will react to different input signals. Figure 12-10 shows periodic step functions of water input at the same time as the water flows out of the container. In this case the average input rate is greater than the average output rate so that the volume of water tends to increase with each filling and the beginning rate of outflow is increasing with each filling.

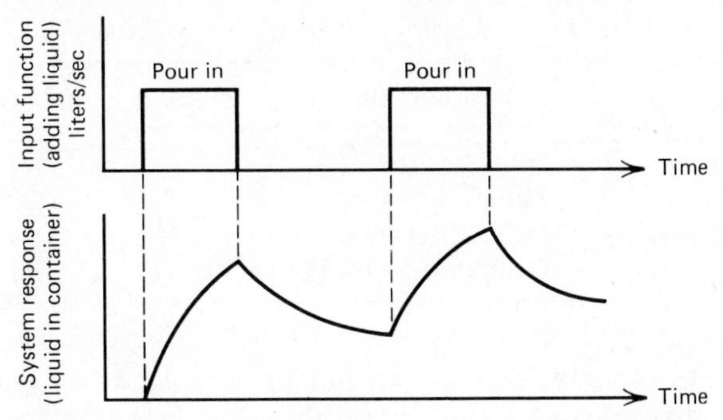

Figure 12-10. A time-history of the container-water system.

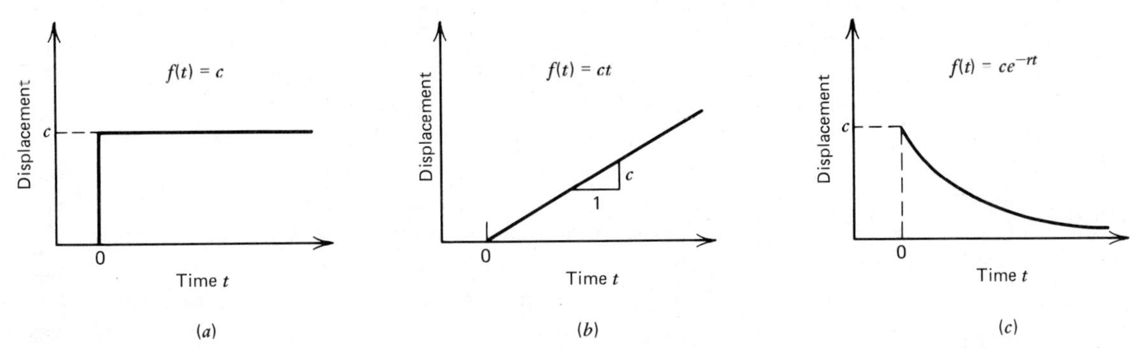

Figure 12-11. Different displacement-time functions. (*a*) A step function. (*b*) A ramp function. (*c*) A decay function.

Laplace Transforms

Consider a target moving over time as given by the function $f(t)$ where the vertical axis denotes the displacement or movement in a given direction and the horizontal axis is time t. A step function, as shown in Figure 12-11*a*, shows that the target made a single move of c units initially and then stayed there, whereas the ramp function shown in Figure 12-11*b* shows a target moving at a constant rate of ct over time. Almost any functional form of $f(t)$ can be used including a decay function (Figure 12-11*c*) where the target moves toward the reference at a constantly decreasing rate of ce^{-rt} where r is the exponential rate of decrease. Notice that these descriptions are all functions of time.

The Laplace transform is a conversion of a *time* function to a function of the frequency variable s. At this point, don't worry about the variable s as we will cover that later. A transform is a mathematical operation where an operator acts on a function of one variable and changes it to another variable. With the Laplace transform the operator is e^{-st}, which operates on the function $f(t)$ and the transform is:

$$\mathscr{L}(s) = \int_0^{+\infty} e^{-st} f(t)\,dt \qquad (12\text{-}1)$$

TABLE 12-1 *Laplace Transforms of Some Time Functions*

Time Function	Laplace Transform
Step, $f(t) = c$	c/s
Ramp, $f(t) = ct$	c/s^2
Decay, $f(t) = ce^{-rt}$	$c/(s + r)$
Growth, $f(t) = c(1 - e^{-rt})$	$c/[1/s - 1/(s + r)]$

In the case of the step function, replace $f(t)$ with the parameter c (the displacement) and the resulting Laplace transform is c/s (Figure 12-12). Table 12-1 also shows the transforms of some other selected time functions. Note that the step function and those others shown in this table all start at $t = 0$. If the time function is the same functional form as the step function but the displacement occurs k time units later, then the Laplace transform is equal to that of the time form multiplied by the factor e^{-ks}, shown in Figure 12-13. Another point is illustrated by a time function where there is first a displacement of c units of movement at k_1 time units in the future and a displacement of c movement units back again after k_2 time units. This movement over time is shown in Figure 12-14*a* whereas Figure 12-14*b* shows that time movement pattern is equivalent to the difference between the two step functions that occur at different points in future time. The Laplace transform of this time displacement function is $c(e^{-k_1s} - e^{-k_2s})/s$; the difference between the Laplace transforms of the two

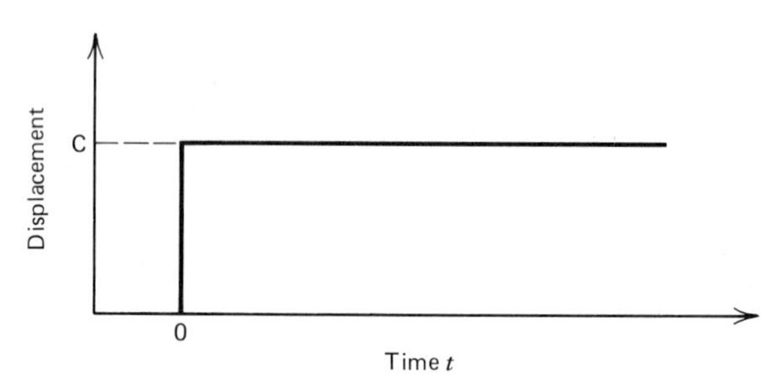

Figure 12-12. A step time function beginning at t = 0.

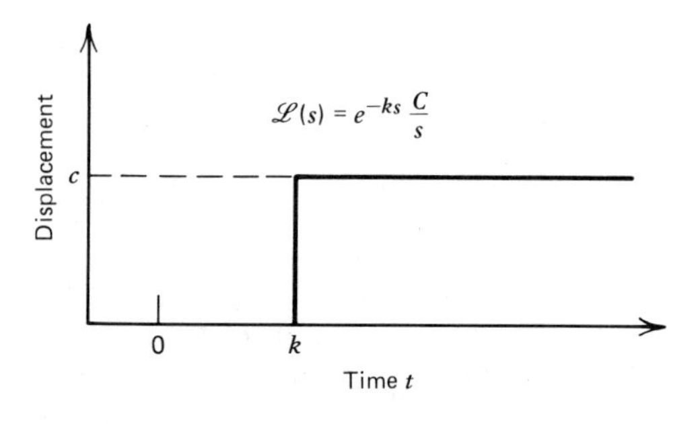

Figure 12-13. Time delayed step function.

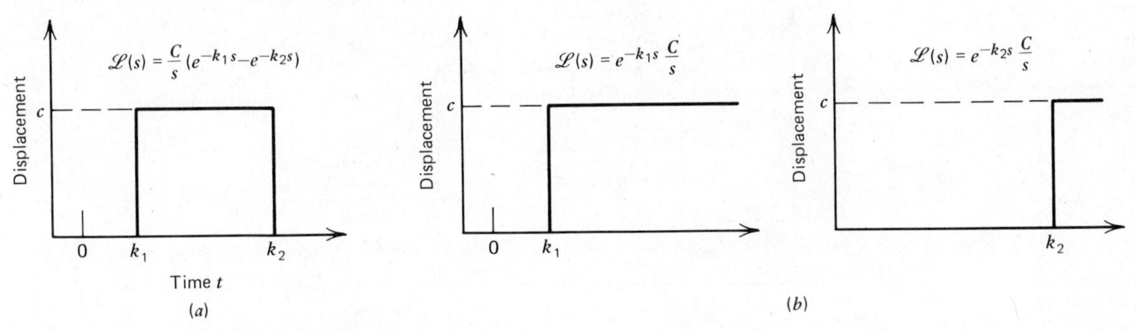

Figure 12-14. Illustration of linearity property.

step functions. It is this *linearity* property that makes Laplace transform so useful in control theory. Linearity means, in part, that the transform of the sum (or difference) of two time functions is equal to the sum (or difference) of the two transforms. Laplace transforms of a constant times a time function is simply the constant times the transform of the time function; linearity under multiplication is by a constant. These properties provide a form of algebra for control.

Transfer Functions in Tracking

Suppose that you are holding a lever that activates one of those electronic games. As the lever is moved forward and back, a flashing dot on the screen moves up and down. The object to the game is to move the lever so that the flashing dot follows a short dash in the center of the screen. The inventor of the game has programmed into the game a change in displacement of this dash over time, which is unknown to the player. Once the game is started the dash moves rightward and another dash appears in the screen center following the inventor's program. Also the location of the past flashing dot moves from left to right at the same speed as the dash. When you position the lever correctly, the dots appear between the dashes to the right of the screen center so that you have a history of your tracking ability. The dots appeared even farther to the right earlier in this brief game history. Errors occur in your tracking when the dotted path separates from the dashed line. The greater the separation, the greater the error. Let the time path of the dashed line be called $w(t)$ and the dotted path $y(t)$ so that the tracking error over time is $e(t) = y(t) - w(t)$. When $e(t)$ is positive, you have pushed the lever too far forward. Notice how the $e(t)$ value serves as feedback in this game.

You are holding the lever constant in perfect tracking as indicated by the dash-dot line to the right. All of a sudden the leftward dash moves up c_1 units on the screen and stays at that upward location. Since it takes you a few moments to see the location change in the dash and react to it, there is a momentary delay before your reaction. Then you start pushing the lever forward trying to bring the flashing dot up to the same location as

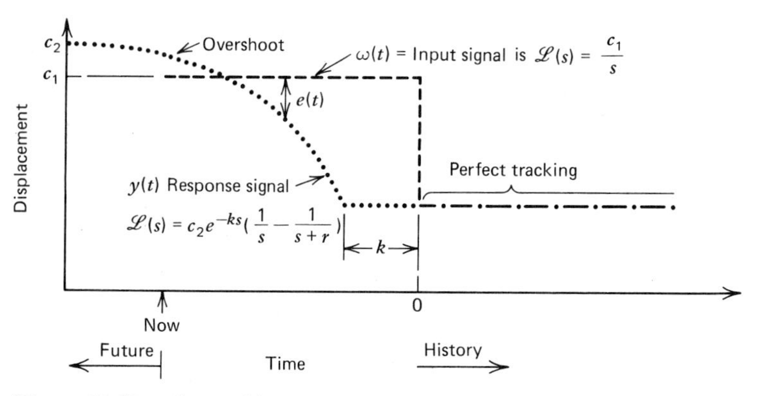

Figure 12-15. Control in an electronic game.

the dash. The gap between the two paths now starts to close, where Figure 12-15 shows $e(t)$ getting smaller and smaller but currently overshooting the intended displacement by $c_2 - c_1$. In this case the time function $w(t)$ is a step function with the Laplace transform of C_1/s. However your response function is a growth function or $y(t) = C_2(1 - e^{-rt})$. The Laplace transform of your response relative to the start of change of the dashed line is $C_2e^{-ks}[s^{-1} - (s+r)^{-1}]$ where k is the response delay time and r is the rate of error closure over time. These two Laplace transforms are the input and output, respectively. The ratio of the output-to-input transforms is known as the transfer function or

$$H(s) = \left\{C_2e^{-ks}\left(\frac{1}{s} - \frac{1}{s+r}\right)\right\} \div \left\{\frac{C_1}{s}\right\} = \frac{C_2}{C_1}e^{-ks}\left(1 - \frac{s}{s+r}\right)$$

where the three resulting factors are known as the gain (C_2/C_1), the delay lag (e^{-ks}), and compensation correction $[1 - s/(s + r)]$. Gain is the ultimate output-to-input signal ratio. Your hi-fi amplifier is primarily a gain device that takes a weak signal from the turntable and makes it louder. If the amplifier takes all tones coming to it from the very low to very high frequencies and makes these tones uniformly louder, then the transfer function of the amplifier is a constant to show that it is purely a gain device. But the constant becomes greater as you turn up the volume control of the amplifier. The delay lag factor, as the name implies, describes the time period wait before there is a beginning response. Finally the compensatory factor shows how the response is made over time relative to the error. In the foregoing case, the amount of correction decreased proportionately with the error. If the response, once started, has been a step function of exactly the same magnitude as the input as shown in Figure 12-16, then the gain would be the constant one and so would the compensation factor; leaving only the delay lag factor. Transfer functions to specific input functions describe many operating characteristics of your ability to perform various tracking tasks. While the Laplace transform algebra allows the combining of these fre-

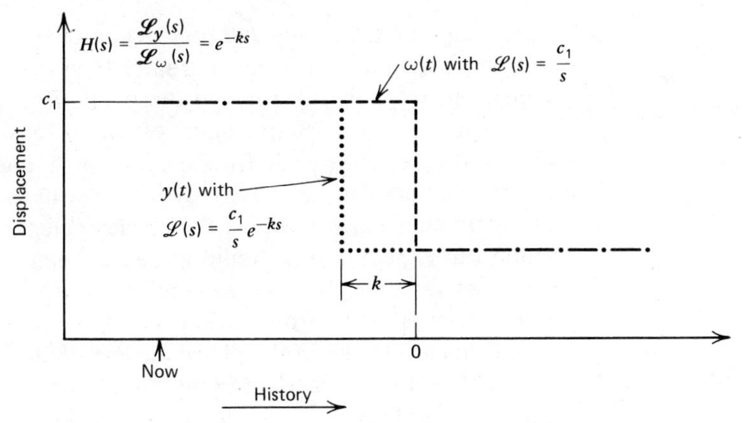

Figure 12-16. Electronic game with only a time delay.

quency functions and the transfer function shows the output response relative to an input signal, the resulting time properties of the system can not be shown directly from these transforms or the transfer function. An inverse Laplace transform is needed, reversing the operation of Equation 12-1, to find the $f(t)$. In this inverse transform, the time function can be described using a term denoting the position of the error over time.

A classical experiment by Ellson and Wheeler (1949) gave subjects a series of input step functions of differing heights in random height sequences. Results from their experiments showed that the average output magnitude was equal to the input magnitude. However, there was a small amount of overshooting small input signals and undershooting the large input magnitudes of a set of the range of input magnitudes. This over- and undershooting is known as the *range effect* because it depends on the range of magnitudes and not on the average or absolute magnitudes. This study and others obtained results that show a high degree of linearity in human tracking responses to simple input functions.

When the input signal to the electronic game is a sine wave, then the input and output traces may look like Figure 12-17. In this case there was

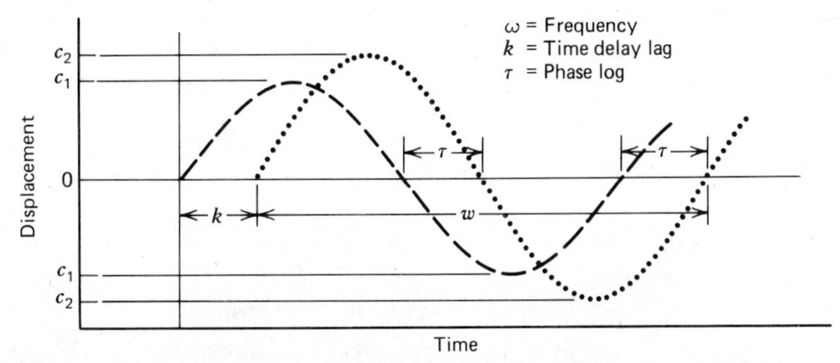

Figure 12-17. Electronic game with a sine wave path.

a time lag of k time units and then accurate signal tracking except for the output peaks and valley being greater than the signal and the output was consistent by time units behind the input. The difference in C_2/C_1 is the gain whereas the constant lagging of output from the input by τ time units relative to the time period for the cycle of W time units is the *phase angle* ρ. Since a complete sine wave cycle takes W time units, the frequency s of the input signal is $1/W$ and the phase angle $\rho = 360\tau/W$ in degrees. It is natural to expect that ρ should increase when the oscillating input signal becomes faster and so phase angles were examined for various input frequencies. Results from various experiments (e.g., Elkind, 1956) show a near-linear change in the phase angle with an increase in the frequency s. These results to random input stimuli indicate further linearity in human control. However, as the person tracked the sine waves longer and longer, errors tended to decrease; indicating that people no longer simply followed the input, as a mechanical mechanism would do, but rather they began anticipating the input form and started rhythmatically leading what they perceived to be the signal. This anticipation is similar to aiming a rifle in front of a moving target to compensate for the time the projectile takes to move from the rifle to the target. Anticipation effects create nonlinearities in human response patterns where τ in Figure 12-17 reduces over time.

Suppose that the electronic game was modified so that the input signal (the dashed trace) was shown starting well to the left of the flashing dot on the tube that you control. This modification gives you the advantage of previewing changes in the input well ahead of having to react to them. As you would expect, people can utilize this preview information and anticipate changes to make in the control lever much better and the input-output diagrams change from the one shown in Figure 12-18*a*, without preview information, to that in Figure 12-18*b* with the preview information. The error in control with preview information tends to be considerably less with better anticipation and there is less linearity in the sense that people behave less like mechanical servocontrol mechanisms.

General Human Transfer Functions

A typical human transfer function tends to have a linear part that looks somewhat like the following (Kelley, 1968).

$$H(s) = \frac{Ke^{-ks}(1 + T_L s)}{(1 + T_N s)(1 + T_I s)} = \frac{\text{(Gain) (Delay lag) (Lead compensation)}}{\text{(Neuromuscular lag) (Compensatory lag)}}$$

It also follows that the inverse Laplace transform for this general human transfer function is

$$\Theta_0(t) + (T_N + T_I)\frac{d\Theta_0(t)}{dt} + T_N T_I \frac{d^2\Theta_0(t)}{dt^2} = K\left[\Theta_i(t-k) + T_L \frac{d\Theta_i(t-k)}{dt}\right]$$

$$\begin{matrix}\text{Output} \\ \text{position}\end{matrix} + \begin{matrix}\text{Output} \\ \text{velocity}\end{matrix} + \begin{matrix}\text{Output} \\ \text{acceleration}\end{matrix} = \text{Gain}\left(\begin{matrix}\text{Input} \\ \text{position}\end{matrix} + \begin{matrix}\text{Input} \\ \text{velocity}\end{matrix}\right)$$

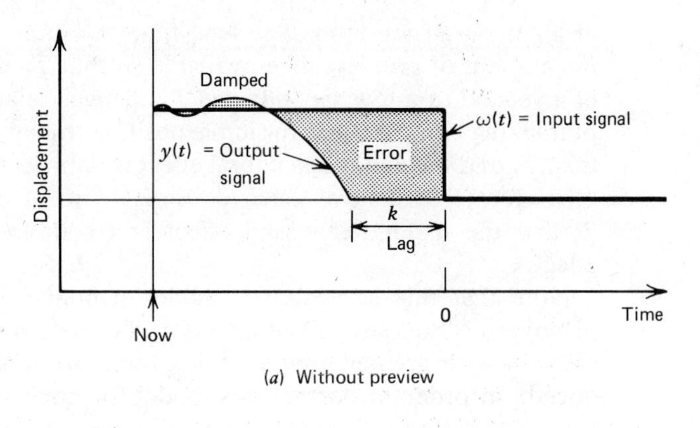

(a) Without preview

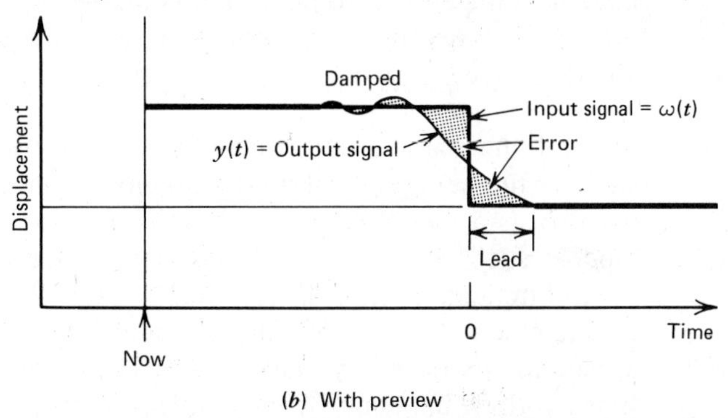

(b) With preview

Figure 12-18. Input-output diagrams without and with preview information. (*a*) No preview information. (*b*) With preview information.

A number of studies confirm that the delay lag k in compensatory tracking is around 0.15 sec, that is, when the human operator is only shown the positioning error. With a sine wave at about $1\frac{2}{3}$ Hz, a 0.15-sec delay lag amounts to a phase angle of about 90° behind the input signal. The neuromuscular lag T_N tends to be more variable in different system dynamics and greater when the human body member that operates the control has greater mass. There are three other factors in the general human transfer function (gain, lead, and compensatory lag constants) that tend to vary in related ways depending on the nature of the input function, the skill of the human operator, and whether there is noise (random variation) present. The gain K, lead constant T_L, and lag constant T_I tend to be near optimum for minimizing the root mean squared error for highly skilled operators. When there is noise in the error signal observed, then most human operators decrease the gain K and increase the lag constant T_I, so the ratio K/T_I is nearly constant at around 10 and behavior tends to be more conservative. This same conservative behavior is typically observed when the system being controlled is unstable with high-frequency responses; such as sudden acceleration or jerky braking of a car on an icy road. The lead time constant T_L depends heavily on the amount of preview time available so that T_L changes from a fraction of a second to a few seconds and T_L changes relative to T_I with the type of tracking control used. In simple position tracking T_I tends to be greater than T_L and the major behavioral effect is noticed in the gain change. With other forms of control, such as direct control of the acceleration, T_L T_I and the major behavioral effect is a combined gain and frequency effect.

Note that this servocontrol model of human behavior excludes any planning on the part of the operator. To the extent operators can plan, such as with greater preview time, there are more nonlinearities introduced. In order to correct this model for nonlinearities, experimenters have introduced a correction that is known as *remnant terms*. However, there are many different proposed remmant terms reported (Shinner, 1967).

Other Models of Human Control

It is not unusual to study human behavior in contrast to a physical model such as the servocontrol model previously discussed. Also it is not unusual to find model variations proposed because of deficiencies in other models such as human nonlinearities. There are many other forms of control models. One of these is Bekey's (1962) *Sample and Hold Model*. In this model it is assumed that discrete flows of information occur within a person about every half second. Between these clocked-reference times nothing happens. Since people generally require a fixed time period for a minimum reaction time, this sample-data or sample-and-hold model has considerable appeal. However, it is a difficult model to test against the continuous models presented earlier.

Some of the other model forms presented deal with the strategy of the human operator and optimization. In some of these models it is assumed that human operators act in some optimum way. Ornstein (1961), Adams (1963), and Bekey, Meissinger, and Rose (1965) have used forms of steepest ascent optimization to set parameter values in models similar to the general human transfer function. Others have used dynamic programming and other forms of optimization to set the model parameters so that the model best mimics observed human performance. More recently the theory of modern control has been used as a basis for control models. Modern control employs time variables rather than frequency variables and this theory describes multivariate control. These modern control models (Kleinman, Baron & Levison, 1971; Obermayer & Muckler, 1965) also assume that the human operator observes the controls, his or her own dynamics (as well as system dynamics), and the constraints of control, and then acts to optimize some performance criterion. In essence, such models of human performance are based on this assumption of optimal behavior and that the criterion

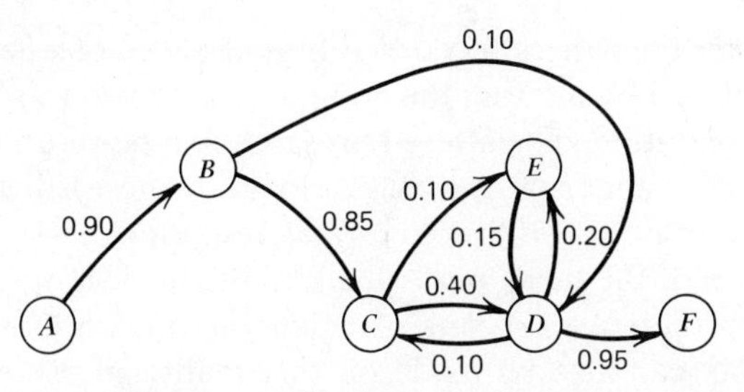

Figure 12-19. A simple activity sequence generator.

assumed is that used by people. Many variations of these models are summarized by McRuer (1980), Sheridan and Ferrel (1974), and Pew (1974).

Another aspect of control strategy is the form and sequence of control actions. Many processes including vehicle control require more than a single control dimension, such as with aircraft and submarines where altitudes (or depths) are controlled as well as the horizontal direction. A microstrategy is the sequence of control actions between horizontal and vertical correction as well as the velocity control devices on these dimensions. *Activity sequence generators* (Miller, Jagacinski, Nalavade, & Johnson, 1981) may be used to describe these microstrategies exhibited by people during the control activities. Figure 12-19 shows an activity-sequence generator where the nodes show the activities and the directed branches denote the next activity with branch probability values. Commonalities between operators' activity sequences can be found from such generators and these commonalities serve as important design information in control layouts or for control-aiding purposes.

Simulation in Process Control

One of the contemporary forms of modeling control actions is through the use of computer simulation. This technique is a form of mathematical model building but instead of using the continuous or sampling forms of mathematical models, computer programs are written that describes the tasks that the humans perform, task sequences, time requirements of the tasks, and other features of logic required in this description. Resulting models do not show a transfer function but rather a detailed computer program or a general computer program and a network of task nodes and branches such as the activity-sequence generator shown in Figure 12-19 and, of course, a lot of data on the model parameter values. When these models are put into a large digital computer, the program may be rerun a large number of times in order to determine statistical measures of system or individual operator's performance. In cases where there are a number of operators in a crew that operates a system or a process, and some operators do some tasks while other operators or machines do other tasks, then the effects of different ways of operating the system can be explored by altering the computer program to reflect the different manner of operation. The altered program can be rerun to find out the differences in the manner of operation. Some of these effects are greater work loads on certain crew operators and less on others. As a result of this effect, some of the tasks performed by the busiest operators can be allocated to the others as a means of balancing human jobs (Buck & Maltas, 1979). Tests can also be made on the quality of system performance with different operating procedures in emergency conditions (Siegel & Wolf, 1969). Other applications are possible as well. The feature that makes computer simulation particularly appealing is that the computer model development can be made in decomposed parts, which are logically connected without concern with all of the interconnecting details of some of the more complex models just described. The computer performs much of the model integration under the imposed logic. A shortcoming of this approach, however, is that subtle assumptions of system operation or individual operator performance can be included in the model that are very difficult to detect and validate.

Advantages and Criticisms of Mathematical Models of Human Operators

Mathematical models of human operators provide such advantages as showing the integration of a human and a machine working together in a way that performance measures can be predicted. In this way various design aids can be investigated for better or worse performance. Also, the models serve as a known reference point for describing relevant features of behavior. However, there is a danger in believing a model that has been inadequately tested because precise results are no substitute for an accurate description. Many of the manual control models, particularly the earlier ones, do not account for how humans filter, identify, and interpret potential information about them. Due to this inadequacy, control models often predict identical performance regardless of the types of visual and auditory display used, whereas human performance typically varies greatly with alternative forms of displays and display formatting of the same available data. Also a large number of the manual control models do not allow for the effects of human memory of similar past situations. An integration of the immediate past is no substitute for a past memory. Human interpretation of previewed information is not fully achieved from the current derivatives of control conditions. Part of this interpretation is affected by the internal representation of the operating system that can only be vaguely mimicked by a mathematical model. There are also times where people display shifts in criteria and behavioral discontinuities that are very difficult to model mathematically. As progress continues, however, some of these criticisms will fade. Even if many criticisms don't fully disappear, an imperfect model is better than none.

TASK MANAGEMENT AND SUPERVISORY CONTROL

The principal forms of human control shown above are basically that of continuously steering a vehicle rather than managing discrete tasks. However, many situations in controlling industrial processes, nuclear or conventional power generation plants, and supervising robots require people to monitor the process, plan, instruct the process, and intervene when necessary. In these cases the human operator performs discrete tasks, some of which are done while a computer keeps the moment-to-moment process operations going. Only when the operator intervenes with the process directly are the human and machine tasks truly synchronous. Figure 12-20 illustrates this situation. During normal operations the process is controlled by the computer as indicated by loop B. An important human task during normal operations is *monitoring* where the operator calls up informational displays, through loop A, of operations going on in loop B. Another human task is *planning,* which uses loop A and the computer offline of direct process control to anticipate future events and to select planned responses to these events. Once this plan is chosen, the operator must *instruct* the computer about the planned response by selecting the appropriate computer program and the event that will trigger that program through loop A. One form of triggering event is by direct operator instruction for the computer to start the program now. In nonnormal operation, the operator can directly *intervene* with the process through loop C. Intervention may occur during emergencies, planned maintenance and repair, or just for a human ego trip. These are modes of human operations where various discrete tasks are selected by the operator from a time stream of tasks that includes the directing of sensors to acquire the needed information and the mechanisms for effecting the chosen actions.

There are numerous roles for the computer in this paradigm as well. The computer selects sensed data, operates on it, and presents it to the human operator. During the operator/computer communications, the computer evaluates commands and suggests alternative responses or plans to the operator. In fact, the computer should request information

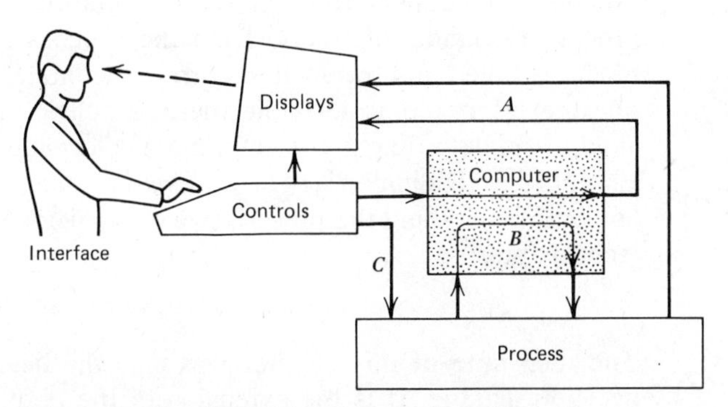

Figure 12-20. Supervisory control paradigm.

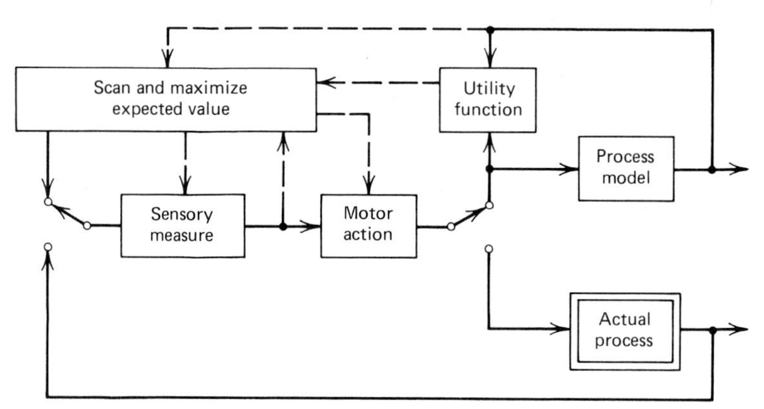

Figure 12-21. Generalized model of expected value seeker.

from the operator at times and tell the operator when it cannot implement an action. At times the computer may need to override an operator's command. In these ways the computer and the human operator become in ways partners in the process control.

The operations of this operator/computer partnership may be viewed under Sheridan's (1976) model of *supervisory control* as shown in Figure 12-21. A decision must be made about the form of sensed information needed to assess the true state of the process. In this model it is assumed that information is selected to maximize the expected value of the information. Alternative motor actions need to also be assessed in light of the state of the process. These alternative actions are evaluated either directly by a utility function (i.e., a combined model of criteria with relative importance measures) or by a model of the process under control and the utility function. The process model serves to describe the implications of alternative actions on the process. In the selection of motor actions it is assumed that an action that maximizes the expected utility will be selected. This selected action is then implemented in the actual process and feedback from this implementation is monitored. If the actual process responds differently from the process model, then the process model is tuned and trimmed to correct the model. Thus goes the concept of supervisory control. While there is considerable appeal to this and similar models (e.g., the Optimum Decision Model—Krishna-Rao, Ephrath, & Kleinman, 1979), there is some question about the behavioral fidelity and the prescriptive compulsion for an elusive optimality (see Chapter 14).

SUMMARY

The final note of this chapter was also the beginning: control is really decision making. It is the extension of the body and mind of people to the systems that they devise. Sometimes this decision making is continuous over time and sometimes it is discrete. But in either case good control requires feedback. In many practical contexts, there are control considerations required at different control levels, for the order of control, the needs in control aiding or automation, and for adaptation of the environment. To address these concerns one must have some behavioral theory of control. As the discussion shows, many such theories exist from servo theory to modern control theory to supervisory control theory. There are useful aspects of these various theories and limitations. Special mathematical techniques are required to use these theories. While it appears that these theories and models are in transition, perhaps they always will be. Human factors engineering must use the best available theories and the most appropriate models at hand to solve problems and design systems while researchers seek to improve the existing theories and models.

13 HUMAN FACTORS IN COMPUTER PROGRAMMING

H. E. DUNSMORE
Department of Computer Sciences
Purdue University
West Lafayette, Indiana 47907

EXPERIMENTAL PROBLEMS

We would like to say that we understand how people interact with computers, how programmers construct computer programs, and how people use computer systems. We would like to say these things, but we can't. Out understanding of the human factors of computer programming is very naive; in this regard we still think the world is flat and the moon is made of green cheese. But, in recent years, we have started to sail to the edges of that flat world and have begun to take some nibbles out of the green cheese.

Since the beginning of computing in the 1940s, most of our efforts have been concentrated in creating powerful hardware, complex operating systems, and versatile programming languages to be used by a *typical* programmer. But the problem has been that we have completely overlooked what the typical programmer is capable of doing. We have envisioned a programmer as an automaton able to use any computer system or computer language regardless of how complicated or how confusing it might be (Figure 13-1). For nearly 30 years few paid much attention to a critical element in the use of computers: computers are used by people—all kinds of people from the cleverest programmers to the simplest of data enterers. What if people are unable to use these facilities intended for them?

In the late 1960s and early 1970s a few psychologists and computer scientists became concerned about the human factors of computing. These researchers began to perform experiments using human subjects to try to discover the best ways to design computer systems and program-

Figure 13-1. The *typical* programmer? Most designers of programming languages and computing systems have failed to consider that human beings have frailties and limitations. (Courtesy of American Telephone & Telegraph.)

ming languages. They were also concerned with the best ways to construct and modify computer programs. This chapter concerns those efforts. First, we discuss briefly some problems that experimenters working in this area have encountered. In general the experimentation problems discussed in Chapter 1 are as much a problem for researchers concerned with human factors in programming as for anyone else working in the human factors area. However, a few problems are especially difficult for experimenters in this area. So, we will take a side trip to discuss them before considering some of the findings of the research.

But, before the side trip, a comment is in order about the state of this research and of our knowledge concerning the human factors of programming. The careful reader will notice that most of the experiments reported in this chapter seem quite naive—rather like placing a raw egg in front of a garbage truck to verify that the egg will, indeed, be squashed. The value of such an experiment may seem meaningless. It will only serve to support a contention about which there is no disagreement, anyway. And, if the egg survives unscathed while wrecking the garbage truck, no one will believe the result.

But, to date, the research concerning the human factors of computer programming has been primarily from the "egg and garbage truck" paradigm. Shortly before this chapter went to press the first Human Factors in Computer Systems Conference was held in March 1982. The research results presented at this conference reflected this tendency. There were papers on command names, documentation, and text editors. But there was clearly no unified effort to attack the most significant problems concerning human use of the computer. In early 1982 computer scientists are just beginning to enlist the aid of cognitive psychologists to help in constructing models, deriving hypotheses, and interpreting results.

Experimental Approaches

The approach taken by researchers considering the human factors of programming has been almost totally empirical in nature. There are no universally accepted theories of programming. There is usually only an interesting question and a desire to gather data to try to answer that question. Once an experiment is conducted, it is not clear how to generalize the findings to other computer systems, other tasks, and other programmers. This empirical experimentation continues primarily because it is easy to do and does not require a model of programming or of computer use (Moran, 1981).

ESTABLISHING SYSTEM GOALS

The Lack of a Standard Software Metric

Several metrics have been proposed for "measuring" software but none has gained universal acceptance as a standard software metric. Some metrics concern time and cost (effort) of software production, some concern effort involved only in debugging, and some are ostensibly only useful for testing, modification, or determining the reliability of software. Computer scientists are still very far from any kind of agreement on what kinds of metrics we need and which (if any) of the current candidates should be used universally.

The world-famous, best-known (and perhaps least useful) software metric is "lines of code" (or LOC) (Jones, 1978). A computer program is a set of instructions. Generally each instruction (something like "Add A to B and put the result in C") is a code written on one line. Some instructions are very long and have to be continued to additional lines. Some lines are not instructions at all, but comments describing what the instructions do. And sometimes programmers insert blank lines to set groups of instructions apart from others. There are many suggestions about how to compute the measure, but most would agree that a reasonable LOC would be a count of all the lines in a program excluding comments, statement continuations, and blank lines. The hypothesis is that LOC is related to the cost (time) of software production, but no one has ever been able to offer conclusive evidence to support this. In fact, anecdotal evidence thrives to the contrary that nearly equal-sized (in terms of LOC) programs may be wildly different in programming times and costs.

The "cyclomatic complexity" [referred to as $v(G)$] of a program was originally proposed as a software metric related to testing complexity (McCabe, 1976). This measure requires a graphical representation of a program and is a function of vertices, edges, and connected components. In theory, $v(G)$ is the number of distinct paths in a program (i.e., all the unique ways one can proceed from the start to the end of the program). Experimental results have not led computer scientists to the conclusion that this metric is a good way of predicting effort.

Software science is a theory and family of metrics proposed by Halstead (1977). From it comes an effort measure E, which is a function of the operators (+, −, etc.) and operands (variables, statement numbers, etc.) used in the program. It is claimed that E can be converted to an estimate of programming time (in seconds) by dividing by the number 18. Several experiments have shown software science measures to work moderately well in some cases in explaining time and effort (Shen, Conte, & Dunsmore, 1982). But the theory behind several of the measures troubles computer scientists. Thus, software science has not become the standard software metric.

But, if experimentation is conducted entirely this way there are some serious problems: What experiments should we do? How can we (or even, can we) generalize the results? Does any one experiment tell us anything in general about the human factors of programming? Empirical work is better if it grows out of some models of human-computer interaction and if there is a real effort on the part of the experimenter to set up an experiment with tasks and subjects that are representative of a broader class of situations. Experimentation concerning the human factors of computer programming is one of the newest horses in the human factors stable. To date the major thing lacking are some theoretical models to guide experimentation and to help interpret the results we have gotten.

Subjects

In addition to the usual problems acquiring representative subjects as discussed in Chapter 6, there is a potentially more serious problem in the human-computer area. The empirical process works best when subjects are all about the same. (We're back to the automata, again. They would make great experimental subjects and we wouldn't have to break for lunch.) But programmers are not all the same. They come in all levels of ability from neophytes who can spend weeks constructing the simplest computer programs to grizzled veteran programmers who can knock out a tough program before their first coffee break. So, if we're to use representative subjects, whom should they represent? And if we try to represent everybody, there's no hope for using subjects who are nearly equivalent.

The compromise has been to use college students in most of the research. They represent a group of fairly equivalent subjects. But in some cases findings based on experiments with them may not say very much about what professional programmers do on the job (Brooks, 1980, and Figure 13-2). Results with college programmers probably do tell us something about novice programmers on the job and can be useful for understanding nonprogrammers (especially if we grab subjects out of college majors other than computer science).

Performance measure	Poorest score	Best score	Ratio
1. Debug hours Algebra	170	6	28:1
2. Dubug hours Maze	26	1	26:1
3. CPU time Algebra (sec)	3075	370	8:1
4. CPU time Maze (sec)	541	50	11:1
5. Code hours Algebra	111	7	16:1
6. Code hours Maze	50	2	25:1
7. Program size Algebra	6137	1050	6:1
8. Program size Maze	3287	651	5:1
9. Run time Algebra (sec)	7.9	1.6	5:1
10. Run time Maze (sec)	8.0	.6	13:1

Figure 13-2. It is fairly well-known that programmers can be vastly different in abilities. This table which appeared in one of the earliest Human-Computer papers (Sackman et al., 1968) shows some incredible ratios of "best" and "poorest" performances of a group of programmers. It is because of this table that computer professionals generally claim a 20:1 difference in their best and worst employees.

Tasks

Even if we could pull our subjects out of a hat and have them be perfectly representative of the people to whom we want to generalize, we face another problem. What should we have them do? Should they construct a 10,000-statement computer program? (That could take weeks and some of them may not want to stay with us that long.) Should they answer questions about computer programs that have already been constructed by someone else? (What would our results tell us about what they would do if they had to *construct* that program? Anything at all?)

There are different task types that we can require of our subjects when we experiment in the programming arena. A (not necessarily exhaustive) list appears as follows.

Construction—construct a computer program.
Comprehension—answer questions about an existing program.
Debugging—find errors in an existing program.
Maintenance—make modifications in an existing program.

These tasks can be used in experiments that either mimic the programming process with subjects using a computer and trying to get something to work, or they can be used in a "classroom" situation where computers are not present. Classroom experiments may tell us something about how people interact with computers, but probably don't tell us as much as experiments in which the subjects really have to use computers.

Furthermore, within each of these task types there are questions about how large and complex the software should be. Remember the question about whether our subjects should construct a 10,000-statement computer program? If that takes too long, why not just have each construct a 1000-statement program? But, we suspect that a 10,000-statement program is not just 10 times harder than a 1000-statement program (Brooks, 1980). We aren't ever sure if the findings will generalize.

Practice Effects

Practice effects are generally not considered in programming experiments (Sheil, 1981). For one thing, finding out about them would be much more expensive than what we find from the experiments that are done. Also, people seem to be able to adapt to almost anything . . . regardless of how awkward it may seem at first. Thus, a programmer forced to use a computer terminal at which all characters appear in Old English font would probably be annoyed (and less productive) at first, but sometime later would probably be oblivious to the whole matter (Figure 13-3). On the other hand there are some effects that don't change with practice (e.g., a computer terminal at which the characters don't print at all). In such cases the effect on performance should be nearly the same regardless of when the empirical measurement takes place.

In the remainder of this chapter you will find that we are learning that the programming world is not flat (but we still don't know what shape

```
IF HOURS ARE GREATER THAN 40
THEN
SUBTRACT 40 FROM HOURS
MULTIPLY OVER-HOURS BY 1.5
ELSE
MOVE 0 TO OVER-TIME
END
```

Figure 13-3. Programming work would seem to be extremely difficult with output in Old English. The intriguing thing is that most programmers can adapt to almost anything after some initial grousing. The real question is "Will their performance be appreciably affected?"

Figure 13-4. Imagine trying to describe a beach to someone who has no idea what sand is like. The empirical results that we have concerning human-computer interaction are not complete enough yet, but they're much better than no results at all. (Joel Gordon.)

it is exactly) and that the programming moon is not made of green cheese (but things still seem a little moldy). In short, what we have learned is like taking home a bucket of sand scooped up from a beach. We can't spread it out on a table and have the homefolks understand exactly what the beach is like, but it's a lot better than trying to understand it with no sand at all (Figure 13-4).

PROGRAMMING TECHNIQUES

An important facet of human-computer interaction is the way programmers attack the task of constructing computer software. In this section we consider some of the different ways of accomplishing this task.

On-line or Off-line Processing

The earliest use of computers was done necessarily via something called the *off-line* mode in which programmers wrote their programs at their desks, punched them into cards, and submitted the cards to the computer. Sometime later (often much later) the programmer was able to receive a printout listing the program and what the computer system had done with it. Often the computer might have balked at the syntax of the program statements or, if these were all right, the program might have been run and fallen apart due to more serious logic errors. For example, "ADD 1 TO X" is not the way to request that X be incremented by 1 in the Fortran programming language. "X = X + 1" is the way to do it. So "ADD 1 TO X" is a *syntax* error, but "X = X + 1" (which is all right syntactically) is a *logic* error if you really mean to add 2. In the off-line mode, programmers retreated to their desks with their printouts, decided what was wrong with their programs, changed them, and resubmitted them.

Most programmers welcomed the new world of the late 1960s and early 1970s of *on-line* processing in which the programmer sits at a computer terminal, enters programs, views output nearly immediately, and continues to work without interruption until programs are completed. But, this new world was not without its detractors. Some claimed that programmers would be encouraged to be sloppy (most needed no encouragement; they were sloppy anyway). The claim was that instead of carefully considering all probable errors, the programmer might only correct one and resubmit to see if that one was corrected before tackling the next. The interactive debugging process (which is fairly typical these days) seemed a bit revolutionary in the late 1960s and terrified some people because they foresaw a lot of computer time being eaten up doing those things that programmers might do at their desks by hand. But, in the interim, the cost of computing has decreased and the expense of paying programmers has skyrocketed. Thus, we are currently concerned with doing anything we can (within reason) to save programmer time at the expense of computer time.

Experiments were conducted (Sackman, Erikson, & Grant, 1968) in which programmers were each required to develop one program in the off-line and one program in the on-line mode. These experiments involved some experienced programmers and some programmer trainees (Figure 13-5). The results for the two groups were similar. On-line program con-

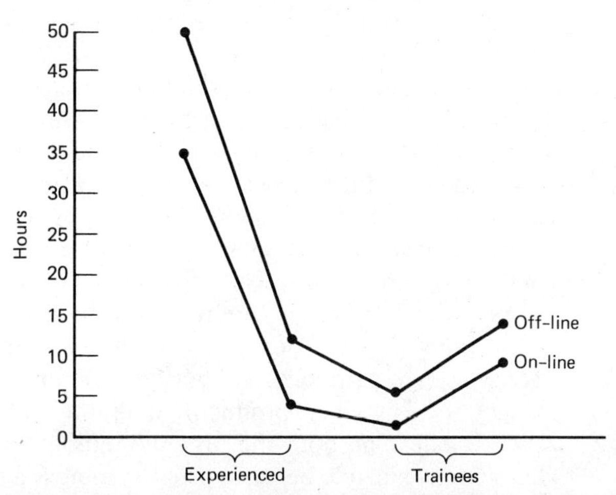

Figure 13-5. In the studies conducted by Sackman et al. (1968) the actual hours of work were less for on-line processing than off-line processing. This result occurred for both experienced and trainee programmers.

struction took no more than two-thirds the time that off-line construction took. Furthermore, this increased performance was accomplished at little (or no) increase in machine processing time.

On-line processing is probably more efficient because the programmer is able to continue to work continuously on the same task until it is completed or until some natural stopping point is achieved. With off-line processing the programmer is forced to take long gaps between thinking (and doing something about) the programming task. This is inefficient because of the "startup" overhead that has to be expended each time before getting into a period of productive work.

Specifications

To guide programmers in the production of computer software they are generally given a set of *specifications* outlining *what* the software is to do (although usually not *how* it is to do it). The way these specifications are written can have a major effect on the software produced. Obviously, this is what we want to happen. We don't want to give a programmer specifications for a payroll system and get an accounts receivable system.

But specifications can convey some other notions as well concerning the mode of constructing the software or the way we want the program to act when it is in operation. Furthermore, specifications are not always confined to paper. Directives given verbally by one's supervisor (e.g., "Be sure to check for employee numbers that are not valid.") take on the same degree of (or maybe more) importance as anything printed on paper.

One way to investigate this phenomenon is to give programmers nearly identical tasks and to observe whether the differences in specifications lead to differences in performance. In such a study (Weinberg & Schulman, 1974) two groups of programmers were told to construct a program to prepare a book index using several sorting and formatting rules. Both received identical specifications of about 25 pages except for the last page. One group ("Fast Programming" treatment) was told on that page "to get a fully debugged program in as short a time as possible." The other group ("Efficient Program" treatment) was directed "to get a fully debugged program which is as efficient as possible." The subjects were able to meet the requirements imposed on them (Figure 13-6). Notice that the "Fast Programming" group took far fewer runs than the "Efficient Program" group, while the latter produced programs that in general ran with test data in one tenth the time as those programs produced by the other group. Thus, our model of programming as a very human activity seems strongly supported. Different considerations can be brought to bear when producing computer programs. They can have a dramatic effect on both the programming activity and the final product. This seems plausible because programming is a process in which different methods can lead to largely different behaviors and products that try to pass for "equivalent."

	Mean Number of Runs	Mean Scaled Execution Time
Fast programming	30.5	10
Efficient program	78	1

Figure 13-6. Weinberg and Schulman (1974) found that the "Fast Programming" group required fewer runs and the "Efficient Program" group produced more efficient programs.

In a follow-up study reported by the same experimenters, they found that in addition to "program this as quickly as possible" and "write a program that runs as efficiently as possible," programmers seem to be able to satisfy other requests to do such things as "write a program that produces very easily readable output" and "write a program that takes as little memory space as possible." You may be thinking, "Why not just require all those kinds of specifications for every program?" The problem is that some of them are nearly opposites of others. Weinberg and Schulman found that a group whose primary objective was "minimum programming hours" and secondary objective was "program readability" beat all the other groups in the time to produce a program but had the *most* unreadable program of the lot. And, the group whose primary goal was "minimum execution time" and who tried also to satisfy the secondary goal of "minimum programming hours" took almost the longest time to produce their fast-running program. Thus, supervisors must be very careful that the specifications their programmers are acting under are the ones they want them to use—taking into consideration both the written and verbal specifications that they have (or think they have) given.

Flowcharts

One programming technique that enjoyed a good deal of favor in the early days of computing is the *flowchart* (Figure 13-7). The flowchart is a graphical representation of program structure. It shows how parts of the program are related to other parts, how some parts may be executed conditionally, and how some parts are to be reiterated until some condition is met. A flowchart can be constructed at quite a high level showing only how large parts of a program are related to other large parts, or it can be very detailed reflecting nearly a statement-by-statement account of the program.

Until about 1975, most programmers were taught to construct a flowchart from the program specification and to use it to guide program construction. The benefits they were told were many—making program construction easier via a flowchart, making it easier to understand a program written by someone else (a so-called "alien" program) if that

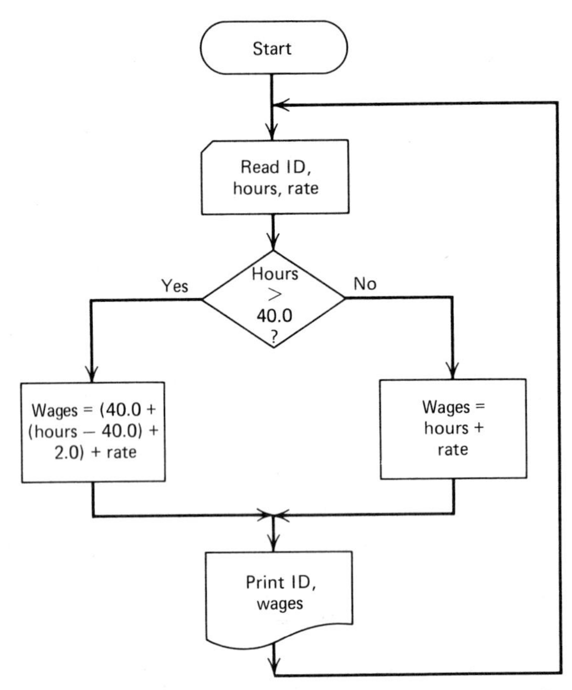

Figure 13-7. This is an example of a fairly detailed flowchart. Notice that the card-shaped object represents input to the program. The diamond is a decision point. Arrows show the way the program executes its various parts. The curly bottom box indicates output from the program. (From Kreitzderg, C. R. and Shneiderman, B., 1975. Copyright Harcourt Brace Jovanovich used with permission.)

someone else had provided a flowchart, and making it easier to modify a program that they or someone else had constructed if a flowchart could be referred to before and during modification. On the other hand, programmers privately questioned the utility of detailed flowcharts—almost never using them to guide program construction, and cursing the fact that before submitting a completed program a flowchart had to be constructed.

Shneiderman, Mayer, McKay, and Heller (1977) investigated flowcharting in a series of experiments involving hundreds of university students and faculty. As Figure 13-8 shows there were no instances to suggest that *detailed* flowcharts are of any value for construction, debugging, comprehension, or modification. The mean scores for subjects doing any of those tasks seemed to be no better or worse with a flowchart than without it. Brooke and Duncan (1980) noted that the subjects in the Shneiderman experiment were not *required* to use flowcharts and suggested that perhaps the results were equivalent because the subjects with the flowcharts simply ignored them. So, Brooke and Duncan replicated the flowchart experiment with the twist that subjects who received a flowchart received nothing else. They could not "cheat" and ignore the flowchart in favor of the program. Their results were nearly

		Mean Scores	
Experiment	Activity	Flowchart	No Flowchart
I	Composition	94	95
II	Comprehension	94	93
III	Comprehension and debugging	39	38
IV	Modification	77	75
V	Comprehension	52	60

Figure 13-8. Shneiderman et al. (1977) could find no activity in which their subjects performed better with flowcharts. The mean scores were nearly the same in every instance.

identical to Shneiderman et al.'s; the flowcharts led to no improved performance in finding errors by either naive or experienced subjects.

These results occur probably because a detailed flowchart is simply another way of stating what is in the program code itself. This is roughly equivalent to studying a book in English and a translation in French. If one understands the English book, the translation is of very little benefit. On the other hand, since the French translation simply repeats what is in the original, it probably is of no help to someone who cannot understand the English version.

Thus, by about the mid 1970s we were no longer teaching the use of flowcharts with the missionary zeal that we had and the government and private sector were gradually dropping their insistence on the use of flowcharts. No one has attempted to determine experimentally whether higher-level flowcharts are valuable in planning or understanding a very large program or a system of programs. This is generally where and how flowcharts are being used today.

Pretty-Printing

The programming technique *pretty-printing* is sometimes known as indenting or paragraphing (Figure 13-9). Whatever one calls it, this is the use of blank lines, spacing, and indentation to make programs easier to read (and, it is assumed, easier to construct, debug, comprehend, and modify). Consider the relationship between flowcharting and pretty-printing. Flowcharts are separate representations of programs that were thought to make it easier to work with the programs, while pretty-printing

```
IF HOURS ARE GREATER THAN 40 THEN
SUBTRACT 40 FROM HOURS GIVING OVER-HOURS
MULTIPLY OVER-HOURS BY 1.5 GIVING OVER-TIME
ELSE MOVE 0 TO OVER-TIME END
```

Figure 13-9. This program segment is the same as the one that appears in Figure 13-15. But, that one is pretty-printed and this one is not. Don't you think it's easier to understand the pretty-printed version?

	Mean Scores for Finding Errors	
	Binary Search	Merge
Pretty-printed	13.6	9.2
Not pretty-printed	13.5	7.6

Figure 13-10. Shneiderman (1980) asked people to find errors in pretty-printed programs. Did they do better than those using "ugly"-printed programs? If so, only barely.

is a technique used right within the program itself to try to achieve the same goal.

To investigate the usefulness of pretty-printing, Shneiderman (1980) required subjects to find errors in two programs—one of which was presented with nice indentation and the other with no indentation whatsoever. The results were inconclusive (Figure 13-10). Although the means slightly favored pretty-printing (very, very slightly in the case of the binary search), the differences were by no means significant. Similar results occurred in experiments conducted by Weissman (1974) and Love (1977). However, in Weissman's work, subjects indicated that they preferred the pretty-printed versions to the others. Norcio (1982) asked subjects to study programs and then to supply "missing" statements (i.e., where their programs contained blank lines). He found that the combination of pretty-printing and comments led to significantly more correct guesses.

Comments are statements in a program that are not instructions but describe what some instruction statements do. Mnemonic terms are words that sound and look like what they stand for (e.g., SALARY as a name for someone's salary). It may be that pretty-printing, flowcharts, comments, and mnemonic terms are all providing basically the same type of information and that in the absence of any of the others, any one might be useful. But, all together there may be instances when they create confusion rather than aid in comprehension. For example, Weissman found that performance was the *worst* for programs that were both pretty-printed and commented.

Structured Walkthroughs

A programming technique that is becoming increasingly popular is the *structured walkthrough.* This is a by-product of the "egoless" programming movement (Weinberg, 1971) and team programming concept in which several programmers work together to construct software. Before this movement, programming was a very personal activity akin to alchemy in which programmers went off into their private cubicles to conjure up software. In such times programming was rather a mystical art which, when it worked, was viewed with awe, and when it didn't, was viewed with disdain. "Programming must cease being an art and become an engineering discipline" was the frequent anguished cry, and computer scientists responded by creating the discipline of "software engineering." This is a set of rules and techniques that attempt to standardize the programming process.

During the program construction process (or modification process, for that matter) when programmers are organized in groups (the "team" concept), they can help each other via structured walkthroughs. In such cases one programmer describes to the others in her team what a particular piece of software is supposed to do, and the other team members read through the software trying to find loopholes, unanticipated problems, or more serious logic errors. Why can't the programmer who constructed the software find those things herself? Because the more intimate one becomes with a computer program the more easy it is for one to overlook subtle or obvious problems that someone unfamiliar with the software (and trying to understand it) would question. In the course of answering those questions, the originator of the program often finds that the other person is having trouble understanding it because it doesn't do what it's supposed to do.

Myers (1978) required some professional programmers to debug a small program into which he had seeded some errors. Some subjects used a structured walkthrough. Others received a program specification (i.e., a written description of what the program was supposed to do). Others received both the specification and the program code (i.e., the list of program statements). The mean number of errors found (5.7) was better using the structured walkthrough technique than the two other techniques: computer-based testing using only a specification (4.5) and computer-based testing using both specification and code (5.4). Furthermore, Lemos (1979) found that training in the structured walkthrough methodology leads to better performance by the people involved (Figure 13-11). Programmers who had been trained to develop Cobol programs

	Adjusted Mean Exam Scores	
	Reading	Writing
Structured walkthrough subjects	115	132
No training in structured walkthroughs	86	107

Figure 13-11. Those people who had learned Cobol programming by the structured walkthrough technique did much better reading and writing programs on their final exam in Lemos' (1979) study.

using the structured walkthrough technique scored about 25% higher on a final exam testing their ability to read (and comprehend) and to write Cobol programs than programmers in a control group. Programmers trained in this technique probably perform better because they become accustomed to looking at code in a critical light—trying to understand what a program really will do rather than what it purports to do.

Debugging Aids

The debugging process (i.e., removing errors from software) is one of the hardest tasks a programmer faces. Designing and coding software are pleasant tasks (something akin to writing a symphony or painting a pastoral scene). But, if the computer program doesn't work correctly, then the programmer is just as crestfallen as the composer who learns that his orchestra does not have trumpets or the painter who finds that her oils have run into little pools at the bottom of the canvas. So, it would be good to be able to provide debugging aids that would help remove some of the drudgery associated with this phase of the programming process.

In a study of the difficulty of finding different types of bugs and the usefulness of several debugging aids (Gould & Drongowski, 1974), experienced programmers were asked to find errors in several single-page Fortran programs. The programs were seeded with three classes of bugs: (1) array bugs—in which references are made to members of arrays that cannot exist (an example would be referring to X(11) when X is an array with only 10 members), (2) iteration bugs—in which loops are executed an inappropriate number of times (e.g., going through a set of statements N times when it should really be N-1), and (3) assignment statement bugs—in which variables are given incorrect values (the problem we discussed earlier of having the statement "X = X + 1" if you really mean to add 2).

THE BUG

A popular term used in programming is the word "bug." This refers to an error in computer software. It can be a "syntactic" bug (a violation of the rules for writing statements in the programming language), or a "semantic" bug (a syntactically correct statement that doesn't do what the programmer intended).

Some bugs are found during program development, others during testing, a few after the software goes into production, and some may never be found at all. (As an example of the latter, suppose a credit card program produces gibberish whenever anyone owes $9999.99. But if no one ever owes exactly $9999.99, then the bug will never be found.)

The origin of the term "bug" is debated, but it is believed to have originated with Dr. Grace Murray Hopper, one of the pioneers of computing and the first recipient of the Data Processing Management Association's "Man of the Year" Award (which she graciously accepted in spite of the title). Working on one of the first large-scale computers (the MARK II) in the summer of 1945, she found a real insect (a large moth) that had been beaten to death in a relay and that had caused the Mark II to stop. She removed the victim with a pair of tweezers and carefully taped it into the logbook. Later, when asked why the computer was not running, she announced that she had been removing a "bug." "Debugging" is a term that has been used by millions of programmers ever since.

As a footnote to history, that original bug may still be found in the logbook in the Naval Museum in the Naval Surface Weapons Center, Dahlgren, Virginia (Hopper, 1981).

Subjects were randomly assigned to one of five debugging aid groups:

1. Control—no debugging aid beyond the program listing.
2. Input/output—in addition to the program listing a listing of input data and program output.
3. Input/output + correct—same as (2) plus the output that would have resulted if the program had run correctly.
4. Class of bug—these subjects were given the program listing only and told whether it contained an array, iteration, or assignment statement bug.
5. Line number—these subjects were given the program listing only and told in which line the error was.

Figure 13-12 indicates that the most difficult type of error to find was the assignment statement bug. It took about four times as long to find them and the number of assignment bugs not found at all was more than four

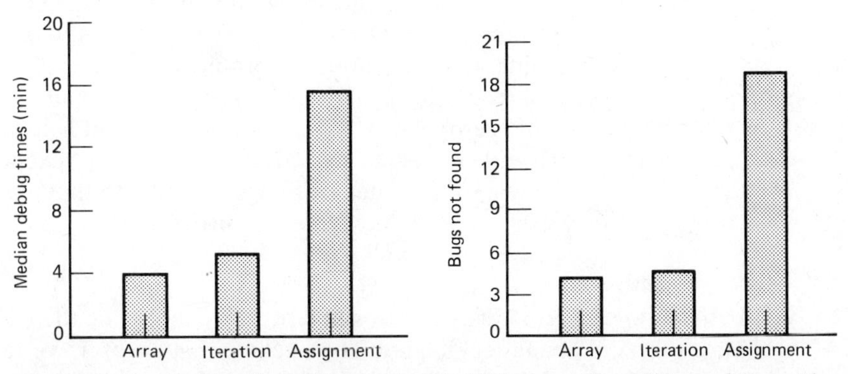

Figure 13-12. In Gould and Drongowski's (1974) experiment the assignment bug was more resistant to being discovered than the array or iteration bug.

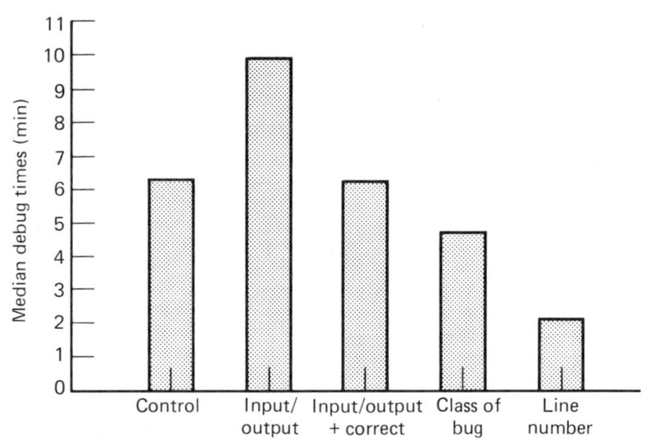

Figure 13-13. Gould and Drongowski (1974) found only one debugging aid that really helped—identifying the line number in which the bug occurred.

times the number of array or iteration bugs. This is probably because comments and redundancy don't help much in finding assignment bugs (you'll read about comments and redundancy later in this chapter) and the effect of assignment bugs seems to be more localized to a single statement. Finding array bugs can be greatly aided by considering statements where array sizes are declared and iteration bugs frequently make a large segment of code seem suspect.

Figure 13-13 has some striking results. Notice that in terms of median debug times that the control group did as well (or better) than three of the treatment groups. Only the line number group (which was told in which line the error resided) performed significantly better than the group with only a listing. Thus, this study found no debugging aids superior to pointing out the statement in which the error occurs.

The problem is that "pointing out the statement in which the error occurs" is sometimes very difficult to do, especially for logic errors. For example, the fact that a program attempts to divide by X, which is zero in line 475, may not be a problem with line 475. Instead, it may reflect a problem with line 452 where X was computed (and given the insidious value zero). Thus, any debugging aids that help programmers isolate such information (such as aids that trace the "execution" of a program by showing the values of things every time they change) might be valuable in the program development process.

PROGRAMMING LANGUAGES

Now, we've seen some results concerning the human factors of programming techniques. To conclude this chapter, we're going to consider some of the features of programming languages that experiments suggest affect the programming process.

Structured Transfer of Control

One of the fallouts of the software engineering movement mentioned above is a programming technique known as *structured programming.* The major characteristic of this technique is that it places some very strict rules on how control may be transferred within a computer program. Rather than allowing pieces of software to bring other pieces into execution willy-nilly, structured programming requires that statements be executed sequentially, or when nonsequentially by carefully constructed if-then-else or while-do constructs. Figure 13-14 graphically shows the allowable sequence, choice (if-then-else), and repetition (while-do). The diamond-shaped boxes represent decision points. In the "choice" construct "if" the decision question is true "then" the statements on the left are executed "else" the statements on the right branch are executed. In the "repetition" construct "while" the decision question is true, the program continues to "do" (i.e., execute) the statements directly below it.

One aspect of structured programming is its requirement that forward branching be constrained to the very strict if-then-else process. That is, at any point in a program if a section of the code is to be skipped, it is done by examining the truth or falsehood of a Boolean expression (e.g., HOURS ARE GREATER THAN 40). If the expression is true given the current value of HOURS, then a set of predetermined statements (known as the THEN clause) are executed—probably those that compute one's pay with overtime. But, if the expression is false, then another set of statements (the ELSE clause)—probably having

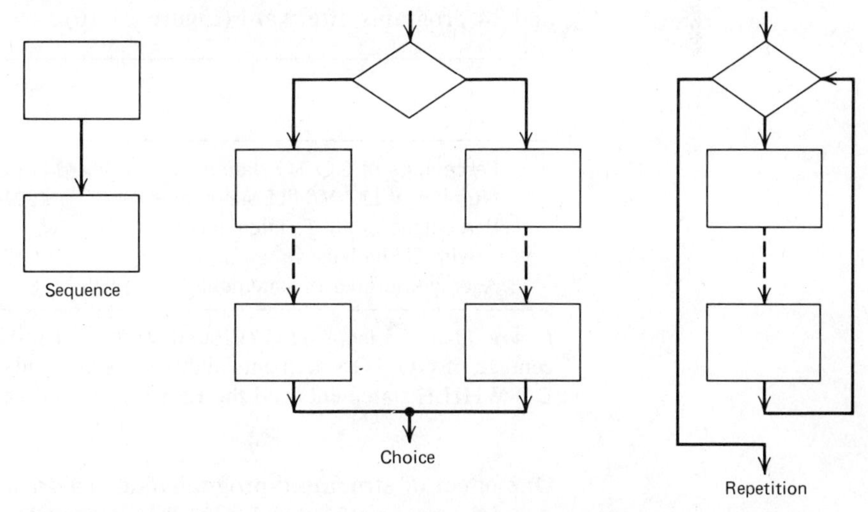

Figure 13-14. Structured programming allows only three basic control constructs: sequence (one statement always follows another), choice (one group of statements or another are executed dependent on the testing of a condition), and repetition (a group of statements are executed over and over until a condition is met).

to do with pay without overtime – are executed. After either case the program is resumed at some point that is independent of which branch was taken. This means of transfer of control is thus known as "branch and join." (Figure 13-15)

```
IF HOURS ARE GREATER THAN 40
  THEN
    SUBTRACT 40 FROM HOURS GIVING OVER-HOURS
    MULTIPLY OVER-HOURS BY 1.5 GIVING OVER-TIME
  ELSE
    MOVE 0 TO OVER-TIME
  END
```

Figure 13-15. This is an example of a structured IF-THEN-ELSE. Depending upon the current value of HOURS either the THEN or the ELSE part will be executed.

But, programmers do not always program like this. Some programming languages allow much more creative (and dangerous!) transfers of control. These are called "test and branch" situations because if a Boolean expression is true, the programmer may transfer control to any other point in the program without regard to whether (or how-ever) it will ever get to the rest of the program.

Elshoff (1977) collected 120 programs written by professional programmers before structured programming concepts were taught to them and 34 programs afterward (Figure 13-16).

	Nonstructured Programs	Structured Programs
Percentage of GO TO statements	11.7	2.8
Number of DO-WHILE statements	11	109
Percentage of all IF statements with ELSE clauses	17.0	36.4
Average number of statements	853	593

Figure 13-16. Elshoff (1977) found structured programs led to a smaller percentage of GO TO statements and fewer statements while increasing the use of DO-WHILE statements and the ELSE clause in the IF statement.

One effect of structured programming is a drastic decrease in the unconditional transfer of control (GO TO) statement that is often used for producing convoluted control structures; their usage drops by 75%. There is also a big increase in the use of the DO-WHILE construct and the ELSE clause. The former is used for establishing structured loops. Finally, the structured programs are more succinct than the nonstructured ones. On the other hand, these were different programs and cannot be compared directly. But, they were representative of work being done by professional programmers. There is no reason to believe that the structured programs had fewer statements because they were inherently simpler. Instead, the conclusion is that structured programming techniques have led to more succinct programs with less-convoluted control structure.

Furthermore, in an experiment in which 36 professional programmers attempted to understand programs written in both structured and nonstructured Fortran (Sheppard, Borst, & Curtis, 1978), the mean comprehension score for the structured programs (56%) exceeded that of the nonstructured ones (42%).

In experiments concerned with only the transfer of control aspect of structured programming (Sime, Green, & Guest, 1973), 18 subjects were asked to write programs using two languages.

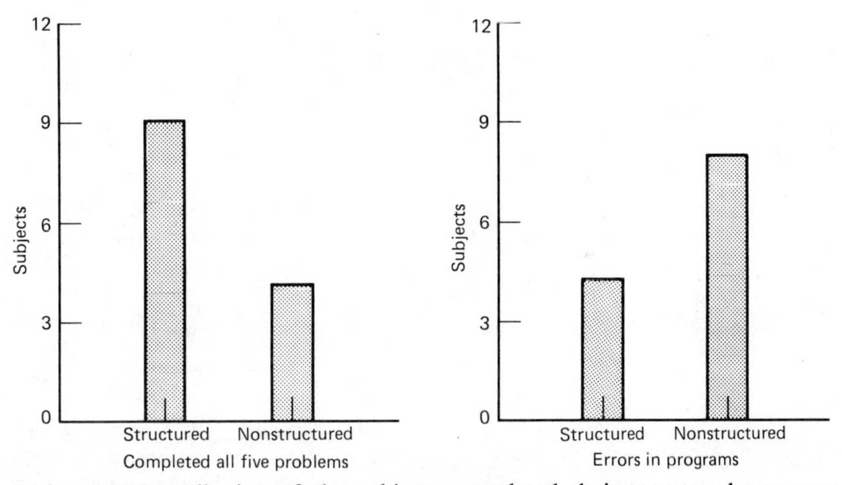

Figure 13-17. All nine of the subjects completed their structured programs. Also, subjects committed fewer errors with the structured language in the study conducted by Sime et al. (1973).

All 9 of the subjects who used the structured language solved all five problems assigned to them, and only 4 subjects had any problem with a program not running as originally written (Figure 13-17). On the other hand, only 4 (of the 9) subjects using the nonstructured language were able to complete all five assigned problems, and 8 of these subjects had problems with program errors. In addition (considering only completed, correct solutions), structured programs were written more quickly than nonstructured ones. In the second week the 18 subjects used the language they had *not* used the week before. Those who used the structured language the second week were faster solving the same five problems, while those who used the nonstructured language were slower. Since the first week could be considered as practice for the second week (and people should naturally get better at a task with practice), this latter result really points out the inefficiency of the nonstructured language. Programming is complicated enough without adding layers of complication to those that already exist, so it is not surprising that programmers typically perform better when using structured conditionals.

In later research Green (1977) investigated a syntactical variation of the IF-THEN-ELSE construction (see Figure 13-18) in which the predicate appears more than once.

```
IF HOURS ARE GREATER THAN 40:
    SUBTRACT 40 FROM HOURS GIVING OVER-HOURS
    MULTIPLY OVER-HOURS BY 1.5 GIVING OVER-TIME
NOT HOURS ARE GREATER THAN 40:
    MOVE 0 TO OVER-TIME
END HOURS ARE GREATER THAN 40
```

Figure 13-18. This is the IF-NOT-END version of the IF-THEN-ELSE as proposed by Green (1977). Notice that the predicate (HOURS ARE GREATER THAN 40) appears three times.

He found that when figuring out the conditions to make a program perform a designated action, mean response times were about 6% shorter with the IF-NOT-END variation than with the standard IF-THEN-ELSE form.

Modularity

Most programming languages (Cobol is a notable exception) provide easy facilities for breaking up a complex program into several subprograms called *modules*. Modularization actually began as a response to small computer memories. Only a subset of the entire program could be kept in memory at a time. But, now that computer memories are (virtually) unlimited, this behavior persists because it has been found to be an effective way of organizing a large computer program. Generally, what is included in each module is a segment of code that deals with one function (e.g., invert a matrix) or one collection of data (e.g., a large airline reservation table). The process of modularizing allows the original programmer to segment his or her thinking so that each module deals only with some mentally manageable subset of the problem the entire program is trying to solve. Furthermore, when a program is modularized, it is easier to comprehend segments of it later. This is particularly useful when performing modifications and not wanting to have to understand the entire program just to change some small part of it.

There is general agreement that modularization is useful for constructing and comprehending software. However, it is possible that overmodularization may actually be detrimental. That is, a 1000-statement program divided into four 250-statement routines may be easier to understand. But, if it's divided into 1000 one-statement routines (a ridiculous extreme), then comprehension couldn't possibly be any easier.

The usefulness of modularity and the potential problem with overmodularity are supported by two experiments. Meldrum, Bowles, Brown, and Szanto (1979) produced three versions of a Fortran program (each about 100 statements long).

Unmodularized—the entire program was written as one routine.
Partially modularized—the program was broken up into a "moderate" number of subroutines.
Super modularized—the program was broken up into more than twice the number of subroutines as the partially modularized version. An attempt was made to make this "super" decomposition as natural (albeit overdone) as possible.

Twenty-six subjects were assigned randomly to each of the three treatments. The median score for a comprehension test for the partially modularized version (31.5) was significantly better than the medians (both were 19) for the unmodularized and super-modularized versions (Figure 13-19).

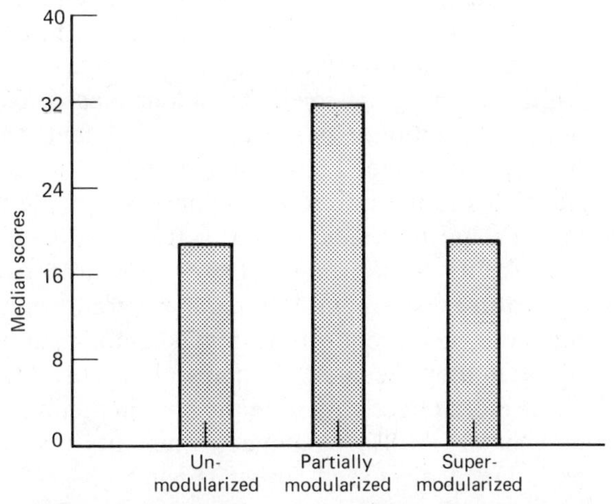

Figure 13-19. Meldrum et al. (1979) found that comprehension was the best for a program broken up into a "moderate" number of subroutines.

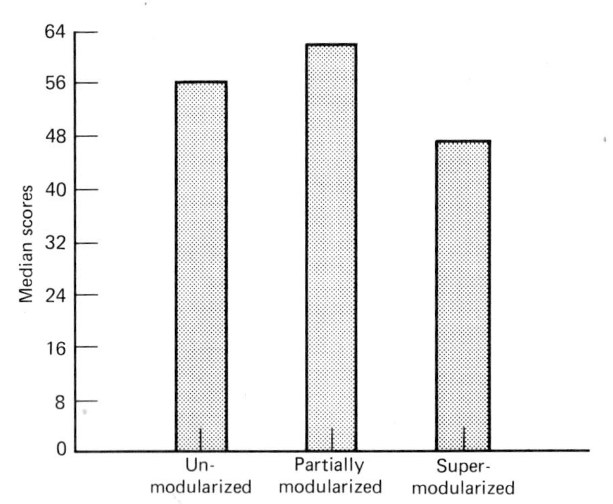

Figure 13-20. Working with larger programs and more subjects than Meldrum et al., Woodfield et al. (1981) also found comprehension of the program was better for the partially modularized versions.

Woodfield, Dunsmore, and Shen (1981) conducted a similar comprehension experiment with a program nearly twice the size and with nearly twice the number of subjects as the Meldrum et al. experiment. (*Note.* Woodfield actually used two types of partial modularization. The median reported here is the median of the two types.) His research led to similar results—with a 62 median for partial modularization, 56 for unmodularized, and 47 for super modularization (Figure 13-20).

Comments

When we discussed flowcharts and pretty-printing it was pointed out that these are two ways of attempting to aid in comprehension of what the program does. Pretty-printing is a technique that may be employed right within the program itself. Most languages provide a facility that can aid in pretty-printing. *Comments* are statements that are interspersed within the program code. These are ignored by the compiler (i.e., the thing that translates a program into machine-executable form), but comments may be seen by the reader of the software (Figure 13-21). Hubka, Baechle, Miller, and Parthasarathy (1979) prepared two versions of a small (27 executable statements) Fortran program with and without comments. Subjects were requested to make modifications to the version of the program they received. As shown in Figure 13-22 the modifications were made more effectively with the commented version. Furthermore, Woodfield et al. (1981) examined comments (as well as modularity) in the experiment discussed earlier (Figure 13-23). Half the subjects received commented versions with short comments describing the function of each module. Half received uncommented programs. The comprehension scores were significantly higher for the commented programs.

Fortran comment—

```
C  COMPUTE OVER-TIME AS TIME AND A HALF OF ALL THE HOURS
   OVER 40
C  AND ZERO IF THE EMPLOYEE WORKS 40 HOURS OR LESS
```

Cobol comment—

```
NOTE COMPUTE OVER-TIME AS TIME AND A HALF OF ALL THE HOURS
OVER 40 AND ZERO IF THE EMPLOYEE WORKS 40 HOURS OR LESS.
```

PL/1 comment—

```
/*  COMPUTE OVER-TIME AS TIME AND A HALF OF ALL THE HOURS
    OVER 40 AND ZERO IF THE EMPLOYEE WORKS 40 HOURS OR LESS  */
```

Figure 13-21. **These comments could be placed right into Fortran, Cobol, and PL/1 programs (respectively) to aid the reader in understanding what the adjacent program statements are trying to accomplish.**

	Mean Modification Scores
Comments	52.0
No comments	42.7

Figure 13-22. **In two small Fortran programs, subjects did a better job of making modifications in a commented version (Hubka et al., 1979)**

	Mean Comprehension Scores
Comments	60.3
No comments	42.8

Figure 13-23. **In the Woodfield et al. (1981) experiment comprehension scores were much higher for the commented versions of programs.**

Two notes of caution are in order here. Misleading comments may be more detrimental than no comments at all. Imagine trying to find an error in a program in which the comment reads

INCREMENT THE COUNTER BY 1

but in which the next statement in the language really adds two to the counter. Furthermore, no compiler checks the program against the comments to ensure that the comments really reflect what is going on within. Comments are purely a psychological tool for the programmer.

Mnemonic Terms

Most programmers are taught to name variables things that make sense. Thus, above we used the name "HOURS" to refer to the number of hours worked by an employee. This is usually an easier term to remember that "X" or "J12R4." The use of mnemonic variables should aid in program construction, comprehension, and maintenance (Figure 13-24).

Mnemonic

salary = SALARY rate = RATE

Non-mnemonic

salary = X rate = Y

Anti-mnemonic

salary = RATE rate = SALARY

Figure 13-24. Names selected for things in computer programs can either be potentially helpful (mnemonic), probably neutral (non-mnemonic), or potentially harmful (anti-mnemonic). Why would anyone use an anti-mnemonic variable? Usually because the original purpose for a variable changes drastically during the construction of a program. But, sometimes just because programmers are weird.

Strich (1979) conducted a series of experiments concerning the use of mnemonic terms. First, through a pilot study, some lists of data names were compiled. One list consisted of the 20 most-frequently mentioned names (called the "mnemonic" list) for the items and its opposite consisted of 20 of the least-frequently mentioned names (the "nonmnemonic" list). Interestingly, the mean length of the data items in the two lists was about 7 characters, even though some subjects were allowed to use up to 12 characters. Three other lists were compiled consisting of the most mnemonic names of 1 to 4 characters, 5 to 8 characters, and 9 to 12 characters (Figure 13-25). In a pair of experiments subjects were requested to answer questions by recalling item names that they had seen earlier. If they did not recall the names unaided, then they were allowed to look them up in a large data dictionary of more than one hundred related names. In the first experiment recall was clearly better (both in terms of time and errors) for the mnemonic names. In the second experiment, recall time slightly favored the 5- to 8-character range. Performance was obviously worse (both time and errors) with names in the 9 to 12 range (Figure 13-25).

Booth, Austin, Lape, and Roman (1981) replicated some of the above work using mnemonic names in the 5- to 9-character range and the 10 to 30 range. This was done because several programming languages allow a

	Mnemonic Names	Nonmnemonic Names	
Mean response time	10.6 sec	18.4 sec	
Number of wrong responses	7	39	
	1 to 4 characters	5 to 8 characters	9 to 12 characters
Mean response time	10.8 sec.	10.3 sec.	12.1 sec.
Number of wrong responses	13	13	19
	5 to 9 characters	10 to 30 characters	
Mean number recalled correctly	17.0	14.3	

Figure 13-25. Strich (1979) found that mnemonic terms were clearly better than non-mnemonic ones and that 5-8 character terms were a little easier to use than shorter or longer ones. Booth et al. (1981) found that very long variable names were not better than 5-9 character names.

virtually unlimited name length. The question was whether longer mnemonic names are better than relatively short mnemonic ones. Subjects were directed to recall as many of 20 terms as possible in a 15-minute interval. Performance was about 20% better for the shorter terms than the longer ones.

Very short names may not be rich enough to help memory or comprehension. For example, "HRS" might mean "hours worked" or it could mean "number of horses" or worse. Longer names invite excessive creativity that (in the absence of guidelines concerning their formation) may lead to names that are hard to remember exactly (and certainly take longer to enter at a keyboard). Thus, "HOURS-WORKED" might be misremembered as "HOURS-WORK," "WORK-HOURS," or even "EMPLOYEE-HOURS-WORKED" if there is no limit on the number of characters or guidelines on how names are constructed.

Error Messages

When a programmer makes an error that the programming language compiler can catch, an error message is generated. The simplest of these is a syntax error. Good syntax error messages should help programmers find out what is wrong fairly quickly. But, what does "good" mean?

Brown, Dickman, Grosso, Kosor, and Seekamp (1979) constructed two versions of a Pascal compiler. One gave "near and complete" error messages (Figure 13-26), that is, diagnostics for as many errors as possible in each statement and, for each, indication of the type and point of

```
                PROGRAM TEST(INPUT,OUTPUT) ;

                CONST
                  MAXNUMBERS := 23;
*************                ^16   '=' EXPECTED
*************                ^50   ERROR IN CONSTANT

                VAR
                  J : INTERGER;
*************            ^104  IDENTIFIER NOT DECLARED

                BEGIN
                  WRITELN(' THE QUICK RED FOX JUMPED ') ;

                  J := 1; / ; I := I+1;
*************          ^6  ILLEGAL SYMBOL
*************            ^104  IDENTIFIER NOT DECLARED
*************                ^104  IDENTIFIER NOT DECLARED
                END.
```

Figure 13-26. An example of "near and complete" diagnostics. Notice that all syntax errors are explained at the point of occurrence.

```
                  PROGRAM TEST(INPUT,OUTPUT) ;

                  CONST
****ERROR  16       MAXNUMBERS := 23;

                  VAR
****ERROR 104       J : INTERGER;

                  BEGIN
                    WRITELN (' THE QUICK RED FOX JUMPED ') ;

****ERROR   6       J := 1; / ; I := I+1;
                  END.

      ERROR SUMMARY
      ******* ***********

  6:  ILLEGAL SYMBOL
 16:  '=' EXPECTED
104:  IDENTIFIER NOT DECLARED
```

Figure 13-27. An example of "far and incomplete" diagnostics. Only the first syntax error found in each statement is indicated and these are not explained until the ERROR SUMMARY at the end of the listing.

error. The other gave "far and incomplete" diagnostics (Figure 13-27), that is, a message for at most one error in any statement, no indication of the point of error, and postponement of description of the type of error until an error summary at the end of the program.

Thirty-four subjects constructed two programs using the two Pascal "compilers." Notice that if there were no errors in a program, then the program listing was identical for the two compilers. If "near and complete" error messages aid programming effort, then subjects should be able to reach their first syntactic error-free compile earlier with them. Figure 13-28 shows that this is exactly what happened. Another thing that error messages should affect is the persistence of errors as characterized by the number of runs between successful compilations. Errors were more persistent with the "far and incomplete" diagnostics—requiring about 20% more submissions between successful compiles. Finally, error messages should be related to programming time. If the "far and incomplete" messages are hindering the program development process, then the intervals between submissions should be longer for programs developed using that compiler. Excluding all intervals of more than 90 minutes, the subjects with "near and complete" messages took about 35% less time to resubmit programs following runs with errors. Thus, it appears that error summaries at the end of a program listing are not as useful as error messages interleaved with the statements of the program.

Logic errors are another matter. These *cannot* be detected by the compiler and only a handful of them cause exceptional conditions that the computer can complain about; "division by zero" and "subscript out of range" are examples of errors for which messages can be provided. But, no computer system can detect that you added 2 when you really meant to add 1.

	Near and Complete	Far and Incomplete
Median submissions until first successful compilation	4	6
Mean submissions between successful compilations	2.57	3.18
Mean time between submissions following a run with an error	12.5 min	16.9 min

Figure 13-28. Brown et al. (1979) found that "near and complete" diagnostics helped subjects attain their first successful compile more quickly. They also helped reduce the number of and time between submissions following a syntax error.

Redundancy

Redundancy is the process of repeating information. Generally, redundancy is time consuming and annoying. Generally, redundancy is time consuming and annoying. Isn't it? But, there are instances when redundancy can be valuable. In a study comparing two programming languages (Gannon & Horning, 1975), a language TOPPSII (pronounced "TOPS 2") was created that in general required more redundancy than an original language TOPPS. For example, in TOPPSII, a programmer had to declare in each procedure those variables that he or she wished to use from other procedures. In TOPPS any variable from another procedure was available without declaration. The study consisted of having 25 subjects write two programs (one in each of the two languages). Errors, runs submitted, and persistence of errors (consecutive runs without being able to correct a problem) were recorded for each subject. Errors in TOPPSII (the language that required more redundancy) were only about half as persistent as in TOPPS.

Most programming languages subtly incorporate various types of redundancy to aid in error detection and correction. For example, if a variable is declared to be a singly subscripted array (like a vector X of length 50), it will generally have to appear in the program followed by a single subscript [e.g., X(23)]. (There are situational exceptions to this rule, of course; but this is typical.) Thus, if R is declared to be an integer array of 10 values, then the compiler or run-time system can complain if you refer to R(3,6), or R(15), or even try to put a real value into one of R's locations. None of these complaints would be possible without redundancy.

On the other hand, redundancy can be detrimental when it requires that duplicate information be provided. The classic case is when the same variable used in different subprograms must be redeclared each time when all the programmer really wants to say is "use the same R as I declared the first time I used it." However, there is a thin line between being detrimental and helping to catch errors. If the programmer gave slightly different declarations (e.g., declaring R as size 11 instead of 10) because of a misunderstanding, then the requesting of "duplicate" information has served a purpose.

There is controversy over how much to try to correct errors by considering redundant information. In the example above if the programmer requests that R be set equal to zero, this can be pointed out as an error (using the previously provided fact that R is an array) or the "forgiving" stance can be taken that the programmer is really asking to place zeros in each element of the R array. The latter action could lead the program to exhibit strange behavior if the programmer really meant to place a zero in the first element [i.e., R(1)] but simply forgot the subscript.

Statement Delimiters

Every language has *statement delimiters*. This is the way we tell where one statement ends and the next begins. In written English we use the period ("."). In spoken English it is the short pause. Programming languages also employ statement delimiters. Pascal uses the semicolon and Cobol the period.

Another facet of the Gannon and Horning (1975) studies (referred to in the last section) was that TOPPS and TOPPSII had differing philosophies concerning statement delimiters. TOPPS really had no statement delimiter but used the semicolon as a statement *separator* in a manner that made it appear to operate a lot like a delimiter. In TOPPSII the semicolon was really employed as a statement delimiter. Errors concerning the semicolon were 10 times more prevalent in TOPPS than in TOPPSII.

You may think that some languages (e.g., Fortran and Basic) use no statement delimiter since their statements have no explicit characters to conclude them. However, these languages have a more subtle statement delimiter. It is the "end-of-line." Unlike English (or some programming languages) in which statements may follow each other freely (like the sentences do in this book) some programming languages require that there be exactly one statement per line. Thus, the end-of-line is also the end-of-statement. Everything's fine with these languages until you have to use a statement that is longer than a line. Then there is generally a special continuation indicator to signify that the current statement goes on to the next line. In Fortran, for example, a character in column 6 indicates that this is not a new statement, but simply more of the preceding statement.

Inconsistencies with statement delimiters can cause problems for programmers. The worst cases are programming languages in which statement delimiters are not what they seem to be. For example, in Cobol the statement delimiter "." is optional EXCEPT in some very important situations. The range of an IF statement is completely determined by the first delimiter encountered. Thus, the two program segments in Figure 13-29 are completely different even though they appear to be about the same.

Notice the difference between segments 1 and 2. (A lot of Cobol pro-

Segment 1

```
IF HOURS ARE GREATER THAN 40
   SUBTRACT 40 FROM HOURS GIVING OVER-HOURS
   MULTIPLY OVER-HOURS BY 1.5 GIVING OVER-TIME
   ADD OVER-TIME TO HOURS.
```

Segment 2

```
IF HOURS ARE GREATER THAN 40
   SUBTRACT 40 FROM HOURS GIVING OVER-HOURS.
   MULTIPLY OVER-HOURS BY 1.5 GIVING OVER-TIME.
   ADD OVER-TIME TO HOURS.
```

Figure 13-29. Can you detect the subtle differences in these two program segments?

Segment 3

```
IF HOURS ARE GREATER THAN 40
   SUBTRACT 40 FROM HOURS GIVING OVER-HOURS.
   MULTIPLY OVER-HOURS BY 1.5 GIVING OVER-TIME.
   ADD OVER-TIME TO HOURS.
   ENDIF.
```

Figure 13-30. Notice that an IF statement terminator (ENDIF) makes errors like those in segment 2 much less probable.

grammers have had a lot of trouble because they didn't.) Segment 1 will do what we want; it increases the HOURS worked by the appropriate number so that when multiplied by the hourly rate later it will lead to pay that reflects normal salary for all hours up to 40 and time-and-a-half for all hours over 40. Unfortunately, segment 2 looks very similar but has some fatal flaws. First, since the range of the IF statement is controlled by the *first* delimiter, this means that the two statements "MULTIPLY OVER-HOURS BY 1.5 GIVING OVER-TIME." and "ADD OVER-TIME TO HOURS." are executed *regardless* of whether the number of hours is more than 40. So, what happens then? An error? Nope. Since OVER-HOURS has some value in it (the last legitimate value computed by the IF statement), everybody gets some overtime (even those who have worked only 40 hours or less).

The statement delimiter is a problem in this case because it serves different purposes in different situations. The period at the end of "SUBTRACT 40 FROM HOURS GIVING OVER-HOURS." ends the IF statement as well as the SUBTRACT statement. It would be far better to add to this language an IF statement terminator (such as ENDIF) leading to something like segment 3 in Figure 13-30. Then "what you see is what you get." This points out another problem with segment 2. Notice that the pretty-printing is, in fact, misleading—making us think that the MULTIPLY and ADD statements are part of the IF statement. So, pretty-printing can be a two-edged sword when it represents what we want instead of what we have really said.

SUMMARY

In this chapter we have considered the research concerning human factors in computer programming. We noted that there are no universally accepted theories of programming and that the approach taken by researchers has been almost totally empirical in nature.

But, some experimental results are beginning to appear that may lead to theories of programming that could affect language design and methods of teaching programming. We outlined research that suggested that (1) online program construction is superior to off-line construction, (2) programmers can perform quite differently with different specifications for the same program, (3) detailed flowcharts may be of no value in typical programming tasks, (4) "pretty-printing" may lead to more comprehensible programs, (5) the "walkthrough" mode of programming may lead to better programs, and (6) certain debugging aids may be of limited value.

Considering programming language features, there is some experimental evidence to suggest that (1) structured transfer of control may be superior to nonstructured, (2) proper modularization, good comments, and mnemonic terms may aid in program comprehension, and (3) the placement and amount of detail in error messages may affect the programming process.

14 DECISION MAKING WITH APPLICATIONS TO INSPECTION AND MAINTAINABILITY

J. R. BUCK and B. H. KANTOWITZ

Two peddlers meet in the village square on market day to barter their wares. One peddler has a basket full of shleems and the other peasant a cart piled high with blechs. Although it is general knowledge that shleems are more valuable than blechs, the two peasants at first are unable to come to an agreement. Finally, one peasant suggests tossing a coin to determine the exchange. "If the coin is heads, I'll trade you four of my blechs for each of your shleems," says the first merchant. "If it comes up tails, you trade me one of your shleems for four of my blechs." "OK," says the second peasant, "but I want to flip the coin to make sure you don't cheat me!"

Making decisions, like trading shleems for blechs, is part and parcel of everyday life. Some decisions are quite important—What job should I take?—while others are relatively unimportant—Should I have apple or cherry pie for dessert? But all decisions require thought and people try to make the best decisions they can. Mathematicians have developed several techniques for making rational decisions that we will discuss in this chapter. Alas, people are not always rational decision makers and we will discuss that too. One problem is finding a value or metric that will allow us to compare shleems and blechs. While we may suspect that the second peddler got ripped off, we can't be sure until shleems and blechs are related to a common unit of value.

Decision making is especially important if you are troubleshooting some piece of equipment that is not working properly. You must go through some systematic investigation to diagnose the fault, and then must decide on corrective action. Human detection and diagnosis of system failures has obvious practical importance and a recent NATO symposium (Rasmussen & Rouse, 1981) was devoted entirely to this topic. Keeping equipment up and running is as much a part of human factors as designing displays and controls. In fact, the human factors specialist is responsible for ensuring that the final design of equipment makes it easy to service in the field. A piece of equipment that cannot be maintained not only is itself useless, but also may cause an entire system to go down at great expense to those operators and/or customers relying on it.

The importance of maintenance as a contributor to operating cost is hard to overestimate. For example, such maintenance errors as returning equipment that could have been repaired in the field to a central maintenance depot require costly inventory of replacement equipment if the system is to be kept on line. In general, as systems become more complex, they are more difficult to maintain. This is especially true for weapon systems and one forceful argument against supersophisticated military hardware is that the excessive downtime compared to simpler weapons tremendously inflates the effective cost of the equipment. Not only are these supersystems more expensive to begin with, they require more maintenance and have a higher proportion of downtime. Some shocking statistics on field maintenance (Feineman, 1978) reveal that 70% of U.S. Army electronic equipment was not working, 30% of U.S. Navy equipment was inoperative while on maneuvers, maintenance costs for the Army air corps electronic equipment were 10 times the original cost over a 5-year period, and first-year maintenance costs for naval aviation equipment were double the original price. Improper maintenance can have more than economic cost. The crash of a DC-10 in Chicago in 1979 that killed 272 people was due to faulty maintenance procedures. Other examples of defective field maintenance (Christensen & Howard, 1981) emphasize the need for human factors in maintainability design. Maintainability is considered in the second half of this chapter.

DECISION MAKING

In real life, decision making consists of choosing between alternative courses of action available and enjoying or suffering the consequences of the chosen actions. One can also avoid decisions as Figure 14-1 shows. The theories about decision making are either *normative* in that the theory tells one how to behave to consistently realize rationality or *descriptive* in that the theory describes how people do behave. Could it be that the theories are the same? Well, the answer to this mystery is found in the following discussion.

Normative Decision Making

Decisions can be described in simple theoretical terms as consisting of four component parts: (1) a set of actions, (2) one or more events that describe the uncontrolled future, (3) the outcomes and consequences of

Figure 14-1. Avoiding decisions. (Al Isler, *Good Housekeeping.*)

each action given each event, and (4) the chances favoring each event. For example, a decision one can face before coming to work is whether or not to carry an umbrella. Thus the two actions: a_1 carry an umbrella, and a_2 do not carry it. In considering the decision further there are two possible events connected with this decision, either e_1 that it would rain or e_2 it wouldn't. Obviously if it rained and one doesn't bring an umbrella, then the physical outcome would be a wet, soggy you, an undesired prospect. There are also physical outcomes associated with rain and bringing the umbrella or no rain and both actions. In order to keep these factors straight, they can be written down as follows.

	Events	
Actions	e_1: Rain	e_2: No Rain
a_1: carry an umbrella	Get slightly damp, lug around the umbrella all day, and perhaps lose it	Lug around the umbrella, perhaps lose it, and have to explain to your friends why your're carrying an umbrella on a sunny day
a_2: do not carry an umbrella	Get very wet and uncomfortable; drip all over the papers	Nothing

The only remaining component of this decision is to evaluate the chances of rain by looking out the window, reading the morning paper, or turning on the radio or TV to see what the experts predict.

The Notion of Utility

In normative or prescriptive decision theory, there are statements about how people ought to make decisions, such as the great umbrella decision, and remain consistent with their true feelings of value. Most of these theories start with a measure called *utility,* which is a numerical measure of the relative worth of all physical outcomes associated with a decision. Utility, however, starts with an ordering of the physical outcomes from least to most desired. In the umbrella decision, for example, getting wet and soggy is likely the least desirable outcome, the "nothing" outcome is the most desirable, and the other two are in the middle somewhere, depending on a personal evaluation.

After ordering the consequences, the next job in measuring the utility of these outcomes is to assign values to these physical outcomes that truly reflect their relative worth. According to the theory of von Neumann and Morgenstern (1953), the least and most desired outcomes can be arbitrarily assigned the respective values of 0 and 1 and the utility values of the others can be found through a comparison between each of the other outcomes individually as a certainty and the *standard gamble* consisting of a risky situation. Suppose we had a wheel of fortune (Figure 14-2) where the circle was divided into two colors, light and dark, but where the amount of light color could be increased or decreased, inversely changing the dark color portion proportionately. Now to describe this work, consider yourself to be the subject with an experimenter there to show you how this concept works. The wheel of fortune is set up as shown in Figure 14-2 and the pointer will be spun on the wheel of fortune. If the pointer stops on the dark color, you get the most *undesirable* outcome but if it lands on the light color you get the most *desirable* outcome. That is the standard gamble where the proportion of the wheel of fortune circumference in light color is denoted as p. Now let us take the outcome where you carry an umbrella and it rains and the experimenter asks, "Would you prefer that outcome for certain or the standard gam-

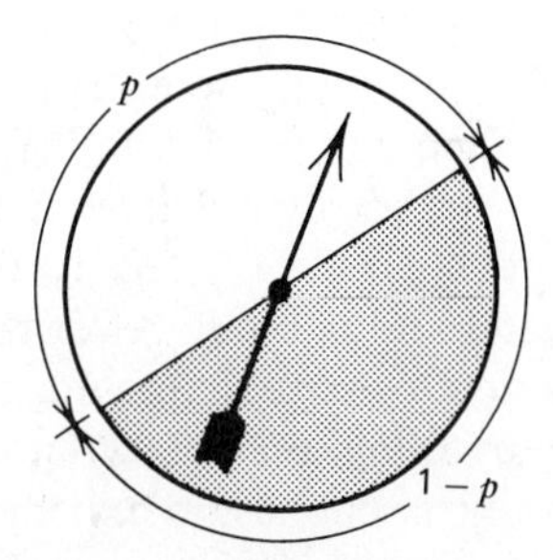

Figure 14-2. Wheel of fortune.

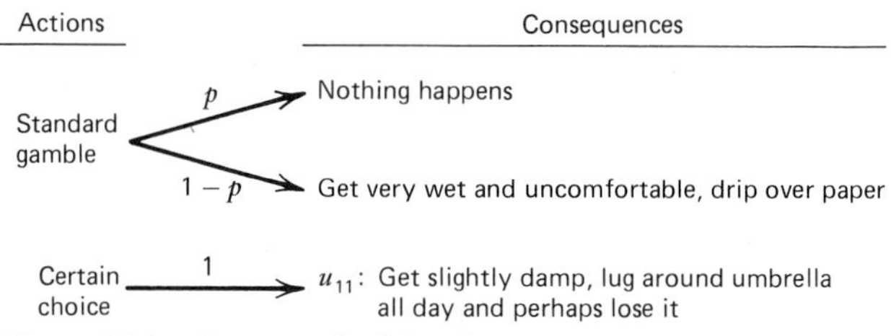

Figure 14-3. Lottery decision in utility measurement.

ble?" (See Figure 14-3.) If you say the certain outcome, then the experimenter will *increase p* and repeat the question with the revised wheel of fortune. If you say the standard gamble, then the experimenter will *decrease p* and repeat. Eventually we will find a value of p where you will be indifferent. That is, a value of p where any increase always causes you to select the standard gamble and any decrease causes you to select the certain outcome. It is this indifference value of p that reflects the utility of that certain outcome according to this normative theory. By repeating this procedure with the final outcome and finding a new value of p where indifference occurs, then we would have a measure of the utility of each outcome in this decision. It is interesting to note that measuring one's utility values is similar to the psychophysical techniques that psychologists have used to determine the perceived intensity of physical variables, such as the perceived loudness of a tonal frequency (see Chapter 3).

SOME NECESSARY MATHEMATICS

Before you can understand how humans make decisions and how the decision process is incorporated into maintainability, we must first lay an elementary mathematical foundation. The concepts of expected value, conditional probability, and Bayes' theorem, which follows from the definition of conditional probability, are necessary. Readers who already are fluent with these concepts can skim over the next two sections. The rest of you need not panic since our treatment will be easy to understand.

Expected Value

Imagine you are sitting in your friendly neighborhood tavern and a stranger approaches you with a proposition for a game of chance: "If you flip a coin and it comes up heads I'll pay you \$2.00. But if it comes up tails you pay me \$0.50. However, the entry fee for this game is \$1 that you must pay me every time you flip a coin." Would you play this game?

In order to analyze whether or not it would be profitable for you to play, we will first ignore the \$1 entry fee. Let's go through two successive coin tosses, one head and one tail. After the first toss you have won \$2. On the second toss you lose \$0.50, leaving a net profit of \$1.50. Since it took two tosses for you to make \$1.50, your average profit is \$0.75 for each coin toss. This profit is called the expected value of the coin toss.

Expected value can be calculated by taking the expected gain or loss for each outcome (heads or tails), multiplying by the probability of the particular outcome, and then adding up gains and losses. For our coin example, the gain associated with heads ($2) is multiplied by the probability of heads (.5) to get a value of $1. To this must be added the other outcome, tails, and its loss of $0.50 × (.5) or $0.25. Adding plus $1 and minus $0.25 yields an expected value of $0.75. Mathematically, this is expressed as follows.

$$EV = O_1 \times p_1 + O_2 \times p_2 + \ldots$$

where EV is expected value, O is an outcome such as win $2 and p is the probability of that outcome.

Our discussion of the game is incomplete because, until now, we have neglected the $1 entry fee. Since the expected value is less than $1, a rational decision maker would refuse to play the game. Anyone who played could expect to lose $0.25 for each coin toss. If the entry fee were lowered to $0.75, you would, in the long run, neither win nor lose money. Thus, this price is called your *indifference point*. A rational decision maker will be happy to play at any price less than the indifference point and will refuse to play at prices greater than the indifference point.

Let's say that the price of the game has been lowered to $0.70. You should be eager to play. But now let's up the ante. If the coin comes up heads you get paid $2000. If it comes up tails, you must pay $500. Will you pay $700 as an entry fee? Since the expected value of the game is $50 ($750 average profit per toss minus $700 entry fee), a rational decision maker would decide to play. But most people would be hesitant. This illustrates the difference between the economic value of a bet and its psychological *utility*. For small sums of money, value and utility are about equal. But, unless you are rich and accustomed to dealing with large sums of money, utility goes down because of the psychological risk associated with large sums. A rational decision maker will play any game with an expected positive return regardless of the sums involved. People who focus on subjective utility rather than expected value will not play any game even if it can be expected to be profitable.

Bayes' Theorem

What is the probability that you will receive A's in all your courses this semester? At the beginning of the term, you were able to guess a probability, but might not have been very confident about your guess. As the term progresses and you receive grades for exams and projects, your estimate should improve. The process of revising probability estimates as new data are received is the subject of Bayes' theorem. Bayes' theorem provides an optimal model of how such probability revisions should be carried out. Optimal models are not models of human behavior (Kantowitz & Hanson, 1981). Nevertheless, they are extremely useful since they provide an upper limit for behavior; that is, we would not expect a person to perform better than an optimal model. Since people usually perform worse than an optimal model, the optimal model tells us how much room there is for improvement.

In order to understand Bayes' theorem, you must first know what a conditional probability is. We will explain conditional probability first by a simple example taken from an excellent introductory statistics text (Mosteller, Rourke, & Thomas, 1961) and then more formally. You have two dice, one red and one green. Each die can have a value of from 1 to 6. The sum of the values of the two dice is less than (<) 4. What is the probability that the red die has a value of 1? If you answered 2/3 you can skip this section. If you answered anything else, read on. There are two ways to solve this problem. First, we can count (or enumerate, as mathematicians would say) outcomes. How many combinations of dice satisfy the condition that red + green < 4? There are three such combinations: both dice having a value of 1, red die = 1 and green die = 2, red die = 2 and green die = 1. Any other combination (e.g., both = 2) will add up to more than 3. For two of these three combinations, the red die = 1. Thus, the probability that $r = 1$ given that $r + g < 4$ is two out of three or 2/3. (See Table 14-1.)

This method of enumeration works fine when the number of outcomes is small. But it soon becomes tedious for problems of any size. So mathematicians have invented a better way to calculate conditional probabilities. By the way, the 2/3 we calculated above is a conditional probability since it was based on the condition that $r + g < 4$. The condition in conditional probability is usually expressed by the words "given that." The terse mathematical notation for a conditional probability is $P(A|B)$, which is read as "the probability of A given B;" the vertical line between A and B stands for "given." A and B could be anything. For example, what is the probability of class meeting (A) given that the professor is 10 minutes late (B)? In the dice example, A stands for the outcome that the red die has a value of 1, and B stands for the total $r + g$ being < 4. There are 36 possible outcomes for throwing two dice ranging from both dice showing 1 to both dice showing 6. This set of 36 events is called the sample space (Table 14-1). Outcome B ($r + g < 4$) consists of three

TABLE 14-1 *Sample Space for Dice Example*

		1	2	3	4	5	6	Red Die
	1	2	3	4	5	6	7	
	2	3	4	5	6	7	8	
Green	3	4	5	6	7	8	9	
Die	4	5	6	7	8	9	10	Sum
	5	6	7	8	9	10	11	
	6	7	8	9	10	11	12	

events as listed in the foregoing paragraph. Outcome A $(r=1)$ consists of six events: (1,1), (1,2), (1,3), (1,4), (1,5), (1,6) where the first number in each pair represents the red die and the second number the green die, (r, g). In order to calculate conditional probability we must know the intersection of outcomes A and B. The intersection is those events common to both sets. In this case there are two events in the intersection of A and B: $A \cap B =$ (1,1), (1,2) where $\cap$ stands for intersection. Now we have all the raw material necessary and only require a formula for conditional probability.

$$P(A|B) = \frac{P(A \cap B)}{P(B)}$$

Since there are 36 events in the sample space, $P(A \cap B) = 2/36$. Similarly, $P(B) = 3/36$. Putting these probabilities into the formula we find that $P(A|B) = 2/3$. Note that this is the same answer we got by using the method of enumeration.

Starting with the formula for conditional probability, we can easily derive Bayes' theorem.

$$P(A \cap B) = P(B \cap A) = P(A|B)P(B)$$

Similarly,

$$P(B|A) = \frac{P(B \cap A)}{P(A)}$$

Substituting the topmost equation into the one just above produces Bayes' theorem.

$$\boxed{P(B|A) = \frac{P(A|B)P(B)}{P(A)}}$$

$P(B|A)$ is called the posterior probability because it is the probability after data have been obtained. $P(B)$ is sometimes called the prior probability because it is the probability before getting any new data.

How Bayes' theorem is used is best explained through an example. Assume there are two urns filled with poker chips. Urn A contains 70% red chips and 30% black chips. Urn B has 30% red chips and 70% black chips. However, the labels have peeled off the urns and you don't know which urn is which. You reach into one urn and pull out a red chip. What is the probability you have reached into urn A? Write down your guess before continuing. The probability of getting a red chip given that you have reached into urn A is .70, since urn A has 70% red chips. Since you have no information about which urn is which, the probability that you have drawn from urn A is .5. Therefore, the numerator in Bayes' theorem is $(.7)(.5) = .35$. The denominator is the probability of getting a red chip. This can happen in two ways. First, you could have drawn from urn A with a .5 probability. This is multiplied by the probability of getting a red chip from urn A (.7) to once again get .35. Second, you could have drawn from urn B. This gives a probability of a red chip of (.5)(.3) = .15. Thus, the total probability of getting a red chip is .35 + .15 = .50. Therefore, according to Bayes' theorem the probability that you have drawn from urn A given that a red chip was selected equals .35/.50 = .70. Your guess was probably less than .70.

Now let's put back the red chip and draw another chip from the same urn. (This is called sampling with replacement.) The second chip is also red. Now what is the probability that urn A is the source of the two red chips? After a second red chip the numerator of Bayes' theorem is (.7)(.7). How did the .5 get changed to .7? Drawing the first red chip, we had no information about the urns. But now we know (because we just finished calculating it) that there is a .7 probability we are drawing from urn A. The information gained by drawing the first red chip has modified the probability that we are drawing from urn A. This is called the Bayesian revision of probability. So the numerator is now .49. The denominator will also change. Although there are still two ways to get a red chip, the probabilities of the urns are now different. The probability of getting a red chip from urn A is now (.7) (.7) while getting a red chip from urn B is now (.3) (.3) since if urn A has a .70 probability then urn B must have a probability of .30. (These two probabilities must sum to 1.0 since there only two urns.) So the revised probability of sampling from urn A given two red chips is .49/.58 = .84. Again, your guess was probably too low. (Additional practice with Bayes' theorem is given in the Workbook.) Of course, drawing poker chips out of urns has little to do with human factors. Human factors specialists are too busy to play poker. But later in the chapter you will see how Bayes' theorem applies directly to troubleshooting and maintainability.

HUMAN DECISION MAKING

Most people would find it hard to believe that sampling only two red chips is sufficient to overwhelmingly implicate urn A as the source. Their lower probability estimates reveal this innate conservatism. People are conservative decision makers (Figure 14-4). They consistently underestimate Bayesian probability revisions. Possible reasons for this conservatism were discussed by Edwards, Lindman, and Phillips (1965) who concluded that "the finding of human conservatism raises some problems for the design of man-machine systems intended to perform information processing in a more or less optimal way."

Although modern computers make a flood of information available to the decision maker, people generally make decisions the same way they did tens and hundreds of years ago: by guess and by gosh. People rely on intuition and "gut feelings" to make decisions. They do not use rational means to integrate and evaluate information that often is unreliable and

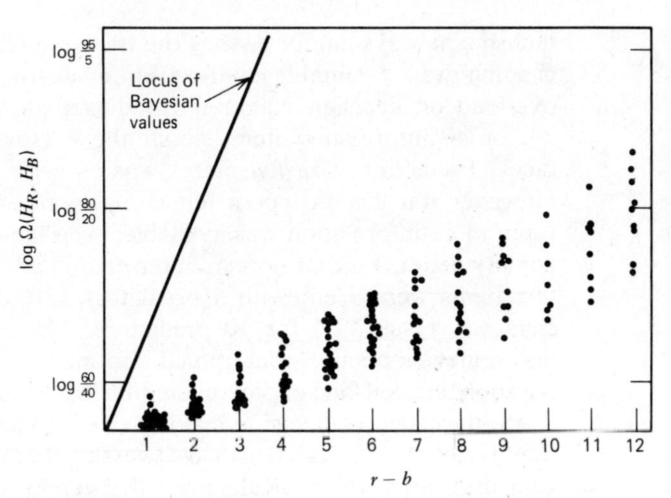

Figure 14-4. People are conservative decision makers. An optimal decision maker would follow the line labeled "locus of Bayesian values"; however, the slope of the human data is far less steep. (From Edwards, Lindam & Phillips, 1965)

of unknown validity. The full effects of an important decision are not always foreseen. For example, a Zero Defects program to improve quality control was adopted by over 12,000 firms (Birkin & Ford, 1973). Management believed it was of utmost importance to get the job done right the first time. But a direct consequence of the Zero Defects program was decreased productivity and missed deadlines. Many companies reconsidered and decided they would do better without zero defects.

You are probably saying to yourself that this outcome was entirely predictable. You would have known better. This is called hindsight bias (Fischoff & Beyth, 1975). People tend to believe that they "knew it all along." The knowledge that some event has occurred increases our feeling that the event had to happen. Furthermore, to make matters worse we also believe that others should have foreseen the event. This human tendency towards hindsight bias—which most people are unaware of—makes it difficult to learn from past mistakes.

The concept of random error has been discussed in the following chapters: for example, the communication model and information theory (Chapter 5), psychomotor skills (Chapter 5), signal detection theory (Chapter 3), and reliability (Chapter 2). Any human-machine system must cope with random error. Human decision making is equally plagued. One of the classic studies that demonstrated random error in expert judgments was conducted with radiologists who examined X rays for the presence of lung disease (Garland, 1960). When they evaluated the same X ray on two different occasions, the radiologists changed their mind about 20% of the time. A related study (Slovic, 1981) used expert handicappers at race tracks. (Slovic was not trying to establish a new system for beating the track but felt that horseracing handicapping was a suitable context for evaluating effects of information overload on decision making.) Handicappers were given either 5, 10, 20, or 40 information items about the horses' performance in earlier races. Prediction with five items was as good as with more than five. However, the handicappers felt more confident about their judgments when more information was available, even though their judgments were not any better. Furthermore, random error increased when more information items were used: with 5 predictors 22% of the first-place choices changed versus 39% for 40 predictors. These kind of results suggest that scurrying about for additional information before making a decision is a sophisticated kind of procrastination that will not improve the decision.

Another type of decision-making error is called the base-rate fallacy (Bar-Hillel, 1980; Kahneman & Tversky, 1973). We can illustrate this with the Cab problem (Kahneman & Tversky 1972 cited by Bar-Hillel, 1980).

> *Two cab companies operate in Metro City. Eighty-five percent of the cabs are from the Me-First Taxi Company and the rest from the Running Meter Cab Company. One night a cab was involved in a hit-and-run accident. An eyewitness testified that the cab was from the Running Meter Cab Co. The court conducted a nighttime visibility test and discovered that the witness was able to correctly identify a cab at night 80% of the time. What is the probability that the witness correctly identified the cab at the scene of the accident?*

Most people would answer .80. If you answered .80 you have just committed the dreaded base-rate fallacy. This problem is really another form of the urn and poker chip problem discussed earlier. You can apply Bayes' theorem to obtain an optimal answer. There are two kinds of data in this problem. The data about the distribution of cabs in Metro city is called base-rate information. The witness report is called diagnostic (or indicant) information. Using Bayes' theorem to combine these two forms of data gives a posterior probability of .41. Most people find this estimate hard to believe because they allow diagnostic information to overwhelm base-rate information. If we disregard the eyewitness report, there was a .15 probability that the cab was from the Running Meter Cab Co. So the best estimate must be between .15 and .80. Bayes' theorem lets us calculate the optimal answer of .41. The base-rate fallacy can be overcome by presenting base-rate information in a format that stresses its relevance (Bar-Hillel, 1980). Note, however, that this Bayesian analysis of the Cab problem has been criticized as being incomplete (Birnbaum, 1983). Birnbaum employed a signal detection theory analysis and some general assumptions about the witnesses' response criteria (see Chapter 3), to predict answers much closer to the .80 value. His analysis suggests that there may be no evidence for a base-rate fallacy. Simple normative models are not necessarily good descriptions of human behavior (Kantowitz & Hanson, 1981).

These are only some of the ways in which people make poor decisions. A detailed review can be found in Slovic (1981). Until we understand how people make decisions—that is, the cognitive processes that are involved—it will be quite difficult to devise techniques that improve the decision making cabability of human-machine systems. But new techniques are being studied and offer hope (Slovic, 1981).

A SLICE OF LIFE

Making Rational Decisions

One of the simplest techniques for making rational decisions is to organize a set of outcomes according to expected utility. Then we select the outcome that has the greatest expected utility and our decision is made. People who violate the mathematical principles associated with maximization of utility can look awfully silly. For example, if a person prefers outcome A to outcome B, and outcome B to outcome C, outcome A must be preferred to outcome C. (This is called transitivity by mathematicians.) Any person who does not follow this ordering can be turned into the infamous "money pump." Take the example of a person who prefers A to B, B to C but irrationally also prefers C to A. A, B, and C can be anything but it might help to think of them as Christmas presents under a tree to be traded among family or friends. Present C is picked up from under the tree. You have already grabbed presents A and B. But, being generous you offer to trade B for C if the person will give you $1 for your trouble. Now you have A and C and they have B. Since they prefer A to B, you offer to trade again demanding another $1. Now they have A and you have B and C (and $2). Since they prefer C to A, you are happy to organize another trade. Now they have C, you have A and B and $3, and the cycle is back at the beginning. This process can be repeated until the "money pump" is out of funds.

Let's see how maximization of expected utility can help you decide which of three job offers to accept after graduation. Job BC is at a big corporation, Job TN is at a tiny new company just getting started, and Job R is located in a pleasant resort area. First, we make a list of important job characteristics such as salary, interesting work, etc. (See Table 14-2). We then assign a subjective utility, on an arbitrary scale of from 1 to 10 where 10 is best, to each job characteristic. For example, in Table 14-2 the characteristic "high salary" was given a utility of 5. (Remember, professors are filling this out!) Now for each characteristic and each job we estimate the probability that the characteristic will be obtained. (Since we live in an uncertain world where even weather reports are probabilistic, we can't know exactly what these probabilities are so we enter subjective probabilities based on knowledge of the companies from interviews, etc.) Now we can calculate expected utility for each job by multiplying the assigned utility by the subjective probability and adding up these products. These calculations have already been done in Table 14-2, which shows that the job that maximizes expected utility is with the tiny new company. You might want to try this with your own set of assigned utilities since this might lead to a different decision.

The other side of the coin on rationality is that the normative theory can also be suspect. Most of the normative theories neglect the human effort needed to complete their sense of rationality. For example, suppose that the ice cream truck in Figure 14-1 had 200 flavors. Few people would consider it rational to go through the entire list to be absolutely sure that one's expected utility was maximized. Perhaps people are *satisficers* (Simon 1957) rather than optimizers; they would stop scanning the ice cream list when they found a satisfactory choice rather than keep looking for the very best choice. So you might not want to rely completely on maximizing expected utility to select your life's work.

TABLE 14-2 *Job Offers and Expected Values*

Job Characteristic	Assigned Value	Estimated Probabilities: Job BC	Job TN	Job R
1. High salary	5	.8	.9	.3
2. Opportunity for rapid promotion	6	.5	.8	.2
3. Nice place to live	8	.2	.2	.9
4. Interesting work	10	.8	.8	.5
5. Friendly co-workers	4	.5	.5	.7

Calculating Total Expected Values
Job Characteristic

	1		2		3		4		5		Total
Job BC	(.8 × 5)	+	(.5 × 6)	+	(.2 × 8)	+	(.8 × 10)	+	(.5 × 4)	=	18.6
Job TN	(.9 × 5)	+	(.8 × 6)	+	(.2 × 8)	+	(.8 × 10)	+	(.5 × 4)	=	20.9
Job R	(.3 × 5)	+	(.2 × 6)	+	(.9 × 8)	+	(.5 × 10)	+	(.7 × 4)	=	17.7

How to calculate expected values for a set of three job offers. Note that the expected value for Job TN is higher than that for Jobs BC and R. Thus, Job TN would be the best choice.

DECISION MAKING IN INSPECTION

Sam, who works as an inspector at an apple-packing plant, says that he recently quit his job because he couldn't stand all that decision making. That may seem silly because the decisions are so simple, but there are a lot of decisions to be made. With each apple examined the inspector must decide if all of the examined items of specification are met. If any specification item was not met, then Sam was supposed to reject the apple. However, Sam would forget to consider a specified item, thereby accepting the apple, and sometimes this forgetfulness resulted in a type-2 error (i.e., accepting a defective apple). On other occasions he would reject an apple that did meet specifications (creating a type-1 error). Sometimes, when the supervisor was away, he would eat one (an unclassified error). Inspection performance improves as both types of errors grow smaller. In effect a decision of the following type must be made by the inspector for every apple presented.

	Events	
Actions	e_1: Defective Apple	e_2: Nondefective
a_1: accept the apple	Type-2 error	No error
a_2: reject the apple	No error	Type-1 error

With most inspection situations the inspector must physically remove or mark an item to signify rejection and therefore ignored or overlooked items are automatically accepted. It is obvious that either type of inspection error causes production costs to rise so it is important for human factors specialists to try to reduce both error types to the lowest economic limit. Ballou and Pazer (1982) showed that in serial multistage operations the costs are more sensitive to increases in type-1 errors. Therefore the inspection errors can be viewed as a case of *signal detectibility,* as indicated in Figure 14-5, in order to separate the detectibility strength d' for improving the inspection performance from the inspectors' criterion β. A number of researchers have investigated inspection decision making in this context (Buck, 1975; Drury, 1974; Smith & Barany, 1971; Wallace & Adams, 1969). The role of human factors specialists is then to maximize d' up to an economic limit and then to

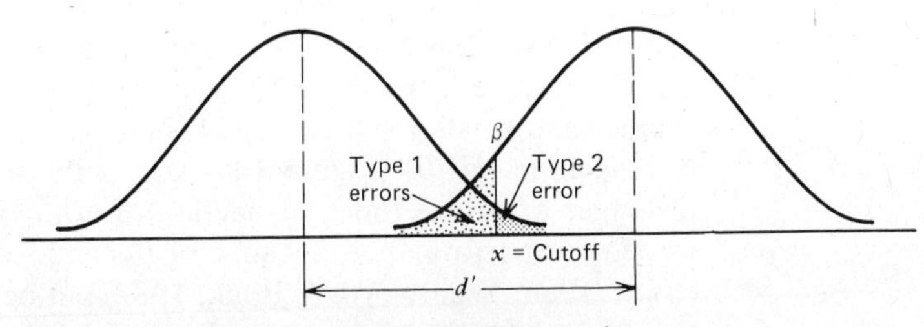

Figure 14-5. TSD model of visual inspection.

calibrate the inspectors' criterion β to approximate the relative cost situation.

A natural question arises: How can human factors specialists help inspection decision making? One clear way toward this objective is to enhance the inspector's sensory process to the maximum so that accept-reject discrimination features are highlighted. Another way is to provide sufficient inspection time per item and/or provide adequate time breaks or relief time to overcome fatigue. To be most economic with time breaks the inspection time should be such that the added benefits for an additional unit of time should equal the added costs for a unit less time. Another source of inspection aiding is through providing means for maintaining consistency in the inspector's judgments of acceptability of specifications and reducing social pressures in making those judgments. Finally, there would seem to be personal characteristics that good but not poor inspectors possess and that tests could serve to select the appropriate personnel. However, Wiener (1975) made a summary of studies about this conjecture but he found no variables that had consistent correlations with better inspection (e.g., intelligence, extroversion, or sex). The only positive characteristic with visual inspection that appears to discriminate out poor inspection performance is visual testing (Nelson & Barany, 1969). Some details on other forms of human factors aiding is discussed as follows.

Improving Inspection Decisions by Improving the Senses

Inspection, as with most tasks, involves the senses and motor responses along with decision making. Vision is the typical sensory modality that brings information to the inspection decision maker. Since the decision making can not be any better than the informational input, one of the more important ways to aid the inspector is to optimize the visual task (Blackwell, 1970). Sometimes the simple positioning of lights or a viewing angle can be found that enhances flaws of a particular type. Figure 14-6 shows three lighting and viewing conditions of scratched acrylic sheet where the combined lighting and viewing dramatically brought out flaws (Faulkner & Murphy, 1973). Colored lights have also proved to be helpful in color matching inspection. It is relatively easy and inexpensive to experiment with lighting and viewing arrangements that will create the desired effects. Chapter 17, Microenvironments, considers factors relating to illumination in greater detail.

Object crowding or pacing in inspection is another cause of reduced inspection performance. Some investigators have shown that inspection errors increase exponentially when the time available per object gets smaller and smaller (Drury, 1974; Nelson & Barany, 1969; Rizzi, Buck, & Anderson, 1979). Figure 14-7 shows the total inspection error effect of longer exposure time. Inspection of a moving object further reduces inspection performance because of dynamic visual acuity requirements (Burg, 1966; Wentworth & Buck, 1982) but better visual search patterns can improve this performance (Bloomfield, 1975; Buck & Rizzi, 1974).

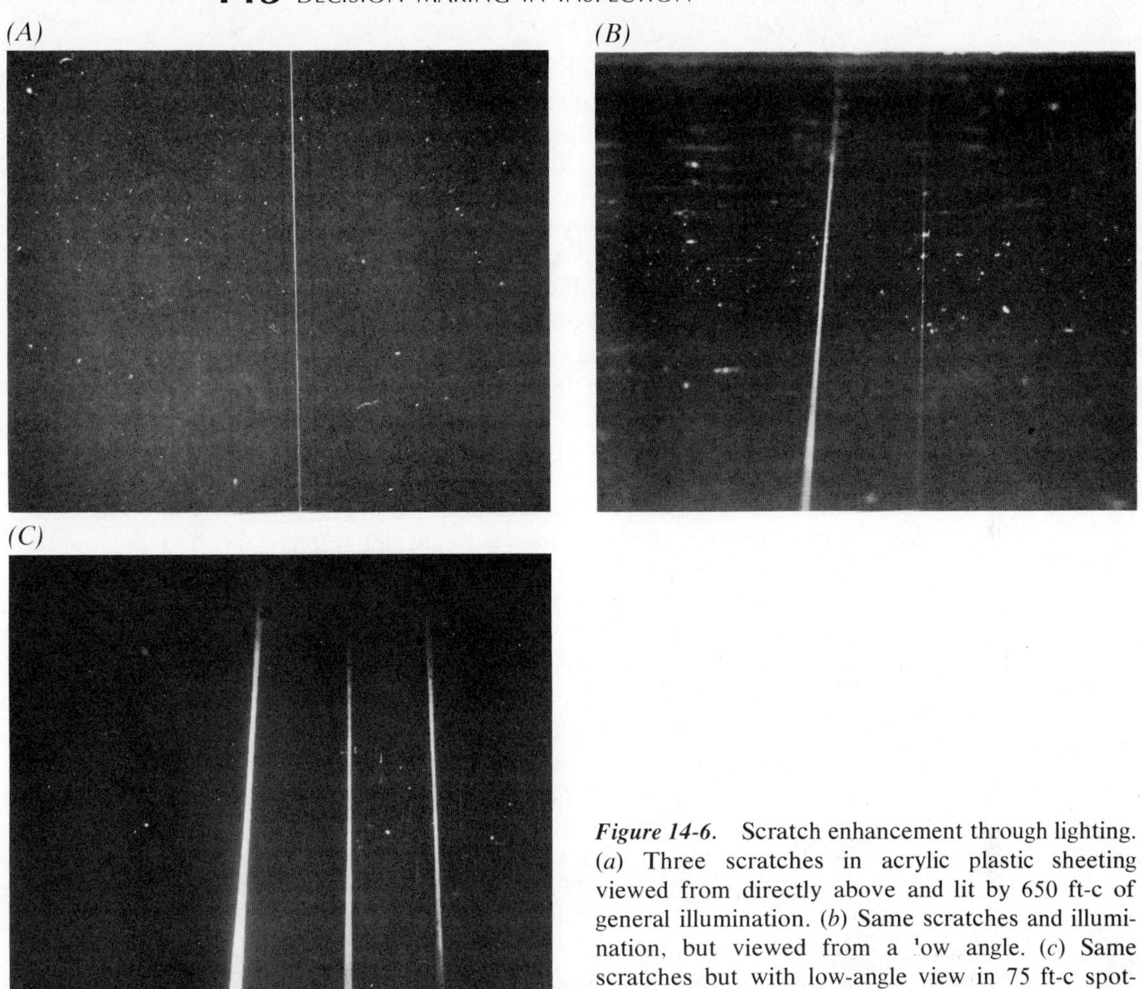

Figure 14-6. Scratch enhancement through lighting. (*a*) Three scratches in acrylic plastic sheeting viewed from directly above and lit by 650 ft-c of general illumination. (*b*) Same scratches and illumination, but viewed from a low angle. (*c*) Same scratches but with low-angle view in 75 ft-c spotlighting (From Faulkner & Murphy, 1973).

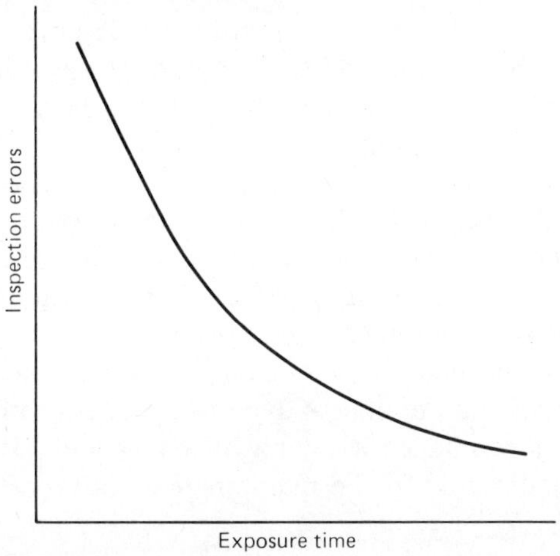

Figure 14-7. Inspection errors as a function of the time each item was exposed to view. (From Buck, 1975.)

While greater inspection speeds cause reduced inspection performance, greater speeds reduce quality control costs and so tradeoffs are usually necessary in the design of inspection tasks. In any case, removing any unnecessary visual constraints helps.

Improving Inspection Decisions by Improving Judgments and Social Pressures

A source of directly aiding the decision making part of inspection is achieved by clarifying the specifications that the inspector is checking, reducing the number of these specifications, and setting up a better check sequence. With fuzzy specifications, inspector judgments become more subjective. Training inspectors on the specifications helps. Since relative judgments are typically more precise and consistent than absolute judgments, photographs of minimally defective specifications can serve as an anchoring aid at the inspection station to prevent a drift in standards over time. Also some types of faults are more likely to result than others and if the inspectors check for faults in decreasing order of uncertainty (see Chapter 5), then less time is required per inspected item. Since some faults in manufacturing frequently occur in conjunction with other faults, some of those highly correlated faults can be removed from the list to be checked in order to reduce the inspectors' memory requirements.

Some of the problems with inspection are social. If Sam's friend operates the apple polishing machine, then it is much harder for Sam to reject apples that aren't adequately polished. A number of ways have been advocated to help solve this problem. One approach is *peer inspection*. When two people are assembling or making components, then each can be assigned the additional task of inspecting the other's work (Rigby & Swain, 1975). In this way peer pressure helps to improve the assembly operations. An independent inspector later in the process can sample check the possibility of peer collusion. Another suggested solution is *quality circles* where the production and inspection workers hold regular meetings together to iron out the causes in the loss of quality products. This is one form of participative management. The ill-fated attempt to improve quality through the zero-defects program was socially oriented but it failed to account for needed production and quality tradeoffs (Birkin & Ford, 1973). Needless to say, the 12,000 firms that had this program are now exhibiting hindsight bias.

TROUBLE-SHOOTING AND MAINTAINABILITY

Troubleshooting and maintenance activities are centuries old. But their importance was not generally recognized until World War II. There were a number of reasons supporting this recognition. The complexity of aircraft and war equipment rose abruptly in World War II compared to the past. Both military and civilian personnel saw the magnitude of the problem and the military advantage of better maintenance so that more equipment would be available when needed. It was then that research was first directed to the maintenance activity. Since then maintainability has been an important area of human factors and systems research and engineering. Some of the troubleshooting, maintenance, and repair procedures devised by NASA for needs in space exploration have reached great heights of sophistication (no pun intended).

Much of the research and many engineering studies have focused on the actual repair activities. The purpose was to see how much better systems could be designed so that quicker and more accurate tests and repairs could be made by the maintenance personnel. Also there was concern for less required training and fewer specialized tools. This area of study has been referred to by many as maintainability ergonomics.

While some of the maintenance activity occurs on a regular time basis, other repair activity occurs after a system fault is clearly identified or during the search for the fault. The activity of searching to isolate the fault and diagnose its cause is a task known as *troubleshooting*. Sometimes the troubleshooting task also includes the detection and correction of a fault and verification of the correction.

Fault Isolation and Diagnosis

Once a system is known to contain a fault, then the challenge is to locate the system component where the fault exists. Contemporary aircraft, power generating stations, computers, ships, or industrial plants are complex systems consisting of a hierarchy of subsystems, subsubsystems, and so on, down to components and elements. The degree of fault isolation within this hierarchy down to the finest detail is known as *resolution*. Fault location of higher resolution means that the troubleshooter has located the fault to a finer detail. The needed resolution in troubleshooting dictates to a large degree the complexity of the troubleshooting task. For this reason, many systems are designed with replacement components so that troubleshooting in the field need only be done in modest resolution to a replaceable component and then finer-resolution troubleshooting is done at specialized facilities. Therefore, an important design decision for systems maintainability is the resolution level required for field (site) troubleshooting compared to special-location final troubleshooting.

The Split-half Approach

In the early days of troubleshooting research this task was likened to a soothsayer who only told yes and no answers to queries of satisfactory operation. Putting a probe at a system location and turning on the test equipment was viewed as analogous to an inquiry. As one carries this analogy farther under the assumption that all components are equally likely to contain the fault before the search starts, then the obvious strategy for most quickly isolating the fault is to progressively split the system in half. Information theory (see Chapter 5) shows that a unique component out of 16 can be isolated in exactly *four* yes-no inquiries if each successive question asks if the fault is in a specific half of the remaining system. Accordingly this split-half strategy was advocated by many as the proper approach for troubleshooting. However, several practical difficulties intervened. Many systems could not be designed to follow this strategy because of the physics of the system and test points could not be planned for a middle split. Also faults were not necessarily unique (e.g., two or more system faults) so that the fact that one-half contained a fault did not preclude an additional fault in the other. Reliability engineers further complicated the story by showing economic limits to the design of components so that some components had a lower chance of containing a fault than others. While this observation did not deny the use of the information theory analogy, it did complicate the elegant simplicity of the split-half approach. However, some Sherlock Holmes buff pointed out that clues from test devices give more information to a troubleshooter than the presence or absence of a fault within a subsystem. This buff may have said, "My car's engine does not exhibit smooth running, good acceleration, and continued idling while parked. Why should I be concerned about the bumpers, tires, radio, and hood ornament?" These clues, along with knowledge of the system operation, indicate that the cause of the improper auto operation may be in the spark subsystem, the fuel subsystem, and/or the internal engine operations (e.g., a valve, seal, or piston). Each specific clue made a particular cause of fault more or less likely and the logic of these clues further contributed to these changing probabilities.

The Bayesian Approach

This detective concept led thinking away from the split-half strategy toward a Bayesian approach to diagnosis and trouble shooting. Under this Bayesian approach there are symptoms that one could observe and there are causes of the malfunction. If one could document the symptoms associated with a *known cause* over a series of malfunctions and find relative frequencies of each class of causes, then the probability of a cause given a set of symptoms could be found through Bayes' equation as $p(C|S) = p(S|C)p(C) \div [\Sigma p(S/C)p(C)]$ where $p(\ )$ is probability, C is a specific cause, S a specific set of symptoms, and the summation in the denominator is over all causes. Note that $p(C)$ is the probability of a cause *prior* to knowledge of the symptom set S and the $p(C|S)$ is the cause probability *after* this knowledge. When a particular symptom is unrelated to the cause (i.e., statistically independent or with low correlation), then the product $p(S|C)p(C)$ is approximately equal to $p(S)p(C)$ and $p(C|S)$ is approximately equal to $p(C)$. Therefore, the probability of a cause given a symptom is unchanged by the presence of the symptom. On the other hand, a highly related cause and symptom will result in $p(C|S)$ being vastly different from $p(C)$, denoting that one ought to disregard causes with low in differences between $p(C|S)$ and $p(C)$ and to concentrate on those with high differences. Example 14.1 illustrates a case of Bayesian diagnosis. The simple mathematical beauty of this approach gave it great appeal and the diagnosis problem was thought to be solved so long as good estimates could be made on $p(S|C)$ and $p(C)$. In some cases contemporary troubleshooting is enhanced through this concept. However, as investigations went on, it was discovered that there were problems with this naive statistical approach. Sometimes data toward the estimate of $p(S|C)$ was entered where the specific cause was uncertain because the operation of tearing down the equipment made it difficult to be sure of the cause. Also it took a lot of time to get good estimates of the relative frequency of causes; particularly when there were only a few systems built. More importantly, however, the large number of possible sets of symptoms made the data collection exceedingly difficult for any reasonable resolution. Therefore, this naive Bayesian approach to diagnosis never was highly successful unless system knowledge was sufficient to reduce the symptom set to a small manageable collection.

Example 14.1 A special electronic device has been investigated by reliability experts and it has been estimated that the probability of a cause when failure occurs is: .25 due to loss of a ground connection, .20 due to an inadequate voltage supply, and .55 for all other causes. There were failure and repair records also available where various symptoms were reported with failure incidents that included a red warning light, an oscillating meter reading, or neither. In these 91 failure-repair records with symptoms and verified causes, the frequency of incidents was as follows:

Symptoms	C_1 Ground	C_2 Voltage Supply	C_3 Other	Sum
S_1: red light	19	1	1	21
S_2: oscillating meter	2	8	0	10
S_3: neither	7	0	53	60
Sum	28	9	54	91

These records of symptom-cause frequency could be used to compute the probability of a symptom given a cause but with the limited number of records, it was decided to add the constant 1 to every cell in the above table for smoothing these data and then estimate the probability of the symptom-given-a-cause as the revised table entry divided by 100. For example, $p(S_1|C_2) = 2/100 = .02$. Cause probability estimates could be made from the reliability studies. With these two forms of probability estimates, Bayes' equation can be used to determine probability estimates of causes given specific symptoms. For example, the probability of a faulty ground when the red warning light is on is

$$p(C_1|S_1) = \frac{p(S_1|C_1)p(C_1)}{p(S_1|C_1)p(C_1) + p(S_1|C_2)p(C_2) + p(S_1|C_3)p(C_3)}$$

$$p(C_1|S_1) = \frac{(20/31)\ (.25)}{(20/31)\ (.25) + (2/12)\ (.20) + (2/57)\ (.55)} = .754$$

On the other hand the probability of an inadequate voltage supply being associated with a red warning light is

$$p(C_2|S_1) = \frac{(2/12)\ (.20)}{(20/31)\ (.25) + (2/12)\ (.20) + (2/57)\ (.55)} = .156$$

In contrast to the above, the probability of an inadequate voltage supply being associated with an oscillating meter is:

$$p(C_2|S_2) = \frac{(9/12)\ (.20)}{(3/31)\ (.25) + (9/12)\ (.20) + (1/57)\ (.55)} = .816$$

Where the probability of an inadequate voltage supply with an oscillating meter (.831) compared to the red warning light symptom (.156) is 5.33 times more likely (.831/.156). The full table of cause probabilities without knowledge of the symptom relationship and with knowledge of the three classes of symptoms is

	Probabilities			
Cause	Without Knowledge	With the Warning Light	With the Oscillating Meter	With Neither Symptom
C_1: ground connection	.250	.754	.132	.107
C_2: voltage supply	..200	.156	.816	.028
C_3: other	.550	.090	.052	.865

Input-Output Matching and Patterns

Investigations toward split-half and Bayesian diagnosis pointed toward the need to embed system knowledge into the troubleshooting process. One basis for doing this is to take a component of the system and provide a test pattern of input signals. Knowledge of the component operations would then specify what the output signals should be if the component were functioning properly. As one gets a bit more sophisticated in this approach, two or more connected components could be tested together as a macro component or a small subsystem. The concept here is that of working from inside the system outward as opposed to a top-down approach from the overall system toward details. Another aspect of sophistication with this approach is to first identify the collection of components where the fault existed. Then a pattern of input signals is specified so that the output signal pattern can isolate the specific component or components where the fault existed, thereby reversing the approach once a collection of components was known to contain a fault. Sometimes the Bayesian approach was employed with the collection of components in order to isolate the specific component and cause. However, this conceptual approach did not denote the sequence of attack on component collections. In practice troubleshooters frequently select the sequence of component collections based on the highest relative frequencies of past observed failures or collections with the lowest reliabilities.

Higher sophistication in the input-output approach has led to input-output pattern matching with signal timing and output patterns as *signatures* to specific fault causes. Some of these patterns or signatures have been empirically identified while many are a result of theoretical predictions from expected failure causes. In many cases the fine visual pattern recognition of people is an important asset to troubleshooting. However, the result of much of this sophistication has led to automatic forms of troubleshooting where humans serve computer-programming needs by placing probes as told and in doing other tasks as directed by the computer program. The notion behind these human-assisted methods of automatic troubleshooting is that systems operations are sufficiently complex that experts in the component designs are needed and these experts are represented by the programs they devise to do the troubleshooting. Onsite people are then used as supplemental devices (e.g., moving probes and checking things that the computer program cannot, such as seeing if the electricity is on).

Network Diagnostics

Electronic systems and others as well are often described as nodes and connecting arcs analogously to a transistor and a connecting wire. Troubleshooting tests are made at the output side of the nodes, which show if the node element and the inputs are working by a test result that either denotes a satisfactory result or an unsatisfactory result. A particular node may only operate satisfactory if *all* elements leading to it are satisfactory (i.e., as an *and* gate) or if at least one of the elements leading to it is satisfactory (i.e., as an *or* gate). In both cases the node itself must also operate satisfactorily if the test equipment so indicates. In this concept of troubleshooting all the subjective judgments about the qualitative and quantitative signals from the node are handled automatically but the human troubleshooter must decide the sequence of nodes to be tested. The human troubleshooter must also use logic in these judgments as modified by test results (Rouse, 1978, 1981). Obviously this logic depends on the network configuration (Rouse & Rouse, 1979). For example the two networks of nodes in Figure 14-8 would generate different test sequences as the example below shows. This example shows that the average number of trials needed to find a solution to the troubleshooting

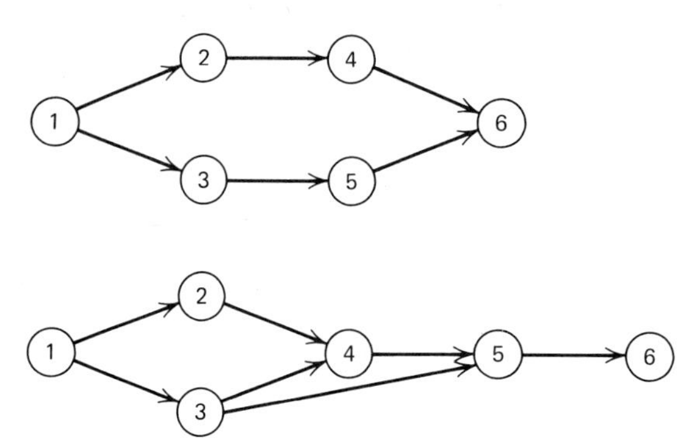

Figure 14-8. Two networks of connected components as **And** nodes.

problem changes both with the configuration and with the probability of a failure at any of these six nodes.

Example 14.2 Table 14.3 shows the probable test sequences for both networks shown in Figure 14-8 given the fault location in each node. The final test sequence number is the number of tests required. If the fault were equally likely to be in any of the six nodes, then the average number of tests and the standard deviation of this number for Network A is respectively $4\frac{1}{6}$ trials and 0.687 trials. For Network B these two statistics are respectively $4\frac{1}{3}$ trials and 1.312 trials. Accordingly Network B is more time consuming on the average and more variable in time than Network A under these conditions. However, if the probability of failure is different for these components, then the average number of tests needed differs for these test sequences. Some examples are illustrated as follows.

Components						Average Number of Tests Needed	
1	2	3	4	5	6	Network A	Network B
.25	.25	.125	.125	.125	.125	4.250	4.625
.125	.125	.125	.125	.25	.25	4.000	3.875
.25	.125	.125	.125	.125	.25	4.000	4.250
Probabilities of component failure							

Empirical tests on humans performing network troubleshooting tasks have shown that people can perform well on smaller network problems of this type but human performance tends to deteriorate with larger and more intricate networks. Memory limitations appear to be a substantive cause of this deterioration. These empirical data, as Example 14.2 indi-

TABLE 14-3 *Test Sequences for the Figure 14-8 Networks with Faults*

Network	If Fault Is in	Components Tested						
		1	2	3	4	5	6	
	1	4−	3−		2−		1−	Probable test *sequence* and test results (− indicates a failed test and + a satisfactory test)
	2	5+	4−		3−	2+	1−	
	3	5+		4−	3+	2−	1−	
A	4		4+		3−	2+	1−	
	5			4+	3+	2−	1−	
	6				3+	2+	1−	

Network	If Fault Is in	Components Tested						
		1	2	3	4	5	6	
	1	6−	5−	4−	3−	2−	1−	Probable test *sequence* and test results (− indicates a failed test and + a satisfactory test)
	2	5+		4+	3−	2−	1−	
	3		5+	4−	3−	2−	1−	
B	4		5+	4+	3−	2−	1−	
	5				3+	2−	1−	
	6					2+	1−	

cates, show that the nature of the network arrangements as well as the problem size affect the number of trials and the time needed to locate the fault. Human pattern recognition ability was also a factor in better performance, but there are large individual differences other than pattern recognition that affect the apparent strategy of troubleshooting and hence performance.

Automanual Troubleshooting

Another troubleshooting approach is a composite of automatic pattern matching methods with allowed human interaction and control. In this concept human controllers can call on the given mathematical techniques of analysis, perform intermediate search manually, or any mixture which they choose. Based on recent field tests (Mitchell & Buck, 1981), this highly flexible approach appears to provide excellent promise because it allows for high individual differences in abilities and strategies as well as supplementing human deficiencies in memory and computing abilities. Less experienced maintenance people have aids available to help them but people with detailed knowledge of the system can devise their own tests and go directly to components suspected. In addition, automanual troubleshooting allows for a modification of strategy to fit the complexity of the apparent problem complexity and the clues found along the way rather than giving a rigid approach to meet the worst suspected conditions.

Test Noise in Troubleshooting

Almost all approaches to troubleshooting carry the implicit assumption that information derived from test devices is without error. The fact is that the test equipment can contain faults in its design, manufacturing, operation, or through use. Obvious faults with test equipment are readily discovered but subtle faults pose particular difficulties to detection and to the locating of faults in the systems that the test equipment is used on. One of the principal difficulties with test equipment is when the test result states that a condition isn't present just because it has a low probability. The troubleshooter may be highly misled and make a lot of hypotheses built on the misinformation. Such noise in test results creates a spread of doubt in the troubleshooter. If this noise occurs in the Bayesian approach, the whole analysis can blow up because the conditional statement must be true. However, noise in test results is still an unresolved question.

Fault Detection

Part of the job of pilots, process controllers, and ship controllers, and many others is to detect the presence of a fault, in addition to diagnosing an appropriate response and acting on it. In many contemporary systems a computer is employed to assist in this task. When this task is to be completely automated, the designer must precisely define all malfunctions in advance. Since this design task is usually an impossible one, there typically are malfunction detections assigned to the process monitoring computer in various forms of automatic protection systems and a residual of malfunction detections are assigned to operating personnel. However, the philosophical basis for this assignment remains a strong controversy.

An important basis of human detection of faults of process situations that could readily lead to a fault is the operator's prior expectations based on past experience and his or her model of the process (Anyakora & Lees, 1972). Information that shows deviations from these expectations is usually delivered to the operator via visual displays (see Chapter 7). Numerous situations exist where only current status information is available, requiring operator memory and performing signal integration over time. Other situations exist where operators have historical displays for fault detection. Figure 14-9 shows some typical strip chart recordings of conditions that imply the existence of some fault. Such cases as signal drift, erratic noise, or other types of signal variations indicate a system malfunction and human visual pattern recognition is well developed and a distinct advantage. While it is often uncertain that the fault is in the process equipment, sensors, or other parts of the information reporting system, most designs incorporate sufficient informational redundancies between different information displays that checks can be made to ascertain and correct the problem. When displays do not carry the necessary

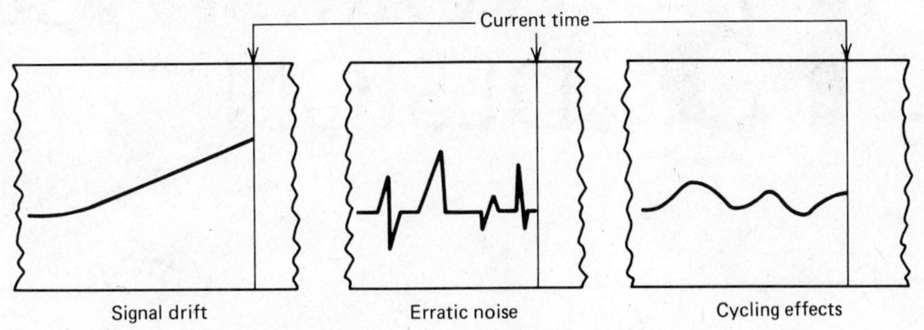

Figure 14-9. Examples of strip-chart recordings of system fault effects.

check information, other personnel can often be dispatched to make a direct check on the equipment in question.

SUMMARY

Human decision making is an important part of many human tasks. This is as it should be because human goals are needed for our systems. There is concern, however, whether human decision making is rational. When we compare human decision making as it is currently done to how some experts say it ought to be done, there appear to be some discrepancies. Part of these discrepancies are human biases and part are deficiencies of normative theories. In practice, there are a number of things that human factors people can do to improve system performance. Such improvements are directly related to the decision making, providing aiding displays and analytical techniques. In other cases, system performance can be improved by enhancing the necessary sensory processes and reducing interferences. Some of these aiding features are illustrated in the two areas of application just described.

15 WORKSPACE DESIGN

A well-designed human-machine system must have more than good or even optimum displays and controls. These system components must also be located in the right places. The world's most perfect visual display does no good if the operator cannot see it because it is located behind him or her. Similarly, the best control is useless if it is beyond the operator's reach. The essence of proper workspace design is to locate components so that the operator is able to utilize them easily.

In a system of any complexity, it is impossible to locate all components in the best possible place, that is, where a component would be put if it were the only display or control in the entire system. As is often the case in human factors design, the specialist will try to satisfice (see Chapter 14) since the optimal solution may either be impossible due to conflicting system demands or to limitations of time and resources allocated to finding an optimal solution. In order to achieve a workable design, the human factors specialist must have detailed knowledge about the operators as well as various relations among system components. In this chapter, you will learn the rudiments of engineering anthropometry and design layout. Anthropometry refers to the physical dimensions of people and answers such questions as how far can the human arm reach and how much pressure can the human foot exert on a pedal. Design layout takes into account not only the position of components relative to the operator but also various functional relationships among the components themselves. The chapter ends by using specific design applications, such as office equipment and vehicles, to illustrate workspace design.

ENGINEERING ANTHROPOMETRY

There has always been a close relationship between measurement and the human body. Even today, linear distances are measured in feet and horses are measured in hands. Since measurement started with parts of the body, it is rather appropriate that modern application of scientific measurement techniques is today used to better fit equipment to the human body. Since the human body comes in only two models, male and female, this knowledge is easily standardized. Engineering anthropometry is defined as the application of scientific physical measurement methods to the human body in order to optimize the interface between humans and machines and other manufactured products. Such body measurements assure that manufactured goods will be suitable for intended user populations (Roebuck, Kroemer, & Thomson, 1975). Our discussion of engineering anthropometry starts with a review of necessary statistical concepts that can be skimmed or omitted by readers with background in descriptive statistics. Some results and methods in static measurements, dynamic measurements and muscular strength capabilities are then presented.

Statistical Concepts

The goal of anthropometry is to obtain a set of physical measurements that accurately describe some population of users. The key word here is population. A population is an entire set of individuals: for example, all U.S. citizens of voting age, all individuals with valid driving licenses, all airline passengers during 1983. It is seldom feasible to obtain measurements from every member of a population. Therefore, some available fraction of the population, called a *sample,* is measured instead. Note that if the entire population is small and available, sampling is unnecessary and statistics are much simplified, if used at all. Thus, when NASA was designing the Mercury capsule there were only seven astronauts in the entire population and all were carefully measured. This is a rare situation and sampling is almost always used. It is essential that the sample be truly representative of the population. Statisticians have developed several strategies to generate unbiased samples from a population. The simplest is random selection where every individual in the population has an equal chance of being picked. A more complex procedure is stratified sampling. Here the population is first segmented based on one or more distinctive characteristics such as sex or age. Sampling is usually random within each segment. Even more complex sampling schemes can be found in statistics texts. However, all share the same goal of achieving an unbiased representative sample.

It is absolutely essential that the human factors specialist using anthropometric data be fully aware of the population from which the data are drawn. If the data are to be applied to a different population, great caution is required. Let us take the example of a car manufactured in Japan that was designed for the Japanese population. Japanese people are smaller than Americans. This, it makes sense for the manufacturer to place the brake and clutch pedals closer together than in customary in American cars. But once this same car is exported to the United States a serious problem emerges. A certain proportion of larger American males will be unable to step on the brake without simultaneously stepping on the clutch pedal. Americans have larger feet than Japanese. Using anthropometric data derived from one population (Japanese) caused the cars to be unusable for members of a different population (Americans). Some of the most detailed and comprehensive anthropometric data available are based on a population of military personnel. These people tend to be younger and in better physical condition than the population of civilians. Thus, these data cannot be used directly as the design basis for civilians. For example, a maintenance passageway that would allow two 20-year old naval technicians to just pass one another might be a very tight squeeze for two portly 55-year old civilian technicians. Thus, the human factors designer must try to obtain anthropometric data tables based upon the actual user population. In critical cases, this may mean that data collection will be required if no handbook contains usable data.

A (descriptive) statistic is a number that describes some characteristic of a population. Since the statistic is based on a sample, not a population, it is subject to error. The statistic is not a perfect predictor of the population value. There are several ways to make statistics predict better. The simplest is to increase the number of observations in the sample. Larger sample sizes yield better predictions. Tables of anthropometric data contain numbers that tell how well the sample data predict population values. These will be discussed shortly. First, you must discover what kinds of statistical data are presented in the typical anthropometric data table.

The basic statistic used is the percentile. Although the percentile can theoretically fall between 0 and 100, in practice tables almost always have 99 as the highest percentile and 1 as the lowest. The percentile is the percentage of the population having a body dimension equal to or less than that indicated in the table. For example, the 95th percentile for overhead reach (the distance from the floor to the fingertip of a standing man whose arm is pointing straight up) is 87.6 inches for Air Force officer flying personnel. An officer whose overhead reach is 87.6 inches can reach higher than 95% of his fellow officers. The 5th percentile for overhead reach is 78.6 inches. A control located on the ceiling at a height of 78.6 inches could not be reached (without jumping) by 5% of Air Force officers.

There are some common errors you should avoid when using percentiles. It is dangerous to add percentile values. For example, the 50th percentile value for fingertip to elbow length plus the 50th percentile value for elbow to shoulder length will not equal the 50th percentile value for arm length. The reason for this is simple: there is no such animal as the 50th percentile person who exhibits precisely the values of the 50th percentiles for all body dimensions. These values represent a statistical average similar to the average taxpayer or the average family with 2.3 children and not any specific person. Real (nonstatistical) people exhibit considerable variation in body dimension as can be seen in the profile analysis of Figure 15-1. This plots the percentile values for twelve body dimensions for three real people. A horizontal line would indicate a person whose various body dimensions remain fixed at the same percentile.

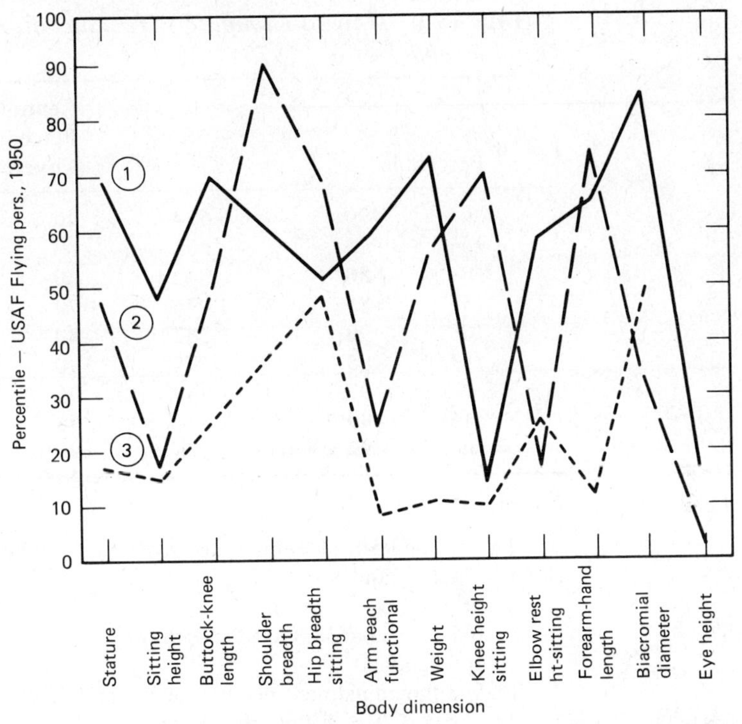

Figure 15-1. A profile of anthropometric measurements from individual subjects. Each line represents one person. Note that none of the lines are flat. (From Roebuck et al., 1975, Wiley.)

None of the lines are flat. The **n**th percentile person is only a statistical abstraction.

Repeated sampling of some body dimension from a set of individuals tends to form a normal (Gaussian) distribution (Figure 15-2). The 50th percentile, also called the median, is at the peak of this distribution. Normal distributions can have values that remain close to the peak (a tall and skinny-looking distribution) or values that are far from the peak (a fat and

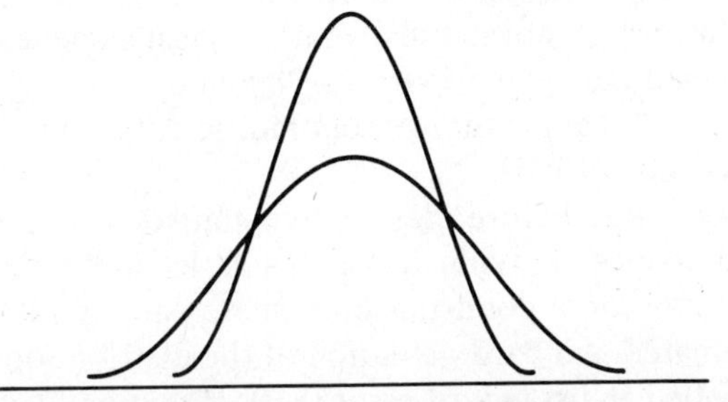

Figure 15-2. Two normal distributions with different variances.

TABLE 15-1 *How to Compute Percentiles from Standard Deviations (from Roebuck et al., 1975)*

Percentile		K_1	Central Percent Covered	$K_2 = 2K_1$
30	70	0.524	40	1.045
25	75	.674	50	1.349
20	80	.842	60	1.683
15	85	1.036	70	2.073
10	90	1.282	80	2.563
5	95	1.645	90	3.290
2.5	97.5	1.960	95	3.920
1.0	99.0	2.326	98	4.653
0.5	99.5	2.576	99	5.152

Examples:

1. To find the 95th percentile, use $K_1 = 1.645$. When $\bar{X} = 35.1$ in. and $S = 1.5$ in.;
   $1.5 \times 1.645 = 2.5$ in.
   $35.1 + 2.5 = 37.6$ in., the 95th percentile.
2. To find the adjustment needed for the middle 90% of the same group:
   $1.5 \times 3.29 = 4.9$ in., the range of adjustment.

wide-looking distribution). The "spread" of a distribution is called its dispersion. A common index of dispersion is the standard deviation (S) of a distribution. As the standard deviation gets larger, the distribution becomes more dispersed (more spread out or fatter). Tables of anthropometric data often state the standard deviation of the sampling distribution. In a normal distribution the median and mean (arithmetic average) are equal. If the mean ($\bar{X}$) and standard deviation of a normal distribution are known, percentile values can be calculated using Table 15-1.

Sometimes distributions depart from normality (Figure 15-3). If a distribution is not symmetrical, this departure from normality is called skewness. A normal distribution has a skewness of 0. Another departure is called kurtosis (Figure 15-3*c* and *d*). For a normal distribution kurtosis equals 3. Tables of anthropometric data sometime have values for skewness and kurtosis.

As stated before, descriptive statistics are merely estimates of population values. Having learned a little about distributions, you can now calculate how good these estimates are. To do this, you must know the estimated standard deviation of the distribution of sampling errors, which is called the standard error (SE). If you know the standard deviation (S)

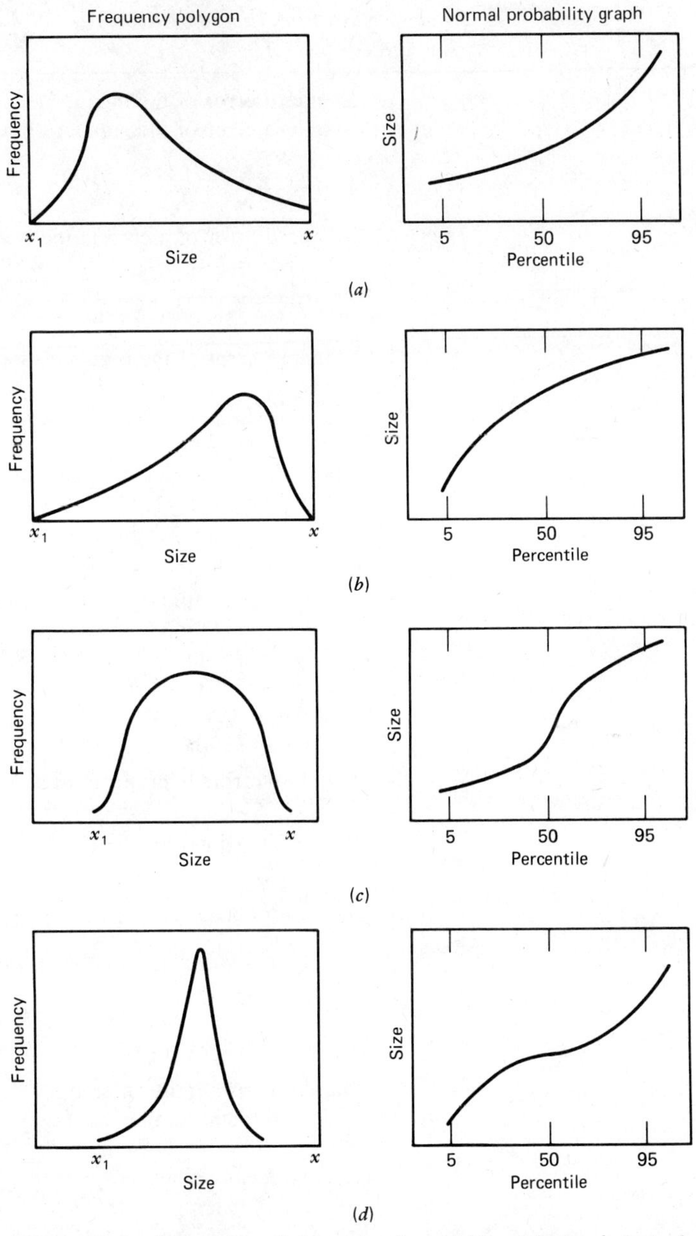

Figure 15-3. Examples of skewness (a and b) and kurtosis (c and d). (a) Skewed to the right (negative a_3). (b) Skewed to the left (positive a_3). (c) Platykurtic (negative $a_4 - 3$). (d) Leptokurtic (positive $a_4 - 3$). (From Roebuck et al., 1975, Wiley.)

TABLE 15-2 *Standard Error Formulas for Common Statistics (from Roebuck et al., 1975)*

1. Standard error of the mean: $S/\sqrt{N}$
2. Standard error of the standard deviation:

$$\frac{S}{\sqrt{2N}} = 0.71 \times \text{SE Mean}$$

3. Standard error of a correlation coefficient:

$$\frac{1-r^2}{\sqrt{N-3}}$$

4. Standard error of the percentiles:

$$\frac{k(100-k)S}{100 f_k N}$$

where f_k is the ordinate of the normal curve at the kth percentile. Or

30th through 70th	$1.3 \times S/\sqrt{N}$
20th, 25th, 75th, 80th	$1.4 \times S/\sqrt{N}$
15th and 85th	$1.5 \times S/\sqrt{N}$
10th and 90th	$1.7 \times S/\sqrt{N}$
5th and 95th	$2.1 \times S/\sqrt{N}$
3rd and 97th	$2.5 \times S/\sqrt{N}$
2nd and 98th	$2.9 \times S/\sqrt{N}$
1st and 99th	$3.7 \times S/\sqrt{N}$

5. Standard errors of proportions and percentages:

(a) a proportion p: $\sqrt{\frac{p(1-p)}{N}}$

(b) a percentage $P\%$: $\sqrt{\frac{P(10\ -P)}{N}}$

6. Standard error of the coefficient of variation:

$$\frac{V}{\sqrt{2N}}$$

7. Standard error of the difference between the values of the same statistic computed on two independent samples:

(a) two mean values: $\sqrt{\frac{S_1^2}{N_1} + \frac{S_2^2}{N_2}}$

(b) two standard deviations: $\sqrt{\frac{S_1^2}{2N_1} + \frac{S_2^2}{2N_2}}$

(c) generally: $\sqrt{\text{SE}_1^2 + \text{SE}_2^2}$

and the number of observations or samples (N) you can calculate the standard error for various statistics by using Table 15-2. Knowing the standard error will allow you to improve product design as the following example shows.

You are using anthropometric weight data to design a ladder. It is important that this ladder be as light as possible so it can be carried by one person but it still must be strong enough to support people safely. You find that the weight of the 50th percentile user is 180 pounds with a standard error of 10 pounds. Designing the ladder to hold 180 pounds on each rung would be a mistake. "Aha," you say. "I'll add the standard error and make each rung hold 190 pounds." This is better but you would be wiser to consult Table 15-1. Since weight is normally distributed you can determine the weight of a 99th percentile user. Multiply the standard error (10) by 2.326 (from Table 15-1), which equals 23.26. Add this to the mean obtaining $180 + 23.26 = 203.26$ pounds. The ladder should be designed to hold 203.26 pounds to be safe for 99% of the users. This technique of using the standard error is called finding a confidence limit. While the ladder example used a 99% confidence limit since the relative cost of strengthening the ladder was low, most of the time a 95% confidence limit is used since it is more economical. To obtain 95% confidence limits merely multiply the standard error by 1.645 (from Table 15-1) if you wish to establish only an upper (or only a lower) limit. (Statisticians call this a one-sided estimate.) If you need to establish an upper and lower limit simultaneously, (a two-sided estimate), multiply the standard error by 1.96 to obtain a 95% confidence limit.

Static Measurements

Static measurements are obtained when the body is stationary and at rest. For example, your height measured during a routine physical examination is a static measurement. A complete set of static measurements of your own body would allow construction of an accurate manikin—a three-dimensional model such as found in clothing store windows—that would contain all the anthropometric information to duplicate the various parts of your body, provided that the manikin was stationary. As we will see in the next section, the requirement for no motion limits the use of static measurements. However, static measurements are easy to obtain and many tables of static data are available to the human factors designer.

Static measurements include linear dimensions such as arm length as well as circumferential dimensions such as waist circumference. Mechanical devices such as anthropometers, calipers, compasses and tapes are used to make static measurements (see Figure 15-4). It is also possible to use photographic methods to obtain static measurements. Static measurements are often related to the three body planes—sagittal, coronal, and transverse—illustrated in Figure 15-5.

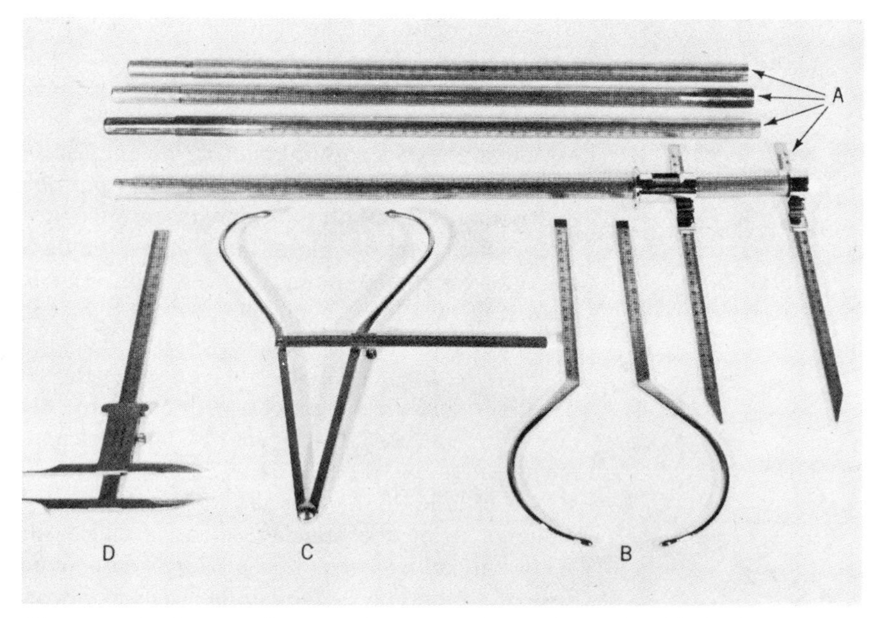

Figure 15-4. Some anthropometric measuring instruments: (*a*) Anthropometer, disassembled and put together. (*b*) Curved branches for anthropometer. (*c*) Calipers. (*d*) Sliding compass. (Photo by James Abbott; from Roebuck et al., 1975, Wiley.)

As an example of a typical static measurement, we have selected the buttock-knee length of a female member of the U.S. Army from a survey conducted by the Clothing, Equipment & Materials Engineering Laboratory in 1977. Instructions for obtaining this measurement from the sample of Army women are contained in Figure 15-6. Such detailed instructions are aimed at obtaining the same posture from all sample members that are tested. This is very important because uncontrolled changes in posture can increase the variability of measurement. For example, chest circumference changes depending upon whether you have just inhaled or exhaled. Standardized instructions for chest measurement will specify respiration conditions.

The results of the survey are shown in Figure 15-7 along with certain summary statistics. The buttock-knee length is used to design all kinds of chairs and seating equipment, from office furniture to vehicle cockpits. Examining Figure 15-7 shows a range of values from 15.1 cm for the 1st percentile to 65.3 cm for the 99th percentile; this total range is thus 14.2 cm, the 99th percentile value minus the 1st percentile value (65.3 − 51.1 = 14.2 cm). If however, we select a range of values that goes from the 5th to the 95th percentile, the range is smaller: 63.33 − 53.09 = 10.13 cm. The median value is found by checking the 50th percentile, in this case equal to 57.67 cm.

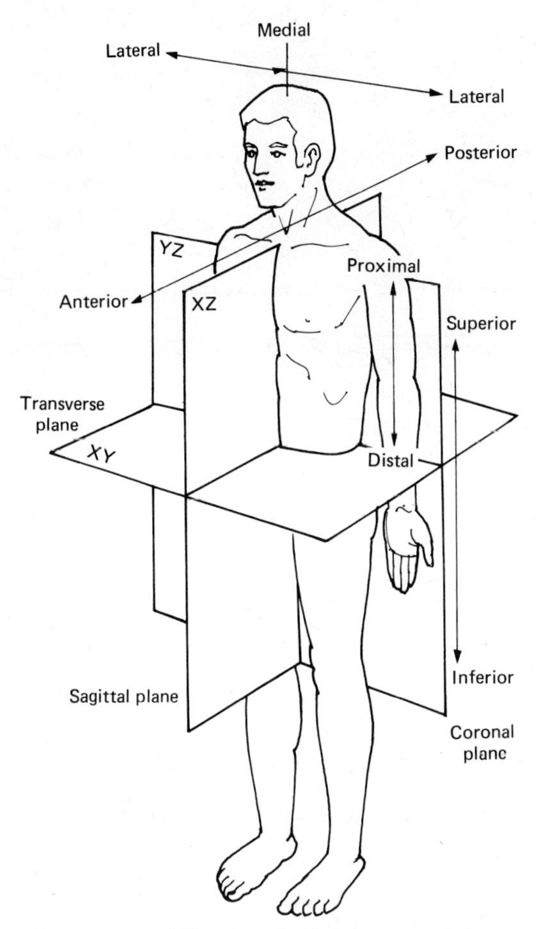

Figure 15-5. The three body planes and the names for location relative to them.

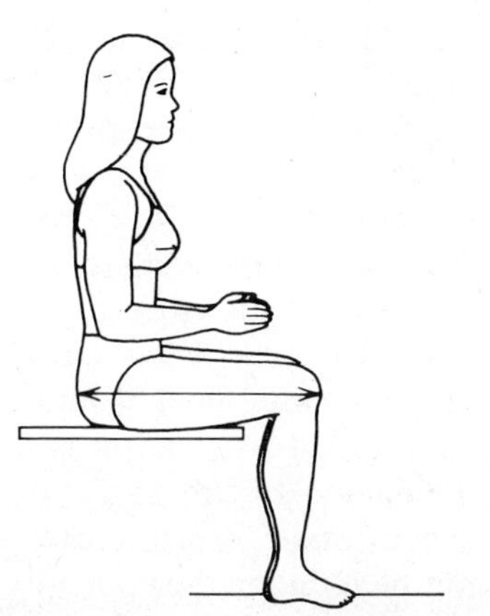

Landmark: Buttock, maximum protrusion.

Instrument: Beam caliper.

Position of Subject: Subject sits erect on a flat surface, looking straight ahead, feet on a platform which is adjusted so that the knees are together and flexed 90°.

Procedure: With a beam caliper held parallel to the long axis of the thigh, measure the horizontal distance from the most posterior aspect of the right buttock to the most anterior aspect of the right knee.

Figure 15-6. Instructions for measuring buttock-knee length. (From Churchill et al., 1977)

The Percentiles

Centimeters		Inches
65.30	99th	25.71
64.53	98th	25.41
63.99	97th	25.19
63.22	95th	24.89
61.98	90th	24.40
61.13	85th	24.07
60.45	80th	23.80
59.88	75th	23.57
59.37	70th	23.37
58.91	65th	23.19
58.48	60th	23.02
58.07	55th	22.86
57.67	50th	22.71
57.29	45th	22.55
56.90	40th	22.40
56.51	35th	22.25
56.12	30th	22.09
55.69	25th	21.93
55.23	20th	21.74
54.71	15th	21.54
54.05	10th	21.28
53.09	5th	20.90
52.44	3rd	20.64
51.94	2nd	20.45
51.10	1st	20.12

The horizontal distance from the most posterior protrusion of the buttock to the most anterior point of the kneecap

The Summary Statistics

Centimeters		Inches
57.85	Mean	22.78
.08	SE (M)	.03
3.06	ST DEV	1.21
.06	SE (SD)	.02
	****	
Coef. of variation		5.3%
Symmetry—VETA I		.24
Kurtosis—VETA II		2.95
	****	
Number of Subjects		1331
	****	

Figure 15-7. Results for buttock-knee length measurements. (From Churchill et al, 1977)

The summary statistics tell us that the distribution of buttock-knee lengths is very close to a normal distribution. This is not surprising since the sample included 1331 subjects as noted in Figure 15-7. As sample size increases, the distribution of values is more likely to be normal. With over 1000 subjects we would expect a near normal distribution. There are several ways for you to check on the normality of the distribution. First, the mean of 57.85 is quite close to the median of 57.67 cm. You will recall that mean and median are identical in a normal distribution. Second, symmetry of .24 is close to the value of 0.0 for a normal distribution. Third, kurtosis of 2.95 is close to the value of 3.0 for a normal distribution. Thus, a designer would be safe in treating this distribution as if it were a normal distribution as shown in Figure 15-2.

Once you read the data in Figure 15-7, you will be able to understand just about any table of static measurements. Similar tables for other body dimensions can be found in the appendix at the end of this chapter.

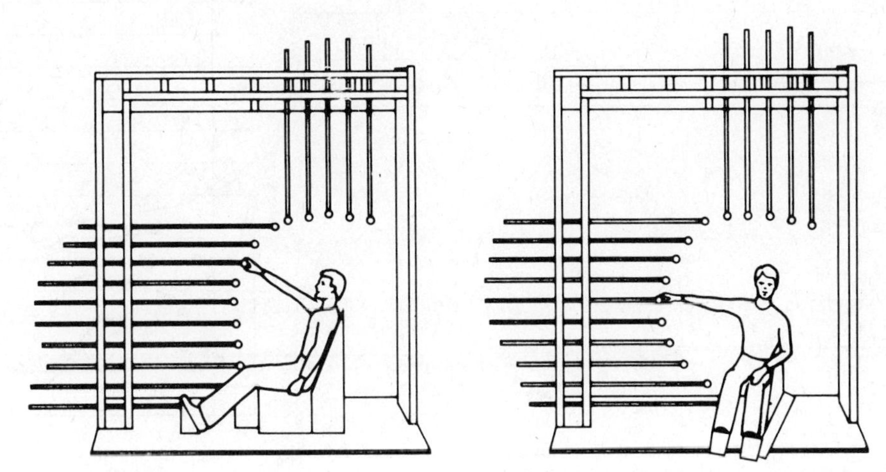

Figure 15-8. Device used to measure functional arm reach. (From Roebuck et al., 1975, Wiley.)

Dynamic Measurements

Few tasks require the human to be rigid and motionless. Often the designer is more concerned with how far you can reach rather than the length of your arm. Thus, the static measurement of arm length is of limited use for this purpose. What is needed is a dynamic measurement defining a volume of space that can be reached by a population. This spatial volume is called a reach envelope (Figure 15-8). Dynamic measurements involve a human in motion who is performing some task or function. They require more complex instrumentation than static measurements but often yield data that are more helpful to the equipment designer.

Reach envelopes are obtained relative to the seat reference level (Kennedy & Bates, 1965) as shown in Figure 15-9. For example, subjects are required to grasp knobs at set distances (e.g., 10, 20, or more inches etc) above the seat reference level (SRL). Each distance creates one graph of angular reach. For example, Figure 15-8 shows results at the 20-inch level. Other levels can be found in the appendix at the end of the chapter. Boundaries are usually shown for 5th, 50th and 95th percentiles.

Muscular Strength

Strength is the maximal force a muscle can exert isometrically in a single effort for a very short time period. This definition has some qualifications that need to be discussed. Maximal force means the greatest or peak force but this depends on precise instructions. For example, Figure 15-10 shows that instructions to hold a maximal force for five seconds yield a lower estimate of strength than instructions to apply gradually increasing force until your maximum is reached.. Isometrically means that the length of the muscle remains constant while the force is exerted. This

Reach: Horizontal Boundaries, 20-in. Level

Angle (deg)	N	Min	Percentiles (in.) 5th	50th	95th
L165					
L150					
L135					
L120					
L105					
L 90	11			14.00	18.75
L 75	16			18.00	21.50
L 60	20	17.00	17.50	20.50	24.50
L 45	20	18.25	19.50	22.75	26.75
L 30	20	20.25	21.50	24.75	28.25
L 15	20	22.50	23.50	26.75	29.75
0	20	25.00	25.50	28.75	31.75
R 15	20	27.25	28.00	30.50	34.00
R 30	20	29.00	30.00	32.00	35.75
R 45	20	30.50	31.00	33.50	36.25
R 60	20	31.50	32.00	33.75	36.25
R 75	20	31.50	32.25	34.00	36.50
R 90	20	31.75	32.25	34.00	36.00
R105	20	31.50	31.75	33.50	35.75
R120	19		30.50	33.00	35.50
R135	9				34.50
R150					
R165					
R180					

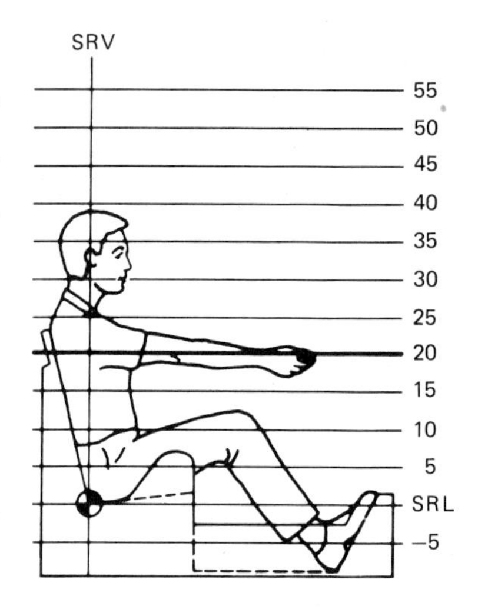

Angular Reach from SRV at the 20-inch Level

Minimum	L65° to R110°
5th %ile	L66° to R122°
50th %ile	L90° to R134°
95th %ile	L90° to R146°

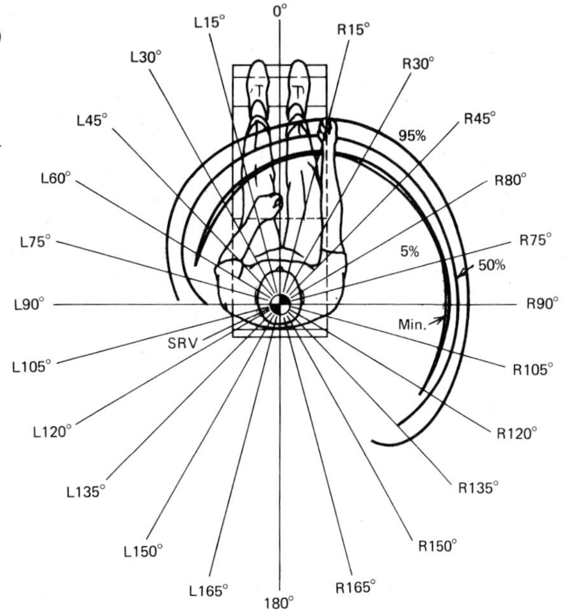

Figure 15-9. Reach envelope at 20-inch level. SRV means seat reference vertical. (From Kennedy, 1964.)

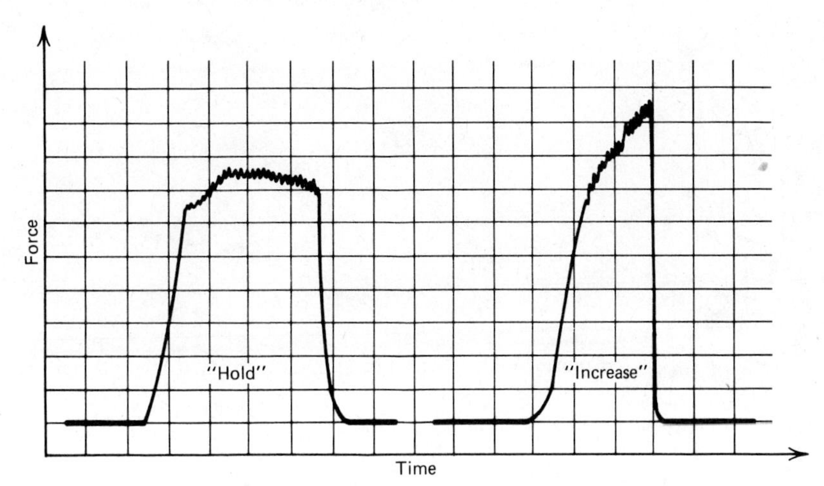

Figure 15-10. Different instructions generate difference maximum force levels. (From Roebuck et al., 1975, Wiley.)

means that static strength is being measured. (Sometimes the term "isotonic" is used in strength measurements. This means that muscle tension is constant. Isotonic is not the same as isometric.) Finally, a very short time period means only a few seconds. All of these qualifications mean that when a handbook or other source is consulted for strength data, one must carefully consider how the measurements were made and what instructions were given.

Some representative strength data for Army women are shown in Figure 15-11. Note that the pull is from a point 38 cm above the ground; lifting an object on the ground is more difficult, especially for women since the socket of the female hip joint is in front of the body's center of gravity (Tichauer, 1978). The median woman can exert a peak force of about 140 pounds using arms, shoulders, and legs. Men in their thirties can exert approximately half again as much strength as women the same age (Hertzberg, 1972). Additional strength data for men and women can be found in the end-of-chapter appendix.

The load that a person can lift depends on age, sex, and method of lifting. There are detailed methods for calculating biomechanical lifting equivalents for various lifting procedures and geometric loads (Tichauer, 1978). However, for occasional lifting by any method the values in Figure 15-12 can be used to establish safe limits.

Endurance is the ability to continue to exert force over time. The amount of physical work a person can accomplish is limited much more by endurance than by strength. Endurance depends on the percent of strength required by a task (Figure 15-13). This implies that for a particular task requiring some particular strength—say, lift a 35-pound package and toss it into a bin 5 feet away—people with greater strength will

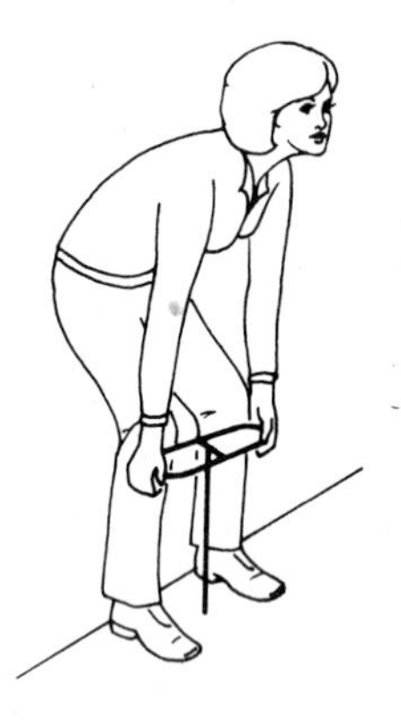

Figure 15-11. Strength statistics for two-handed pull. (From Churchill et al., 1977.)

Standing Two-Handed Pull: 38 Centimeter Level

The subject stands with her feet 45 centimeters apart and her knees bent. She bends at the waist and grasps both sides of the long handle which is attached 38 centimeters above the platform and directly in front of her. She is instructed to minimize pull with her back to lessen the chance of injury. She attempts to lift the handle, primarily using the arms and shoulders but also using her legs by extending them upwards.

Summary-Statistics

	kg lb	kg lb			
Mean force—1	M = 56.6/126.8	SD = 15.2/33.6	V = 26.9%	V-I = .2	V-II = 2.8
Mean force—2	M = 58.3/128.5	SD = 15.1/33.2	V = 25.8%	V-I = .3	V-II = 3.2
Peak force—1	M = 63.3/139.6	SD = 15.9/35.0	V = 25.1%	V-I = .1	V-II = 2.7
Peak force—2	M = 65.1/141.5	SD = 15.2/33.5	V = 23.4%	V-I = .2	V-II = 3.2

Percentiles

Kilograms

	5th	10th	15th	25th	35th	50th	65th	75th	85th	90th	95th
Mean force—1	32.3	36.8	40.2	45.4	49.8	55.8	62.1	66.8	72.8	76.9	83.0
Mean force—2	33.8	38.9	42.5	48.0	52.4	58.2	63.7	67.9	73.2	77.1	83.4
Peak force—1	37.6	42.7	46.3	52.0	56.7	63.1	69.5	74.3	80.3	84.4	90.3
Peak force—2	40.5	45.6	49.2	54.7	59.0	64.9	70.7	75.0	80.6	84.5	90.6

Pounds

	5th	10th	15th	25th	35th	50th	65th	75th	85th	90th	95th
Mean force—1	71.2	81.2	88.6	100.1	109.7	123.1	136.8	147.3	160.6	169.6	183.0
Mean force—2	74.4	85.5	93.6	105.8	115.6	128.2	140.5	149.7	161.4	169.9	183.8
Peak force—1	82.9	94.1	102.2	114.7	125.0	139.0	153.2	163.8	177.0	186.0	199.2
Peak force—2	89.2	100.6	108.6	120.5	130.1	143.0	155.8	165.4	177.8	186.4	199.7

N—349

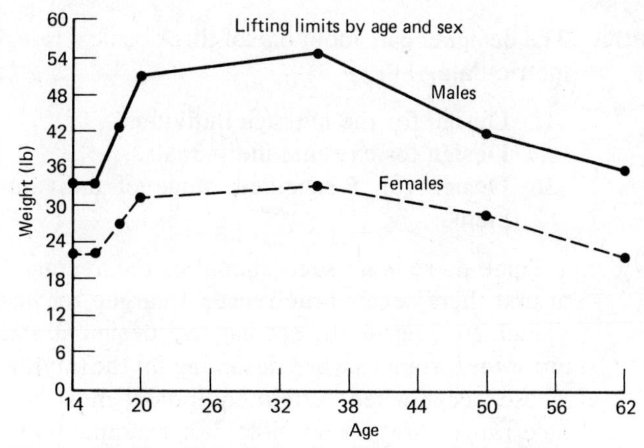

Figure 15-12. Safe limits for lifting weight by sex and age group. (Data from International Occupational Safety and Health Information Center, 1962.)

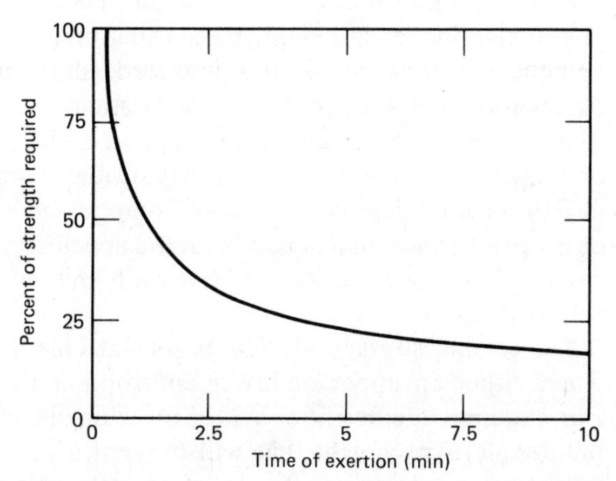

Figure 15-13. Endurance as a function of time (Kroemer, 1970). (From Roebuck et al., 1975.)

exhibit greater endurance since a lower pecentage of their strength will be required by the task.

DESIGN LAYOUT

One of the major design tasks is deciding where components should be located. The human factors specialist must correctly locate displays and controls (1) relative to each other, and (2) relative to the operator. Link analysis (see Chapter 7) is one important technique to achieve the first goal. Use of anthropometric data is needed to accomplish the second goal.

Using Anthropometric Data

The designer can adopt one of three basic strategies when using anthropometric data.

1. Design for the average individual.
2. Design for extreme individuals.
3. Design for a specified range of individuals by providing adjustments.

Since there is no such animal as the average person (see Figure 15-1), at first there seems little reason to argue for design for the average individual. In general, the second two design strategies are better. But there are some instances when designing for the (mythical) average is not foolish. These occur when certain equipment must be standardized and adjustable ranges are not possible. An example will clarify this kind of situation. Countertop height in most American kitchens has been standardized at 3 feet. (You can verify this by measuring the kitchen counter in your own home and in homes you visit.) There is no simple way to provide an adjustable counter height, especially since manufacturers of kitchen cabinets cannot economically produce sets of cabinets that differ by only one or two inches in height. By settling on an average height, the inconvenience to most people is minimized, although this is probably small consolation to short people who must stand on telephone books or boxes to use a counter efficiently or to tall people who must add telephone books or boxes to the counter top to elevate the workspace. (The workspace design for a kitchen is even more complicated since the optimum height of a work surface also depends on the specific task being done. Precision work—decorating a cake—requires a higher surface than heavy work—rolling dough.)

The second strategy, designing for extreme individuals, is widely used when either an upper or lower anthropometric limit must be specified. For example, the height of a door or ceiling is obviously more critical for tall people. Any height that will prevent a tall person from hitting their head will automatically be satisfactory for shorter people. So the designer might select a height that would accommodate the 95th percentile person without at all being concerned about the 5th percentile person. Similarly, a ladder or platform that would successfully hold up a 99th percentile individual would not collapse under the weight of a 1st percentile individual. When we are concerned with minimum distances, lower percentile limits are used. For example, if a 5th percentile person can reach all the controls in a workspace, then so can anyone with a higher percentile reach.

The third strategy, design for an adjustable range, should be used whenever economically feasible. Seating is a common example of this design strategy. Automobile seats can be moved toward or away from the steering wheel to accommodate a range of individuals. Seats for typists and data entry personnel have height adjustments. Many seats

TABLE 15-3 *Pilot Requirements for Armed Services (from Roebuck et al., 1975).*

Service	Age (yr)	Height (in.)	Weight (lb)	Miscellaneous
U.S. Air Force	$10\frac{1}{2}-17\frac{1}{2}$	64–76 (maximum sitting height of 39)	113–230 (depends on age and height)	Vision – 20/20 uncorrected College graduate Physically and mentally fit.
U.S. Army	18–30	62–75	103–200 (depends on age and height)	Vision – 20/50 uncorrected High school graduate Physically and mentally fit.
U.S. Navy and Marine Corps	20–28	60–78	120–280 (depends on age and height)	Vision – 20/20 uncorrected College graduate Physically and mentally fit.

also have assorted kinds of back adjustments. It is common to design adjustments to accommodate people from the 5th to the 95th percentiles. But if the cost in a particular design is low, this range could be extended.

Where adjustment costs are high, constraints are sometimes placed on worker populations. Thus, the military will not accept people for pilot training unless their height is within limits that airplane cockpits are designed for (see Table 15-3). However, it is important to realize that such population constraints very rapidly diminish the percentage of the population that remains eligible. Since there is no average person, each successive anthropometric selection standard eliminates part of the remaining population (see Figure 15-14). As Figure 15-14 shows, even a wide range of 5th to 95th percentile eliminates almost half of the potential pilot population when several body dimensions must be restricted.

There are many instances in which such population restriction is unacceptable. For example, laws governing equal employment opportunity have made it necessary for women to work at jobs that once were restricted to men. Since women lack the physical size and strength of men, it has become necessary to redesign the jobs so that they can be accomplished by women. Thus, hand tools originally designed for the male hand, have been redesigned for the smaller female hand. Chapter 20 contains an interesting example of redesign of a heavy ladder used by telephone installers in order to make this job accessible to women.

Design Procedure

The minimum goal of any anthropometric design procedure is to ensure that the human gets the same care and consideration as would a piece of equipment. No engineer would generate installation specifications for a piece of machinery such as a motor without checking that there was enough room to mount the motor, enough clearance for moving parts and sufficient space to allow maintenance as required. Yet many systems

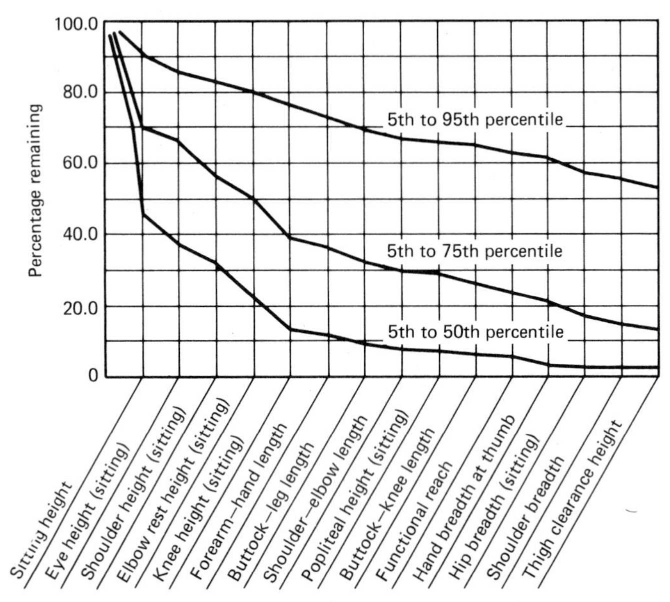

Figure 15-14. Reduction of work force by application of successive anthropometric standards. Three different percentile ranges are shown. (From Roebuck et al., 1975.)

have been built where the human operator is cramped, uncomfortable, and unable to reach all controls easily. This is not only unpleasant for the operator but also degrades the performance of the entire system of which the operator is a part.

There are several ways to characterize the formal aspects of workspace design procedures. Rather than present most of them in outline form, we will go through one procedure (Roebuck, Kroemer, & Thomson, 1975) in detail. The 14 steps illustrated in Figure 15-15 should be viewed as representative, and not as the only correct way to achieve good anthropometric design. The important point for you to grasp as you work your way through the many boxes and arrows in the flowchart is that sound design procedure requires far more than just looking up some numbers in tables of anthropometric data.

Step 1. *Establish requirements.* This step has been discussed in Chapter 1. Before any analysis can proceed, the designer must know system goals.

Step 2. *Define population.* Once the population is specified its anthropometric description is needed. With luck an appropriate description can be found in available tables such as those in the appendix at the end of the chapter. If not, it may be necessary to take measurements directly.

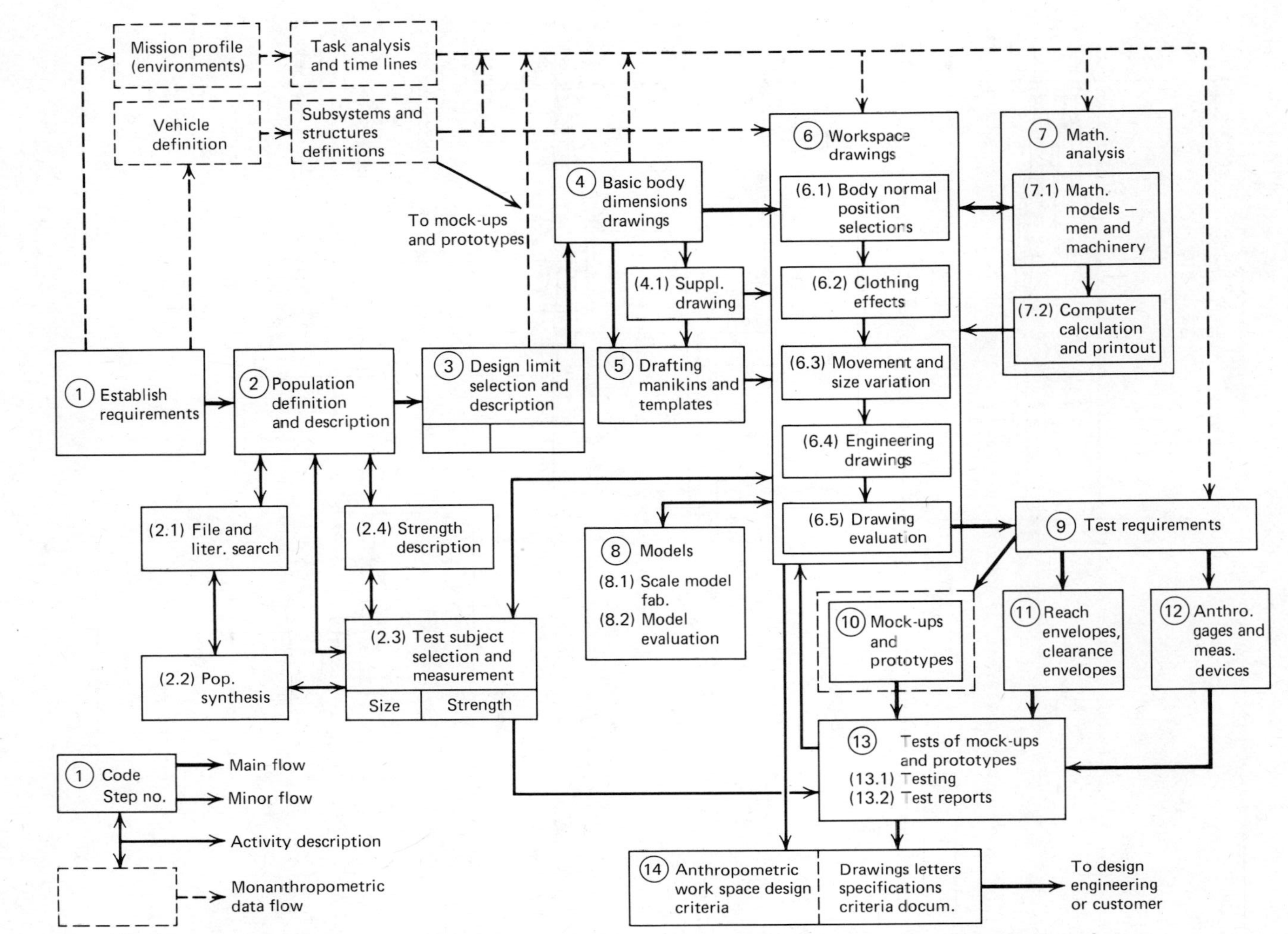

Figure 15-15. Flow chart for work space design procedure. See text for explanation of numbered boxes. (From Roebuck et al., 1975.)

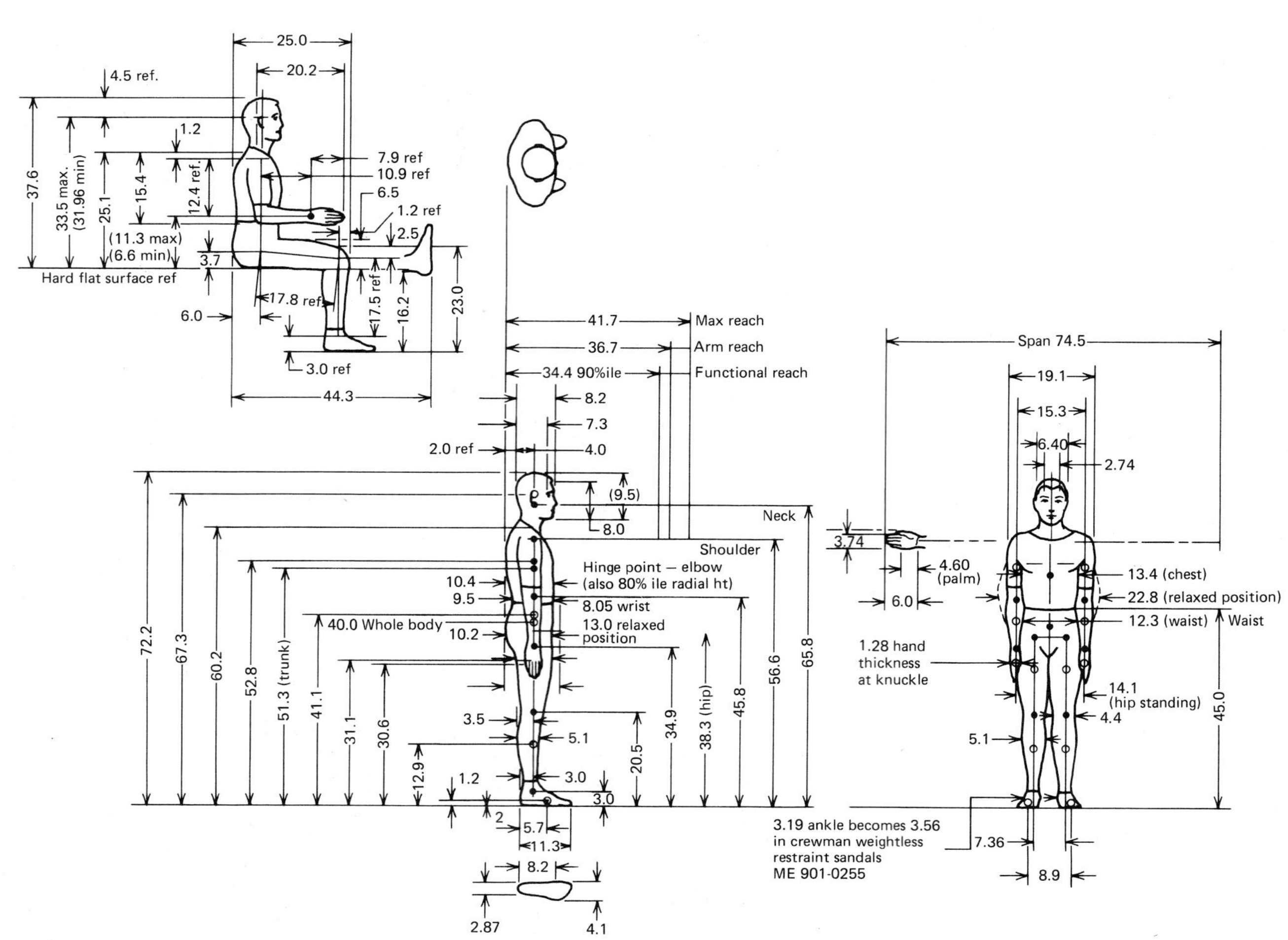

Figure 15-16. Sample basic body dimensions for work space design. (From Roebuck et al., 1975.)

Step 3. *Design limits selection.* The essential activity here is the selection of design percentiles. Should the range include the 99th percentile or the 95th percentile? What specific body dimensions must fall within the range? How much restriction upon the available population (see Figure 15-14) can be tolerated? Note that at this point in the design process the economic trade-offs between different ranges may not be known. Thus, the designer may not be able to state how much more it would cost to include 99th percentile individuals. Therefore, although this is not shown in Figure 15-15, it may be necessary to return to this step again once this information is available. This may require redoing later steps. Any design process is an iterative procedure (see Chapter 1) and the need to go back and start again is not represented in Figure 15-14.

Step 4. *Basic body dimensions.* The data are collected and drawn as a stylized extreme percentile individual (see Figure 15-16). This also makes it easier to detect any missing data.

Step 5. *Drafting aids.* These will help carry out later steps. Plastic overlays are usually drawn to scale. Sometimes scale manikins are built.

Step 6. *Workspace layout.* This is a critical step. The designer must use descriptive geometry to determine functional layout based on both static and dynamic anthropometric data. Clearance limits for body parts must be established. Reach envelopes determine panel location and control positions (see Figure 15-17). Computed-aided graphics are becoming more popular for this important step.

Step 7. *Mathematical analysis.* Several biomechanical models are available to aid graphic representation of body positions. These models can be superimposed on the planned layout to generate movement envelopes.

Step 8. *Small-scale models.* Physical models are made from cardboard and wood. These models may reveal design flaws.

Step 9. *Prepare test requirements.* In order to evaluate the proposed design, explicit requirements must be formulated to determine if system criteria have been met.

Step 10. *Mock-ups.* This is another key step. Full size mock-ups are built. These mock-ups may have special features to aid evaluation. For example, one Apollo Command Module mock-up could be split in half to aid photography of crew movements, others had removable windows and sections, plumbing for space suits, etc. (Roebuck et al., 1975). Mock-ups are built from Fome-Cor, a foam and paper sandwich material, wood, metal and plastic. These are relatively low cost materials, especially Fome-Cor, that can be ripped apart and rebuilt when design flaws are discovered. Despite their temporary construc-

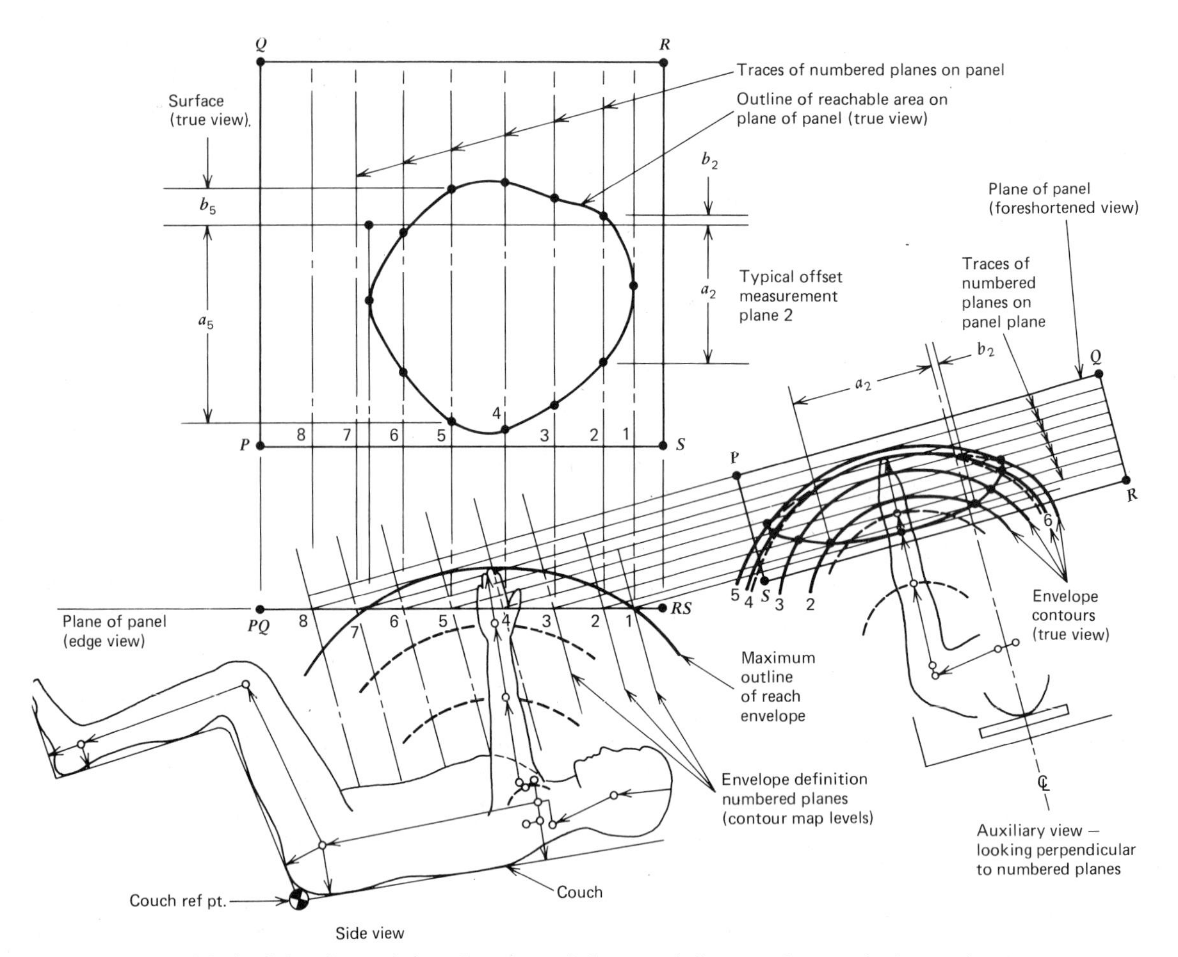

Figure 15-17. Method for determining the area of the panel that can be reached. (Roebuck et al., 1975.)

tion, mock-ups can be very realistic (see Figure 15-18). It is impossible to design a system of any complexity without extensive use of mock-ups.

At one time it was believed that a simulator could not provide useful data unless there was a high degree of physical similarity between it and the system that was simulated. For example, one very expensive Air Force jet simulator has a complex computer-generated display, motion cues, and even has seat cushions that can be inflated to simulate gravity forces on the buttocks of the pilot. Recent research has shown that such physical fidelity is not always required to obtain valid results. Fixed-base vehicle simulators that do not move can still provide very useful data (Blaauw, 1982). For example, a study of cockpit traffic displays in a fixed-base, relatively unrealistic looking cockpit mock-up found that it

Figure 15-18. Fome-Cor mock-up of a space shuttle orbiter cockpit. (Courtesy of Space Division, Rockwell International Corporation.)

was more important to simulate a realistic communication flow with messages similar to those that occur during actual flying than to have a sophisticated high-fidelity moving-base simulated cockpit (Hart, 1982).

Step 11. *Reach and clearance envelopes.* These are tested with a human inside the mock-up. Physical envelope mock-ups are constructed and placed inside the overall mock-up.

Step 12. *Preparation of special measuring devices.* Each workspace evaluation may require unique devices. For example, a special acrylic device was used to provide visual cues on the docking window of the Apollo mock-up.

Step 13. *Testing.* This can be quite extensive. For example, mock-ups have been placed in large planes that perform acrobatic maneuvers to simulate weightlessness. The performance of test subjects is recorded for a wide range of conditions.

Step 14. *Documentation.* Once the design is set, this information must be communicated. Since such information is really a recommendation for production, it must be well documented. Even then, not all of the recommendations will be accepted by higher management or by the customer.

Since the human factors designer who has completed the previous 13 steps can make better predictions about the use (and misuse) of the system than can someone less familiar with the system, clear and complete communication is essential. Nothing is more frustrating and wasteful than to have carefully gathered and tested design data ignored because it was not presented properly. Some would argue that this step is the most crucial of all, since a poor job here can negate all the good work accomplished in prior steps.

DESIGN APPLICATIONS

The best way to illustrate workspace design is with actual examples. Thus, the rest of this chapter is devoted to specific examples of design applications currently in use across the street and across the world. While the data presented in this section will allow you to evaluate designs you are now using, for example, the chair you are sitting on, it is also important for you to understand how these results and suggested designs relate to the preceding parts of this chapter. Workspace design could not be accomplished without engineering anthropometry and the techniques of design layout discussed previously.

Seating

The human race spends a great deal of time sitting down, whether working in an office, studying in a library, commuting by bus, car, or airplane, or eating in a restaurant. Some seats are far more comfortable than others. Indeed, in some fast-food restaurants, seats are intentionally designed to be uncomfortable after a short period so that customers will not tarry too long. The support provided by seating is especially crucial when the seat must be occupied for several hours, as in classroom and office spaces. Improper seat design can lead to back pain and cardiovascular ailments.

Seating research has confirmed some general design guidelines on which most experts agree. (There is, however, disagreement about details of seating design for specific applications, some of which will be discussed later. There is no ideal seat for all people and all applications (Pile, 1979, Appendix 1).

- The main weight of the body should be carried by the bony protuberances of the buttocks, more technically known as the ischial tuberosities.
- The thighs should exert as little pressure as possible on the seat or on the front edge of the seat.
- The lumbar (lower) portion of the back must be supported.
- The feet must be able to be placed firmly on the floor or, if this is not possible, on a footrest.
- The seated person should be able to change posture (without getting up).

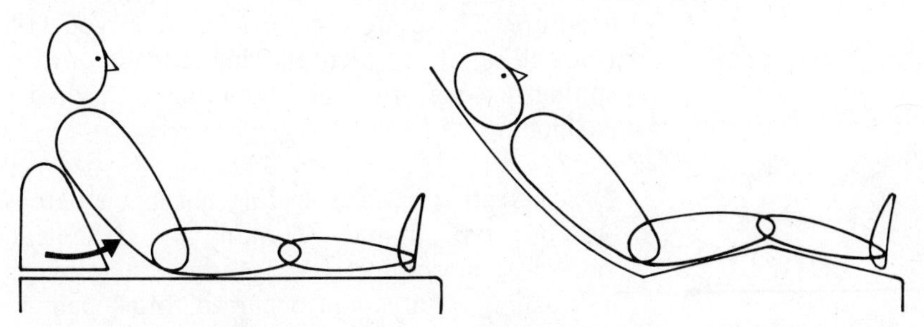

Figure 15-19. Problems in sitting on a flat surface can be corrected by an inclined backrest. (From Pile, 1979.)

Figure 15-20. Using an inclined seat and footrest to correct seating problems. (From Pile, 1979.)

Some common seating problems, due to violation of one or more of the above guidelines, are shown in Figures 15-19 and 15-20. Sitting on a flat surface, such as a bed, causes a slump because the lumbar region is not supported (Figure 15-19). Providing a slanted back solves this problem but also dictates a particular posture. You can now buy beds for home use that have electric motors to raise and lower the head and foot. These are versions of hospital beds that have been designed to fit in the home. Some even have vibrators and/or heat pads built in. Since many people use their bed for sitting, such a device may well be worth its extra cost. When a seat back is tilted away from vertical (Figure 15-20) the upper part of the body tends to push the lower part forward. If the seat is horizontal, you will slide forward and generate a slump. Tilting the seat so that the thighs are no longer horizontal solves this problem, but only at the expense of violating another rule by placing too much pressure on the thighs and behind the knees. This new problem can be eliminated by providing an adjustable leg rest.

Padding and springs are commonly used in seats to reduce the pressure by spreading out the contact area between body and chair. They also aid postural adjustment and reduce vibration in vehicle seats. However, there can be too much of a good thing as in the case of soft, overstuffed furniture. These are comfortable when you first sit down, but soon become tiring. Too much padding spreads out the weight excessively, resulting in squeezing flesh between the ischial tuberosities and the soft cushioning.

Office Seating Office seating has been extensively studied. The first step of a typical study (Hunting and Grandjean, 1976) was to evaluate causes of discomfort for 246 office employees by observing postures and asking where pain occurred. Most pain was reported in the back (57%), followed by knees and feet (29%), neck and shoulders (24%), thighs (19%), bottom (16%), and head (14%). The reasons for discomfort in the upper and lower parts of the body were due to improper (or no) use of anthropometric data. Tables and desks that were too high caused people to scrunch their shoulders and neck. Aching feet were a problem in short people who were unable to reach the floor and thus perched upon the edge of their chairs. A chair whose seat was between 27 and 30 cm below the work surface would eliminate such problems for most workers. Since leaning back was a very common posture, Hunting and Grandjean designed a chair with a high back. This chair has a slight concave dip at the top and a pronounced convex bulge to support the lumbar region (see Figure 15-21). This chair provides support when the worker is leaning forward and also when leaning back to rest the dorsal muscles.

Figure 15-21. Office chair designed with ergonomic factors taken into account. (Courtesy E. Grandjean & Grioflex Company.)

Summary of Mandal theory right-angled chairs which dictate backwards-inclined rest position make reading almost impossible (eyes are 500–600mm from desk surface) and strain hip joints and lumbar region

To read or write, occupant may bend the lumbar region (about 20° in this case) which places stress on 3rd, 4th and 5th vertebral discs (those most likely to slip). Back begins to hunch

Alternatively, occupant may sit forward, sloping the thighs about 20°. Back is straight but seat edge cuts into soft tissue of thighs, interferes with blood supply and may push on iscia nerve

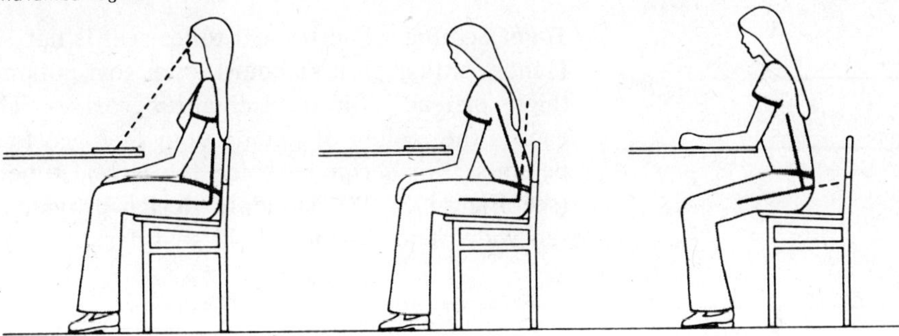

During lectures, meetings or periods of rest, occupant can sit back and push the bottom forward. Bend in hips is reduced from 90° to about 70° (Mandal considers 45° ideal) but position is impractical for most work

Children's universal tendency to tip chairs forward onto the front legs allows thighs to slope as much as 30° below horizontal; the hip joints bend about 60° and they can keep their backs straight

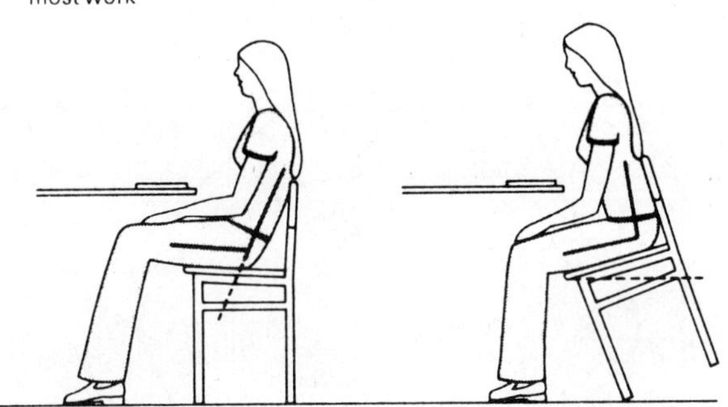

Figure 15-22. The theory behind the Mandal chair. (Courtesy Design Council; drawing by Carol Preedy, from *Design, 333*:33.)

A somewhat controversial design has been created by Aage Mandal, a Danish surgeon. This chair slopes forward so that the occupant does not touch the backrest when working. The rationale for this novel design is presented in Figure 15-22. When people lean forward in a standard chair with a horizontal seat, strain on the lumbar region is increased. Sitting forward as an alternative to leaning gets the worker closer to the work but increases pressure on the thighs cutting off blood circulation. The solution used by the Mandal chair is to tip the seat forward, just as children "instinctively" tip their chairs forward to work at their desks. This new design has been criticized (Ward, 1976) because it violates some of the general rules listed above, such as not supporting the lumbar region, exerting too much pressure on the thighs, and not allowing for postural changes while working.

Toilet Seating The typical toilet seat is flat. Your buttocks are curved. Hence, sitting on a standard toilet seat puts most of your weight on the thighs instead of on the ischial tuberosities. This cuts off circulation and causes the feeling of having your feet "go to sleep," especially in older persons. Furthermore, since the ischial tuberosities are close together (see Figure 15-23), standard design provides a seat opening that is far too wide. The Posture Mold seat designed by architect Alexander Kira

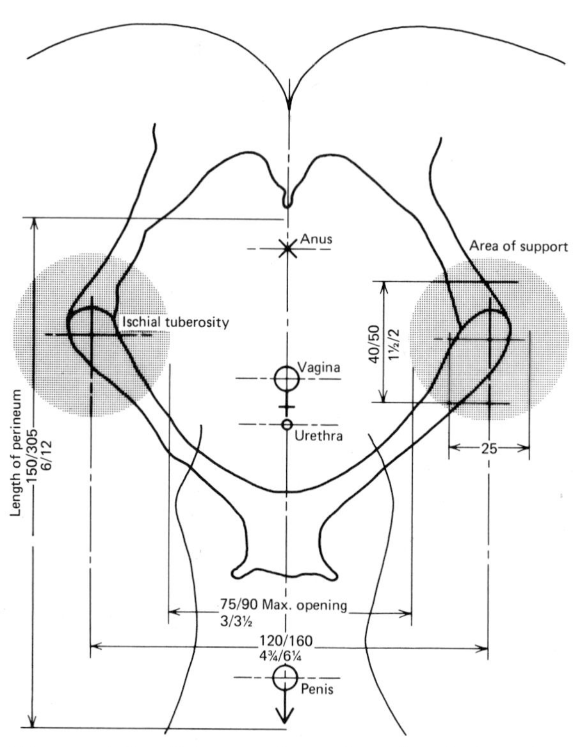

Figure 15-23. Relevant dimensions for design of a toilet seat in English and metric units. (From Kira, 1976.)

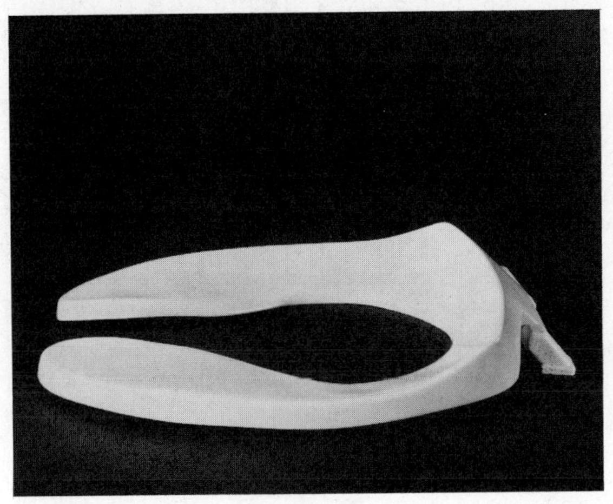

Figure 15-24. The Posture Mold toilet seat. Note contours and shape. (Courtesy Forbes Wright, Church Products.)

(Figure 15-24) is contoured and provides proper support for the thighs. This seat was selected for the design study collection of the Museum of Modern Art showing that good human factors can be esthetically as well as functionally attractive. The Posture Mold seat is manufactured by Forbes Wright, primarily for institutional use in hospitals, nursing homes, etc. Your local plumber will not carry it, although it can be ordered for use in a residence.

Vehicular Seating Layout

Once a seat has been designed to provide proper support for the human body, the human factors specialist must then decide where to locate the seat in relation to the rest of the system. For office seating as discussed previously, the major anthropometric location is the vertical distance between seat and work surface. For vehicular seating in cars, buses, and airplanes the seat is one component of a three-dimensional workspace. Vehicle drivers must be able to have proper lines of sight and to reach all controls in the vehicle cab. Vehicle passengers require seating layouts that promote safety and comfort while maximizing passenger density for economic reasons.

Design for Vehicle Operators Layout of most vehicle cabs begins from a theoretical design eye point. This is an imaginary point in space from which lines of sight are calculated. Its correspondence to the physical location of the operator's eye is not exact—for example, most operators have two eyes so a single point is incorrect—but can be loosely viewed as roughly being the midpoint between the pupils of the two eyes. An idealized graph of optimum lines of sight is drawn (Figure 15-25). This graph is then compared to what is physically possible given assorted con-

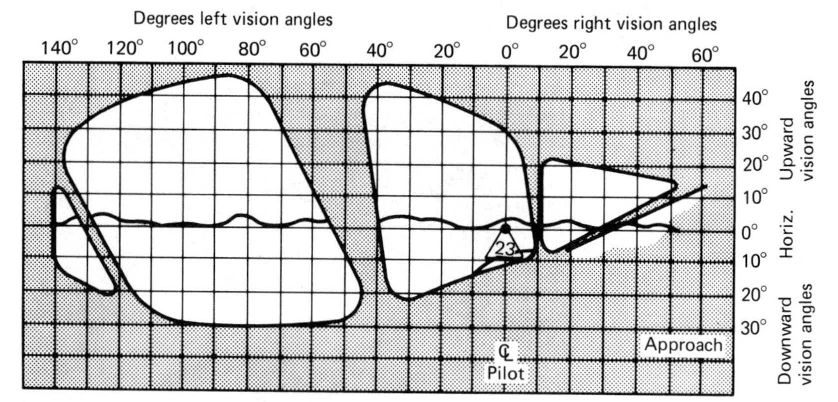

Figure 15-25. Window visibility diagram for airplane cockpit. (From Roebuck et al., 1975.)

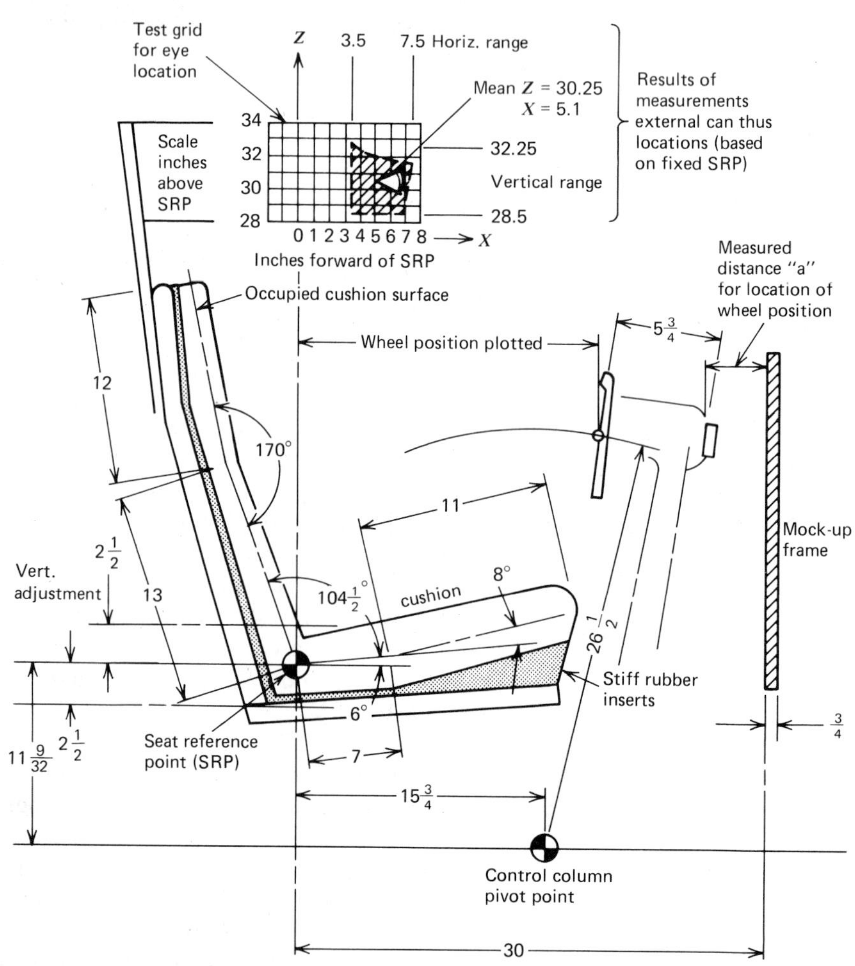

Figure 15-26. Mock-up configuration and grid for eye location measurements. (From Roebuck et al., 1975.)

straints such as structural members that need be placed around the windows. Once a mock-up has been built, better data can be obtained about eye location. For example, the anthropometric measurement called Eye Height—distance from the seat surface to the eye—overestimates actual eye height because people do not sit erect when working. Instead they slump over so that actual eye height in the mock-up may be as much as two inches shorter than the value found in anthropometric tables. The amount of slump depends on the seat as well as the vehicle cab layout. So it is impossible to design a perfect cab only from tables of data. These tables allow a first-approximation mock-up to be built, but then additional data about eye location must be obtained using people or manikins if sufficient resources for data acquisition are lacking. Typical mock-up results for eye location in the DC-8 commercial aircraft are shown in Figure 15-26.

A similar procedure is used for automobile cab layout (Babbs, 1979). Manikins representing the 95th percentile male torso and the 5th percentile female torso are placed in the mock-up. Seat position is then adjusted to account for eye location and the intrusion of pedals and other controls into the workspace (Figure 15-27); this allows the designer to determine front and rear mounting points for the seat and the range of adjustment that is required to accommodate both manikins. This range is represented as the lower rectangles in Figure 15-27.

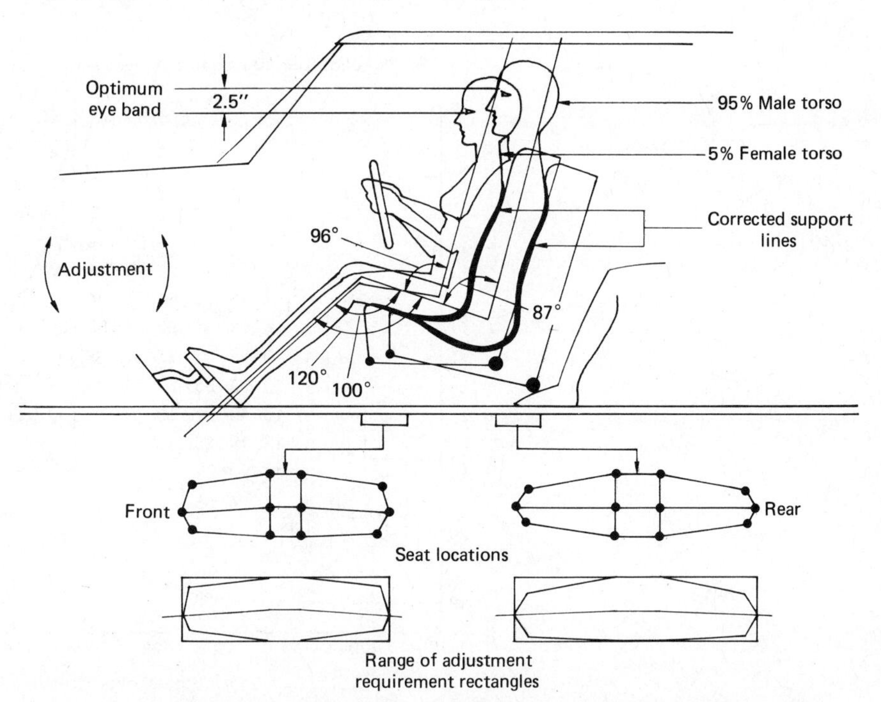

Figure 15-27. Design for automobile seat adjustments. (From Babbs, 1979.)

Leg clearance determination

1. Determine graded series of leg positions for study.

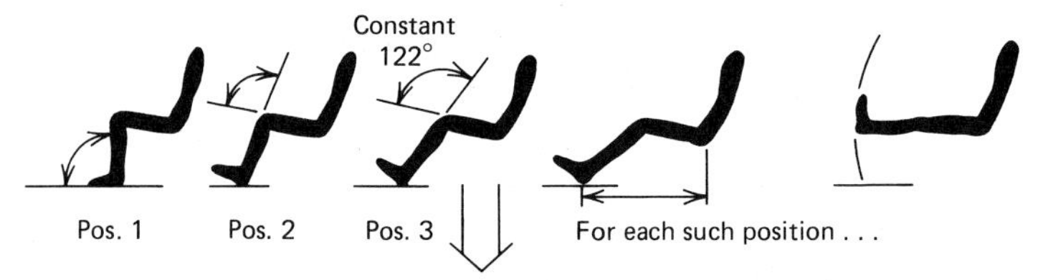

2. Layout allowable variations of leg component lengths in given space.

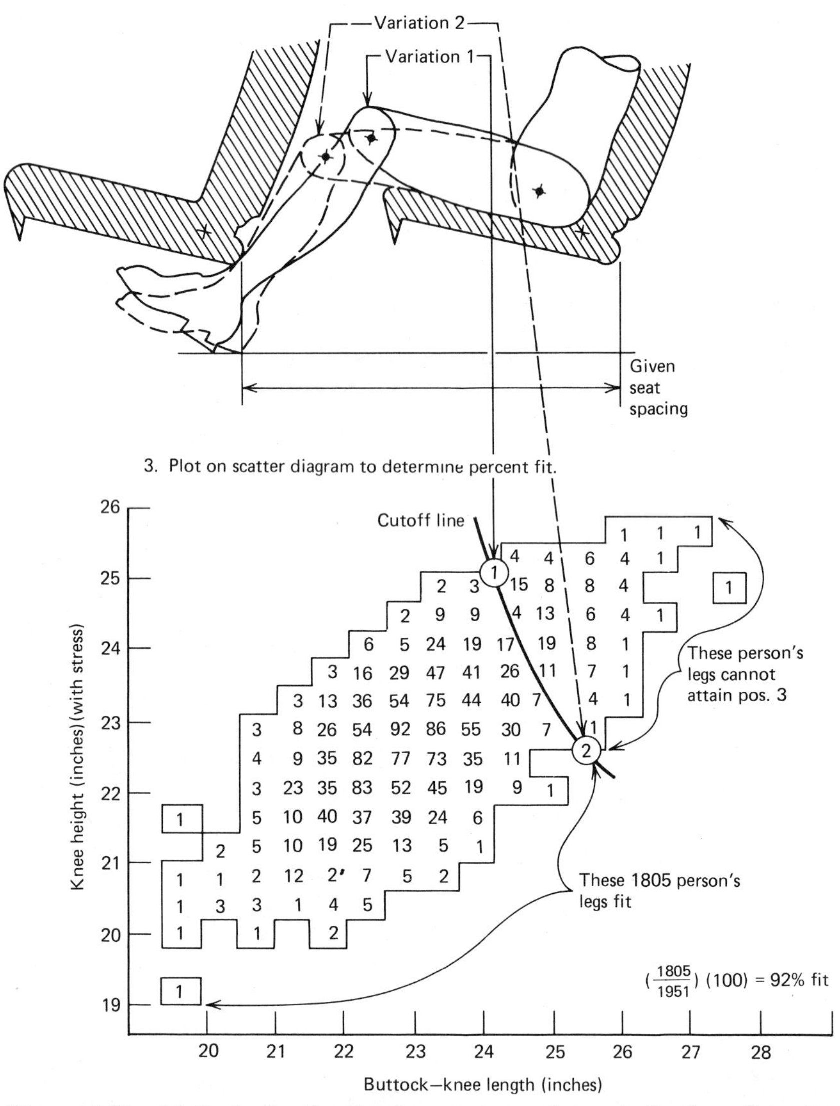

Figure 15-28. Method of estimating leg space requirements for fore-aft seating layout. (From Roebuck et al., 1975.)

Design for Vehicle Passengers Anyone who has traveled extensively by air knows full well that the same airplane can carry different numbers of passengers depending on how many seats the airline has decided to fit. There is a rational method for making this decision based on anthropometric data. First, we discuss fore-aft seat spacing, that is, how many rows of seats should be placed in the aircraft. The designer must select a seat and a posture (upper panel, Figure 15-28). Then the permissible variations of leg positions for that seat must be laid out (middle panel, Figure 15-28). Finally, a scatter diagram based on tabled anthropometric data for knee height and buttock-knee length is plotted to determine the percent of passengers who will not fit (lower panel, Figure 15-28). It is, of course, impossible to fit 100%, and each airline will make its own decision as to what is an acceptable fit.

Anthropometric data also can determine side-by-side seat spacing, that is, how many seats will fit in each row. The crucial dimension is called shoulder breadth. If your shoulders fit, so will your hips. However, an analysis based on shoulder breadth does not guarantee you will have much room to move your elbows. Figure 15-29 is based on statistical analyses of shoulder breadth. The bench criterion assumes that a distribution of three shoulder breadths is all that is needed. Since this is a statistical average, it implies that people with broad shoulders will be sitting next to people with narrow shoulders and the three passengers sitting abreast can lean or wiggle around to optimize the available space. The seat criterion is more generous and is based on probabilities of overlapping shoulders when passengers sit in the center of each seat. There is little to be gained by having seats more than 19 inches wide.

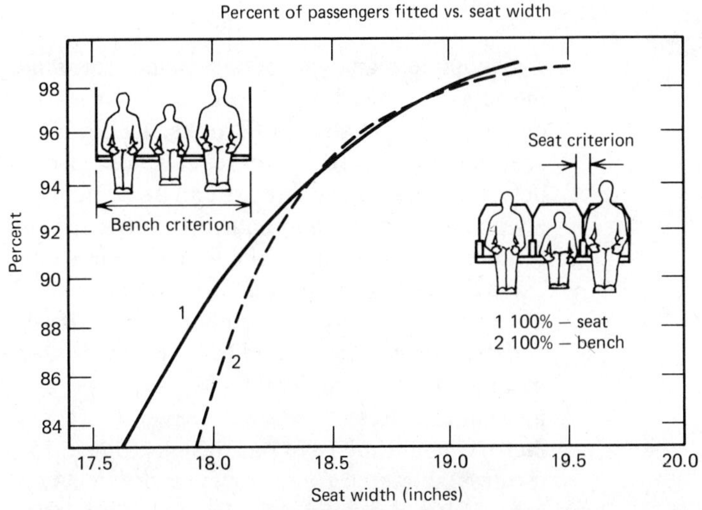

Figure 15-29. Analysis of seat width for three abreast seating. (From Lippert, 1958.)

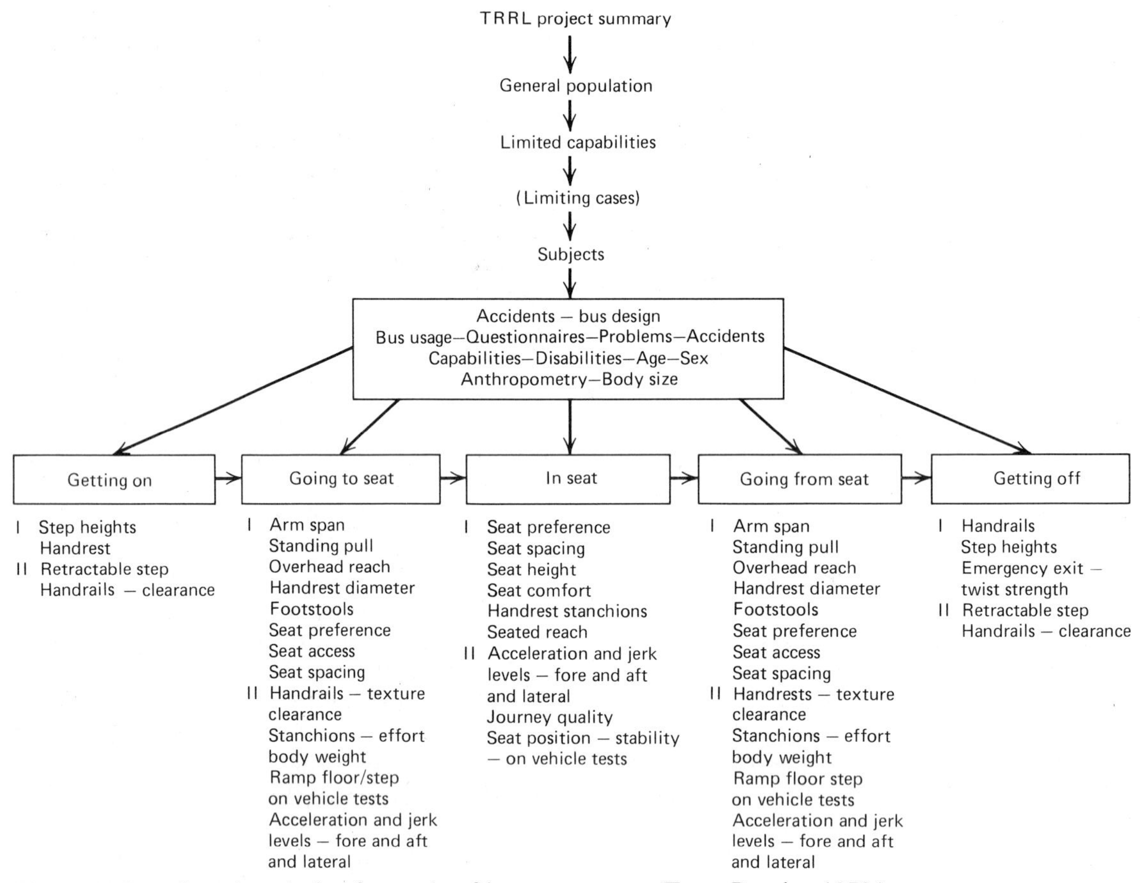

Figure 15-30. Complete design for study of bus passengers. (From Brooks, 1979.)

A complete analysis of passenger requirements goes well beyond seating layout. This can be seen in a thorough analysis of bus design conducted in Great Britain (Brooks, 1979). The flow chart of the entire research plan (Figure 15-30) reveals how many details must be investigated before design improvements can be achieved. Phase I of the project used a wooden mock-up and anthropometric measurements of 200 elderly and disabled persons, since they were representative of the population that experienced the most difficulty in bus travel. Phase II used a new mock-up and a specially instrumented bus. It also conducted an historical accident survey that found that 56% of passenger injuries were sustained in noncollision accidents; 43% of these accidents occurred to elderly passengers. Results showed several areas where ergonomics (human factors) could improve bus design (Figure 15-31). These included adding a retractable step to aid entry and exit (A), recommendations for doorway width and handrails (B), headroom (C), and problems on moving buses due to acceleration and jerk (changes in acceleration) levels.

Figure 15-31. Improvements in bus design. (From work undertaken by Leyland Vehicle under contract to the Transport and Road Research Labortory.)

TABLE 15-4 *Dimensions for Standard Consoles. See Figure 15-32. (from Roebuck et al., 1975)*

	Maximum Console Height from Standing Surface	Console Depth	Vertical Dimension of Panel Including Sills	Console Panel Angle – from Vertical	Minimum Pencil-Shelf Depth	Minimum Writing Surface Depth – Including Pencil Shelf	Minimum Knee Clearance	Foot Support to Seat	Seat Adjustability	Minimum Thigh Clearance at Midpoint of "I"	Writing Surface Height from Standing Surface	Seat Height at Midpoint of "I"	Maximum Console Panel Breadth
Type of Console	A	B	C	D	E	F	G	H	I	J	K	L	M
1. Sit-stand	62.0	Opt.	26	15[a]	4	16	18	18	4	6.5	36.0	28.5	36
2. Sit (with vision over top)	47.5[a] to 58.0	Opt.	22	15[a]	4	16	18	18	4	6.5	25.5 to 36.0	18.0 to 28.5	36
3. Sit (without vision over top)	51.5[b] to 62.0	Opt.	26	15[a]	4	16	18	18	4	6.5	25.5 to 36.0	18.0 to 28.5	36
4. Stand (with vision over top)	62.0	Opt.	26	15[a]	4	16	–	–	–	–	36.0	–	36
5. Stand (without vision over top)	72.0	Opt.	36	15[a]	4	16	–	–	–	–	36.0	–	36

[a] "A" must never be more than 29.5 in. greater than "L."
[b] "A" must never be more than 33.5 in. greater than "L."

Console Design

A console is a unit that contains displays and controls. It can be as small as a portable typewriter or large enough to fill a room. The human factors specialist must make sure that the console operator can see all necessary displays, reach all controls, and assume a posture that is comfortable for long periods of time. Consoles can be designed for seated or for standing operators. It may be necessary for the operator to see over the top of the console. Once this is decided, standardized console dimensions can be found in Figure 15-32 and Table 15-4. These dimensions are based on mock-ups and will accommodate 95% of male and 60% of female U.S. operators (Kennedy & Bates, 1965).

Although designing a console from Table 15-4 will give satisfactory results, it is based only on anthropometric considerations. More recent research has combined performance measures with anthropometric standards to obtain panel width recommendations for superior eye-hand coordination. This optimal panel width has been termed the visual comfort zone (Mital, Halcomb, & Asfour, 1979). We describe the procedures used to determine the visual comfort zone in detail, because this research is quite typical for human factors.

Any experimenter must compromise between laboratory control and

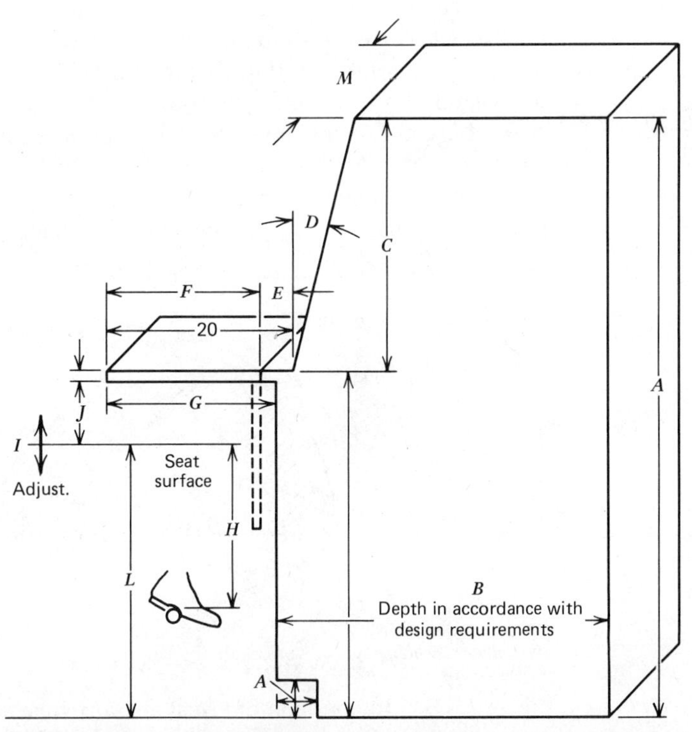

Figure 15-32. Dimensions for standardization console design. See Table 15-4. (From Roebuck et al., 1975.)

realism in the experimental situation. Basic researchers usually choose a high degree of control, even at the expense of having a very artificial laboratory situation. Applied researchers often prefer realistic situations. While the human factors researcher would like both realism and control, this is seldom possible in a single experiment. When laboratory experiments are used, they are apt to be simple. The use of sophisticated computers to control on-line data acquisition requires a great deal of time and effort. The human factors researcher working under a time deadline may not be able to establish a sophisticated laboratory setup. But such a sophisticated laboratory is not always required to obtain useful human factors data. There are many instances in human factors research where it is foolish to obtain data to three decimal places and undue precision may not improve results. The following experiment shows that simple laboratory techniques have a place in human factors research.

This experiment used 30 college undergraduates as participants: half male and half female. An ordinary office desk 76 cm high was the work surface. An adjustable biomechanical chair allowed each participant to sit with feet flat on the floor and thighs horizontal. A grid was superimposed on the desk surface and five-letter nonsense words, made from the letters O and C, were placed at various locations on this grid. In the first experimental session black letters appeared on a white background. In the second session, letters were blue on a white background. Participants had to spell out the five-letter words at each location and give a subjective comfort rating on a four-point scale. Based upon this method, the visual comfort zone was found to be within two semicircles having radii of 8 and 54 cms. Results were the same for black and for blue letters and contour boundaries were the same for both sexes. When this visual comfort zone is superimposed on anthropometric reach data for the 2.5th percentile U.S. female population, the shaded area in Figure 15-33 is

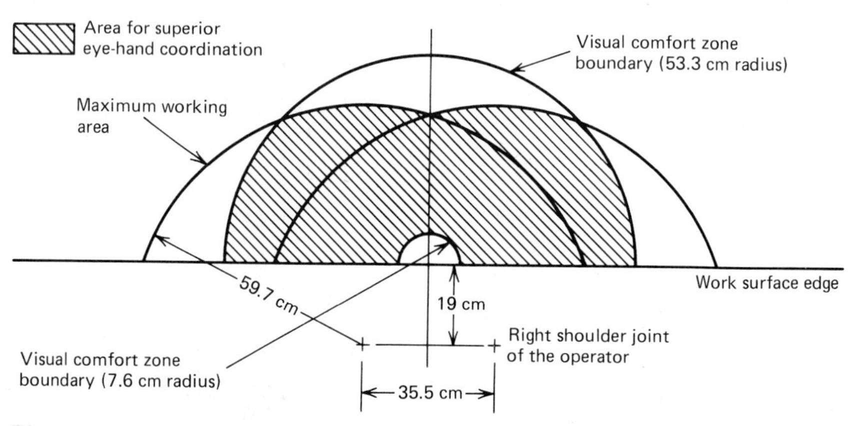

Figure 15-33. Intersection of visual comfort zone and reach envelope for 2.5th percentile female worker. (From Mital et al., 1979.)

obtained. Locations within this region will give best eye-hand coordination. While future research of a more sophisticated nature would probably improve our understanding of both the visual comfort zone and certain puzzling findings obtained in this study, the results shown in Figure 15-33 are still sufficient to aid the console designer.

SUMMARY

Engineering anthropometry is the application of scientific physical measurement methods to the human body in order to optimize the human-machine interface. Anthropometric data cannot be used correctly without a good understanding of sampling and descriptive statistics. Static measurements are obtained when the body is at rest. Dynamic measurements define a volume of space, such as a reach envelope.

Anthropometric data can be used in three ways by the designer: design for the average individual, design for extreme individuals, or design for a range of individuals by providing adjustments. Sound design procedure involves much more than just looking up data in tables. Some important applications of workspace design are seating, vehicle cab layout, and console design.

APPENDIX

Representative Anthropometric Data

From *Anthropometric Source Book, Volume I: Anthropometry for Designers*. NASA Reference Publication 1024, July 1978. Available from National Technical Information Service, Springfield Virginia 22161.

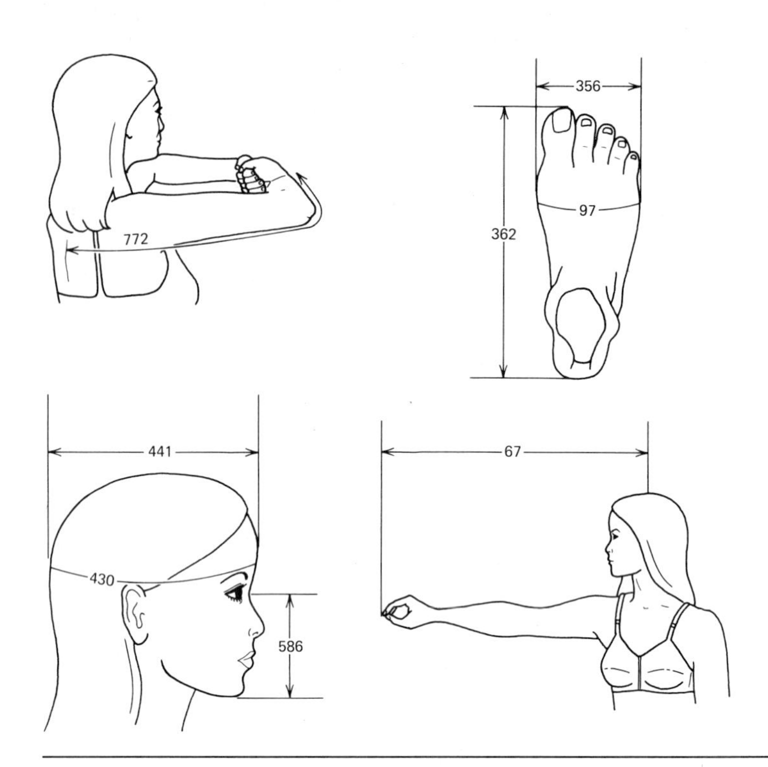

	1985 Female[a]			
No.	Dimension	5%ile	Mean	95%ile
67	Thumbtip reach	67.7 (26.7)	74.3 (29.3)	80.6 (31.7)
772	Sleeve length	74.2 (29.2)	80.0 (31.5)	85.2 (33.5)
441	Head length	17.5 (6.9)	18.6 (7.3)	19.7 (7.8)
430	Head circumference	52.6 (20.7)	55.2 (21.7)	57.9 (22.8)
586	Menton-sellion (face) length	12.6 (9.8)	14.0 (10.8)	15.6 (11.8)
362	Foot length	22.2 (8.7)	24.1 (9.5)	26.1 (10.3)
356	Foot breadth	8.0 (3.1)	8.8 (3.5)	9.7 (3.8)
97	Ball of foot circumference[b]	21.3 (8.4)	23.3 (9.2)	25.3 (10.0)

[a] Data given in centimeters with inches in parentheses.
[b] Estimated from regression equations.

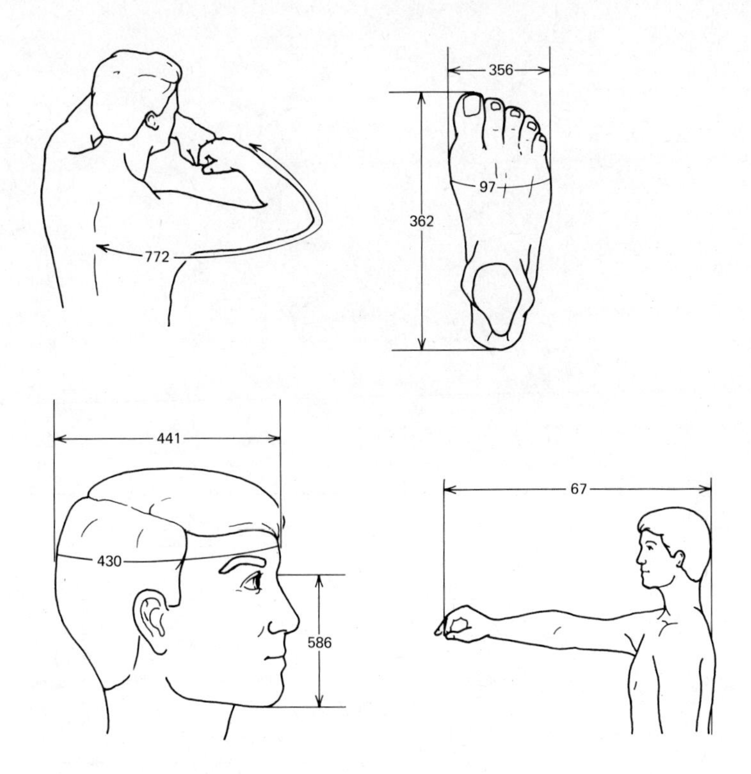

1985 Male[a]				
No.	Dimension	5%ile	Mean	95%ile
67	Thumbtip reach	74.3 (29.3)	80.7 (31.8)	87.4 (34.4)
772	Sleeve length	85.7 (33.7)	91.3 (35.9)	97.3 (38.3)
441	Head length	18.8 (7.4)	19.9 (7.8)	21.0 (8.3)
430	Head circumference	55.3 (21.8)	57.6 (22.7)	60.0 (23.6)
586	Menton-sellion (face) length	11.1 (4.4)	12.0 (4.7)	13.0 (5.1)
362	Foot length	25.3 (10.0)	27.2 (10.7)	29.2 (11.5)
356	Foot breadth	9.0 (3.5)	9.8 (3.9)	10.7 (4.2)
97	Ball of foot circumference	23.0 (9.1)	25.0 (9.8)	27.0 (10.6)

[a] Data given in centimeters with inches in parentheses.

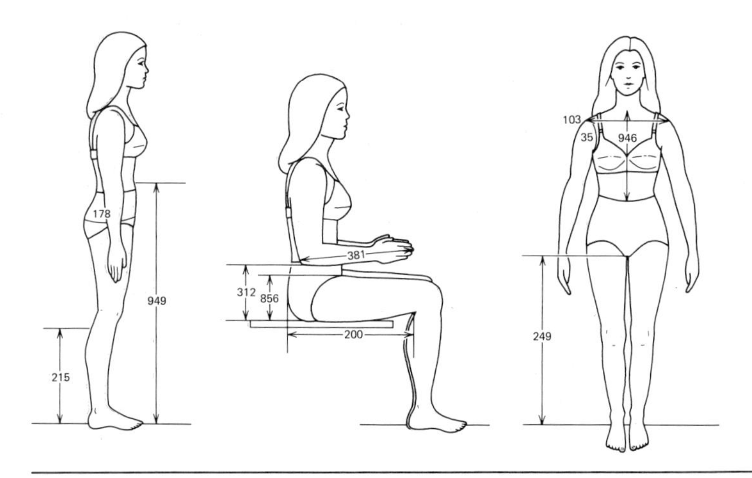

No.	Dimension	5%ile	Mean	95%ile
	1985 Female[a]			
949	Waist height	93.1 (36.7)	100.7 (39.6)	108.1 (42.6)
249	Crotch height	67.7 (26.7)	74.4 (29.3)	81.3 (32.0)
215	Calf height	28.7 (11.3)	33.1 (13.0)	37.5 (14.8)
103	Biacromial breadth	33.4 (13.1)	36.1 (14.2)	38.8 (15.3)
946	Waist front	30.4 (12.0)	33.7 (13.3)	37.1 (14.6)
735	Scye circumference	34.1 (13.4)	37.8 (14.9)	41.9 (16.5)
178	Buttock circumference	86.0 (33.9)	95.1 (37.4)	106.6 (42.0)
312	Elbow rest height	19.2 (7.6)	22.9 (9.0)	27.1 (10.7)
856	Thigh clearance	10.4 (4.1)	12.5 (4.9)	14.9 (5.9)
381	Forearm-hand length[b]	39.7 (15.6)	42.8 (16.9)	45.9 (18.1)
200	Buttock-popliteal length	43.7 (17.2)	47.9 (18.9)	52.7 (20.7)

[a] Data given in centimeters with inches in parentheses.
[b] Estimated from regression equations.

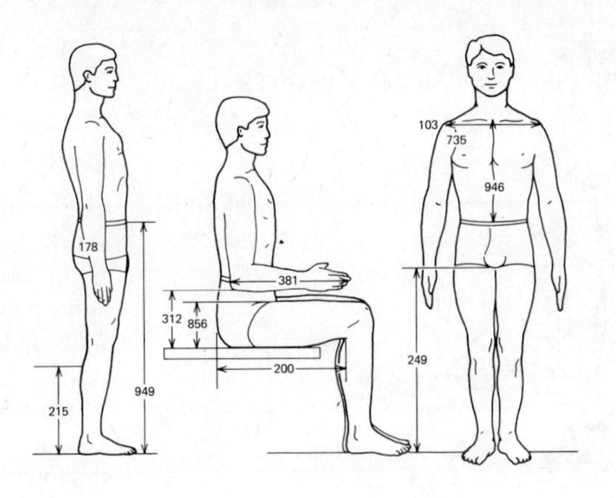

		1985 Male[a]		
No.	Dimension	5%ile	Mean	95%ile
949	Waist height	99.4 (39.1)	107.2 (42.2)	114.8 (45.2)
249	Crotch height	78.9 (31.1)	85.7 (33.7)	92.6 (36.5)
215	Calf height	32.3 (12.7)	35.8 (14.1)	39.6 (15.6)
103	Biacromial breadth	37.6 (14.8)	40.9 (16.1)	44.0 (17.3)
946	Waist front	37.1 (14.6)	40.6 (16.0)	44.2 (17.4)
735	Scye circumference	44.2 (17.4)	48.7 (19.2)	53.3 (21.0)
178	Buttock circumference	90.3 (35.6)	99.5 (39.2)	108.9 (42.9)
312	Elbow rest height	21.0 (8.3)	25.3 (10.0)	29.7 (11.7)
856	Thigh clearance	14.5 (5.7)	16.8 (6.6)	19.1 (7.5)
381	Forearm-hand length[b]	45.7 (18.0)	49.1 (19.3)	52.6 (20.7)
200	Buttock-popliteal length	46.4 (18.3)	50.8 (20.0)	55.1 (21.7)

[a] Data given in centimeters with inches in parentheses.

[b] Estimated from regression equations.

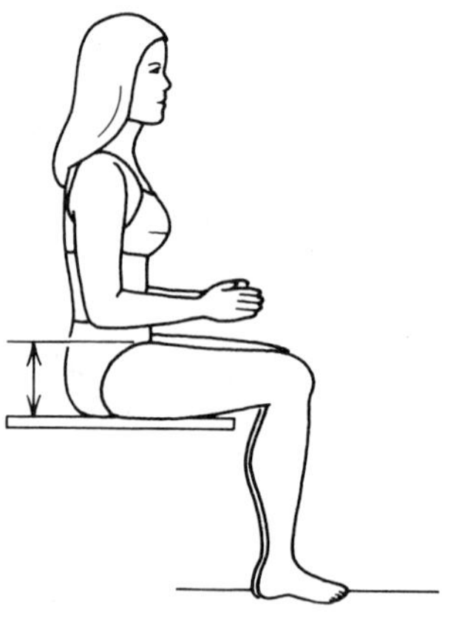

Thigh Clearance

Definition: The vertical distance from the sitting surface to the highest point on the right thigh. The subject sits erect with his knees and ankles at right angles.

Application: Workspace layout. Equipment design: vertical clearance from the top of the seat surface to the underside of work tables and consoles for the seated operator.

Sample and Reference	Survey Date	Number of Subjects	Age Range	Descriptive Statistics[a]			
				$\overline{X}$	SD	5%ile	95%ile
Females							
USAF Women	1968	1905	18–56	12.4 (4.9)	1.3 (0.5)	10.4 (4.1)	14.6 (5.7)
U.S. HEW civilians	1960–1962	1165	25–40	13.9 (5.5)	1.9 (0.7)	10.7 (4.2)	17.8 (7.0)
Swedish civilians	1968	214	20–49	15.4 (6.1)	1.3 (0.5)	13.2 (5.2)	17.5 (6.9)
Males							
USAF flying personnel	1967	2420	21–50	16.5 (6.5)	1.4 (0.6)	14.3 (5.6)	18.8 (7.4)
RAF flying personnel	1970–1971	588	18–45	15.8 (6.2)	1.2 (0.5)	13.9 (5.5)	17.8 (7.0)
Italian military	1960	1342	18–59	16.1 (6.3)	1.1 (0.4)	14.4 (5.7)	18.0 (7.1)
French fliers	1973	65	27–32	14.5 (5.7)	1.1 (0.4)	12.7 (5.0)	16.4 (6.5)
German AF	1975	1004	Not reported	15.5 (6.1)	1.5 (0.6)	13.2 (5.2)	18.0 (7.1)

[a] Data given in centimeters with inches in parentheses.

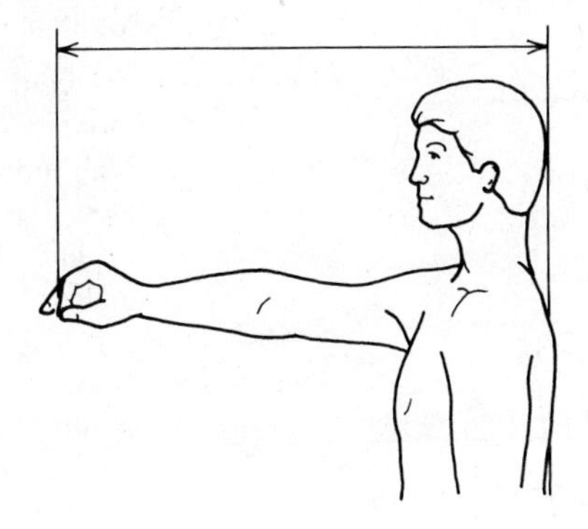

Thumb-tip Reach

Definition: The horizontal distance from the wall to the tip of the thumb, measured with the subject's back against the wall, his arm extended forward, and his index finger touching the tip of his thumb.

Application: Workspace layout. Equipment design: a minimum forward thumbtip reach distance with shoulder and torso restrained.

Sample and Reference	Survey Date	Number of Subjects	Age Range	Descriptive Statistics[a]			
				$\overline{X}$	SD	5%ile	95%ile
Females							
USAF women	1968	1905	18–56	74.1 (29.2)	3.9 (1.5)	67.7 (26.7)	80.5 (31.7)
Males							
USAF flying personnel	1967	2420	21–50	80.3 (31.6)	4.0 (1.6)	73.9 (29.1)	87.0 (34.3)
RAF flying personnel	1970–1971	1997	18–45	80.2 (31.6)	3.6 (1.4)	74.4 (29.3)	85.1 (33.5)
Italian military	1960	1342	18–59	75.3 (29.6)	3.7 (1.5)	69.3 (27.3)	81.6 (32.1)
German AF	1975	1004	Not reported	80.0 (31.5)	4.3 (1.7)	73.1 (28.8)	87.1 (34.3)

[a] Data given in centimeters with inches in parentheses.

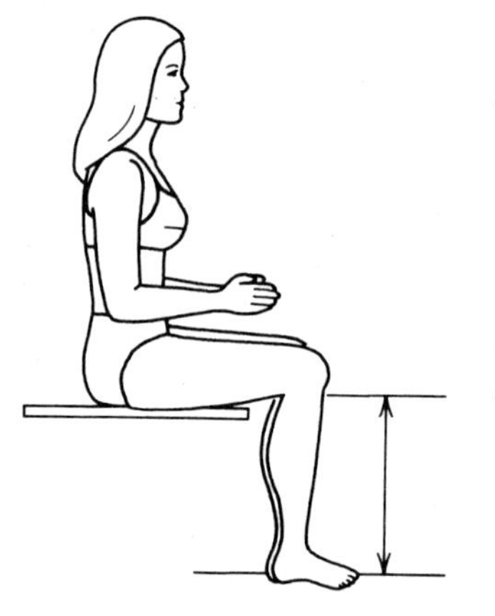

Popliteal Height

Definition: The vertical distance from the floor to the underside of the thigh immediately behind the knee. The subject sits erect with his knees and ankles at right angles.

Application: Workspace layout; Equipment design: vertical distance from the floor to the top forward edge of the seat pan for the seated operator.

Sample and Reference	Survey Date	Number of Subjects	Age Range	Descriptive Statistics[a]			
				$\overline{X}$	SD	5%ile	95%ile
Females							
USAF women	1968	1905	18–56	41.1 (16.2)	1.9 (0.7)	38.0 (15.0)	44.1 (17.0)
U.S. HEW civilians	1960–1962	1165	25–40	40.0 (15.7)	2.6 (1.0)	35.8 (14.1)	44.3 (17.4)
Males							
USAF flying personnel	1967	2420	21–50	43.7 (17.2)	2.3 (0.9)	40.1 (15.8)	47.5 (18.7)
Italian military	1960	1342	18–59	40.3 (15.9)	2.3 (0.9)	36.6 (14.4)	44.2 (17.4)
French fliers	1973	65	27–32	45.6 (18.0)	1.5 (0.6)	42.6 (16.8)	47.7 (18.8)
German AF	1975	1004	Not reported	43.8 (17.2)	2.1 (0.8)	40.4 (15.9)	47.4 (18.7)

[a] Data given in centimeters with inches in parentheses.

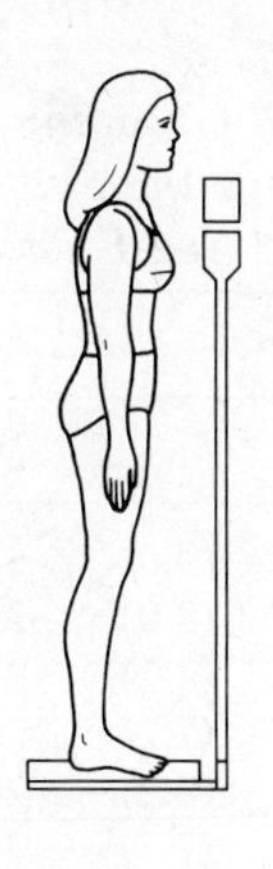

Weight

Definition: Nude body weight as measued on physician's scales.

Application: General body description.
Sizing of clothing and personal protective equipment.
Workspace layout.
Body linkage and models.
Equipment design: structural support for seats, platforms, couches, and body-restraint systems and harness rigging.

Sample and Reference	Survey Date	Number of Subjects	Age Range	Descriptive Statistics[a]			
				$\overline{X}$	SD	5%ile	95%ile
Females							
USAF women	1968	1905	18–56	57.73 (127.27)	7.52 (16.58)	46.4 (102.3)	70.9 (156.3)
U.S. HEW civilians	1960–1962	1165	25–40	62.38 (137.52)	14.26 (31.44)	46.0 (101.4)	89.4 (197.1)
British civilians	1957	4989	18–55+	60.40 (133.15)	10.00 (22.05)	46.6 (102.7)	79.4 (175.0)
Swedish civilians	1968	210	20–49	59.26 (130.64)	6.65 (14.66)	48.3 (106.5)	70.2 (154.8)
Japanese civilians	1967–1968	1622	25–39	51.30 (113.09)	7.00 (15.43)	39.8 (87.7)	62.8 (138.4)
Males							
USAF flying personnel	1967	2420	21–50	78.74 (173.58)	9.72 (21.43)	63.6 (140.2)	95.6 (210.8)
NASA astronauts	Dates vary	59	28–43	74.51 (164.26)	6.92 (15.26)	65.1 (143.5)	87.3 (192.5)
RAF flying personnel	1970–1971	1998	18–45	75.04 (165.43)	8.81 (19.42)	61.4 (135.4)	90.3 (199.1)
Italian military	1960	1342	18–59	70.25 (154.87)	8.42 (18.56)	57.6 (127.0)	85.1 (187.6)
French fliers	1973	65	27–32	74.0 (163.1)	8.10 (17.9)	60.6 (133.6)	88.3 (194.7)
German AF	1975	1004	Not reported	74.73 (164.74)	8.10 (17.86)	62.2 (137.1)	88.8 (195.8)
Japanese civilians	1967–1968 1972–1973	1870	25–39	60.20 (132.71)	8.60 (18.96)	46.1 (101.6)	74.3 (163.8)

[a] Data given in kilograms with pounds in parentheses.

TABLE 15-5 *Women's Right Hand Grasping Reach to a Horizontal Plane 91.4 Centimeters (36 in.) above the Seat Reference Point. Horizontal Distance from the SRV.*[a] *See Figure 15.34.*

Angle to Left or Right	Percentiles 5		50		95	
L 165	22.9	(9.0)	33.0	(13.0)	49.5	(19.5)
L 150	20.3	(8.0)	29.2	(11.5)	45.0	(17.7)
L 135	18.3	(7.2)	25.9	(10.2)	40.6	(16.0)
L 120	18.3	(7.2)	25.4	(10.0)	39.4	(15.5)
L 105	18.3	(7.2)	26.7	(10.5)	38.6	(15.2)
L 90	19.6	(7.7)	29.2	(11.5)	40.6	(16.0)
L 75	20.8	(8.2)	33.0	(13.0)	43.7	(17.2)
L 60	25.4	(10.0)	36.1	(14.2)	45.7	(18.0)
L 45	29.2	(11.5)	39.4	(15.5)	49.5	(19.5)
L 30	33.5	(13.2)	43.7	(17.2)	54.6	(21.5)
L 15	36.1	(14.2)	48.3	(19.0)	47.7	(22.7)
0	41.1	(16.2)	52.1	(20.5)	61.0	(24.0)
R 15	44.5	(17.5)	54.6	(21.5)	62.7	(24.7)
R 30	47.0	(18.5)	57.2	(22.5)	66.0	(26.0)
R 45	48.8	(19.2)	61.0	(24.0)	68.6	(27.0)
R 60	52.6	(20.7)	63.5	(25.0)	70.4	(27.7)
R 75	53.3	(21.0)	64.8	(25.5)	71.1	(28.0)
R 90	56.4	(22.2)	66.5	(26.2)	72.9	(28.7)
R 105	53.8	(21.2)	66.5	(26.2)	72.9	(28.7)
R 120	46.2	(18.2)	63.5	(25.0)	70.4	(27.7)
R 135	31.8	(12.5)	48.3	(19.0)	65.3	(25.7)
R 150	25.4	(10.0)	43.7	(17.2)	59.7	(23.5)
R 165	25.9	(10.2)	40.6	(16.0)	55.9	(22.0)
180	24.1	(9.5)	38.6	(15.2)	53.8	(21.2)

[a] Data given in centimeters with inches in parentheses.
The original data were measured to the nearest $\frac{1}{4}$ inch and are reported here rounded down to the nearest tenth of an inch.

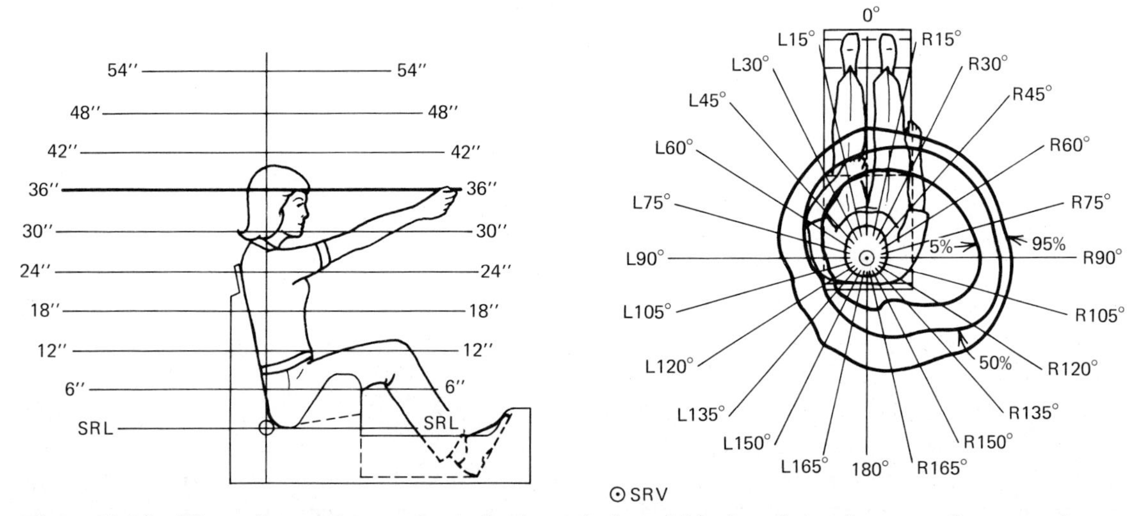

Figure 15-34. Women's grasping reach to a horizontal plane 36 inches above the seat reference point.

TABLE 15.6 *Men's Right Hand Grasping Reach to a Horizontal Plane 88.9 Centimeters (35 in.) above the Seat Reference Point. Horizontal Distance from the SRV.[a] See Figure 15-35.*

Angle to Left or Right	Minimum		Percentiles 5		50		95	
L 165					37.3	(14.7)	53.3	(21.0)
L 150					34.8	(13.7)	50.8	(20.0)
L 135					33.5	(13.2)	48.3	(19.0)
L 120			27.2	(10.7)	33.5	(13.2)	47.5	(18.7)
L 105			31.0	(12.2)	35.6	(14.0)	47.5	(18.7)
L 90	32.3	(12.7)	34.8	(13.7)	39.4	(15.5)	50.8	(20.0)
L 75	36.1	(14.2)	38.1	(15.0)	43.7	(17.2)	53.3	(21.0)
L 60	38.6	(15.2)	40.6	(16.0)	47.5	(18.7)	54.6	(21.5)
L 45	41.1	(16.2)	43.7	(17.2)	52.1	(20.5)	62.7	(24.7)
L 30	45.7	(18.0)	48.8	(19.2)	57.2	(22.5)	66.5	(26.2)
L 15	48.8	(19.2)	53.3	(21.0)	62.7	(24.7)	68.6	(27.0)
0	52.6	(20.7)	56.4	(22.2)	67.3	(26.5)	72.4	(28.5)
R 15	57.7	(22.7)	62.7	(24.7)	70.4	(27.7)	78.7	(31.0)
R 30	62.2	(24.5)	67.8	(26.7)	74.2	(29.2)	83.1	(32.7)
R 45	67.8	(26.7)	71.6	(28.2)	77.5	(30.5)	85.6	(33.7)
R 60	71.1	(28.0)	73.7	(29.0)	78.7	(31.0)	85.6	(33.7)
R 75	72.9	(28.7)	74.9	(29.5)	79.2	(31.2)	86.4	(34.0)
R 90	73.7	(29.0)	75.4	(29.7)	79.2	(31.2)	85.1	(33.5)
R 105	73.7	(29.0)	75.4	(29.7)	80.0	(31.5)	85.1	(33.5)
R 120	72.4	(28.5)	73.7	(29.0)	78.7	(31.0)	85.1	(33.5)
R 135					72.39	(28.5)	85.1	(33.5)
R 150							80.0	(31.5)
R 165							55.1	(21.7)
180					41.9	(16.5)	56.4	(22.2)

[a] Data given in centimeters with inches in parentheses.
The original data were measured to the nearest ¼ inch and are reported here rounded down to the nearest tenth of an inch.

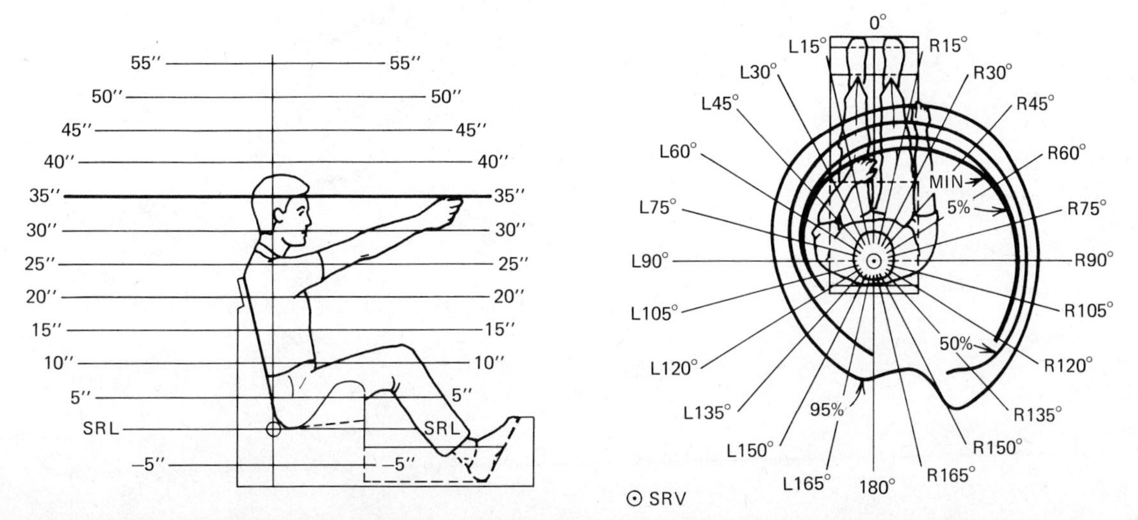

Figure 15-35. Men's grasping reach to a horizontal plane 35 inches above the seat reference point.

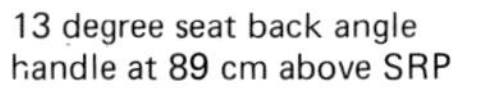

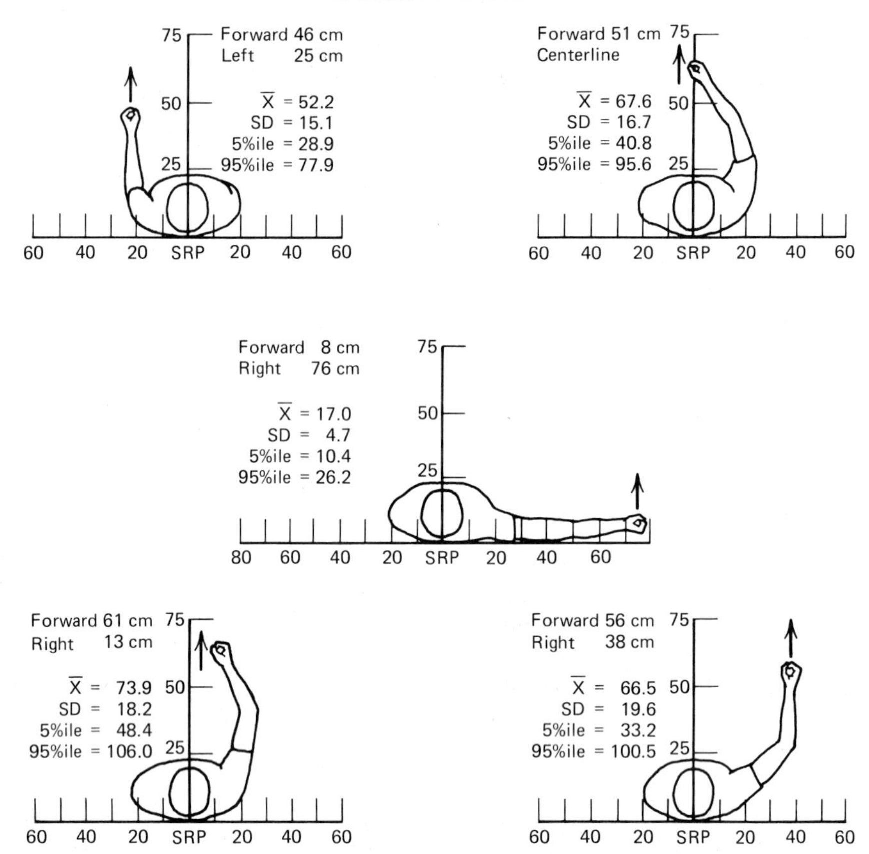

Figure 15-36. Force exerted on handle assembly at various locations relative to the seat reference point and seat centerline (values in kiloponds).

PART FIVE

ENVIRONMENT

16 NOISE

Did you ever have an argument about the sound level of your stereo with somebody in your dorm? "Turn down that ----- racket, it's driving me up the wall!" You may have wondered why your estimate of the sound level of the stereo was so different from that of your angry neighbor. Why can a faint sound appear terribly noisy in certain situations? Can exposure to noise affect our behavior after the noise has been turned off? How dangerous to our hearing is ordinary environmental noise? These are a few examples of some frequently asked questions about the effect of noise on people. We will try to answer these and other questions in this chapter. To do so, two important aspects of the noise problem must be considered. The first is the physical character of the noise. The second is the psychological condition of the person being exposed to the noise. You will see that we can only predict the effect of the noise when both these aspects of the noise situation have been considered.

Obtaining a complete physical description of noise can be a complex job, requiring the use of expensive recording and analyzing equipment. We touched on this problem in Chapter 3, in the discussion of how sound is generated and measured. Noise is defined as any unwanted sound. Noise in our work and home environment is usually composed of many different frequencies. The intensity and phase of each of these frequencies may be changing with time. It would be nice if there were a way to take the complex physical description of a noise, and boil that down to a single number that would predict its effect on people. Several measures of this type will be described in this chapter.

We also have to specify the psychological state of the people that are being stimulated by the noise. The kind of psychological factors that are important include whether the noise is expected or surprising, whether the noise makes the person's task more difficult, and whether the person is in a relaxed or alert condition. We will see that it is the interaction between the physical and psychological aspects of noise that creates some of the most interesting and varied effects of noise on people.

PHYSICAL DESCRIPTION OF NOISE

Noise is usually a complex sound, composed of many different frequencies of varying level and phase. A rough idea of the complexity of a noise can be obtained from the sound spectrogram shown in Figure 16-1. This is a type of three-dimensional record of level, frequency, and time for some short duration of the noise. The particular noise shown in the figure is the sound output of an automobile muffler while the engine was briefly accelerated. The relative darkness of the lines indicates the level

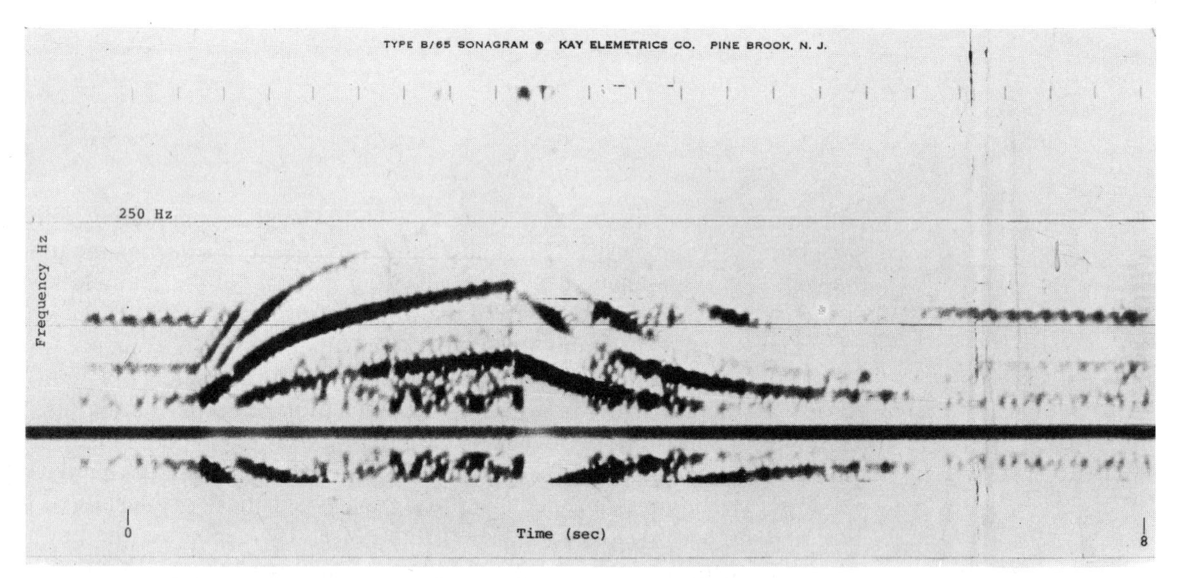

Figure 16-1. The sound spectrogram of sound from an automobile muffler. (From L. Pande & J. W. Sullivan. Herrick Laboratories Report No. 4HL 78-42, August, 1978.)

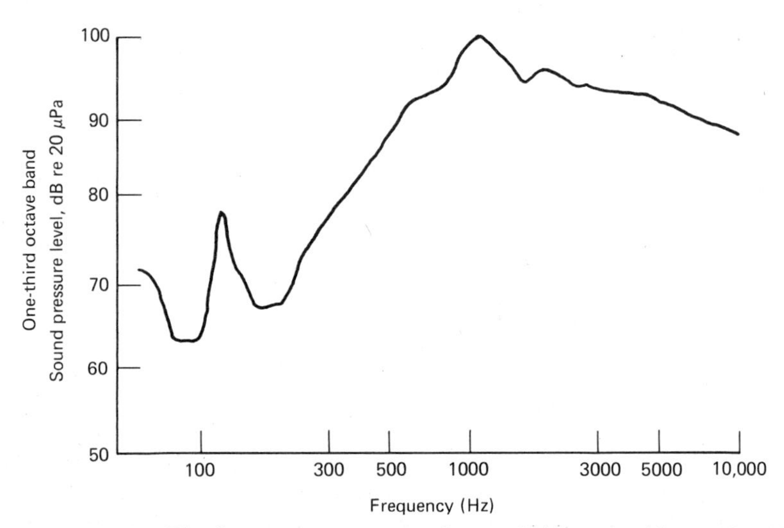

Figure 16-2. The frequency spectrum of a wood planer machine. (After NIOSH, 1975.)

of the noise at each frequency. Most of the energy in this noise is at frequencies below 250 Hz.

Frequency Analyzers If the levels of the different frequency components of the noise are approximately constant over time, we can ignore the time dimension and plot the frequency spectrum of the noise. Figures 16-2 through 16-4 illustrate such spectral plots of sound pressure level versus frequency for

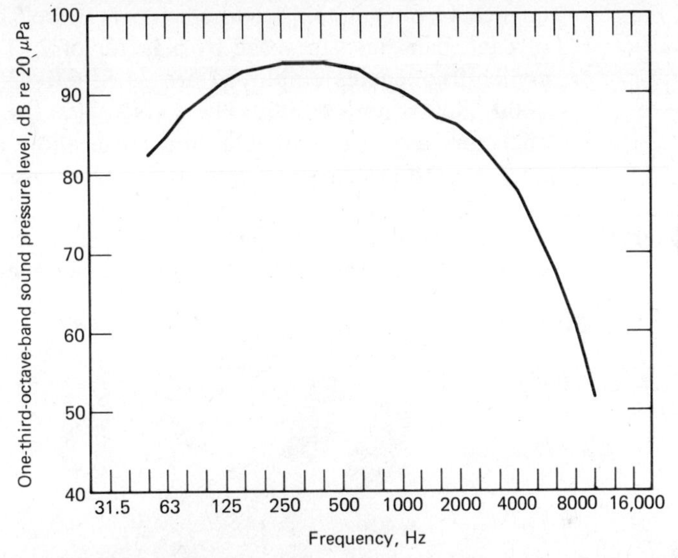

Figure 16-3. Representative spectrum of noise produced by jet-transport airplane flying overhead at a distance of approximately 300 meters at takeoff thrust. (Courtesy A. H. Marsh.)

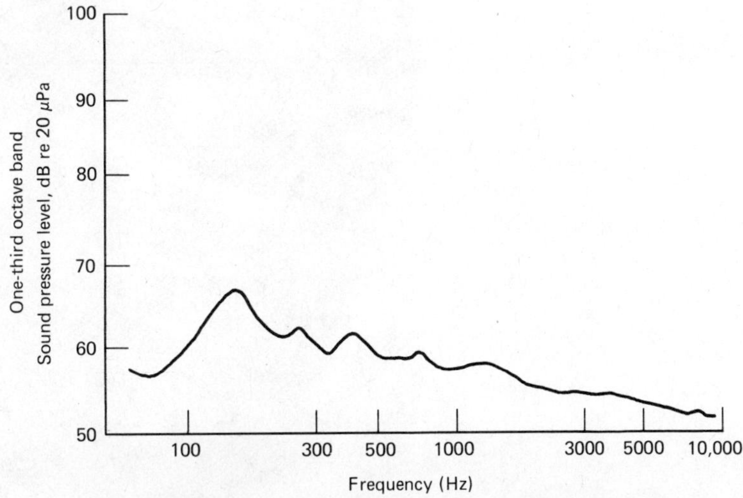

Figure 16-4. The frequency spectrum of a lawnmower. (From J. A. Molino, G. A. Zerdy, N. D. Lerner, & D. L. Harwood, *Journal of the Acoustical Society of America*, 1979, *66*, 1437, and reproduced by permission.)

some typical noise sources in factory, transportation, and community environments. These kinds of spectral plots can be obtained in a variety of ways. One rather precise way is to employ a frequency analyzer, which scans the noise level in each of a large number of very narrow frequency bands. Another type of analyzer commonly employed in noise measurement is an octave band analyzer, which determines the sound level within a limited number of wide frequency bands. Each band includes frequencies from f_{low} to f_{high}, where f_{high} is twice f_{low}. The center frequencies of each band also increase by a factor of 2. The standard octave band center frequencies employed are 31.5, 63, 125, 250, 500, 1000, 2000, 4000, 8000, and 16,000 Hz (ANSI 1971b). One-third octave band analyzers are also available, and these allow a finer description of the noise spectrum than the octave band analyzers.

Sound Level Meter

The simplest type of sound measuring instrument is the sound level meter (Figures 16-5 and 16-6). This device measures sound level across

Figure 16-5. Standard type of sound level meter (Courtesy Gen Rad).

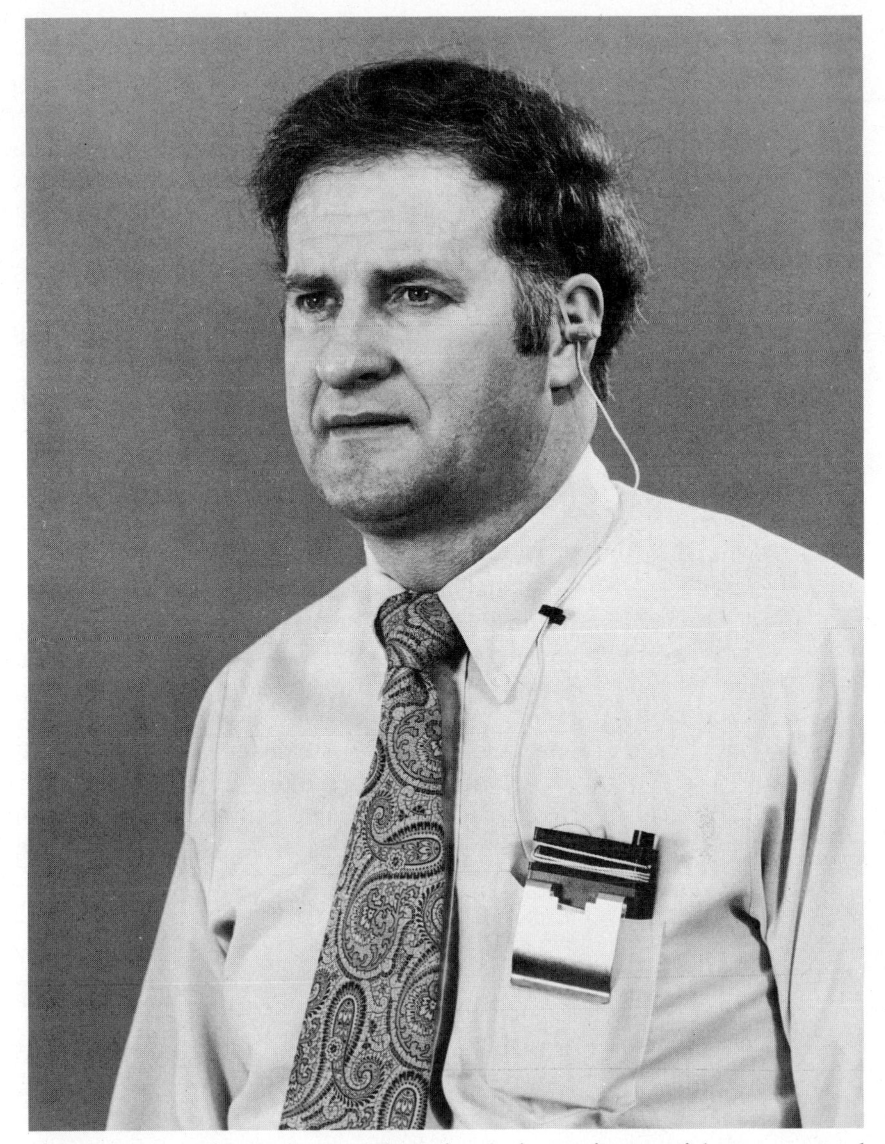

Figure 16-6. Noise dosimeter. The noise dosimeter is a special purpose sound level meter that can measure the total noise exposure of a worker throughout the work day (Courtesy Gen Rad).

the entire range of audible frequencies, and hence yields much less precision than a frequency or octave band analyzer. There are two settings on this instrument that concern us: the meter response setting, and the frequency weighting setting.

The meter response setting of FAST or SLOW determines the dynamic characteristics of the meter. A FAST setting will allow the meter needle or display to more easily follow changes in the level of the noise being measured, and to respond more accurately to peak sound levels (the response time of the meter is about 200 msec on the fast setting). In fact, it is difficult to measure the peak levels of very rapidly changing sounds with a standard sound level meter. Some instruments include special circuits to allow the measurement of very intense brief peaks in the noise level (meter response of 50 μsec or less). The SLOW meter setting averages the sound level over a longer duration (about 500 msec), and thus allows the measurement of fluctuating sound levels. This is the standard setting for most measurements.

The other important setting on a sound level meter is the frequency response of its internal weighting circuit; there are several different weighting curves such as A, B, C, D and E. These curves are defined by international and American standards (ANSI, 1971a). The A, B, and C weighting functions are shown in Figure 16-7. The A setting (with the slow meter setting) is the legal setting used to measure noise to evaluate its effect on people and to indicate the degree of danger to the human auditory system. You can see that the weighting network will cause the meter to be much less responsive to low-frequency components in the noise being measured. The A function was derived from the 40 dB equal loudness contour discussed in Chapter 3. Because the A scale roughly resembles our sensitivity to sound of different frequencies, it is an effective way to generate a single number that will characterize the effect of a complex noise on humans (NIOSH, 1975).

By switching between the A and C scales, you can get a relative idea of the presence of low- and high-frequency components in the noise. Since

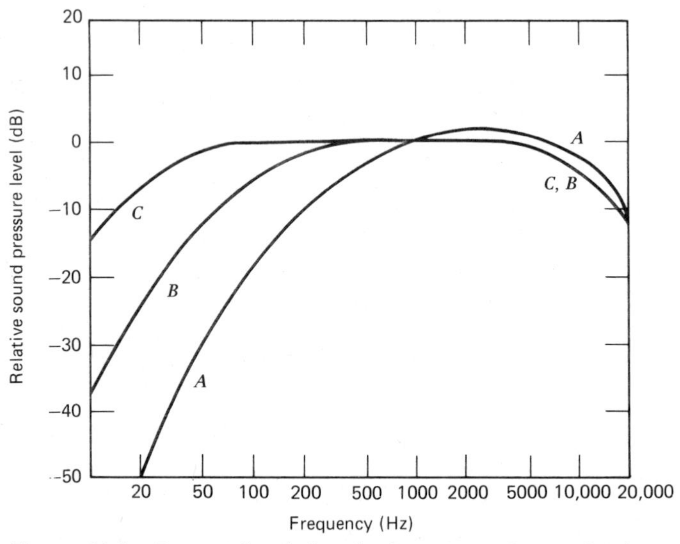

Figure 16-7. International Standard *A, B,* and *C* weighting curves for sound level meters.

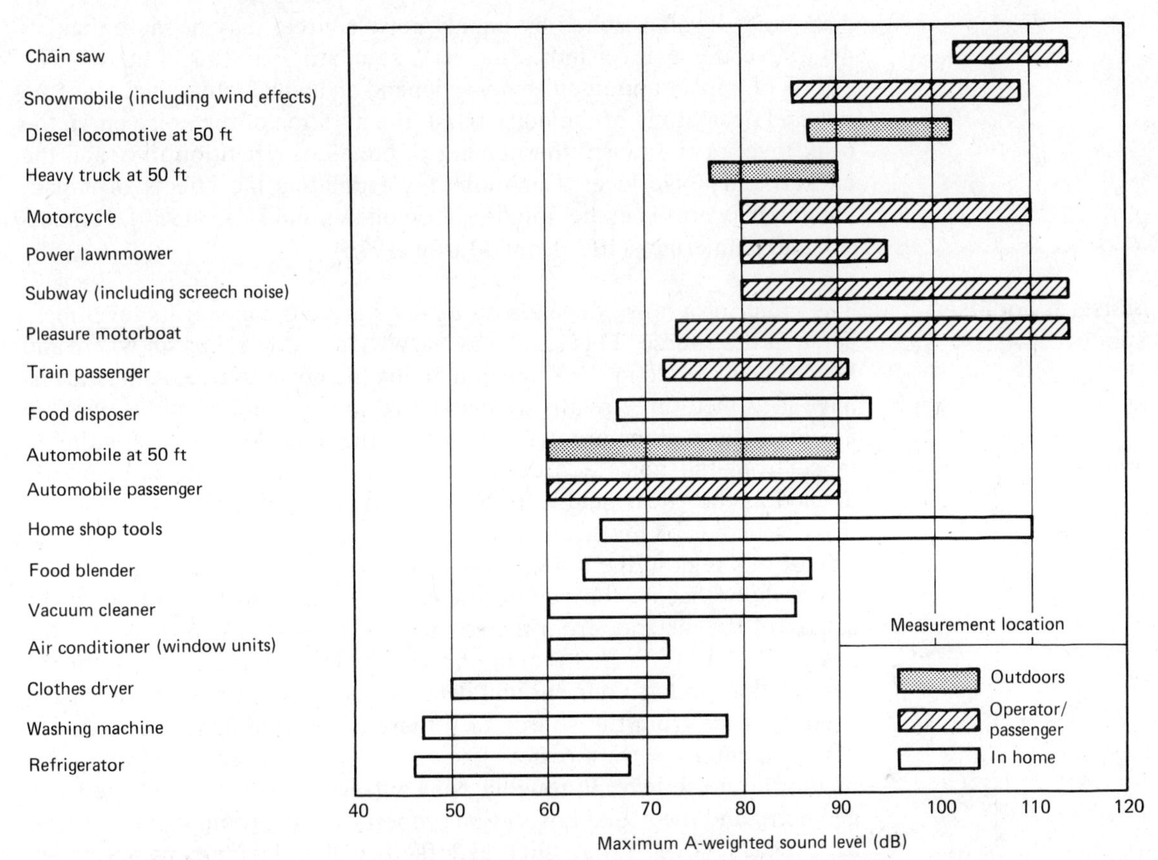

Figure 16-8. The typical range of common sounds, measured in dB(A). (After EPA, 1978.)

the C scale is more sensitive to low frequencies and slightly less sensitive to high frequencies than the A, a higher reading on the C scale will indicate the presence of relatively more low-frequency energy in the noise that you are measuring. A noise level measurement on the A (and slow) setting is indicated by the letter A after the dB, as for example, 80dB(A). Figure 16-8 illustrates the range of some typical noises in A-weighted levels.

Impact and Impulsive Noise

For certain types of noise, the sound level is not constant over short periods of time. Impact and impulsive noises fall into this category. Impact noise arises out of object to object collisions like metal parts falling against metal surfaces. Dump six empty cans into an empty metal trash container and you'll get impact noise. Impulse noise can arise out of the rapid acceleration of an object or the explosive discharge of gases. The sound of rifle fire is a good example. These types of noise can't be measured with a basic sound level meter since such meters will not respond to the extremely short duration peaks of sound intensity that occur. The peak noise levels reached by impact noise sources may be more than 30 dB above the A-level indication on a standard sound level meter! The effects of impact and impulse noise depend on many factors, including the rise and decay time of the noise burst, the duration of the noise burst, the peak level of the noise, the number of bursts or repetition rate, and the background noise level. Equations for estimating the effects of impact and impulse noise can be found in Trémolières and Hétu (1980), Henderson and Hamernick (1978), and Martin (1976).

Noise in Enclosed Spaces

The effect of a noise depends on its level at your ear, not its level measured at the source. This level may vary widely, depending on where and how it is measured in the room containing the noise source. Acousticians have identified three relatively distinct regions around a sound source in a room. The regions are defined by how the noise level varies with distance from the noise source. The region close to the noise source is defined as the "near field." In this region the sound level is variable and may be higher or lower as distance from the source increases. Just outside of this region, the noise intensity decreases according to the inverse square law, that is, the sound intensity decreases in proportion to the square of the distance from the source. This region is known as the "far field." If you are in this region and you double your distance from the source, the level at your ear will decrease by 6 dB. If you are at a sufficient distance from the source or if there are reflecting surfaces near to you, you enter the third region, known as the "reverberant field." In this region the noise level is roughly constant and is a function of the room geometry and the sound absorption properties of the room surfaces. In an acoustically "live" room, such as a factory shop lacking any sound absorption and reduction materials, the noise level when you are far away from the source may be nearly the same as at the source. This is an important aspect of noise in enclosed areas.

Combined Noises

If we know the noise level generated by two or more independent noise sources, we can compute the total amount of noise that will result from their simultaneous combination. We can't simply add the dB levels of the sources together. The total level is given by

$$L_t = 10 \log(10^{L1/10} + 10^{L2/10} + 10^{L3/10} + \ldots) \qquad (16\text{-}1)$$

where $L1$ is the level of source number 1, $L2$ is the level of source number 2, etc. Suppose we had three independent sources of noise levels 75 dB(A), 86 dB(A), and 87 dB(A), what would the combined noise level be if all sources were on at the same time?

$$L_t = 10 \log(10^{7.5} + 10^{8.6} + 10^{8.7})$$

$$L_t = 10 \log(3.16 \times 10^7 + 3.98 \times 10^8 + 5.01 \times 10^8) = 89.7 \text{ dB(A)}$$

If the noise sources include strong pure tone components, this method will overestimate the combined noise level.

NOISE-INDUCED HEARING LOSS

Although it is possible to damage a person's eardrum by exposure to an intense blast of noise, the usual place of injury due to noise exposure is the auditory sense organ of the inner ear, the organ of Corti (see Chapter 3). This is the exquisitely sensitive receptor organ for sound, where very small mechanical movements of the hair cells cause electrical impulses to be transmitted to the brain via the auditory nerve. Very intense, brief noises such as produced by an explosion can cause severe mechanical deformation in the organ of Corti and result in immediate and extensive structural damage. Long-term exposure to moderate noise levels can result in the partial degeneration of the hair cells along the basilar membrane of this organ. The hearing loss associated with an injury to the inner ear depends on the degree of damage to a particular region of the organ of Corti and its distribution over the length of the organ. For example, damage to the lower region of the cochlea might be evident as increases in the threshold for high-frequency sounds (Moody, Stebbins, Johnson, & Hawkins, 1976).

Many people don't appreciate the fact that such damage to the auditory nervous system, once suffered, is permanent and nonreversible. The losses in hearing ability that result will have different implications for a person, depending on his or her life-style and occupation. It is important to note that the type of hearing losses produced by noise are usually not simply a decrease in the effective amplification or gain of the auditory system. A hearing aid, for example, will probably be relatively ineffective in compensating for the loss in your ability to distinguish among different frequencies. If that loss in frequency resolution has spread to the speech frequencies, there may be a large decrease in your ability to understand speech—particularly under noisy conditions. This kind of problem can exist even though only moderate changes have occurred in your thresholds for pure tones (Hodge & Price, 1978). In subsequent sections we consider the government restrictions on occupational noise exposure and indicate some general ways to comply with those regulations. Noise exposure in the home and in places of entertainment can sometimes exceed those occupational risk levels.

Temporary Threshold Shift

Noise induced changes in the measured sensitivity of the auditory system can be temporary or permanent. When a noise is sufficiently intense or the exposure is of sufficient duration, the shift in a person's threshold will contain both temporary and permanent components. The measured temporary increase in a person's threshold for tones is defined as the temporary threshold shift, or TTS. Much of what is believed about the specific effects of noise on human hearing has been generalized from ex-

TABLE 16-1 *Summary of TTS Characteristics (after Miller, 1974 and Ward, 1976)*

1. Factors that will increase the magnitude of TTS caused by a noise:
   (a) Noise level (must exceed 60 to 80 dB(A).
   (b) Exposure duration (up to a maximum of 10 to 12 hours, depending on level).
   (c) Concentration of noise frequencies (the more concentrated is the noise to a few frequencies, the greater the TTS).
   (d) Frequencies in the range 2000 to 6000 Hz.
2. The largest TTS will be for test tones that are one-half to one octave above the frequency of the highest component of the noise.
3. There are large individual differences in a person's susceptibility to TTS.
4. The TTS will be reduced if there are frequent interruptions to the noise exposure.

periments on TTS. This is because it is unethical to intentionally cause permanent hearing loss to human subjects. Suppose that you are exposed to a noise that only produces a temporary shift; if the exposure is repeated enough times or is sufficiently long in duration, you will probably have a permanent threshold shift (NIOSH, 1972). Table 16-1 summarizes some of the important facts about TTS.

Effects of Workplace Noise

There have been a number of studies of worker hearing loss as a function of the workplace noise levels and the number of years working. In general, the amount of permanent threshold shift increases as a function of the level of noise. Figure 16-9 illustrates some data from a study of female workers in a jute weaving mill in which the noise levels were about 98 dB(A) (Taylor, Pearson, Mair, & Burns, 1965). The worker thres-

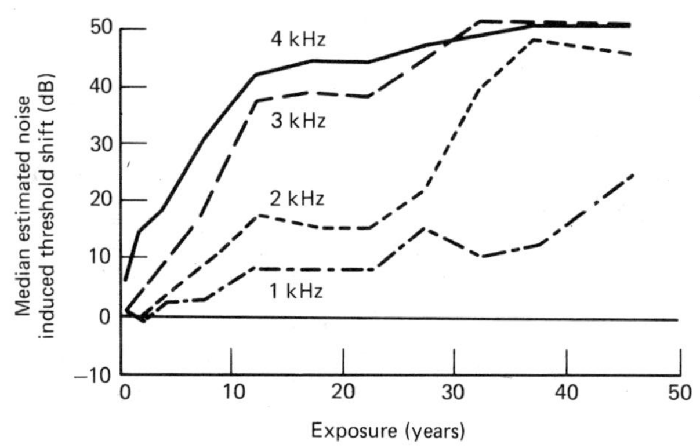

Figure 16-9. Noise-induced threshold shift for jute weavers as a function of duration of exposure to noise of approximately 98 db(A). (From W. Taylor, J. Pearson, A. Mair, & W. Burns, *Journal of the Acoustical Society of America,* 1965, *38,* 118, and reproduced by permission.)

holds for several different test tone frequencies can be seen to rise as a function of the duration of exposure to the noise. The shift at 4000 Hz is relatively rapid and then levels off after about 10 years. After longer periods of time, the loss spreads to other frequencies. These threshold measurements were made on the workers after a weekend away from the noise. This was to allow their thresholds to recover from any temporary threshold shifts prior to measurement. This last point is worth thinking about. This level of noise [98 dB(A)] is sufficient to produce a huge TTS to your hearing. If you were one of these workers, from your very first day of work you would only have normal hearing for a few hours each week (Miller, 1974).

Criteria for Risk of Hearing Damage: OSHA Regulations

The Occupational Safety and Health Act of 1970 (OSHA) applied noise exposure standards to all businesses affecting interstate commerce. In addition, the Noise Control Act of 1972 designated the Environmental Protection Agency with the responsibility for controlling noise in the environment, including the establishment of noise emission standards for new products distributed in interstate commerce, the manufacturer's labeling of products with respect to noise, and many other aspects of noise. Authority for aircraft noise remained with the Federal Aviation Administration.

Permissible noise exposures under OSHA regulation are given in Table 16-2. The limit for average exposure during an 8-hour period is 90 dB(A). Reducing the duration of exposure by one-half allows you to increase the level by 5 dB(A). For example, an exposure of 90 dB(A) for 8 hours is considered equivalent to an exposure of 95 dB(A) for 4 hours. Exposure to impulse or impact noise is not supposed to exceed 140 dB peak sound pressure level.

If the daily exposure is composed of two or more periods of noise at different levels, you must consider their combined effect. If the sum of the fractions: $(C_1/T_1) + (C_2/T_2) + \ldots (C_n/T_n)$ is greater than 1, then the

TABLE 16-2 *OSHA Permissible Noise Exposures*

Hours of Exposure	Sound Level, dB(A)
8	90
6	92
4	95
3	97
2	100
$1\frac{1}{2}$	102
1	105
$\frac{1}{2}$	110
$\frac{1}{4}$ or less	115

mixed exposure would be considered to exceed the allowable limit for a daily exposure. C_n is the total time of exposure at a specified noise level, and T_n is the total time of exposure allowed at that level. For example, suppose that you are exposed to a 100-dB(A) noise for 1.5 hours and to a 92-dB(A) noise for 4 hours. Your combined dosage would be: 1.5/2 plus 4/6 or 1.42; this combined dosage would be 1.42 times as large as the maximum permissible.

The Act specifies that feasible engineering or administrative controls must be applied to protect the hearing of employees. This means that either the noise level at each work area must achieve a net exposure rating of unity or less, or the exposure time of each worker must be controlled by rotating personnel or requiring them to wear hearing protection devices. Civil fines and criminal penalties can result from violation of the Act.

Some scientists believe that these risk criteria are not sufficiently conservative (EPA 1978; NIOSH, 1972; VonGierke & Johnson, 1976). Look at the problem from a personal viewpoint. First, if you have normal hearing before exposure, why should you accept *any* possibility of damage to your hearing when you wouldn't tolerate any damage to your sight? Second, your hearing might be more sensitive than the average. In that case, your risk of hearing loss could be much greater. A legal level of noise that might produce only slight changes (5 dB) to my hearing might really foul yours up. The National Institute of Occupational Safety and Health (1972) has recommended a daily (8-hour) occupational exposure level of 85 dB(A) rather than the 90 dB(A) level currently required by OSHA, and a more conservative trade-off of level with exposure duration. The damage risk criterion for impact-impulse noise is also not sufficiently conservative and is much too simple. Work is continuing in many laboratories on the specification of precise damage risk criteria for exposure to impact and impulsive noise (Martin, 1976; Henderson & Hamernick, 1978; Trémolières & Hétu, 1980).

BEFORE AND AFTER

Noise Control

The low-frequency noise from roof fans on a factory caused considerable disturbance to residents of a housing development about a quarter of a mile away. (See Figure 16-10.) The fans were replaced by new ones with the same capacity but with a larger number of fan blades. The new fans produced less low-frequency noise but more high-frequency noise. Because high-frequency noise is attenuated much more by distance than low-frequency noise, the increase in the high range caused no problem. The reduction in level of the low frequency noise was sufficient to eliminate the community's problem (OSHA, 1980).

Perhaps the noise energy can be moved downward in frequency, rather than upward, to solve a noise problem. The large engine shown in Figure 16-11 was designed to operate at 125 revolutions per minute with a direct drive connection to the ship's propeller. Noise from the propeller would be extremely disturbing to the crew. The solution was to add a differential gear between the engine and propeller so as to gear down the propeller speed to 75 revolutions per minute. A larger propeller was also required. Shifting the noise to the lower frequency results in a much less disturbing situation (OSHA, 1980).

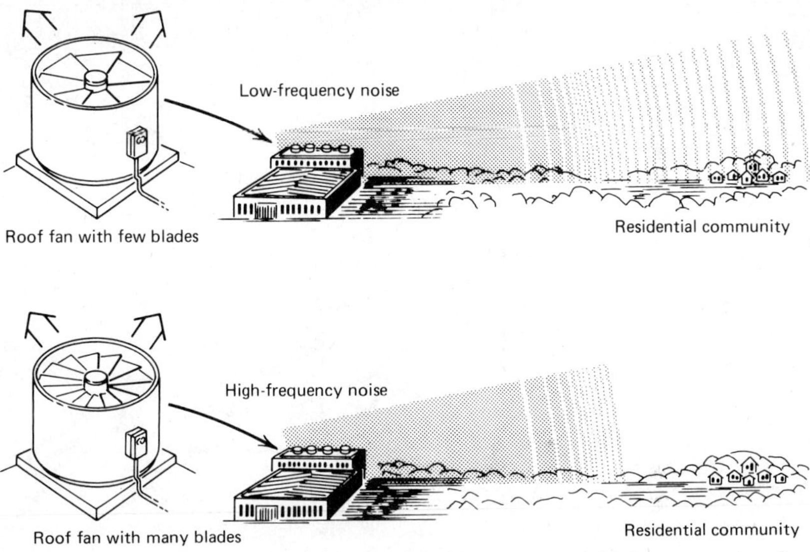

Figure 16-10. The fan design is modified to work at a higher rpm, thereby shifting the noise energy up in frequency, and reducing noise annoyance to the community (see text). (After OSHA, 1980.)

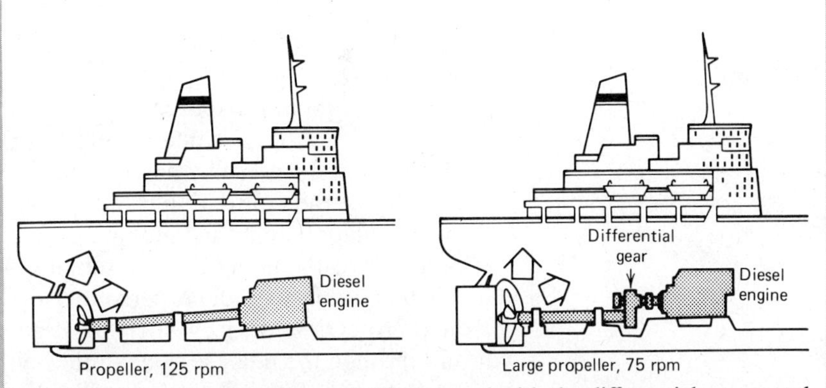

Figure 16-11. The propeller speed is reduced with the differential gear; resultant noise energy is reduced in frequency, causing less disturbance to the crew (see text). (After OSHA, 1980.)

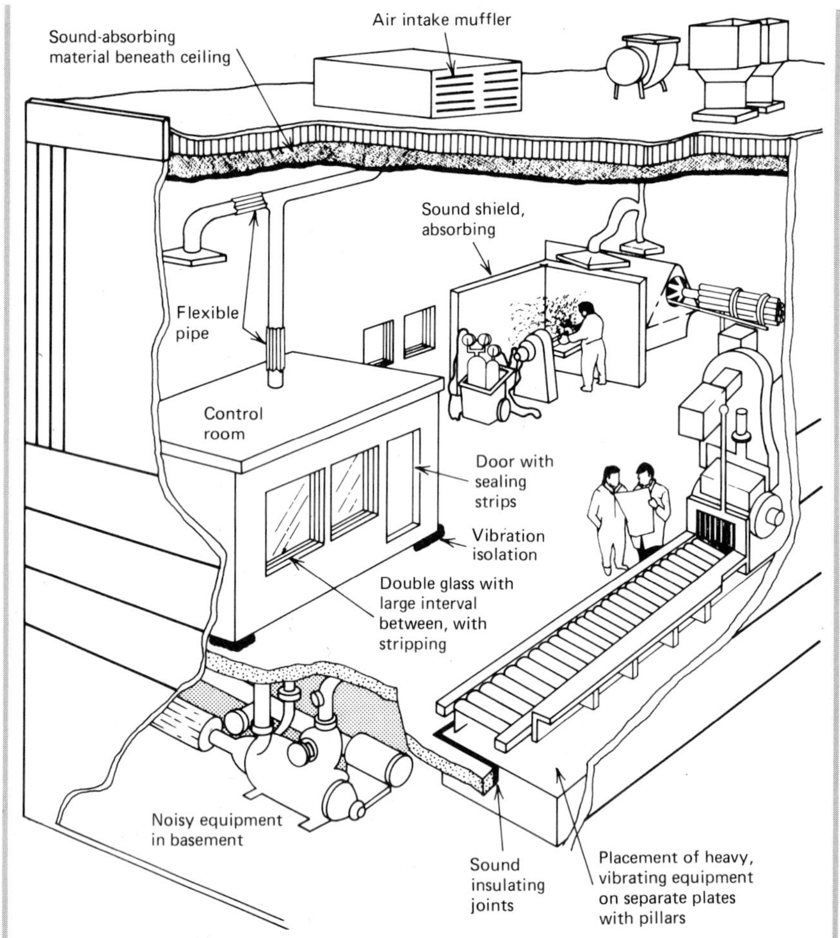

Figure 16-12. Example of noise control measures that can be carried out in an industrial building to avoid the spread of noise. (After OSHA, 1980.)

In some cases a noise source can be controlled with sound-absorbing material, as in the machine pictured in Figure 16-12. Workers can be isolated from the noise by placing them in a shielded room as shown in the center of the figure. Other techniques include the use vibration and sound isolation joints and mountings.

If redesign of the equipment or sound environment is not possible, you might be able to reduce personnel exposure through the use of personal protection devices such as ear protectors, or by rotating workers through the noisy areas. Neither of these techniques is a very good long-term solution to a serious noise problem. Earplugs may be modified by workers for greater comfort and with large decrease in noise protection; they may be worn carelessly or simply removed. It may be very difficult to supervise the use of plugs, particularly if the workers have long hair.

The best solution usually involves controlling the noise at the source, such as by shielding or isolation, by shifting the noise frequencies, or by minimizing turbulence in gas flows, rapid pressure changes, or metal-to-metal impacts. One generalization about noise control is also true of many human factors problems: it's usually cheaper to consider the noise problem before the product, plant, or process is designed, rather than afterward.

EFFECTS OF NOISE ON PERFORMANCE

In the next sections of this chapter we consider two general ways that noise can affect performance. First, noise can prevent us from hearing signals that are important to the performance of a task. Second, noise can have an effect on our mental and physical condition. Sound is such a natural part of our lives that we probably don't realize how important auditory information is to many tasks. Suppose that you had to start your automobile without being able to hear the engine – imagine the starter and gear damage that could result. Auditory feedback may be necessary for the performance of many skilled jobs, such as the operation of a milling machine or lathe. The section on signal detection deals with predicting how well we can hear such signals in noise. Can exposure to noise produce physical and mental changes to our bodies? We discuss this question, as well as worker stress and arousal, when we discuss physiological and psychological effects. We'll also outline some interesting effects that noise can have on our performance after the noise has been turned off. Finally, we'll discuss some aspects of our attitudes about noise.

Interference with Signal Detection

You are in command of a Coast Guard lifeboat on a search and rescue mission on Lake Michigan. A 36-foot motor yacht has struck some rocks in the heavy fog. The yacht's skipper managed to make a distress call on the marine radio, but their signal is fading fast because their engine area has been flooded. Of course they don't know exactly where they are (if they did, they wouldn't have run up on the rocks). Your radar is of no help; the yacht is fiberglass and didn't carry a radar reflector. You told them to continue to sound their air horn for 1 sec every 2 minutes. Assuming that you arrive at their approximate location, will you be able to hear their sound signal? Will it be heard over the noise of your own boat? Should you post one of your crew on your bow or stop your boat's engine to improve detection of the yacht's signal?

The example points up some of the problems of detecting weak signals in a background of noise. Our ability to hear the signal is drastically reduced by the presence of the noise. We discussed the problem of trying to understand speech in interfering noise in Chapter 9. In this section we are concerned with detecting simple signals such as the air horn in the example, or a warning signal in your car or at your job. The problem arises because the auditory system can't separate the sensations that are caused by two tones that are close together in frequency. Not only can't we hear the desired signal, but if the interfering tones are intense enough, we will also hear all sorts of additional tones.

These aspects of our ability to detect signals were clearly demonstrated in a classical experiment by Wegel and Lane (1924). In that experiment subjects were asked to detect tones of different frequency in the presence of an interfering tone, which was at 1200 Hz. Figure 16-13 shows how the threshold for the signal tone varied, starting from signal frequencies below 1200 Hz, and increasing to frequencies well above that of the interfering tone. The *interfering* tone was set at a level of about 80 dB. The ordinate of the graph indicates how large the signal had to be made in order for the subject to hear it along with the interfering tone.

You can see that the signal was very difficult to hear when its frequency was near to 1200 Hz. A good deal of the interference is also evident when the signal was higher than 1200 Hz. The lower curves on the figure show the signal threshold when the interfering tone was reduced in level; the

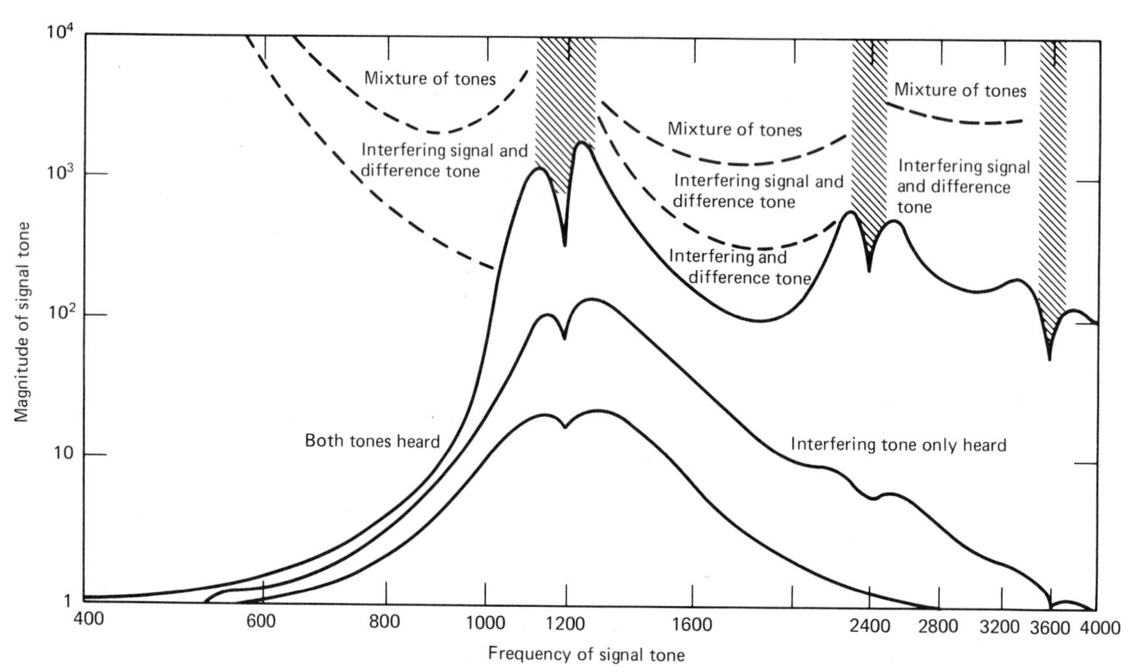

Figure 16-13. The effects of an interfering tone on the threshold of another (the signal). The interfering tone is fixed at 1200 Hz. The upper curve is the threshold level of a signal tone in the presence of an intense interfering tone. The lower curves show signal threshold in the presence of less intense interfering tones. (From R. L. Wegel & C. E. Lane, Physical Review, 1924, *23,* 272, reproduced by permission.)

TABLE 16-3 *Summary of the Effects of Interfering Tones on Signal Detection*

1. Most interference occurs when the signal tone is close to the frequency of the interfering tone.
2. As the interfering tone increases in intensity, the interference spreads to additional signal frequencies.
3. The effect of the interfering tone is worse when it is below the signal frequency, rather than above.
4. The interaction of intense tones results in the generation of additional tones, which include harmonics of the original tones, and tones at summation and difference frequencies.

interference now occurs over a narrower range of frequency. The figure also shows phenomena that result when intense tones are presented to your ear: difference tones, summation tones, and aural harmonics. Difference tones and summation tones are tones of various frequencies including f_1+f_2, f_2-f_1, $2f_1-f_2$, etc., where f_1 and f_2 are the frequencies of the original tones. Aural harmonics are tones at multiples of f_1 and f_2. These phenomena occur when signals are passed through systems that distort the signals in certain ways. Such systems are said to respond in a nonlinear fashion. (This process was described in Chapter 3.) Generally, if you input a signal that is larger than your system was designed to handle, you will run into a problem with distortion of this sort. It doesn't matter if the system is your stereo or your ear—feed two very intense signals in and you'll end up hearing the original signals plus harmonics and summation and difference tones. These noisy new tones are known as distortion products. The effects of interfering tones are summarized in Table 16-3.

Noise Spectrum Level Usually you'll be interested in detecting a signal in a complex noise background, rather than in one or two interfering pure tones. The sea and engine noise of the example was a case of an essentially continuous noise background. Knowing that the overall level of this noise is say, 80 dB, may not be very useful in predicting whether you can hear the distress signal. This is because the measurement of the overall level of a noise includes frequency components that may be far in frequency from components of the signal involved in your task. The total number of gallons of water in your pool is clearly irrelevant to whether you can stand up with your head above water. What matters is the depth where you are standing. An analogous measure is the noise spectrum level, which is defined as the noise level per unit bandwidth, L_{spectrum}. In *white* noise, the noise spectrum level is the same at all frequencies; we say it is "flat." For such flat noise the spectrum level can be determined by subtracting 10 times the log of the noise bandwidth, BW, from the overall noise level;

$$L_{\text{spectrum}} = L_{\text{noise}} - 10 \log BW \tag{16-2}$$

Suppose that the measured overall noise level of a 5000-Hz-wide white noise is 85 dB; then the spectrum level would be

$$L_{\text{spectrum}} = 85 - 10 \log (5000) = 48 \text{ dB}$$

The spectrum level measure allows us to specify the interfering or *masking* effect of noise on a signal. The section on signal detection theory in Chapter 3 discussed how signal and noise factors influence performance in a detection task. Chapter 3 included an equation for the detectability measure, d', of an ideal detection system as a function of signal energy and noise power:

$$d'_{\text{ideal}} = \sqrt{\frac{2E}{N_0}} \tag{16-3}$$

where E is the signal energy (signal power $\times$ signal duration) and N_0 is the noise power per cycle (L_{spectrum} is equal to $10 \log N_0$). The detectability of brief tonal signals (less than 250 msec duration) by human observers follows equation 16-3 but is degraded by various factors such as noise within the auditory nervous system, and observer uncertainty about the signal and noise characteristics. Generally,

$$d'_{\text{human}} = \sqrt{\frac{kE}{N_0}} \tag{16-4}$$

The value of the constant, k, in equation 16-4 depends on the particular task conditions and observer training; we will assume a value of 0.4 to illustrate the calculations involved. Then,

$$\frac{d'_{\text{human}}}{\sqrt{0.4}} = \sqrt{\frac{E}{N_0}}$$

Values of d' equal to 4 or larger should ensure satisfactory detection performance in most applications.

Solving for $\sqrt{E/N_0}$,

$$\sqrt{\frac{E}{N_0}} = \frac{4}{\sqrt{0.4}} = 6.3$$

The $\sqrt{E/N_0}$ ratio can be expressed in dB terms:

$$10 \log \left(\frac{E}{N_0}\right) = 20 \log \sqrt{\frac{E}{N_0}}$$

$$20 \log \sqrt{\frac{E}{N_0}} = 20 \log (6.3) = 16 \text{ dB} \tag{16-5}$$

Thus we need a signal-energy-to-noise-spectrum-level ratio of approximately 16 dB for satisfactory detection performance. Values of 10 log (E/N_0) ranging from about 7 to 16 dB have been obtained in laboratory detection tasks (Green, 1976, Sorkin, Woods, & Boggs, 1979; Tanner, 1961). We can use the 16-dB value to determine the signal level required to detect a 200 msec signal in a noise of 48 dB spectrum level. Note that

$$10 \log \left(\frac{E}{N_0}\right) = L_{\text{signal}} + 10 \log t - L_{\text{spectrum}} \qquad (16\text{-}6)$$

where t is the signal duration and L_{signal} is the signal level expressed in dB. Then in our example,

$$10 \log \left(\frac{E}{N_0}\right) = 16 \text{ dB} = L_{\text{signal}} + 10 \log (0.2) - 48$$

and,

$$L_{\text{signal}} = 16 - 10 \log (0.2) + 48 = 71 \text{ dB}$$

For signal durations greater than 200 msec, signal detectability depends on signal power rather than energy. This is because the auditory system cannot integrate energy for long durations; some energy begins to "leak out" of the detection mechanism as additional energy is acquired from the input signal (cf. Penner, 1980). Under these conditions, detectability in noise depends on the ratio of the signal power to the total noise power of the band of effective interfering frequencies. In order to compute this noise power, we need to know the effective width of this band.

Critical Band If the noise is narrow in bandwidth, and far away from the signal frequency, we would not expect to get much interference or masking from the noise. Harvey Fletcher (1940) demonstrated that when you try to detect a pure tone signal in wide-band noise, only noise frequencies within a relatively narrow band around the signal are effective in masking. Fletcher called this band of frequencies the critical band, since they were the ones critical for interfering with the signal. His view was that the system employed an internal filter through which the external signal and noise were passed; only noise components passing through the filter were *effective* in interfering with the signal. To some extent, you can vary the center frequency of your effective listening band depending on the frequency of the signal you are trying to detect.

Fletcher found that for center frequencies greater than about 500 Hz, the critical bandwidth was approximately proportional to the center fre-

$$BW_c = 0.06 f_c \qquad (16\text{-}7)$$

quency, as in Equation 16-7. Figure 16-14 summarizes some data showing this relationship. Roughly speaking, we can say that a signal will be

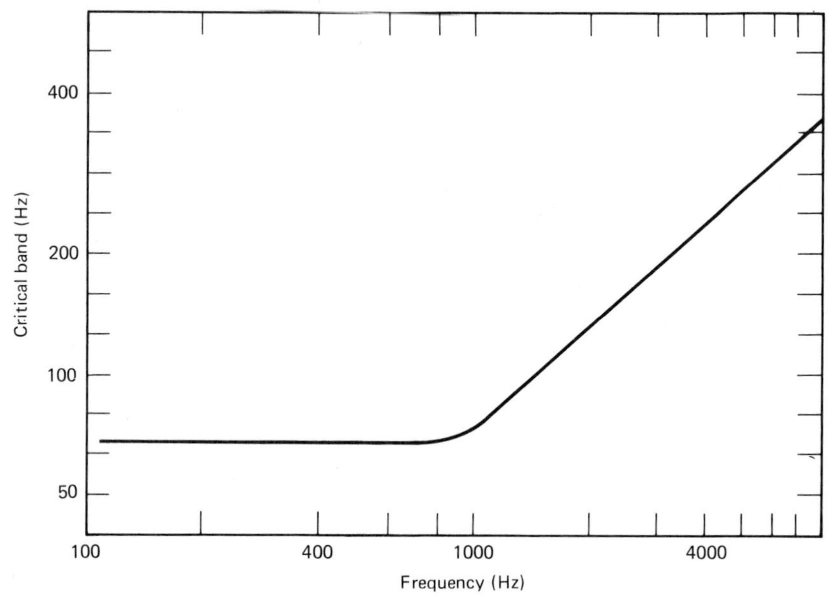

Figure 16-14. Critical Band width plotted as a function of signal frequency. (After R. D. Patterson, *Journal of the Acoustical Society of America,* 1973, *55,* 803, and reproduced by permission.)

detectable in a wide-band noise if the signal power is greater than the total power of the noise coming through the critical band. In dB, the level of noise power in a noise band of width, BW_c, will be

$$L_{\text{noise}} = L_{\text{spectrum}} + 10 \log BW_c \tag{16-8}$$

The effective noise level of a critical band 60 Hz wide, at a center frequency of 1000 Hz, for a noise of 48 dB spectrum level, will be

$$L_{\text{noise}} = 48 + 10 \log(60) = 66 \text{ dB}$$

It will take a signal of 66 dB at 1000 Hz, to be just detectable in this noise.

This rule of thumb should be qualified. The effective width of your listening band will depend on many task variables, such as the particular characteristics of the signal and the noise. To a great extent you can optimize the position and width of the listening band for the detection of specific signals (Green & Swets, 1966). In the laboratory situation, efforts may be directed toward minimizing the subject's uncertainty about the signal and noise characteristics. Furthermore, the subject's attention can be directed toward the detection task and minimal distractions will be present. It's probably a mistake to make comparable assumptions about the uncertainty and attentional factors present in the applied setting. In the field, the signal-to-noise ratios that you need for consistent detection performance could be much higher than the values we have suggested.

Our purpose has been to allow you to estimate the minimal signal levels required for detection.

Suppose that the spectrum level of the sea and boat noise in the lifeboat example were about 48 dB in the frequency range of the yacht's air horn (we assume about 1000 Hz). Although the air horn produces a very high-intensity sound, the inverse square law and the uncertain orientation of the horn relative to the lifeboat would result in a large (and unpredictable) drop in the horn's sound level as a function of the distance between the yacht and lifeboat. From our last calculation, the horn level at the lifeboat would have to exceed 66 dB for detection. Would reducing the lifeboat engine noise improve matters? That would depend on the relative contribution of sea and engine noise to the total noise level. Recall equation 16-1. If the sea noise were uncorrelated with the engine noise and about equal in level, the advantage in cutting off the boat's engines would be 3 dB. We will leave the decision of whether to cut the engines to the lifeboat skipper and to his or her estimate of the distance to the endangered craft.

Physiological and Psychological Effects

Earlier in the chapter we considered noise effects that are primarily acoustic in nature, such as threat of hearing loss and interference with signal detection. In this section we consider effects that are more indirect, and that involve the interaction between the physical variables of the noise stimulus and the physiological and psychological state of the person stimulated. The reaction of the person to noise may be quite different in two situations where the physical parameters of the noise are identical.

Stress Noise has been observed to induce temporary changes in a person's physiological state, including neurological, endocrinological, and cardiovascular changes. These changes are temporary, and apparently do not result in permanent damage to the human exposed (Kryter, 1970). Some investigators have reported a correlation between exposure to noise and the occurrence of circulatory and neurological problems, but it is not clear in these studies whether noise was the cause of the disorder, or simply one contributing or correlated factor (NIOSH, 1972).

One hypothesis is that intense noise is a stressor and can cause permament changes like those resulting from chronic exposure to stress. Miller (1974) has pointed out that if this hypothesis were true, some important factors associated with the effect would be the noise intensity, the amount of fear and annoyance produced by the noise, and the individual's susceptibility to noise. In recent articles, Kryter (1976, 1980) has stated that the net result of many human and animal studies so far does not support the presence of harmful autonomic nervous system reactions to noise—except when the noise is "psychologically meaningful" to the organism. He has suggested that it is the psychological annoyance resulting from the noise, rather than the body's autonomic system response, that generates general negative effects on an organism's health.

As long as the noise is not intense enough to damage our hearing, the major effect of the noise on our health is probably via this indirect psychological route, rather than via a direct effect on our nervous system. It is the meaning of the noise, such as its indication of danger, its interference with desired auditory input or speech, or its caused irritation or annoyance, that may induce stress; this stress, in turn, may be a potent source of physiological harm to the individual.

Case 1 You are speaking to workers in a factory while machines are making a loud din around you. Although you can make yourself understood, you are uneasy about being in this unfamiliar workplace. Perhapy you will enter a dangerous area before someone can direct you clear ($L = 85$ dB(A), high frequencies, continuous).

Case 2 You are in a restaurant with your date attempting to discuss a serious, personal problem. A group of noisy children are enjoying themselves in the booth behind you. You find yourself getting angry and unable to concentrate on what your date is saying ($L = 70$ dB(A), mid to high frequencies, intermittent).

In Cases 1 and 2 we can observe the interaction of some of these psychological and physical parameters. The increased effort necessary for efficient speech communication, the perceived difficulty in hearing potentially important information, the presence of distracting sounds, and the initial psychological state of the persons involved all interact with the physical noise conditions. This reaction influences the psychological state of the person stimulated. In one case a salient aspect of the situation is the potential danger of the workplace; in the second case, there is the serious nature of a potential personality conflict. In these cases the judged disruption and annoyance from the noise would no doubt be very high, and the negative effects on performance might also be significant. Note that the overall noise levels, and the temporal character of the noise in these cases are quite different.

Case 3 After a strenuous climb, you lie down in a shady spot near a mountain waterfall. The sound of the falling water is very soothing and relaxing; the ache in your muscles seems to fade away ($L = 75$ dB(A), white, continuous).

Case 4 You are cruising on a large sailboat. This morning the sea was dead calm and you were sleepy and lethargic. Now the wind has come up and the boat is moving fast in increasing seas and strong gusts of wind. You find the roar of wind and sea exhilarating; all of your senses are alert and at a fine edge ($L = 75$ dB(A), white, continuous).

Arousal Cases 3 and 4 illustrate some additional effects related to the psychological state of the people in the situation: the state of arousal of the person being stimulated by the noise. At any moment our nervous systems are at some level of general sensitivity ranging from an extreme level of alertness (I just drank three cups of black coffee), to very deep sleep (5 minutes after dozing off in Human Factors class). To some extent, any increase in the stimulation of our senses can increase this level of arousal. (Your classmate hisses at you to wake up.) The increase in arousal may be small, however, and it may only be temporary (there goes your head nodding, again). If we are initially at a low level of arousal, a moderate amount of noise stimulation can raise our alertness and produce an improvement in performance (Poulton, 1979). On the other hand, an intense noise could increase our level of arousal enough to induce a general muscular tension that would degrade our performance (your classmate shouts "Wake up!" and you drop all of your books off the arm of your chair). In this respect, noise is not very different from any other source of stress or potentially arousing sensory stimulus.

One thing to remember is that the prior arousal state of the individual will play on important part in determining whether the noise stimulation results in an increase or a decrease in the quality of the person's performance. Before you advocate adding noise to your work environment, we would also point out that there is controversy about the significance of the arousal hypothesis and the possible effects that increased arousal (with noise) will have on performance (Broadbent, 1978, 1980; Hartley, 1981; Poultion, 1977, 1978, 1979, 1981). There may be far better ways to produce the same effective increase in worker arousal. For example, it may be possible to introduce some other stimulus into the job situation, or to revise the manner in which the worker is rewarded. Monetary payoffs can be very arousing. Stress and arousal are considered more fully in Chapter 19.

Case 5 The model shop supervisor in your plant has come to you with a problem about a newly hired machinist. It seems that the new worker tends to give up on difficult jobs much too quickly. In reviewing the file you see that the worker worked for a number of years in a factory with high frequency, short duration noises, that occurred randomly throughout the workday. Can there be any connection between the worker's work history and present behavior? (L = 110 dB peak, high frequencies, short duration, intermittent).

Noise Aftereffects Case 5, and to some extent Case 2, both involve stimulation by noise that is noncontinuous over time; the noise is intermittent and not predictable in occurrence. This type of noise seems to be much more irritating and disrupting to a person's activities than continuous or predictable noise. Another key aspect of a person's sensitivity

TABLE 16-4 *Possible Aftereffects of Noise*

1. During the noise exposure, the worker's learning of certain parts of his or her job is degraded. This decreased job skill becomes evident after some time has passed.
2. The noise exposure causes mental fatigue, which persists for some time after the noise has been turned off.
3. Turning off the noise causes a decrease in the worker's arousal level; this decrease results in a drop in performance.
4. The noise stimulation is so unpleasant and unavoidable that it causes the person to feel unable to handle hard tasks. The person feels inadequate and unable to cope with difficult problems.

to noise stimulation is whether the person believes that he or she has some degree of control over the noise level or its occurrence.

Several studies have reported that certain types of noise exposure results in poorer performance on tasks after the noise exposure has been terminated (Glass & Singer, 1972; Percival & Loeb, 1980). There is some controversy about the origin of these aftereffects, but a number of reasonable hypotheses have been generated, summarized in Table 16-4.

You probably would like some additional explanation of the last point in Table 16-4. This point relates to the psychological phenomenon of "learned helplessness." This refers to what happens when an animal is repeatedly exposed to an unavoidable, very unpleasant stimulus. After this treatment, the animal seems much less able to cope with difficult tasks (Maier & Seligman, 1976). In a typical experiment, people are exposed to noise while performing a task. After this exposure, they work on a second task in quiet conditions. The second task includes work that is designed to assess the person's perseverance, such as proofreading text or trying to solve hard puzzles. Certain kinds of noise during the first task have a negative effect on performance in the second. If the noise was highly unpredictable and unpleasant, people do poorer and persevere less on the difficult problems in the second task. The psychological explanation is that these people learn that they cannot control their environment; they extend this attitude to the problems facing them in the second task. The key aspect of the noise seems to be its highly unpredictable nature, and whether the people believe that they have any control over it. These factors are much more important than the frequency distribution of the noise (Percival & Loeb, 1980).

Attitudes

A variety of attitudinal factors have been found to be involved in our responses to noise and to noise sources. The degree of measured annoyance to noise in the environment has been found to be correlated with people's attitudes toward the noise source, such as (1) whether the noise source is perceived to be a worthwhile activity for society, (2) whether it is perceived as being a necessary outcome of the process which produces

Figure 16-15. The typical startle response to an impulsive sound includes eye blink, facial gesture, and usually an inward and forward head or body movement (Miller, 1974). No doubt Garfield's level of annoyance is partly due to the noise source. (© 1978 United Feature Syndicate, Inc.)

the noise, (3) whether the people in authority over the noise are perceived as being sensitive to the welfare of the persons exposed to it, and (4) whether there is a fear of possible danger from the process producing the noise (Miller, 1974). It is difficult to tell whether these attitudes are the result or the cause of the noise being annoying (see Figure 16-15). In any case, the correlation is useful in assessing the extent of annoyance from a given noise source.

This relationship between a person's attitudes and values and their response to different noise stimuli can be significant when the noise is desirable, rather than unwanted. Preferences and attitudes about noise can play an important role in the area of consumer behavior. A vacuum cleaner or motor cycle manufacturer may be somewhat cautious about reducing the noise level of the product, if it is believed that this action will cause potential customers to downrate the perceived power of the product. On the other hand, it may be possible to design a vacuum cleaner that *sounds* very powerful, even though it actually emits a greatly reduced amount of noise energy (Miller, 1974).

Subjective factors involving spectral noise distributions that are judged to be "good" have been found in the perception of exhaust system noise from automobiles. The judgment of acceptable-sounding mufflers seems to depend on complex aspects of the muffler noise spectra; automotive engineers describe this noise in such subjective terms as "trucky," "hollow," "husky," "blatty," "sharp," "chirpy," "tinny," etc. These characteristics are related to the phase spectrum of the muffler's sound, the existence of beats at different frequency ranges, and the overall level. The muffler must also sound good under different load and acceleration conditions. The point is that the precise acoustical definition of a good muffler is so far not available; muffler acceptability is a complex function of the noise spectrum of the muffler, and not simply a function of the level of sound produced (Pande & Sullivan, 1978).

Finally, we offer one last example of the subjective aspect of noise in the marketplace. During the past few years we have seen the widespread entry of the microcomputer-based game into the family game and toy market. A commonly heard remark from salespeople who market these games is: "the more noise the thing makes, the better it sells!" Whether such auditory inputs are to be considered as carrying information or as noise, depends on your own orientation! Table 16-5 summarizes some of the physiological and psychological effects of noise on performance.

TABLE 16-5 *Summary of the Physiological and Psychological Effects of Noise on Performance*

1. Continuous noise won't have much effect on performance unless it is very intense.
2. High-frequency components (> 2000 Hz) produce more interference than low-frequency components.
3. Short-term decrements in performance may occur after brief noise inputs due to transient physiological reactions to the noise.
4. Complex, high-capacity-demanding tasks are more likely to be affected by noise; the effects will probably be on the accuracy of the performance measured.
5. Increase in worker stress, fatigue, or training time, may be consequences of noise in situations where there is partial interference with speech or other auditory information.
6. Irregular, unpredictable noise causes the most disruption of tasks, and may produce performance aftereffects.
7. Much of the stress and annoyance of a particular noise may be related to the "psychological message" the noise sends to the exposed person, and to that person's attitudes about the noise source.

MEASUREMENT OF ENVIRONMENTAL NOISE

A number of measures have been devised for assessing the effects of different types of environmental noise, such as aircraft noise and transportation noise. Computation of these measures involves various procedures that attempt to boil down the effects of different frequencies and levels to a single number. This single number is supposed to describe the effect of the noise, such as its noisiness, annoyance value, etc. In this section we describe the most commonly employed measures; Goldstein (1978) has a good summary of additional noise ratings. In Chapter 9 we describe measures that relate primarily to the interference of noise with speech and to the design of rooms and offices.

PNL, Perceived Noisiness Level

Kryter (1970) developed a measure of perceived noisiness level, PNL (or PNdB), which is computed in a similar way to the measure of loudness level in phons, discussed in Chapter 3. The first step is to measure the sound level, L_i, in each of eight octave bands (or one-third octave bands).

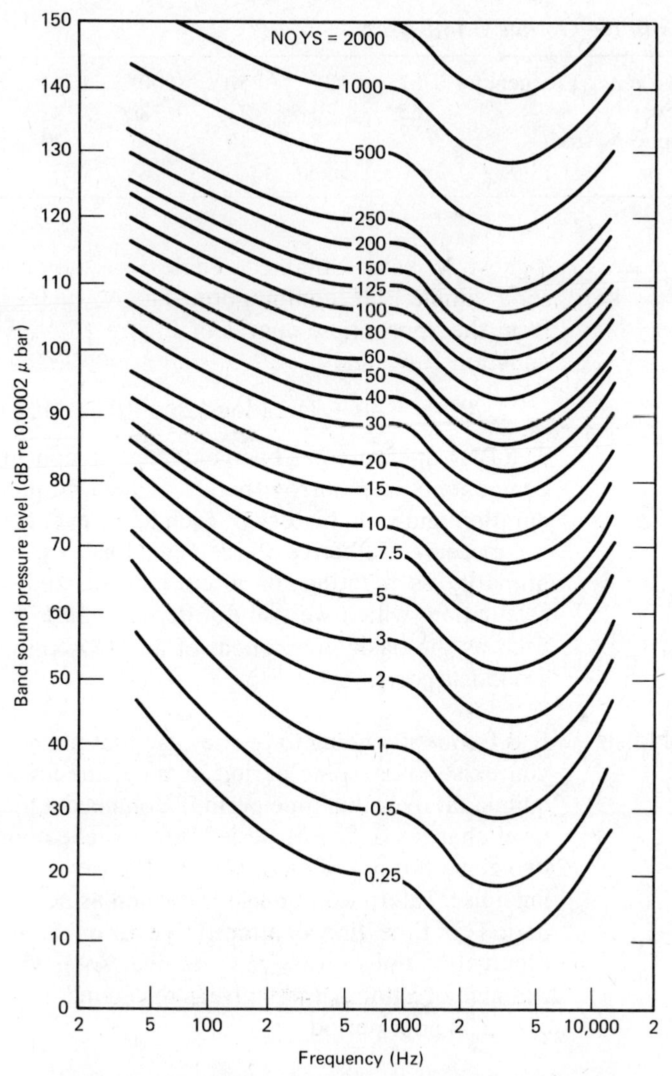

Figure 16-16. Contours of perceived noisiness. (From K. D. Kryter, & K. S. Pearsons, *Journal of the Acoustical Society of America,* 1963, *35,* 871, and reproduced by permission.)

Figure 16-16 is then employed to determine the perceived noisiness value, N_i in noys, for each of the octave band contributions. These values are then entered into the equation

$$\text{PNL} = 40 + \frac{10}{0.301} \log \left[N_{\max} + k \left(N_1 + N_2 + \ldots N_8 - N_{\max}\right)\right] \quad (16\text{-}9)$$

where k is equal to 0.3 for octave band measurements, and $N_{\max}$ is the

TABLE 16-6 *Sample Octave Band Data*

Octave Band Center Frequency	63	125	250	500	1000	2000	4000	8000
Measured Level L_i	65	70	73	78	82	75	74	74
Noisiness Value N_i from Figure 16-16	2	5	10	15	20	20	20	15

highest N_i value obtained. Table 16-6 contains some octave band data for a sample PNL computation. The N_i values in the table were obtained from the appropriate curves of Figure 16-16. Entering these values into equation 16-9 yields

$$\text{PNL} = 40 + 33.22 \log [20 + (0.3)\ (107 - 20)] = 95 \text{ PNdB}$$

The PNL measure has been extended to consider the effects of pure tone components and narrow bands of high-frequency noise, as well as the duration and time history of each noise event (Kryter, 1970). The resulting measure, Effective Perceived Noise Level, EPNL, was developed primarily as a rating for aircraft noise, and involves a fairly complex calculation, which we will not describe here. The EPNL measure also is used as the basic component of an FAA rating for evaluating land use around airports.

Time-varying Noise

It is frequently going to be the case that the noise situation that concerns you exists over some period of time; the level of the noise may change quite a bit over this time period. You need a measure that boils down the level changes over this period into a single number. This is the case with two general types of environmental noise: intermittent noise and fluctuating noise. Intermittent noise is defined as noise that is produced for short periods of time, like an aircraft flyover or the ringing of your alarm clock. Fluctuating noises vary in level over somewhat longer time periods; an example of a fluctuating noise is the sound at an expressway interchange, over a 24 hour period.

Equivalent Sound Level One of the most important general measures of time-varying noise is the equivalent sound level, L_{eq}. This measure is based on equating the energy of the time-varying environmental noise with the energy of a fixed level noise. Record the time-varying noise for some time period, and then compute the total energy in this noise. (We will show you how, shortly.) Then determine what the A-weighted level would be of a fixed noise having the same total energy for this same time period. The resulting fixed noise level is L_{eq}. The idea is that the *effects* of a given noise are mainly a function of the total noise energy over some period—not the particular time pattern—and the L_{eq} is a measure of this energy.

L_{eq} is computed by sampling the noise level (A-weighted) over the time

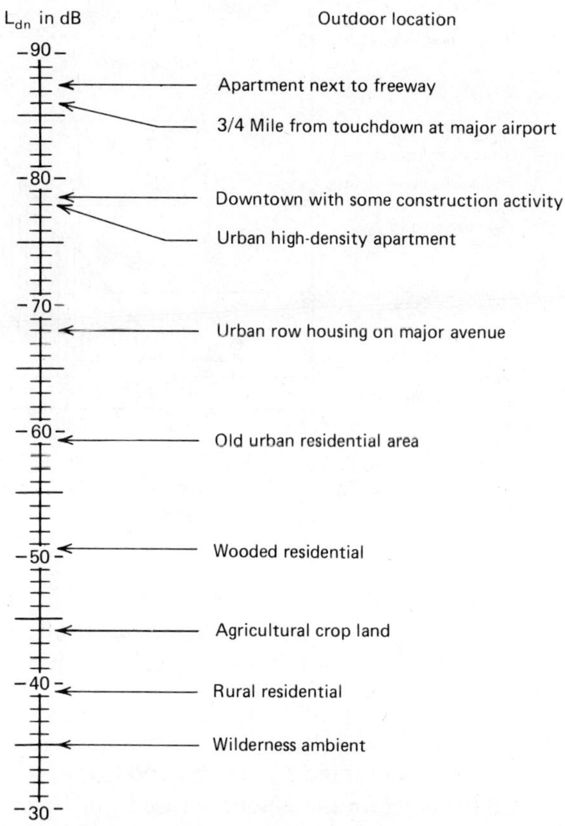

Figure 16-17. Examples of outdoor day-night average sound levels in dB measured at various locations. (From EPA, 1978.)

period of interest in order to get a mean level, L_i, for each time interval, t_i. Then,

$$L_{eq} = 10 \log \left[T^{-1} \sum_{i=1}^{n} (t_i 10^{0.1L_i}) \right] \tag{16-10}$$

where T is the total time period of observation and n is the number of time intervals. For example, suppose that you are interested in L_{eq} for a 2-hour time period, and the average levels recorded every 30 minutes for this period are $L_1 = 68$, $L_2 = 70$, $L_3 = 74$, and $L_4 = 70$ dB(A). Then, $T = 2$ hours, $t_i; = 0.5$ hours,

$$L_{eq} = 10 \log\{(\tfrac{1}{2})[(0.5)10^{0.1L_1} + (0.5)10^{0.1L_2} + (0.5)10^{0.1L_3} + (0.5)10^{0.1L_4}]\}$$

$$= 10 \log \tfrac{1}{4}[10^{6.8} + 10^{7.0} + 10^{7.4} + 10^{7.0}]$$

$$= 10 \log \tfrac{1}{4}[6.31 \times 10^{6} + 2 \times 10^{7} + 2.51 \times 10^{7}]$$

$$= 10 \log \tfrac{1}{4}[5.14 \times 10^{7}] = 71.1 \text{ dB(A)}$$

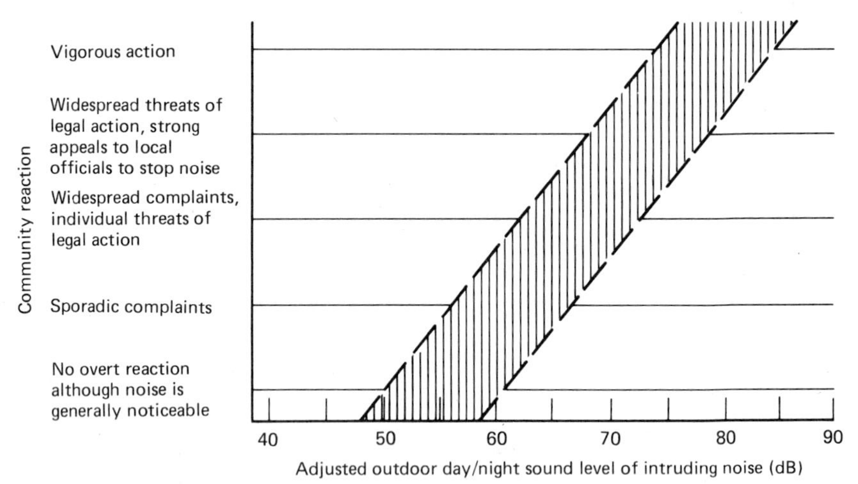

Figure 16-18. Trend of data from community case studies adjusted for conditions of exposure. (From EPA, 1978.)

One very nice feature of this measure is that you can define the time period over which the energy calculation is made, according to the particular interest at the time. Useful versions of the L_{eq} measure include

$L_{eq(24)}$, the A-weighted level for an equivalent 24-hour period.
$L_{eq(8)}$, the level for an 8-hour workday.
L_d, the level for the 15-hour period from 7 A.M. through 10 P.M.
L_n, the level for the 9-hour period from 10 P.M. to 7 A.M.

A widely used descriptor of the total outdoor noise environment is L_{dn}, the Day-Night Sound Level. This measure is computed on the basis of both L_d and L_n, but an additional 10-dB weight is given to the night rating, L_n.

$$L_{dn} = 10 \log(\tfrac{1}{24}) \{[15 \times 10^{(L_d/10)}] + [9 \times 10^{(L_n+10)/10}]\} \quad (16\text{-}11)$$

Figure 16-17 shows some L_{dn} values for various types of typical outdoor locations. Figure 16-18 illustrates some data showing how community reaction varies as a function of the L_{dn} levels of the intruding noise.

SUMMARY

The possible harmful effects of noise include (1) hearing loss, (2) interference with task performance, and (3) increase in worker stress, fatigue, or training time. Tables 16-1 and 16-3 summarize the physical aspects of noise important for predicting hearing loss and interference with auditory tasks. The third category of effects depends on the disruptive nature of the noise, its temporal predictability, and the person's physiological state and perceived control over the noise. These aspects are summarized in Table 16-5.

17 MICRO-ENVIRONMENTS

Did you know that:

- Men and women react differently to physical closeness and invasion of their personal space?
- Color of the carpet and background walls alters the way people move around an art gallery?
- Crowding in a men's room can speed up some biological processes and slow down others?
- Inefficient incandescent lamps are best for lighting people's faces but people with black or yellow faces dislike efficient high-pressure sodium lamps that people with white faces find acceptable?
- Warm wall colors (reds, oranges) do not make you feel thermally warmer?
- Your bathroom is the most dangerous room in your house?

All these items are findings from experiments studying the micro-environment that will be discussed in this chapter. But before getting down to the nitty-gritty details, we need to define the microenvironment, as well as its kissing cousin, the macroenvironment.

Put down this book (but only for a few seconds!) and look around you. The odds are quite good that you are surrounded by glass, steel, plaster, ceilings and walls rather than by green grass, fluffy clouds, and fragrant pine trees. Human beings spend most of their time indoors in what is sometimes termed the built or constructed environment. Until recently, the built environment was not considered to be an area where the human factors specialist had much to contribute. We are happy to report that this situation is changing rapidly; we hope that this and the following chapter convince you that the built environment is yet another complex system where people-system relationships can be improved and thus is legitimate turf for human factors efforts. The relationship between human factors and the built environment is implied in Figure 17-1.

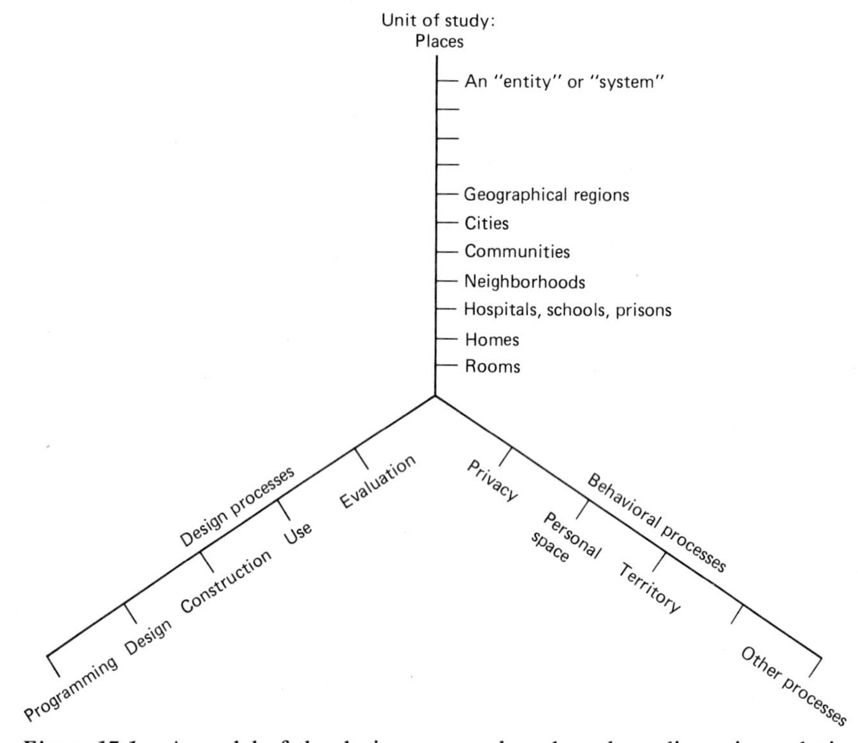

Figure 17-1. A model of the design process based on three dimensions: design processes, behavioral processes, and places. (From Altman, 1973.)

In order to build a place, be it a room, a city, or a highway system, so that people can function well in that place, we must combine design and behavioral processes (Figure 17-1). Architects and engineers are interested primarily in the design processes. Psychologists and sociologists care about behavioral processes. It is the human factors specialist who has the professional commitment and methodologies that can blend design and behavioral processes, with the help of these other specialists, to create better places that try to optimize the relationships between people and the constructed environment.

The microenvironment is that part of the built environment that is immediately accessible to the human. Certainly a room is part of the microenvironment. Small buildings such as a grade school or a home are also part of the microenvironment since people can easily and immediately go from one room to another. Very large buildings such as giant shopping malls and major airport terminals are not immediately accessible and so form part of the macroenvironment. Assemblies of buildings such as neighborhoods are clearly part of the macroenvironment. So we will in this chapter consider parts of the constructed environment up to the level of rooms and one-family residences. Larger elements are discussed in Chapter 18, Macroenvironments.

We start our discussion with the leg of the triad (Figure 17-1) labeled behavioral processes, in particular personal space and territoriality. From there, we progress to actual design examples for offices and residences. The chapter concludes with the topic of illumination of the built environment.

PERSONAL SPACE

The alarm clock rings. Time to wake up and go jogging. But today you have a new piece of apparel to wear over your jogging suit. It is a large, transparent, plastic dome with holes for your arms and legs. Now fully clothed you trot out to face the world. Seems odd, doesn't it? Why wear a plastic dome unless you habitually run in the rain? But every day you wear an invisible psychological bubble that is designed to protect you from spatial assaults by people around you. We use the term "personal space" to refer to the psychological bubble that each of us carries about to aid in regulating social interaction with other people.

How do behavioral scientists know that you are surrounded by such a mysterious bubble, since it is invisible, odorless, and silent? How large is your bubble and does its size depend on your sex? Are the walls of your bubble flexible like a balloon, or rigid? The answers to these questions are part of *proxemics,* the scientific study of how people (and animals) use the physical space around them. Proxemics was invented by Edward Hall (1966), an anthropologist who was interested in the cultural differences associated with personal space. But it was not long before psychologists realized that proxemics had important implications for the design of spaces used by people (Sommer, 1969; 1974). Indeed, a new area called environmental psychology has developed and its workers are deeply concerned with using knowledge about proxemics to improve the built environment (Aiello & Baum, 1979; Bell, Fisher, & Loomis, 1978).

Territoriality

Personal space and territoriality are closely related concepts that are often confused. So before going on to discuss personal space invasion, we must first explain territoriality and show how it differs from personal space. Territoriality is a behavior—that is, a pattern of responses designed to achieve a goal. Specifically, it is the behavior used by an individual to defend a specific physical space or area against members of its own species. When a dog urinates on a fire hydrant it is marking its territory with its own scent. The behavioral pattern differs from ordinary urination since the dog's leg is lifted much higher when the dog is creating a scent-post marker. Of course, this is a subtle difference that is obvious only to other dogs and ethologists who are careful observers of animal behavior. We also mark territory in ways that seem obvious to us but probably go unnoticed by dogs (Figure 17-2), although it is not entirely clear that humans are territorial in precisely the same way as animals (Edney,

Figure 17-2. Some examples of markers used to define territory. (Yan Lukas/ Photo Researchers.)

1976). Street gangs in cities spray graffiti with the gang's name to mark their turf and the first signs of invasion of territory by another gang are the appearance of competing gang graffiti (Ley & Cybriwsky, 1974). Territoriality and fear of property loss are correlated in elderly people. One study (Patterson, 1978) classified elderly citizens into Low and High Territoriality groups by counting the number of markers—fences, surveillance devices, warning signs—on their property. The elderly were then questioned about their fear of crime to establish a scale for fear of property loss. Citizens who showed high territoriality exhibited decreased fear of property loss. Since this was a correlational study, the obvious conclusion that erecting territorial markers makes people feel more secure is not the only explanation of this finding. Perhaps those who display more territorial markers are people who feel more secure to begin with. Although scientists are just starting to explore the role of territoriality in human behavior, it is clear that people are territorial to some extent.

Territory has visible markers or boundaries. Its location is always fixed. Territories do not move. Its size is constant whether the territory is large or small. In contrast, personal space is portable. You carry it with you wherever you go. Its boundaries are invisible and its size will vary with changes in the psychological environment. Personal space is measured by invading it. An early study of the personal-space bubbles of violent prisoners (Kinzel, 1970) used a simple measurement procedure.

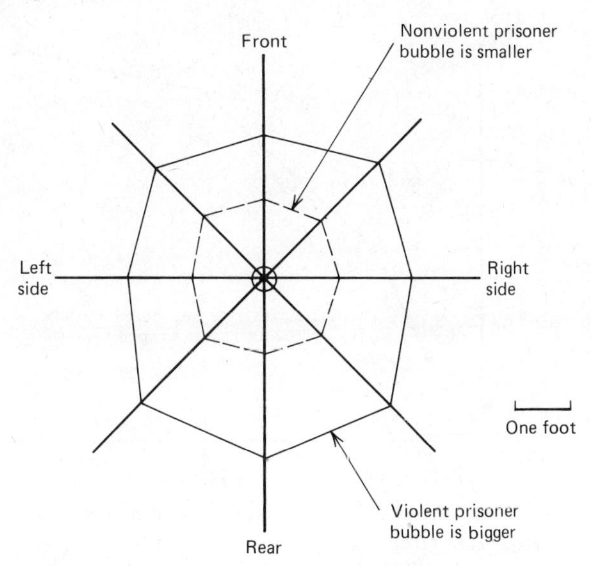

Figure 17-3. The small circle in the center represents the prisoner's head. The dotted circle is the personal-space bubble for nonviolent prisoners and the solid large circle is the bubble for violent prisoners. (Adapted from Kinzel, 1970.)

The prisoner stood in the center of an empty room. The experimenter approached the prisoner until the prisoner felt uncomfortable. By repeating this from several different angles, a map of the prisoner's bubble was drawn. Figure 17-3 shows that the personal-space bubbles for violent and nonviolent prisoners are different sizes. Men have smaller bubbles than women (Evans & Howard, 1973). This is probably due to American cultural norms that have discouraged women from interacting with strangers. This finding offers hope for evaluating the impact of the women's liberation movement. We would expect that liberated women should have smaller bubbles than traditional women and that the truly emancipated woman would have a bubble identical in size to a man's. However, this speculation has yet to be tested.

Invasion of Personal Space

All of us have an arsenal of defensive techniques designed to prevent others from encroaching on our personal space. But because we use these defensive measures automatically, we generally are not aware of them and of which measures are most effective. So let's create an imaginary situation for you to defend. Imagine you are seated at a table in the student union and wish to prevent others from joining you. It is lunchtime and the room is quite crowded. Furthermore, you must use purely nonverbal means to defend your space. What would you do?

First, you might spread personal items over the table and chairs. You

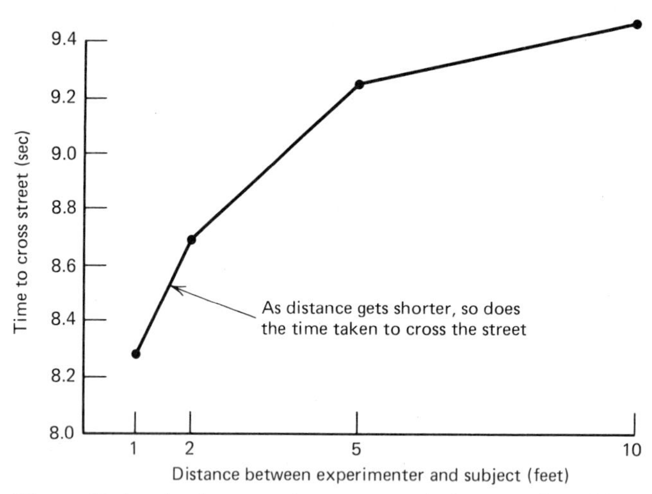

Figure 17-4. As the experimenter stood closer to the subject, who was waiting to cross the street, the subject escaped by walking more quickly. (Data from Konecni et al., 1975.)

could fill up the space with books, your coat, your purse, etc. In doing so you are marking territory and these items function as markers. You might glare at oncoming strangers and otherwise appear hostile. You might feign illness and start coughing violently when someone approaches. And there are several other defensive measures you could devise. One of the authors tried to test these techniques on a crowded airplane flight. He and a friend sat in a row of three seats with the empty seat between them. They piled markers on the empty seat and began an intense conversation about a graph placed on the empty seat. This conversation was interrupted by a stewardess who asked them to move their things to make room for another passenger. The plane had been completely booked and theirs was the last empty seat. Even so, the passenger was unwilling to ask us to make room and resorted to invoking a higher authority. It is clear that most people are tuned in to the social cues involved in defending personal space.

But sometimes even the best defense fails and your personal space is invaded. People usually react to such invasion by escaping as soon as they can. The worse the invasion, the faster the flight from it (Figure 17-4). In one study (Konecni, Libuser, Morton, & Ebbesen, 1975) the experimenter invaded the bubble of a pedestrian standing at a street corner. The independent variable was how close the experimenter stood to the pedestrian. The dependent variable was how fast the pedestrian crossed the street. As the experimenter stood closer, pedestrians decreased the time it took to cross the street.

Of course, there is something unusual about a silent stranger standing quite close to you. What happens if the stranger asks permission to invade your bubble? If you think this will eliminate flight, you are half correct. When males seated alone are invaded by a male who asks permission to sit down, the men do not flee. But when women seated alone are invaded by another woman they flee more rapidly if the invading woman asks permission than if she remains silent (Sundstrom & Sundstrom, 1977). The communication and protection functions of the bubble work differently in men and women. (See Baron, Byrne, & Kantowitz, 1980, Chapter 14 and Bell, Fisher, & Loomis, 1978 for an introduction to these issues.) This complicates the work of the human factors designer who not only must consider personal space as a design element but also the sex of the people present. For example, whether chairs in a waiting area should be arranged side to side or face to face to minimize personal space invasion depends on the sex of the people sitting in these chairs. Men feel less invaded when approached from the side whereas females prefer to be approached from the front (Fisher & Byrne, 1975). This sex difference can lead to unintended, although perhaps amusing, outcomes. A male wishing to approach a strange female in a friendly fashion will often sit or stand to one side of her since, for males, this is a less threatening invasion. But females prefer face-to-face contact so that the man's good intentions go unnoticed as she departs!

FOCUS ON RESEARCH

No Place to Flee: Psychologists Invade the Men's Room

Violation of personal space usually results in flight and escape. This makes it difficult for scientists to study invasion since the experiment often ends as soon as the invasion begins. But clever selection of the location of an experiment can overcome this problem. A man standing in front of a urinal has limited mobility to escape from invasion until he must no longer attend to matters at hand. He cannot even turn away or adopt a defensive posture to minimize invasion. This reasoning, rather than a weak bladder, led psychologist Eric Knowles and his colleagues to use a men's room as the environment for a test of spatial invasion (Middlemist, Knowles, & Matter, 1976).

Once the locale was selected, several problems remained that might block the study. First, there was a need for unobstrusive observation. An experimenter washing his hands for several hours might be suspicious. A new piece of equipment (Figure 17-5) consisting of a periscope concealed in a pile of books solved this problem. It was placed in a closed toilet stall with a good view of the urinals, allowing the experimenter to record data. Second, what would be the dependent

a

b

Figure 17-5. The periscope used to observe inside a men's room. (*a*) When installed inside a stall (*b*), the device looks like an ordinary pile of books. (Photo courtesy of Eric Knowles.)

variable? Knowles' first choice, collecting the urine for chemical analysis, was foiled when he was denied permission to tap into the plumbing. So instead he recorded two time intervals. One was the time between a man stepping up to the urinal and the onset of urination (Z to P interval). The other was the time to complete urination after onset. Why should spatial invasion be expected to affect these two time measures? Current models of personal space (e.g., Sherrod & Cohen, 1979) relate invasion to stress. Stress has biological effects on the body (see Chapter 19). In particular, stress should make it more difficult to relax the muscles that must be relaxed before urination can start. This should increase the Z to P interval when personal space is invaded. But once urination is under way, increased muscle tension should shorten the time needed to finish.

Spatial invasion was manipulated by placing a bucket and an out-of-order sign in either the middle or end urinal in a row of three. A confederate occupied one urinal, leaving no choice for the subject who had to take the only empty urinal. The confederate could either be adjacent to the subject or one urinal away, depending on where the bucket was placed. No confederate was present in a control condition. Results are shown in Figure 17-6. As predicted, it took longer to get started when

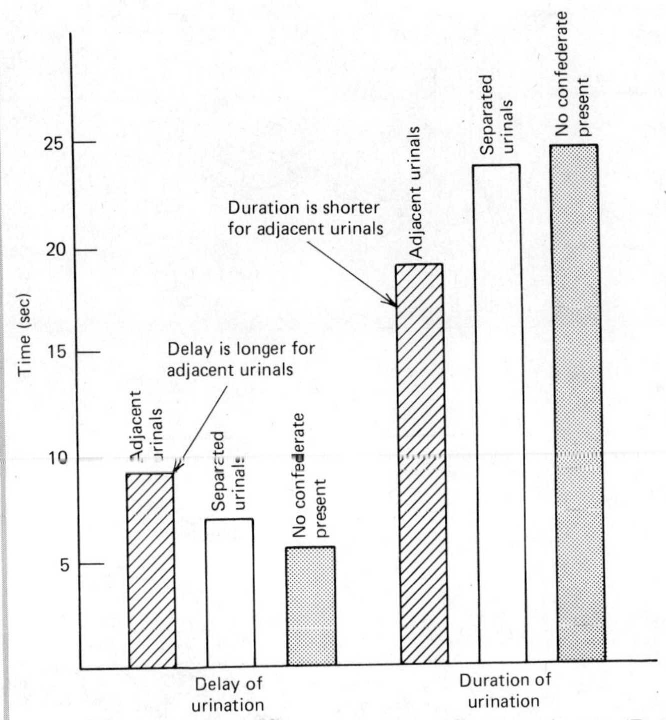

Figure 17-6. Results of the spatial invasion experiment. (Data from Middlemist et al., 1976.)

the confederate was closer. But once started, performance was faster. This experiment shows that personal-space invasion has immediate biological effects that the human factors designer must consider.

ROOM DESIGN

In this section we discuss how the human factors specialist can improve the design of individual rooms. Although many factors such as furniture (see Chapter 15) affect how well a room functions, here we concentrate on the placement and layout of furnishings and fixtures, rather than the design of individual items of furniture. Although this approach can be taken to improve anything from a classroom to a barroom (Sommer, 1969), our discussion of commercial room design will be limited to offices. Then we turn to residential design focusing on the one-family home. Finally, the effects of color on room design will be sketched.

Offices: Design for Working

One of the earliest attempts at applying human factors to office design was the Action Office (Propst, 1966). The arrangement of the furniture and fixtures was designed to improve efficiency as can be seen in Figure 17-7.

a

b

Figure 17-7. Action offices for a plant manager (*a*) and a research specialist (*b*). (From Propst, 1966. Copyright 1966 by the Human Factors Society and used by permission.)

Evaluation of the Action Office noted several advantages over conventional office design (Fucigna, 1967, p. 593).

1. Workspaces were provided for sitting and standing. This relieves the fatigue associated with only one working posture.
2. Flip-up display panels and integral filing systems provide for convenient storage and retrieval of information.
3. Telephone, dictation equipment, etc. were all located together in a communications center.
4. A variety of coded information-storage racks prevented messy accumulation of files and data.

As part of his evaluation, Fucigna had workers keep logs on various activities such as receiving information, retrieving data, communicating, etc. Workers were first studied for a month in their conventional offices. Then, after allowing some time for them to become familiar with the Action Office, similar logs were kept. Suprisingly, when these logs were compared there were no differences in the amounts of time spent on the various activities. However, fewer phone calls and conferences were required in the Action Office, although they were for longer time periods. Presumably, the information available in the Action Office allowed more to be accomplished in a single phone call or conference. Workers did prefer the Action Office, even though there was no objective evidence for improved efficiency.

Later work on the human factors of offices has provided checklists of the type used in workspace layout. These lists cover such items as ambient conditions, spatial arrangement, and office equipment (Parsons, 1976). There has also been an emphasis upon effects of personal space in the office environment (Fong & Bennett, 1981; Zweigenhaft, 1976). These studies have examined the location of desks in offices as related to interactions with visitors to the office.

Zweigenhaft (1976) studied desk placement in faculty offices in a small college where a large part of the faculty's job is interacting with student visitors. Offices were classified into two types: those where the desk was a barrier between the professor and the student and those where student and professor sat side by side without a desk between them. Student evaluations were reliably more positive for those faculty members who did not use their desk as a barrier. Furthermore, senior faculty were significantly more likely to use the desk as a barrier than were junior faculty. Since these data are correlational, we cannot conclude that desk position determines student evaluation, although our knowledge of personal space makes it more likely that certain furniture arrangements will encourage or discourage student-faculty interaction (Sommer, 1969). The most reasonable interpretation of these findings is that senior faculty have a more formal style and that desk orientation is only one way in which this style is implemented. Student evaluations probably reflect this entire style constellation. But it is interesting that desk orientation predicts style so well. Later research (Fong & Bennett, 1981) has confirmed that the desk can function as a barrier.

Whither Windows No discussion of office design would be complete without commenting on the question of whether or not windows are necessary. There has been a tendency to eliminate or reduce window size in new construction in order to save energy. Passions run high on this issue, and in addition to the status role of windows, there are some who claim windows are necessary for maintaining the morale and mental health of office workers. One careful study (Young & Berry, 1979) compared an experimental office that could have either no window, a real window, or an artificial window that was really a rear-projection movie screen with an endless loop of movie film that could portray assorted outdoor/nature scenes. The dependent variable was subjective comparisons using a statistical analysis known as paired comparisons. The no-window condition was liked least of all. However, the artificial window was almost as liked as the real window and the authors concluded that artificial windows were viable substitutes for the real thing. But it is important to realize that no objective performance measures were taken in this study. There is as yet no substantial evidence that worker productivity is influenced by the presence or absence of windows.

Open and Closed Offices One of the hottest new concepts in office design is the landscaped or open office. This term landscaped is unfortunate since it conjures up an image of clerical workers in a tropical jungle. Actually all it means is that the office is a large open space divided by movable partitions (Figure 17-8). Partitions can be moved around whenever changes are desirable. However, there is less privacy and often more noise than in conventional closed offices with fixed walls. The little research that has evaluated landscaped offices is ambiguous. Brookes and Kaplan (1972) found that workers regarded it as a more physically attractive environment but also one that made it harder to work efficiently whereas Nemecek and Grandjean (1973) found many workers who did not object to the open plan. We speculate that landscaped offices will work where workers have some say about the arrangement of modules and are given special instruction in the use of a landscaped office. For example, in a traditional office someone wanting privacy can close their door. More subtle cues must be noticed by workers in a landscaped office. Thus, one boss instructs his workers that he can be interrupted when seated at his desk but must be left alone when seated at his worktable. But even landscaped offices require closed cubicles for confidential conversations.

Figure 17-8. Landscaped office. (K. Bendo.)

Residential Dwellings

The typical American family lives in a house or condominium. This living environment is designed to permit, and hopefully facilitate, a number of activities and behaviors. People prepare and eat food, wash dishes and themselves, sleep, read, watch television, and use the bathroom. A well-designed house lets these activities occur efficiently and pleasantly. Indeed, it is the function of the system we call a house to allow such behaviors. Few people in America bathe under fire hydrants, sleep on park benches, make love in alleys, and urinate in the streets.

Until recently, the optimal design of rooms in a house has been left to architects and builders who often place concern for structural and economic aspects of a building ahead of the constraints a building can place on the behavior of its occupants. While residential design is hardly a dominant theme in human factors, there has been some work relating rooms and behavior. We briefly examine design criteria for three rooms and conclude with a discussion of floor plans.

The Bedroom The bedroom is where we spend more time than any other room of the house, close to 40% of our lives (Parsons, 1972). A partial listing of activities commonly performed in the bedroom includes sleeping, sexual intercourse, dressing, watching television, reading, and housekeeping chores. Yet the average bedroom is not designed to enhance all of these diverse behaviors. Over the years, the bedroom has tended to get larger, taking on many of the functions of the old-fashioned sitting room; it has become a place to read, watch television, and, especially for parents, a refuge from the hustle and bustle of family life. The dominant fixture is, of course, the bed. Yet a review of design characteristics of beds (Parsons, 1972) shows little firm evidence for specifying such important parameters as size, firmness, height, etc. This lack of data is appalling considering that the human race spends more time in contact with this single piece of equipment than any other. For example, how wide should an optimal bed be? This question cannot be answered until the number of people sharing the bed is first specified. The American cultural norm is the double bed, designed to sleep two people in comfort. Larger double beds are readily available in queen and king sizes for those who might feel unduly crowded in the standard double bed. The choice between single and double beds has important sexual implications. The Opinion Research Corporation (1966 cited by Parsons, 1972) found that single beds imply sexual rejection for young married women. Women commented that "There must be something wrong with married people who sleep in twin beds" and "If my husband suggested getting twin beds, I would think we were on the road to divorce." However, some married men preferred to sleep in a single bed with comments like "After the bloom is off the rose, I think it is more important to get a good night's sleep than it is to sleep in the same bed as your wife." Some couples solve this problem by sleeping in a large double bed with separate blankets to avoid midnight tussles. Others put a single large cover over twin beds so visitors will not suspect that a divorce is imminent.

The bed, while perhaps not even optimally designed for sleeping and sex, is quite inadequate for reading and watching television, activities that require more of an upright posture. Stacking pillows to raise one's head is a common solution, but leads to aches and discomfort if the posture need be maintained for any length of time. Beds with electric motors to elevate the head and feet (Figure 17-9), similar to hospital beds, are now available for residential use. From a human factors perspective, the additional cost of such beds is money well spent for people who use their beds for activities performed in nonhorizontal positions. However, this new bed technology creates a problem for married couples who conform to the norm of sleeping in a double bed. What happens when one person wishes to remain awake with the bed upright, while the other wants to go to sleep on a horizontal surface? Twin electric beds are the rational solution, but rationality sometimes yields to cultural stereotypes. And perhaps twin beds may decrease marital satisfaction. It should be clear that the bedroom is a design problem just waiting for human factors solutions. Our discussion has merely scratched the surface (see Parsons, 1972 for more detail).

The Kitchen Homemaking is among the largest occupations in the United States and much of homemaking involves food preparation. It is clear that the kitchen should benefit from the same kinds of workspace

Figure 17-9. An example of good human factors design for multiuse furniture. (Courtesy Sealy, Inc.)

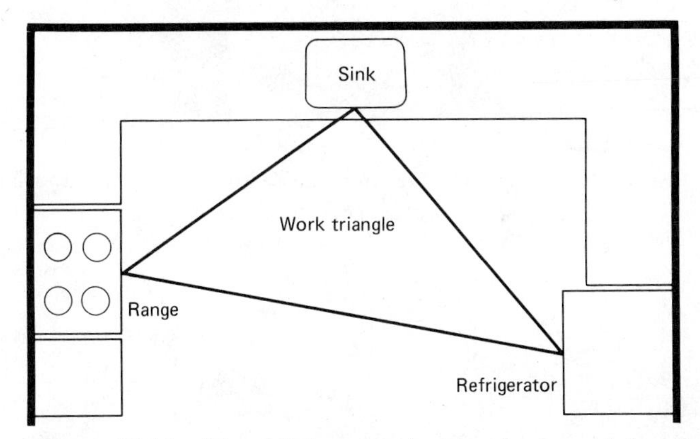

Figure 17-10. The kitchen work triangle formed by the range, sink, and refrigerator.

design principles used in industrial work stations (Steidl, 1972). One well-known principle is the work triangle (Figure 17-10). Stove, refrigerator, and sink should be located in a triangle so that the flow of work can easily go among these three critical pieces of kitchen equipment. Sufficient counter space and storage cabinets are essential. (See Falcone, 1978 for specific design suggestions for kitchens and other rooms.)

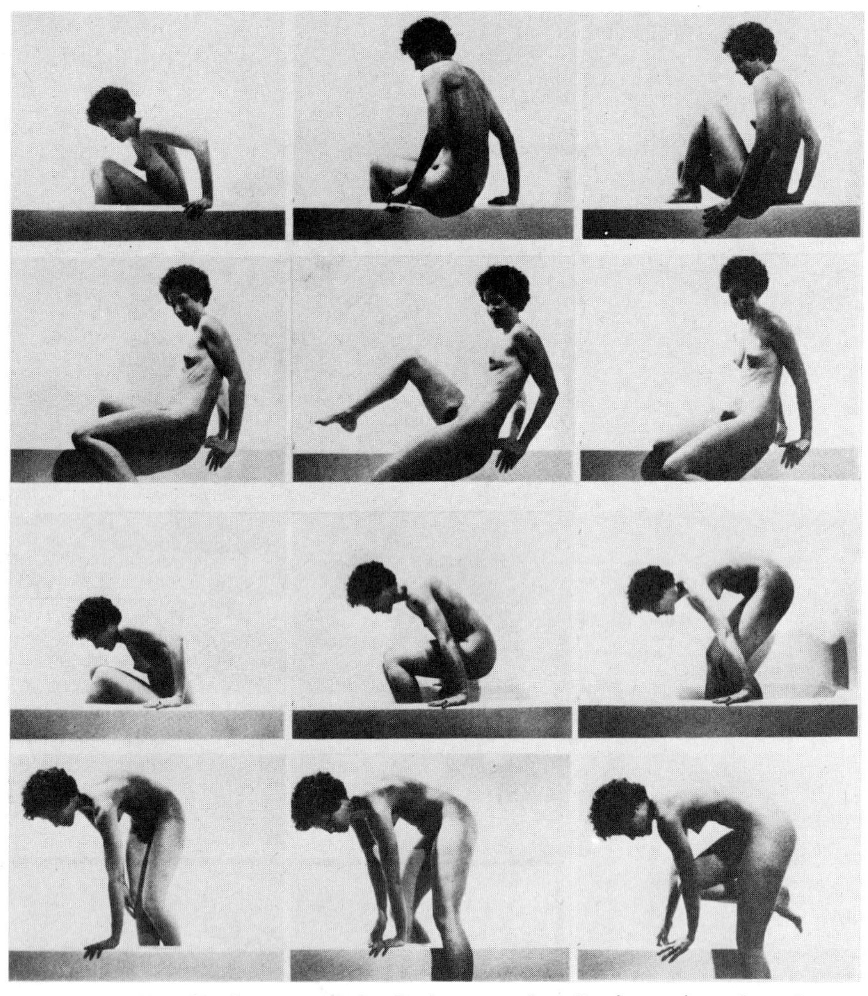

Figure 17-11. Getting out of the bathroom tub calls for awkward, and even dangerous, postures. (From Kira, 1976.)

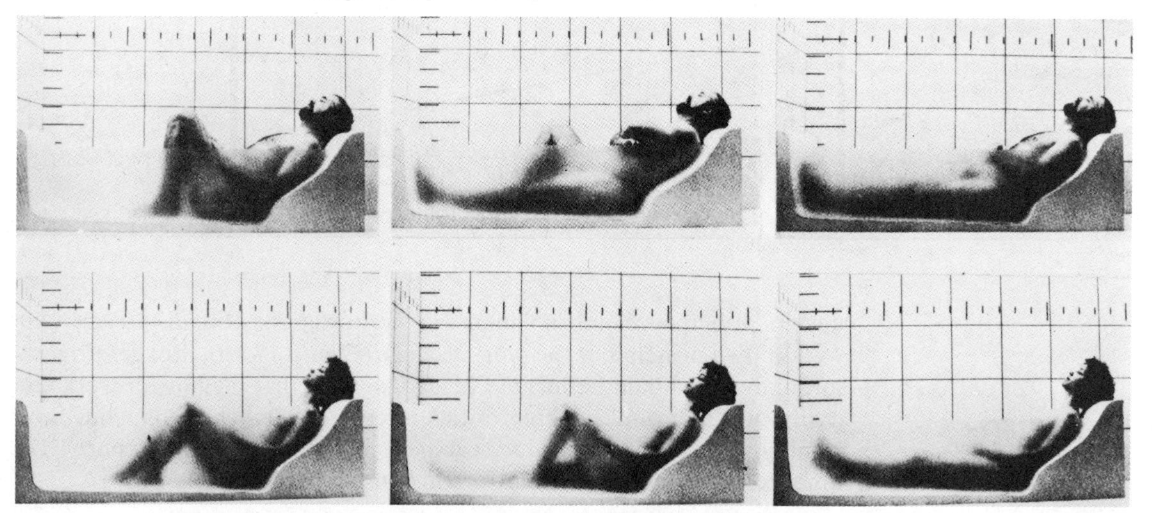

Figure 17-12. Postures assumed for relaxation in a tub with a backrest angle of 32.5°. (From Kira, 1976.)

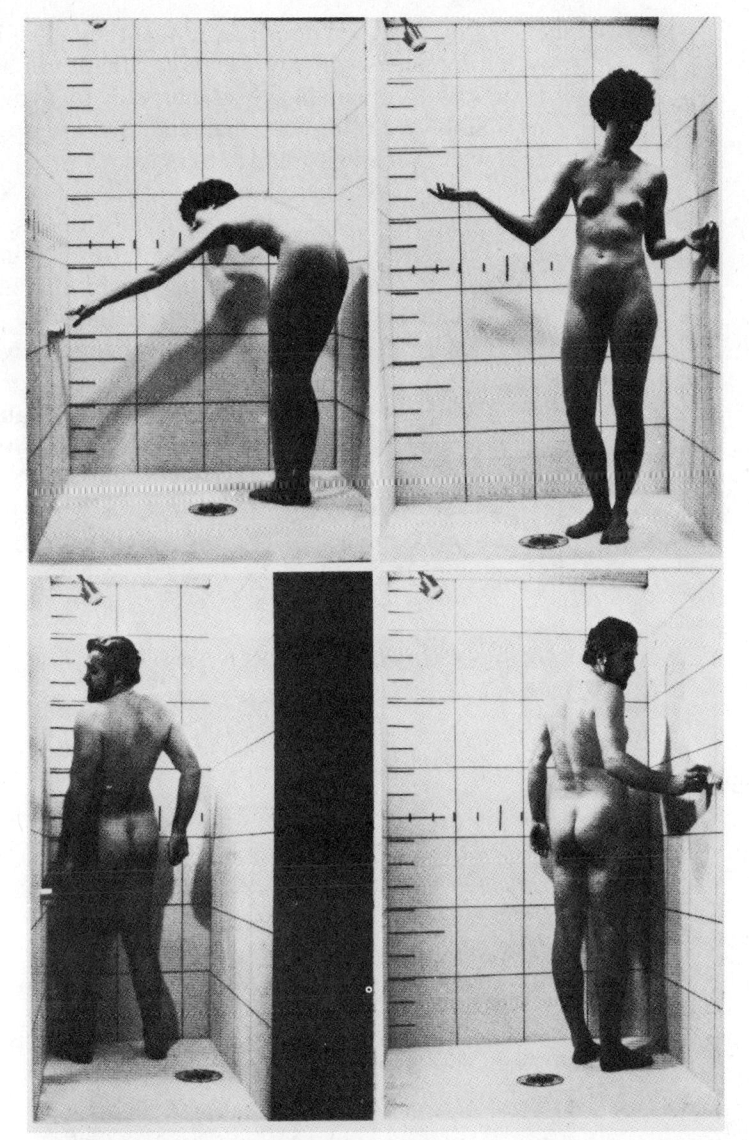

Figure 17-13. Postures assumed to avoid the water spray (left panels) when the control valve is incorrectly located under the shower head. Proper location (right panels) allows the user to adjust water temperature without getting hit by the spray. (From Kira, 1976.)

The Bathroom American cultural norms discourage research about the bathroom (Kira, 1976). The poor design of the bathroom is most often suffered in silence. Approximately 20,000 people are injured every year in their bathrooms (Willis, 1975). We have already discussed seating problems in the bathroom (Chapter 15). Toilets are hard to clean and inefficiently designed considering that they must be used from a standing position by males who are forced to aim at too small a target, especially if they wish to maintain acoustic privacy by avoiding the standing pan of water. Sinks are often too small and at the wrong height. Medicine cabinets placed over sinks are hard to reach.

The bathtub is dangerous and poorly designed. Getting in and out of the tub is awkward and hazardous, especially for elderly people. Figure 17-11 shows the contortions required to enter/exit a tub with high sides. Once in the tub, the average person has some difficulty achieving a relaxed posture. This is particularly annoying since few people take a bath only for cleansing; this purpose is most often achieved by taking a shower (Kira, 1976). However, a proposed new design provides a contoured back making the tub far more comfortable (Figure 17-12). The optimal back rest angle should be between 125° and 140° (Zabcik, 1981) rather than the 105° to 120° used in standard tubs. Finally, it should be possible to reach the water temperature controls while taking a shower without having to encounter the stream of water (Figures 17-13, 17-14). For a full discussion of the many design defects in bathrooms see Kira (1976).

Figure 17-14. The controls for both showers are located at the center of the six-foot tub. The user can safely adjust water temperature. (© B. Kantowitz, TEXT-EYE, 1983.)

Floor Plans In addition to proper design of individual rooms, the floor plan linking the rooms must be considered (Steidl, 1972). Houses should be divided into zones according to the activities that dominate each area. Thus, the bedrooms should be located away from the living room or family room and should have adequate acoustic isolation. Kitchens should be near the garage so that groceries need not be carried far. The dining room should be near the kitchen. Living rooms should be a dead end so that one need not walk through the living room to get to any other room. There is a growing trend to locate utility rooms (bathrooms, laundry rooms, storage) against the north wall of a house to save energy by minimizing the need for windows that face north. The accompanying Workbook contains floor plans for you to evaluate.

Room Color

It is well known that people have affective (emotional) responses to different colors. You may just love shocking pink while your best friend prefers aqua. Interior designers and decorators, to say nothing of clothing designers, spend a great deal of time and effort trying to coordinate colors in pleasing combinations. There are substantial individual differences in color preferences and it is quite difficult to find a color that would be universally admired (or hated). Some systems for cataloging colors are discussed in Chapter 3. Here we discuss the effects of room color on behavior.

Although many people have speculated that color will change behavior, most published research has not supported this claim. Since red-orange colors are considered warm, and blue-green colors cool, the speculation that thermal comfort (see Chapter 19) can be affected by room color has been unduly dignified by terming it the "hue-heat hypothesis." If it is true, then by painting rooms warm colors we should be able to lower the thermostat and save energy without having people feel cold. A recent study (Greene & Bell, 1980) that carefully controlled temperature in rooms painted different colors found absolutely no evidence to support the hue-heat hypothesis. While you may find one color more pleasant than another, color does not change your perception of thermal comfort.

There is one classic study (Srivastava & Peel, 1968) that did find effects of color on behavior. These researchers used a device called a hodometer—hodos is the Greek word for path—that consisted of many switches concealed under a carpet. This allowed them to record the movement patterns of visitors to an art gallery. When the walls and carpet were light beige, people explored less, that is, covered less area, than when walls and carpet were chocolate brown. In the brown gallery, people took more footsteps but spent less time in the room. While this study is often cited as the example showing how color affects behavior, we should point out that no convincing explanation for this result has emerged even though the study is over a dozen years old.

Thus, although the notion that color affects non-visual behavior is intuitively appealing to many, there is little scientific support for this idea. The color of lighting does affect feelings. It also affects visual performance and we will return to this topic in a discussion of color temperature in the following section on illumination.

ILLUMINATION

In Chapter 4, we mentioned that different light sources can have quite different distributions of light energy over wavelength (see Figures 4-2 and 4-3). Because the sensitivity of the human eye varies as a function of the wavelength of light, useful units of light measurement (*photometric* units) incorporate a factor for weighting this differential sensitivity (shown in Table 4-1). The total amount of light given off by a light source in all directions in measured is *lumens* (lm). A new 100-W incandescent lamp gives off about 1700 lm, and a new 40-W fluorescent lamp approximately 3200 lm. The amount of light given off in a specific direction is measured in *candelas*. The term *illuminance* refers to the amount of light reaching a surface, and is measured in *footcandles* (fc) or *lux* (lx). A surface that is one square foot in area receiving 1000 lm of light would have an illuminance of 1000 fc; a 1 m^2 meter surface receiving 1000 lm would have an illuminance of 1000 lx.

Luminance is a measure of the amount of light coming off a surface. A surface radiating one lumen per square foot of surface area has a luminance of one *footlambert*. Luminance can also be measured in *candelas per square meter* or in *nits*. Generally the luminance of a surface will depend on how much light it receives (illuminance) and how much of that light it reflects rather than absorbs. *Reflectance* is the ratio of reflected light to received light. Most surfaces have reflectances of from 5 to 95%; this page probably has a reflectance of 85% or more. The Workbook will give you some practice using these measures.

Wavelength Distribution of Illumination

A number of parameters describing light sources will affect our estimates of their usefulness, comfort, and adequacy in different environments. The actual distribution of light energy over wavelength (the spectral power distribution or spectrum of the light) will be an important factor in many situations. The spectral distribution will determine the apparent color of the light source, and will have a major effect on the color of surfaces illuminated by that source. Chapter 4 discussed the CIE system for defining the color of any spectral distribution of light. A simple but crude way to describe the color of a given light spectrum is to specify its *correlated color temperature*. To define correlated color temperature we need to describe radiation from a blackbody. A blackbody is a physical device that, when heated, radiates a broad and continuous spectrum of energy according to a known spectral distribution. Both the total power and the relative power of short wavelengths in this spectrum will increase greatly as the temperature of the blackbody radiator is raised. Because the spectrum changes with increasing temperature, the color of the light radiated also changes, and the blackbody temperature thus can provide an approximate way to define the color of other spectral distributions.

Remember from our discussion of color in Chapter 4 that many different spectra of light will have roughly the same color as a blackbody at a given temperature. So we can say that some colored light, *x*, is the same color as the light from blackbody heated to, say, 3500 degrees Kelvin(K). This is the correlated color temperature. (Around 1000 K is red, 3000 K is yellow, 5000 K is white, 8000 to 10,000 K is pale blue.) If the spectrum of a light source has a very different shape than a blackbody or possesses a discrete rather than a continuous distribution, then the correlated color temperature specification will not be a very useful way to describe the light source. Incandescent sources such as tungsten filament lamps have spectra that approximately match those of blackbodies; most other light sources such as fluorescent lamps, do not. Incandescent lamps have rather warm spectra, corresponding to color temperatures of below 3000 K.

One important and complex aspect of a light source is its ability to preserve the color of objects viewed under its light as compared with viewing in normal sunlight. Objects viewed under a pair of color matched lamps such as a yellow sodium lamp and yellow fluorescent lamp will appear very different. Their appearance under those lamps will also be very different from the colors seen under normal daylight conditions. One approach to the problem of specifying the color-rendering properties of light sources is a color rendering index (CRI) that was developed by the CIE (see Boyce, 1981). This index measures how well colors will match when viewed under the light source and a reference source of the same color temperature. An index of 100 indicates that the colors of objects viewed under that lamp matches those under the reference source. The reference sources used are either the blackbody spectrum of the appropriate color temperature (below 5000 K) or a spectrum approximating that of daylight. A standard fluorescent white lamp at 3000 K has a color rendering rating of about 60. Some typical CRI's are given in Table 17-1.

If you desired light with a color temperature approximating noon daylight, you would look for high CRI's at a reference temperature of 5000 K. A CRI of 90 or better will usually mean a reasonably constant apparent color under the source and reference conditions. The choice of the reference light would depend on the quality of light desired and the nature of the activity in the environment. For graphics art work, color references of 5000 K and 7500 K have been recommended; the 7500 K reference is an approximation of the light entering a north-facing skylight on a moderately overcast day. Light references possessing this latter classification arose out of their use in raw cotton classification tasks in the cotton industry. In spite of the romantic aspect of a north-facing garret studio in Paris, some artists may prefer the warmth of lower temperature references, per-

TABLE 17-1 *Color and Color Rendering Characteristics of Common Light Sources (IES Lighting Handbook, 1981 Reference Volume)*

Test Lamp Designation	CIE Chromaticity Coordinates x	y	Correlated Color Temperature (Kelvin)	CIE General Color Rendering Index CRI
Fluorescent lamps				
Warm white	.436	.406	3020	52
Warm white deluxe	.440	.403	2940	73
White	.410	.398	3450	57
Cool white	.373	.385	4250	62
Cool white deluxe	.376	.368	4050	89
Daylight	.316	.345	6250	74
Three-component A	.376	.374	4100	83
Three-component B	.370	.381	4310	82
Simulated D_{50}	.342	.359	5150	95
Simulated D_{55}	.333	.352	5480	98
Simulated D_{65}	.313	.325	6520	91
Simulated D_{70}	.307	.314	6980	93
Simulated D_{75}	.299	.315	7500	93
Mercury, clear	.326	.390	5710	15
Mercury, improved color	.373	.415	4430	32
Metal halide, clear	.396	.390	3720	60
Xenon, high pressure arc	.324	.324	5920	94
High-pressure sodium	.519	.418	2100	21
Low-pressure sodium	.569	.421	1740	−44
DXW tungsten halogen	.424	.399	3190	100

haps as low as those approaching the color temperature of incandescent sources (see Nuckolls, 1981).

Color temperature references of 5000 K are recommended for tasks requiring the accurate judgment of color quality; while a 7500-K reference is recommended for tasks involving accurate judgments of color uniformity. Suppose that you purchase a print or painting and intend to hang it in your living room, illuminated by incandescent light of a color temperature of around 3000 K. You'd better hope that the gallery has incandescent or fluorescent lamps which have CRI's of better than 90, referenced to a color temperature of 3000 K. If not, the colors of the print in your room might not be what you expect. Some tasks will require the very fine discrimination of color shades, such as in the grading of subtle color differences in tobacco, fruit or textiles. In these environments, certain light sources may be used to emphasize color differences. For example, a light source rich in short wavelengths will enable small differences to be discriminated in yellow specimens. However, this special light source might greatly distort the normal appearance of objects in that environment.

Special fluorescent lamps are available that can illuminate an environment with relatively full spectrum light with color rendering indices of 90 or better. Many scientists believe that use of a high CRI, full-spectrum light source results in greater apparent brightness, visual clarity, and satisfaction, than possible with a typical lower CRI, cool white fluorescent light source (see Hughes & Neer, 1981). The difference in task performance under different light sources may be especially evident for people over 55 years old. This may be due in part to an effect of the light spectrum on the mechanism of accommodation (Hughes & Neer, 1981). In a recent study, Chao and Bennett (1981) had 60 subjects judge six different light sources on the basis of their acceptability in illuminating faces (their own faces viewed in mirrors). Incandescent (CRI = 100) and warm white fluorescent (CRI = 52) were judged best for viewing faces, three other sources (cool white fluorescent, CRI = 62; improved color mercury, CRI = 32; and metal halide, CRI = 60) were seen as less acceptable, and two sodium sources (high-pressure sodium, CRI = 21 and low-pressure sodium, CRI = 44) were worst. The authors reported an effect attributable to the race of the subjects; the acceptability of the sodium sources was dependent on the skin color of the viewer.

Quantity of Illumination

In today's energy- and cost-conscious economy, there may be pressure to design work environments with the least illumination levels that one can justify as necessary for task performance. In tasks that involve low-contrast material (such as discriminating fine differences in dark shades of gray), the resolution of small details, or viewing under time pressure, performance will increase as the level of illumination increases. The range of illumination for acceptable task performance in such demanding tasks as compared to nondemanding visual tasks, is very great. Recommended levels (*IES Lighting Handbook*) exist for illumination in different work environments, but these should be considered not as minimum requirements, but as guidelines for design. Table 17-2 contains excerpts from recommended illuminance tables developed by the Illuminating Engineering Society. Part I gives general categories of activity and the range of illuminances needed. Part II and III give values for specific types of tasks. The determination of the low, middle, or high value in each category range depends on factors such as worker age, requirement for task speed and/or accuracy, and the reflectances of background (nontarget) surfaces in the room or work area. Anyone specifying illuminance values for a specific environment or work area should read the material in Chapter 2 of the *Application Volume of* the *IES Lighting Handbook;* this chapter describes how to consider many factors in making lighting recommendations. The Handbook also contains a table for defining the illuminance

TABLE 17-2 *Currently Recommended Illuminance Categories and Illuminance Values for Lighting Design—Target Maintained Levels (IES Lighting Handbook 1981 Application Volume)*

I. Illuminance Categories and Illuminance Values for Generic Types of Activities in Interiors

Type of Activity	Illuminance Category	Ranges of Illuminances		Reference Work-Plane
		Lux	Footcandles	
Public spaces with dark surroundings	A	20–30–50	2–3–5	
Simple orientation for short temporary visits	B	50–75–100	5–7.5–10	General lighting throughout spaces
Working spaces where visual tasks are only occasionally performed	C	100–150–200	10–15–20	
Performance of visual tasks of high contrast or large size	D	200–300–500	20–30–50	
Performance of visual tasks of medium contrast or small size	E	500–750–1000	50–75–100	Illuminance on task
Performance of visual tasks of low contrast or very small size	F	1000–1500–2000	100–150–200	
Performance of visual tasks of low contrast and very small size over a prolonged period	G	2000–3000–5000	200–300–500	
Performance of very prolonged and exacting visual tasks	H	5000–7500–10000	500–750–1000	Illuminance on task, obtained by a combination of general and local (supplementary lighting)
Performance of very special visual tasks of extremely low contrast and small size	I	10000–15000–20000	1000–1500–2000	

II. Commercial, Institutional, Residential and Public Assembly Interiors

Area/Activity	Illuminance Category	Area/Activity	Illuminance Category
Auditoriums		**Barber shops and beauty parlors**	E
Assembly	C	**Club and lodge rooms**	
Social activity	B	Lounge and reading	D
Banks		**Conference rooms**	
Lobby		Conferring	D
General	C	Critical seeing (refer to individual task)	
Writing area	D	**Court rooms**	
Tellers' stations	E	Seating area	C
		Court activity area	E
		Dance halls and discotheques	B

category based on the required equivalent task contrast (defined in the next paragraph), if that is known. Remember that the effectiveness of light sources may be reduced due to the presence of glare, lamp failures, temperature and voltage variations, decreased light output of lamps with age, and dirt on lamp and work surfaces and in the air. So the illumination levels must encompass these expected reductions.

One approach to quantifying the illumination level required for various

TABLE 17-2 *(Continued)*

III. Industrial Group

Area/Activity	Illuminance Category	Area/Activity	Illuminance Category
Aircraft manufacturing		**Book binding**	
Assembly		Folding, assembling, pasting	D
Simple	D	Cutting, punching, stitching	E
Moderately difficult	E	Embossing and inspection	F
Difficult	F	**Breweries**	
Very difficult	G	Brew house	D
Exacting	H	Boiling and key washing	D
Bakeries		Filling (bottles, cans, kegs)	D
Mixing room	D	**Candy making**	
Face of shelves	D	Box department	D
Inside of mixing bowl	D	Chocolate department	
Fermentation room	D	Husking, winnowing, fat extraction, crushing and refining, feeding	D
Make-up room			
Bread	D	Bean cleaning, sorting, dipping, packing, wrapping	D
Sweet yeast-raised products	D		
Proofing room	D	Milling	E
Oven room	D	Cream making	
Fillings and other ingredients	D	Mixing, cooking, molding	D
Decorating and icing		Gum drops and jellied forms	D
Mechanical	D	Hand decorating	D
Hand	E	Hard candy	
Scales and thermometers	D	Mixing, cooking, molding	D
Wrapping	D		

tasks is called the *visibility level.* To define the visibility level, we first have to define a quantity called the *threshold contrast.* Threshold contrast is the ratio of target luminance to background luminance (the increment in the luminance of a target divided by the background luminance) that is just at the point where an observer can see the target. The target is a 4 minute in arc luminous disc presented for 200 msec against a uniform background. Figure 17-15 shows how threshold contrast decreases as the background or adaptation luminance increases (*IES Lighting Handbook,* 1981). However you can see that at high levels of background luminance there are increasingly smaller gains to be made in the threshold.

A related term is *equivalent contrast.* Equivalent contrast is the same type of measure but it describes the target contrast ratio in the actual task of interest, rather than the reference task with the disc. The observer background adaptation is the same as in the disc reference task. Visibility level is defined as the ratio of equivalent contrast to threshold contrast. Suppose the visibility level were 8; that would imply that the equivalent contrast value was eight times the threshold contrast value, at any background luminance. The 8 times factor is an estimate of the increased contrast that would be needed in realistic viewing tasks. It approximates the

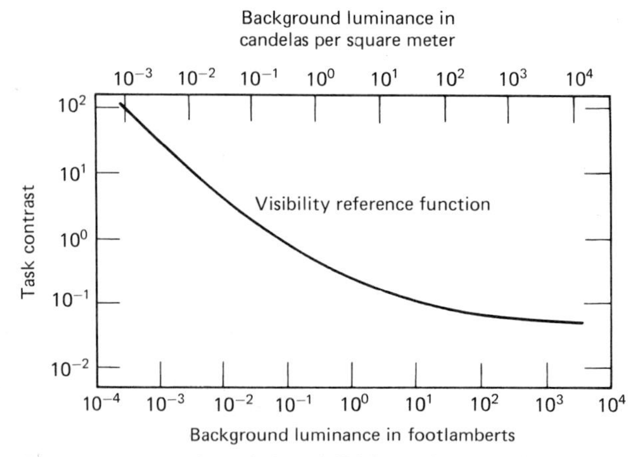

Figure 17-15. Plot of the visibility reference function representing task contrast required at different levels of task background luminance to achieve threshold visibility for a 4-minute luminous disk exposed for $\frac{1}{5}$ sec. (From *IES Handbook,* 1981 Reference Volume.)

effect of factors such as the presence of target spatial and temporal uncertainty, dynamic versus static presentation, and the requirement for a much higher level of detection performance.

We have indicated that the advantage to increasing the luminance is reduced at very high levels of luminance. However there is some controversy about whether increasing the luminance for high contrast targets may eventually cause a reduction in their visibility (Boyce, 1981). This question is probably not of much practical importance, since the luminance levels involved are much higher than those we normally encounter. There has, however, been a clear upward trend in recommended illuminances over the past 30 years. Boyce (1981) has pointed this out in his recent book (see Figure 17-16). Boyce suggests that many changes have led to this result, including changes in the nature of office work, in the efficiency and availability of lighting, and in our expectations of an efficient and comfortable lighting environment.

An extremely important factor in establishing the amount of light required for a task is the age of the worker. As we age there are significant decreases in our visual acuity, our ability to focus and adapt rapidly to different light levels, and our resistance to glare. The probability of visual defects also increases sharply as we pass through our sixties. The increase in lens opacity, light scatter in the aging eye, and limitations in accommodation, can result in serious problems in the performance of visual tasks. Blackwell and Blackwell (1971) determined the elevation in the contrast threshold for people in several older age groups than used to obtain the curve of Figure 17-15. For people in the 50 to 60 and 60 to 70 year groups, the curve was elevated by a factor of 1.86 and 2.51, respectively.

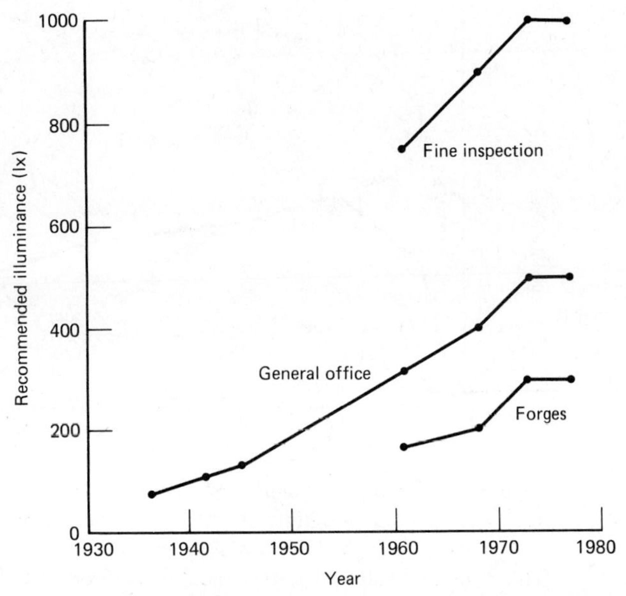

Figure 17-16. Illuminances recommended for fine inspection work, general offices, and forges, in successive IES codes. (From P. R. Boyce, *Human Factors in Lighting,* 1981, p 382, Macmillan Publishing Co., Inc. Copyright 1981 by Applied Science Publishers, Ltd., England, and reproduced by permission.)

A study by Hughes and McNelis (1978) studied the performance of 12 workers in a simulation of an office work situation under three levels of illuminance: 538, 1076, and 1614 lx. Six workers were between 19 and 27 years, and six were between 46 and 57. The lighting levels were varied during the work sessions to prevent worker awareness of how many levels were involved in the study. For both groups, the average time to perform the task was a decreasing function of the level of illumination. But the advantage of the highest over the lowest lighting level was much greater for the older workers (9% versus 6.7%). All of the workers rated the higher illumination levels as preferred on a number of scales including work effort, desirability, and satisfaction.

A quite different kind of test situation was employed by Sivak, Olson, and Pastalan (1981), in a study of the effects of age on the legibility of highway signs at night. Different colors, background luminances and luminance contrast ratios (of target letter and background) were employed. Subjects were either between 18 to 24 years or 62 to 74 years old. The subjects rode in or drove an automobile and watched for a small sign along the road; they pressed a push button switch to indicate the orientation of the target letter on the road sign. Figure 17-17 shows how the distance to the sign (for essentially 100% correct responses) varied as a function of the subject age group and contrast ratio of the sign. Over a wide range of sign materials, older subjects perform significantly worse

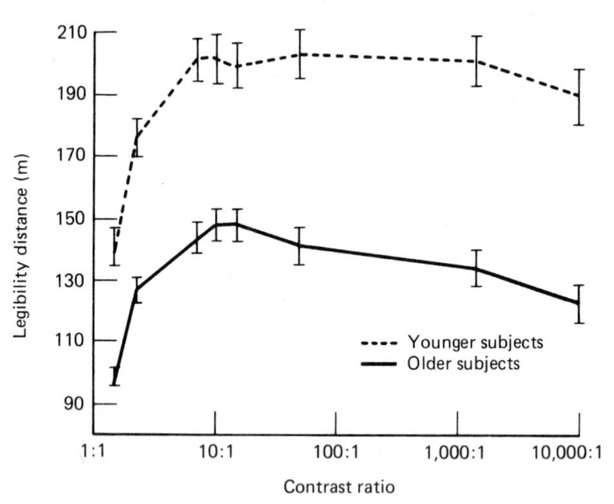

Figure 17-17. Sign legibility as a function of luminance contrast and observer age. The vertical bars indicate ± 1 standard error of each mean. (From M. Sivak, P. L. Olson, & L. A. Pastalan, *Human Factors,* 1981, *23,* 63. Copyright 1981 by The Human Factors Society, Inc., and reproduced by permission.)

than younger subjects, and hence have much less time to act on any information displayed.

Quality of Illumination

Luminance Distribution When performing a visual task, the visual system reaches a state of adaptation that depends on the average luminance of the work. This adaptation shifts the operating range of the system (see Chapter 4) so that small differences in light intensity will result in discriminable differences in brightness. Figure 17-18 illustrates how a family of operating curves can together provide a good contrast sensitivity over a large range of intensities. The eye can rapidly shift its operating curve over a small range as it looks at different parts of a scene. However if the required adaptation shift is too great, comfort and work efficiency will be affected. Excessive differences in the luminance of different parts of the work area and the surroundings should be avoided. These luminance ratios between primary and surrounding areas should be less than 1 to 3 (see (see the *IES Lighting Handbook*). The distribution and diffusion of light sources in the environment will determine these luminance ratios. Some tasks may require directional, nondiffuse illumination, so as to improve the contrast ratios between parts of the visual task. Examples are visual tasks of a three-dimensional nature such as examining imperfections in metal surfaces or sculptured objects.

As people age, the speed and extent of adaptation decreases, and the negative effects of a poor luminance distribution increase. A study by Fozard and Popkin (1978) cited evidence that most accidental falls on stairs by older people occur on the first step of a landing. In this region there is the greatest decrease in luminance, and the visual system cannot

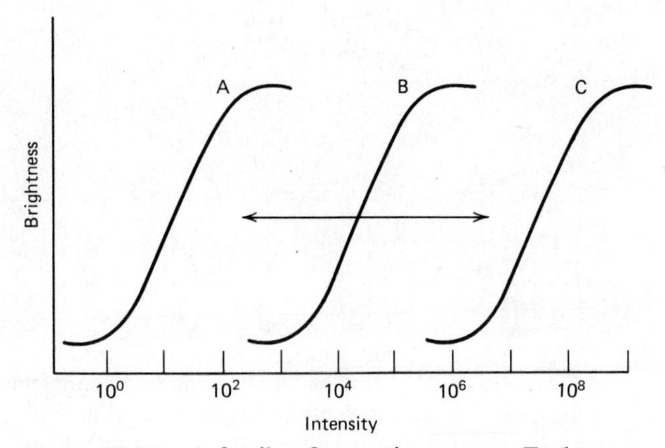

Figure 17-18. A family of operating curves. Each curve enables good contrast discrimination over a narrow range of intensity. Adaptation shifts the observer to higher or lower (arrows) operating curves. The net result is good contrast discrimination over a very wide range of intensity. (From *Sensation and Perception* by E. Bruce Goldstein. © 1980 by Wadsworth Inc. Reprinted by permission of Wadsworth Publishing Company, Belmont, California, 94002.)

make the rapid change in adaptation that is required. The situation can be greatly improved by adding lighting to the first several stairs of a stairway. This problem is not limited to the elderly. A study by Rinalducci, Hardwick, and Beare (1979) indicated that the excessive ratio of luminances at the entrances to some long vehicular tunnels has a definite effect on performance. Speed of traffic flow, driver braking pattern, and visibility, are all influenced adversely by these luminance factors and the requirement for rapid eye adaptation immediately after entering a tunnel.

Glare Bright light sources and bright reflections off environmental surfaces can cause glare. The term *discomfort glare* refers to glare situations producing annoyance or discomfort but not necessarily interference with visual performance. *Disability glare* is a type of glare where there is a reduction in the visibility of objects and a resulting effect upon performance. The types of reflection off surfaces are shown in Figure 17-19.

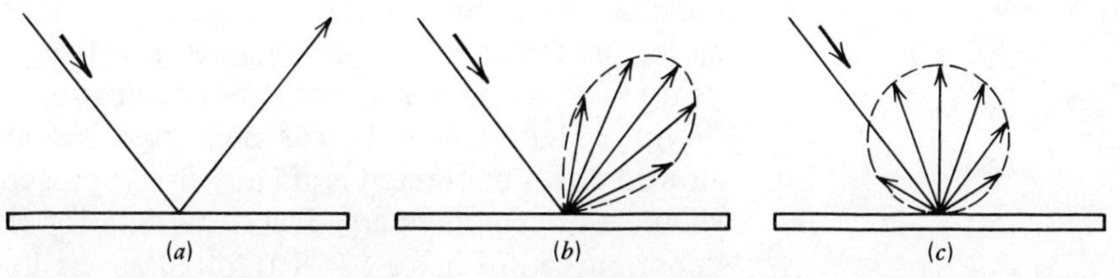

Figure 17-19. The general type of reflection depends on properties of the surface. (*a*) Polished surface (specular). (*b*) Rough surface (spread). (*c*) Matte surface (diffuse). (From *IES Lighting Handbook,* 1981 Reference Volume.)

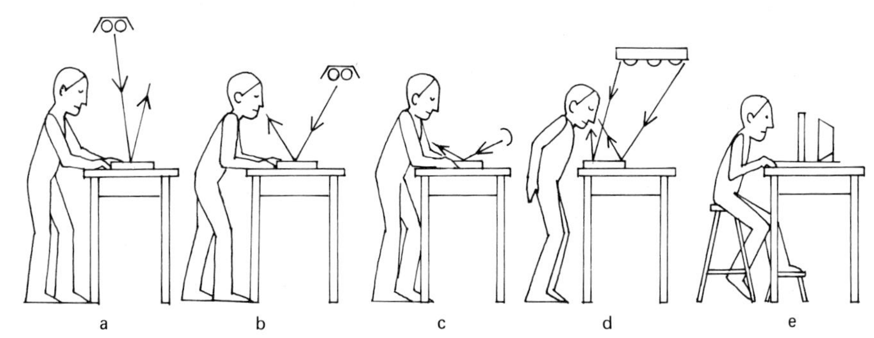

Figure 17-20. Examples of placement of supplementary lighting. (*a*) Source located to prevent reflected glare—reflected light does not coincide with angle of view. (*b*) Reflected light coincides with angle of view. (*c*) Low-angle lighting to emphasize surface irregularities. (*d*) Large-area surface source and pattern are reflected toward the eye. (*e*) Transillumination from diffuse sources. (From *IES Lighting Handbook,* 1981 Application Volume.)

Many glare situations involve combinations of these general types of reflection. Reflected glare is said to reduce visibility by producing a *veiling reflection,* because it obscures or veils part of the visual task. The losses in contrast produced by veiling reflections may not always be obvious. Their effects can be reduced by employing work space arrangements and light sources that limit the amount of light reflected from the visual task toward the worker's eyes. Figure 17-20 illustrates some of these principles. The light sources should be mounted above and away from the normal lines of sight and the amount of light emitted or reflected toward the eyes should be minimized.

In an interesting study of the relationships between headlight glare effects and driver age, Pulling, Wolf, Sturgis, Vaillancourt, and Dolliver (1980) compared physiological glare threshold with a measure of headlight glare resistance. Physiological glare thresholds were assessed in an apparatus that determined the threshold for recognizing a target's orientation in a glare test close to that of the headlight glare situation. (The measure was the log of the ratio of glare illuminance to target illuminance at threshold.) The relationships between glare threshold and age for 148 subjects is shown as the lower curve in Figure 17-21. It's possible to conclude from this result that there is a pronounced drop in most people's glare thresholds with age, equal to a 0.1 log unit deterioration every 4 years. The authors also evaluated each subject's resistance to glare in a driving simulation. This was measured by observing when the subject slowed down or steered eratically in the presence of headlight glare from simulated oncoming cars. The fit to data for 30 subjects is shown as the upper curve in Figure 17-21 (plotted as the log of the ratio of oncoming headlight (glare) illuminance to ambient illumination).

Assuming that the simulator task provides an appropriate estimate of

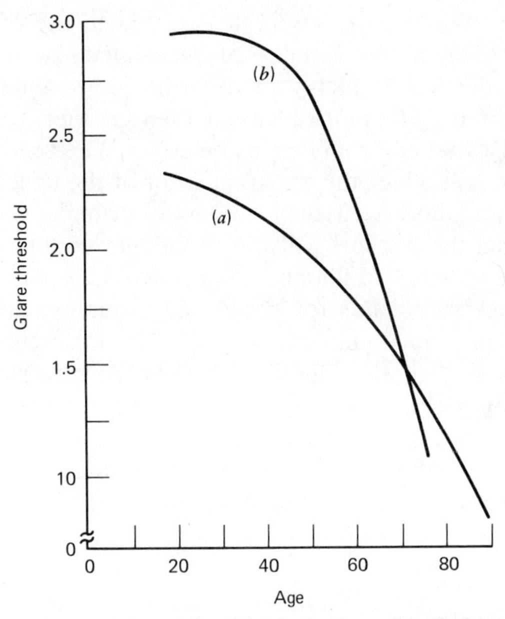

Figure 17-21. Physiological glare threshold (*a*), and headlight glare resistance (*b*), as a function of age. (After N. H. Pulling, E. Wolf, S. P. Sturgis, D. R. Vaillancourt, & J. J. Dolliver, *Human Factors,* 1980, *22,* 107, 8. Copyright 1980 by The Human Factors Society, Inc., and reproduced by permission.)

driver behavior, one might wonder why the headlight glare resistance ratios are much higher than those for the physiological glare situation. The authors suggest that the difference is attributable to each driver's *subjective glare tolerance,* which is a measure of the driver's overreach or conservatism in the night driving task. A driver's subjective tolerance for glare might be much smaller in the riskier real-life night driving situation. However we cannot conclude from this study that drivers will not ordinarily overreach their glare thresholds in driving. Do you know any elderly people who continue to drive in spite of known degradation in their vision? Techniques to reduce the effects of headlight glare include roadway screens to block the glare of oncoming traffic, and improved roadway striping, marking, and lighting.

SUMMARY

The microenvironment is that part of the built environment that is immediately accessible to people. The construct of a personal-space bubble helps us to understand behavior in the microenvironment. Personal space is portable and of varying size in contrast to territoriality, which explains behavior regarding fixed locations of constant size. Invasion of personal space has psychological and physiological effects on people.

Human factors principles have been applied to the design of individual rooms. While most research has studied the office environment, there have been a few studies of residential dwellings. The bathroom in particular leaves plenty of room for improvement. While you may find one room color more pleasant than another, there is little evidence to suggest that color influences behavior. The spectrum of the room illumination will affect the apparent color of the light source, the color of objects and how well they will be discriminated from one another, and whether those colors will appear the same in other rooms.

The quantity of illumination needed depends on the nature of the task, the task's demands for speed and accuracy, the worker's age, and the environmental luminance distribution. Glare is an important aspect of the quality of the illuminated environment. The design of illuminated environments is a complex task that involves physical, engineering, psychological, and aesthetic factors. The human factors designer involved in lighting questions should become familiar with the *IES Lighting Handbook* and a recent book by Boyce (1981).

18 MACRO-ENVIRONMENTS

Human factors has traditionally dealt with people-systems relationships that focus primarily on a small number of individuals such as a single operator working at a particular machine or a small crew interacting at a workstation. The larger view of the interaction between society and systems has been left for sociologists and other behavioral scientists until quite recently. But since society is composed of individuals, it seems reasonable that the human factors specialist should have something to say about improving the relationship between people and their molar environments. Many of the human factors principles discussed in earlier chapters can be successfully applied to some of the more general problems of society. Human factors has grown well beyond the aerospace applications that gave it its initial impetus and the field now also includes such environmental problems as crime prevention, design of airports and highways, and mass transportation. As the techniques and methods of human factors are directed towards these applications, we can expect an increasing demand for human factors specialists in these new areas. Although all of these areas represent an increased scope for human factors, some are newer than others. Transportation is the most established of the areas covered in this chapter. Crime prevention and urban issues are currently being expanded as outlets for human factors activity. In this chapter we try to demonstrate that the system in people-system relationships can be as large as one's imagination will allow. Cities, transportation corridors connecting cities, and even architectural design are all legitimate topics that can benefit from human factors analyses.

TRANSPORTATION

It is a geographic fact of life that people are spatially distributed. Modern society uses an extremely complex system to transport goods and people throughout the country and throughout the world. Even the most primitive transportation system, the human back, requires some equipment: a backpack. This allows a human to carry 100 pounds at a cost of about 20 cents per ton-mile. More sophisticated systems require greater capital investment to lower the cost of transportation. A pack horse brings the cost down to 16 cents per ton-mile, still quite expensive. A team and wagon lowers the cost to 3 cents per ton-mile. A truck operating on paved roads costs about 2 cents per ton-mile, and a railroad train only 1 cent per ton-mile (Morlok, 1978, Table 2-1). We have already discussed the human factors requirements for some of this equipment such as vehicles (Chapter 15). Here we focus upon some larger aspects of transportation systems: airports, highways, and mass transportation. The transportation network in the United States is vast, consisting of 3,700,000 miles of roads, 41,200 miles of interstate freeways, 219,000 miles of railroad freight lines, 162,000 miles of scheduled domestic airline service, 26,000 miles of inland waterways, and 436,000 miles of pipelines (U.S. Dept. of Transportation, 1972). The expenditures for transportation are equally vast as Figure 18-1 reveals. The human factors problems are also vast ranging from design of turnstiles in mass transportation, to terminal design in airports, to optimization of existing systems by operating decisions, such as lower fares in off-peak times that require little or no capital expenditure, to network analysis using complex computer models and simulations. One of the newest branches of engineering, transportation engineering (Morlok, 1978), draws on experts in many diverse fields to solve complex system problems including "the many aspects (that) cannot be quantified (p. 29)." Human factors is especially needed to deal with these less quantitative aspects since, as we have already discovered, it is seldom easy to generate accurate numerical predictions for human behavior.

Airports

Any comprehensive discussion of airport design must deal with many topics that cut across several disciplines: forecasting air traffic demands, land use, air traffic control, air cargo facilities, terminal design, airport access roads, parking facilities, runway pavement, navigation aids, drainage, fire fighting, public relations, and environmental impact. You can find several texts that offer comprehensive coverage (American Society of Civil Engineers, 1971; Ashford & Wright, 1979; de Neufville, 1976). Here we limit discussion to only two topics out of many that are of interest to the human factors specialist: visual displays that guide aircraft pilots before and after landing, and design of passenger terminals.

Visual Displays A pilot emerging from the cloud cover must quickly locate the runway where the plane will touch down. The approach lighting system (ALS) must provide cues that inform the pilot of the plane's position in three-dimensional space relative to the runway. The ALS must take into account glide path angle, range, aircraft landing speed, and cockpit visibility. The Federal Aviation Agency has standardized ALS to prevent confusion due to unique systems at particular

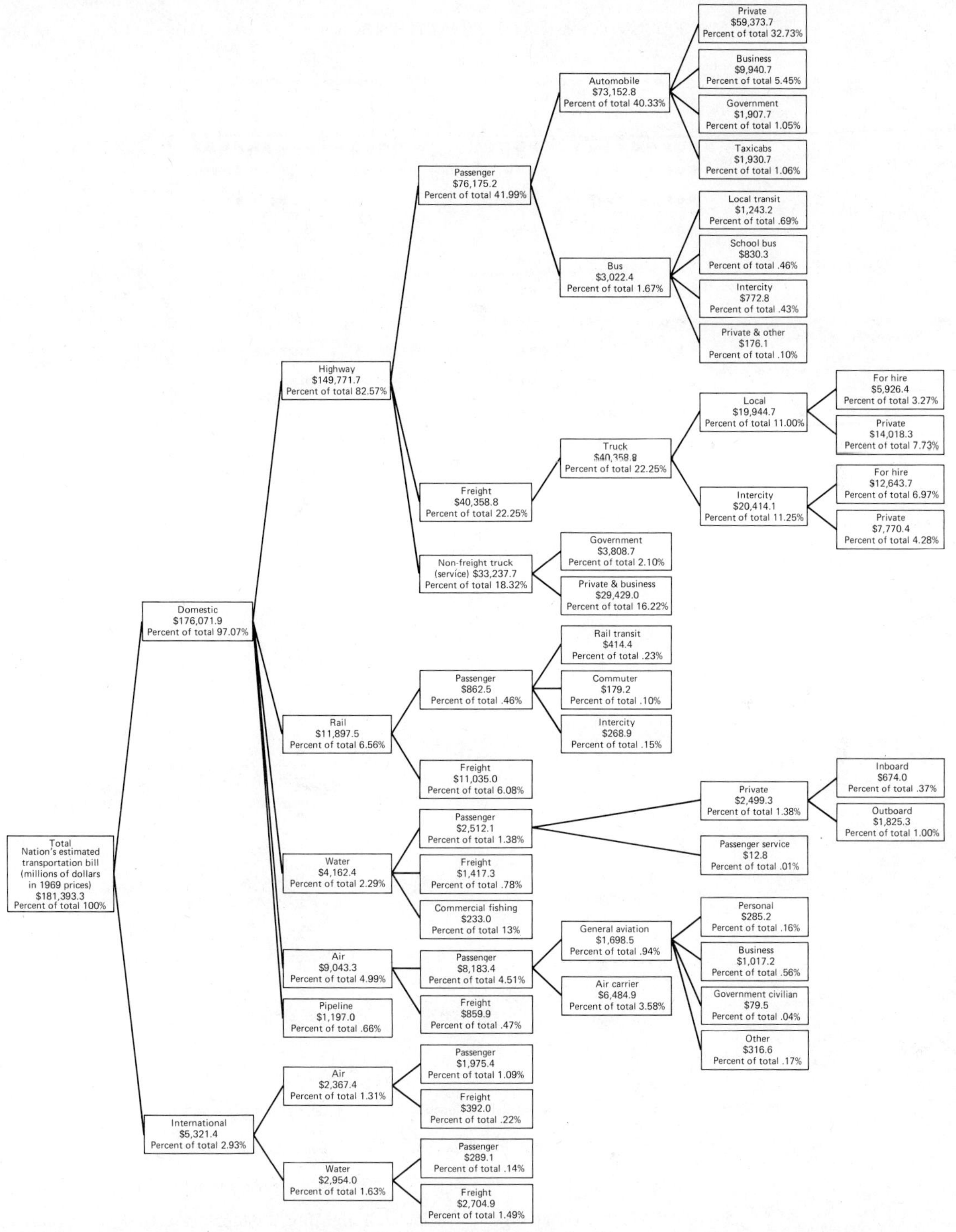

Figure 18-1. Transportation expenditures in the United States in 1970. (From U.S. Department of Transportation.)

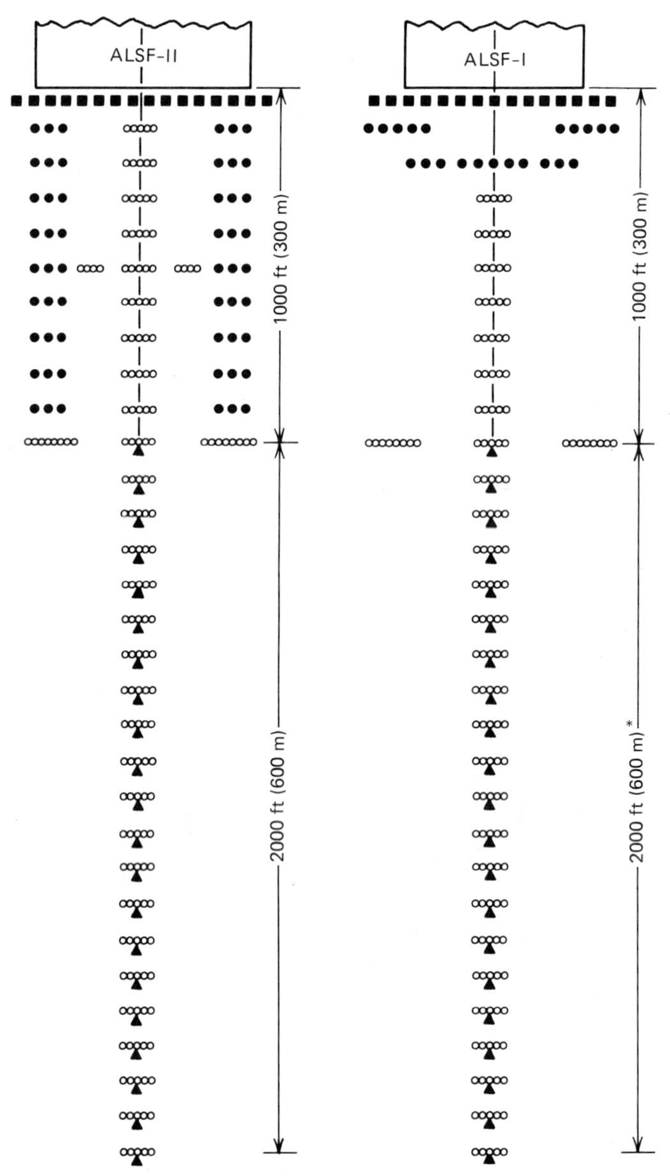

Legend

○ High-intensity steady burning white lights

□ Medium-intensity steady burning white lights

● Steady burning red lights

▲ Sequenced flashing lights

* See paragraphs 42 & 43

■ ALS threshold light bar

Figure 18-2. Airport runway approach lighting system. Asterisk ("see paragraphs 42 & 43") designates FAA Advisory Circular AC/150/5300-2C. September 21, 1973. (From Ashford & Wright, 1979.)

airports (see Figure 18-2). One such standardized system (termed ALSF-II by the FAA) has high-intensity white lights arranged in a bar for 2000 feet along the extended runway. These lights flash in sequence to give the effect of a bar of light moving towards the beginning of the actual runway. The apparent motion of this light source helps the pilot orient toward the runway. Furthermore, since perceived brightness is a product of intensity multiplied by time (see Chapter 4) it is possible to use very intense lights, that would momentarily blind the pilot if they stayed on continuously, since each light flash is of short duration. The inner 1000 feet of the approach has red light bars on each side of the centerline to warn the pilot that touchdown is imminent. Finally a bar of green lights marks the threshold of the runway itself (see Figure 18-2). Thus, the pilot has perceptual cues based upon color coding, spatial location, and apparent motion to guide the plane to a safe touchdown.

Once the pilot has landed and taxied to the terminal, the plane must be aligned with the pier so passengers can deplane. This maneuver is called docking. In many airports an agent runs out and signals to the pilot. This is a poor example of allocation of function since a machine can provide the necessary signals. One such automatic docking system, at Schipol Airport, Amsterdam, is shown in Figure 18-3. It is called the Burroughs optical lens docking system (BOLD). The pilot first aligns the aircraft nose with the left vertical bar. When the two components form a straight

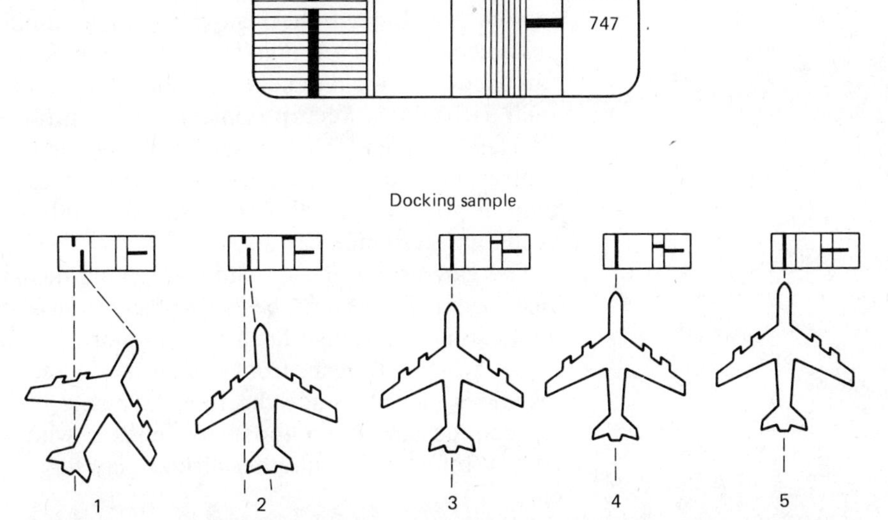

Figure 18-3. The BOLD aircraft docking system. (From Ashford & Wright, 1979.)

line, the aircraft is properly located on the docking centerline. The pilot then switches attention to the right horizontal markers. When they form a straight line, the plane is at the correct distance from the display. The BOLD display is a good human factors example since it separates information on horizontal and vertical axes. It is easier for a pilot to first adjust location on the centerline, by steering the plane right or left, and then to adjust distance. A two-dimensional display would require more attention (see Chapter 6) and be more difficult to use since the position adjustments are made sequentially rather than simultaneously; that is, the pilot cannot adjust distance until the plane is on the centerline.

Passenger Terminal Design Unless you are a seasoned air traveler, you may not be aware of the variety of terminal designs currently in use in the United States and around the world. There are three "pure" design arrangements although many hybrid combinations are possible (de Neufville, 1976). As can be seen in Figure 18-4 there are three basic types.

- Centralized with finger piers or satellites.
- Linear or gate arrival.
- Transporter.

Each type has distinctive advantages and disadvantages. System goals determine which type to use. For example, one system goal might be minimum turnaround time for aircraft. Another might be maximum convenience for passengers but this would have to be broken into clearer subgoals such as minimizing walking distances, having attractive restaurant, shopping, and other customer facilities, having rapid baggage service, etc. Indeed, these subgoals might conflict.

The centralized terminal has a common central hall where passengers can check in, obtain tickets, find shops and restaurants and other services such as banking. Security officers like centralized terminals since it decreases the need for many security checkpoints. On the other hand, the central hall can be a confusing place for passengers since so much activity is going on. Furthermore, this design has the longest distance from check-in to airplane gate.

The gate-arrival design eliminates long distances between planes and passengers. However, the lack of centralization means there is much duplication of terminal facilities and more operating staff to pay. Passengers who must transfer from one airline to another, can have even longer distances to walk since the total layout is not as compact as the centralized design. Even those passengers who are not transferring but merely returning on another airline may have a very long walk to their car.

The transporter system minimizes the need for terminal structures by having vehicles, called mobile lounges although there is no room to lounge or transporters, convey passengers from a central location di-

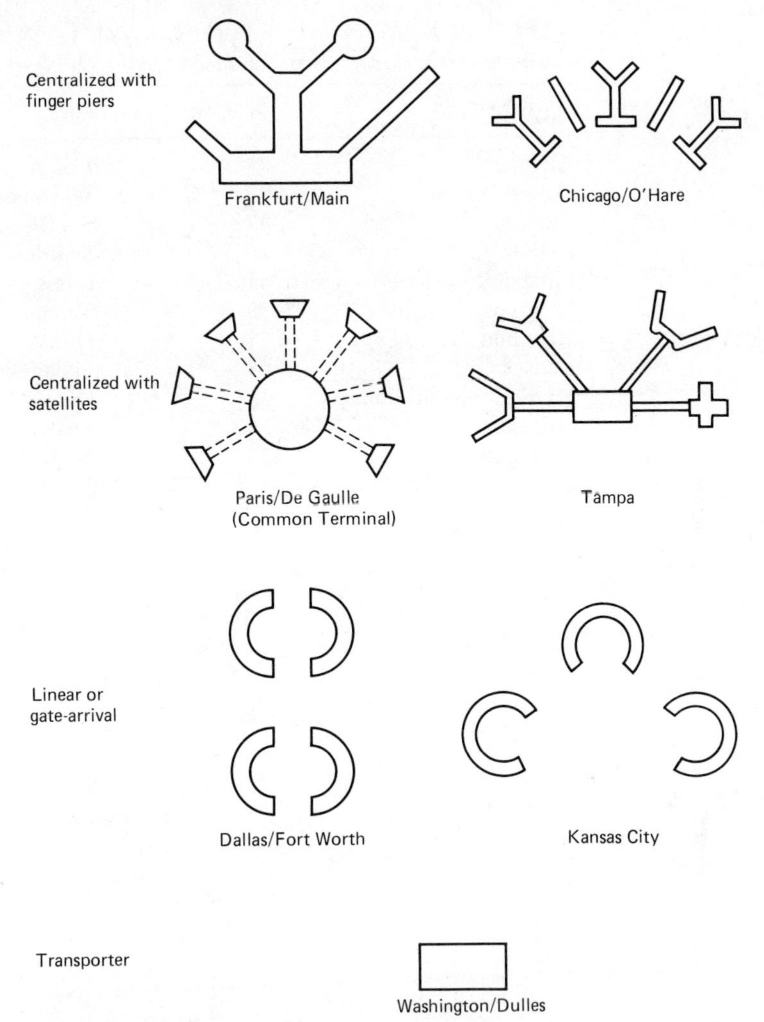

Figure 18-4. Basic design configurations for airport terminals. (From DeNeufville, 1976.)

rectly to the aircraft. The transporter substitutes vehicles for buildings. It can be expensive to operate since vehicles require operators and maintenance, as well as replacement, but on the other hand vehicles can be parked when not needed. This makes the transporter potentially very useful for dealing with peak traffic loads.

How might the designer choose among these types? The best answer is to consider the needs of the passengers using the terminal. In particular, we must determine how many passengers use the terminal to transfer from one plane to another and how many use it as an origin or destination. Some airports function primarily in one of these two modes.

TABLE 18-1 *Percentage of Passengers Transferring between Flights (domestic interstate transfer 1974; from DeNeufville, 1976)*

City	Percent	City	Percent
Atlanta	73	Los Angeles	22
Dallas/Fort Worth	55	Washington, D.C.	22
Chicago	47	San Francisco	21
Denver	46	Seattle	20
Pittsburgh	41	Las Vegas	18
St. Louis	38	Tampa	18
Honolulu	31	Houston	17
Kansas City	28	Philadelphia	15
Minneapolis–St. Paul	26	Detroit	15
Cleveland	25	Boston	10
New Orleans	24	New York	6

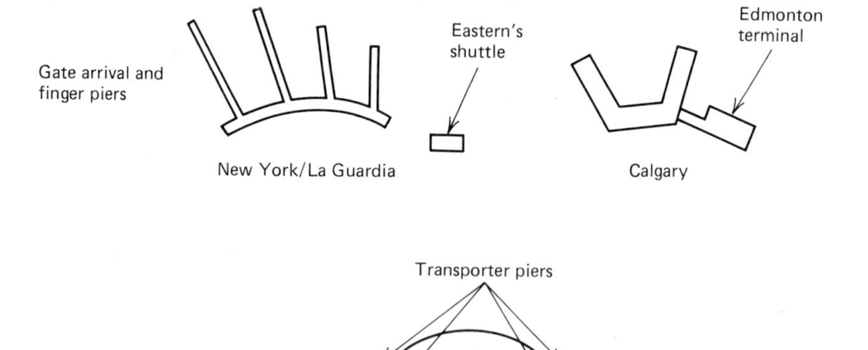

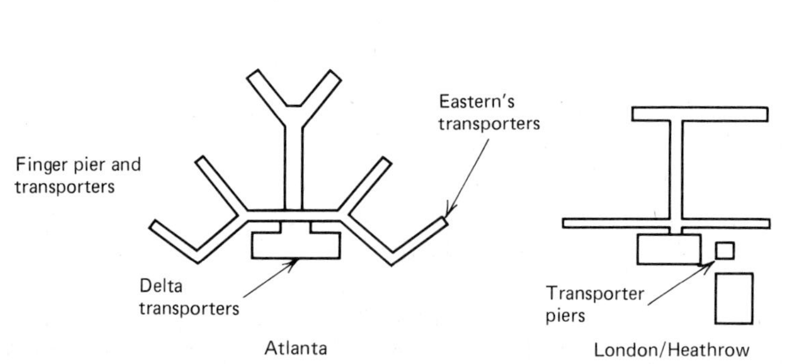

Figure 18-5. Some examples of hybrid terminal design. (From DeNeufville, 1976.)

For example, New York passengers almost always end or start their flight and do not transfer. But Atlanta is such a hub for plane travel to southern states that passengers have a saying about it: "You can't even get to Heaven without changing in Atlanta!" Table 18-1 shows the percent of transfer passengers for several important U.S. airports. A gate-arrival design is best for airports with low transfers. Centralized designs are better for airports with many transfers.

In practice hybrid combinations often work better than any single pure design. Transporters can be used in combination with fixed gates. A rough rule of thumb assigned transporters to one-third of aircraft in hybrid designs (de Neufville, 1976, p. 118). This allows good utilization of fixed gates but still permits efficient operation during peak periods. Figure 18-5 shows some hybrid terminal designs currently in use.

One of the most important factors that determines how many gates are needed is the time required for a plane to unload passengers and baggage, be serviced, load, and depart. If this time can be decreased, then fewer gates will be required. Figure 18-6 shows the activities that must be completed during a 30 minute turnaround time. The longest is fueling, which requires 23 minutes. However, decreasing fueling time would not allow the plane to leave the gate any sooner since refueling is not on the critical path. Cabin service, requiring 16 minutes, is the longest task on

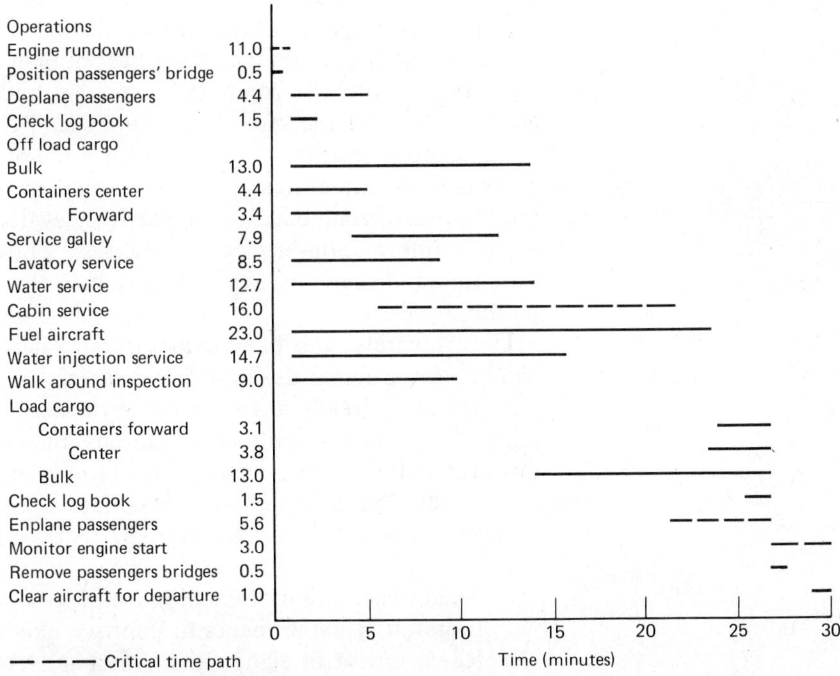

Figure 18-6. Aircraft gate service times showing critical time path. (From Morlok, 1978 and Ralph M. Parsons Co., 1973.)

the critical path. Therefore, to decrease turnaround time, the human factors specialist would have to find a way, perhaps by improving procedures, equipment or adding personnel, to speed up cabin service.

Getting to the Airport Terminal

Major airports are seldom located in the heart of the business district. Airports require too much land, especially if proper buffer zones are maintained around the airport, to make a convenient downtown location feasible. (Sometimes there are suggestions to create convenient new land as in the proposed new airport for Chicago to be built on Lake Michigan.) So the air passenger faces the serious problem of getting to the airport. This trip can sometimes take longer than the actual flying time in the air.

Large numbers of passengers arrive at the airport via automobile. Trying to find the right terminal and parking at a major airport can be a bewildering experience. In the midst of heavy traffic the driver must locate and quickly interpret road signs that contain the information needed to find the correct terminal. Passengers transferring between flights must be able to get from their arrival terminal to their departure terminal. As we will see, human factors experts have been able to apply their techniques to help passengers find the correct airport terminal.

The major human factors efforts in this area have concentrated upon improving guide signs. A sign is, after all, another type of visual display, and the general principles of display design (see Chapter 7) are relevant. As is almost always the case, the best human factors suggestions can be generated before the roadway system is built. Dunn (1973) made an animated film of the most important visual aspects of vehicular travel along a proposed roadway system for a major airport. This film showed the roadway guide signs a driver would encounter while approaching the airport. Certain parts of the roadway system severely overloaded the driver's information-processing abilities. Thus, as a result of this film, the original design was revised to reduce the information load imposed on the driver.

Unfortunately, the human factors specialist is not always a participant in the original design effort and instead can be called in to fix a problem that already has been cast in concrete (Lewis, 1979). But even under such less than optimal conditions, the specialist can often improve the situation. One such successful example improved the complex roadway system that drivers had to negotiate to reach Toronto International Airport (Dewar, Ells, & Cooper, 1977). The study had three phases.

1. Field observation of existing traffic flow and an opinion survey.
2. Laboratory experiments to improve existing signs.
3. Replacement of eight signs with a subsequent evaluation similar to the first phase.

The field observation and interviews determined which signs caused

Figure 18-7. Original (left) and improved (right) traffic signs. (Courtesy of Robert Dewar.)

difficulties for drivers. Then laboratory studies used reaction time (see Chapter 5) as a dependent variable with participants viewing slides of the highway and signs and having to select the lane that would get them to their desired destination. Additional laboratory studies presented the slides for a brief time (700 msec) and used accuracy of responding as the dependent variable. Revised signs were designed and the laboratory studies were repeated. For some (but not all) of the new signs, performance improved. Finally, improved signs were placed at the airport approaches and then evaluated (see Figure 18-7). The new simplified messages with larger arrows were effective in improving traffic flow in one location but not in another. This second location suffered from complex roadway geometry so that the new signs could not compensate for the poor original roadway design. Retrofit analyses can correct some existing errors but seldom are as good as incorporating human factors analyses into the original design process.

Passengers who are changing planes from one airline to another can have many of the same difficulties in locating the correct terminal as do automobile drivers. The Dallas-Fort Worth International Airport uses an automated ground transportation system to connect four terminals and two parking locations, with more planned and/or under construction. The 51 electric vehicles are divided into five routes, each identified by a separate color. The passenger thus has a good chance of getting on the wrong vehicle, especially if their plane connection is tight causing them to rush. Information about which vehicle to take is divided among five sources. While the passenger is consulting these at least one vehicle will arrive, and some passengers will be tempted to enter it even though it might be going to the wrong destination. The information format is confusing since for some airlines it tells where to get off but not which vehicle to enter; for others it tells which vehicle but not where to get off; for major airlines it provides two route colors and instructions to use a nearby TV display to find which station is closest to the departing flight (Bateman, 1979). The operational sequence diagram (see Chapter 1) for the system is shown in Figure 18-8. There are four shaded elements indicating opportunity for error. An improved scheme is shown in Figure 18-9. The scheme consolidates static information, eliminates one display, and improves information format. The operational sequence diagram has only two shaded elements instead of the four contained in the original system. However, the system is still complex. The optimal system (Figure 18-10) uses a computer to store information about flights and is used interactively (see Chapter 13). It has the simplest operational sequence diagram and can be expected to minimize user error.

Highway Location

The bulk of human factors research on highways and railroads has dealt with visual displays like traffic signs (see Chapter 7) and vehicle design (see Chapter 15). Here we consider the molar aspects of the de-

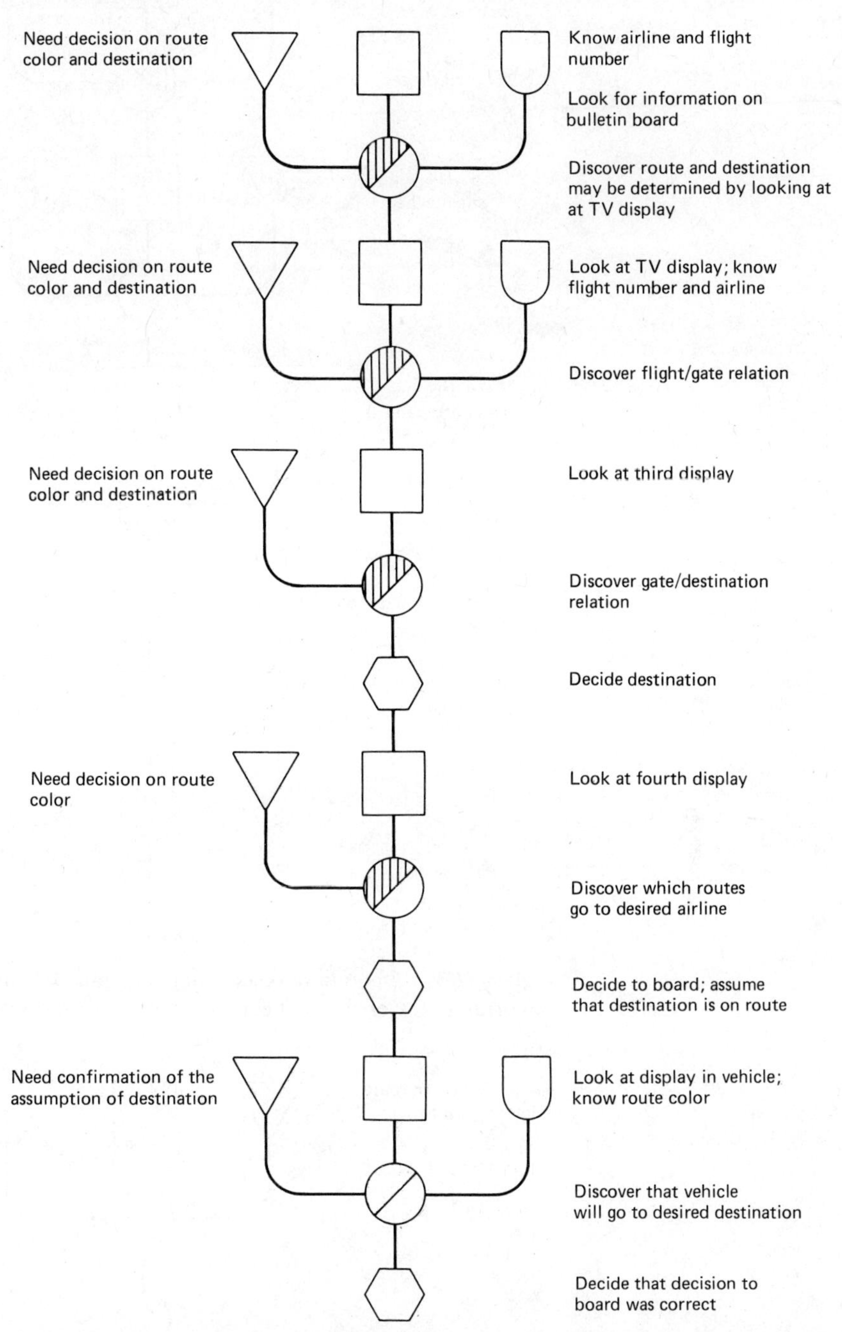

Figure 18-8. Operational sequence diagram for the original system. Shaded circles indicate possibility of errors. Note that five displays are required to obtain two pieces of information. (From Bateman, 1979.)

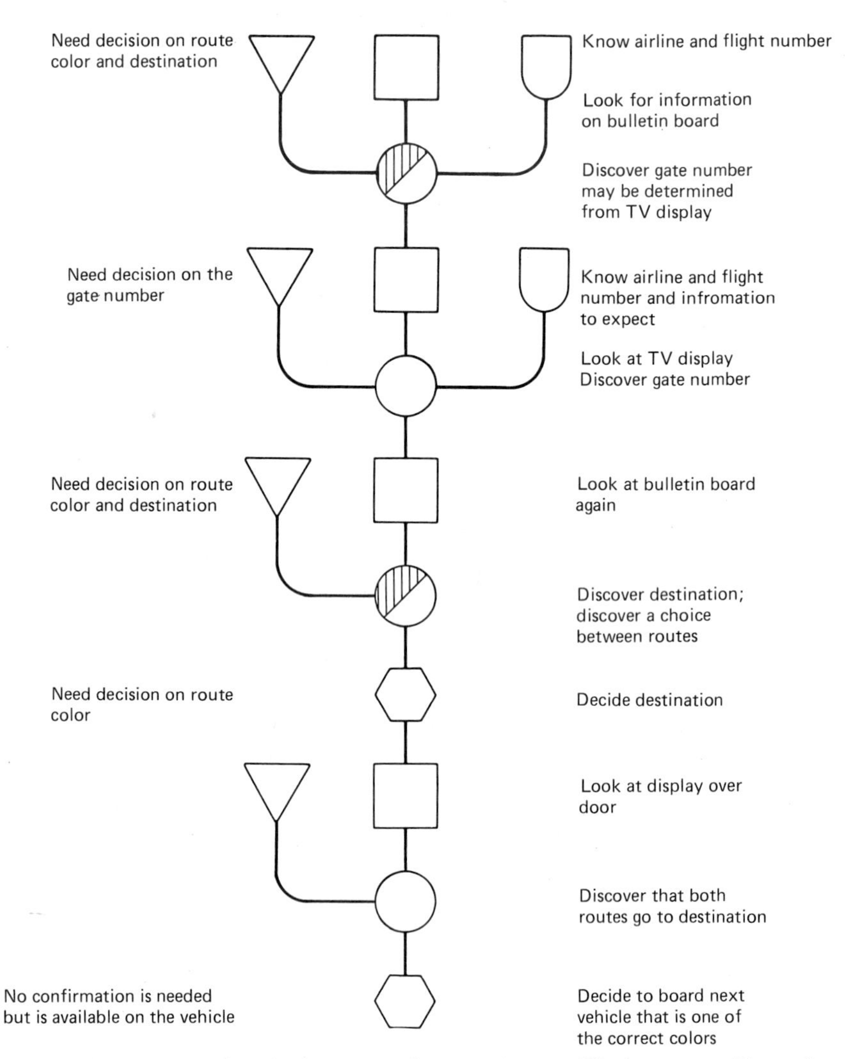

Figure 18-9. Operational sequence diagram for modified system. Note that opportunity for error has been decreased. (From Bateman, 1979.)

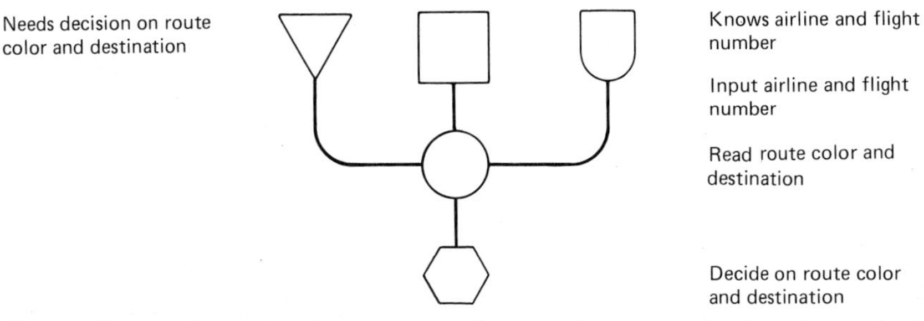

Figure 18-10. Operational sequence diagram for computer-based terminal. Note how much simpler this diagram is when compared to the preceding two figures. (From Bateman, 1979.)

cision to build a highway or railroad, or indeed any other fixed-location transportation facility such as a pipeline or canal, in some particular location. There are five major criteria to be considered when alternative locations are being evaluated (Morlok, 1978).

- Construction costs.
- User costs.
- Impacts on natural environments.
- Impacts on human activities.
- Acceptance by relevant interest groups.

Impacts on human activities comprise the area of greatest human factors application. This category includes creation of boundaries within neighborhoods, consistency with established residential and business activities, aesthetics, noise control, and impact on historical structures or archaeological sites. Acceptance by interest groups is a political process involving governmental agencies at local, state, and federal levels as well as private firms with strong local interests at stake.

An example of a railroad relocation project is shown in Figure 18-11. There were four alternatives: do nothing plus the three shown in the map. The community had several railroad lines running through the middle of the central business district. This incurred social and economic costs including delays to 13,000 motorists per day adding up to 314,000 hours per year of lost time. Ambulances experienced 985 delays per year. An average of 16 auto/train accidents occurred each year with a total cost of almost $1 million. Residents also objected to noise, reduction of property values near the railroad tracks, and loss of taxes due to depressed values. It was clear that the do-nothing alternative could not be tolerated. In the period 1969 to 1981 several relocation studies were conducted and there was much community discussion. Finally, the railroads and the community agreed that the proposed corridor along the riverfront would best eliminate the railroad-highway conflict, improve railroad operations and decrease their cost, and would minimize the number of residents displaced by relocation.

This analysis was performed without human factors specialists. A great deal of time and effort was expended to find the best route without formalizing the basic problem of comparing apples and oranges: for example, what is the trade-off between hours of delay for motorists and cost of new routes. While the community agreed on the need for railroad relocation, there was no objective method for combining the costs and benefits of alternative routes into some understandable metric. A solution was eventually reached by political means and consensus. However, we believe that some of the decision-making techniques (see Chapter 14) used by human factors specialists could have helped the community reach agreement more efficiently. This is, of course, speculation since these techniques were not tried. But it is important to suggest new areas

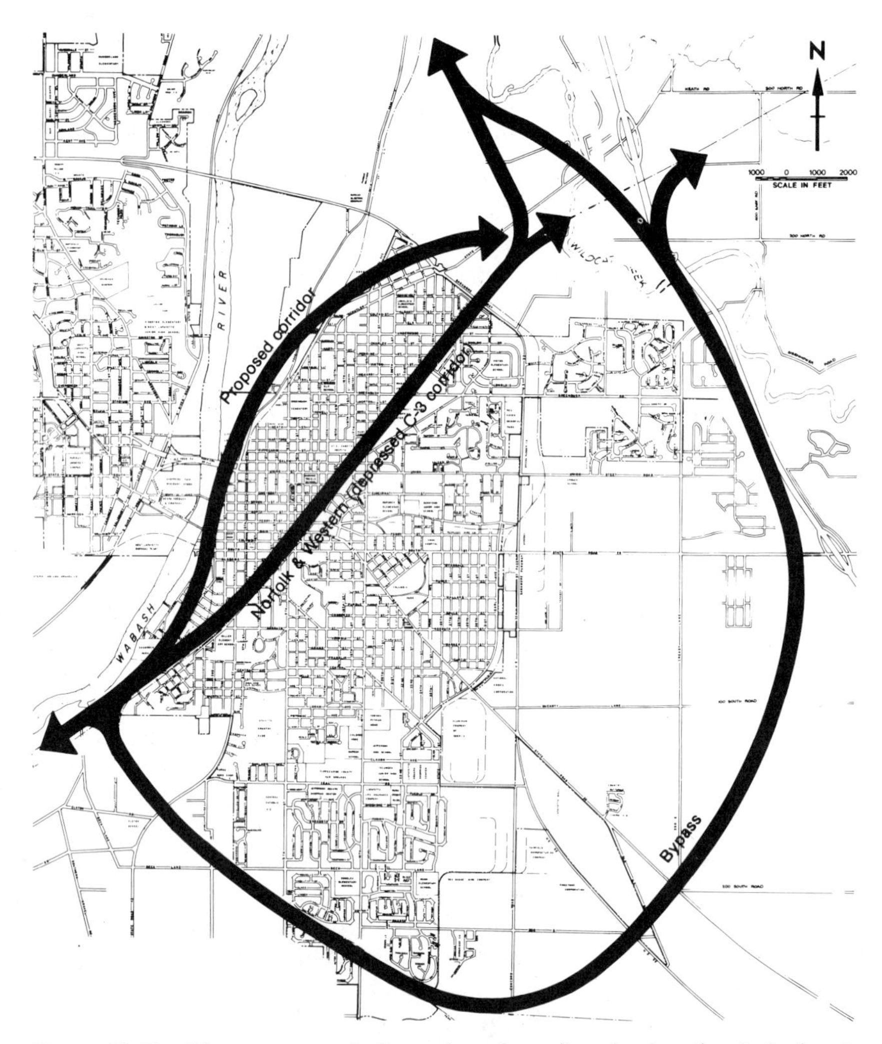

Figure 18-11. Three proposed alternatives for railroad relocation in Lafayette Indiana. (Courtesy Lafayette Railroad Relocation Commission, 1981.)

where it seems worthwhile to attempt human factors analyses. We suspect that macroenvironments offers many new problems that can benefit from human factors.

Mass Transportation

The problems of moving large numbers of people increase as these numbers get larger. Thus, the most critical transit problems are in large urban areas where millions of people must travel to and from work at roughly the same times. It is clear that the automobile is the vehicle of choice for most Americans. We can list some of its advantages and disadvantages (Weigelt, Gotz, & Weiss, 1977).

Advantages	Disadvantages
Door-to-door transportation.	Extremely sensitive to any impediments in traffic flow.
Always available.	Pollutes environment.
High operating speed since no intermediate stops need be made.	Requires parking space.
Pride of ownership.	Driver can't relax or read.
Convenience relative to public transport.	High initial cost and operating costs.

It is clear that Americans are willing to suffer the disadvantages and will prefer the automobile to mass transit, even though the risk of accident is much higher for private vehicles. However, local government policy can have a strong effect on this preference by either making mass transit more attractive or conversely by making automobile use less attractive. Both of these approaches have been tried. For example, it is virtually impossible to park a car on public streets in Manhattan and the city has a vigorous policy of towing away illegally parked vehicles. There is also a shortage of parking garages so that the cost of parking is high. A substantial portion of Manhattan would have to be rebuilt as parking lots in order to accommodate the potential demand for space. This discourages citizens from using automobiles, as does the congestion during rush hours. Building more highways and parking garages only increases automobile traffic without solving congestion problems since the private-car transportation system acts as a positive feedback loop (see Chapter 12). Furthermore, some citizens cannot use cars because they are too young or too old to drive or too poor to afford them. So mass transit is absolutely necessary in urban environments.

Mass transit works best when there is a high density of people going to the same place. The central business district (CBD) of a large city meets this requirement. The designer must provide a system that has low transit times—hopefully faster than private automobiles during rush hour—yet is cost effective and reliable. This is a difficult task and many solutions have been tried. Express buses using reserved lanes for buses only have worked in some cities. Various kinds of tracked vehicles—subways, trolleys, monorails—are also common. New technologies for the design of small-cabin systems called Personal Rapid Transit (Weigelt et al., 1977) are being tested throughout the world. In this section we discuss operations plans for rapid transit and a new development called paratransit that combines features of traditional public transport with many of the conveniences of private transport.

Rapid Transit Human factors analysis does not end once the equipment design is completed. In order to optimize human-machine systems, operations plans that take advantage of system capabilities are also needed. Different managements operating virtually the same equipment experience widely different results depending on operating rules, maintenance procedures, and other prerogatives selected by management. One of the most important management decisions concerns how to match system capacity and user needs. (This, of course, is a very general problem that applies to industries outside the transportation area. For example, electric utilities cannot afford to build enough standard generating plants to handle peak loads and must resort to special "peaking" turbines, pumped-water storage systems, and as a last resort, planned brownouts to economically match system capacity and user needs.) In a rapid transit line, there is heavy demand in the CBD but traffic volume decreases in the outer zones of service. Figure 18-12 shows that having trains stop at all stations results in much wasted capacity outside the CBD. In fact, if traffic volume decreases uniformly as illustrated in Figure 18-12, the average capacity is twice the demand. The load factor *f*, which equals passenger-miles traveled divided by set-miles provided, equals 0.5. The skip-stop service shown in Figure 18-12*b* uses two trains with each stopping at half the stations. While this provides faster service, the load factor remains unchanged since full capacity is still provided for the entire length of the line. Going to an express-local system (Figure 18-12*c*) gives an even better speed improvement but does not alter load factor. Furthermore, this requires a double-track line so that expresses are not "stuck" behind locals. This extra cost is partially balanced by the need for fewer express trains, with savings in operating costs. (See preceding discussion of construction versus other costs in this chapter.) In order to match capacity and need better, a zonal operations plan is used (Figure 18-12*d*). Here service is concentrated in the CBD. Only certain trains travel the entire line, and indeed, if passengers are forced to change trains at zone boundaries even this requirement can be dropped with yet additional savings in capacity. This kind of zone operation reduces the number of cars required while saving time for passengers traveling to the CBD from the outer zone. However, other passengers will experience greater travel times due to increased waiting times for trains. The zone operation plan allows the frequency of trains to be matched with zones, rather than being the same across the entire line. The point is that the same equipment will produce different results according to how it is operated.

Paratransit Another example of using standard equipment to provide alternative forms of service is paratransit. These include Dial-A-Ride services, jitneys, and minibuses. While paratransit was at first limited to

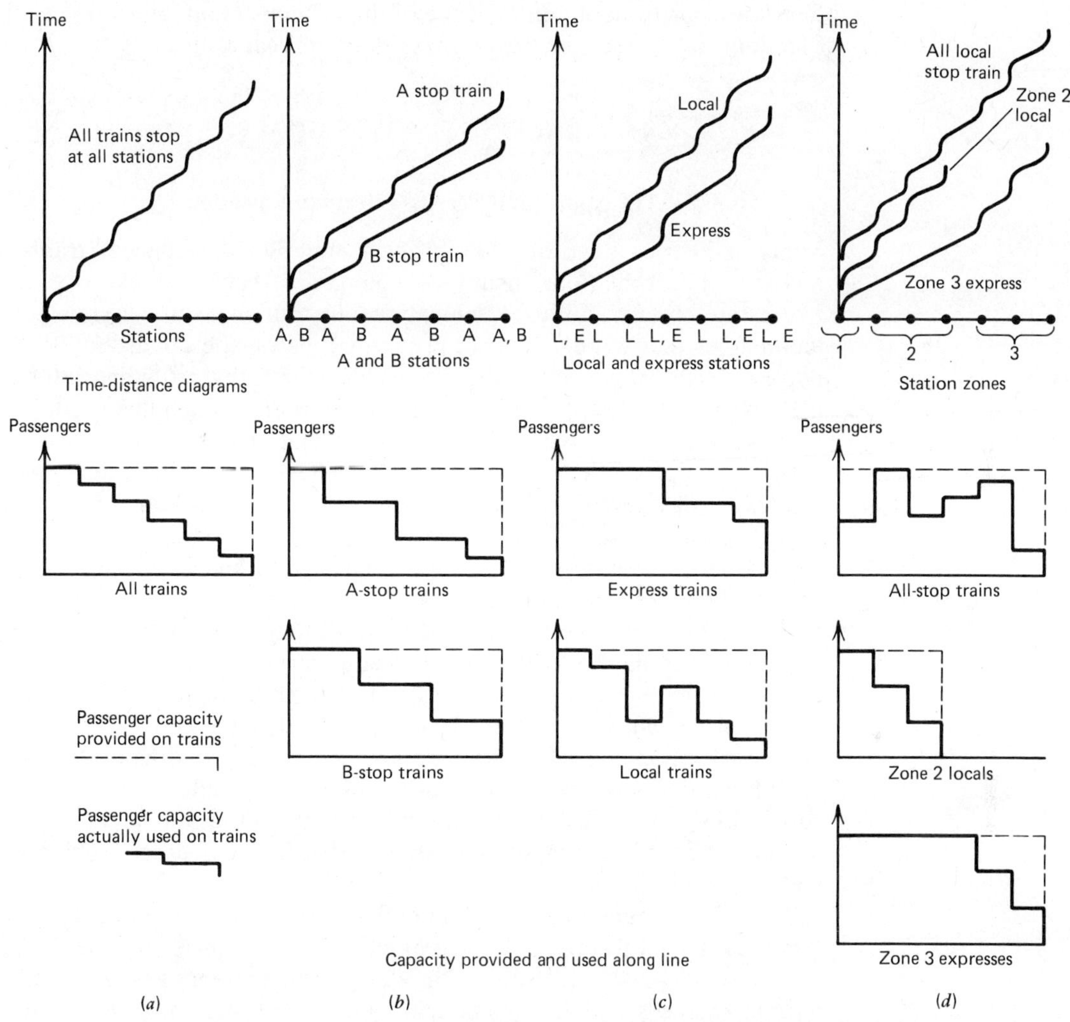

Figure 18-12. Four operations plans for a rapid transit line. (*a*) All stops. (*b*) Skip stops. (*c*) Local and express trains. (*d*) Zone express and locals. (From Morlok, 1978.)

special populations like the elderly and the handicapped the concept has been extended to the general population making it even more cost effective (Teal & Giuliano, 1980). There are many operations plans that can be used for paratransit. For example, jitneys or minibuses can run on set routes but not depart until full or until *n* minutes have elapsed, whichever comes first. Similarly, Dial-A-Ride systems can take reservations or can inform prospective passengers of the delay until a vehicle will arrive. Since paratransit systems use small vehicles, it seems that they do not benefit from economies of scale (Pagano, McKnight, Johnson, & Robins, 1980) so it is not yet clear how these systems will expand. Paratransit offers the door-to-door convenience of the private automobile at the cost of having each passenger travel a less than optimal route.

ACHIEVING SYSTEM GOALS

Transportation versus Telecommunication

Many industries have as their primary function the efficient transmission of information. Insurance companies, banks, stockbrokers, real estate brokers, libraries, educational institutions, software houses (companies that write computer programs), newspaper, magazine and book publishers, and clerical services all operate by providing information to customers. Even businesses that produce a tangible product must keep track of inventories, order raw material and communicate with customers. The vast number of business letters and telegrams sent every day reveals that no business can function without a reliable flow of information.

This need for information, both within a firm and between a firm and its clients and suppliers, is a major cause of transportation demand. If people did not have to be physically present at the office, if salespeople did not have to physically travel to service customers, society could eliminate many of the transportation problems and costs discussed earlier in this chapter. Rush hours and traffic congestion, the need for parking spaces and automobiles, would be drastically reduced. This box raises two questions for you to ponder. First, is the substitution of telecommunication for transportation technically feasible? Second, if it is feasible what kind of human factors problems will have to be solved?

The most commonplace telecommunication device is the telephone. Americans tend to take this for granted and most households have at least one telephone. In Europe, however, while telephones are common in business establishments, many households do not have private telephones. Thus, the concepts discussed here may not be practical in Europe since we assume that telephones are plentiful in homes in order to disburse work from the office to the home. The ideal situation in terms of minimizing transportation demand is to have work performed at home. (Of course, demand can be substantially decreased by decentralization with satellite workstations located outside the CBD closer to homes (Nilles, Carlson, Gray, & Hanneman, 1976) but we will consider only the extreme case of working at home.) Telephone lines are not limited to voice communication. Standard interfaces (called modems) allow digital data to be transmitted as well. Even graphics like blueprints can be transmitted over ordinary telephone lines. Computer terminals and even small but powerful microcomputers are available now at modest to reasonable costs ($1K to

$5K) for home and office use. The technology for working at home is here now at practical prices. Widespread use of terminals at home would lower prices and increase feasibility.

To see how the transportation-telecommunication trade-off might work, let us select a common information task required in virtually every business: typing letters. At present, an executive dictates a letter either to a tape recorder or to a secretary directly. Many organizations have typing pools to increase efficiency. All this happens, of course, at the office. Now we substitute a new service, the Secrepool system (Nilles et al., 1976), to perform this same function at home. The information flow diagram of the Secrepool system is shown in Figure 18-13. As before, the letter is dictated into a tape recorder. Then the executive sends the vocal recording over the telephone to the Secrepool distribution center where the message is given a job code and sent to the typist. The letter is typed on a computer terminal or an electronic typewriter and transmitted, in digital form using a modem, back to the distribution center. When the originator of the

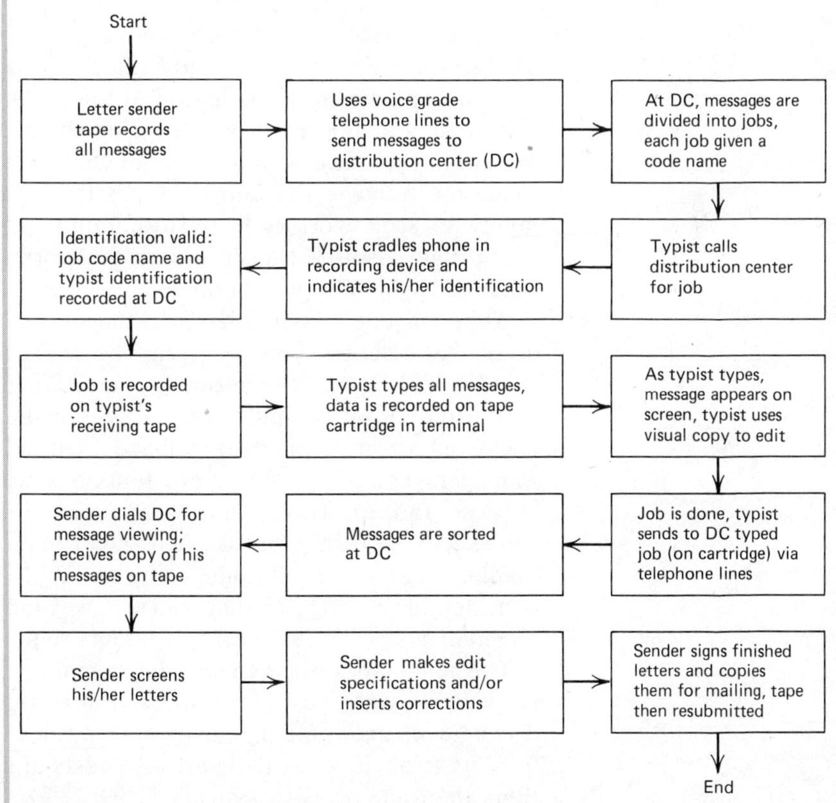

Figure 18-13. Flow diagram for the Secrepool clerical system. (From Nilles et al., 1978.)

letter or document is ready, they telephone the distribution center, which sends it for review on the originator's terminal screen. The turnaround time is selected by the sender who must pay a higher price for faster service. The sender corrects or makes changes and then signs the document using a graphic tablet. (Or, each organization using Secrepool might prefer having only a few tablets with signatures kept on file at the distribution center. In this case the sender would enter a private code that would affix the signature.)

What kinds of problems might such a diffuse system encounter? There are human factors issues associated with data entry and terminals but these have already been discussed in Chapter 11. There are also motivational and social problems. Some executives might fear a loss of control over the secretaries as well as a loss of the other functions some secretaries perform in addition to typing. Some secretaries may enjoy commuting and "getting out of the house." For many people the workplace is a support structure that provides friendship and an important social network. Some secretaries might lack the self-discipline required to turn on the terminal and report for work. But since the terminals can easily monitor productivity, such problems are readily discovered. Unions may object to having the productivity of their members so easily quantified. Goofing off at work may be more acceptable than goofing off at home. On the other hand, fewer distractions at home—it is often difficult to politely get rid of a co-worker in the office who drops in to chat—may increase productivity. Since the trend in the United States for both adults in a two-parent family to seek work is likely to continue, the option of working at home may ease the transition into the work force for many parents, thus increasing the pool of qualified workers.

Although our example has focused on clerical tasks, it should be clear that telecommunication can be substituted for transportation through all levels of the business organization. Many companies right now use internal computer networks to disburse memos, documents, and even to have executives make their own travel arrangements. Managers could perform these functions using computer terminals at home. Indeed, with modern word-processing packages (see Chapter 11) executives can even do their own typing. The text you are now reading was written at home using an Apple II microcomputer with considerable savings in time since it was calculated that work completed at home in 3 uninterrupted hours required 5 hours at the office.

The technology to exchange telecommunication for transportation is available right now. Implementation of this technology will cause dramatic changes in organizational and social life. There are clear benefits to society but there are also costs of new technology. Do you think the trade-off is worthwhile?

CITIES

For some people, cities represent the very best of modern civilization. Here the high concentration of people permits and encourages commerce and culture to flourish. A wealth of goods and services are available in the city. People, especially young people, flock to the cities to advance their careers and to participate in the excitement of city life (Figure 18-14). Professional sports, music, and the other lively arts reach a height in the city that cannot be matched in rural areas.

For other people, cities represent the worst of modern civilization. Here the high concentration of people permits and encourages crime, filth, pollution, poverty, slums, and a callous indifference to the welfare of one's fellow citizens. Nowhere else in our society is the line between haves and have-nots so clearly drawn. Vast areas in our cities look like battlegrounds littered with the remains of pitted and scarred burnt-out buildings no longer safe or suitable for human habitation (Figure 18-15). Those who have the income to afford to it have fled to the relative safety of the suburbs where they must spend 2 or more hours per day commuting to and from work. The heart of the city is deserted after 5 P.M. while the departing workers fume and curse in rush-hour traffic jams.

Regardless of which of these descriptions fits your own image of the city, it is clear that cities are here to stay. Millions of people will be city

Figure 18-14. New York's Lincoln Center with its four major theaters illustrates the cultural diversity a large city can offer. (Lincoln Center for the Performing Arts. Photo by Bob Serating.)

Figure 18-15. New York's South Bronx illustrates the devastation a large city can offer. (Owen Franken/Stock, Boston.)

residents during your lifetime. You may be one of them. What can human factors do to minimize the harmful aspects of city life while maintaining and perhaps even improving the benefits? This section discusses three aspects of city life—defensible space, crime prevention, and information overload—in order to answer this question.

Defensible Space

As the preceding chapter showed, your behavior is influenced by the kinds of rooms you live and work in. It is a small step from this to realize that behavior is also influenced by the size, shape, configuration and other architectural features of the assembly of rooms we call a building. This section focuses on high-rise residential buildings where the concept of defensible space was conceived by Oscar Newman, an urban housing specialist, in a landmark study of public housing projects in New York City (Newman, 1972). However, this concept applies equally well to more expensive private high-rise apartments (Gates, 1977).

In its simplest form, the concept of defensible space refers to the creation of a secure residential environment by permitting physical building features to encourage residents to take an active interest in the behaviors of people, both residents and strangers, in and around the building. Thus, conditions that permit good surveillance are one essential aspect of defensible space. Although this may sound simple at first, guidelines for establishing defensible space are complex and proceed in a hierarchy that extends from the street to the apartment (Newman, 1975). Therefore, to make the concept easier for you to follow, we start with an example of a housing project that lacks defensible space. Then we give some design principles for establishing defensible space. Finally, we compare two housing projects that are similar except for defensible space.

Space in and around high-rise apartments can be divided into three categories. Public space, such as streets and parking lots, is available to residents and nonresidents alike. Semipublic space, such as the building lobby and elevators, is open to residents and visitors. Private space, such as individual apartments, is restricted to residents and their guests. In a typical high-rise apartment the resident can control only the private space. When a building has a dozen or more stories with from 5 to 10 apartments on each floor, it is almost impossible for a resident to know and recognize all the other residents. Therefore, when a stranger appears in the semipublic space, there is no simple way to tell if the stranger is a resident from another floor or is a potential burglar. Even worse, the physical design of most high-rise apartments prevents any kind of visual surveillance once a stranger leaves the lobby. Stairways are often lacking windows and are soon taken over by teen-age gangs as "private" spaces for indulging in drugs and sex. Any resident who uses the stairs, especially at night, risks an attack. Elevators too are convenient places for muggers and rapists to await victims. Even a person trapped alone in a stuck elevator may have to wait an hour or more for rescue since the residents typically ignore alarm bells so that the victim need await the normal rounds of the housing police to be discovered. Residents may not even be aware of crimes committed on their own floor, since it is usually unwise to monitor activity in the hallways by keeping one's door open. Each resident is an island unto himself or herself.

By following guidelines for defensible space, the semipublic and public spaces can be controlled by the building residents. Newman lists four design elements that promote secure environments.

- Subdivide the residential environment into small zones so that adjacent residents can exercise territoriality.
- Locate apartment windows so that residents can maintain surveillance of the approaches to the building as well as building interior.
- Adopt building forms that prevent outsiders from perceiving the vulnerability and isolation of the inhabitants.
- Locate residential developments adjacent to activities that do not provide continued threat. (Don't have bars and massage parlors across the street.)

These are general design principles. Here are some additional more specific examples of design that either improves or diminishes defensible space:

Improves	Diminishes
Illuminate exterior areas.	Sliding glass door or vulnerable windows in isolated areas.
Stairwells open to public view.	Blind corners in public pathways.
Number apartments in logical order so police can respond quickly.	Elevator stop buttons that do not sound alarm.

Two Comparable Housing Projects Newman was able to document the effects of defensible space by comparing two public housing projects that were across the street from each other. Both have residents with similar economic and social backgrounds, both house about 6000 people, and both have a density of 288 persons per acre. However, the Van Dyke (Figure 18-16) houses are mostly 14-story buildings with empty lawns between each building. The Brownsville project (Figure 18-17) consists of low-rise buildings (3 to 6 stories) with less empty space. The Van Dyke buildings all have entrances that are inside the project whereas the Brownsville entrances tend to be located directly off public streets. Thus, observation and surveillance is much easier in the Brownsville project.

Figure 18-16. The Van Dyke Housing project. Note that the building entrances are far from the street making surveillance difficult by police and pedestrians. (N.Y.C. Housing Authority.)

The Van Dyke corridors serve eight families and are public spaces where it is unsafe for children to play. The Brownsville interiors are grouped into two sets of three families, each with a common vestibule area; children are allowed to play in the interior halls. Residents in Brownsville take more responsibility for their territory whereas those in Van Dyke will not go out of their way to cooperate with police officers. These characteristic differences result in a much greater crime rate in Van Dyke: 264% more robberies, 60% more felonies and misdemeanors, and 72% more repair maintenance. It is quite clear that defensible space reduces crime and vandalism. But until guidelines that promote defensible space are written into building codes, this knowledge will go unused (Gates, 1977).

At present, the concept of defensible space is fuzzy and poorly defined. There is no metric that will tell us precisely how much more defensible space is present in Building A versus Building B. Designers who wish to provide defensible space must be guided by intuition. We see this vacuum as an opportunity for the human factors specialist. At one time control panels were developed by guess and by gosh—that is, the designer used intuition to determine control placement. Now there are mathematical metrics that allow us to evaluate the "goodness" of the design (see Chapter 10). We look forward to similar advances in measuring defensible space.

Figure 18-17. The Brownsville housing project. These low-rise buildings have defined areas near them that promote defensible space. (N.Y.C. Housing Authority.)

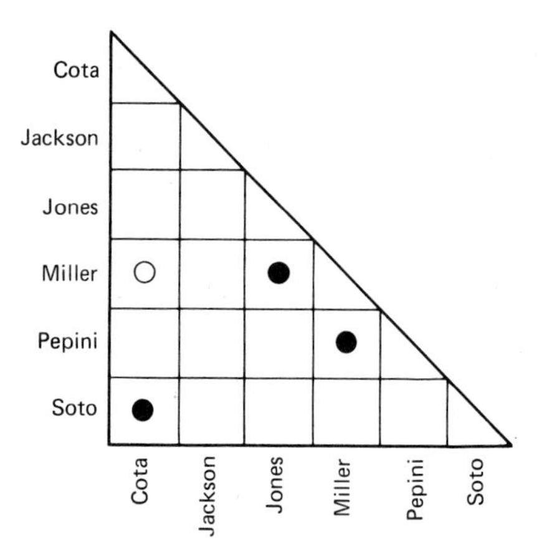

Figure 18-18. A simple association matrix. Solid circles represent strong links and open circles weak links. (From Harper & Harris, 1975.)

Crime Prevention

Human factors specialists have started to fight crime with studies that suggested improvements for police patrol vehicles (Clark & Ludwig, 1970) and evaluated the relationship between street lighting and burglaries (Krauss, 1977). In this section instead of reviewing several studies, we discuss only one in detail because it illustrates an innovative use of an important human factors technique. We will see how link analysis, discussed in Chapter 7 with regard to visual displays, was used to combat crime.

The most important component of the police intelligence process where criminal activity is uncovered is data analysis (Harper & Harris, 1975). Until link analysis was suggested, police had no systematic method for combining bits and pieces of isolated information into a coherent picture. Instead, this vital process was dependent on the intuitive abilities of the police investigator. Link analysis can reveal the interlocking structure in a combination of legal and illegitimate organizations as well as hierarchical features of criminal organizations.

A link analysis of criminal activity starts by classifying relationships among individuals into one of three categories: strong link, weak link, and no link. An example of a strong link would be a father and son who are often seen together. An example of a weak link would be frequent telephone calls by an individual to a restaurant operated by a suspected criminal. This link information is assembled into an association matrix (Figure 18-18). The association matrix is then used to create a link diagram (Figure 18-19). Individuals are represented by circles in the link attempts or iterations to produce a neat link diagram that represents the information in the association matrix. The link diagram offers two major advantages for law enforcement personnel. First, large amounts of data can be summarized and integrated in a format that is easy to understand. Second, new hypotheses about individuals and organizations can be developed from the link diagram.

Successful applications of link analysis have been summarized by

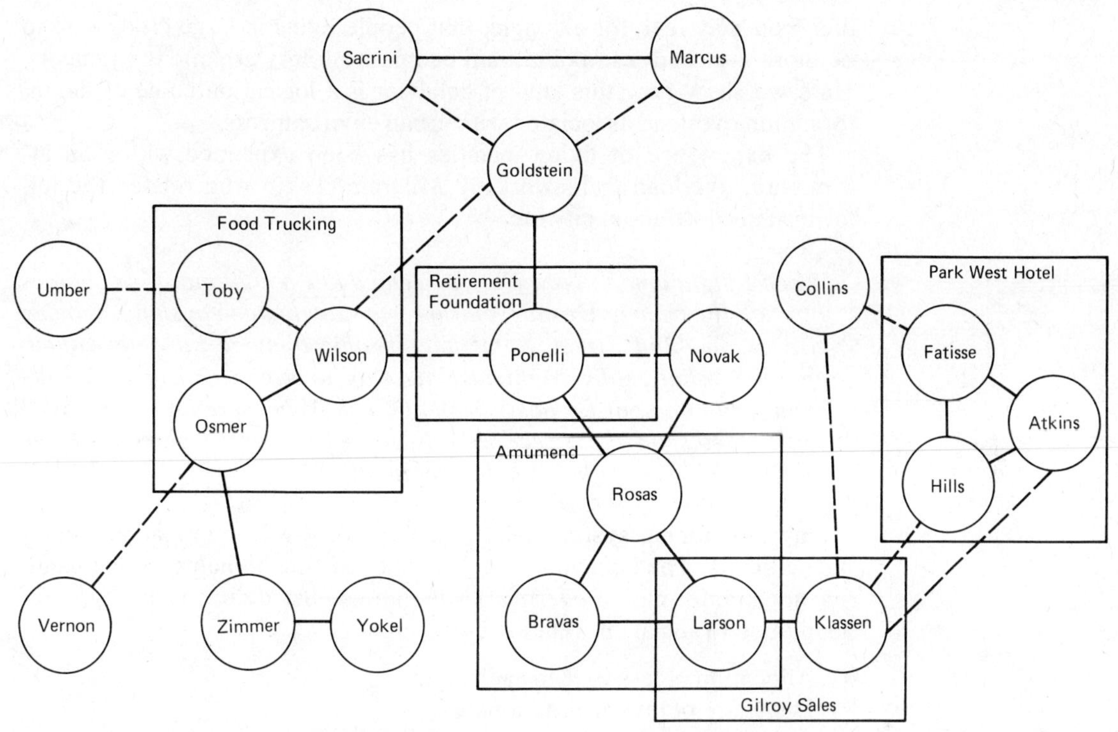

Figure 18-19. Final link diagram. Solid lines represent strong links and dotted lines weak links. Rectangles represent organizations. (From Harper & Harris, 1975.)

Harper and Harris (1975). In Northern California, link analysis identified a criminal organization engaged in robbery, prostitution, and gambling worth $2 million a year. In Southern California, a multiagency task force identified a criminal organization engaged in narcotics, real estate fraud, and pornography to the tune of $10 million per year. Furthermore, link analysis identified several owners of legitimate hotels and restaurants who were part of this organization.

At first glance, design of visual displays and law enforcement may appear to have little in common. But, as this example of link analysis reveals, human factors methods can be applied in many ways. We believe that the human factors profession is capable of making major contributions not only to crime prevention, but to other problems of the macro-environment as well.

Information Overload

People have characteristic ways of handling excessive stimulation. This general topic of attention (see Chapter 6) and its implications for design of displays (see Part 3, Human-Machine Interfaces) have already been discussed. In this section we cover the implications of human information processing limitations as applied to the social aspects of urban life. You may feel, for example, that people living in large cities tend to be more curt and less polite than people from less urban environments. Here we show how this kind of behavior is a logical outcome of the information overload associated with urban environments.

The experience of living in cities has been explained within an information overload framework by Milgram (1970) who related the following anecdote about city life.

> *When I first came to New York it seemed like a nightmare. As soon as I got off the train at Grand Central I was caught up in pushing, shoving crowds on 42nd Street. Sometimes people bumped into me without apology; what really frightened me was to see two people literally engaged in combat for possession of a cab. Why were they so rushed? Even drunks on the street were bypassed without a glance. People didn't seem to care about each other at all.*

Any functioning system must be able to deal with overload and competing inputs. The behavior described above is the human's way of handling social information overload. Milgram has listed six specific types of adaptation to social information overload.

- Allocation of less time to each input.
- Disregard of low-priority inputs.
- Redrawing boundaries in social exchanges.
- Ignoring all inputs.
- Filtering to diminish the intensity of inputs.
- Creating specialized social institutions to absorb inputs that would overwhelm individuals.

In a small town, neighbors might chat about the weather or sports for 5 or 10 minutes. In New York City this same conversation might take only 1 or 2 minutes. Since the urbanite has many more people to deal with, he or she cannot afford to spend 10 minutes with each. The urbanite copes with this overload by allocating less time to each input. Similarly, New Yorkers will casually step over a prostrate drunk in the street if the street is too crowded to prevent walking around the drunk. Such behavior shocks out-of-towners. Yet if the urbanite stopped to help every drunk, he or she would be always late to work. Disregarding low-priority inputs helps reduce overload (Figure 18-20). Workers in cities have the same problem. Once upon a time in New York bus drivers made change for riders and U.S. embassy employees in London would process requests for visas by themselves. Now bus riders must present exact change or tokens and London residents are forced to personally visit the embassy to sort through cartons of passports in order to get a visa in less than several weeks. The boundaries of social responsibilities have been redrawn to allow harried workers to transfer some of the burden

Figure 18-20. This kind of attire takes longer to be noticed in cities because of information overload. (Wide World Photos.)

of information overload. (Of course, such a transfer does not reduce overload in general; it merely dumps it elsewhere.) A surefire way of reducing overload is to ignore all inputs. Thus, city dwellers have proportionately more unlisted telephone numbers than rural environments. (Since the phone company charges extra for the service of not listing one's telephone number some people merely leave their phone off the hook.) New Yorkers riding on public transportation keep their head buried in newspapers and magazines to better ignore inputs like elderly people who might want their seats. Filtering allows inputs to be handled in a superficial way that requires little mental effort. Few people expect a medical report when they ask you how you are feeling today. Finally, specialized institutions like welfare departments handle poverty-stricken people who might otherwise form an army of beggars, as for example is the case in India where begging is a recognized career, that would further overload more prosperous citizens.

Milgram (1970) has listed many other examples of apparently antisocial behavior that can be explained by information overload: refusing to help someone being murdered by so much as telephoning the police, not saying "excuse me" when bumping someone in a crowd, being unwilling to make a telephone call for a stranger, failing to return a stranger's smile, etc. All of these behaviors have been shown by experiment to be more likely in urban settings. While the information overload framework seems to explain such results, it remains for human factors specialists to try to devise ways of minimizing the socially harmful effects these results document.

We have deliberately saved the most speculative topic for last. It is not at all clear how the human factors specialist can solve the problems of urban overload. But at one time airplane pilots were faced with tremendous information overloads. These problems, and their successful solutions, lead to the formation of human factors as an effective discipline during World War II. We are confident that if human factors can solve information overload in the cockpit, it can also reduce overload in the cities. Milgram, a social psychologist, has taken the first step by borrowing the concept of information overload, originally developed and refined by human factors researchers (e.g., Broadbent, 1958). It is time for human factors to take methodologies that were refined in the testbed of traditional person-machine systems analysis and to apply them to urgent problems of the macroenvironment that face our society. Certainly, human factors cannot solve this problem alone. The human factors expert needs to function as part of an interdisciplinary team focused on urban problems. If the human factors specialist is reluctant to propose solutions, others, perhaps less qualified, will step forward to address these issues.

SUMMARY

Complexity of transportation systems varies from one person carrying a backpack to international networks of airways and airports. There is need for human factors analysis at each level of complexity. Visual displays that guide the pilot to a landing and in docking are traditional examples of human factors contributions to transportation systems. The design of airport passenger terminals, and in particular vehicular approachways and automated ground transport, are newer areas where human factors is improving system design by analysis of human information needs.

Mass transportation is another new area where human factors can help. People will not use mass transit unless it satisfies a substantial list of criteria. Analyzing system goals, especially with relation to people, is a component of human factors that could well be applied to mass transportation. Indeed, new technologies such as telecommunication and inexpensive microcomputers and word processors may reduce the need for transportation as work is moved to the home and to surburban satellite centers.

Cities present the greatest challenge to the human factors specialist since little work has been accomplished. Link analysis has been applied successfully to crime prevention. Defensible space is a useful concept that badly needs quantification and measurement techniques. The effects of information overload upon city dwellers have been documented, but little has been offered in the way of solutions to minimize such stress.

19 ENVIRONMENTAL STRESSORS

The jungle is hot and humid. Your arm hurts from swinging your machete to clear a path through the dense underbrush. Great rivers of sweat run down your body. You would trade your best friend for a cold beer, or maybe even just an iced tea. You are not sure how much more of this your body can take but you continue to forge your way through the twisted, overgrown trail. Surely, the lost ark can't be far away.

The air is clear and bright. You are returning from a solo patrol, heading West toward the English Channel. You listen to the thrum of the giant Rolls-Royce engine propelling your Spitfire home and think of a cozy evening at the pub. Suddenly, from out of the sun, you see a flash of red. Tracers stitch an uneven path, smashing your engine. Your plane banks and then spins toward the earth below. Flames burst out. It is time to bail out but your canopy is jammed. You struggle frantically to free it as dive speed increases. Curse you, Red Baron!

You have obtained a summer job as a quality-control inspector for Coca Cola. Endless rows of full Coca Cola bottles pass before your inspection station. A strong light behind them lets you see any suspended material inside the bottles. You must reject any defective bottle. This job is so boring you find it hard to keep alert. In fact, the only thing that keeps you from going to sleep is the occasional 7-UP bottle that goes by to test your alertness. If you fail to reject the 7-UP bottles, you will be fired.

Fate has finally caught up with you. Tomorrow is the Human Factors final exam and you are unprepared. You were unfairly penalized for doing your workbook projects in crayon and now must get an A in the exam to get a B in the course. If you don't get a B, you don't graduate. Your professor is immune to begging and has adapted to a low salary so that bribery will not work either. So you pull an all-nighter cramming your brain with 24 continuous hours of study. By the time you reach the examination room you are exhausted from lack of sleep. But when the exam is placed before you, you become more alert and manage to complete the 4-hour exam in the allotted 2-hour exam period. Then you stumble home and crash, dead to the world for the next 10 hours.

STRESS

The foregoing vignettes are all examples of stress, stressors, and stressful situations. Although the human race has more control over its environment than any other species, we must still tolerate a varied assortment of environment stressors in our daily lives. Laws that protect the worker (see Chapter 20) are based on the assumption that stress can lead to illness, but careful research is required to lay out the details of possible relationships between stress and illness (Dohrenwend & Dohrenwend, 1978). While there is at least preliminary evidence that stress and health are related (Cohen, Glass, & Phillips, 1977), there is still much to be learned about the biological and behavioral effects of stress, both at home and in the workplace. In this chapter we begin by comparing some definitions of stress, stressors, and stressful situations. We then examine some theoretical explanations of the effects of stress on people. From there we progress to effects of specific stressors such as heat and cold. We also discuss effects of stress on mental skills and thought. The chapter concludes with a look at aftereffects once a stressor has been removed.

Definitions

Hans Selye, a physician who is well-known for his studies of stress since 1936, defined stress as "the non-specific response of the body to any demand (Selye, 1979 p. 12)." This is a rather broad statement that implies that stress can be defined in terms of biological effects of exposure to various stimuli. The key word in Selye's definition is "nonspecific." It means that regardless of the particular nature of the stimuli the body responds with a set of biological events that together form a pattern indicative of stress. Selye termed this biological pattern the General Adaptation Syndrome (GAS).

There are three specific stages to the syndrome (see Figure 19-1). First, is the alarm reaction. The shock phase of the alarm reaction is the immediate response of the body to a noxious stimulus. The alarm reaction's characteristic symptoms are depressed blood pressure, lowered body temperature, loss of muscle tone, and tachycardia (abnormally fast heartbeat). The countershock phase of the alarm reaction occurs as the body rebounds and begins to fight. Adrenocorticoid hormones (ACTH) are secreted and the adrenal cortex is enlarged. The second stage of the syndrome is called resistance. Symptoms improve or disappear as the body adapts to the noxious stimulus, also called the stressor. At the same time, however, the body exhibits a decrease in resistance to other stimuli. The third stage of the syndrome is called exhaustion. A stressor that is severe and applied for a prolonged time

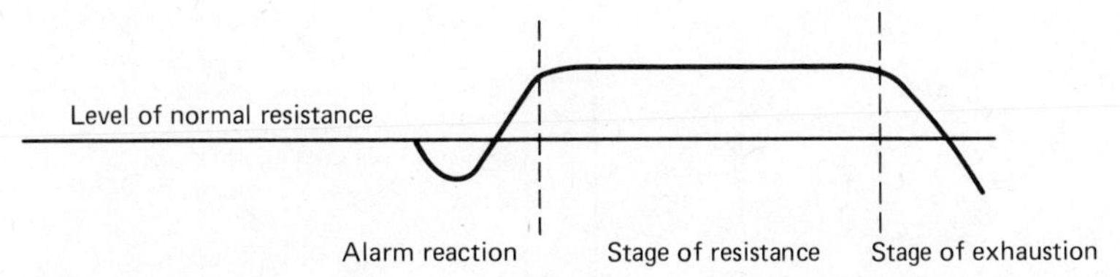

Figure 19-1. The three stages of the General Adaptation Syndrome. (From Selye, 1979.)

will eventually cause the symptoms to reappear. This can cause the ultimate response of the body: death.

Although physicians use this biological pattern to define stress—that is, where there's GAS there's stress—such a definition ignores psychological aspects of stress. Many different situations can evoke similar biological response patterns. For example, heart rate will increase when you are digging ditches and also when you are making love. Despite the similarity of physiological symptoms associated with each form of physical exercise, most people would prefer engaging in sexual intercourse to digging ditches. Furthermore, some researchers argue that the physiological patterns exhibit too much variance so that it is impossible to draw firm conclusions from physiological data (McGrath, 1970). There is considerable evidence that the psychological aspects of a situation are more important in determining stress than the physical aspects (Cohen, Glass, & Phillips, 1979).

One example of a situation where psychological determinants are more important than physical determinants can be found in a study that subjected rats to electrical shock (Weiss, 1972). There were three experimental conditions, with different rats in each condition. First, a rat heard a tone 10 sec before the electrical shock occurred. This is called shock with signal. Second, a rat received the same shock as in the shock with signal but the tone occurred randomly. This is called shock without signal since the random tone did not give the rat any information about when the next shock will happen. Third, a rat heard the tone but received no shock. This is called the control condition. After the experiment had gone on for some time, the rats were killed and examined for gastric ulcers. If stress were determined primarily by physical stimuli, in this case the electric shock, rats in both the shock conditions should show identical amounts of ulceration as measured by the length of any lesions. The control condition with no shock should have very little ulceration. Results of this experiment are shown in Figure 19-2. Ulceration was small in the control condition as expected. But shock without signal produced far more ulceration than the same shock with signal. Furthermore, shock with signal was close to the control condition in amount of ulceration. What can we conclude from these findings? Severe gastric ulceration, a traditional index of stress, was controlled much more by

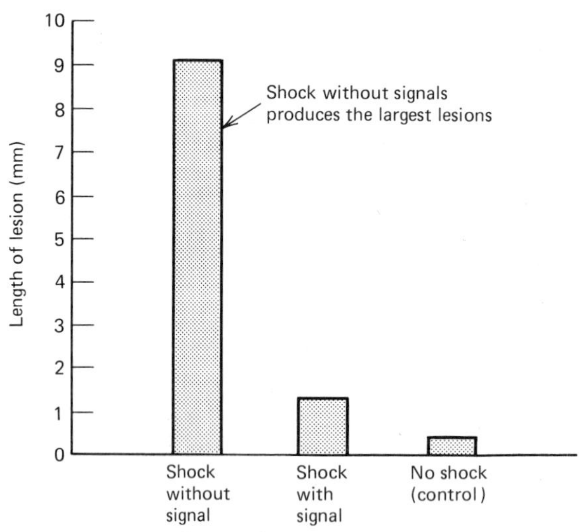

Figure 19-2. Effects of shock and signal on ulceration in rats. Note that shock with signal is much closer to the no shock control than to shock without signal. (Adapted from Weiss, 1972.)

the psychological variable of predictability. Rats that knew the shock was coming had less ulceration. (This was not due to the rats jumping up to avoid the shock since the electrode was attached to their tails.) Predictability, and not the physical shock itself, controlled ulcer formation.

Predictability is only one psychological factor that can affect stress. Since it is difficult to define psychological stress precisely, many researchers instead concentrate on stressful situations (McGrath, 1970). A stressful situation occurs when there is a substantial imbalance between the demands imposed on an organism by the environment and the organism's capability to successfully handle those demands. This means that stressful situations can come about in two ways. First, environmental demands can exceed the organism's ability to cope. The vignette about the Spitfire descending in flames is an example of this kind of environmental overload. Second, environmental demands can be too low. The vignette about the Coca Cola inspector is an example of environmental underload. Underload can be every bit as stressful as overload. In fact, gastric ulceration was once considered the mark of an ambitious, driving, high-status executive. But low socioeconomic status individuals suffer from even more gastric ulceration than do executives (Nugent, 1978, p. 182).

Yerkes-Dodson Law

The traditional view of stress links stress to arousal. Arousal is a concept that is related to nonspecific changes in the body such as hormone secretion and brain activity due to external stimulation (see Figure 19-3). Stress increases arousal. Increasing arousal in the traditional view is

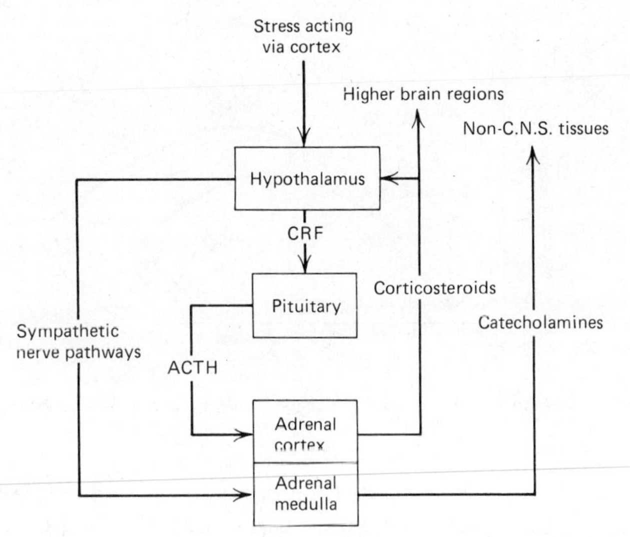

Figure 19-3. The adrenal systems that are involved in the hormonal stress response. (From Warburton, 1979.)

like turning up the volume control of the body (Hockey, 1979). The Yerkes-Dodson law (Yerkes & Dodson, 1908) relates performance and arousal.

Yerkes and Dodson trained mice to make a visual discrimination between two boxes. An incorrect choice resulted in an electric shock. Up to a point, increasing the amount of shock caused the mice to learn faster. But too much shock retarded learning. There was an optimum amount of shock and shocks that were greater or less than this optimum resulted in poorer performance (see Figure 19-4). This inverted U-shaped function is regarded as one of the strongest and most replicated findings in stress research (Mandler, 1979). It implies that to do your best you must reach the optimal arousal level: if you are taking a test and you are too loose or too uptight you will not do your best. The big question, then, is exactly where is this optimal level? This depends on the task difficulty. An easy task requires more arousal than a hard task. So what is optimal arousal for a simple task would be excessive arousal for a complex task. Performance on the complex task would not be best with this much arousal (Figure 19-4).

Although the empirical validity of the Yerkes-Dodson law is unquestionable, it offers only the most general guidance to the human factors designer. There are two major problems with using the Yerkes-Dodson law to predict what performance will be in some task. First, the law is an empirical result and not a theoretical statement. When used as a theoretical statement, it is circular. If performance is too poor we might increase arousal by adding stimulation (white noise, incentive pay, etc.). If

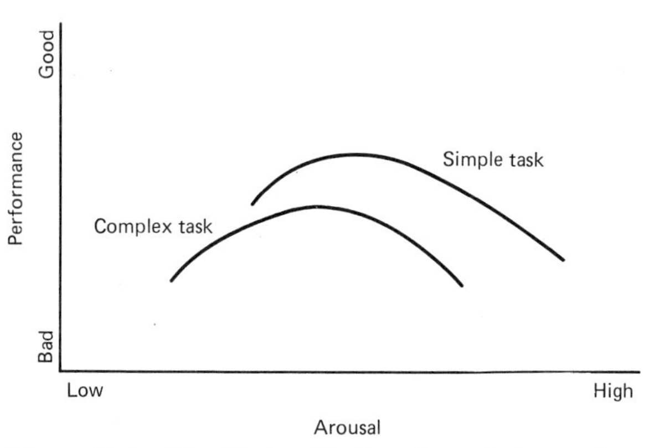

Figure 19-4. The Yerkes-Dodson law relating performance and arousal.

performance fails to improve, or indeed gets worse, we merely state that too much stimulation was added so that the worker is too aroused. We can always "explain" any level of performance by arguing that arousal is too high or too low. Arousal effects are inferred from the very performance that arousal manipulations were intended to control. This is circular. Second, the meaning of the term "arousal" is hard to pin down (Lacey, 1967). There are many physiological indices of arousal that have been used. Although the abscissa of the Yerkes-Dodson law is labeled arousal, the concept of arousal is not unitary as the law implies. Different components of arousal may be affected differently by the same external situation. It is extremely difficult, perhaps impossible outside of the laboratory, to measure arousal in concrete units and have those units solidly linked to changes in physical stimulation and workload. There is no simple way to plot the Yerkes-Dodson function for a particular factory or a particular worker.

Fortunately, there have been some theoretical advances that offer hope for planning workloads and human-machine systems (Broadbent, 1971; Easterbrook, 1959; Kahneman, 1973). These advances explain the mechanisms that cause the Yerkes-Dodson law. One important theoretical statement relates stress and a person's ability to pay attention (see Chapter 6) to cues in the environment. As stress increases, the range of cues that are processed becomes restricted (Easterbrook, 1959). This concept explains many puzzling findings in human factors. For example, airplane pilots crash by "locking on" to some particular instrument that is creating a problem, even if the instrument is of minor importance, and failing to attend to crucial information like a rapidly descending altimeter reading. Stress caused by a defective instrument or indicator—for example, a defective light that fails to come on indicating that the wheels are locked—has caused the pilot to narrow attention and to ignore other instruments (cues). While this example shows performance to be greatly worsened under stress, it is incorrect to conclude that stress always causes performance decrements. Stress always causes qualitative differences in a person's ability to allocate attention to environmental cues. In a simple situation with few cues, stress will improve performance by causing attention to be focused. In a complex situation with many cues, stress will decrease performance because many cues will go unattended. This view of stress will be elaborated later in this chapter in a section on information overload.

The Yerkes-Dodson law is a valid description of the effects of arousal and stress in a general way. It is difficult to apply the law to practical situations where specific design decisions are required. This requires a more detailed theoretical analysis of stress.

TEMPERATURE

In the course of a year, all of us experience wide variations in temperature. Atmospheric conditions are the most common environmental stressor. The current emphasis on energy conservation makes it essential to know precisely how people will react to turning down the thermostat in winter and turning it up in summer.

Nude humans are most comfortable over a narrow range of temperatures. Since the natural environment in most places presents temperatures that are above or below this range, the human has resorted to technology to keep comfortable. Such technology maintains human comfort by providing clothing in cold climates and air conditioning in hot climates. In order to design clothing and heating/cooling systems we must know how well the human functions at different temperatures. Then we can answer questions like how hot must it be before it is necessary or advisable to provide air conditioning in offices and how long can people function efficiently under heat or cold stress.

Heat

The human body generates heat even when stationary. As mechanical work is performed, more heat is generated. The body will try to give off excess heat to the environment, primarily by sweating, in order to maintain a constant core temperature of 37°C (98.6°F). A rise in core temperature of more than 2°C (3.6°F) will impair physiological efficiencies and the World Health Organization has recommended limiting increases to only 1°C (Buck & McAlpine, 1981).

The unit for heat generated by the body is the *met*. A sedentary male produces about 1 met. In physical units a met equals 50 kilogram calories per square meter per hour, which equals 58 watts per square meter. The met unit is scaled by skin area. An average sedentary male produces about 90 kilogram calories per hour. Table 19-1 shows metabolic rates for different human activities in met units. The maximum energy generated by a young male is about 12 mets versus about 8.5 mets for a young female.

TABLE 19-1 *Metabolic Rates of Typical Human Activities and Classifications of Physical Effort in Terms of Met Units (Multiply mets by 90 to get kcal/hr for average male)*

Activity	Mets
Seated quitely	1.0
Walking on the level 2 mph	2.0
3 mph	2.6
Using a table saw to cut wood	1.8 to 2.2
Hand sawing wood	4.0 to 4.8
Using a pneumatic hammer	3.0 to 3.4
Teaching	1.6
Driving a car	1.5
Driving a motorcycle	2.0
Driving a heavy vehicle	3.2
Typing in an office	1.2 to 1.4
Miscellaneous office work	1.1 to 1.3
Filing papers while sitting (standing)	1.2 (1.4)
Light assembly work	2.0 to 2.4
Heavy machine work	3.5 to 4.5
Light lifting and packing	2.1
Very light work	1.6
Light work	1.6 to 3.3
Moderate work	3.3 to 5.0
Heavy work	5.0 to 6.7
Very heavy work	6.7 to 8.3

The body's success in exchanging heat with the environment depends on four important environmental factors. The first factor is temperature, or more technically, dry-bulb temperature. This is the temperature you observe when reading a standard thermometer. The second factor is humidity, which is the amount of moisture contained in the air. One way to measure humidity requires a wet-bulb thermometer that is merely a normal (dry-bulb) thermometer surrounded by a wet wick. Evaporation from the wet wick cools the thermometer and lowers the effective temperature. As the air becomes more moist, this evaporation is less effective. Humidity can be measured by the difference between dry- and wet-bulb temperatures. Humidity can also be measured by finding the dew point—the temperature at which moisture condenses as the object is cooled in air. Dew point is directly proportional to humidity. The third factor is air movement velocity. Moving air helps to remove heat. This is why fans make you feel cooler. The fourth factor is mean radiant temperature. You will feel warmer standing near a wood stove than near a forced-air heating vent, even though both situations have identical temperature, humidity, and air velocity because of the radiant heat of the stove. Radiant heat is measured by a dry-bulb thermometer placed into a flat-black spherical globe (Berglund, 1977).

Several indices have been proposed to predict thermal comfort and thermal sensations from physical environmental parameters (see Buck & McAlpine, 1981 for a review). For example, a new effective temperature standard (Figure 19-5) is based on 50% relative humidity but ignores radiant heat transfer effects (ASHRAE, 1977). The wet-bulb globe temperature is another index but it too has limitations (Azer & Hsu, 1977). A more complete index proposed by Fanger (1972) is based on activity level, thermal resistance of clothing, dry-bulb temperature, radiant temperature, humidity, and air velocity. The Fanger model uses equations to predict thermal comfort based on these variables. All of these indices suffer from the same problem discussed earlier in this chapter—ignoring psychological aspects of stress (see Figure 19-2). Physical variables alone are insufficient to predict human response to heat stress.

Howell and Kennedy (1979) made this point quite clearly when they conducted a field validation of the Fanger model. Although Fanger's physical predictors were statistically reliable thus confirming the model, they accounted for only 6% of the variance of thermal judgments. This means that the Fanger model has limited practical application in real-world situations. The range of acceptable thermal comfort was shown to extend beyond that predicted by the Fanger model. Incidentally, the most comfortable temperature is the same for both sexes and does not change with the season of the year (Osada, 1978). It is clear that as-yet-unknown psychological variables exert powerful control over thermal comfort. One might speculate that a technique used by a local indoor tennis court operator to keep down complaints about lack of heat has considerable psychological validity. The operator has numerous thermostats located around the courts. Players can adjust them as they wish. However, unknown to the players, none of these thermostats are connected. Yet hardly any players complain of lack of heat whereas many complaints were registered before the phony thermostats were installed. This technique would be even more effective with the installation of thermometers that gave incorrect readings indicating higher temperatures than actually present. Of course, the human factors designer must stop and consider if these techniques, although possibly quite effective, are ethical and proper.

Heat and Performance

Although the effects of temperature on physiological functioning of the body are well known (Burse, 1979), effects of heat stress on human performance are more complex. Although there is a general consensus that prolonged exposure to heat will cause performance decrement, studies do not agree about how much decrement will occur and exactly how long it takes the decrement to set in. One recent review (Buck & McAlpine, 1981) has resolved these apparently conflicting results by concluding that performance will be degraded when new effective temperature begins to exceed 30°C (86°F) and exposure durations exceed 3 hours per

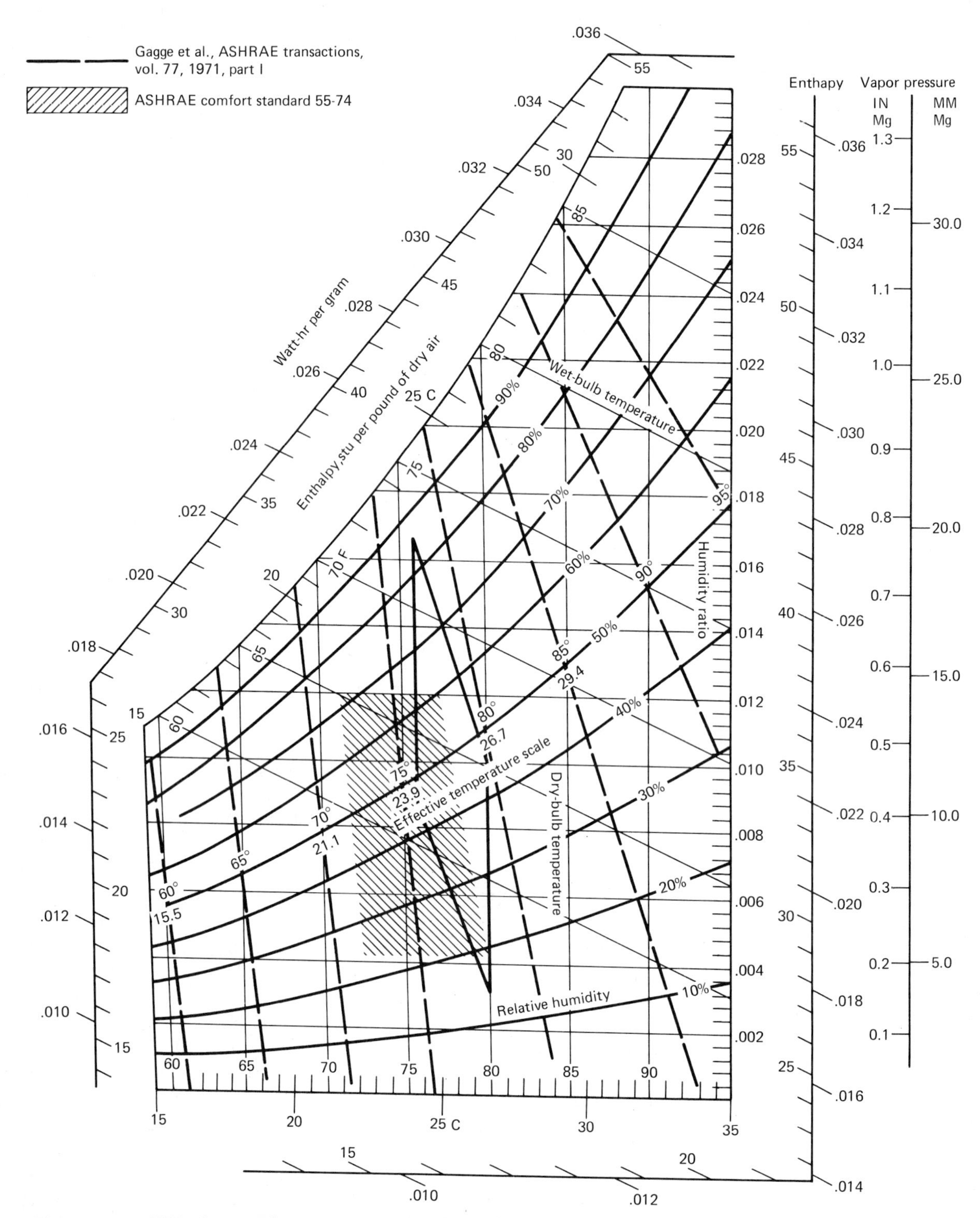

Figure 19-5. ET* chart. **The envelope applies for lightly clothed, sedentary individuals in spaces with low air movement, where the MRT equals air temperature.* (From ASHRAE, 1977.)

day. For shorter durations, people have higher tolerance levels before performance deteriorates.

There is currently strong disagreement about the effects of heat stress on the mental efficiency of sedentary workers. Wing (1965) performed the classic study of this problem. Wing found that mental tasks such as arithmetic and memory tasks were not affected for very short exposures (6 minutes) but that exposures of about 43 minutes to effective temperatures over 100°F did cause impairment. Wing proposed an exponential function relating exposure time, and an effective temperature (see Figure 19-6) later adopted by the National Institute for Occupational Safety and Health (NIOSH, 1972) as the recommended standard for the lower limit of heat-impaired mental performance. These limits are reached before the physiological tolerance limit of the human. However, this standard has been strongly criticized as being far too conservative (Hancock, 1980). A detailed reanalysis of Wing's data has led to the line in Figure 19-6 labeled Hancock. It is close to the physiological limits, shown as the line labeled Taylor in Figure 19-6. Until this debate is resolved, cautious human factors specialists will base decisions on a curve falling between the Hancock and the Wing stress tolerance functions; extremely cautious designers may prefer to use the more conservative Wing tolerance curve.

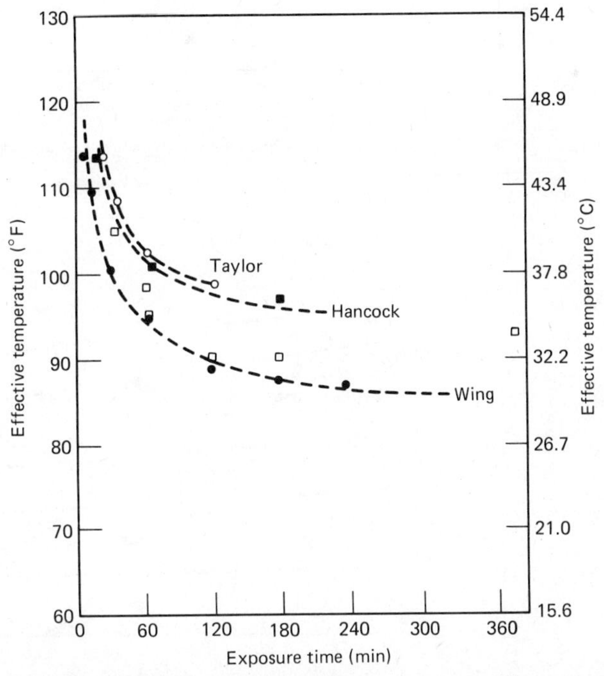

Figure 19-6. Heat stress tolerance curves. The curve labeled Taylor shows physiological limits while the other two curves show limits for mental performance. (From Hancock, 1980.)

Although there is disagreement about the safe limit for unimpaired mental performance in heat, we can specify isodecrement curves for heat stress (Ramsey & Morrissey, 1978). An isodecrement curve is a function where performance decrement is constant. It displays the trade-off between temperature and exposure time. Ramsey and Morrissey computed isodecrement curves for a wide variety of mental tasks including tracking, reaction time, vigilance, eye-hand coordination, and complex tasks combining two or more elements of the simpler tasks. Their final results (Figure 19-7) integrated this large amount of data into only two iso-

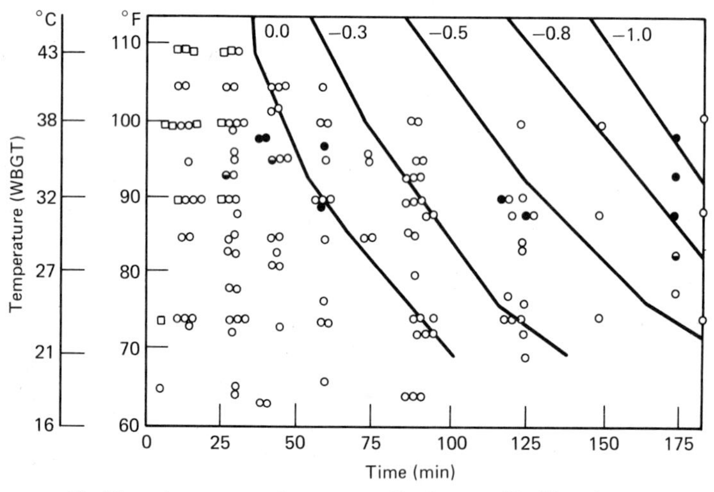

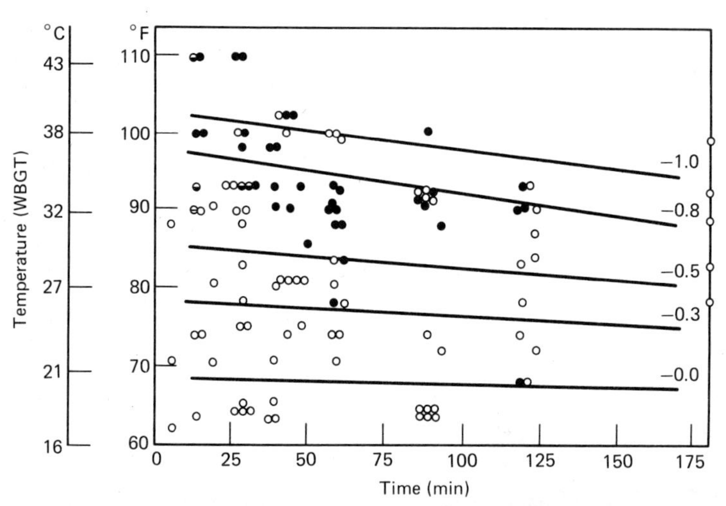

Figure 19-7. Isodecrement curves for mental reaction time performance (above) and for vigilance and complex tasks (below). (From Ramsey & Morrisey, 1978.)

decrement curves: one for mental reaction time performance and the other for vigilance and complex tasks. There are five functions in each plot. The function labeled zero means no performance decrement and that labeled −1.0 corresponds to the greatest decrement.

Cold

The body attempts to maintain a core temperature in cold environments by shivering and by restricting blood flow to the extremities. In addition to heat loss through the skin, about one-fifth of the total body heat loss occurs by evaporation from the lungs and by heating the cold air that is breathed. Core temperatures below 35°C (95°F) are very dangerous. Temporary amnesia can develop at 34°C and cardiac irregularities and possible unconsciousness occur between 30 and 32°C (NOAA, 1975). Humidity is not an important factor in cold temperatures since cold air is unable to hold large amounts of moisture. However, air velocity is quite important and the wind chill index (Table 19-2) is often given in weather reports. It is important to realize that wind chill results in a physical lowering of effective temperature so that when you feel colder in a wind this is not just a psychological effect.

Manual dexterity is impaired when the hand-skin temperature falls below 60°F. This decrement can be partially avoided by providing spot heating (Lockhart & Kiess, 1971). Cold is particularly threatening to underwater divers since water has a thermal conductivity about 25 times that of air. This is why the body loses heat much faster in water than in air of equal temperature. Figure 19-8 shows that an unprotected diver can tolerate less than one hour underwater when temperature is less than 60°F.

A recent laboratory study (Please, Ludwig, Green, & Millslage, 1980) showed that cold increases both simple reaction time and movement time. Male college athletes had their arms packed in ice for 30 minutes. Reaction time increased from 199 msec (room temperature pretest) to 218

TABLE 19-2 *Table entries are effective temperature equivalent to no-wind conditions. For example, a 40 MPH wind and a temperature of 0°F, is equivalent to a temperature of −53°F and no wind*

	Wind Speed (Miles Per Hour)							
Temperature (°F)	5	10	15	20	25	30	35	40
0	−5	−22	−31	−39	−44	−49	−52	−53
5	0	−15	−25	−31	−36	−41	−43	−45
10	6	−9	−18	−24	−29	−33	−35	−37
15	11	−3	−11	−17	−22	−25	−27	−29
20	16	3	−5	−10	−15	−18	−20	−21
25	22	10	2	−3	−7	−10	−12	−13
30	27	16	9	4	1	−2	−4	−5
35	32	22	16	12	8	6	4	3

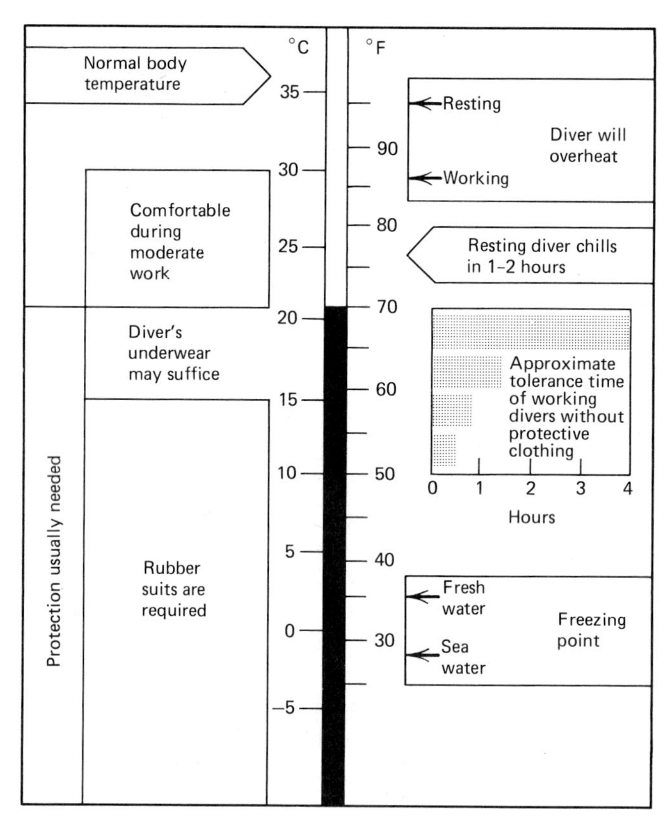

Figure 19-8. Temperature and clothing requirements for underwater divers. (From NOAA, 1975.)

msec and movement time increased from 149 to 161 msec. Since even an elementary laboratory task shows cold deficit, it is not surprising that workers cannot perform optimally in cold environments.

Clothing

People wear clothing as protection against cold stress. (They also wear clothing as decoration and to indicate status but these uses of fashion and apparel are beyond the scope of this text.) Clothing provides insulation that slows down the loss of body heat. The amount of insulation is measured by *clo* units. One clo is that amount of insulation required to keep a nude sedentary man comfortable at 70°F, 50% relative humidity in a normally ventilated room. The clo unit is defined mathematically as

$$\text{clo} = {}^\circ\text{F}/\ (\text{Btu/h})\ (\text{body area in square feet})$$

Clo values for individual items of clothing range from 0.05 for bra and panties to 0.49 for a man's heavy jacket (Figure 19-9).

A linear model has been developed to predict how much clothing is necessary for thermal comfort (Rohles, Konz, & Munson, 1980). First,

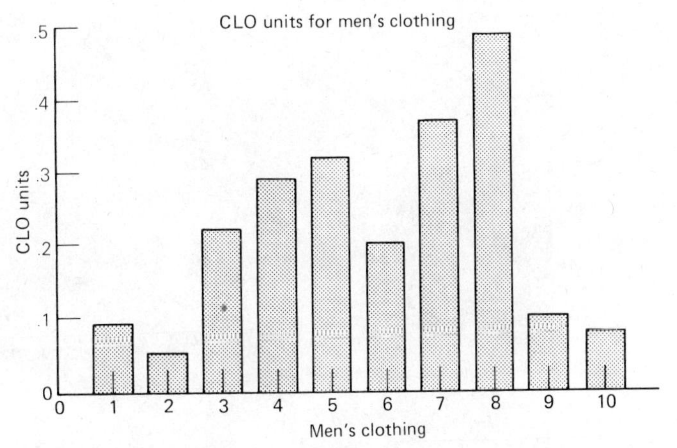

Figure 19-9a. (1) T Shirt. (2) Briefs. (3) Light, long-sleeve shirt. (4) Heavy, long-sleeve shirt. (5) Heavy trousers. (6) Light sweater. (7) Heavy sweater. (8) Heavy jacket. (9) Ankle-length socks. (10) Oxfords shoes. (Data from Rohlens, et al., 1980.)

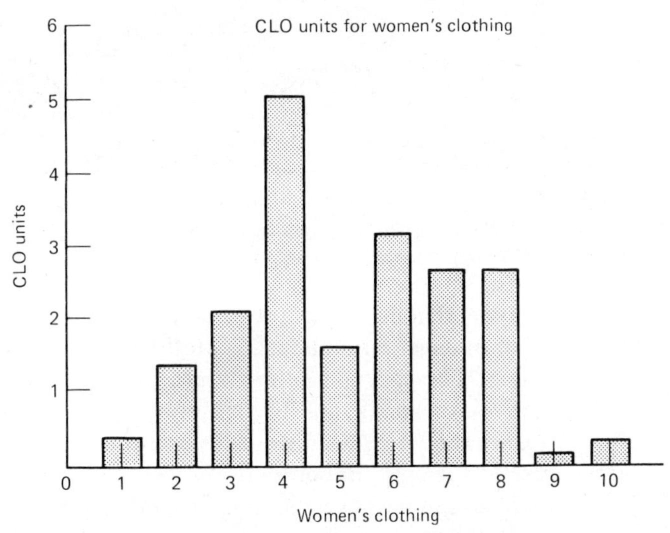

Figure 19-9b. (1) Bra and Panties. (2) Full Slip. (3) Heavy Blouse. (4) Heavy dress. (5) Heavy skirt. (6) Heavy slacks. (7) Heavy sweater. (8) Heavy jacket. (9) Panty hose. (10) Pumps, shoes. (Data from Rohlens, et al., 1980.)

the individual clo values for each item of clothing are added and then this total is multiplied by 0.82 to yield the clo value of the ensemble, called Icl. Then the following equation is used:

$$CET^* = 29.75 - 7.28 \text{ Icl}$$

where CET* is new effective temperature (°C). Since it is impossible to

Figure 19-10. Protective clothing permits firefighters to work in high-temperature environments. (Courtesy Fyrepel TM.)

satisfy 100% of the population (see Chapter 15), this equation is based on a model that has 6% of the population unhappy.

In unusual circumstances clothing can be worn to protect against heat stress. The special clothing worn by race drivers, geologists investigating active volcanoes, and firefighters (Figure 19-10) insulates the body from external heat. But unless active refrigeration is provided, these suits are useful for only limited periods.

In addition to its insulating properties, clothing must conform to anthropometric measurements of the population wearing it (Roebuck, Kroemer, & Thomson, 1975, Chapter 9). Thus, design of clothing should follow the same procedures as specified in Chapter 15, Workspace Design. Specialized clothing such as space suits and underwater suits for divers must be checked to ensure adequate reach envelopes since the protective clothing can restrict movement. It is particularly important to distinguish between static and functional measurements. For example, a glove designed from static measurements will have the top and bottom of each finger of equal length. This makes it difficult for the wearer to curve his or her fingers. Yet, the natural resting state of fingers is arched rather than straight. Jobs or environments that require the worker to wear gloves for several hours should be tailored so that the bottom of each finger is shorter, allowing the glove to mimic the arch in the resting human hand. Gloves with straight fingers will prove uncomfortable when worn for more than short periods of time.

STRESSORS

Information Load

In our technological society, more and more jobs require cognitive effort from the worker as opposed to the mechanical efforts of simpler societies (see Chapter 1). But humans have a limited capacity for mental work (see Chapter 6). Having to process too much information simultaneously, or too high a rate of information, can result in highly stressful situations. Many modern jobs—for example, nuclear power plant operator, airplane pilot—have been characterized by workers as "hours of boredom relieved by minutes of sheer terror." When an automatic plant or system is functioning normally there is little for the operator to do, aside from monitoring the system. Even that can be performed in a perfunctory fashion since the operator knows that alarms will be energized if something goes wrong. Therefore, many companies require operators to log entries periodically, even though the computer could do this automatically, just to keep the operator awake and involved. The minutes of sheer terror occur when an alarm goes off. The operator must react quickly within minutes, or even seconds for pilots, to prevent disaster. Thus, many jobs require the worker to alternate between two stressful situations: boredom where information load is insufficient to keep the operator aroused, and an alarm condition where disaster is imminent and information load is excessive.

The human's ability to cope with the stress imposed by information overload (Broadbent, 1971) has been discussed extensively in Chapter 6 in the section devoted to attention and mental workload. Effects of overload can be summarized by stating that systematic performance decrements, especially in tasks that require selection and execution of responses, occur when capacity is exceeded.

Effects of information underload have been less studied than effects of overload. A recent survey of boredom research since 1926 found an average of less than one paper per year (Smith, 1981). Smith decided that the most robust finding in boredom research was related to individual differences. People who are extroverts cannot perform boring tasks as well as introverts. There is evidence that repetitive tasks, such as placing ball bearings into a container, create physiological symptoms consistent with stress and arousal. Repetitive tasks caused elevated heart rate, depressed alpha activity, and elevated adrenaline excretion rate (Weber, Fussler, O'Hanlon, Gierer, & Grandjean 1980). Although industrial robots are taking over some of the more boring assembly line jobs (see Chapter 5), there still remain many occupations, such as driving a truck cross country, where information underload is a problem. Truck drivers with good safety records cope with boredom successfully by engaging in mental games, counting objects they pass, etc. to make time appear to pass more quickly (McBain, 1970). These drivers do not report being troubled by boredom on the road. Unfortunately, the human factors designer, lacking a substantial data base on boredom, is not in a position to suggest effective ways of avoiding boredom for dull jobs occupied by individuals who are less inventive than the safe truck drivers. Boredom remains one area where human factors is primitive and only slightly better than common sense. However, a recent theoretical integration proposed by O'Hanlon (1981) to explain boredom offers hope that rapid progress can be made by combining practical and theoretical efforts.

Danger

The ultimate stressful situation occurs when a person fears that his or her own life is threatened. Most jobs are not this stressful. For example, an air traffic controller who has the power of life and death over the passengers and crew of the planes in her airspace will not die if she errs causing a midair collision unless one of the planes strikes the control facility. But some jobs and sports such as underwater diver, parachute jumper, mountain climber, construction worker on high-rise buildings and bridges, astronaut, etc., are risky and can threaten life and limb. If the previous analysis of stress as leading to attentional deficit is correct, we would expect that danger should also cause a narrowing of attention. This is especially important since danger is not a physical stimulus like noise, or heat, or electric shock. Instead, danger is a psychological problem resulting from an individual's assessment of inability to cope with an environmental situation. It is closely related to emotional states like panic and anxiety (Mandler, 1979).

It would be unethical to create dangerous situations in the laboratory merely to observe how people react to danger. Therefore, most studies of danger are field studies that observe workers, primarily divers, on the job. A convenient way to vary danger underwater is to control the depth at which divers work. Deep dives are more dangerous than shallow dives. An unassisted diver can make it to the surface alone in an emergency far more easily from a shallow depth. Divers are well aware of this, even under nonemergency conditions. If a diver is deep enough for some time—these combinations of time and depth are found in standard dive tables (NOAA, 1975)—a direct ascent to the surface is inadvisable without a decompression stop underwater. Several studies that compared diver performance at different depths and in pressure chambers on land where physical conditions were similar to those underwater but not nearly as dangerous have revealed narrowing of attention (Baddeley, 1971, 1972) and changes in arousal indicating anxiety (Mears & Cleary, 1980). A study that simulated danger in a pressure chamber found increased arousal and impaired attention to peripheral stimuli, although a central task was performed without decrement (Weltman, Smith, & Egstrom, 1971).

Figure 19-11. Blood pressure before and after a jump differs for novice versus experienced parachutists. (Ellis Herwig/Stock, Boston.)

The perception of danger depends on one's ability to cope with the environment. Novice divers and parachute jumpers perceive danger in situations that are not stressful for experienced divers and parachutists (Figure 19-11). While novice jumpers have high pulse rates before a drop and normal rates after landing, experienced jumpers have the opposite: low pulse rates before jumping and high rates on landing (Epstein & Fenz, 1965). This interesting reversal can be explained by noting that emergencies and arousal occur when the individual has ongoing behavior and plans interrupted (Mandler, 1979). A stressful situation results when the individual has no immediate plan of action to deal with the situation. But experienced and highly trained personnel have immediate plans and response sequences to deal with the interruption. For example, simulators are widely used to practice dealing with assorted emergencies. Thus, as a result of training and experience, emergencies become routine situations. If a plan of action is readily available there is no emergency and no stressful situation. Before the novice jumps, he can imagine many possibilities that could be beyond his control and arousal goes up. Once the jumper goes down and lands the original goal has been achieved and arousal diminishes. The experienced parachutist knows she can deal with the jump and is not aroused. Upon landing the pleasant and planned jump experience is interrupted and arousal occurs. While the explanation of arousal as due to interruption of plans and action is too recent to be universally accepted, it offers a useful theoretical tool for the designer who must plan for danger. Providing hardware and training for emergencies will minimize arousal and narrowing of attention.

Crowding

All of us have experienced physical crowding. Being pushed and shoved in line at a rock concert or at a sale in a department store is not pleasant. Is crowding a stressor? Since it would be unethical to crowd people for very long periods of time, much crowding research has been done with animals. The results of a long series of studies conducted at the National Institute for Mental Health (Calhoun, 1966, 1971) showed appalling behavior disorders developing in successive generations of crowded rats. Implications of these results for human behavior were frightening.

Most crowding research with humans has been demographic field studies where assorted measures of density are correlated with social pathologies like crime rate, admissions to mental hospitals, and death rate. While early studies found moderate correlations (Galle, Gove, & McPherson, 1972), later work found little or no relationship between density and pathology (Freedman, Heshka, & Levy, 1975; McPherson, 1975; Ward, 1975). We cannot conclude that crowding is necessarily harmful (Cohen & Sherrod, 1978) and people who are crowded do not develop the pathologies of crowded rats (Freedman, 1975). But there is one important exception to this conclusion.

People living in institutional settings—prisons, college dormitories, nursing homes, etc.—do suffer from high density. Prisoners develop more psychiatric illness, higher death rates, and higher blood pressure when crowded (Paulus, McCain, & Cox, 1978). Students in dormitories react to crowding by withdrawal and decreased social interaction (Baum & Valins, 1977; McCarthy & Saegert, 1979; Paulus, 1979).

Crowding, like danger, is a psychological problem (Figure 19-12). High physical density by itself does not ensure that people will feel crowded (Stokols, 1976). A person must believe that his or her freedom of movement is constricted before density can cause a stressful situation. Crowding is an intensifier rather than just a stressor. It amplifies feelings and emotion. Attending a crowded disco or football game adds to the fun. But studying in a crowded space is unpleasant (Cozby, 1973).

Figure 19-12. Crowding is a fact of urban life. (*Funny Business,* © 1976 by Newspaper Enterprise Association.)

The human factors designer working on institutional settings should minimize physical density. The designer working on fun places should plan for some crowding to heighten pleasurable feelings.

CONFLICTING SYSTEM GOALS

The Tragedy of the Commons

It has long been a guiding philosophical principle of democratic society to strive for "the greatest good for the greatest number." Not only do we try to increase the average standard of living, we also try to improve the condition of even the poorest members of society as well as the richest. A mathematical expression of the system goal of the greatest good for the greatest number implies that society is attempting to maximize two variables, good and number, at the same time. In general, this is mathematically impossible. Calculus can only maximize one variable at a time. Attempts to maximize more than a single variable result in some sort of compromise whereby no variable is optimized.

Garrett Hardin (1968), an ecologist, has made this point quite clear by using the English commons as a graphic example. The commons is a pasture shared by all herders. Each has an equal right to graze animals on the commons. Since each herder is a rational decision maker (see Chapter 14) he or she decides to add an additional animal to the flock. This is a rational decision for each individual, since the benefits of an extra animal outweigh the costs of increasing the load on the commons which is divided among all herders. But once enough herders make this rational decision, the result is catastrophic. Too many animals deplete the commons. There is not enough food and most, if not all, of the animals die of starvation. The rational decisions of individuals have brought tragedy to the group.

The dilemma of the commons is not limited to grazing animals. It occurs wherever finite resources must be shared. Another example is the smokestack found in every town. The purpose of the smokestack is to carry waste gases and pollution away from the generating source. The higher the smokestack, the farther will this pollution be carried by air currents. Since few towns want their own soot and wastes deposited in their own backyards, smokestacks are built higher and higher. But there is only one atmosphere on spaceship earth. Pure air is a vast but still finite resource. Dumping pollution into the atmosphere does not get rid of it. Instead, it lands somewhere else. One of many examples of this problem is the acid rain falling on the Northeast coastal region of the United States (Bohrmann, 1982). It results from the smokestack gases of Midwestern factories and power plants carried east by the prevailing winds. A smokestack does not control pollution: it displaces pollution. When we realize that the earth is a single ecological system we appreciate that nothing can be thrown away. Displacement is not pollution control. Hardin (1972) used wry humor to accentuate this point by prophesizing a Museum of Antiquities 100 years in the future. Busloads of children are brought to view a tall cylindrical object with the following bronze plaque at its base.

SMOKESTACK
QUAINT DEVICE USED BY OUR PREECOLOGIC ANCESTORS, IN THE SUPERSTITIOUS BELIEF THAT THERE WAS AN "AWAY" TO THROW THINGS TO. LAST USED IN 1987.

Is there a solution to the tragedy of the commons? Technology is often suggested. Improved fertilizer or hybrid grass can increase the carrying capacity of the commons. But this only postpones the problem. Eventually, the expanded herd will reach the new carrying capacity and the problem returns. Only an unending stream of technological improvements can forestall the tragedy. Some people believe that such continual innovation is not only possible but likely. Any linear extrapolation of societal problems will be incorrect since new technology creates in quantum leaps. According to this view, the major problem of cities 100 years ago that threatened urban civilization was removing tons of horse manure from the streets. Of course, new technology prevented this problem from smothering urban life. Other people believe that a new morality is the solution. Since it is impossible to maximize good and number we should voluntarily decide to limit population growth in order to achieve the greatest good instead of the greatest number. But such voluntary constraints will fail in the long run. For example, those parents who decide to limit themselves to only two children will in a few generations become a tiny minority if other parents decide to have more than two offspring. This difficulty with voluntary altruistic solutions has lead some to propose a "lifeboat ecology." For example, Hardin has argued with perfect logic that it is wrong for the United States to send surplus food to countries where there are famines due to excessive population. This only encourages more excessive population and increases the need for external food supplies. Yet, most citizens believe in charity and feeding starving nations. As you have seen in other chapters, conflicting system goals lead to inconsistent behaviors. Would you prefer the greatest good, the greatest number, or do you believe that calculus does not apply to human social groups and that both can be achieved?

Air Pollution

Society believes that breathing polluted air is harmful and has passed many laws to control automobile exhausts, smokestack effluent, coal dust, and other pollutants that can escape into the atmosphere. Even the breathing space of nonsmokers has legal protection. Detailed models calculate the trade-off between the costs of air pollution and the gains to society of the technology that produces the pollution (Small, 1977). However, while it is generally conceded that pollution has harmful physiological effects on people, relatively little is known about the behavioral effects of breathing polluted air (Evans & Jacobs, 1981).

One of the first behavioral studies (Lewis, Baddeley, Bonham, & Lovett, 1970) compared a group of people breathing pure air to a group breathing polluted air. The polluted air was obtained 15 inches above the ground—approximately the location of the air intake vent on a car—on a highway in England that was not very busy by "central London standards" since only 830 vehicles passed each hour. Neither the experimenter nor the participants knew which source of air was being used during the experiment. This is called a double blind procedure and prevents the results from being influenced by anyone's expectations about polluted air. Of course, after the experiment was completed, the coding could be deciphered to identify the air supply. Both groups were tested on an assortment of information-processing tasks similar to those discussed in Chapter 6. On three of the four tasks used, people breathing pure air did better. Similar results were obtained in a more recent American experiment where participants breathed small amounts of carbon monoxide, a pollutant found in automobile exhausts (Putz, 1979). This experiment used a time-sharing task to measure attention. You will recall from Chapter 6 that such attentional tasks are often more sensitive indicants of human performance than are single tasks performed in isolation. Breathing carbon monoxide made it more difficult to pay attention to two tasks at the same time.

Although the number of studies of air pollution and behavior is small, the conclusions are consistent. Air pollution decreases mental efficiency. This suggests that automobile accident rates are higher in congested areas not only because of greater numbers of vehicles (Figure 19-13) but also

Figure 19-13. The air pollution caused by this traffic jam may decrease the mental efficiency of drivers thus increasing the chances for accidents. (United Press International.)

because traffic pollution itself contributes directly to accidents by making drivers less able to process information. Similar effects might be expected in polluted industrial environments.

STRESS AND WORK

Studies of stress in work settings customarily distinguish between stressors and strain. Strain is the effect of stressors on the worker. It depends on the stressors that are present, the duration of work, and the individual's capacity to tolerate stressors. The term "strain" is derived from engineering where stress is applied to some structural member, for example a steel beam, and the effects of stress – that is, strain – are measured, for example, by observing how much the steel deforms or bends. It is much harder to measure strain in people and this makes the distinction between stressors and strain less valuable when applied to humans rather than to pieces of metal. While strain in physical work can be related to physiological measures such as heart rate and oxygen consumption (Rohmert & Luczak, 1979), strain associated with mental work load is more difficult to pin down. (See discussion of mental work load in Chapter 6.)

There is evidence that changes in physiological measures are related to mental work load rather than to the risk and anxiety that often occur when work load is increased. One study using experienced test pilots (Roscoe, 1978) monitored the heart rates of both pilot and copilot. Both pilot and copilot shared the same risk but only the pilot had to handle the workload. The copilot was a "safety" pilot who normally would not fly the airplane. Heart rate increased during periods of increased mental work load such as take-off and landing for the pilot but not for the copilot. However, heart rate as an index of strain has a serious deficiency. As yet, researchers have not determined how much of an increase is required before the pilot is unable to perform the job satisfactorily. While structural engineers can predict the breaking strain for a steel beam, human factors specialists cannot yet predict on the basis of physiological measures how much mental strain is unsafe for a worker.

Another profession where mental work load is widely regarded as stressful is that of air traffic control. The recent strike where controllers were willing to lose their jobs rather than return to stressful work conditions is proof that the controllers view their job as highly stressful. This point is strongly made by a letter to the editor from the wife of an air traffic controller written during the strike (Figure 19-14). Physiological and biochemical studies of air traffic controllers have had mixed results with some studies finding no more strain than in normal populations and others finding greater long-term risk of hypertension and peptic ulcers. However, the lack of a consistent pattern on medical and physiological problems may indicate that these measures of strain are insensitive rather than that air traffic control is not a stressful occupation (Crump, 1979).

A more promising approach may be assessment of spare mental capac-

Letters

Controllers' life filled with stress

I am the wife of an Air Traffic Controller and very proud of the fact. During the last number of days I have been living through a "work stoppage" with my husband, a government employee for 25 years, with only two more years before he would take a reduced retirement.

Are these men, highly trained in their profession, really criminals like the government and media make them out to be? No! They are some of the most conscientious, law abiding men in our country! They have families, work in the community on may community projects, and volunteer their time to schools and other organizations, in order to promote their profession and general aviation. They are a breed of men proud of being air traffic controllers! That sounds contradictory, considering 12,000 plus controllers would walk off of their jobs and get fired — doesn't it? Why did these men prefer to give up their jobs instead of going back to work?

I WILL SPEAK ONLY only as the wife of an Air Traffic Controller! At first, I was shocked and disappointed, until, I really sat down and began to think of the realities of the past years. Would retirement make sense if he wasn't around to enjoy it? I think not!

In 1979 I was called from my job to Home Hospital where my husband was taken for an apparent heart attack! He was three days in intensive care and, Thank God, no damage was apparent. What was it? The doctors could only say — STRESS!

July 3, 1980 I was called to the control tower because my husband was vomiting and did not know quite where he was. Again to the hospital and he went through a series of tests. Diagnosis — STRESS.

This past spring, during intense fog a plane was given directions from the controller to land on a certain runway. The pilot became disoriented in the fog and came directly at the tower. All controllers working hit the floor thinking the end had come! The plane swerved, missing the tower! The controllers came up and directed the plane to a safe landing.

Another day, Air Wisconsin, was given clearance to land from a distance near Montmorenci over the outer marker, when suddenly a small aircraft came out of "nowhere." The pilot was fortunate enough to see it and avert a collision which would have killed over 30 people. Purdue Airport does not have radar and small aircraft who invade the airspace without contacting the tower by radio cannot be seen by the controller from such a distance.

I would like to relate controller stress to an example! If you were in your automobile traveling at 55 m.p.h. and another car was coming straight at you at the same rate of speed; and suddenly when you thought you would collide, at the last moment the car swerved and missed you. How would you feel? These kinds of situations happen more frequently than the general public realizes. Why doesn't the public know? Because that is a controllers life!

He must learn to react in split seconds, the decision must be made instantly because his decision is LIFE and DEATH! As time wears on so does the human body, and yet through it all the record of air traffic control is phenomenal in the United States!

Where is the stress seen? After the work day when the man comes home and you see in his face the strain. Perhaps if you are lucky he might tell you of the near miss — but more likely he will not say anything because he cannot relate his feelings because you were not there to feel it with him. It is difficult to relate to something you yourself have not experienced!

Do I want my husband to go back? That decision is up to him, if given the opportunity. If he doesn't I'll understand! What else does he love? What else is he trained for? Who will take a man 45 years old and train him for another occupation? I don't know — but God give us strength we will find a way!

The system can be changed and it must be changed!

LOIS WARTOLEC
Lafayette

Figure 19-14. A letter from an air traffic controller's wife. (Courtesy of the *Journal and Courier,* August 19, 1981.)

ity using techniques discussed in chapter 6 (Kalsbeek, 1971). Another approach has used information theory to evaluate stress from patterns of radio telephone communication (Reiche, Kirschner, & Laurig, 1971). Reducing uncertainly by communication reduces strain. However, despite many hopeful statements (e.g., Finkelman & Kirschner, 1980) about using information-processing interpretations of air traffic control stress, an index that can actually be usefully applied inside the control tower on a daily basis has yet to be achieved. Progress would undoubtedly be accelerated if some of the human factors specialists working in this area became more familiar with the theoretical models of mental work load formulated in laboratory research.

AFTEREFFECTS

It is hardly surprising that stressors will affect performance while the stressor is present. But there is a considerable body of research that shows exposure to a stressor can harm performance after the stressor has been removed (Cohen, 1980). In fact, even stressors that are mild enough not to hurt performance while they are being presented can, under certain circumstances, depress performance once they have been removed (Percival & Loeb, 1980). The impact of a stressor that was presented earlier and is no longer physically present is termed an *aftereffect*.

The classic study of aftereffects of noise (Glass & Singer, 1972) found that people exposed to intermittent and unpredictable noise did worse, once the noise was removed, on tasks like proofreading and attempting to solve puzzles that actually could not be solved. These people made fewer attempts to solve the puzzles and made more proofreading errors than people exposed to predictable noise. Glass and Singer tried to explain these results by postulating that unpredictable noise was more aversive than predictable noise. This argument should remind you of the rats receiving electric shock that were discussed at the beginning of this chapter (Figure 19-2): rats with predictable shock had less ulceration. Glass and Singer believed that people exposed to unpredictable stressors felt more at the mercy of their environment and less able to control their world. This is called learned helplessness.

Further tests of the learned helpless idea were carried out by telling some people that they could control the noise with a switch that would turn it off if it became too unpleasant. However, the experimenters asked the people not to use the switch unless it was truly necessary. Most of these people did not switch off the noise, although they believed they could if they wanted to. This is termed perceived control. Perceived control should decrease learned helplessness. Results showed that people in the perceived-control situations suffered less from aftereffects of noise (Glass & Singer, 1972; Sherrod, Hage, Halpern, & Moore, 1977).

While there is general agreement that predictability and controllability of a stressor minimizes aftereffects (Cohen & Weinstein, 1981), a rival to the explanation based on learned helplessness soon emerged. This rival explanation (Cohen, 1978) was based on the attentional framework previously discussed in this chapter. Basically, a stressor causes a narrowing of attention and uses up some of the human's limited capacity (see Chapter 6). It is postulated that unpredictable stressors cause a greater shift of attention to the stressor, and therefore a greater capacity drain and a greater likelihood of not attending to cues unrelated to the stressor, than do predictable stressors. This explanation predicts that a situation with high mental work load, even with no physical stressor present, will produce an aftereffect similar to those already discussed. Results supported this prediction (Cohen & Spacapan, 1978), which is difficult to explain in terms of learned helplessness.

It is too early to state all the practical implications of these findings for the human factors designer. However, it is quite clear that noise stress outside the laboratory has such harmful effects as lowering children's reading scores, school achievement and cognitive skills (Cohen, 1980; (Cohen, Evans, Krantz, Stokols, & Kelly, 1981). The findings that brief exposures to stressors that do not cause impairment during exposure nevertheless cause measurable aftereffects also is ominous. The human factors specialist will need to be aware of new research findings in this area in order to cope with subtle aftereffects that can do serious harm to workers in the long run. What little evidence is present suggests that adaptation to noise over a one-year period is minimal (Cohen et al., 1981).

SUMMARY

Stress is the nonspecific response of the body to any demand. However, this medical definition is not as useful to the human factors specialist as the more psychologically oriented definition of a stressful situation. This is defined as a situation where there is a substantial imbalance between the demands imposed by the environment and the human's capability to successfully handle those demands. Stressful situations are created by overload and also by underload. The Yerkes-Dodson law relates stress and arousal. New theoretical conceptions link stress and attention.

Human core temperature can vary over only a narrow range before performance, and even life itself, are impaired. The body's success in exchanging heat with the environment depends on dry-bulb temperature, humidity, air movement velocity, and radiant temperature. There is currently a dispute over how close the heat tolerance level for mental performance approaches the physiological limit for heat tolerance. However, isodecrement curves for heat stress are available.

Cold stress also is undesirable. Protective clothing is worn to prevent cold stress and clo values for individual items of clothing are available.

Information load is becoming a more important stressor as society replaces physical work with mental work. Information overload will cause systematic performance decrements. Information underload leads to boredom, which is also a stressful situation. The ultimate stressful situation, danger to life and limb, obeys the same behavioral principles as do lesser stressors. Crowding acts to intensify feelings and is not necessarily a stressor unless the person being crowded is in an institutional setting.

Evaluating long-term effects of stress in the workplace is difficult. Aftereffects of stress can alter behavior after the stressor is no longer present.

20 LEGAL ASPECTS OF HUMAN FACTORS

In this chapter we discuss some legal issues that the human factors specialist must be aware of. While firms employing the human factors specialist have attorneys to handle legal affairs, an understanding of some basic legal ideas will help the specialist communicate with these attorneys. As you will see, there is a great deal of difference in the framework of legal concepts and the framework of science. This often makes it difficult for scientists and lawyers to understand each others' point of view. Yet there are very practical reasons for becoming familiar with some points of law.

The most important legal arena for the human factors specialist is that of product liability. Human factors tries to create products that are safe and reliable by taking the human's propensities for error into account. But proving a product is safe in a court of law is very different from designing a safe product! In this chapter we discuss legal terminology, legal descriptions of product defects, the use of warnings, and government standards. We also briefly discuss the Occupational Safety and Health Act and the litigation it has caused.

PRODUCT LIABILITY

How would you define a safe product? "A perfectly safe product is one that can do no objectionable harm, at any time, under any circumstances, to people (McAllister, 1975, p. 181)." How many perfectly safe products can you list? None. There is no such thing as a perfectly safe product. A pencil can poke someone in the eye, a fire extinguisher can fall from its mounting bracket and break someone's toe, and a button can be swallowed by a child causing death by choking.

Any product, especially if used in a manner not intended by the manufacturer, can inflict harm. Thus, if society decided that an absolute standard of perfect safety was required before any product could be distributed, all manufacturers would close their doors. Instead, there must be a balance between the potential harm a product may cause and the benefits to society of a plentiful supply of products. This balance is achieved through the laws and regulations, promulgated by government and interpreted by the judicial system, that control product liability. Such balance is never a static process. As laws and judicial interpretations evolve, the definition of a reasonably safe product changes. Therefore, any discussion of product liability must be preceded by a review of some basic concepts in law.

Legal Terminology and Concepts

The human factors specialist accustomed to the precision of technology and the clear logic of scientific method often finds the legal process fuzzy, frustrating, and even fearsome. To the uninitiated, legal decisions may seem to devolve from fine points of legal technology rather than from the obvious common-sense merits of substantive issues. A theory in the legal sense bears scant resemblance to theory as used by scientists. Even the most interested and concerned layperson is tempted to leave legal issues solely to lawyers. But such a strategy is akin to locking the barn door after the horse has escaped. The best time for the human factors specialist to consider legal issues is before any litigation develops. Early consultation with lawyers allows the specialist to discover the legal adequacy of a product before it is delivered to users. For example, products must comply with statutory standards. Lawyers know the relevant statutes better than most human factors specialists. It is true that law is complex and written in a dialect of English that is designed more to confuse than to communicate. Since most governmental bodies are composed primarily of lawyers, this sort of obfuscation is to be expected since the legal profession gains substantial financial benefits from jargon. If the average citizen understood the law, demand for legal services would diminish. So it is clear that the human factors specialist who attempts to understand law is tackling a tough job. This chapter will not make you a legal expert. But it will explain enough of legal technology so that you can converse with a product-liability lawyer without being overwhelmed. (See Jobe, 1981; Walkowiak, 1980 for more technical discussion of legal issues.) In a product liability lawsuit an injured party—the plaintiff—brings suit against a manufacturer or seller—the defendant—who has provided the allegedly defective product. In order to win the case the plaintiff must prove that the product was defective. The mere fact of injury is not sufficient to establish liability. For example, a plaintiff who was injured by a knife when peeling carrots would not carry the day by claiming the knife was defective because it was sharp enough to cut human flesh. In addition to the injury, a legal theory of liability must be established by the plaintiff. Current law allows several legal principles under which the plaintiff can bring a product-liability action: negligence, strict liability, implied warranty, express warranty, and misrepresentation (Weinstein, Twerski, Piehler, & Donaher, 1978).

Negligence To establish negligence, the plaintiff must prove that the conduct of the defendant involved an unreasonably great risk of causing damage. The plaintiff must prove that the defendant failed to exercise ordinary or reasonable care. A manufacturer has a legal obligation to use new developments that will decrease the level of risk associated with his or her product. Failure to keep abreast of new technology can be grounds for negligence. For example, one case involved an exploding boiler that had injured the plaintiff (*Marsh* v. *Babcock and Wilcox,* 1932). The defendant had continued to use an obsolete hydrostatic test of boiler integrity when an improved test was available. The jury found the defendant negligent for failing to use the more sophisticated test that had a higher probability of discovering flaws in the boiler.

There is no absolute standard for "unreasonably great risk." It is quite possible that the same defendant would be judged negligent in one court and innocent in another court. This kind of variability is disturbing for human factors specialists who are accustomed to consistency in behavior of people and systems. Although judges tend to decrease variability, there can still be differences of opinion among judges. It is not unheard of for an appellate judge to overturn the decision of a lower court. Uncertainty is the handmaiden of litigation.

Strict Liability and Implied Warranty Strict liability is another theory of product liability. A tort in law is a wrongful act or injury for which a plaintiff can bring civil (as opposed to criminal) charges. Examples of torts are malpractice and defamation of character. The theory of strict liability is quite different from that of negligence even though both are torts. Negligence focuses on the conduct of the manufacturer. Strict liability focuses on the quality of the product, regardless of whether or not the manufacturer acted reasonably. Strict liability applies when the product is of an inherently dangerous nature. Strict liability represents a social policy that shifts the costs of the inevitable injuries associated with a dangerous product onto those who stand to profit from the product. Thus, arguments that the defendant exercised all possible care, while important for negligence cases, are irrelevant in a strict liability case. Another advantage of the strict liability theory for the plaintiff is that liability attaches to the entire distributive chain, rather than only to the manufacturer as in negligence. Thus, the retailer, wholesaler, and distributor can all be liable. This is true even if they have no duty to inspect the product and indeed, even if they have no practical way to inspect. Thus, while failure to inspect might be reasonable—therefore removing any grounds for liability under a negligence action—it offers no protection against strict liability. Reasonable conduct is no defense against strict liability claims.

The Uniform Commercial Code states that goods must be fit for the ordinary purposes for which such goods are used. This implied warranty of merchantability offers an avenue similar to strict liability. However, a plaintiff proceeding under the Uniform Commercial Code, which was designed for commercial enterprises rather than specifically for consumers, must take care to comply with several complex provisions of the Code. For example, the injured party must give the manufacturer notice of the product defect within a reasonable time after the injury occurred. Manufacturers often try to get around implied warrantability by issuing disclaimers. Courts have disagreed about whether such disclaimers are valid when a product caused a personal injury (Weinstein et al., 1978). However, it is clear that disclaimers are no defense against claims of strict liability.

Express Warranty and Misrepresentation The two legal theories discussed above both require the plaintiff to establish a product defect. This step is not necessary under a theory of express warranty. A plaintiff can establish liability under a theory of express warranty if he or she can prove that injury was due to the failure of the product to meet the warranty given by the seller. A good example is the case of *Crocker* v. *Winthrop Laboratories* (1974). Winthrop Laboratories sold a painkiller that was advertised as nonaddictive. The plaintiff took this medicine as prescribed by his physician and became addicted. The plaintiff died from an overdose of the drug. The court found the defendant liable even though it truly believed the drug was nonaddictive at the time the drug was advertised.

An implied warranty occurs under the Uniform Commercial Code (Section 2-315) when the seller knows that the buyer is relying on the seller's judgment to select suitable goods. Even though the product has no defect, and no express warranty was made by the seller, the seller can still be liable. An implied warranty arises whenever there is reliance by the buyer. For example, a buyer purchases wood from a lumber yard and asks the yard to select material suitable for an outdoor deck. If some of the wood has a crack so that it cannot bear the weight of a person standing on the deck, the yard can be held liable if an injury occurs because of this. Note that the crack in the wood is not necessarily a defect since lumber yards normally sell grades of wood that have cracks.

Product Defects

The first step in establishing product liability in cases where no express warranty or misrepresentation is involved is to prove that the product was defective. If a plaintiff cannot establish this, his or her cause will fail. Since there is no perfectly safe product, it is not sufficient for the plaintiff to demonstrate that a product was dangerous. A sharp knife is dangerous but that does not mean it is defective. Product defects arise from two sources. First, a flaw in the manufacturing process may cause a defect. This results in a defective product that does not meet the manufacturer's own quality-control standards. The classic example of a manufacturing defect is an exploding soda-pop bottle. Second, a product may have a design defect. Human factors specialists are far more likely to be involved in product liability cases based on design defects than on manufacturing defects.

How does one determine that a design defect has occurred? A manufacturer may have used reasonable care in designing and even testing a product before it was released to the public but such reasonable care is no protection against the theory of strict liability. The manufacturer is liable for any defects in a product, including those that were unknown at the time of manufacture. A defective product is "unreasonably dangerous." All products present some risk. The law must balance the risks against the usefulness of the product and the cost of providing greater safety in order to determine if a product is only reasonably dangerous. A reasonably dangerous product means that "a reasonable person who had actual knowledge of the product's potential for harm would conclude that it was proper to market it in that condition (Weinstein et al., 1978, p. 32)."

There is no absolute standard for unreasonable danger. Expert witness testimony is often used to establish the degree of danger associated with a product. But even experts disagree. It is not unusual to find the human factors specialist testifying for the plaintiff at odds with the specialist for the defendant. However, the following checklist (Wade, 1965) gives seven factors that need be weighed in product liability cases.

1. The usefulness and desirability of the product.
2. The availability of other and safer products to meet the same needs.
3. The likelihood of injury and its probable seriousness.
4. The obviousness of the danger.
5. Common knowledge and normal public expectation of the danger (particularly for established products).
6. The avoidability of injury by care in use of the product (including the effect of instructions or warnings).
7. The ability to eliminate the danger without seriously impairing the usefulness of the product or making it unduly expensive.

Legal precedent has not always considered all of these points. One especially important case (*Campo* v. *Scofield,* 1950) ignored all but number 4. The plaintiff was feeding onions into a machine that cut off the tops. He caught his hands in the revolving rollers and was severely injured. The plaintiff claimed that the absence of safety guards on the machine showed negligent design. The court held for the defendant on the grounds that the danger was obvious. Only a "concealed danger" or a "latent defect" would have been grounds for establishing liability according to the court. This opinion is most unfortunate for the human factors specialist. It implies that a manufacturer is better off not providing any safety devices and instead should make machinery that is dangerous in an obvious way instead of designing to reduce danger. In fact, according to the logic of this decision, attempts at reducing danger might make the manufacturer more liable by reducing an obvious danger to a concealed danger or latent defect.

The unfortunate Campo doctrine was rejected by the New York Court of Appeals (*Metcallef* v. *Miehle Co.,* 1976). The plaintiff operated a large photo-offset printing press. A foreign object was found on the plate; this would cause a blemish known as a "hickie" in the printing trade. The plaintiff tried to remove the hickie while the press was running. This was called "chasing the hickie on the run" and was a common practice since stopping the press and restarting it would take about 3 hours. While chasing the hickie the plaintiff's hand was drawn into the printing press. No safety guard was built into the machine and the emergency shutoff switch could not be reached quickly from the plaintiff's location. The testimony of an expert witness revealed that a safety guard would have been good engineering practice since the danger of a hand being caught in the rollers was rather predictable. Since the danger was obvious, the Campo doctrine would find for the defendant. The court rejected the Campo doctrine and ruled that obviousness of danger should be only one factor, rather than the only factor, in determining a product defect. From a human factors perspective, this more recent decision is far more reasonable since it encourages attempts to increase product safety by removing the legal incentive to keep dangers obvious. However, some courts in other states still adhere to the Campo doctrine.

Warnings

The general position of courts regarding warnings is clear: always use warnings. The logic behind this approach is based on the low cost of supplying warnings. Usually a warning can be added by merely printing a few extra words on a label or painting some words on a piece of machinery. Therefore, many cases have found manufacturers liable for failure to warn, even when the (missing) warning would have been of dubious value (Weinstein et al., 1978). From a human factors perspective, such court decisions are unsound. While human factors can contribute to making warnings more effective, the legal issue ignores this at present and concentrates on the presence or absence of a warning rather than its effectiveness.

A representative case involves a bottle of Faberge's Tigress Cologne, a product with a 27-year history without accident (*Moran* v *Faberge,* 1975). Two teenage girls were spending the evening watching a candle burn down and had decided that the candle was not scented. One of them decided to add scent and began to pour the cologne onto the candle at a point below the flame. A fire burst out immediately and the 17-year-old plaintiff was injured. During the trial, it was determined that cologne has a high percentage of alcohol. Thus, cologne is a dangerous combustible with a flash point of 73°F. The court found for the plaintiff on the grounds that the cologne bottle failed to warn of flammability.

From a human factors perspective, excessive warnings are as bad as insufficient warnings. People become accustomed to the warnings and tend to ignore them. Warnings should be reserved for high-probability events. Even then, it is difficult to get people to pay attention to them.

One obvious example of this difficulty occurred when a human factors specialist was observing workers in an explosives plant. Workers were supposed to carry only one detonator at a time. The human factors specialist decided to terminate his observations when a worker carrying a handful of detonators dropped one, but caught it on his shoe before it struck the concrete floor (Kolb & Ross, 1980, p. 151). It is doubtful that a sign warning "Dropping Detonators Can Be Harmful to Your Health" would have altered worker behavior. Since people tend to ignore warnings of even high-probability risks, there is some likelihood that adding product warnings about low-probability risks will only increase the odds that all warnings will be ignored. In order to guard against all possibilities of misusing a bottle of cologne, the warning label would have to be much larger than the bottle itself. A separate warning sheet would be thrown away and is far less useful than a limited warning that was concise enough to remain attached to the product. It is poor design to have to include warnings against such low-probability dangers as pouring cologne onto a candle that occur once in 27 years.

Yet there have been cases where courts have demanded warnings that would have failed to increase safety at all. In one case a boy was injured when jumping on a trampoline. He missed the canvas bed, which was only 16 inches in diameter, and caught one leg on the supporting springs. The leg had to be amputated. The court found the product to be defective solely because of a failure to warn that the user's foot could be trapped in the supporting cables (*Nissen Trampoline Co.* v. *Terre Haute National Bank*). It is hard to imagine how such a warning would have made the product safe. The real problem is the design of a trampoline that requires the user to land with both feet in a 6-inch diameter area. The solution is to make the canvas larger and to cover the support cables. Basing a decision on an omitted useless warning diverts attention from the true problem. (To be fair we note that a dissenting opinion of the Indiana Supreme Court, 358 N.E. 2d 974 at 980, made many of these points.)

All the human factors expert can do is continue to improve warnings and wait for the law to catch up. Human factors specialists have been working on improving written messages for some time (Chapanis, 1967). Warnings like "The batteries in the AN/MSQ-55 could be a lethal source of electrical power under certain conditions" have been changed to "LOOK OUT!—This can kill you!" Eventually the legal community will take the utility of a warning into account.

Governmental Standards

The various regulatory agencies of the federal government often issue standards that control manufactured products. For example, the Environmental Protection Agency controls the amount of permissible pollutants in automobile emissions and the Nuclear Regulatory Commission must approve the design and construction of nuclear power plants. In this chapter we do not consider whether it is more prudent for a manufacturer to comply with these regulations, to lobby for reduced standards as did the automotive industry, to shut down plants, or to operate without compliance and regard any fines as additional operating costs. We assume that the manufacturer is in full compliance with governmental standards. Does following these standards reduce or eliminate product liability?

An interesting case concerns a helicopter crash (*Berkebile* v. *Brantly Helicopter Corp.*, 1975). The pilot took off with almost no fuel and the helicopter crashed while climbing. The plaintiff claimed that the helicopter was defective because it allowed the pilot only one second to engage the autorotation system, a safety device used in case of engine failure. However, this short time was in full compliance with Federal Aviation Agency standards for helicopter autorotation systems. The FAA regulation stated that "In taking corrective action, the time delay for all flight conditions shall be based on the normal pilot reaction time, except that for the cruise condition the time delay shall not be less than one second." The defendant claimed that taking off with an almost empty gasoline tank put the product to an "abnormal use" that was the true cause of the accident. The court decided for the plaintiff on the grounds that the autorotation system was designed to prevent a crash in case of engine failure for any reason. Running out of gas was one of several events that could cause engine failure so that the reason for engine failure was deemed irrelevant by the court.

The court ruled that failure to comply with the regulation would have been negligence. So compliance is protection against lawsuits based on the theory of negligence. However, governmental agencies set minimum standards. Therefore, compliance is no protection against lawsuits based on the theory of strict liability. Each manufacturer must decide if compliance with minimum standards will be sufficient to produce a reasonably safe product. A plaintiff cannot be precluded from recovery in a strict liability case because of his or her own negligence—in this case, taking off without sufficient fuel. Some readers may feel that this decision was incorrect because contributory negligence should be taken into account. However, the court ruled that the existence of due care on the part of the consumer is irrelevant. This means that in strict liability cases the product must be safe even if used in ways not intended by the manufacturer.

From a human factors perspective, one second is inadequate to engage an autorotation system. It may take a pilot more than one second to determine that the engine has failed, let alone to plan and execute corrective action. Thus, the helicopter was defective. But there is another human factors issue in this case. The normal takeoff procedures—checklist, etc.—failed since there was insufficient fuel. Should helicopters have interlocks that prevent the engine from being started if the fuel tank is not at least half full? Should the human factors definition of a reasonably safe product be stricter than the legal definition? Human factors cannot be expected to protect a system from any and all human error. But should human factors be expected to offer protection from human error that has predictable drastic consequences such as loss of life? If your answer to this question is "no," why spend money on human factors? If your answer is "yes," why did you disconnect the buzzer on your automobile seat belts on your old car and why are you opposed to paying for air bags for your new car?

There is still plenty of room for additional human factors research on warnings. For example, most human factors experts would agree that the effective life of a warning is short. But there is a shortage of scientific evidence that empirically demonstrates how brief the effective life of a warning on a power press really is. Would changing the wording of the warning every month increase safety? Many similar questions are unanswered at present. The human factors profession could provide relevant data that would be useful in court whenever warnings are an important factor.

A SLICE OF LIFE

Product Liability Case Histories: The Expert Human Factors Witness

The preceding discussions of case histories have concentrated on the rulings issued by the court. Here we consider the point of view of the human factors expert who is often called to testify in such proceedings. The first report we consider (Samuels, 1980) represents the typical experience of an expert who is new to the legal arena. It illustrates some of the difficulties that an expert witness can encounter, especially when his or her own common sense and professional experience lead to different legal conclusions than those reached by the jury.

Samuels, an industrial designer, was contacted by a lawyer in 1975 as an expert witness in a product liability case. This case concerned an accident that occurred in 1973 when a liquid drain cleaner, placed in an upright position with its top on, was bumped by a woman getting up from the toilet. The drain cleaner fell into her lap, the cap dropped off, and a few ounces of the liquid contents severely burned her stomach, genitals, thighs, and legs. The lawyer believed that if it could be demonstrated that the product did not use technology available at the time of the accident to protect users from the contents—93% sulfuric acid—he could win the case. It is not clear from Samuels' report whether the lawyer was proceding under a theory of negligence or of strict liability.

Samuels' first job was to analyze the container. He decided it was inadequate for several reasons. It was a tall cylinder that was easily tipped over. It was constructed of polypropyrene, a plastic that interacts with sulfuric acid in a manner that renders the container brittle and potentially dangerous if shelf life exceeded 2 weeks. The internal threads of the "child resistant" top were corroded from contact with the acid contents. The warning on the label had been partially eroded from acid dripping down the side of the bottle.

The next step was to find a bottle available in 1973 that lacked these defects. Samuels was unable to accomplish this and so designed a bottle with a spring-loaded top that was permanently attached to the bottle. Opening the bottle required two hands pushing on the screw-on closure. If the bottle fell accidentally it would close itself and only a few drops would get out.

In 1978 Samuels gave his deposition to six lawyers representing the defendant: "For six hours these lawyers questioned me on every aspect of my professional qualifications, experience, clients, patents, philosophy and my analysis of the container in question. Every word, phrase, idea and conclusion was dealt with. It was this phase of the project that made me think that my fee was much too low for what I was going through." This was a milestone in Samuels' legal education. He discovered that many of the points he considered crucial as a designer were considered irrelevant by the lawyers. The fading of the instructions and warnings had no direct causal-effect relationship to the accident and so did not matter. The lawyers also felt that the material used to make the container was also irrelevant since it didn't directly cause the accident. This conclusion was reached despite Samuels' research about the corroded threads.

The trial took place 5 years after the accident, 3 years after Samuels was hired, 2 years after he built his new bottle, and 1 year after giving his deposition. Speed is not a characteristic of our legal system. After some "clever manuevering by the opposing lawyers" the judge decided that Samuels could not testify about acids since he was not a chemist. Instead, he was to comment on the original container as if it were empty. Since this instruction makes little sense to a designer—few people design empty containers—the trial was interrupted seven times by the judge who invited Samuels and the lawyers into chambers so he could instruct Samuels about what was proper testimony. Samuels was now positive his fee was too low. Testimony lasted for 5 hours. At the conclusion the plaintiff's lawyer took the model Samuels had constructed and placed it on the rail before the jury. The bottle was filled with water. The lawyer dramatically smacked the 3-year old model, knocking it down to demonstrate that it would not leak. Of course, Samuels had not been informed of this test and expected the bottle to burst apart. Fortunately, the spring held and no water leaked out.

To Samuels' surprise and dismay the jury found for the defendant. It ruled that the woman had been careless. The manufacturer was not required to anticipate misuse of sulfuric acid in this case. Samuels concluded his narration by stating he would be reluctant to serve as an expert witness again.

This case illustrates the inconsistency of legal practice. Common sense and expert testimony of a human factors witness indicates that the cap was defective. According to the legal principles discussed earlier in this chapter, the manufacturer should have been liable. Of course, the plaintiff has the right to appeal but Samuels did not state if such an appeal was forthcoming.

Human factors experts sometimes win cases too. The defective ring guard on the autopilot discussed in Chapter 10 is one such favorable example (Bowen & Mauro, 1980). Recall that the aircraft crashed because the third pilot had leaned against the test button that caused a climbing turn. Here we discuss the legal, rather than the design, aspects of this product liability case.

Since all three pilots were killed in the crash, there was no possibility of direct testimony to establish that the test button had been pushed accidentally. Therefore the legal doctrine of *res ipsa loquitor*—Latin for the thing speaks for itself—was used to establish circumstantial evidence of product liability. The crucial legal aspect of the case was the judge's decision to allow a mock-up to be introduced into court. The defense spent a day arguing unsuccessfully that the mock-up was not admissible. However, the plaintiff armed with records of how the model had been built was able to show that the model was a true copy. Using the model, the plaintiff convinced the jury that the test button was the cause of the accident.

The plaintiff used large models of the indicator instruments to prove that the pilot would not have noticed the change of course until it was too late to take corrective action. Thus, the jury was convinced that the product was not protected against inadvertent activation of the test button. The jury awarded $3.5 millions for the plaintiff. After the verdict was given, members of the jury told the human factors experts for the plaintiff that they would not have been able to understand the issues without the working models and visual aids. A thing may speak for itself, but nonverbal presentations help the thing to speak more clearly.

OCCUPATIONAL SAFETY AND HEALTH ACT

The Occupational Safety and Health Administration (OSHA) was created by an Act of Congress that was signed into law on December 29, 1970 and took effect on April 28, 1971. In the slightly more than one decade that OSHA has existed, it has been cursed and reviled by most business critics as a prime example of government intervention that drastically increased the cost of doing business while adding little to health and safety in the workplace. Proof of considerable dissatisfaction with the priorities and effectiveness of OSHA can be found in the 46 bills introduced in the Ninety-sixth Congress to amend OSHA; 6 of these bills were intended to repeal the original act. Yet when Congress passed the Act it expected OSHA to improve health and safety in the workplace, a goal that few can claim is undesirable. What went wrong?

The Act had several parts (see Rothstein, 1981 for a more complete description and evaluation), some of which are not directly related to human factors. However, the part of the Act that is most closely related to human factors, the setting of safety standards in the workplace, also was the part responsible for much of the criticism of OSHA. The Secretary of Labor was given the power to set and enforce occupational safety and health standards. Employers who failed to comply with these standards could be fined. In order to speed the creation of such standards, a special authority good only for the first 2 years of OSHA permitted "national consensus standards" and "established federal standards" that did not have to go through the more elaborate rulemaking procedures specified by the Act. These original standards were hastily constructed without benefit of human factors research. Hence, some of these original standards were incorrect, trivial, and even outright foolish. For example, OSHA required that all toilet seats used in workplaces be "open-front" and prohibited the use of ice in drinking water. The basic problem with these standards was that they mandated specific construction standards rather than more flexible performance standards. For example, a performance standard might state that a ladder be capable of holding 400 pounds. Many kinds of ladder would satisfy this performance standard. However, a specification standard that ladder rungs must be made of wood and must be one inch in diameter gives the employer little choice about what kind of ladder must be provided in the workplace. Finally, in 1978 OSHA revoked 607 general industry standards and 321 special industry standards but many regarded this action as too little too late.

The legal issues raised by industries protesting OSHA action were, and still remain, formidable and complex. The best-known is the Supreme Court decision in *Industrial Union Department, AFL-CIO* v. *American Petroleum Institute;* this case is often called the billion-dollar benzene blunder (Grove, 1980). At question was OSHA's benzene health standard that had been invalidated by the Fifth Circuit Court on the grounds that OSHA neglected to provide a quantitative estimate of the health benefits resulting from lowering the permissible exposure limit. The standard applied by federal courts in reviewing cases involving protection of the public against hazardous substances is called the "substantial evidence" test. If the Secretary of Labor has formulated a standard for which factual verification is available, all the courts must do is to ensure that the standard is not arbitrary or capricious (Grove, 1980). But many standards lack precise empirical data because it might take years to establish firm scientific grounds for effects of some hazard. Here the issue becomes cloudy and the courts have demanded that the Department of Labor must produce evidence justifying its actions. Several cases have rejected standards based on tests of laboratory animals (Grove, 1980) because this did not provide substantial evidence that humans would necessarily be at hazard. Thus, extrapolation from animal data is legally not sufficient to establish limits for human tolerance.

In 1977 OSHA reduced the permissible level for benzene from 10 ppm to 1 ppm because of the hazard of leukemia. The rationale for this lowered standard was an assumption that any level of benzene could be harmful and therefore the reduction would protect workers. In a 5 to 4 decision the Supreme Court affirmed the decision of the Fifth Circuit and eliminated the 1 ppm standard. The Court held that OSHA had failed to establish a "significant risk of harm" under the old 10 ppm standard. But legal matters are seldom this clear. The Fifth Circuit had held that the Secretary had to determine "whether the benefits expected from a standard bear a reasonable relationship to the costs imposed by the standard." This was a major victory for industry. However, the Supreme Court made its ruling on narrow grounds—that is, whether there was a risk of harm—and avoided the cost-benefit issue. Thus, the Supreme Court did not go on record as requiring OSHA to conduct a cost-benefit analysis as ruled by the Fifth Circuit. Furthermore, since the Supreme Court was sharply divided—five separate opinions were issued by the Justices—it is clear that there will be many more legal tests of OSHA standards.

EQUAL EMPLOYMENT OPPORTUNITY

The legal basis for equal employment opportunities for women was Title VII of the Civil Rights Act of 1964. Two subsequent Executive Orders (11375 in 1968 and 11246 in 1965) made it clear that sex discrimination would be prohibited at work. Stereotypes about women being the "weaker" sex kept women out of many skilled trades, especially those which required mechanical aptitude (Safilios-Rothschild, 1981). Even when women were able to obtain such jobs, for example, as construction workers, they were placed in a "no-win" situation. A successful female worker was regarded as highly unusual whereas any unsuccessful worker was considered typical with her inadequate job performance generalized to all women. Women often were compelled to use tools designed for men and this made their jobs even more difficult. (See Chapter 10, Controls and Tools and Chapter 15, Workspace Design for discussion of this point.) However, companies slowly learned that jobs could be redesigned for women through the use of human factors.

Raising the Equal Opportunity Ladder

An outstanding example of job redesign to achieve affirmative action concerns the outside craft jobs in the Bell Telephone system (Sheridan, 1975). These jobs require physical exertion with tasks like climbing

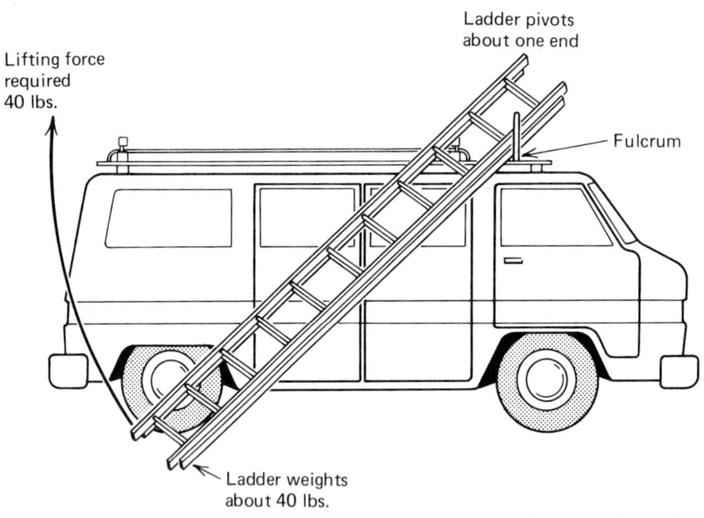

Figure 20-1. The original telephone truck and ladder. (Adapted from Sheridan, 1975.)

telephone poles, using 3-pound hammers, and carrying 24- or 28-foot ladders. The Bell System had agreed with the federal government to have 19% of these outside craft jobs filled by women. But the company had great difficulty in finding women who wanted such jobs. Only half of those women completed job training. Those women who successfully completed training usually lasted for less than a year on the job. Furthermore it appeared that women had a higher accident rate than men. Ironically, this could lead to problems with OSHA, yet another instance of governmental agencies acting in ways that industry regards as conflicting. How could the Bell System meet its obligation to hire and keep female outside craft workers without compromising occupational safety?

A major stumbling block for female craft workers was handling ladders. The first problem was getting the long ladder on and off the truck (Figure 20-1). This required a lifting force of about 40 pounds, an unsafe task for women (see Figure 15-12). Redesigning the rack holding the ladder by moving the fulcrum to the middle of the ladder decreased the lifting force to about 10 pounds (Figure 20-2). This was a great improvement, but not a complete solution since many women are too short to get the ladder on top of the $6\frac{1}{2}$-foot-high van. Indeed, one 4-foot 11-inch trainee required a stepladder to accomplish this. A stepladder will be unstable in some situations and so may violate OSHA standards. It may be necessary to mount the ladder lower or to bolt a vertical ladder to the side of the truck to solve this problem.

Another problem for women was extending the top section of the ladder. A standard ladder has a single pulley and a rope to accomplish this. There is no mechanical advantage and the worker must lift the full weight of the extension section, about 40 pounds. By adding a second

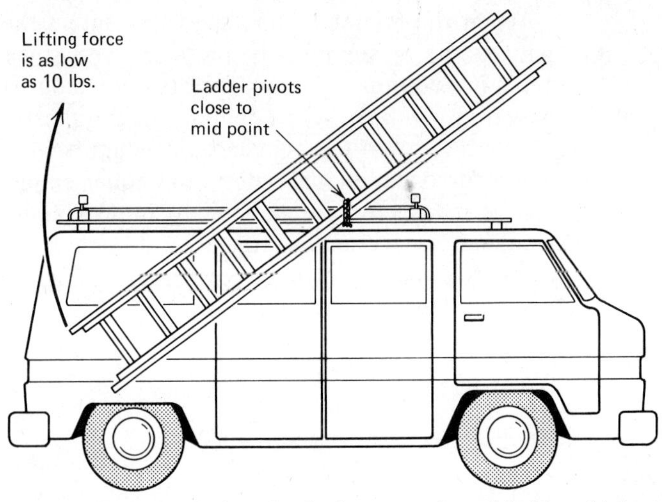

Figure 20-2. Redesigned telephone truck and ladder. (Adapted from Sheridan, 1975.)

pulley to provide mechanical advantage, this force was cut in half. (Adding a third pulley to reduce force even more didn't work since the rope tended to get tangled.) Another modification made it easier for women to raise the extended ladder. The last problem to be overcome was carrying the heavy ladder. Men usually carried it on their shoulder but this was impossible for women. (It also was a poor design for men.) This problem was solved by adding a wheelbarrow to one end of the ladder, drastically reducing the need for physical strength.

Thus, redesign of the ladder and associated tools made a job that was extremely difficult for women into one that could be accomplished. It also made the job easier for men. So affirmative action can help both sexes.

SUMMARY

Human factors specialists are becoming more involved with the legal aspects of human factors. Employees may meet with attorneys to discuss product safety and government standards. Consultants may serve as expert witnesses in court.

No product is 100% safe. Society must balance the potential harm a product may cause against the benefits of a plentiful supply of products. Legal principles under which a plaintiff can bring a product-liability action include negligence, strict liability, implied warranty, and misrepresentation. The human factors specialist can improve product safety by designing effective warnings. Compliance with governmental standards does not necessarily protect a manufacturer from legal action in strict liability cases.

When the human factors specialist enters the legal arena as an expert witness, he or she must be prepared with some knowledge of legal concepts in addition to human factors knowledge. Opposing attorneys often may try to limit the scope of human factors testimony.

There is room for human factors inputs in meeting government standards for occupational safety and equal employment opportunity. Jobs can be redesigned, for example by improving tools, so that both men and women can accomplish them. Although such redesign has been performed primarily to meet government standards, it can also make jobs easier for the people who perform them.

REFERENCES

Acton, M. B. At home on the range? *Human Factors Society Bulletin,* 1976, *19*(7), 2. Also corrected 1976, *19*(8), 2.

Adams, J. A. A closed-loop theory of motor learning. In G. E. Stelmach (Ed.), *Motor control: Issues and trends.* New York: Academic Press, 1976.

Adams, J. A. Issues in human reliability. *Human Factors,* 1982, *24,* 1–10.

Adams, J. J. *A simplified method for measuring transfer functions.* Technical Note D–1782, NASA, 1963.

Adkins, J. M., Boss, D. W., Driscoll, P. A., & Michtom, G. C. The ease of learning and use of line editors and screen editors. CS 590D Project Report, Department of Computer Sciences, Purdue University, 1980.

Aiello, J. R., & Baum, A. (Eds.). *Residential crowding and design.* New York: Plenum, 1979.

Albanese, R. A. Mathematical analysis and computer simulation in military mission workload assessment. *Proceedings of the AGARD conference on methods to assess workload.* AGARD-CPP-216, April 1977, A13-1-A13-6.

Allport, D. A., Antonis, B., & Reynolds, P. On the division of attention: A disproof of the single-channel hypothesis. *Quarterly Journal of Experimental Psychology,* 1972, *24,* 225–235.

Altman, I. Some perspectives on the study of man-environment phenomena. *Representative Research in Social Psychology,* 1973, 4, 109-126.

American National Standards Institute. *Methods for the calculation of the Articulation Index.* ANSI S3.5–1969.

American National Standards Institute. *American National Standard specification for sound level meters.* ANSI S1.4–1971. (a)

American National Standards Institute. *American National Standard specification for octave, half-octave, and third-octave band filter sets.* ANSI S1.11–1966 (R1971). (b)

American National Standards Institute. *American National Standard immediate evacuation signal for use in industrial installations.* ANSI/ANS N2.3–1979.

American Society of Civil Engineers. *Airports, key to the air transportation system.* New York: ASCE, 1971.

Anyakora, S. N., & Lees, F. P. Detection of instrument malfunction by the process operator. *The Chemical Engineer,* August, 1972, 304–309.

Ashford, N., & Wright, P. H. *Airport engineering.* New York: Wiley, 1979.

ASHRAE. *ASHRAE Handbook and product directory.* New York: ASHRAE, 1977.

Atal, B. S., & Hanauer, S. Speech analysis and synthesis by linear prediction of the speech wave. *Journal of the Acoustical Society of America,* 1971, *50,* 637–655.

Attneave, F. *Applications of information theory to psychology.* New York: Holt, Rinehart & Winston, 1959.

Ayoub, M., & LoPresti, P. The determination of an optimum size cylindrical handle by use of electromyography. *Ergonomics,* 1971, *14,* 509–518.

Azer, N.Z. & Hsu, S. OSHA heat stress standards and the WBGT index. ASHRAE Transactions, 1977, 83, part 2, 30-40.

Babbs, F. W. A design layout method for relating seating to the occupant and vehicle. *Ergonomics,* 1979, *22,* 227–234.

Backstrom, J., Cole, R., & Duffley, G. The effect of menu type on user efficiency. CS 590D Project Report, Department of Computer Sciences, Purdue University, 1981.

Baddeley, A. D. Diver performance. In J. D. Woods & J. N. Lythgoe (Eds.), *Underwater science.* London: Oxford University Press, 1971.

Baddeley, A. D. Selective attention and performance in dangerous environments. *British Journal of Psychology,* 1972, *63,* 537–546.

Baddeley, A. D., & Hitch, G. Working memory. In G. H. Bower (Ed.), *The psychology of learning and motivation,* Vol. 8. New York: Academic Press, 1974.

Ballou, D. P., & Pazer, H. L., The impact of inspector fallibility on the inspection process in serial production systems. *The Institute of Management Science,* April 1982.

Banks, W. W. & Boone, M. P. A method for quantifying control accessibility. *Human Factors,* 1981, 23, 299–303.

Banks, W. W., & Goehring, G. S. The effects of degraded visual and tactile information on diver work performance. *Human Factors,* 1979, *21,* 409–415.

Bar-Hillel, M. The base-rate fallacy in probability judgments. *Acta Psychologica,* 1980, *44,* 211–233.

Barnwell, T. P., III. Objective measures for speech quality testing. *Journal of the Acoustical Society of America,* 1979, *66,* 1656–1663.

Baron, R. A., Byrne, D., & Kantowitz, B. H. *Psychology: Understanding behavior.* New York: Holt, Rinehart and Winston, 1980.

Bateman, R. P. Design evaluation of an interface between passengers and an automated ground transportation system. *Proceedings of the Human Factors Society,* 1979, *23,* 119–123.

Baum, A., & Valins, S. Architecture and social behavior. Hillsdale, N.J.: Lawrence Erlbaum Associates, 1977.

Baxa, D. E., & Seireg, A. The use of quantitative criteria for the optimum design of concert halls. *Journal of the Acoustical Society of America,* 1980, *67,* 2045–2054.

Beach, L. R., & Peterson, C. R. Subjective probabilities for unions of events. *Psychonomic Science,* 1966, *5,* 307–308.

Bejczy, A. K. Sensors, controls, and man-machine interface for advanced teleoperation. *Science,* 1980, *208,* 1327–1335.

Bekesy, G. von. *Sensory inhibition.* Princeton, N.J.: Princeton University Press, 1967.

Bekey, G. A. The human operator as a sample data system. *IRE Transactions Human Factors in Electronics,* 1962, HFE–3, 43–51.

Bekey, G. A., Meissinger, H. F., & Rose, R. E. *A study of model matching techniques for the determination of parameters in human pilot models.* Tech. Report CR-143, NASA, 1965.

Bell, P., Fisher, J. D., & Loomis, R. J. *Environmental psychology.* Philadelphia: W.B. Saunders, 1978.

Beranek, L. L. The design of speech communication systems. *Institute of Radio Engineers, Proceedings,* 1947, *35,* 880–890.

Beranek, L. L., Blazier, W. E., & Figwer, J. J. Preferred Noise Criterion (PNC) curves and their application to rooms. *Journal of the Acoustical Society of America,* 1971, *50,* 1223–1228.

Berglund, L. G. Radiation measurement for thermal comfort in the built environment. In B. W. Magnum & J. E. Hill (Eds.), *Thermal analysis—Human comfort—Indoor environments.* NBS Special Publication 491, 1977.

Beringer, D. B., Willeges, R. C., & Roscoe, S. N. The transition of experienced pilots to a frequency-separated aircraft altitude display. *Human Factors,* 1975, *17,* 401–414.

Biederman, I. On the semantics of a glance at a scene. In M. Kubovy, & J. R. Pomerantz, (Eds.), *Perceptual organization.* Hillsdale, N.J.: Lawrence Erlbaum Associates, 1981, pp. 213–254.

Birkin, S. J., & Ford, J. S. The quantity/quality dilemma: The impact of a zero defects program. In J. L. Cochrane & M. Zeleny (Eds.), *Multiple criteria decision making.* Columbia, S.C.: University of South Carolina Press, 1973.

Birmingham, H. P., & Taylor, F. V. A design philosophy for man-machine control systems, *Proc. IRE,* 1954, 42, 1748–1758.

Birnbaum, M. Base rates in Bayesian inference: Signal detection analysis of the cab problem. *American Journal of Psychology,* 1983, in press.

Blaauw, G. J. Driving Experience and task demands in stimulator and instrumented car: A validation study. *Human Factors,* 1982, *24,* 473–486.

Blackwell, H. R. Development of procedures and instruments for visual task evaluation. *Illuminating Engineering,* 1970, *65,* 267–291.

Blackwell, O. M., & Blackwell, H. R. Visual performance data for 156 normal observers of various ages. *Journal of Illuminating Engineering Society,* 1971, *1,* 3-13.

Bloomfield, J. R. Studies on visual search. In C. G. Drury, & J. G. Fox (Eds.), *Human reliability in quality control.* London: Taylor and Francis, Ltd., 1975, pp. 31–44.

Bolt, R. H., Cooper, F. S., David, E. E., Jr., Denes, P. B., Pickett, J. M., & Stevens, K. N. Speaker identification by speech spectrograms: A scientist's view of its reliability for legal purposes. *Journal of the Acoustical Society of America,* 1970, *47,* 597–612.

Bolt, R. H., Cooper, F. S., David, E. E., Jr., Denes, P. B., Pickett, J. M., & Stevens, K. N. Speaker identification by voice spectrograms: Some further observations. *Journal of the Acoustical Society of America,* 1974, *54,* 531–534.

Booth, S., Austin, M., Lape, J., & Roman, D. *An experimental evaluation of long and short data names.* CS 590D Project Report, Department of Computer Sciences, Purdue University, 1981.

Bormann, F. H. The New England landscape: Air pollution stress and energy policy. *Ambio,* 1982, *11,* 188–194.

Bowen, H. & Mauro, C. L. Product liability: The case of the ten-cent ring guard. In H. R. Poydar (Ed.), *Human factors and industrial design in consumer products.* Medford, Mass.: Tufts University, Department of Engineering Design, 1980.

Bower, G. H. Mental imagery and associative learning. In L. Gregg (Ed.), *Cognition in learning and memory.* New York: Wiley, 1972.

Boyce, P. R. *Human factors in lighting.* New York: Macmillan, 1981.

Bradley, J. V. Tactual coding of cylindrical knobs. *Human Factors,* 1967, *9,* 483–496.

Brebner, J., & Sandow B. The effect of scale side on population stereotype. *Ergonomics,* 1976, *19,* 571–580.

Bregman, A. S. Asking the "What For" question in auditory perception. Chapter 4 in M. Kubovy & J. R. Pomerantz (Eds.), *Perceptual organization.* Hillsdale, N.J.: Lawrence Erlbaum Associates, 1981.

Brickman, P. Optional stopping on ascending and descending series. *Organizational Behavior and Human Performance,* 1972, *7,* 53–62.

Briggs, G. E. On the predictor variable for choice reaction time. *Memory & Cognition,* 1974, *2,* 575–580.

Broadbent, D.E. *Perception and communication.* London: Pergamon, 1958.

Broadbent, D. E. Application of information theory and decision theory to human perception and reaction. In N. Wiener & J. P. Schade (Eds.), *Cybernetics of the nervous system.* Amsterdam: Elsvier, 1965.

Broadbent, D. E. *Decision and stress.* New York: Academic Press, 1971.

Broadbent, D. E. The current status of noise research: A reply to Poulton. *Psychological Bulletin,* 1978, *85,* 1052–1067.

Broadbent, D. E. Noise in relation to annoyance, performance, and mental health. *Journal of the Acoustical Society of America,* 1980, *68,* 15–17.

Broadbent, D. E. Task combination and selective intake of information. *Acta Psychologica,* 1982, *50,* 253–290.

Brook, R. O. *New towns and communal values.* New York: Praeger, 1974.

Brooke, J. D., & Duncan, K. D. An experimental study of flowcharts as an aid to identification of procedural faults. *Ergonomics,* 1980, *23*(4), 387–399.

Brookes, M. J., & Kaplan, A. The office environment: Space planning and effective behavior. *Human Factors,* 1972, *14,* 373–392.

Brooks, R. E. Studying programmer behavior experimentally: The problems of proper methodology. *Communications of the ACM,* April 1980, *23*(4), 207–213.

Brooks, R. M. An investigation into aspects of bus design and passenger requirements. *Ergonomics,* 1979, *22,* 175–188.

Brown, R. L., Dickman, K. J., Grosso, R. P., Kosor, C. D. & Seekamp, C. R. *The effect of error messages on the debugging process.* CS 590D Project Report, Department of Computer Sciences, Purdue University, 1979.

Buck, J. R. Dynamic visual inspection. In C. G. Drury, & J. G. Fox (Eds), *Human reliability in quality control.* London: Taylor and Francis, Ltd., 1975, pp. 165–187.

Buck, J. R., and Maltas, K. L. Stimulation of industrial man-machine systems, *Ergonomics,* 1979, *22*(7), 785–797.

Buck, J. R., & McAlpine, D. B. The effects of atmospheric conditions on people. 1981. Technical report. School of Industrial Engineering, Purdue University.

Buck, J. R., & Rizzi, A. M. Viewing strategies and window effects on dynamic visual inspection. *American Institute of Industrial Engineers Transactions,* 1974, *6,* 196-205.

Buck, J. R., Tanchoco, J. M. A., & Sweet, A. L. Parameter estimation for discrete exponential learning curves. *American Institute of Industrial Engineers Trans.,* 1976, *8*(2), 184–194.

Buck, L. Motor performance in relation to control-display gain and target width. *Ergonomics,* 1980, *23,* 579–589.

Buckner, D. N., & McGrath, J. J. (Eds.). *Vigilance: A symposium.* New York: McGraw-Hill, 1963.

Burg, A. Visual acuity as measured by dynamic and static tests: A comparative evaluation. *Journal of Applied Psychology,* 1966, *50,* 460–466.

Burse, R. L. Sex differences in human thermoregulatory response to heat and cold stress. *Human Factors,* 1979, *21,* 687–699.

Caelli, T., & Porter, D. On difficulties in localizing ambulance sirens. *Human Factors,* 1980, *22,* 719–724.

Cakir, A., Hart, D. J., & Stewart, T. F. M. *Visual display terminal.* New York: Wiley, 1980.

Calhoun, J. B. The role of space in animal sociology. *Journal of Social Issues,* 1966, *22,* 46-58.

Calhoun, J. B. Space and strategy of life. In S. H. Esser (Ed.), *Behavior and environment.* New York: Plenum, 1971.

Campbell, F. W., & Wurtz, R. Saccadic omission—Why we do not see a grey-out during a saccadic eye movement. *Vision Research,* 1978, *18,* 1297–1303.

Carpenter, B. E., & Lavington, S. H. The influence of human factors on the performance of a real-time speech recognition system. *Journal of the Acoustical Society of America,* 1973, *53,* 42–45.

Carr, R. M. The effects of color coding indicator displays on dark adaptation. *Human Factors,* 1967, *9,* 175–179.

Carroll, R. F. (Chairman). *Guidelines for the design of man/machine interfaces for process control,* Purdue Laboratory for Applied Industrial Control. W. Lafayette, Ind.: Purdue University, 1976.

Carson, D. H. Human factors and elements of urban housing. In T. K.

Sen (Ed.), *Human factors applications in urban development.* New York: Riverside Research Institute, 1970.

Carter, R. C., Jr. Knobology underwater. *Human Factors,* 1978, *20,* 641–647.

Carter, R. C., Jr. Visual search with color. *Journal of Experimental Psychology: Human Perception and Performance,* 1982, *8,* 127–136.

Carter, R. C., Jr. Visual search and color coding. *Proceedings of the Human Factor Society,* 1979, *23,* 369–373.

Carter, R. C., Jr., & Cahill, M. C. Regression models of search time for color-coded information displays. *Human Factors,* 1979, *21*(3), 293–302.

Chao, A., & Bennett, C. A. Lamps for lighting people. *Proceedings of the Human Factors Society,* 1981, *25,* 485–487.

Chapanis, A. *Research techniques in human engineering.* Baltimore: The Johns Hopkins Press, 1959.

Chapanis, A. Words, words, words. *Human Factors,* 1967, *7,* 1–17.

Chapanis, A. Human factors in systems engineering. In K. B. DeGreene (Ed.), *Systems psychology.* New York: McGraw-Hill, 1970.

Chapanis, A., & Halsey, R. M. Luminance of equally bright colors. *Journal of Optical Society of America,* 1955, *45,* 1–6.

Chapanis, A., & Kinkade, R. G. Design of controls. In H. P. Van Cott & R. G. Kinkade (Eds.), *Human engineering guide to equipment design.* Washington, D.C.: U.S. Government Printing Office, 1972.

Chapanis, A., & Lindenbaum, L. E. A reaction time study of four control-display linkages. *Human Factors,* 1959, *1,* 1–7.

Christ, R. E. Review and analysis of color coding research for visual displays. *Human Factors,* 1975, *17,* 542–570.

Christensen, J. M., & Howard, J. M. Field experience in maintenance. In J. Rasmussen & W. B. Rouse (Eds.), *Human detection and diagnosis of system failures.* New York: Plenum, 1981.

Christensen, J. M., & Mills, R. G. What operators do in complex systems. *Human Factors,* 1967, *9,* 328–340.

Clark, G. E., & Ludwig, H. G. Police patrol vehicles. *Human Factors,* 1970, *12,* 69–74.

Clark, H., & Clark, E. *Psychology and language.* New York: Harcourt Brace Jovanovich, 1977.

Cochran, D. J., Riley, M. W., & Douglas, E. I. An investigation of shapes for warning labels. *Proceedings of the Human Factor Society,* 1981, 395–399.

Coffey, J. L. A comparison of vertical and horizontal arrangements of alpha-numeric material—experiment I. *Human Factors,* 1961, 93–98.

Cohen, S. Environmental load and the allocation of attention. In A. Baum, J. E. Singer, & S. Valins (Eds.), *Advances in environmental psychology* (Vol. 1) Hillsdale, N.J.: Lawrence Erlbaum Associates, 1978.

Cohen, S. The aftereffects of stress on human performance and social behavior: A review of research and theory. *Psychological Bulletin,* 1980, *88,* 82–108.

Cohen, S., Evans, G. W., Krantz, D. S., Stokols, D., & Kelly, S. Aircraft noise and children: Longitudinal and cross-sectional evidence of adaptation to noise and the effectiveness of noise abatement. *Journal of Personality and Social Psychology,* 1981, *40,* 331–345.

Cohen, S., Glass, D. C., & Phillips, S. Environment and health. In H. E. Freeman, S. Levine, & J. G. Reeder (Eds.), *Handbook of medical sociology.* Englewood Cliffs, N.J.: Prentice-Hall, 1977.

Cohen, S., Glass, D. C., & Singer, J. E. Apartment noise, auditory discrimination, and reading ability in children. *Journal of Experimental Social Psychology,* 1973, *9,* 407–422.

Cohen, S., & Sherrod, D. R. When density matters: Environmental control as a determinant of crowding: Effects in a laboratory and residential settings. *Journal of Population,* 1978, *1,* 189–202.

Cohen, S., & Spacapan, S. The aftereffects of stress: An attentional interpretation. *Environmental Psychology and Nonverbal Behavior,* 1978, *3,* 45–57.

Cohen, S., & Weinstein, N. Nonauditory effects of noise on behavior and health. *Journal of Social Issues,* 1981, *37,* 36–70.

Cole, E. L., Milton, J. L., & McIntosh, B. B. *Routine maneuvers under day and night conditions, using an experimental panel: The ninth of a series of reports on eye fixations of aircraft pilots.* U.S.A.F. WADC Technical Report 53-220, 1954.

Cole, R. A., & Scott, B. Toward a theory of speech perception. *Psychological Review,* 1974, *81,* 348–374.

Colquhoun, W. P. Evaluation of auditory, visual, and dual-mode displays for prolonged sonar monitoring in repeated sessions. *Human Factors,* 1975, *17,* 425–437.

Commission International de l'Éclairage. National Physical Laboratory, Teddington. Cambridge: University Press, 1932.

Conrad, R. Short-term memory factor in the design of data-entry keyboards: An interface between short-term memory and S-R compatibility. *Journal of Applied Psychology,* October 1966, *50*(5), 353–356.

Cooper, G. E. A survey of the status of and philosophies relating to Cockpit Warning Systems. NASA-CR-152071, June 1977, A03/MF A01 CSCL d5E.

Cooper, G. E., & Harper, R. P. *The use of pilot rating in the evaluation of aircraft handling qualities.* Moffett Field, Calif.: NASA, Ames Research Center, TN-D-5153, April 1969.

Coover, J. E. A method of teaching typewriting based upon a psychological analysis of expert typing. *National Education Association Addresses and Proceedings,* 1923, *61,* 561–567.

Cornog, D. Y., & Rose, F. C. *Legibility of alphanumeric characters and other symbols: II. A reference handbook.* National Bureau of Standards Miscellaneous Publication 262-2. Washington, D.C.: U.S. Government Printing Office, 1967.

Cornsweet, T. N. *Visual perception.* New York: Academic Press, 1970.

Cozby, P. Effects of density, activity and personality on environmental preferences. *Journal of Research in Personality,* 1973, *1,* 45–60.

Craig, A. Nonparametric measures of sensory efficiency for sustained monitoring tasks. *Human Factors,* 1979, *21,* 69–78.

Craig, A. Effect of prior knowledge of signal probabilities on vigilance performance at a two-signal task. *Human Factors,* 1980, *22,* 361–371.

Craig, A. Monitoring for one kind of signal in the presence of another: The effects of signal mix on detectability. *Human Factors,* 1981, *23,* 191-197.

Craig, J. C. Pictorial & abstract cutaneous displays. In F. A. Geldard, (Ed.), *Cutaneous communication systems and devices.* Austin, Texas: The Psychonomic Society, 1974, pp. 78–83.

Craig, J. C. Vibrotactile pattern perception: Extraordinary observers. *Science,* 1977, *196,* 450–452.

Craig, J. C., & Sherrick, C. E. Dynamic tactile displays. In W. Schiff, & E. Foulke, (Eds.), *Handbook on haptic perception.* Oxford Press, 1982 in press.

Craig, J. D. Modes of vibrotactile pattern generation. *Journal of Experimental Psychology: Human Perception & Performance,* 1980, *6,* 151–166.

Craik, F.I.M. & Lockhart, R.S. Levels of processing: A framework for memory research. *Journal of Verbal Learning and Verbal Behavior*, 1972, 11, 671-684.

Crowder, R. G. The demise of short-term memory. *Acta Psychologica,* 1982, *51,* 291–323.

Crump, J. H. Review of stress in air traffic control: Its measurement and effects. *Aviation, Space, and Environmental Medicine,* 1979, *50,* 243–248.

Curley, M. D., & Bachrach, A. J. Tactile sensitivity in the one-atmosphere diving system JIM. *Human Factors,* 1981, *23,* 291–297.

Dainoff, M. J., Happ, A., & Crane, P. Visual fatigue and occupational stress in VDT operators. *Human Factors,* August 1981, *23*(4), 421–438.

Dale, H. D. A. Fault finding in electronic equipment. *Ergonomics,* 1958, *1,* 356.

Danaher, J. W. Human error in ATC system operations. *Human Factors,* 1980, *22,* 535–545.

Darwin, C. J., Turvey, M. T., & Crowder, R. G. An auditory analogue of the Sperling partial report procedure: Evidence for brief auditory storage. *Cognitive Psychology,* 1972, *3,* 255–267.

de Neufville, R. *Airport systems planning.* Cambridge, Mass.: MIT Press, 1976.

Derrick, W. L. The relationship between processing resource and subjective dimensions of operator workload. *Proceedings of the Human Factors Society,* 1981, *25,* 532–536.

Deutsch, D., & Feroe, J. The internal representation of pitch sequences in tonal music. *Psychological Review,* 1981, *88,* 503–522.

Deutsch, M. Trust and suspicion. *Journal of Conflict Resolution,* 1958, *1,* 558–568.

Dewar, R. E., Ells, J. G., & Cooper, P. J. Evaluation of roadway guide signs at a large airport. *Transportation Engineering,* 1977, *47*(6), 19–23.

Dohrenwend, B. S., & Dohrenwend, B. P. Some issues on research on stressful life events. *The Journal of Nervous and Mental Disease,* 1978, *166,* 7–15.

Drury, C. G. The effect of speed of working on industrial inspection accuracy. *Applied Ergonomics,* 1974, *4,* 2–7.

Drury, C. G. Application of Fitts' law to foot-pedal design. *Human Factors,* 1975, *17,* 368–373.

Drury, C. G., & Clement, M. R. The effect of area, density, and number of background characters on visual search. *Human Factors,* 1978, *20,* 597–602.

Drury, C. G., & Fox, J. G. *Human reliability in quality control.* London: Taylor & Francis, Ltd.

Duncan, J. Response selection rules in spatial choice reaction tasks. In S. Dornic (Ed.), *Attention and performance VI.* Hillsdale, N.J.: Lawrence Erlbaum Associates, 1977.

Duncan, J. The demonstration of capacity limitation. *Cognitive Psychology,* 1980, *12,* 75–96.

Eargle, J. *Sound recording* (2nd ed.). New York: Van Nostrand, 1980.

Earl, W. K., & Goff, J. D. Comparison of two data entry methods. *Perceptual and Motor Skills,* 1965, *20,* 369–384.

Easterbrook, J. A. The effect of emotion on cue utilization and the organization of behavior. *Psychological review,* 1959, *66,* 183–201.

Easterby, R., Kroemer, K. H., & Chaffin, D. B. (Eds.) *Anthropometry and biomechanics.* New York: Plenum Press, 1982.

Eckert, S., & Woods, D. D. *Recommended human engineering approach to auditory alarms for the Clinch River fast breeder reactor control room.* Westinghouse Research Report 80-8C57-CRFBR-R2, January 20, 1981.

Edney, J. J. Human territoriality. *Psychological Bulletin,* 1974, *81,* 959–975.

Edney, J. J. Human territories: Comment on functional properties. *Environment and Behavior,* 1976, *8,* 31–48.

Edwards, W., Lindam, H., & Phillips, L. D. Emerging technology for making decisions. *New Directions in Psychology,* Vol. II, New York: Holt, Rinehart & Winston, 1965.

Egan, J. P. Articulation testing methods. *Laryngoscope,* 1948, *58,* 955–991.

Eggemeier, F. T. Current issues in subjective assessment of workload. *Proceedings of the Human Factors Society,* 1981, *25,* 513–517.

Elkind, J. I. *Characteristics of simple manual control systems.* Lincoln Park Lab. Tech., Report 111, MIT, 1956.

Elliot, P. B. Tables of *d'*. In J. A. Swets (Ed.), *Signal detection and recognition by human observers.* New York: Wiley, 1964.

Ellis, J. G., & Dewar, R. E. Rapid comprehension of verbal and symbolic traffic sign messages. *Human Factors,* 1979, *21*(2), 161–168.

Ellson, D. G., & Wheeler, L., *The Range Effect.* Tech. Report 5813, Air Material Command, USAG, Wright-Patterson AFB, 1949.

Elmes, D., Kantowitz, B. H., & Roediger, H. L. *Methods in experimental psychology.* Boston: Houghton Mifflin, 1980.

Elshoff, J. L. The influence of structured programming on PL/I program profiles. *IEEE Transactions on Software Engineering,* September 1977, *3*(5), 364–368.

Emanuel, J. T., Mills, S. J., & Bennett, J. F. In search of a better handle. In H. R. Poydar (Ed.), *Human factors and industrial design in consumer products.* Medford, Mass.: Tufts University, Dept. of Engineering Design, 1980.

Engel, S. E., & Granda, R. E. *Guidelines for man/display interfaces.* IBM Technical Report TR 00.2720. New York: Poughkeepsie Laboratory, December, 1975.

Engelberger, J. F. Designing robots for industrial environments. *Mechanism and Machine Theory,* 1977, *12,* 403–412.

English, W. K., Engelbart, D. C., & Berman, M. L. Display-selection techniques for text manipulation. *IEEE Transactions on Human Factors in Electronics,* March 1967, *8*(1), 5–15.

Environmental Protection Agency. *Protective noise levels. Condensed version of EPA levels document.* EPA 550 9-79-100, November 1978.

Epstein, S., & Fenz, W. D. Steepness of approach and avoidance gradients in humans as a function of experience: theory and experiment. *Journal of Experimental Psychology,* 1965, *70,* 1–13.

Eriksen, C.W. & Collins, J.F. Sensory traces versus the psychological moment in the temporal organization of form. *Journal of Experimental Psychology,* 1968, 77, 376-382.

Erickson, R. A. Line criteria in target acquisition with television. *Human Factors,* 1978, *20,* 573–588.

Evans, G. H., & Jacobs, S. V. Air pollution and human behavior. *Journal of Social Issues,* 1981, *37,* 95–125.

Evans, G. W., & Howard, R. B. Personal space. *Psychological Bulletin,* 1973, *80,* 334–344.

Evans, S. W. Design implications of spatial research. In J. R. Aiello & A. Baum (Eds.), *Residential crowding and design.* New York: Plenum, 1979.

Evarts, E. V. *Science,* 1973, *179,* 501–503.

FAA-RD-76-222-I. Aircraft alerting systems criteria study. Volume I. Collation and analysis of aircraft alerting system data. D6-44199 May 1977. U.S. Department of Transportation, *Federal Aviation Administration,* Systems Research & Development Service, Washington, D.C. 20590. (a)

FAA-RD-76-222-II. Aircraft alerting systems criteria study. Volume II. Human factors guidelines for aircraft alerting systems. D6-44200 May 1977. U.S. Department of Transportation, *Federal Aviation Administration,* Systems Research & Development Service, Washington, D.C. 20590. (b)

Falcone, J. D. *How to design, build, remodel and maintain your home.* New York: Wiley, 1978.

Fanger, P. O. *Thermal comfort: Analysis and applications in environment engineering.* New York: McGraw-Hill, 1972.

Faulkner, T. W., & Murphy, T. J. Lighting for difficult visual tasks. In C. G. Drury & J. G. Fox, (Eds.), *Human reliability in quality control.* London: Taylor and Francis, Ltd., 1975, pp. 133–147.

Feallock, J. B., Southard, J. F., Kobayashi, M., & Howell, W. C. Absolute judgment of colors in the federal standards system. *Journal of Applied Psychology,* 1966, *50,* 266–272.

Feineman, G. How to live with reliability engineers. *Spectrum,* Spring 1978.

Fidell, S. Effectiveness of audible warning signals for emergency vehicles. *Human Factors,* 1978, *20,* 19–26.

Finkelman, J. M., & Kirschner, C. An information-processing interpretation of air traffic control stress. *Human Factors,* 1980, *22,* 561–567.

Fiorentini, A., & Maffei, L. Spatial contrast sensitivity of myopic subjects. *Vision Research,* 1976, *16,* 437–438.

Fischoff, B., & Beyth, R. I knew it would happen—Remembered probabilities of once-future things. *Organizational Behavior and Human Performance,* 1975, *13,* 1–16.

Fisher, D. F., Monty, R. A., & Senders, J. W. (Eds.). *Eye movements: Cognition and visual perception.* Hillsdale, N.J.: Lawrence Erlbaum Associates, 1981.

Fisher, J. D., & Byrne, D. Too close for comfort: Sex differences in response to invasions of personal space. *Journal of Personality and Social Psychology,* 1975, *32,* 15-21.

Fitts, P. M. Engineering psychology in equipment design. In S. S. Sevens, (Ed.), *Handbook of Experimental Psychology.* New York: Wiley, 1951.

Fitts, P. M. The information capacity of the human motor system in controlling the amplitude of movement. *Journal of Experimental Psychology,* 1954, *47,* 381–391.

Fitts, P. M. & Jones, R. E. Analysis of factors contributing to 460 "pilot-error" experiences in operating aircraft controls. Report TSEAA-694-12, Air Material Command, Wright-Patterson Air Force Base, 1947. Reprinted in H. W. Sinaiko (Ed.), *Selected papers on human factors in the design and use of control systems.* New York: Dover, 1961.

Fitts, P. M., Jones, R. E., & Milton, J. L. Eye movements of aircraft pilots during instrument-landing approaches. *Aeronautical Engineering Review,* 1950, *9*(2), 1–6.

Fitts, P. M., & Peterson, J. R. Information capacity of discrete motor responses. *Journal of Experimental Psychology,* 1964, *67,* 103–112.

Fitts, P. M. & Seeger, C. M. S-R compatibility: Spatial characteristics of stimulus and response codes. *Journal of Experimental Psychology,* 1953, *46,* 199-210.

Flanagan, J. L., Coker, C. H., Rabiner, L. R., Schafer, R. W., & Umeda, N. Synthetic voices for computers *IEEE Spectrum,* 1970, *7,* 22–45.

Flanagan, J. L., Rabiner, L. R., Schafer, R. W., & Denman, J. D. Wiring telephone apparatus from computer generated speech. *Bell System Technical Journal,* 1972, *51,* 391–397.

Fletcher, H. Auditory patterns. *Review of Modern Physics,* 1940, *12,* 47-65.

Fong, Y., & Bennett, C. A. The desk as a barrier. *Proceedings of the Human Factors Society,* 1981, *25,* 482–484.

Forrin, B., Kumler, M. L., & Morin, R. E. The effects of response code and signal probability in a numeral naming task. *Canadian Journal of Psychology,* 1966, *20,* 115-124.

Fozard, J. L., & Popkin, S. J. Optimizing adult development: Ends and means of an applied psychology of aging. *American Psychologist,* 1978, *33,* 975–989.

Freedman, J. L. *Crowding and behavior.* San Francisco: Freeman, 1975.

Freedman, J. L., Heshka, S., & Levy, A. Population density and pathology: Is there a relationship? *Journal of Experimental Social Psychology,* 1975, *11,* 539–552.

French, N. R., & Steinberg, J. C. Factors governing the intelligibility of speech sounds. *Journal of Acoustical Society of America,* 1947, *19,* 90–119.

Frijters, J. E. R. Variations of the triangular method and the relationship of its unidimensional probabilistic models to three-alternative forced-choice signal detection models. *British Journal of Mathematical and Statistical Psychology,* 1979, *32,* 229–241.

Frijters, J. E. R., Kooistra, A., & Vereijken, P. F. G. Tables of *d'* for the triangular method and the 3-AFC signal detection procedure. *Perception & Psychophysics,* 1980, *27,* 176-178.

Fucigna, J. T. The ergonomics of offices. *Ergonomics,* 1967, *10,* 589-604.

Gaertner, K. P., & Holzhausen, K. P. Controlling air traffic with a touch sensitive screen. *Applied Ergonomics,* 1980, *11,* 17–22.

Galle, O. R., Gove, W. R., & McPherson, J. M. Population density and pathology: What are the relationships for man? *Science,* 1972, *176,* 23–30.

Gannon, J. D., & Horning, J. J. Language design for programming reliability. *IEEE Transactions on Software Engineering,* June 1975, 1(2), 179-191.

Gardner, M. B. Historical background of the Haas and/or Precedence effect. *Journal of the Acoustical Society of America,* 1968, *43,* 1243–1248.

Garland, L. H. The problem of observer error. *Bulletin of the New York Academy of Medicine,* 1960, *36,* 569–384.

Garner, W. R. Information and structure as psychological concepts. New York: Wiley, 1962.

Gates, D. F. Honeycomb projects: An architectural crime problem. *The Police Chief,* 1977, 38–42.

Geldard, F. A. (Ed.) *Cutaneous communication systems and devices.* Austin, Texas: Psychonomic Society, 1974.

Geldard, F. A. *Sensory saltation: Metastability in the perceptual world.* Hillsdale, N.J.: Lawrence Erlbaum Associates, 1975.

Giarretto, H. The effects of stereoscopy on the recognition of patterns in visual noise. *Human Factors,* 1968, *10,* 403–412.

Gibson, J. J. Observations on active touch. *Psychological Review,* 1962, *69,* 477–491.

Gibson, J. J. What gives rise to the perception of motion? *Psychological Review,* 1968, *75,* 335–346.

Gilson, R. D. Vibrotactile masking: Some spatial and temporal aspects. *Perception & Psychophysics,* 1969, *5,* 176–180. (a)

Gilson, R. D. Vibrotactile masking: Effects of multiple maskers. *Perception & Psychophysics,* 1969, *5,* 181–182. (b)

Glass, D. C., & Singer, J. E. *Urban stress: Experiments on noise and social stressors.* New York: Academic Press, 1972.

Glass, S. W., & Suggs, C. W. Optimization of vehicle accelerator-brake pedal foot travel time. *Applied Ergonomics,* 1977, *8,* 215–218.

Goldberg, A. J. Practical implementations of speech waveform coders for the present day and for the mid 1980s. *Journal of the Acoustical Society of America,* 1979, *66,* 1653–1657.

Goldstein, E. B. *Sensation and perception.* Belmont, Calif.: Wadsworth, 1980.

Goldstein, I. L., & Dorfman, P. W. Speed and load stress as determinants of performance in a time sharing task. *Human Factors,* October 1978, *20*(5), 603–609.

Goldstein, J. Fundamental concepts in sound measurement. In D. M. Lipscomb, (Ed.), *Noise and audiology.* Baltimore: University Park Press, 1978, pp. 3–58.

Goldwater, B. C. Psychological significance of pupillary movements. *Psychological Bulletin,* 1972, *77,* 340–355.

Gordon, I. E., & Cooper, C. Improving one's touch. *Nature,* 1975, *256,* 203–204.

Gould, J. D. Visual factors in the design of computer controlled CRT displays. *Human Factors,* 1968, *10*(4), 359–376.

Gould, J. D., & Drongowski, P. An exploratory study of computer program debugging. *Human Factors,* 1974, *16*(3), 258–277.

Gravely, M. L., & Hitchcock, L. The use of dynamic mock-ups in the design of advanced systems. *Proceedings of the Human Factors Society,* 1980, *24,* 5–8.

Green, D. M. *An introduction to hearing.* Hillsdale, N.J.: Lawrence Erlbaum Associates, 1976.

Green, D. M., & Birdsall, T. G. Detection and recognition. *Psychological Review,* 1978, *85,* 192–206.

Green, D. M., & Swets, J. A. *Signal detection theory and psychophysics.* New York: Wiley, 1966.

Green, P., & Pew, R. W. Evaluating pictographic symbols: An automotive application. *Human Factors,* 1978, *20*(1), 103–114.

Green, T. R. G. Conditional program statements and their comprehensibility to professional programmers. *Journal of Occupational Psychology,* 1977, *50,* 93–109.

Greene, T. C., & Bell, P. A. Additional considerations concerning the effects of "warm" and "cool" colours on energy conservation. *Ergonomics,* 1980, *23,* 949–954.

Gregory, R. L. *Eye & brain* (2nd ed.). New York: McGraw-Hill, 1973.

Grice, R. G., Nullmeyer, R., & Spiker, A. V. Human reaction time: Toward a general theory. *Journal of Experimental Psychology: General,* 1982, *111,* 135–153.

Grossman, J. D., & Whitehurst, H. O. The relative effects of multiple factors on target acquisition. *Human Factors,* 1979, *21,* 423–432.

Grove, P. L. The billion dollar benzene blunder. *Tulsa Law Journal,* 1980, *16,* 252–285.

Guilford, J. P. *Psychometric methods.* 2nd ed. New York: McGraw-Hill, 1954.

Haber, R. N., & Hershenson, M. *The psychology of visual perception.* New York: Holt, Rinehart and Winston, 1973.

Hacker, M. J., & Ratcliff, R. A revised table of d′ for M-alternative forced choice. *Perception & Psychophysics,* 1979, *26,* 168–170.

Hall, A. D., & Fagen, R. E. Definition of system. In W. Buckley (Ed.), *Modern systems research for the behavioral scientist.* Chicago: Aldine, 1968.

Hall, E. T. *The hidden dimension.* New York: Doubleday, 1966.

Halstead, M. H. *Elements of software science.* New York: Elsevier North-Holland, 1977.

Hammerton, M. The use of same or different sensory modalities for information and instructions. *Ergonomics,* 1975, *18*(6), 683–686.

Hamming, R. W. Coding and information theory. Englewood Cliffs, N.J.: Prentice-Hall, 1980.

Hancock, P. A. Mental performance impairment in heat stress. *Proceedings of the Human Factors Society,* 1980, *24,* 363–366.

Hanson, R. H., Payne, D. G., Shively, R. J., & Kantowitz, B. H. Process control simulation research in monitoring analog and digital displays. *Proceedings of the Human Factors Society,* 1981, 154–158.

Hardin, G. The tragedy of the commons. *Science,* 1968, *162,* 1243–1248.

Hartzell, E. J., Dunbar, S., Beveridge, R., & Cortilla, R. Helicopter pilot response latency as a function of the spatial arrangement of instruments and controls. *Eighteenth Annual Conference on Manual Control, 1982.* Dayton, Ohio: in press.

Hawley, M. E. (Ed.). *Benchmark papers in acoustics, speech intelligibility & speaker recognition.* Stroudsberg, Pa.: Dowden, Hutchinson & Ross, 1977.

Hecht, S., & Hsia, Y. Dark adaptation following light adaptation to red and white lights. *Journal of the Optical Society of America,* 1945, *35,* 261–267.

Henderson, D., & Hamernik, R. P. Impulse noise-induced hearing loss: An overview. In D. M. Lipscomb, (Ed.), *Noise and audiology.* Baltimore: University Park Press, 1978, pp. 143–166.

Hennessy, R. T. Instrument myopia. *Journal of the Optical Society of America,* 1975, *65,* 1114–1120.

Herman, L. M. Study of the single channel hypothesis and input regulation within a continuous simultaneous task situation. *Quarterly Journal of Experimental Psychology,* 1965, *17,* 37–46.

Herman, L. M., & Kantowitz, B. H. The psychological refractory period effect: Only half the double-stimulation story? *Psychological Bulletin,* 1970, *73,* 74–88.

Hertzberg, H. T. E. Engineering anthropology. In H. P. Von Cott, & Kinkade, R. G. (Eds.), *Human engineering guide to equipment design.* (Rev. ed.) Washington, DC: U.S. Government Printing Office, 1972.

Hick, W. E. On the rate of gain of information. *Quarterly Journal of Experimental Psychology,* 1952, *4,* 11–26.

Higbee, K. L. *Your memory: How it works and how to improve it.* Englewood Cliffs, N.J.: Prentice-Hall, 1977.

Hirsch, J. Rate control in man-machine systems. In F. A. Geldard (Ed.), *Cutaneous communication systems and devices.* Austin, Texas: Psychonomic Society, 1974, pp. 65–71.

Hitt, W. D. An evaluation of five different abstract coding methods – Experiment IV. *Human Factors,* 1961, 120–130.

Hockey, R. Stress and the cognitive component of skilled performance. In V. Hamilton & D. M. Warburton (Eds.), *Human stress and cognition.* New York: Wiley, 1979.

Hodge, D. C., & Price, G. R. Hearing damage risk criteria. In D. M. Lipscomb, (Ed.), *Noise and audiology.* Baltimore: University Park Press, 1978, pp. 167–192.

Hopper, G. M. The first bug. *Annals of the History of Computing,* July 1981, *3*(3), 285–286.

Horak, J. Gas lines need deodorant, not fix, jittery callers told. *Lafayette Journal and Courier,* September 12, 1979, 60, 1.

House, A. S., Williams, C. E., Hecker, M. H. L., & Kryter, K. D. Articulation-testing methods: Consonantal differentiation with a closed-response set. *Journal of the Acoustical Society of America,* 1965, *37,* 158–166.

Howell, W. C., & Kennedy, P. A. Field validation of the Fanger thermal comfort model. *Human Factors,* 1979, *21,* 229–239.

Howell, W. C., & Kraft, C. L. *Size, blur, and contrast as variables affecting the legibility of alphanumeric symbols on radar-type displays.* Technical Report USAF:WADC TR 59–536, 1959.

Hubka, D., Baechle, E. M., Miller, R. W., & Parthasarathy, B. *The value of comments in modifying programs: Some empirical evidence.* CS 590D Project Report, Department of Computer Sciences, Purdue University, 1979.

Hughes, P. C., & McNelis, J. F. *Lighting, productivity and the work environment.* Paper presented at the Annual Illuminating Engineering Society Technical Meeting, Denver, August, 1978.

Hughes, P. C., & Neer, R. M. Lighting for the elderly: A psychobiological approach to lighting. *Human Factors,* 1981, *23,* 65–85.

Hull, J. C., Gill, R. T., & Roscoe, S. N. Locus of the stimulus to visual accommodation: Where in the world, or where in the eye? *Human Factors,* 1982, *24,* 311–319.

Hunting, W., & Grandjean, E. Hunting and Grandjean highback. *Design,* 1976, *333,* 34–35.

Hurvich, L. M. *Color vision.* Sunderland, Mass.: Sinauer Associates Inc., 1981.

Hutchinson, T. P. A review of some unusual applications of signal detection theory. *Quality and Quantity,* 1981, *15,* 71–98.

Hyman, R. Stimulus information as a determinant of reaction time. *Journal of Experimental Psychology,* 1953, *45,* 188–196.

IES lighting handbook, 1981 (Application vol.). New York: Illuminating Engineering Society, 1981.

IES lighting handbook 1981 (Reference vol.). New York: Illuminating Engineering Society, 1981.

Ince, F., & Williges, R. C. Detecting slow changes in system dynamics. *Human Factors,* 1974, *16,* 278–285.

Isreal, J. B., Wickens, C. D., Chesney, G. L., & Donchin, E. The event-related brain potential as an index of display-monitoring workload. *Human Factors,* 1980, *22,* 211–224.

Jagacinski, R. J., Miller, D. P., & Gilson, R. D. A comparison of kinesthetic-tactual and visual displays in a critical tracking task. *Human Factors,* 1979, *21,* 79–86.

Jenkins, W. L., & O'Connor, M. B. Some design factors in making settings on a linear scale. *Journal of Applied Psychology,* 1949, *33,* 395.

Jobe, C. K. The model uniform product liability act. *Journal of Air Law and Commerce,* 1981, *46,* 387–447.

Johanssen, G., Moray, N., Pew, R., Rasmussen, J., Sanders, A., & Wickens, C. Final report of experimental psychology group. In N. Moray (Ed.), *Mental Workload.* New York: Plenum, 1979.

Johnson, S. L., & Roscoe, S. N. Symbolic flight displays. *Naval Research Reviews.* 1971, *24*(2), 1–13.

Jones, D. P., Schipper, L. M., & Holtworth, R. J., Effects of the amount of information on decision strategies. *Journal of General Psychology,* 1978, *98,* 281–294.

Jones, M. R., Kidd, G., & Wetzel, R. Evidence for rhythmic attention. *Journal of Experimental Psychology: Human Perception and Performance,* 1981, *7,* 1059–1073.

Jones, T. C. Measuring programming quality and productivity. *IBM Systems Journal,* 1978, *17*(1), 39–63.

Josefowitz, A. J., North, R. A., & Trimble, J. Combined multisensor displays. *Proceedings of the Human Factors Society,* 1980, 17–21.

Kahn, D. *The code breakers.* New York: Macmillan, 1967.

Kahneman, D. *Attention and effort.* Englewood Cliffs, N.J.: Prentice-Hall, 1973.

Kahneman, D., & Tversky, A. Subjective probability: A judgment of representativeness. *Cognitive Psychology,* 1972, *3,* 430–454.

Kahneman, D., & Tversky, A. On the psychology of prediction. *Psychological Review,* 1973, *80,* 237–251.

Kalikow, D. N., Stevens, K. N., & Elliot, L. L. Development of a test of speech intelligibility in noise using sentence materials with controlled word predictability. *Journal of the Acoustical Society of America,* 1977, *61,* 1337–1351.

Kalsbeek, J. W. H. Standards of acceptable load in ATC tasks. *Ergonomics,* 1971, *14,* 641–650.

Kantowitz, B. H. (Ed.) *Human information processing.* Hillsdale, N.J.: Lawrence Erlbaum Associates, 1974.

Kantowitz, B. H.. Double stimulation. In B. H. Kantowitz (Ed.), *Human information processing.* Hillsdale, N.J.: Lawrence Erlbaum Associates, 1974. (a).

Kantowitz, B. H. Double stimulation with varying response requirements. *Journal of Experimental Psychology,* 1974, *103,* 1092–1107. (b)

Kantowitz, B. H. Ergonomics and the design of nuclear power plant control complexes. In T. W. Kvalseth, (Ed.), *Arbeidsplass og miljøbruk av ergonomiske data.* Trondheim, Norway: Tapir, 1977.

Kantowitz, B. H. Interfacing human information processing and engineering psychology. In W. C. Howell & E. A. Fleishman (Eds.), *Human performance and productivity,* Vol. 2. Hillsdale, N.J.: Lawrence Erlbaum Associates, 1981.

Kantowitz, B. H. Interfacing human information processing and engineering psychology. In W. C. Howell & E. A. Fleishman (Eds.), *Human performance and productivity* (Vol. 2). Hillsdale, N.J.: Lawrence Erlbaum Associates, 1982.

Kantowitz, B. H., & Hanson, R. H. Models and experimental results concerning the detection of operator failures in display monitoring. In J. Rasmussen & W. B. Rouse (Eds.), *Human detection and diagnosis of system failures.* New York: Plenum, 1981.

Kantowitz, B. H., & Herman, L. M. Effects of increasing feedback processing requirements on motor system capacity and on Fitts' law. Paper presented to Midwestern Psychological Association, Chicago, 1967.

Kantowitz, B. H., & Knight, J. L. Testing tapping timesharing: II. Auditory secondary task. *Acta Psychologica,* 1976, *40,* 343–362. (a)

Kantowitz, B. H., & Knight, J. L. Testing tapping timesharing. *Journal of Experimental Psychology,* 1974, 103, 331–336. (b)

Kantowitz, B. H., & Roediger, H. L. *Experimental psychology.* Chicago: Rand McNally, 1978.

Kaplan, H. L., Macmillan, N. A., & Creelman, C. D. Tables of *d'* for variable-standard discrimination paradigms. *Behavior Research Methods & Instrumentation,* 1978, *10,* 796–813.

Keele, S. W. Behavioral analysis of movement. In V. Brooks (Ed.), *Handbook of physiology,* Motor Control volume, in press.

Keele, S. W. Movement control in skilled motor performance. *Psychological Bulletin,* 1968, *70,* 387–403.

Keele, S. W., & Posner, M. I. Processing of visual feedback in rapid movements. *Journal of Experimental Psychology,* 1968, *77,* 155–158.

Keele, S. W., & Summers, J. J. The structure of motor programs. In G. Stelmach (Ed.), *Motor control: Issues and trends.* New York: Academic Press, 1976.

Keidel, W. D. The cochlear model in skin stimulation. F. A. Geldard (Ed.), *Cutaneous Communication Systems and Devices.* Austin, Texas: The Psychonomic Society, 1974, pp. 27-32.

Kelley, C. R. *Manual and automatic control.* New York: Wiley, 1968.

Kellogg, R. S., Kennedy, R. S., & Woodruff, R. R. A comparison of color versus black and white visual display as indicated by bombing pefor-mance in the 2B35 TA-45 flight simulator. *Proceedings of the Human Factors Society,* 1981, *25,* 233–234.

Kennedy, K. W., & Bates, C. *Development of design standards for ground support consoles.* WPAFB, Ohio: AMRL-TR-66-27. Aero-space Medical Labs, USAF, 1965.

Kersta, L. G. Voice print identification. *Nature,* 1962, *196,* 1253–1257.

Kinsbourne, M. Single-channel theory. In D. Holding (Ed.), *Human skills.* New York: Wiley, 1981.

Kinzel, A. S. Body buffer zone in violent prisoners. *American Journal of Psychiatry,* 1970, *127,* 59–64.

Kira, A. *The bathroom.* (Rev. ed.). New York: Viking Press, 1976.

Kirman, J. H. Tactile Communication of speech: A review and analysis. *Psychological Bulletin,* 1973, *80,* 54–74.

Kirman, J. H. Tactile perception of computer-derived format patterns from voiced speech. *Journal of the Acoustical Society of America,* 1974, *55,* 163–169.

Klapp, S. T. Reaction time analysis of programmed control. *Exercise and Sport Science Reviews,* 1977, *5,* 231–253.

Klatzky, R. *Human memory,* 2nd ed. San Francisco: W. H. Freeman, 1980.

Klatsky, G. J., Teitelbaum, R. C., Mezzanotte, R. J., & Beiderman, I. Mandatory processing of the background in the detection of objects in scenes. *Proceedings of the Human Factors Society,* 1981, *25,* 272–276.

Klatt, D. H. Review of the ARPA Speech Understanding Project. *Jour-nal of the Acoustical Society of America,* 1977, *62,* 1345–1366.

Kleinman, D. L., Baron, S., & Levison, W. H. A control theoretic approach to manned-vehicle systems analysis, *Institute of Electrical & Electronics Engineering Transactions Automatic Control,* 1971, AC-16, 824–832.

Kolb, J., & Ross, S. S. *Product safety and liability.* New York: McGraw-Hill, 1980.

Kolers, P. A. *Aspects of motion perception.* New York: Pergamon Press, 1972.

Konecni, V. J., Libuser, L., Morton, H., & Ebbesen, E. B. Effects of a violation of personal space on escape and helping responses. *Journal of Experimental Social Psychology,* 1975, *11,* 288–299.

Konz, S. Design of handtools. *Proceedings of the Human Factors Society,* 1974, *18,* 292–300.

Konz, S., Chawla, S., Sathaye, S., & Shah, P. Attractiveness and legibility of various colours when printed on cardboard. *Ergonomics.* 1972, *17,* 189–194.

Kornblum, S. Simple reaction time as a race between signal detection and time estimation: A paradigm and model. *Perception & Psychophysics,* 1973, *13,* 108–112.

Krauss, P. B. The impact of high intensity street lighting on nighttime business burglary. *Human Factors,* 1977, *19,* 235–239.

Kreitzberg, C. R., & Shneiderman, B. *Fortran programming.* New York: Harcourt Brace Jovanovich, 1975.

Krishna-Ras, P., Ephrath, A. R., & Kleinman, D. L. *Analysis of human decision-making in multi-task environments.* Tech. Report No. EECS TR-79-15 under AFOSR 78-3733, 1979.

Kryter, K. D. Validation of the Articulation Index. *Journal of the Acoustical Society of America,* 1962, *34,* 1698–1702.

Kryter, K. D. *The Effects of Noise on Man,* New York: Academic Press, 1970.

Kryter, K. D. Speech communication. In H. P. Van Cott, & R. G. Kinkade (Eds.), *Human engineering guide to equipment design.* Washington, D.C.: American Institutes for Research, 1972.

Kryter, K. D. Extra auditory effects of noise. In D. Henderson, R. P. Hamernik, S. Dosanjh, & J. H. Mills (Ed.), *Effects of noise on hearing.* New York: Raven Press, 1976, pp. 531–546.

Kryter, K. D. Physiological acoustics and health. *Journal of the Acoustical Society of America,* 1980, *68,* 10-14.

Kryter, K. D., & Pearsons, K. S. Some effects of spectral content and duration on perceived noise level. *Journal of the Acoustical Society of America,* 1963, *35,* 866–883.

Kubovy, M. Concurrent pitch segregation and the theory of indispensable attributes. Chapter 3 in M. Kubovy & J. R. Pomerantz (Eds.), *Perceptual organization.* Hillsdale, N.J.: Lawrence Erlbaum Associates, 1981.

Kuller, R. (Ed.). *Architectural psychology.* Stroudsburg, Pa.: Dowden, Hutchinson & Ross, 1973.

Kurke, M. I. Operational sequence diagrams in system design. *Human Factors,* 1961, *3,* 66–73.

Kvaolseth, T. O. An experimental paradigm for analyzing human information processing during motor control tasks. *Proceedings of the Human Factors Society,* 1981, *25,* 581–585.

Lahy, J. M. Motion study in typewriting. *In Studies and Reports Series J (Educational) No. 3.* Geneva: International Labor Office, 1924.

Laming, D. Choice reaction performance following an error. *Acta Psychologica,* 1979, *43,* 199–224.

Lane, D. Attention. In W. C. Howell & E. A. Fleishman (Eds.), *Human performance and productivity,* Vol. 2. Hillsdale, N.J.: Lawrence Erlbaum Associates, 1981.

Langolf, G. D., Chaffin, D. B., & Foulke, J. A. An investigation of Fitts' law using a wide range of movement amplitudes. *Journal of Motor Behavior,* 1976, *8,* 113–128.

Langolf, G., & Hancock, W. M. Human performance times in microscope work. *AIEE Transactions,* 1975, *7,* 110–117.

Lathrop, R. G. Perceived variability. *Journal of Experimental Psychology,* 1967, *73,* 498–502.

Lederman, S. J. "Improving one's touch" . . . and more. *Perception and Psychophysics,* 1978, *24,* 154–160.

Ledgard, H. F., Whiteside, J. A., Seymour, W., & Singer, A. An experiment on human engineering of interactive software. *IEEE Transactions on Software Engineering.* November 1980, *6*(6), 602–604.

Lee, D. R., & Buck, J. R. The effect of screen angle and luminance on microform reading. *Human Factors,* 1975, *17*(5), 461–469.

Lees, F. P., Research on the process operator. In E. Edwards & F. P. Lees (Eds.), *The human operatory in process control.* London: Taylor and Francis Ltd., 1974.

Leibowitz, H. W., & Owens, D. A. Night myopia and the intermediate dark focus of accommodation. *Journal of the Optical Society of America,* 1975, *65,* 1121–1128.

Leibowitz, H. A., & Owens, D. A. Nighttime driving accidents and selective visual degradation. *Science,* 1977, *197,* 422–423.

Lemos, R. S. An implementation of structured walkthroughs in teaching Cobol programming. *Communications of the ACM,* June 1979, *22*(6), 335–340.

Levine, J. M., & Samet, M. G. Information seeking with multiple sources of conflicting and unreliable information. *Human Factors,* 1973, *15,* 407–419.

Lewis, J., Baddeley, A. D., Bonham, K. G., & Lovett, D. Traffic pollution and mental efficiency. *Nature,* 1970, *225,* 96.

Ley, D., & Cybriwsky, R. Urban graffiti as territorial markers. *Annals of the Association of American Geographers,* 1974, *64,* 491–505.

Liberman, A. M., Cooper, F. S., Shankweiler, D. P., & Studdert-Kennedy, M. Perception of the speech code. *Psychological Review,* 1967, *74,* 431–461.

Licklider, J. C. R. The influence of interaural phase relations upon the masking of speech by white noise. *Journal of the Acoustical Society of America,* 1948, *20,* 150–159.

Licklider, J. C. R., Bindra, D., & Pollack, I. The intelligibility of rectangular speech-waves. *American Journal of Psychology,* 1948, *61,* 1–20.

Lindsay, P. H., & Norman, D. A. *Human information processing* (2nd ed.). New York: Academic Press, 1977.

Lockhead, G. R., & Byrd, R. Practically perfect pitch. *Journal of the Acoustical Society of America,* 1981, *70,* 387-389.

Loftus, G. R., Dark, V. J., & Williams, D. Short–term memory factors in ground controller/pilot communication. *Human Factors,* 1979, *21,* 169–181.

Long, G. M., & Waag, W. L. Limitations on the practical applicability of d' and β measures. *Human Factors,* 1981, *23,* 285–290.

Loomis, J. M., & Collins, C. C. Sensitivity to shift of a point stimulus: An instance of tactile hyperacuity. *Perception & Psychophysics,* 1978, *24,* 487–492.

Love, T. Relating individual differences in computer programming performance to human information processing abilities (Doctoral Dissertation, University of Washington, 1977).

Loveless, N. E. Direction of motion stereotypes: A review. *Ergonomics,* 1962, *5,* 357–383.

Lowenstein, W. R., & Skalak, R. Mechanical transmission in a Pacinian corpuscle. An analysis and a theory. *Journal of Physiology,* 1966, *182,* 346–378.

Mackworth, N. H., Morandi, A. J. The gaze selects information details within pictures. *Perception & Psychophysics,* 1967, *2,* 547–552.

Macmillan, N. A., Kaplan, H. L., & Creelman, C. D. The psychophysics of categorical perception. *Psychological Review,* 1977, *84,* 452–471.

Maddox, M. E. Two-dimensional spaciat frequency content and confusion among dot matrix characters. *Proceedings of the Human Factors Society,* 1979, *23,* 384–388.

Maier, S. F., & Seligman, M. E. P. Learned helplessness: Theory and evidence. *Journal of Experimental Psychology: General,* 1976, *105*(1), 3–46.

Martin, A. The equal energy concept applied to impulse noise. In D. Henderson, R. P. Hamernik, S. Dosanjh, & J. H. Mills (Ed.), *Effects of noise on hearing.* New York: Raven Press, 1976, pp. 421–456.

Mandler, G. Thought processes, consciousness, and stress. In V. Hamilton & D. M. Warburton (Eds.), *Human stress and cognition.* New York: Wiley, 1979.

Mathews, M. L., Angus, R. G., & Pearce, D. G. Effectiveness of accommodative aids in reducing empty field myopia in visual search. *Human Factors,* 1978, *20,* 733–740.

Mayer, M. *The builders.* New York: W. W. Norton, 1978.

McAllister, J. F. Exploring the consumer product safety challenge. In S. Steingiser (Ed.), *Safety and product liability.* Westport, Conn.: Technomic Publishing Co., 1975.

McBain, W. N. Arousal, monotony, and accidents in line driving. *Journal of Applied Psychology,* 1970, *54,* 509–519.

McCabe, T. J. A complexity measure. *IEEE Transactions on Software Engineering,* December 1976, *2*(4), 308–320.

McCarthy, D., & Saegert, S. Residential density, activity and personality on environmental preferences. *Journal of Research in Personality,* 1979, *1,* 45–60.

McCormick, E. J., & Sanders, M. S. *Human factors in engineering and design.* New York: McGraw-Hill, 1982.

McGrath, J. E. (Ed.). *Social and psychological factors in stress.* New York: Holt, Rinehart and Winston, 1970.

McHarg, I. *Design with nature.* New York: Doubleday/Natural History Press, 1971.

McPherson, J. M. Population density and social pathology: A reexamination. *Sociological Symposium,* 1975, *13,* 77–90.

McRuer, D. T. Human dynamics in man-machine systems, *Automatica,* 1980, *16,* 237–253.

Meagher, S. W. Designing accident-proof operator controls. *Machine Design,* June 23, 1977, 80–83.

Mears, J. D., & Cleary, P. J. Anxiety as a factor in underwater performance. *Ergonomics,* 1980, *23,* 549–557.

Meister, D. *Human factors: Theory and practice.* New York: Wiley, 1971.

Meldrum, L., Bowles, P., Brown, L., & Szanto, G. *Some experimental evidence on the effect of modularity on program comprehension.* CS 590D Project Report, Department of Computer Sciences, Purdue University, 1979.

Mermelstein, P. Evaluation of a segmental SNR measure as an indicator of the quality of ADPCM coded speech. *Journal of the Acoustical Society of America,* 1979, *66,* 1664–1667.

Meyer, D. E., Smith, J. E. K., & Wright, C. E. Models for the speed and accuracy of aimed movements. *Psychological Review,* 1982, 449–482.

Middlemist, R. D., Knowles, E. S., & Matter, C. F. Personal space invasions in the laboratory: Suggestive evidence for arousal. *Journal of Personality and Social Psychology,* 1976, *33,* 541–546.

Milgram, S. The experience of living in cities. *Science,* 1970, *167,* 1461–1468.

Miller, G. A. The magical number seven, plus or minus two. *Psychological Review,* 1956, 63, 81–97.

Miller, G. A., Heise, G. A., & Lichten, W. The intelligibility of speech as a function of the context of the test materials. *Journal of Experimental Psychology,* 1951, *41,* 329–335.

Miller, G. A., & Isard, S. Some perceptual consequences of linguistic rules. *Journal of Verbal Learning and Verbal Behavior,* 1963, *2,* 217–228.

Miller, J. D. Effects of noise on people. *Journal of the Acoustical Society of America,* 1974, *56,* 729–764.

Miller, L. H. A study in man-machine interaction. *National Computer Conference,* 1977, 409–421.

Miller, R. A., Jagacinski, R. J., Nalavade, R. B., & Johnson, W. W. Plans and the structure of target acquisition behavior. *Proceedings of the Human Factors Society,* 1981, *25,* 571–575.

Mills, A. W. Auditory Localization. Chapter 8 in J. V. Tobias (Ed.), *Foundations of modern auditory theory.* Vol. II. New York: Academic Press, 1972.

Mital, A., Halcomb, C. G., & Asfour, S. S. Visual comfort zone. *Proceedings of the Human Factors Society,* 1979, *23,* 193–195.

Mitchell, D. C., & Buck, J. R. Man-in-the-loop trouble shooting. Technical Report, Purdue School of Industrial Engineering, Purdue University, W. Lafayette, Ind., 1981.

Molino, J. A. Zerdy, G. A., Lerner, N. D., & Harwood, D. L. Use of the "acoustic menu" in assessing human response to audible (corona) noise from electric transmission lines. *Journal of the Acoustical Society of America,* 1979, *66,* 1435–1445.

Moody, D. B., Stebbins, W. C., Johnson, L., & Hawkins, J. E., Jr. Noise induced hearing loss in the monkey. In D. Henderson, R. P. Hamernik, S. Dosanjh, & J. H. Mills (Eds.), *Effects of noise on hearing.* New York: Raven Press, 1976, pp. 309–326.

Moore, B. C. J. *Introduction to the Psychology of Hearing,* London: Macmillan, 1977.

Moore, T. G. Industrial push-buttons. *Applied Ergonomics,* 1975, *6,* 33–38.

Moran, T. P. An applied psychology of the user. *ACM Computing Surveys,* March 1981, *13*(1), 1–11.

Moray, N. (Ed.), *Mental workload: Its theory and measurement.* New York: Plenum Press, 1979.

Morin, R. E., & Grant, D. A. Learning and performance of a key-pressing task as a function of the degree of spatial stimulus-response correspondence. *Journal of Experimental Psychology,* 1955, *49,* 39–47.

Morlok, E. K. *Introduction to transportation engineering.* New York: McGraw-Hill, 1978.

Mosteller, F., Rourke, R. E. K., & Thomas, G. B. *Probability and statistics.* Reading, Mass.: Addison-Wesley, 1961.

Mountford, S. J., & North, R. A. Voice entry for reducing pilot workload. *Proceedings of the Human Factors Society,* 1980, 185–189.

Mountford, S. J., & Somberg, B. Potential uses of two types of stereographic display systems in the airborne fire control environment. *Proceedings of the Human Factors Society,* 1981, *25,* 235–239.

Munson, W. A., & Karlin, J. E. Isopreference method for evaluating speech-transmission circuits. *Journal of the Acoustical Society of America,* 1962, *34,* 762–774.

Murrell, K. F. H. *Human performance in industry.* New York: Reinhold Publishing Corp., 1965.

Myers, G. J. A controlled experiment in program testing and code walkthrough/inspections. *Communications of the ACM,* September 1978, *21*(9), 760–768.

Myers, W. Interactive computer graphics: flying high—Part I. *Computer,* July 1979, *12*(7), 8–17.

Myers, W. Computer graphics: A two-way street. *Computer,* July 1980, *13*(7), 49–58.

National Institute for Occupational Safety and Health. *Criteria for a recommended standard.* HEW No. (NIOSH) 73-11001, 1972.

National Institute for Occupational Safety and Health. *Criteria for a recommended standard-occupational exposure to hot environment.* HEW No. (NIOSH), 1972.

National Institute for Occupational Safety and Health. *Industrial noise control manual.* HEW No. (NIOSH) 75-183, 1975.

Navon, D., & Gopher, D. On the economy of the human processing system. *Psychological Review,* 1979, *86,* 214–255.

Neilson, P. D., & Neilson, M. D. Influence of control-display compatibility on tracking behavior. *Quarterly Journal of Experimental Psychology,* 1980, *32,* 125–135.

Nelson, J. B., & Barany, J. W. A dynamic visual recognition test. *American Institute of Industrial Engineers Transactions,* 1969, *1,* 327–332.

Nelson, M. A., & Halberg, R. L. Visual contrast sensitivity function obtained with colored and achromatic gratings. *Human Factors,* 1979, *21,* 225–228.

Nemecek, J., & Grandjean, E. Results of an ergonomic investigation of large space offices. *Human Factors,* 1973, *15,* 421–450.

Newman, D. G. *Engineering economic analysis,* 3rd ed. San Jose, Calif.: Engineering Press, 1980.

Newman, O. *Defensible space.* New York: Macmillan, 1972.

Newman, O. Design guidelines for creating defensible space, 1975.

Newman, W. M., & Sproull, R. F. *Principles of interactive computer graphics.* New York: McGraw-Hill, 1973.

Nilees, J. M., Carlson, F. R., Gray, P., & Hanneman, G. J. *The telecommunication-transportation trade-off.* New York: Wiley, 1976.

NOAA Diving Manual, U.S. Government Printing Office, Washington, D.C., 1975.

Norcio, A. F. Indentation, documentation and programmer comprehension. *Proceedings of Human Factors in Computer Systems Conference,* March 1982, Gaithersburg, Maryland, 118–120.

Noyes, L. The positioning of type on maps: The effect of the surrounding material on word recognition time. *Human Factors,* 1980, *22*(3), 353–360.

Nuckolls, J. L. Cited by R. Horn. Studio lighting. *American Artist,* 1981, 74–77.

Nugent, N. *How to get along with your stomach.* Boston: Little, Brown, 1978.

Obermayer, R. W., & Muckler, F. A. *On the inverse optimal control problem in manual control systems.* NASA Tech. Report, CR-208, 1965.

Occupational Safety and Health Administration. *Noise control, A guide for workers and employers.* U.S. Department of Labor, OSHA 3048, 1980.

Ogden, G. D., Levine, J. M., & Eisner, E. J. Measurement of workload by secondary tasks. *Human Factors,* 1979, *21,* 529–548.

O'Hanlon, J. F. Boredom: Practical consequences and a theory. *Acta Psychologica,* 1981, *49,* 53–82.

Ohlson, M. System design considerations for graphics input devices. *Computer,* November 1978, *11*(11), 9–18.

O'Keefe, B., Brandes, M., Collins, N., Crow, J., Moore, N., & Notarnicola, R. Performance with screen editor versus line editor by novice users. CS 590D Project Report, Department of Computer Sciences, Purdue University, 1981.

Olson, P. L., & Bernstein, A. The nighttime legibility of highway signs as a function of their luminance characteristics. *Human Factors,* 1979, *21*(2), 145–160.

Ornstein, G. N. *Applications of a technique for the automatic analog determination of human response equation parameters.* Report NA 61H-1, North American Aviation, Columbus, Ohio, 1961.

Osada, Y. Experimental studies on the sexual and seasonal differences of the optimal thermal conditions. *Journal of Human Ergology,* 1978, *7,* 145–155.

Ostberg, O. *Fatigue in clerical work with CRT display terminals.* Goteborg Sweden: Goteborg Psychological Reports, 1974.

Ostry, D. J. Execution-time movement control. In G. E. Stelmach & J. Requin (Eds.), *Tutorials in motor behavior.* Amsterdam: North Holland, 1980.

Owens, D. A., & Leibowitz, H. W. Night myopia: Cause and a possible basis for amelioration. *American Journal of Optometry and Physiological Optics,* 1976, *53,* 709–717.

Pachella, R. G. The interpretation of reaction time in information processing research. In B. H. Kantowitz (Ed.), *Human information processing.* Hillsdale, N.J.: Lawrence Erlbaum Associates, 1974.

Pagano, A. M., McKnight, C., Johnson, C., & Robins, L. The costs of providing paratransit services. *Proceedings of the Transportation Research Forum,* 1980, *21,* 79–85.

Palmer, E. A., Jago, S. J., & O'Connor, S. L. Perception of horizontal aircraft separation on a cockpit display of traffic information. *Human Factors,* 1980, *22*(5), 605–620.

Pande, L., & Sullivan, J. W. *Preliminary study of the subjective assessment of exhaust system noise.* Purdue University, Herrick Laboratories Report No. 4 HL 78-42, August 1978.

Park, W. R. *Cost engineering analysis.* New York: Wiley, 1973.

Parsons, H. M. Life and death. *Human Factors,* 1970, *12,* 1–6.

Parsons, H. M. The bedroom. *Human Factors,* 1972, *14,* 421–450.

Parsons, H. M. Work environments. In I. Altman & J. F. Wohlwill (Eds.), *Human behavior and environment* (Vol. 1). New York: Plenum, 1976.

Patterson, A. H. Territorial behavior and fear of crime in the elderly. *Human Ecology,* 1978, *2,* 131–144.

Patterson, R. D. The effects of relative phase and the number of components on residue pitch. *Journal of the Acoustical Society of America,* 1973, *53,* 1565–1572.

Patterson, R. D. Auditory filter shape. *Journal of the Acoustical Society of America,* 1974, *55,* 802–809.

Patterson, R. D., & Milroy, R. *Existing and recommended levels for auditory warnings on civil aircraft.* MRC Applied Psychology Unit Civil Aviation Authority Contract Number 7D/S/0142, February 1979.

Patterson, R. D., & Milroy, R. *Auditory warnings on civil aircraft: the learning and retention of warnings.* MRC Applied Psychology Unit. Civil Aviation Authority Contract Number 7D/S/0142, February 1980.

Paulus, P. B. Crowding. In P. B. Paulus (Ed.), *Psychology of group influence.* Hillsdale, N.J.: Lawrence Erlbaum Associates, 1979.

Paulus, P. B., McCain, G., & Cox, V. C. Death rates, psychiatric commitments, blood pressure and perceived crowding as a function of institutional crowding. *Environmental Psychology and Nonverbal Behavior,* 1978, *3,* 107–116.

Pease, D. G., Ludwig, D. A., Green, E. B. & Millslage, D. G. Immediate and follow-up effects of cold on performance time and its components, reaction time and movement time. *Perceptual and Motor Skills,* 1980, *50,* 667–675.

Penner, M. J., & Shiffrin, R. M. Nonlinearities in the coding of intensity within the context of a temporal summation model. *Journal of the Acoustical Society of America,* 1980, *67,* 617–627.

Percival, L., & Loeb, M. Influence of noise characteristics on behavioral aftereffects. *Human Factors,* 1980, *22*(3), 341–352.

Peterson, C. R., Schneider, R. J., & Miller, A. J. Sample size and the revisions of subjective probability. *Journal of Experimental Psychology,* 1965, *69,* 522–527.

Pew, R. W. Performance of human operators in a three-state relay control system with velocity-augmented displays, *IEEE Trans. Human Factors,* Electronics, 1966, *7,* 77–83.

Pew, R. W. The speed-accuracy operating characteristic. *Acta Psychologica,* 1969, 30, 16–26.

Pew, R. W. Human perceptual motor performance. In B. H. Kantowitz (Ed.), *Human information processing: Tutorials in performance and cognition.* Hillsdale, N.J.: Lawrence Erlbaum Associates, 1974, pp. 1–39.

Pew, R. W. Secondary tasks and workload measurement. In N. Moray (Ed.), *Mental Workload.* New York: Plenum, 1979.

Pheasant, S., & O'Neill, D. Performance in gripping and turning: A study in hand-handle effectiveness. *Applied Ergonomics,* 1975, *6,* 205–208.

Pickett, J. M. Advances in sensory aids for the hearing-impaired: Visual and vibrotactile aids. In P. E. Brookhauser, & J. E. Bordley (Eds.), *Childhood communication disorders: Present status and future priorities. The annals of otology, rhinology, & laryngology,* 980, *89,* Supplement 74, pp. 74–78.

Pile, J. F. *Modern furniture.* New York: Wiley, 1979.

Pirenne, M. H. *Vision and the eye* (2d ed.). London: Associated Book Publisher, 1967.

Pirn, R. Acoustical variables in open planning. *Journal of the Acoustical Society of America,* 1971, *49,* 1339–1345.

Pisoni, D. B., & Hunnicutt, S. Perceptual evaluation of MITalk: The MIT unrestricted text-to-speech system. International Conference on Acoustics, Speech and Signal Processing, IEEE, April 1980.

Pisoni, D. B. Perceptual evaluation of voice response systems: Intelligibility, recognition & understanding. Workshop on Standardization for Speech I/O Technology, National Bureau of Standards (ICST), Gaithersburg, Maryland, March 18–19, 1982.

Pitz, G. F. Response variables in the estimation of relative frequency. *Perceptual and Motor Skills,* 1965, *21,* 873–876.

Poulton, E. C. Continuous intense noise masks auditory feedback and inner speech. *Psychological Bulletin,* 1977, *84,* 977–1001.

Poulton, E. C. A new look at the effects of noise: A rejoinder. *Psychologial Bulletin,* 1978, *85,* 1068–1079.

Poulton, E. C. Composite model for human performance in continuous noise. *Psychological Review,* 1979, *86,* 361–375.

Poulton, E. C. Not so! Rejoinder to Hartley on masking by continuous noise. *Psychological Review,* 1981, *88,* 90–92.

Propst, R. L. The Action Office. *Human Factors,* 1966, *8,* 299–306.

Porteous, J. D. *Environment & behavior.* Reading, Mass.: Addison-Wesley, 1977.

Proceedings of the Human Factors Society, 23rd Annual Meeting, Boston, Massachusetts, October 29–November 1, 1979.

Proxmire, W. *The fleecing of America.* Boston: Houghton Mifflin, 1980.

Pulat, B. M., & Ayoub, M. A. A computer-aided instrument panel design procedure. *Proceeding of the Human Factors Society,* 1979, *23,* 191–192.

Pulling, N. H., Wolf, E., Sturgis, S. P., Vaillancourt, D. R., & Dolliver, J. J. Headlight glare resistance and driver age. *Human Factors,* 1980, *22,* 103–112.

Putz, V. R. The effects of carbon monoxide on dual-task performance. *Human Factors,* 1979, *21,* 13–24.

Rabbitt, P., & Rodgers, B. What does a man do after he makes an error? An analysis of response programming. *Quarterly Journal of Experimental Psychology,* 1977, *29,* 727–743.

Rabbitt, P., & Rodgers, B. "The last of the old lag?" A reply to Welford's comment. *Quarterly Journal of Experimental Psychology,* 1979, *31,* 543–548.

Rainwater, L. Fear and the house-as-have in the lower class. *Journal of the American Institute of Planners*, 1966, *16*, 293–296.

Ramsey, J. D., & Morrissey, S. J. Isodecrement curves for task performance in hot environments. *Applied Ergonomics*, 1978, *9*, 66–72.

Rasmussen, J., & Rouse, W. B. (Eds.). *Human detection and diagnosis of system failures*. New York: Plenum, 1981.

Ray, R. D., & Ray, W. D. An analysis of domestic cooker control design. *Ergonomics*, 1979, *22*, 1243–1248.

Reiche, D. E., Kirschner, J. H., & Laurig, W. Evaluation of stress factors by analysis of radio telecommunication in ATC. *Ergonomics*, 1971, *14*, 603–609.

Reid, G. B., Shingledecker, C. A., & Eggemeier, F. T. Application of conjoint measurement to workload scale development. *Proceedings of the Human Factors society*, 1981, *25*, 522–526.

Rigby, L. V., & Swain, A. D. Some human factors applications to quality control in a high technology industry. (In C. G. Drury & J. G. Fox (Eds.), *Human Reliability in Quality Control*. London: Taylor and Francis, Ltd., 1975, pp. 201–216.

Rinalducci, E. J., Hardwick, D. A., & Beare, A. N. An assessment of visibility at the entrance of a long vehicular tunnel. *Human Factors*, 1979, *21*, 107–117.

Rizzi, A. M., Buck, J. R., & Anderson, V. L. Performance effects of variables in dynamic visual inspection. *American Institute of Industrial Engineers*, 1979, *11*, 278–285.

Robinson, G. H. Continous estimation of a time-varying probability. *Ergonomics*, 1964, *7*, 7–21.

Robinson, G. H. Dynamics of the eye and head during movement between displays: A qualitative and quantitative guide for designers. *Human Factors*, 1979, *21*, 343–352.

Rochester, N., Bequaert, F. C., & Sharp, E. M. The chord keyboard. *Computer*, December 1978, *11*(12), 57–63.

Rock, I. *An introduction to perception*. New York: Macmillan, 1975.

Roebuck, J. A., Kroemer, K. H. E., & Thomson, W. G. *Engineering anthropometry methods*. New York: Wiley, 1975.

Roederer, J. G. *Introduction to the physics and psychophysics of music*. London: The English Universities Press, 1973.

Rogers, J. G. Peripheral contrast thresholds for moving images. *Human Factors*, 1972, *14*, 199–205.

Rohles, F. H., Konz, S., & Munson, D. Estimating occupant satisfaction from effective temperature. *Proceedings of the Human Factors Society*, 1980, *24*, 223–227.

Rohmert, W., & Luczak, H. Stress, work, and productivity. In V. Hamilton & D. M. Warburton (Eds.), *Human stress and cognition.* New York: Wiley, 1979.

Rolfe, J. M. The secondary task as a measure of mental load. In W. T. Singleton, J. C. Fox, & D. W. Whitfield (Eds.), *Measurement of man at work.* London: Taylor & Francis, 1971, pp. 135–148.

Roscoe, A. H. Stress and workload in pilots. *Aviation, Space, and Environmental Medicine,* 1978, *49,* 630–636.

Roscoe, S. N. When day is done and shadows fall, we miss the airport most of all. *Human Factors,* 1979, *21,* 721–731.

Roscoe, S. N. *Aviation psychology.* Ames, Iowa: The Iowa State University Press, 1981.

Roscoe, S. N., & Bergman, C. A. Flight performance control. In S. N. Roscoe (Ed.), *Aviation psychology.* Ames, Iowa: The Iowa State University Press, 1980.

Roscoe, S. N., & Kraus, E. F. Pilotage error and residual attention: The evaluation of a performance control system in airborne area navigation. *Navigation,* 1973, *20,* 267–279.

Rostron, A. B. Brief auditory storage: Some further observations. *Acta Psychologica,* 1974, *38,* 471–482.

Rothenberg, M., Verillo, R. T., Zaborian, S. A., Brachman, M. L., & Bolanowski, S. J. Jr. Vibrotactile frequency for encoding a speech parameter. *Journal of the Acoustical Society of America,* 1977, *62,* 1003–1012.

Rothstein, M. A. OSHA after ten years: A review and some proposed reforms. *Vanderbilt Law Review,* 1981, *34,* 74–139.

Rouse, W. B., A model of human decision making in a fault diagnosis task. *IEEE Transactions on Systems, Man & Cybernetics,* 1978, *SMC-8*(5), 357–361.

Rouse, W. B. Experimental studies and mathematical models of human problem solving performance in fault diagnosis tasks. *Human detection and diagnosis of system failures.* New York: Plenum Press, 1981.

Rouse, W. B., & Rouse, S. H. Measures of complexity of fault diagnosis tasks. *IEEE Transactions on Systems, Man & Cybernetics,* November 1979, *SMC-9*(11), 720–722.

Rowland, G. E., & Cornog, D. Y. *Selected alphanumeric characters for closed-circuit television displays.* Courtney and Co., New York, TR No. 21, 1 July 1958.

Sackman, H., Erikson, W. J., & Grant, E. E. Exploratory experimental studies comparing online and offline programming performance. *Communications of the ACM,* January 1968, *11*(1), 3–11.

Safilios-Rothschild, C. Women and minorities in construction: The impact of affirmative action and its effects on work productivity. In M. Helander (Ed.), *Human factors/ergonomics for building and construction.* New York: Wiley, 1981.

Samuels, A. J. Being an expert witness in a product liability case. In H. R. Poydar (Ed.), *Human factors and industrial design in consumer products.* Medford, Mass.: Tufts University, Dept. of Engineering Design, 1980.

Sanders, A. F. Some remarks on mental load. In N. Moray (Ed.), *Mental workload.* New York: Plenum, 1979.

Scapin, D. L. Computer commands in restricted natural language: some aspects of memory and experience. *Human Factors,* June 1981 *23*(3), 365–375.

Schiff, W. *Perception: An applied approach.* Boston: Houghton Mifflin, 1980.

Schmidt, R. A., Zelaznik, H. N., & Frank, J. S. Sources of inaccuracy in rapid movement. In G. Stelmach (Ed.), *Information processing in motor control and learning.* New York: Academic Press, 1978.

Schmidt, R. A., Zelaznik, H., Hawkins, B., Frank, J. S., & Quinn, J. T. Motor-output variability: A theory for the accuracy of rapid motor acts. *Psychological Review,* 1979, *86,* 415–451.

Schroeder, M. R. Acoustics in human communications: Room acoustics, music, and speech. *Journal of the Acoustical Society of America,* 1980, *68,* 22–28.

Schroeder, M. R., Atal, B. S., & Hall, J. L. Optimizing digital speech coders by exploiting masking properties of the human ear. *Journal of the Acoustical Society of America,* 1979, *66,* 1647–1652.

Schroeder, M. R., Gottlob, D., & Siebrasse, K. F. Comparative study of European concert halls: Correlation of subjective preference with geometric and acoustic parameters. *Journal of the Acoustical Society of America,* 1974, *56,* 1195–1201.

Schultz, T. J. Noise criterion curves for use with the USASI preferred frequencies. *Journal of the Acoustical Society of America,* 1968, *43,* 637–638.

Scott, B. L., & De Filippo, C. L. Evaluating a two-channel lip-reading aid. *Journal of the Acoustical Society of America,* 1976, *60,* S124–S125(A).

Sears, T. A., & Davis, J. N. 1968, *Ann. NY Acad. Sci., 155,* 183–190.

Selye, H. The stress concept and some of its implications. In V. Hamilton & D. M. Warburton (Eds.), *Human stress and cognition.* New York: Wiley, 1979.

Seminara, J. L., Gonzalez, W. R., & Parsons, S. O. *Human Factors*

Review of Nuclear Power Plant Control Room Design. Electric Power Research Institute Report NP-309, March 1977.

Senders, J. W. The estimation of operator workload in complex systems. In K. B. DeGreene (Ed.), *Systems psychology.* New York: McGraw-Hill, 1970.

Sergeant, L., Atkinson, J. E., & Lacroix, P. G. The NSMRL tri-word test of intelligibility (TTI). *Journal of the Acoustical Society of America* 1979, *65,* 218-222.

Shaffer, L. H. Control processes in typing. *Quarterly Journal of Experimental Psychology,* 1975, *27,* 419–432.

Shaffer, L. H. Intention and performance. *Psychological Review,* 1976, *83,* 375–392.

Shaffer, L. H. Timing in the motor programming of typing. *Quarterly Journal of Experimental Psychology,* 1978, *30,* 333–345.

Shaffer, L. H. Analysing piano performance: A study of concert pianists. In G. E. Stelmach & J. Requin (Eds.), *Tutorials in motor behavior.* Amsterdam: North Holland, 1980.

Sheil, B. A. The psychological study of programming. *ACM Computing Surveys,* March 1981, *13*(1), 101–120.

Shen, V. Y., Conte, S. D., & Dunsmore, H. E. Software science revisited. *IEEE Transactions on Software Engineering,* 1982, to appear.

Sheppard, S. B., Borst, M. A., & Curtis, B. Predicting programmers' ability to understand and modify software. *Symposium Proceedings: Human Factors and Computer Science,* June 1978, Washington, D.C., 115–135.

Sheridan, J. A. *Designing the work environment.* New York: AT&T Company, July 1975.

Sheridan, T. B. Toward a general model of supervisory control. In T. B. Sheridan & G. Johannesen (Eds.), *Monitoring behavior and supervisory control.* London: Plenum Press, 1976, pp. 272–281.

Sheridan, T. B., & Ferrel, W. R. *Man-machine systems: Information control, and decision models of human performance.* Cambridge, Mass.: The MIT Press, 1974.

Sherrick C. E. Cutaneous communication. In W. D. Neff (Ed.), *Contributions to sensory physiology.* New York: Academic Press, 1982 expected.

Sherrick, C. E., & Rogers, R. Apparent haptic movement. *Perception & Psychophysics,* 1966, *1,* 175–180.

Sherrod, D. R., & Cohen, S. Density, personal control, and design. In J. R. Aiello & A. Baum (Eds.), *Residential crowding and design.* New York: Plenum, 1979.

Sherrod, D. R., Hage, J. N., Halpern, P. L., & Moore, B. S. Effects of personal causation and perceived control on responses to an aversive environment: The more control the better. *Journal of Experimental Social Psychology,* 1977, *13,* 14–27.

Shinar, D. Control-display relationships on the four-burner range: Population stereotypes versus standards. *Human Factors,* 1978, *20,* 13–17.

Shinner, S. M., Man-machine control systems. *Electro-Technology,* 1967, 61–76.

Shneiderman, B. *Software psychology: Human factors in computer and information systems.* Cambridge, Mass.: Winthrop Publishing Co., 1980.

Shneiderman, B., Mayer, R., McKay, D., & Heller, P. Experimental investigations of the utility of detailed flowcharts in programming. *Communications of the ACM,* June 1977, *20*(6), 373–381.

Shuford, E. H. A comparison of subjective probabilities for elementary and compound events. Psychometric Laboratory Report No. 20, University of North Carolina, 1959.

Shultz, H. G. An evaluation of methods for presentation of graphic multiple trends—Experiment III. *Human Factors,* 1961, *3*(3), 108–119.

Shultz, H. G. An evaluation of formats for graphic trend displays—Experiment II." *Human Factors,* 1961, *3*(3), 99–107.

Shum, D. A. Inferences on the basis of conditionally nonindependent data. *Journal of Experimental Psychology,* 1966, *72,* 401–409.

Shurtleff, D. A. Studies in television legibility—A review of the literature. *Information Displays,* January–February 1967, *4,* 40–45.

Siegel, A. I., & Wolf, J. J. *Man-machine simulation models.* New York: Wiley, 1969.

Sime, M. E., Green, T. R. G., & Guest, D. J. Psychological evaluation of two conditional constructions used in computer languages. *International Journal of Man-Machine Studies,* January 1973, *5*(1), 105–113.

Simon, H. A. *Models of man.* New York: Wiley, 1957.

Simpson, C. A., & Williams, D. H. Response time effects of alerting tone and semantic context for synthesized voice cockpit warnings. *Human Factors* 1980, *22,* 319–330.

Simpson, W., & Voss, J. F. Psychological judgments of probabilistic stimulus sequences. *Journal of Experimental Psychology,* 1961, *62,* 416–420.

Sivak, M., Olson, P. L., & Pastalan, L. A. Effect of driver's age on nighttime legibility of highway signs. *Human Factors,* 1981, *23,* 59–64.

Slovic, P. Toward, understanding and improving decisions. In W. C. Howell & E. A. Fleishman (Eds.), *Human performance and productivity,* Vol. 2. Hillsdale, N.J.: Lawrence Erlbaum Associates, 1981.

Small, K. A. Estimating the air pollution. *Journal of Transport Economics and Policy,* 1977, *11,* 109–132.

Smith, D. L. Self-service elevator control panels, A survey of user preference. *Proceedings of the Human Factors Society,* 1979, *23,* 124–128.

Smith, J. E. K. Simple algorithms for M-alternative forced-choice calculations. *Perception & Psychophysics,* 1982, *31,* 95–96.

Smith, L. A., & Barany, J. W. An elementary model of human performance on paced visual inspection. *American Institute of Industrial Engineers Transactions,* 1971, *4,* 298–308.

Smith, R. P. Boredom: A review. *Human Factors,* 1981, *23,* 329–340.

Smith, S. L. Letter size and legibility. *Human Factors,* 1979, *21*(6) 661–670.

Smith, S. L. Exploring compatibility with words and pictures. *Human Factors,* 1981, *23,* 305–315.

Smith, S. L., & Thomas, D. W. Color versus shape coding in 1964, information displays. *Journal of Applied Psychology,* 1964, *48,* 137–146.

Smith, S. W. Is there an optimum high level for office tasks? *Journal of the Illuminating Engineering Society,* 1978, *7,* 255–258.

Snyder, H. L., & Taylor, G. B. The sensitivity of response measures of alphanumeric legibility to variations in dot matrix display parameters. *Human Factors,* 1979, *21*(4), 457–471.

Sommer, R. Man's proximate environment. *Journal of Social Issues,* 1966, *22,* 59–70.

Sommer, R. *Personal space: The behavior basis of design.* Englewood Cliffs, N.J.: Prentice-Hall, 1969.

Sommer, R. *Tight spaces: Hard architecture and how to humanize it.* Englewood Cliffs, N.J.: Prentice-Hall, 1974.

Sorkin, R. D. Extension of the theory of signal detectability to matching procedures in psychoacoustics. *Journal of the Acoustical Society of America,* 1962, *43,* 1745–1751.

Sorkin, R. D., & Pohlman, L. D. Some models of observer behavior in two-channel auditory signal detection. *Perception & Psychophysics,* 1973, *14,* 101–109.

Sorkin, R. D., Boggs, G. J., & Brady, S. L. Discrimination of temporal jitter in patterned sequences of tones. *Journal of Experimental Psychology: Human Perception & Performance,* 1982, *8,* 46–57.

Sorkin, R. D., Pohlmann, L. D., & Woods, D. D. Decision interaction between auditory channels. *Perception & Psychophysics,* 1976, *19,* 290–295.

Sorkin, R. D., Woods, D. D., & Boggs, G. J. Signal detection in computer-synthesized noise. *Journal of the Acoustical Society of America,* 1979, *66,* 1351–1355.

Sperling, G. The information available in brief visual presentations. *Psychological Monographs,* 1960, *74,* 11 (Whole No. 498).

Srivastava, R. K., & Peel, T. S. *Human movement as a function of color stimulation.* Topeka, Kansas: Environmental Research Foundation, 1968.

Stammerjohn, L. W., Smith, M. J., & Cohen, B. G. F. Evaluation of work station design factors in VDT operations. *Human Factors,* August 1981, *23*(4), 401–412.

Steeneken, H. J. M., & Houtgast, T. A physical method for measuring speech-transmission quality. *Journal of the Acoustical Society of America,* 1980, *67,* 318–326.

Steidl, R. E. Difficulty factors in homemaking tasks: Implications for environmental design. *Human Factors,* 1972, *14,* 471–482.

Stelmach, G. (Ed.), *Information processing in motor control and learning.* New York: Academic Press, 1978.

Sternberg, S. Two operations in character recognition: Some evidence from reaction time measurements. *Perception & Psychophysics,* 1967, *2,* 45–53.

Sternberg, S. The discovery of processing stages: Extension of Donder's method. *Acta Psychologica,* 1969, *30,* 276–315.

Sternberg, S., Monsell, S., Knoll, R. L., & Wright, C. E. The latency and duration of rapid movement sequences: Comparison of speech and typewriting. In G. E. Stelmach (Ed.), *Information processing in motor control and learning.* New York: Academic Press, 1978.

Stokols, D. The experience of crowding in primary and secondary environments. *Environment and Behavior,* 1976, *8,* 49–86.

Strich, R. Human factors in the assignment of data item names for data bases (Master's thesis, Department of Computer Sciences, Purdue University, 1979).

Sundstrom, E., & Sundstrom, M. G. Personal space invasions: What happens when the invader asks permission? *Environmental Psychology and Nonverbal Behavior,* 1977, *2,* 76–82.

Swain, A. D. A method for performing a human-factors reliability analysis. *Sandia Corporation Monograph SCR-685,* August 1963.

Swain, A. D., & Guttmann, H. E. *Handbook of human reliability analysis with emphasis on nuclear power plant applications*. (Draft report for interim use and comment.) Washington, D.C.: U.S. Nuclear Regulatory Commission, Technical Report NUREG/CR-1278, October 1980.

Swets, J. A. Mathematical models of attention. In R. Parasuraman, R. Davies, & J. Beatty (Eds.), *Varieties of attention*. New York: Academic Press, 1982 expected.

Swets, J. A., & Birdsall, T. G. Repeated observation of an uncertain signal. *Perception & Psychophysics,* 1978, *23,* 269–274.

Swets, J. A., Green, D. M., Fay, T. H., Kryter, K. D., Nixon, C. M., Riney, J. S., Schultz, T. J., Tanner, W. P. Jr., & Whitcomb, M. A. A proposed standard fire alarm signal. *Journal of the Acoustical Society of America,* 1975, *57,* 756–757.

Swets, J. A., Green, D. M., Getty, D. J., & Swets, J. B. Signal detection and identification at successive stages of identification. *Perception & Psychophysics,* 1978, *23,* 275–289.

Tanner, W. P. Jr. Physiological implications of psychophysical data. *Annals of the New York Academy of Science,* 1961, *89,* 752–765.

Tanner, W. P. Jr., & Sorkin, R. D. The theory of signal detectability, Chapter 2 in J. V. Tobias (Ed.), *Foundations of modern auditory theory*. Vol. II. New York: Academic Press, 1972.

Taylor, D. A. Stage analysis of reaction time. *Psychological Bulletin,* 1976, *83,* 161–191.

Taylor, W., Pearson, J., Mair, A., & Burns, W. Study of noise and hearing in jute weaving. *Journal of the Acoustical Society of America,* 1965, *38,* 113–120.

Teal, R. F., & Giuliano, G. Taxi-based community transit: A comparative analysis of system alternatives and outcomes. *Proceedings of the Transportation Research Forum,* 1980, *21,* 86–93.

Teichner, W. H., & Krebs, M. J. Laws of visual choice reaction time. *Psychological Review,* 1974, *81,* 75–98.

Teichner, W. H., & Mocharnuk, J. B. Visual search for complex targets. *Human Factors,* 1979, *21,* 259–275.

Telford, C. W. The refractory phase of voluntary and associative responses. *Journal of Experimental Psychology,* 1931, *14,* 1–36.

Tichauer, E. R. *Biomechanical basis of ergonomics*. New York: Wiley-Interscience, 1978.

Tinker, M. A. Recent studies of eye movements in reading. *Psychological Bulletin,* 1958, *55,* 215–231.

Tinker, M. A. *Legibility of print*. Ames, Iowa: The Iowa State University Press, 1963.

Toates, F. *Control theory in biology and experimental psychology.* London: Hutchinson Educational, 1975.

Toney, E. R. The effects of operator and system response modality on fault diagnosis performance. M.S. Thesis, Psychology Department, Purdue University, 1981.

Topmiller, D. A., & Aume, K. S. M. Computer-graphic design for human performance. *Proceedings of the Annual Reliability and Maintainability Symposium of the Institute of Electrical Engineers,* 1978, 385–388.

Torgerson, Warren S. *Theory and methods of scaling.* New York: Wiley, 1958.

Tosi, O., Oyer, H., Lashbrook, W., Pedrey, C., Nicol, J., & Nash, E. Experiment of voice identification. *Journal of the Acoustical Society of America,* 1972, *51,* 2030–2043.

Townsend, J. T. Issues and models concerning the processing of a finite number of inputs. In B. H. Kantowitz (Ed.), *Human information processing.* Hillsdale, N.J.: Lawrence Erlbaum Associates, 1974.

Trémolières, C., & Hétu, R. A multi-parametric study of impact noise-induced TTS. *Journal of the Acoustical Society of America,* 1980, *68,* 1652–1659.

Treisman, A. M. Strategies and models of selective attention. *Psychological Review,* 1969, *76,* 282–299.

Treisman, A. M. & Gelade, G. A feature-integration theory of attention. *Cognitive Psychology,* 1980, 12, 97–136.

Triggs, T. J., Levison, W. H., & Sanneman, R. Some experiments with flight-related electrocutaneous and vibrotactile displays. In F. A. Geldard (Ed.), *Cutaneous communication systems and devices.* Austin, Texas: The Psychonomic Society, 1974, pp. 57–64.

Tulving, E. Subjective organization in free recall of "unrelated" words. *Psychological Review,* 1962, *69,* 344–354.

Turpin, J. A. Effects of blur and noise on digital imagery interpretability. *Proceedings of the Human Factors Society,* 1981, *25,* 286–289.

Tversky, A. Choice by elimination. *Journal of Mathematical Pschology,* 1972, *9,* 341–367.

Tversky, A., & Kahneman, D. The belief in the law of small numbers. *Psychological Bulletin,* 1971, *76,* 105–110.

Tversky, A., & Kahneman, D. Availability: A heuristic for judging frequency and probability. *Cognitive Psychology,* 1973, *5,* 207–232.

Tversky, A., & Kahneman, D. Judgment under uncertainty: Heuristics and biases. *Science,* 1974, *185,* 1124–1131.

U.S. Department of Transportation, Office of the Secretary. 1971 National Transportation Report, Washington, D.C.: U.S. Government Printing Office, 1972.

Vallerie, L. L., & Link, J. M. Visual detection probability of "sonar" targets is a function of retinal position and brightness contrast. *Human Factors,* 1968, *10,* 403–412.

Van Cott, H. P., & Kinkade, R. G. (Eds.). *Human engineering guide to equipment design.* Washington, D.C.: U.S. Superintendent of Documents, 1972.

Van Nes, F. L., & Bouma, H. On the legibility of segmented numerals. *Human Factors,* 1980, *2*(4), 463–474.

VDT Manual. Darmstadt, West Germany: Ince-Fiej Research Association, 1979.

Verrillo, R. T. Effect of contactor area on the vibrotactile threshold. *Journal of the Acoustical Society of America,* 1966, *35,* 1962–1966.

Von Gierke, H. E., & Johnson, D. L. Summary of present damage risk criteria. In D. Henderson, R. P. Hamernik, S. Dosanjh, & J. H. Mills (Eds.), *Effects of noise on hearing.* New York: Raven Press, 1976, pp. 547–560.

Von Neuman, J., & Morgenstern, O. *Theory of games and economic behavior,* (3rd ed.). Princeton, N.J.: Princeton University Press, 1953.

Walkowiak, V. S. (Ed.). *Uniform products liability act* (Vol. 1). New York: Mathew Bender, 1980.

Wallach, N., Newman, E. B., & Rosenzweig, M. R. The Precedence effect in sound localization. *American Journal of Psychology,* 1949, *52,* 315–336.

Wallack, P. M., & Adams, S. K. The utility of signal detection theory in the analysis of industrial inspector accuracy. *American Institute of Industrial Engineers Transactions,* 1969, *1*(1), 33–44.

Ward, J. Ward reservations. *Design,* 1976, *333,* 33–34.

Ward, S. K. Overcrowding and social pathology: A reexamination of the implication for human population. *Human Ecology,* 1975, *3,* 275-286.

Ward, W. D. A comparison of the effects of continuous, intermittent, and impulse noise. In D. Henderson, R. P. Hamernik, S. Dosanjh, & J. H. Mills (Eds.), *Effects of noise on hearing.* New York: Raven Press, 1976, pp. 407–420.

Warren, R. M. Perceptual restoration of missing speech sounds. *Science,* 1970, *167,* 392–393.

Warren, R. M. Auditory temporal discrimination by trained listeners. *Cognitive Psychology,* 1974, *6,* 237–256.

Wastell, D. G., Brown, I. D., & Copeman, A. K. Evoked potential amplitude as a measure of attention in working environments: A comparative study of telephone switchboard design. *Human Factors,* 1981, *23,* 117–121.

Watson, C. S. Factors in the discrimination of word-length auditory patterns. In S. K. Hirsh, D. H. Eldredge, I. J. Hirsh, & S. R. Silverman (Eds.), *Hearing and Davis: Essays honoring Hallowell Davis.* St. Louis, Mo.: Washington University Press, 1976.

Watson, C. S., & Kelly, W. J. The role of stimulus uncertainty in the discrimination of auditory patterns. Chapter 3 in D. J. Getty & J. H. Howard Jr. (Eds.), *Auditory and visual pattern recognition.* Hillsdale, N.J.: Lawrence Erlbaum Associates, 1981.

Weber, A., Fussler, C., O'Hanlon, J. F., Gierer, R., & Grandjean, E. Psychophysiological effects of repetitive tasks. *Ergonomics,* 1980, *23,* 1033–1046.

Webster, J. C. Speech interference aspects of noise. In D. M. Lipscomb (Ed.), *Noise and audiology.* Baltimore: University Park Press, 1978, pp. 193–228.

Wegel, R. L., & Lane, C. E. The auditory masking of one pure tone by another and its probable relation to the dynamics of the inner ear. *Psychological Review,* 1924, *23,* 266–285.

Weigelt, H. R., Gotz, R. E., & Weiss, H. H. *City traffic.* New York: Van Nostrand Reinhold, 1977.

Weinberg, G. M. *The psychology of computer programming.* New York: Van Nostrand Reinhold, 1971.

Weinberg, G. M., & Schulman, E. L. Goals and performance in computer programming. *Human Factors,* February 1974, *16*(1), 70–77.

Weinstein, A. S., Twerski, A. D., Piehler, H. R., & Donaher, W. A. *Products liability and the reasonably safe product.* New York: Wiley, 1978.

Weinstein, S. Intensive and extensive aspects of tactile sensitivity as a function of body part, sex, and laterality. In D. R. Kenshalo (Ed.), *The skin senses.* Springfield, Ill.: Charles C Thomas, 1968, pp. 195–218.

Weiss, J. M. Psychological factors in stress and disease. *Scientific American,* 1972, *266,* 104–113.

Weissman, L. Psychological complexity of computer programs: an experimental methodology. *ACM SIGPLAN Notices,* June 1974, *9*(6), 25–36.

Welch, J. R. Automatic speech recognition—Putting it to work in industry. *Computer,* May 1980, *13*(5), 65–73.

Welford, A. T. *Skilled performance.* Glenview, Ill.: Scott, Foresman, 1976.

Welford, A. T. Comment on the paper by Rabbitt and Rodgers. *Quarterly Journal of Experimental Psychology,* 1979, *31,* 539–542.

Weltman, G., Smith, J. E., & Egstrom, G. H. Perceptual narrowing during simulated pressure-chamber exposure. *Human Factors,* 1971, *13,* 99–107.

White, B. W., Saunders, F. A., Scadden, L., Bach-Y-Rita, P., & Collins, C. C. Seeing with the skin. *Perception & Psychophysics,* 1970, *7,* 23–27.

White, F. A. *Our acoustic environment.* New York: Wiley, 1975.

Whyte, W. W. *The last landscape.* New York: Doubleday, 1968.

Wiener, E. L. Individual and group differences in inspection. In C. G. Drury, & J. G. Fox (Eds.), *Human reliability in quality control.* London: Taylor and Francis, Ltd., 1975, 7, pp. 101–122.

Wiener, E. L. Midair collisions: The accidents, the systems, and the realpolitik. *Human Factors,* 1980, *22,* 521–533.

Wierwille, W. W. Physiological measures of aircrew mental workload. *Human Factors,* 1979, *21,* 575–593.

Williams, L. G. The effects of target specification on objects fixated during visual search. *Perception and Psychophysics,* 1966, *1,* 315–318.

Williges, R. C., & Wierwille, W. W. Behavioral measures of aircrew mental workload. *Human Factors,* 1979, *21,* 549–574.

Willis, C. L. An empirical study of bathtub and shower accidents. *Proceedings of the symposium on environmental effects on behavior.* Big Sky, Montana: Environmental Design Group of the Human Factors Society, 1975.

Wing, J. F. Upper thermal tolerance limits for unimpaired mental performance. *Aerospace Medicine,* 1965, *36,* 960-964.

Wise, K. D., Chen, K., & Yokely, R. E. *Micro-computers: A technology forecast and assessment to the year 2000.* New York: Wiley, 1980.

Woodfield, S. N., Dunsmore, H. E., & Shen, V. Y. The effect of modularization and comments on program comprehension. *Proceedings of Fifth International Conference on Software Engineering,* March 1981, San Diego, 215–223.

Woodson, W. E., & Conover, E. W. *Human engineering guide for equipment designers* (2nd ed.). Berkeley: University of California Press, 1966.

Woodworth, R. S. The accuracy of voluntary movement. *Psychological Review Monograph,* 1899, *3,* 1–14.

Wright, P. The harassed decision maker: Time pressures, distractions, and the use of evidence. *Journal of Applied Psychology,* 1974, *59,* 555–561.

Yerkes, R. M., & Dodson, J. D. The relative strength of stimulus to rapidity of habit-formation. *Journal of Comparative and Neurological Psychology,* 1908, *18,* 459–482.

Yost, W. A., & Nielsen, D. W. *Fundamentals of hearing, an introduction.* New York: Holt, Rinehart and Winston, 1977.

Young, H. H., & Berry, G. L. The impact of environment on the productivity attitudes of intellectually challenged office workers. *Human Factors,* 1979, *21,* 399–407.

Zabcik, D. A. Anthropometric design of a personal hygiene center. *Proceedings of the Human Factors Society,* 1981, *25,* 488–491.

Zacarias, P., Benham, B., Dreyer, J., & Duffy, C. A comparative study of video and paper terminals. CS 590D Project Report, Department of Computer Sciences, Purdue University, 1981.

Zweigenhaft, R. L. Personal space in the faculty office: Desk placement and the student-faculty interaction. *Journal of Applied Psychology,* 1976, *61*(4), 529–532.

AUTHOR INDEX

Page numbers in *italics* indicate material found in tables and figures

SUBJECT INDEX